Texts in Statistical Science

# Statistical Methods for Spatial Data Analysis

**CHAPMAN & HALL/CRC**
**Texts in Statistical Science Series**



**Analysis of Failure and Survival Data**
Peter J. Smith

**The Analysis and Interpretation of Multivariate Data for Social Scientists**
David J. Bartholomew, Fiona Steele, Irini Moustaki, and Jane Galbraith

**The Analysis of Time Series — An Introduction, Sixth Edition**
Chris Chatfield

**Applied Bayesian Forecasting and Time Series Analysis**
A. Pole, M. West and J. Harrison

**Applied Nonparametric Statistical Methods, Third Edition**
P. Sprent and N.C. Smeeton

**Applied Statistics — Handbook of GENSTAT Analysis**
E.J. Snell and H. Simpson

**Applied Statistics — Principles and Examples**
D.R. Cox and E.J. Snell

**Bayes and Empirical Bayes Methods for Data Analysis, Second Edition**
Bradley P. Carlin and Thomas A. Louis

**Bayesian Data Analysis, Second Edition**
Andrew Gelman, John B. Carlin, Hal S. Stern, and Donald B. Rubin

**Beyond ANOVA — Basics of Applied Statistics**
R.G. Miller, Jr.

**Computer-Aided Multivariate Analysis, Fourth Edition**
A.A. Afifi and V.A. Clark

**A Course in Categorical Data Analysis**
T. Leonard

**A Course in Large Sample Theory**
T.S. Ferguson

**Data Driven Statistical Methods**
P. Sprent

**Decision Analysis — A Bayesian Approach**
J.Q. Smith

**Elementary Applications of Probability Theory, Second Edition**
H.C. Tuckwell

**Elements of Simulation**
B.J.T. Morgan

**Epidemiology — Study Design and Data Analysis, Second Edition**
M. Woodward

**Essential Statistics, Fourth Edition**
D.A.G. Rees

**A First Course in Linear Model Theory**
Nalini Ravishanker and Dipak K. Dey

**Interpreting Data — A First Course in Statistics**
A.J.B. Anderson

**An Introduction to Generalized Linear Models, Second Edition**
A.J. Dobson

**Introduction to Multivariate Analysis**
C. Chatfield and A.J. Collins

**Introduction to Optimization Methods and Their Applications in Statistics**
B.S. Everitt

**Large Sample Methods in Statistics**
P.K. Sen and J. da Motta Singer

**Linear Models with R**
Julian J. Faraway

**Markov Chain Monte Carlo — Stochastic Simulation for Bayesian Inference**
D. Gamerman

**Mathematical Statistics**
K. Knight

**Modeling and Analysis of Stochastic Systems**
V. Kulkarni

**Modelling Binary Data, Second Edition**
D. Collett

**Modelling Survival Data in Medical Research, Second Edition**
D. Collett

**Multivariate Analysis of Variance and Repeated Measures — A Practical Approach for Behavioural Scientists**
D.J. Hand and C.C. Taylor

**Multivariate Statistics — A Practical Approach**
B. Flury and H. Riedwyl

**Practical Data Analysis for Designed Experiments**
B.S. Yandell

**Practical Longitudinal Data Analysis**
D.J. Hand and M. Crowder

**Practical Statistics for Medical Research**
D.G. Altman

**Probability — Methods and Measurement**
A. O'Hagan

**Problem Solving — A Statistician's Guide, Second Edition**
C. Chatfield

**Randomization, Bootstrap and Monte Carlo Methods in Biology, Second Edition**
B.F.J. Manly

**Readings in Decision Analysis**
S. French

**Sampling Methodologies with Applications**
Poduri S.R.S. Rao

**Statistical Analysis of Reliability Data**
M.J. Crowder, A.C. Kimber, T.J. Sweeting, and R.L. Smith

**Statistical Methods for Spatial Data Analysis**
Oliver Schabenberger and Carol A. Gotway

**Statistical Methods for SPC and TQM**
D. Bissell

**Statistical Methods in Agriculture and Experimental Biology, Second Edition**
R. Mead, R.N. Curnow, and A.M. Hasted

**Statistical Process Control — Theory and Practice, Third Edition**
G.B. Wetherill and D.W. Brown

**Statistical Theory, Fourth Edition**
B.W. Lindgren

**Statistics for Accountants**
S. Letchford

**Statistics for Epidemiology**
Nicholas P. Jewell

**Statistics for Technology — A Course in Applied Statistics, Third Edition**
C. Chatfield

**Statistics in Engineering — A Practical Approach**
A.V. Metcalfe

**Statistics in Research and Development, Second Edition**
R. Caulcutt

**Survival Analysis Using S —Analysis of Time-to-Event Data**
Mara Tableman and Jong Sung Kim

**The Theory of Linear Models**
B. Jørgensen

Texts in Statistical Science

# Statistical Methods for Spatial Data Analysis

**Oliver Schabenberger**
**Carol A. Gotway**

Chapman & Hall/CRC
Taylor & Francis Group
Boca Raton London New York

Published in 2005 by
Chapman & Hall/CRC
Taylor & Francis Group
6000 Broken Sound Parkway NW, Suite 300
Boca Raton, FL 33487-2742


Printed in the United States of America on acid-free paper
10 9 8 7 6

International Standard Book Number-10: 1-58488-322-7 (Hardcover)
International Standard Book Number-13: 978-1-58488-322-7 (Hardcover)



**Library of Congress Cataloging-in-Publication Data**

Catalog record is available from the Library of Congress

T&F informa

Taylor & Francis Group
is the Academic Division of T&F Informa plc.

**Visit the Taylor & Francis Web site at**
**http://www.taylorandfrancis.com**

**and the CRC Press Web site at**
**http://www.crcpress.com**

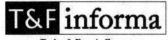

To Lisa and Charlie

# Contents

# Preface

The study of statistical methods for spatial data analysis presents challenges that are fairly unique within the statistical sciences. Like few other areas, spatial statistics draws on and brings together philosophies, methodologies, and techniques that are typically taught separately in statistical curricula. Understanding spatial statistics requires tools from applied statistics, mathematical statistics, linear model theory, regression, time series, and stochastic processes. It also requires a different mindset, one focused on the unique characteristics of spatial data, and additional analytical tools designed explicitly for spatial data analysis.

When preparing graduate level courses in spatial statistics for the first time, we each struggled to pull together all the ingredients necessary to present the material in a cogent manner at an accessible, practical level that did not tread too lightly on theoretical foundations. This book ultimately began with our efforts to resolve this struggle. It has its foundations in our own experience, almost 30 years combined, with the analysis of spatial data in a variety of disciplines, and in our efforts to keep pace with the new tools and techniques in this diverse and rapidly evolving field.

The methods and techniques discussed in this text do by no means provide a complete accounting of statistical approaches in the analysis of spatial data. Weighty monographs are available on any one of the main chapters. Instead, our goal is a comprehensive and illustrative treatment of the basic statistical theory and methods for spatial data analysis. Our approach is mostly model-based and frequentist in nature, with an emphasis on models in the spatial, and not the spectral, domain. Geostatistical methods that developed largely outside of the statistical mainstream, e.g., kriging methods, can be cast easily in terms of prediction theory based on statistical regression models. Focusing on a model formulation allows us to discuss prediction and estimation in the same general framework. But many derivations and results in spatial statistics either arise from representations in the spectral domain or are best tackled in this domain, so spectral representations appear throughout. We added a section on spectral domain estimation (§4.7) that can be incorporated in a course together with the background material in §2.5. While we concentrate on frequentist methods for spatial data analysis, we also recognize the utility of Bayesian hierarchical models. However, since these models are complex and intricate, we leave their discussion until Chapter 6, after much of the foundation of spatial statistics necessary to understand and interpret them has been developed.

The tools and approaches we consider essential comprise Chapters 1–7. Chapter 1, while introductory, also provides a first description of the basic measures of spatial autocorrelation and their role in the analysis of spatial data. Chapter 2 provides the background and theoretical framework of random fields necessary for subsequent chapters, particularly Chapters 4 and 5. We begin the heart of statistical methods for spatial data analysis with mapped point patterns in Chapter 3. Since a good understanding of spatial autocorrelation is necessary for spatial analysis, estimation and modeling of the covariance function and semivariogram are treated in detail in Chapter 4. This leads easily into spatial prediction and kriging in Chapter 5. One of the most important chapters is Chapter 6 on spatial regression. It is unique in its comprehensiveness, beginning with linear models with uncorrelated errors and ending with a succinct discussion of Bayesian hierarchical models for spatial data. It also provides a discussion of model diagnostics for linear and generalized linear spatial models. Chapter 7 is devoted to the simulation of spatial data since we believe simulation is an essential component of statistical methods for spatial data analysis and one that is often overlooked in most other textbooks in spatial statistics. The chapters on non-stationary covariance (Chapter 8) and spatio-temporal models (Chapter 9) are primarily a review of a rapidly evolving and emerging field. These chapters are supplemental to the core of the text, but will be useful to Ph.D. students in statistics and others who desire a brief, concise, but relatively thorough overview of these topics.

The book is intended as a text for a one-semester graduate level course in spatial statistics. It is assumed that the reader/student is familiar with linear model theory, the requisite linear algebra, and a good, working knowledge of matrix algebra. A strong foundation in fundamental statistical theory typically taught in a first course in probability and inference (e.g., probability distributions, conditional expectation, maximum likelihood estimation) is essential. Stochastic process theory is not assumed; the needed ingredients are discussed in Chapter 2. The text does not require prior exposure to spectral methods. There are problems at the end of the main chapters that can be worked into course material.

It is difficult to write a book of this scope without the help and input of many people. First and foremost, we need to express our deepest appreciation to our spouses, Lisa Schabenberger and Charlie Crawford, who graciously allowed us ample time to work on this project. We are also fortunate to have many friends within the greater statistical community and many of them undoubtedly influenced our work. In particular, we are grateful for the insights of Noel Cressie, Linda Young, Konstantin Krivoruchko, Lance Waller, Jay Ver Hoef, and Montserrat Fuentes. We are indebted to colleagues who shared their data with us. Thanks to Thomas Mueller for the C/N ratio data, to Jeffrey Walters for the woodpecker data, to Vaisala Inc. for the lightning strikes data, to Victor De Oliveira for the rainfall data, and to Felix Rogers for the low birth weight data. Thanks to Charlie for being there when shape files go bad; his knowledge of Virginia and his patience with the "dry as toast" material in the book are much appreciated.

Finding time to write was always a challenge, and we are indebted to David Olson for supporting Carol through her struggles with government bureaucracy. Finally, we learned to value our close collaboration and critical thinking ability, enjoying the time we spent discussing and debating recent developments in spatial statistics, and combining our different ideas like pieces of a big puzzle. We hope the book reflects the integration of our knowledge and our common philosophy about statistics and spatial data analysis. A special thanks to our editor at CRC Press, Bob Stern, for his vision, support, and patience.

The material in the book will be supplemented with additional material provided through the CRC Press Web site (`www.crcpress.com`). The site will provide many of the data sets used as examples in the text, software code that can be used to implement many of the principal methods described and illustrated in the text, as well as updates and corrections to the text itself. We welcome additions, corrections, and discussions for this Web page so that it can make statistical methods for spatial data analysis useful to scientists across many disciplines.

Oliver Schabenberger
Cary, North Carolina

Carol A. Gotway Crawford
Atlanta, Georgia

CHAPTER 1

# Introduction

## 1.1 The Need for Spatial Analysis

Statistical methods for spatial data analysis play an ever increasing role in the toolbox of the statistician, scientist, and practitioner. Over the years, these methods have evolved into a self-contained discipline which continues to grow and develop and has produced a specific vocabulary. Characteristic of spatial statistics is its immense methodological diversity. In part, this is due to its many origins. Some of the methods developed outside of mainstream statistics in geology, geography, meteorology, and other subject matter areas. Some are rooted in traditional statistical areas such as linear models and response surface theory. Others are derived from time series approaches or stochastic process theory. Many methods have undergone specific adaptations to cope with the specific challenges presented by, for example, the fact that spatial processes are not equivalent to two-dimensional time series processes. The novice studying spatial statistics is thus challenged to absorb and combine varied tools and concepts, revisit notions of randomness and data generating mechanisms, and to befriend a new vernacular.

Perhaps the foremost reason for studying spatial statistics is that we are often not only interested in answering the "how much" question, but the "how much is where" question. Many empirical data contain not only information about the attribute of interest—the response being studied—but also other variables that denote the geographic location *where* the particular response was observed. In certain instances, the data may consist of location information only. A plant ecologist, for example, records the locations within a particular habitat where a rare plant species can be found. It behooves us to utilize this information in statistical inference provided it contributes meaningfully to the analysis.

Most authors writing about statistical methods for spatial data will argue that one of the key features of spatial data is the autocorrelation of observations in space. Observations in close spatial proximity tend to be more similar than is expected for observations that are more spatially separated. While correlations between observations are not a defining feature of spatial data, there are many instances in which characterizing spatial correlation is of primary analytical interest. It would also be shortsighted to draw a line between "classical" statistical modeling and spatial modeling because of the existence of correlations. Many elementary models exhibit correlations.

**Example 1.1** A drug company is interested in testing the uniformity of its buffered aspirin tablets for quality control purposes. A random sample of three batches of aspirin tablets is collected from each of the company's two manufacturing facilities. Five tablets are then randomly selected from each batch and the diameter of each tablet is measured. This experiment is a nested design with $t = 2$ manufacturing facilities, $r = 3$ batches per facility, $rt = 6$ experimental units (the batches of aspirin), and five sub-samples per experimental unit ($n = 5$), the diameter measurements made on the tablets within each batch. The experimental design can be expressed by the linear model

$$\begin{aligned} Y_{ijk} &= \mu + \tau_i + e_{ij} + \epsilon_{ijk}, \\ i &= 1, t = 2, \qquad j = 1, 2, r = 3 \qquad k = 1, 2, n = 5. \end{aligned}$$

The $e_{ij}$ represent the variability among the batches within a manufacturing facility and the $\epsilon_{ijk}$ terms represent the variability in the diameters of the tablets within the batches, i.e., the sub-sampling errors. By virtue of the random sampling of the batches, the $e_{ij}$ are uncorrelated. The $\epsilon_{ijk}$ are uncorrelated by virtue of randomly sampling the tablets from each batch. How does this experiment induce correlations? Measurements made on different experimental units are uncorrelated, but for two tablets from the same batch:

$$\begin{aligned} \text{Cov}[Y_{ijk}, Y_{ijk'}|e_{ij}] &= \text{Cov}[e_{ij} + \epsilon_{ijk}, e_{ij} + \epsilon_{ijk'}|e_{ij}] = \text{Cov}[\epsilon_{ijk}, \epsilon_{ijk'}] = 0 \\ \text{Cov}[Y_{ijk}, Y_{ijk'}] &= \text{Cov}[e_{ij} + \epsilon_{ijk}, e_{ij} + \epsilon_{ijk'}|e_{ij}] \\ &= \text{Cov}[e_{ij}, e_{ij}] + \text{Cov}[e_{ij}, \epsilon_{ijk'}] + \\ &\quad \text{Cov}[\epsilon_{ijk}, e_{ij}] + \text{Cov}[\epsilon_{ijk}, \epsilon_{ijk'}] \\ &= \text{Var}[e_{ij}] + 0 + 0 + 0. \end{aligned}$$

The data measured on the same batch are correlated because they share the same random effect, the experimental error $e_{ij}$ associated with that unit. □

During the past two decades we witnessed tremendous progress in the analysis of another type of correlated data: longitudinal data and repeated measures. It is commonly assumed that correlations exist among the observations repeatedly collected for the same unit or subject. An important aspect of such data is that the repeated measures are made according to some metric such as time, length, depth, etc. This metric typically plays some role in expressing just how the correlations evolve with time or distance. Models for longitudinal and repeated measures data bring us thus closer to models for spatial data, but important differences remain. Consider, for example, a longitudinal study of high blood pressure patients, where $s$ subjects are selected at random from some (large) population. At certain time intervals $t_1, \cdots, t_{n_i}$, $(i = 1, \cdots, s)$ the blood pressure of the $i^{\text{th}}$ patient is measured along with other variables, e.g., smoking status, exercise and diet habits, medication. Thus, we observe a value of $Y_i(t_j)$, $j = 1, \cdots, n_i$, the blood pressure of patient $i$ at time $t_j$. Although the observations taken on a particular patient are most likely serially correlated, these data fall within the realm of traditional random sampling. We consider the vectors $\mathbf{Y}_i = [Y_i(t_1), \cdots, Y_i(t_{n_i})]'$ as independent random vectors because patients were selected at random. A statistical model for the data from the $i^{\text{th}}$ patient might be

$$\mathbf{Y}_i = \mathbf{X}_i \boldsymbol{\beta} + \mathbf{e}_i, \qquad (1.1)$$

where $\mathbf{X}_i$ is a known $(n_i \times p)$ matrix of regressor and design variables associated with subject $i$. The coefficient vector $\boldsymbol{\beta}$ can be fixed, a random vector, or have elements of both types. Assume for this example that $\boldsymbol{\beta}$ is a vector of fixed effects coefficients and that $\mathbf{e}_i \sim (\mathbf{0}, \mathbf{V}_i(\boldsymbol{\theta}_l))$. The matrix $\mathbf{V}_i(\boldsymbol{\theta}_l)$ contains the variances of and covariances between the observations from the same patient. These are functions of the elements of the vector $\boldsymbol{\theta}_l$. We call $\boldsymbol{\theta}$ the vector of covariance parameters in this text, because of its relationship to the covariance matrix of the observations.

The analyst might be concerned with

1. estimating the parameters $\boldsymbol{\beta}$ and $\boldsymbol{\theta}_l$ of the data-generating mechanism,
2. testing hypotheses about these parameters,
3. estimating the mean vector $\mathrm{E}[\mathbf{Y}_i] = \mathbf{X}_i \boldsymbol{\beta}$,
4. predicting the future blood pressure value of a patient at time $t$.

To contrast this example with a spatial one, consider collecting $n$ observations on the yield of a particular crop from an agricultural field. The analyst may raise questions similar to those in the longitudinal study:

1. about the parameters describing the data-generating mechanism,
2. about the average yield in the northern half of the field compared to the southern-most area where the field surface is sloped,
3. about the average yield on the field,
4. about the crop yield at an unobserved location.

Since the questions are so similar in their reference to estimates, parameters, averages, hypotheses, and unobserved events, maybe a statistical model similar to the one in the longitudinal study can form the basis of inference? One suggestion is to model the yield $Z$ at spatial location $\mathbf{s} = [x, y]'$ as

$$Z(\mathbf{s}) = \mathbf{x}_s' \alpha + \nu,$$

where $\mathbf{x}_s$ is a vector of known covariates, $\alpha$ is the coefficient vector, and $\nu \sim (0, \sigma^2)$. A point $\mathbf{s}$ in the agricultural field is identified by its $x$ and $y$ coordinate in the plane. If we collect the $n$ observations into a single vector, $\mathbf{Z}(\mathbf{s}) = [Z(\mathbf{s}_1), Z(\mathbf{s}_2), \cdots, Z(\mathbf{s}_n)]'$, this spatial model becomes

$$\mathbf{Z}(\mathbf{s}) = \mathbf{X}_s \alpha + \boldsymbol{\nu}, \qquad \boldsymbol{\nu} \sim (\mathbf{0}, \boldsymbol{\Sigma}(\boldsymbol{\theta}_s)), \qquad (1.2)$$

where $\boldsymbol{\Sigma}$ is the covariance matrix of the vector $\mathbf{Z}(\mathbf{s})$. The subscript $s$ is used to identify a component of the spatial model; the subscript $l$ is used for a component of the longitudinal model to avoid confusion.

Similarly, in the longitudinal case, the observations from different subjects can be collected into a single vector, $\mathbf{Y} = [\mathbf{Y}_1', \mathbf{Y}_2', \cdots, \mathbf{Y}_n']'$, and model (1.1) can be written as

$$\mathbf{Y} = \mathbf{X}_l\beta + \mathbf{e}, \tag{1.3}$$

where $\mathbf{X}_l$ is a stacking of $\mathbf{X}_1, \mathbf{X}_2, \cdots, \mathbf{X}_n$, and $\mathbf{e}$ is a random vector with mean $\mathbf{0}$ and covariance matrix

$$\mathrm{Var}[\mathbf{e}] = \mathbf{V}(\boldsymbol{\theta}_l) = \begin{bmatrix} \mathbf{V}_1(\boldsymbol{\theta}_l) & \mathbf{0} & \cdots & \mathbf{0} \\ \mathbf{0} & \mathbf{V}_2(\boldsymbol{\theta}_l) & \cdots & \mathbf{0} \\ \mathbf{0} & \mathbf{0} & \ddots & \vdots \\ \mathbf{0} & \mathbf{0} & \cdots & \mathbf{V}_s(\boldsymbol{\theta}_l) \end{bmatrix}.$$

The models for $\mathbf{Y}$ and $\mathbf{Z}(\mathbf{s})$ are rather similar. Both suggest some form of generalized least squares (GLS) estimation for the fixed effects coefficients,

$$\begin{aligned} \widehat{\beta} &= (\mathbf{X}_l'\mathbf{V}(\boldsymbol{\theta}_l)^{-1}\mathbf{X}_l)^{-1}\mathbf{X}_l'\mathbf{V}(\boldsymbol{\theta}_l)^{-1}\mathbf{Y} \\ \widehat{\alpha} &= (\mathbf{X}_s'\boldsymbol{\Sigma}(\boldsymbol{\theta}_s)^{-1}\mathbf{X}_s)^{-1}\mathbf{X}_s'\boldsymbol{\Sigma}(\boldsymbol{\theta}_s)^{-1}\mathbf{Z}(\mathbf{s}). \end{aligned}$$

Also common to both models is that $\boldsymbol{\theta}$ is unknown and must be estimated. The estimates $\widehat{\beta}$ and $\widehat{\alpha}$ are efficient, but unattainable. In Chapters 4–6 the issue of estimating covariance parameters is visited repeatedly.

However, there are important differences between models (1.3) and (1.2) that imperil the transition from the longitudinal to the spatial application. The differences have technical, statistical, and computational implications. Arguably the most important difference between the longitudinal and the spatial model is the sampling mechanism. In the longitudinal study, we sample $s$ independent random vectors, and each realization can be thought of as that of a temporal process. The sampling of the patients provides the replication mechanism that leads to inferential brawn. The *technical* implication is that the variance-covariance matrix $\mathbf{V}(\boldsymbol{\theta})$ is block-diagonal. The *statistical* implication is that standard multivariate limiting arguments can be applied, for example, to ascertain the asymptotic distribution of $\widehat{\beta}$, because the estimator and its associated estimating equation can be expressed in terms of sums of independent contributions. Solving

$$\sum_{i=1}^{s} \mathbf{X}_i'\mathbf{V}_i(\boldsymbol{\theta}_l)^{-1}\left(\mathbf{Y}_i - \mathbf{X}_i'\beta\right) = \mathbf{0}$$

leads to

$$\widehat{\beta} = \left(\sum_{i=1}^{s} \mathbf{X}_i'\mathbf{V}_i(\boldsymbol{\theta}_l)^{-1}\mathbf{X}_i\right)^{-1} \sum_{i=1}^{s} \mathbf{X}_i'\mathbf{V}_i(\boldsymbol{\theta}_l)^{-1}\mathbf{Y}_i.$$

The *computational* implication is that data processing can proceed on a subject-by-subject basis. The key components of estimators and measures of their precision can be accumulated one subject at a time, allowing fitting of models to large data sets (with many subjects) to be computationally feasible.

In the spatial case, $\boldsymbol{\Sigma}$ is not block-diagonal and there is usually no replication mechanism. The computational ramifications of a dense matrix $\mathbf{\Sigma}$ are formidable in their own right. For example, computing $\widehat{\boldsymbol{\beta}}$ in the subject-by-subject form of a longitudinal model requires inversion of $\mathbf{V}_i$, a matrix of order at most $\max\{n_i\}$. In the spatial case, computing $\widehat{\boldsymbol{\alpha}}$ requires inversion of an $(n \times n)$ matrix. The implications and ramifications of lacking a replication mechanism are even more daunting. We suggested earlier that longitudinal data may be considered as the sampling of $s$ temporal processes. Making the connection to the spatial case suggests a view of spatial data as a single sample ($s \equiv 1$) of a two-dimensional process. This is indeed one way of viewing many types of spatial data (see §2), and it begs the question as to how we can make *any* progress with statistical inference based on a sample of size one.

While there are models, data structures, and analyses in the more traditional, classical tool set of the statistician that are portable to the spatial case, the analysis of spatial data provides some similar and many unique challenges and tasks. These tasks and challenges stimulate the need for spatial models and analysis. Location information is collected in many applications, and failing to use this information can obstruct important characteristics of the data-generating mechanism. The importance of maintaining the spatial context in the analysis of georeferenced data can be illustrated with the following example.

**Example 1.2** Data were generated on a $10 \times 10$ lattice by drawing *iid* variates from a Gaussian distribution with mean 5 and variance 1, denoted here as $G(5, 1)$. The observations were assigned completely at random to the lattice coordinates (Figure 1.1a). Using a simulated annealing algorithm (§7.3), the data points were then rearranged such that a particular value is surrounded by more and more similar values. We define a nearest-neighbor of site $\mathbf{s}_i$, $(i = 1, \cdots, 100)$, to be a lattice site that could be reached from $\mathbf{s}_i$ with a one-site move of a *queen* piece on a chess board. Then, let $\overline{z}_i$ denote the average of the neighboring sites of $\mathbf{s}_i$. The arrangements shown in Figure 1.1b–d exhibit increasing correlations between $Z(\mathbf{s}_i)$, the observation at site $\mathbf{s}_i$, and the average of its nearest-neighbors.

Since the same 100 data values are assigned to the four lattice arrangements, any exploratory statistical method that does not utilize the coordinate information will lead to identical results for the four arrangements. For example, histograms, stem-and-leaf plots, $Q$-$Q$ plots, and sample moments will be identical. However, the spatial *pattern* depicted by the four arrangements is considerably different. This is reflected in the degree to which the four arrangements exhibit spatial autocorrelation (Figure 1.2).

□

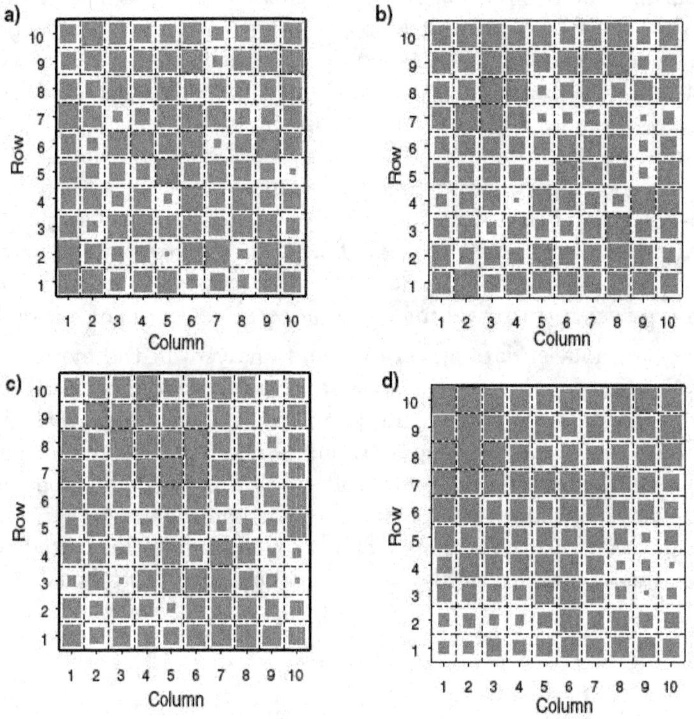


Figure 1.1 *Simulated spatial arrangements on a* $10 \times 10$ *lattice. Independent draws from a* $G(5,1)$ *distribution assigned completely at random are shown in panel a. The same data were then rearranged to create arrangements (b–d) that differ in their degree of spatial autocorrelation.*

## 1.2 Types of Spatial Data

Because spatial data arises in a myriad of fields and applications, there is a myriad of spatial data types, structures, and scenarios. A classification of spatial data invariably is either somewhat coarse or overly detailed. We err on the side of a coarse classification, which allows you, the reader, to subsequently equate a data type with certain important features of the underlying spatial process. These features, in turn, lead to particular classes of models, inferential questions, and applications. For example, for spatial data that is typically observed exhaustively, e.g., statewide data collected for each county, the question of predicting the attribute of interest for an unobserved county in the state is senseless. There are no unobserved counties.

The classification we adopt here follows Cressie (1993) and distinguishes data types by the nature of the spatial domain. To make these matters more precise, we denote a spatial process in $d$ dimensions as

$$\{Z(\mathbf{s}) : \mathbf{s} \in D \subset \mathbb{R}^d\}.$$

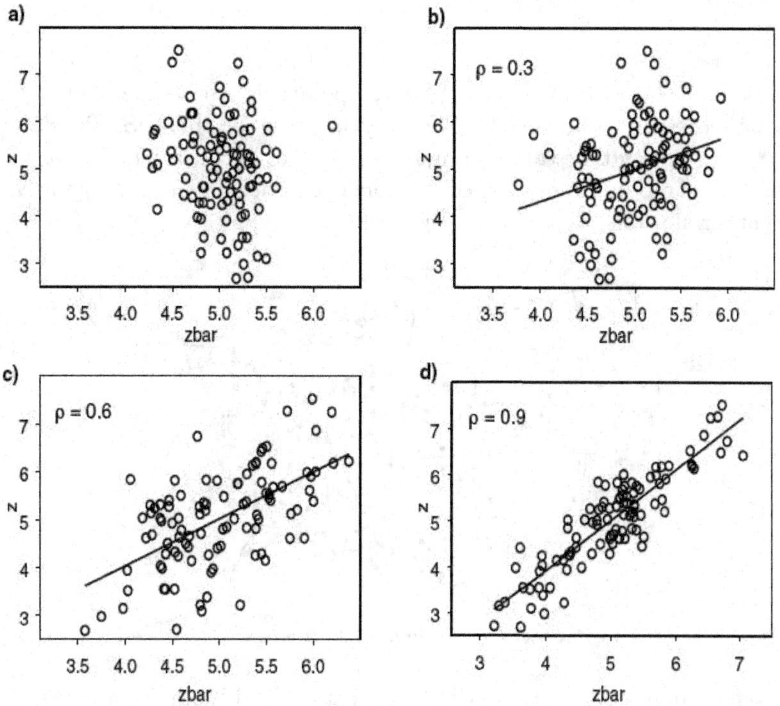


Figure 1.2 *Correlations between lattice observations $Z(s_i)$ and the average of the nearest neighbors for lattice arrangements shown in Figure 1.1.*

Here, $Z$ denotes the attribute we observe, for example, yield, concentration, or the number of sudden infant deaths. The location at which $Z$ is observed is $\mathbf{s}$, a $(d \times 1)$ vector of coordinates. Most of the spatial processes in this book are processes in two-dimensional space, $d = 2$, and $\mathbf{s} = [x, y]'$ are the Cartesian coordinates. The spatial data types are distinguished through characteristics of the domain $D$.

### *1.2.1 Geostatistical Data*

The domain $D$ is a continuous, fixed set. By continuous we mean that $Z(\mathbf{s})$ can be observed everywhere within $D$, i.e., between any two sample locations $\mathbf{s}_i$ and $\mathbf{s}_j$ you can theoretically place an infinite number of other samples. By fixed we mean that the points in $D$ are non-stochastic. Because of the continuity of $D$, geostatistical data is also referred to as "spatial data with continuous variation." It is important to associate the continuity with the domain, not with the attribute being measured. Whether the attribute $Z$ is continuous or discrete, has no bearing on whether the data are geostatistical, or not.

**Example 1.3** Consider measuring air temperature. Air temperature could, at least theoretically, be recorded at any location in the U.S. Practically, however, air temperature values cannot be recorded exhaustively. It is usually recorded at a finite number of specified locations such as those designated as U.S. weather stations, depicted in Figure 1.3.

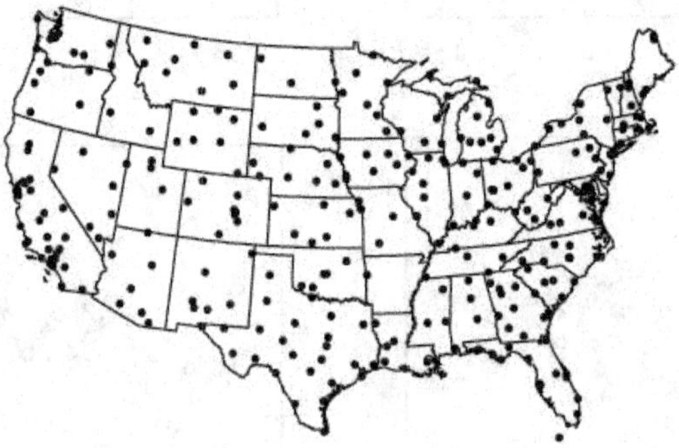

Figure 1.3 *U.S. Weather Stations. Source: National Climatic Data Center.*

It is reasonable to treat these data as geostatistical, defining a continuous temperature surface across the U.S. Our assessment of these data as geostatistical would not change if, instead of the air temperature, we determine an indicator variable, namely whether the air exceeds a specified threshold limit, at each weather station. How we select the locations at which $Z$ is observed also has no bearing on whether the process is geostatistical or not. If instead of a specified network of weather stations, we had measured air temperature at the geographic centroid of each state, we would still observe the same temperature surface, just at different points.

Since the spatial domain $D$ (in this case the entire U.S.) is continuous, it cannot be sampled exhaustively, and an important task in the analysis of geostatistical data is the reconstruction of the surface of the attribute $Z$ over the entire domain, i.e., mapping of $Z(\mathbf{s})$. Thus, we are interested in how the temperature values vary geographically, as in Figure 1.4.

□

### 1.2.2 Lattice Data, Regional Data

Lattice data are spatial data where the domain $D$ is fixed and discrete, in other words, non-random and countable. The number of locations can be infinite; what is important is that they can be enumerated. Examples of lattice data are attributes collected by ZIP code, census tract, or remotely sensed data

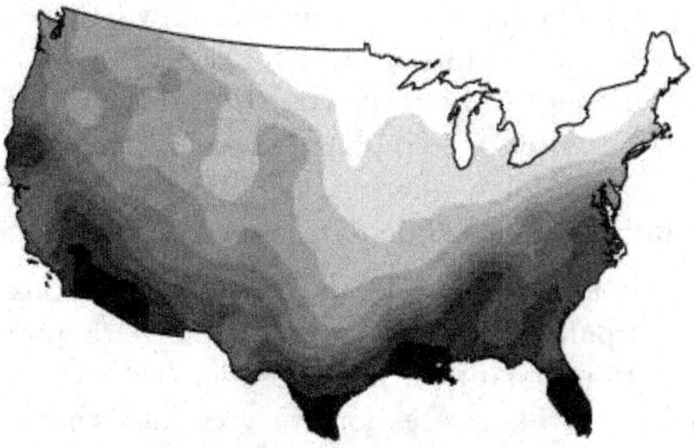

Figure 1.4 *Temperature variations in the U.S.*

reported by pixels. Spatial locations with lattice data are often referred to as **sites** and these sites usually represent not points in space, but areal regions. It is often mathematically convenient or necessary to assign to each site one precise spatial coordinate, a "representative" location. If we refer to Wake County in North Carolina, then this site is a geographic region, a polygon on a map. To proceed statistically, we need to spatially index the sites so we can make measurements between them. For example, to measure the distance between the Wake County site and another county in North Carolina, we need to adopt some convention for measuring the distance between two regions. One possibility would be the Euclidean distance between a representative location within each county, for example, the county centroid, or the seat of the county government.

Unfortunately, the mathematical notation typically used with lattice data, and the term lattice data itself, are both misleading. It seems natural to refer to the observation at the $i$th site as $Z_i$ and then to use $\mathbf{s}_i$ to denote the representative point coordinate within the site. In order to emphasize the spatial nature of the observation, the same notation, $Z(\mathbf{s}_i)$, used for geostatistical data is also routinely used for lattice data. The subscript indexes the lattice site and $\mathbf{s}_i$ denotes its representative location. Unfortunately, this notation and the idea of a "lattice" in general, has encouraged most scientists to treat lattice data as if they were simply a collection of measurements recorded at a finite set of point locations. In reality, most, if not all, lattice data are spatially aggregated over areal regions. For example,

- yield is measured on an agricultural plot,
- remotely sensed observations are associated with pixels that correspond to areas on the ground,
- event counts, e.g., number of deaths, crime statistics, are reported for counties or ZIP codes,
- U.S. Census Bureau data is made available by census tract.

The aggregation relates to integration of a continuous spatial attribute (e.g., average yield per plot), or enumeration (integration with respect to a counting measure). Because the areal units on which the aggregation takes place can be irregularly shaped regions, the name **regional data** is more appropriate in our opinion. If it is important to emphasize the areal nature of a lattice site, we use notation such as $Z(A_i)$, where $A_i$ denotes the $i$th areal unit. Here, $A_i$ is called the **spatial support** of the data $Z(A_i)$ (see §5.7).

Another interesting feature of lattice data is exhaustive observation in many applications. Many spatial data sets with lattice data provide information about every site in the domain. If, for example, state-wide data provide information on cancer mortality rates for every county, the issue of predicting cancer mortality in any county does not arise. The rates are known. An important goal in statistical modeling lattice data is *smoothing* the observed mortality rates and assessing the relationship between cancer rates and other variables.

**Example 1.4 Blood lead levels in children, Virginia 2000.** The mission of the Lead-Safe Virginia Program sponsored by the Virginia Department of Health is to eradicate childhood lead poisoning. As part of this program, children under the age of 72 months were tested for elevated blood lead levels. The number of children tested each year is reported for each county in Virginia, along with the number that had elevated blood lead levels. The percent of children with elevated blood lead levels for each Virginia county in 2000 is shown in Figure 1.5.

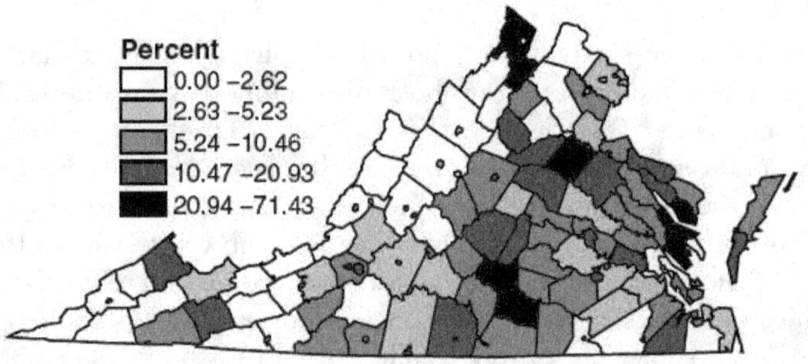


Figure 1.5 *Percent of children under the age of 72 months with elevated blood lead levels in Virginia in 2000. Source: Lead-Safe Virginia Program, Virginia Department of Health.*

This is a very typical example of lattice or regional data. The domain $D$ is the set of 133 counties (the sites) in Virginia which is clearly fixed (not random), discrete, and easily enumerated. The attribute of interest, the percentage of children with elevated blood lead levels, is an aggregate variable, a ratio of two values summed over each county. A common, simple, probability model for this variable is based on the binomial distribution: assume the number of children per county with elevated blood lead levels, $Y(\mathbf{s}_i)$, follows a binomial distribution with mean $\pi n_i$, where $n_i$ is the number of children tested in county $i$ and $\pi$ is the probability of having an elevated blood level. If $Z(\mathbf{s}_i) = Y(\mathbf{s}_i)/n_i$, then $\text{Var}[Z(\mathbf{s}_i)] = \pi(1-\pi)/n_i$, which may change considerably over the domain. Unfortunately, many of the statistical methods used to analyze lattice data assume that the data have a constant mean and a constant variance. This is one of the main reasons we find the term lattice data misleading. It is just too easy to forget the aggregate nature of the data and the heterogeneity (in mean and in variance) that can result when data are aggregated geographically. A plethora of misleading analyses can easily be found in the literature, in statistical textbooks and journals as well as in those within specific subject-matter areas such as geography and epidemiology.

In regional data analysis, counties in close proximity to one another with similar values produce a spatial pattern indicative of positive spatial autocorrelation. Identifying groups of counties in close proximity to one another with *high* values is often of particular interest, suggesting a "cluster" of elevated risk with perhaps a common source. Another goal in regional data analysis is identification of the spatial risk factors for the response of interest. For example, the primary source of elevated blood lead levels in children is dust from lead-based paint in older homes in impoverished areas. Thus, we might seek to correlate the map in Figure 1.5, to one produced from Census data showing the median housing value per county, a surrogate for housing age and maintenance quality (Figure 1.6).

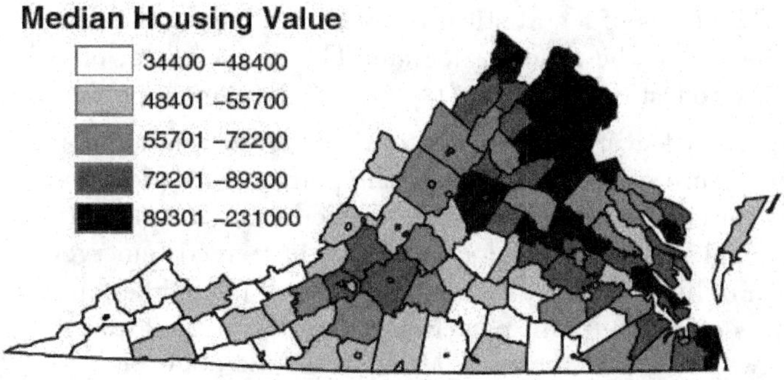


Figure 1.6 *Median housing value per county in Virginia in 2000. Source: U.S. Census Bureau.*

□

### *1.2.3 Point Patterns*

Geostatistical and lattice data have in common the fixed, non-stochastic domain. A domain $D$ is fixed if it does not change from one realization of the spatial process to the next. Consider pouring sand out of a bucket onto a desk and let $Z(\mathbf{s})$ denote the depth of the poured sand at location $\mathbf{s}$. The set

$D$, which comprises all possible locations in this experiment, is the desktop. After having taken the depth measurements at some sample locations on the desk, return the sand to the bucket and pour it again. This creates another realization of the random process $\{Z(\mathbf{s}) : \mathbf{s} \in D \subset \mathbb{R}^2\}$. We now have two realizations of a geostatistical process; the domain has not changed between the first and second pouring, unless someone moved the desk. The objective of a statistical inquiry with such data is obviously the information collected on the attribute $Z$, for example, the construction of an elevation map of the sand on the desk. The objective is not to study the desk, the domain is known and unchanging.

Can we conceive of a random domain in this example, that is, a set of points that changes with each realization of the random process? If $Z(\mathbf{s})$ is the depth of the sand at location $\mathbf{s}$, apply an indicator transform such that

$$I(\mathbf{s}) = \begin{cases} 1 & Z(\mathbf{s}) > c \\ 0 & \text{otherwise.} \end{cases}$$

If $Z(\mathbf{s})$ is geostatistical data, so is the indicator $I(\mathbf{s})$, which returns 1 whenever the depth of the sand exceeds the threshold $c$. Now remove all locations in $D$ for which $I(\mathbf{s}) = 0$ and define the remaining set of points as $D^*$. As the process of pouring the sand is repeated, a different realization of the *random set* $D^*$ is obtained. Both the number of points in the realized set, as well as their spatial configuration, is now the outcome of a random process. The attribute "observed" at every point in $D^*$ is rather un-interesting, $I(\mathbf{s}) \equiv 1$, if $\mathbf{s} \in D^*$. The focus of a statistical investigation is now the set $D^*$ itself, we study properties of the random domain. This is the realm of point pattern analysis. The collection of points $I(\mathbf{s}), \mathbf{s} \in D^*$, is termed a point pattern.

The important feature of point pattern data is the random domain, not the degenerate nature of the attribute at the points of $D$. Many point patterns are of this variety, however. They are referred to as **unmarked** patterns, or simply as point patterns. The locations at which weeds emerge in a garden, the location of lunar craters, the points at which nerve fibers intersect tissue are examples of spatial point patterns. If along with the location of an event of interest we observe a stochastic attribute $Z$, the point pattern is termed a **marked** pattern. For example, the location and depth of lunar craters, the location of weed emergence and the light intensity at that location can be considered as marked point patterns.

**Example 1.5 Lightning strikes.** Once again we turn to the weather for an interesting example. Prompt and reliable information about the location of lightning strikes is critical for any business or industry that may be adversely affected by the weather. Weather forecasters use information about lightning strikes to alert them (and then the public at large) to potentially dangerous storms. Air traffic controllers use such information to re-route airline traffic, and major power companies and forestry officials use it to make efficient use of human resources, anticipating the possibility of power outages and fires.

The National Lightning Detection Network locates lightning strikes across the U.S. instantaneously by detecting the electromagnetic signals given off when lightning strikes the earth's surface. Figure 1.7 shows the lightning flash events within approximately 200 miles of the East coast of the U.S. between April 17 and April 20, 2003.

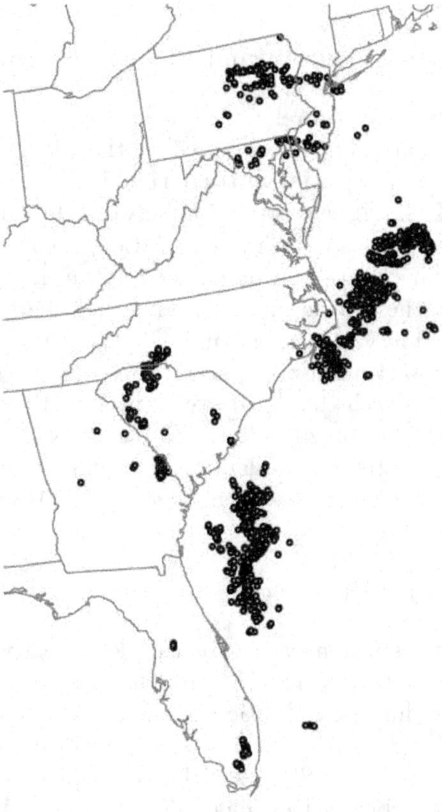

Figure 1.7 *Locations of lightning strikes within approximately 200 miles of the east coast of the U.S. between April 17 and April 20, 2003. Data courtesy of Vaisala Inc., Tucson, AZ.*

This is an example of a spatial point pattern. It consists of the locations of the lightning strikes (often called event locations) recorded within a boundary that is the set of all potential locations within 200 miles of the East Coast of the U.S. It is called a pattern, but this is in fact one of the questions that may be of interest: Do the lightning strikes occur in a spatially random fashion or is there some sort of pattern to their spatial distribution? □

In addition to the locations of the lightning strikes, the National Lightning

Detection Network also records information about the polarity (a negative or positive charge), and peak amplitude of each strike. Thus, these attributes, together with the locations of the lightning strikes, are an example of a marked point pattern. With marked point patterns, we are interested in the spatial relationships among the values of the marking attribute variable, above and beyond any induced by the spatial distribution of the strikes. We will treat such analyses in more detail in Chapter 3.

## 1.3 Autocorrelation—Concept and Elementary Measures

As the name suggests, autocorrelation is the correlation of a variable with *itself.* If $Z(\mathbf{s})$ is the attribute $Z$ observed in the plane at spatial location $\mathbf{s} = [x, y]'$, then the term **spatial autocorrelation** refers to the correlation between $Z(\mathbf{s}_i)$ and $Z(\mathbf{s}_j)$. It is a correlation between the same attribute at two locations. In the absence of spatial autocorrelation, the fact that two points $\mathbf{s}_i$ and $\mathbf{s}_j$ are close or far has no bearing on the relationship between the values $Z(\mathbf{s}_i)$ and $Z(\mathbf{s}_j)$ at those locations. Conversely, if there is positive spatial autocorrelation, proximity in space should couple with similarity in attribute values. Another way to visualize the concept is to consider a three-dimensional coordinate system in which the $(x, y)$-axes represent the spatial coordinates and the vertical axis the magnitude of $Z$. Under positive autocorrelation, points that are close in the $(x$–$y)$-plane tend to have similar values of $Z$. In the three-dimensional coordinate system this creates clusters of points.

### *1.3.1 Mantel's Tests for Clustering*

Some elementary statistical measures of the degree to which data are autocorrelated can be motivated precisely in this way; as methods for detecting clusters in three-dimensional space. Mantel (1967) considered a general procedure to test for disease clustering in a spatio-temporal point process $\{Z(\mathbf{s}, t) : \mathbf{s} \in D \subset \mathbb{R}^2, t \in T^* \subset \mathbb{R}\}$. For example, $(\mathbf{s}, t)$ is a coordinate in space and time at which a leukemia case occurred. This is an unmarked spatio-temporal point pattern, $Z$ is a degenerate random variable. To draw the parallel with studying autocorrelation in a three-dimensional coordinate system, we could also consider the data-generating mechanism as a spatial point process with a mark variable $T$, the time at which the event occurs. Denote this process as $T(\mathbf{s})$ and the observed data as $T(\mathbf{s}_1), T(\mathbf{s}_2), \cdots, T(\mathbf{s}_n)$.

The disease process is said to be clustered if cases that occur close together in space also occur close together in time. In order to develop a statistical measure for this tendency to group in time and space, let $W_{ij}$ denote a measure of spatial proximity between $\mathbf{s}_i$ and $\mathbf{s}_j$ and let $U_{ij}$ denote a measure of the temporal proximity of the cases. For example, we can take for leukemia cases at $\mathbf{s}_i$ and $\mathbf{s}_j$

$$W_{ij} = ||\mathbf{s}_i - \mathbf{s}_j|| \qquad U_{ij} = |T(\mathbf{s}_i) - T(\mathbf{s}_j)|.$$

Mantel (1967) suggested to test for clustering by examining the test statistics

*Mantel Statistics*

$$M_1 = \sum_{i=1}^{n-1} \sum_{j=i+1}^{n} W_{ij} U_{ij} \tag{1.4}$$

$$M_2 = \sum_{i=1}^{n} \sum_{j=1}^{n} W_{ij} U_{ij} \tag{1.5}$$

$$W_{ii} = U_{ii} = 0.$$

Because of the restriction $W_{ii} = U_{ii} = 0$, (1.4) sums the product $W_{ij}U_{ij}$ for the $n(n-1)/2$ unique pairs and $M_2$ is a sum over all $n(n-1)$ pairs. Collect the spatial and temporal distance measures in $(n \times n)$ matrices $\mathbf{W}$ and $\mathbf{U}$ whose diagonal elements are 0. Then $M_2 = \sum_{i=1}^{n} \mathbf{u}_i'\mathbf{w}_i$, where $\mathbf{u}_i'$ is the $i$th row of $\mathbf{U}$ and $\mathbf{w}_i'$ is the $i$th row of $\mathbf{W}$.

Although Mantel (1967) focused on diagnosing spatio-temporal disease clustering, the statistics (1.4) and (1.5) can be applied for any spatial attribute and are not restricted to marked point patterns. Define $W_{ij}$ as a measure of closeness between $\mathbf{s}_i$ and $\mathbf{s}_j$ and let $U_{ij}$ be a measure of closeness of the observed values; for example, $U_{ij} = |Z(\mathbf{s}_i) - Z(\mathbf{s}_j)|$ or $U_{ij} = \{Z(\mathbf{s}_i) - Z(\mathbf{s}_j)\}^2$. Notice that for a geostatistical or lattice data structure the $W_{ij}$ are fixed because the domain is fixed and the $U_{ij}$ are random. We can imagine a regression through the origin with response $U_{ij}$ and regressor $W_{ij}$. If the process exhibits positive autocorrelation, then small $W_{ij}$ should pair with small $U_{ij}$. Points will be distributed randomly beyond some critical $W_{ij}$. The Mantel statistics are related to the slope estimator of the regression $U_{ij} = \beta W_{ij} + e_{ij}$ (Figure 1.8):

$$\widehat{\beta} = \frac{M_2}{\sum_{i=1}^{n} \sum_{j=1}^{n} W_{ij}^2}.$$

The general approach to inference about the clustering, based on statistics such as (1.4) and (1.5), is to test whether the magnitude of the observed value of $M$ is unusual in the absence of spatial aggregation and to reject the hypothesis of no clustering if the $M$ statistic is sufficiently extreme. Four approaches are commonly used to perform this significance test. We focus on $M_2$ here and denote as $M_{2(obs)}$ the observed value of Mantel's statistic based on the data $Z(\mathbf{s}_1), \cdots, Z(\mathbf{s}_n)$.

- **Permutation Test**. Under the hypothesis of no clustering, the observed values $Z(\mathbf{s}_1), \cdots, Z(\mathbf{s}_n)$ can be thought of as a random assignment of values to spatial locations. Alternatively, we can imagine assigning the site coordinates $\mathbf{s}_1, \cdots, \mathbf{s}_n$ completely at random to the observed values. With $n$ sites, the number of possible assignments is finite: $n!$. Enumerating $M_2$ for the $n!$ arrangements yields its null distribution and the probability to exceed $M_{2(obs)}$ can be ascertained. Let $\mathrm{E}_r[M_2]$ and $\mathrm{Var}_r[M_2]$ denote the mean and variance of $M_2$ under this randomization approach.

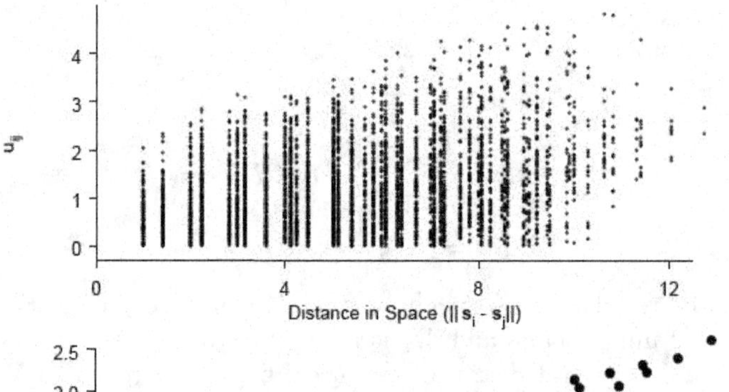

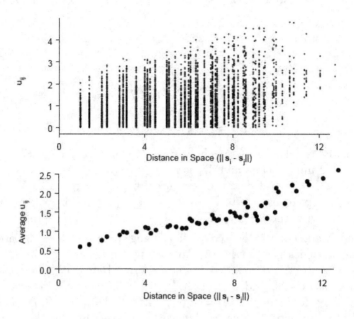

Figure 1.8 *Scatter plots of* $u_{ij} = |Z(s_i) - Z(s_j)|$ *vs.* $w_{ij} = ||s_i - s_j||$ *for the simulated lattice of Figure 1.1d. Bottom panel shows* $\overline{u}_{ij}$ *vs.* $w_{ij}$.

- **Monte Carlo Test.** Even for moderately large $n$, the number of possible random assignments of values to sites is large. Instead of a complete enumeration one can sample independently $k$ random assignments to construct the empirical null distribution of $M_2$. For a 5%-level test at least $k = 99$ samples are warranted and $k = 999$ is recommended for a 1%-level test. The larger $k$, the better the empirical distribution approximates the null distribution. Then, $M_{2(obs)}$ is again combined with the $k$ values for $M_2$ from the simulation and the relative rank of $M_{2(obs)}$ is computed. If this rank is sufficiently extreme, the hypothesis of no autocorrelation is rejected.

- **Asymptotic $Z$-Test With Gaussian Assumption.** The distribution of $M_2$ can also be determined—at least asymptotically—if the distribution of the $Z(\mathbf{s})$ is known. The typical assumption is that of a Gaussian distribution for the $Z(\mathbf{s}_1), \cdots, Z(\mathbf{s}_n)$ with common mean and common variance. Under the null distribution, $\text{Cov}[Z(\mathbf{s}_i), Z(\mathbf{s}_j)] = 0 \, \forall i \neq j$, and the mean and variance of $M_2$ can be derived. Denote these as $\text{E}_g[M_2]$ and $\text{Var}_g[M_2]$, since these moments may differ from those obtained under randomization. A test statistic

$$Z_{obs} = \frac{M_{2(obs)} - \text{E}_g[M_2]}{\sqrt{\text{Var}_g[M_2]}}$$

is formed and, appealing to the large-sample distribution of $M_2$, $Z_{obs}$ is compared against cutoffs from the $G(0,1)$ distribution. The spatial proximity weights $W_{ij}$ are considered fixed in this approach.

- **Asymptotic $Z$-Test Under Randomization.** The mean and variance of $M_2$ under randomization can also be used to formulate a test statistic

$$Z_{obs} = \frac{M_{2(obs)} - \mathrm{E}_r[M_2]}{\sqrt{\mathrm{Var}_r[M_2]}},$$

which follows approximately (for $n$ large) a $G(0,1)$ distribution under the null hypothesis. The advantage over the previous $Z$-test lies in the absence of a distributional assumption for the $Z(\mathbf{s}_i)$. The disadvantage of either $Z$-test is the reliance on an asymptotic distribution for $Z_{obs}$.

The Mantel statistics (1.4) and (1.5) are usually not applied in their *raw* form in practice. Most measures of spatial autocorrelation that are used in exploratory analysis are special cases of $M_1$ and $M_2$. These are obtained by considering particular structures for $W_{ij}$ and $U_{ij}$ and by scaling the statistics.

**Example 1.6** Knox (1964) performed a binary classification of the spatial distances $||\mathbf{h}_{ij}|| = ||\mathbf{s}_i - \mathbf{s}_j||$ and the temporal separation $|t_i - t_j|$ between leukemia cases in the study of the space-time clustering of childhood leukemia,

$$W_{ij} = \begin{cases} 1 & \text{if } ||\mathbf{s}_i - \mathbf{s}_j|| < 1 \text{ km} \\ 0 & \text{otherwise} \end{cases} \qquad U_{ij} = \begin{cases} 1 & \text{if } |t_i - t_j| < 59 \text{ days} \\ 0 & \text{otherwise.} \end{cases}$$

Let $\mathbf{W}^*$ denote the $(n \times n)$ matrix consisting of the upper triangular part of $\mathbf{W}$ with all other elements set to zero (and similarly for $\mathbf{U}^*$). The number of "close" pairs in space and time is then given by $M_1$, the number of pairs close in space is given by $S_w^* = \mathbf{1}'\mathbf{W}^*\mathbf{1}$, and the number of pairs close in time by $S_u^* = \mathbf{1}'\mathbf{U}^*\mathbf{1}$. A $(2 \times 2)$ contingency table can be constructed (Table 1.1). The expected cell frequencies under the hypothesis of no autocorrelation can be computed from the marginal totals and the usual Chi-square test can be deployed.

Table 1.1 $(2 \times 2)$ *Contingency table in Knox's double dichotomy*

| | Proximity in Time | | |
|---|---|---|---|
| In Space | Close | Far | Total |
| Close | $M_1$ | $S_w^* - M_1$ | $S_w^*$ |
| Far | $S_u^* - M_1$ | $\frac{n(n-1)}{2} - S_u^* - S_w^* + M_1$ | $n(n-1)/2 - S_w^*$ |
| Total | $S_u^*$ | $\frac{n(n-1)}{2} - S_u^*$ | $n(n-1)/2$ |

□

Other special cases of the Mantel statistics are the Black-White join count, Moran's $I$, and Geary's $c$ statistic. We now discuss these in the context of lattice data and then provide extensions to a continuous domain.

### *1.3.2 Measures on Lattices*

Lattice or regional data are in some sense the coarsest of the three spatial data types because they can be obtained from other types by spatial accumulation (integration). Counting the number of events in non-overlapping sets $A_1, \cdots, A_m$ of the domain $D$ in a point process creates a lattice structure. A lattice process can be created from a geostatistical process by integrating $Z(\mathbf{s})$ over the sets $A_1, \cdots, A_m$.

Key to analyzing lattice structures is the concept of spatial connectivity. Let $i$ and $j$ index two members of the lattice and imagine that $\mathbf{s}_i$ and $\mathbf{s}_j$ are point locations with which the lattice members are identified. For example, $i$ and $j$ may index two counties and $\mathbf{s}_i$ and $\mathbf{s}_j$ are the spatial locations of the county centroid or the seat of the county government. It is not necessary that each lattice member is associated with a point location, but spatial connectivity between sites is often expressed in terms of distances between "representative" points. With each pair of sites we associate a weight $w_{ij}$ which is zero if $i = j$ or if the two sites are not spatially connected. Otherwise, $w_{ij}$ takes on a non-zero value. (We use lowercase notation for the spatial weights because the domain is fixed for lattice data.) The simplest connectivity structure is obtained if the lattice consists of regular units. It is then natural to consider binary weights

$$w_{ij} = \begin{cases} 1 & \text{if sites } i \text{ and } j \text{ are connected} \\ 0 & \text{if sites } i \text{ and } j \text{ are not connected.} \end{cases} \tag{1.6}$$

Sites that are connected are considered spatial neighbors and you determine what constitutes connectedness. For regular lattices it is customary to draw on the moves that a respective chess piece can perform on a chess board (Figure 1.9a–c). For irregularly shaped areal units spatial neighborhoods can be defined in a number of ways. Two common approaches are shown in Figure 1.10 for counties of North Carolina. Counties are considered connected if they share a common border or if representative points within the county are less than a certain critical distance apart. The weight $w_{ij}$ assigned to county $j$, if it is a neighbor of county $i$, may be a function of other features of the lattice sites; for example, the length of the shared border, the relative sizes of the counties, etc. Symmetry of the weights is not a requirement. If housing prices are being studied and a small, rural county abuts a large, urban county, it is reasonable to assume that changes in the urban county have different effects on the rural county than changes in the rural environment have on the urban situation.

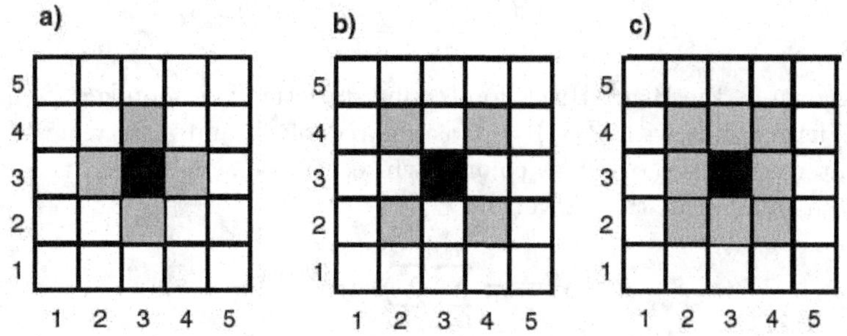


Figure 1.9 *Possible definitions of spatial connectedness (contiguity) for a regular lattice. Edges abut in the rook definition (a), corners touch in the bishop definition (b), edges and corners touch in the queen definition (c). Adapted from Figure 9.26 in Schabenberger and Pierce (2002).*

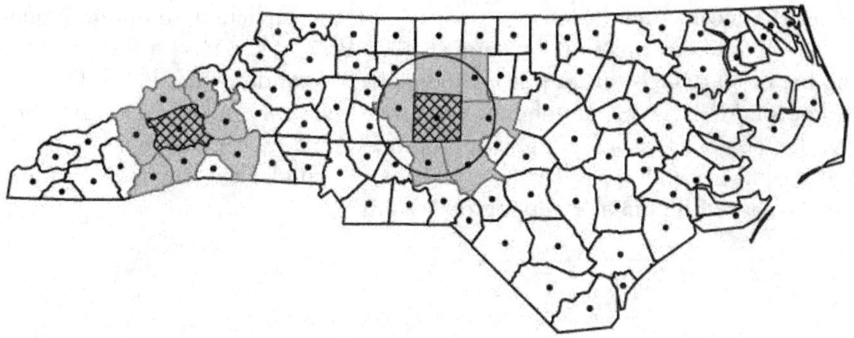

Figure 1.10 *Two definitions of spatial connectedness (contiguity) for an irregular lattice. The figure shows counties of North Carolina and two possible neighborhood definitions. The left neighborhood shows counties with borders adjoining that of the central, cross-hatched county. The neighborhood in the center of the map depicts counties whose centroid is within 35 miles of the central cross-hatched county.*

### 1.3.2.1 Black-White Join Counts

Measures of spatial autocorrelation that are of historical significance—and remain important exploratory tools in spatial data analysis to date—originated from the consideration of data on lattices. If the attribute $Z$ is binary, that is,

$$Z(\mathbf{s}_i) = \begin{cases} 1 & \text{if an event of interest occurred at site } i \\ 0 & \text{if the event did not occur at site } i, \end{cases}$$

a generalization of Mantel's statistic $M_2$ is

$$BB = \frac{1}{2}\sum_{i=1}^{n}\sum_{j=1}^{n} w_{ij}Z(\mathbf{s}_i)Z(\mathbf{s}_j), \tag{1.7}$$

*Black-Black Join Count*

known as the Black-Black join count statistic. The moniker stems from considering sites with $Z(\mathbf{s}_i) = 1$ as colored black and sites where the event of interest did not occur as colored white. By the same token, the Black-White join count statistic is given by

$$BW = \frac{1}{2}\sum_{i=1}^{n}\sum_{j=1}^{n} w_{ij}(Z(\mathbf{s}_i) - Z(\mathbf{s}_j))^2. \tag{1.8}$$

*Black-White Join Count*

A permutation approach to test the hypothesis of no spatial autocorrelation is possible by permuting the assignment of $n_1$ black and $n - n_1$ white cells to the lattice sites. Typically, inference based on $BB$ or $BW$ proceeds by assuming an asymptotic Gaussian distribution of

$$\frac{BB - \mathrm{E}[BB]}{\sqrt{\mathrm{Var}[BB]}} \quad \text{or} \quad \frac{BW - \mathrm{E}[BW]}{\sqrt{\mathrm{Var}[BW]}},$$

where the means and variances are derived for a particular sampling model. Under binomial sampling it is assumed that $\Pr(Z(\mathbf{s}_i) = 1) = \pi$ and that the number of black cells out of $n$ is a Binomial$(n, \pi)$ random variable. If the constraint is added that the number of black cells equals $n_1$, the observed number, then hypergeometric sampling is appropriate. Cliff and Ord (1981, Ch. 2) derive the means and variances for $BB$ and $BW$ under the two assumptions. For the case of binomial sampling one obtains

$$\begin{aligned}
\mathrm{E}[BB] &= \frac{1}{2}\pi^2\sum_{i=1}^{n}\sum_{j=1}^{n} w_{ij} \\
\mathrm{Var}[BB] &= \frac{1}{4}\pi^2(1-\pi)\left\{S_1(1-\pi) + S_2\pi\right\} \\
\mathrm{E}[BW] &= \pi(1-\pi)\sum_{i=1}^{n}\sum_{j=1}^{n} w_{ij} \\
\mathrm{Var}[BW] &= S_1\pi(1-\pi) + \frac{1}{4}S_2\pi(1-\pi)(1-4\pi(1-\pi)) \\
S_1 &= \frac{1}{2}\sum_{i=1}^{n}\sum_{j=1}^{n}(w_{ij} + w_{ji})^2 \\
S_2 &= \sum_{i=1}^{n}\left\{\sum_{j=1}^{n} w_{ij} + \sum_{j=1}^{n} w_{ji}\right\}^2.
\end{aligned}$$

Whereas for binomial sampling we have $\Pr(Z(\mathbf{s}_i) = 1) = \pi$ and $\Pr(Z(\mathbf{s}_i) = 1, Z(\mathbf{s}_j) = 1) = \pi^2$ under the null hypothesis of no autocorrelation, for hypergeometric sampling with $n_1$ black cells the corresponding results are

$\Pr(Z(\mathbf{s}_i) = 1) = n_1/n$ and $\Pr(Z(\mathbf{s}_i) = 1, Z(\mathbf{s}_j) = 1) = n_1(n_1-1)/\{n(n-1)\}$. Using the notation

$$P_{(k)} = \frac{n_1 \times (n_1 - 1) \times \cdots \times (n_1 - k + 1)}{n \times (n-1) \times \cdots \times (n - k + 1)},$$

and putting $w_{..} = \sum_{i,j} w_{ij}$, the mean and variance of the $BB$ statistic under hypergeometric sampling are (Cliff and Ord, 1981, (2.27), p. 39)

$$\begin{aligned} \mathrm{E}[BB] &= \frac{1}{2}P_{(2)}\sum_{i=1}^{n}\sum_{j=1}^{n} w_{ij} \\ \mathrm{Var}[BB] &= \frac{S_1}{4}\left[P_{(2)} - 2P_{(3)} + P_{(4)}\right] + \frac{S_2}{4}\left[P_{(3)} - P_{(4)}\right] \\ &\quad + \frac{w_{..}^2}{4}P_{(4)} - \frac{1}{4}\left[w_{..}P_{(2)}\right]^2. \end{aligned}$$

*1.3.2.2 The Geary and Moran Statistics*

The $BB$ statistic is a special case of the Mantel statistic with $Z(\mathbf{s}_i)$ binary and $U_{ij} = Z(\mathbf{s}_i)Z(\mathbf{s}_j)$. If $Z$ is a continuous attribute whose mean is not spatially varying, i.e., $\mathrm{E}[Z(\mathbf{s})] = \mu$, then closeness of attributes at sites $\mathbf{s}_i$ and $\mathbf{s}_j$ can be expressed by various measures:

$$U_{ij} = (Z(\mathbf{s}_i) - \mu)(Z(\mathbf{s}_j) - \mu) \quad (1.9)$$

$$U_{ij} = (Z(\mathbf{s}_i) - \overline{Z})(Z(\mathbf{s}_j) - \overline{Z}) \quad (1.10)$$

$$U_{ij} = |Z(\mathbf{s}_i) - Z(\mathbf{s}_j)| \quad (1.11)$$

$$U_{ij} = (Z(\mathbf{s}_i) - Z(\mathbf{s}_j))^2. \quad (1.12)$$

The practically important measures are (1.10) and (1.12) because of mathematical tractability and interpretation. If $\overline{Z}$ is consistent for $\mu$, then $\mathrm{E}[(Z(\mathbf{s}_i)-\overline{Z})(Z(\mathbf{s}_j)-\overline{Z})]$ is a consistent estimator of $\mathrm{Cov}[Z(\mathbf{s}_i), Z(\mathbf{s}_j)]$. It is tempting to scale this measure by an estimate of $\sigma^2$. If the $Z(\mathbf{s}_i)$ were a random sample, then $S^2 = (n-1)^{-1}\sum_{i=1}^{n}(Z(\mathbf{s}_i) - \overline{Z})^2$ would be the appropriate estimator. In case of (1.10) this leads us to consider

$$(n-1)\frac{(Z(\mathbf{s}_i) - \overline{Z})(Z(\mathbf{s}_j) - \overline{Z})}{\sum_{i=1}^{n}(Z(\mathbf{s}_i) - \overline{Z})^2}, \quad (1.13)$$

as an "estimator" of the correlation between $Z(\mathbf{s}_i)$ and $Z(\mathbf{s}_j)$. It is not a sensible estimator, however, its statistical properties are poor. Because it combines information about covariation (numerator) with information about variation it is a reasonable estimator. Combining the contributions (1.13) over all pairs of locations, introducing the spatial connectivity weights $w_{ij}$, and scaling, yields the statistic known as Moran's $I$, first discussed in Moran (1950): *Moran's I*

$$I = \frac{n}{(n-1)S^2\, w_{..}}\sum_{i=1}^{n}\sum_{j=1}^{n} w_{ij}(Z(\mathbf{s}_i) - \overline{Z})(Z(\mathbf{s}_j) - \overline{Z}). \quad (1.14)$$

An autocorrelation measure based on (1.12) was constructed by Royaltey, Astrachan, and Sokal (1975). Note that if $\text{Var}[Z(\mathbf{s})] = \sigma^2$, then $\text{E}[(Z(\mathbf{s}_i) - Z(\mathbf{s}_j))^2] = \text{Var}[Z(\mathbf{s}_i) - Z(\mathbf{s}_j)] = 2\sigma^2 - 2\text{Cov}[Z(\mathbf{s}_i), Z(\mathbf{s}_j)]$. Scaling and combining local contributions based on (1.12) leads to the statistic known as Geary's $c$ (Geary, 1954):

*Geary's c*

$$c = \frac{1}{2S^2 w_{..}} \sum_{i=1}^{n} \sum_{j=1}^{n} w_{ij}(Z(\mathbf{s}_i) - Z(\mathbf{s}_j))^2. \tag{1.15}$$

As with the general Mantel statistic, inference for the $I$ and $c$ statistic can proceed via permutation tests, Monte-Carlo tests, or approximate tests based on the asymptotic distribution of $I$ and $c$. To derive the mean and variance of $I$ and $c$ one either proceeds under the assumption that the $Z(\mathbf{s}_i)$ are Gaussian random variables with mean $\mu$ and variance $\sigma^2$ or derives the mean and variance under the assumption of randomizing the attribute values to the lattice sites. Either assumption yields the same mean values,

$$\text{E}_g[I] = \text{E}_r[I] = -\frac{1}{n-1}$$
$$\text{E}_g[c] = \text{E}_r[c] = 1,$$

but expressions for the variances under Gaussianity and randomization differ. The reader is referred to the formulas in Cliff and Ord (1981, p. 21) and to Chapter problem 1.8.

The interpretation of the Moran and Geary statistic is as follows. If $I > \text{E}[I]$, then a site tends to be connected to sites that have similar attribute values. The spatial autocorrelation is positive and increases in strength with $|I - \text{E}[I]|$. If $I < \text{E}[I]$, attribute values of sites connected to a particular site tend to be dissimilar. The interpretation of the Geary coefficient is opposite. If $c > \text{E}[c]$, sites are connected to sites with dissimilar values and vice versa for $c < \text{E}[c]$.

The assumptions of constant mean and constant variance of the $Z(\mathbf{s}_i)$ must not be taken lightly when testing for spatial autocorrelation with these statistics. Values in close spatial proximity may be similar not because of spatial autocorrelation but because the values are independent realizations from distributions with similar mean. Values separated in space may appear dissimilar if the mean of the random field changes.

**Example 1.7** Independent draws from a $G(\mu(x,y), \sigma^2)$ distribution were assigned to the sites of a $10 \times 10$ regular lattice. The mean function of the lattice sites is $\mu(x,y) = 1.4 + 0.1x + 0.2y + 0.002x^2$. These data do not exhibit any spatial autocorrelation. They are equi-dispersed but not mean-stationary. Calculating the moments of the Moran's I statistic with a rook definition for the spatial contiguity weights $w_{ij}$ leads to the following results.

| Assumption | $I$ | $\text{E}[I]$ | $\sqrt{\text{Var}[I]}$ | $Z_{obs}$ | $p$-value |
|---|---|---|---|---|---|
| $Z(\mathbf{s}_i) \sim G(\mu, \sigma^2)$ | 0.2597 | −0.0101 | 0.0731 | 3.690 | 0.00011 |
| Randomization | 0.2597 | −0.0101 | 0.0732 | 3.686 | 0.00011 |

The statistic is sensitive to changes in the mean function, $I$ "detected" spurious autocorrelation.

While this simple example serves to make the point, the impact of heterogeneous means and variances on the interpretation of Moran's $I$ is both widely ignored and completely confused throughout the literature. McMillen (2003) offers perhaps the first correct assessment of this problem (calling it model misspecification). Waller and Gotway (2004) discuss this problem at length and provide several practical illustrations in the context of spatial epidemiology. □

Because it is often not reasonable to assume constancy of the mean over a larger domain, two courses of action come to mind.

- Fit a mean model to the data and examine whether the residuals from the fit exhibit spatial autocorrelation. That is, postulate a linear model $\mathbf{Z}(\mathbf{s}) = \mathbf{X}(\mathbf{s})\boldsymbol{\beta} + \mathbf{e}$ and obtain the least squares estimate of the $(p \times 1)$ vector $\boldsymbol{\beta}$. Since testing for spatial autocorrelation is usually part of the exploratory stages of spatial data analysis, one has to rely on ordinary least squares estimation at this stage. Then compute the Moran or Geary statistic of the residuals $\widehat{e}_i = Z(\mathbf{s}_i) - \mathbf{x}'(\mathbf{s}_i)\widehat{\boldsymbol{\beta}}$. In matrix-vector form the $I$ statistic can be written as

*Moran's I for OLS Residuals*

$$I_{res} = \frac{n}{w_{..}} \frac{\widehat{\mathbf{e}}'\mathbf{W}\widehat{\mathbf{e}}}{\widehat{\mathbf{e}}'\widehat{\mathbf{e}}}, \tag{1.16}$$

  where $\widehat{\mathbf{e}} = \mathbf{M}\mathbf{e}$ and $\mathbf{M} = \mathbf{I} - \mathbf{X}(\mathbf{X}'\mathbf{X})^{-1}\mathbf{X}'$. The mean of Moran's $I$ based on OLS residuals is no longer $-(n-1)^{-1}$. Instead, we have

$$\mathrm{E}_g[I_{res}] = \frac{n}{(n-k)w_{..}}\mathrm{tr}[\mathbf{MW}].$$

- Even if the mean changes globally throughout the domain it may be reasonable to assume that it is locally constant. The calculation of autocorrelation measures can then be localized. This approach gives rise to so-called *LISA*s, local indicators of spatial autocorrelation.

### *1.3.3 Localized Indicators of Spatial Autocorrelation*

Taking another look at Mantel's $M_2$ statistic for matrix association in §1.3.1 reveals that the statistic can be written as a sum of contributions of individual data points:

$$M_2 = \sum_{i=1}^{n}\sum_{j=1}^{n} W_{ij}U_{ij} = \sum_{i=1}^{n} M_i.$$

The contribution $M_i = \sum_{j=1}^{n} W_{ij}U_{ij}$ can be thought of as the autocorrelation measure associated with the $i$th site, a local measure. Since the Black-White join count, Moran's $I$ and Geary's $c$ statistics are special cases of $M_2$, they can be readily localized. The concept of local indicators of spatial autocorrelation

*LISAs*

(LISAs) was advanced by Anselin (1995), who defined a statistic to be a LISA if

1. it measures the extent of spatial autocorrelation around a particular observation,
2. it can be calculated for each observation in the data, and
3. if its sum is proportional to a global measure of spatial autocorrelation such as $BB$, $c$, or $I$.

*Local Moran's I*

A local version of Moran's $I$ that satisfies these requirements is

$$I(\mathbf{s}_i) = \frac{n}{(n-1)S^2}(Z(\mathbf{s}_i) - \overline{Z})\sum_{j=1}^{n} w_{ij}(Z(\mathbf{s}_j) - \overline{Z}), \tag{1.17}$$

so that $\sum_{i=1}^{n} I(\mathbf{s}_i) = w_{..}I$. Anselin (1995) derives the mean and variance of $I(\mathbf{s}_i)$ under the randomization assumption. For example,

$$\mathrm{E}_r[I(\mathbf{s}_i)] = \frac{-1}{n-1}\sum_{j=1}^{n} w_{ij}.$$

The interpretation of $I(\mathbf{s}_i)$ derives directly from the interpretation of the global statistic. If $I(\mathbf{s}_i) > \mathrm{E}[I(\mathbf{s}_i)]$, then sites connected to the $i$th site exhibit attribute values similar to $Z(\mathbf{s}_i)$. There is locally positive autocorrelation among the values; large (small) values tend to be surrounded by large (small) values. If $I(\mathbf{s}_i) < \mathrm{E}[I(\mathbf{s}_i)]$, then sites connected to the $i$th site have dissimilar values. Large (small) values tend to be surrounded by small (large) values. Local indicators of spatial autocorrelation are useful exploratory tools to:

(i) Detect regions in which the autocorrelation pattern is markedly different from other areas in the domain.

(ii) Detect local spatial clusters (hot spots). There are many definitions of hot spots and cold spots. One such definition focuses on the histogram of the data and identifies those sites as hot (cold) spots where $Z(\mathbf{s}_i)$ is extreme. LISAs enable hot spot definitions that take into account the surrounding values of a site. A hot spot can be defined in this context as one or several contiguous sites where the local indicators are unusually large or small.

(iii) Distinguish spatial outliers from distributional outliers. A value unusual in a box plot of the $Z(\mathbf{s}_i)$ is a distributional outlier. Data points that are near the center of the sample distribution do not raise flags or concerns in this box plot. But these values can be unusual relative to the neighboring sites. Anselin (1995) suggests comparing the distribution of $I(\mathbf{s}_i)$ to $I/n$ to determine outliers and/or leverage points in the data.

The exact distributional properties of the autocorrelation statistics are elusive, even in the case of a Gaussian random field. The Gaussian approximation tends to work well, but the same cannot necessarily be said for the local statistics. Anselin (1995) recommends randomization inference as for the global statistic with the added feature that the randomization be conditional; the attribute value at site $i$ is not subject to permutation. The implementation of the permutation approach can be accelerated in the local approach, since only values in the neighborhood need to be permuted. However, Besag and Newell (1991) and Waller and Gotway (2004) note that when the data have heterogeneous means or variances, a common occurrence with count data and proportions, the randomization assumption is inappropriate. In some cases, the entire concept of permutation makes little sense. Instead they recommend the use of Monte Carlo testing.

While localized measures of autocorrelation can be excellent exploratory tools, they can also be difficult to interpret. Moreover, they simply cannot be used as confirmatory tools. On a lattice with $n$ sites one obtains $n$ local measures, and could perform $n$ tests of significant autocorrelation. This is a formidable multiplicity problem. Even in the absence of autocorrelation, the $I(\mathbf{s}_i)$ are correlated if they involve the same sites, and it is not clear how to adjust individual Type-I error levels to maintain a desired overall level.

## 1.4 Autocorrelation Functions

### *1.4.1 The Autocorrelation Function of a Time Series*

Autocorrelation is the correlation among the family of random variables that make up a stochastic process. In time series, this form of correlation is often referred to as serial correlation. Consider a (weakly) stationary time series $Z(t_1), \cdots, Z(t_n)$ with $\mathrm{E}[Z(t_i)] = 0$ and $\mathrm{Var}[Z(t_i)] = \sigma^2$, $i = 1, \cdots, n$. Rather than a single measure, autocorrelation in a time series is measured by a function of the time points. The covariance function of the series at points $t_i$ and $t_j$ is given by

$$\mathrm{Cov}[Z(t_i), Z(t_j)] = \mathrm{E}[Z(t_i)Z(t_j)] = C(t_j - t_i), \tag{1.18}$$

and the (auto)correlation function is then

$$R(t_j - t_i) = \frac{\mathrm{Cov}[Z(t_i), Z(t_j)]}{\sqrt{\mathrm{Var}[Z(t_i)]\mathrm{Var}[Z(t_j)]}} = \frac{C(t_j - t_i)}{C(0)}. \tag{1.19}$$

Figure 1.11 shows the realizations of two stochastic processes. Open circles represent an independent sequence of random variables with mean zero and variance 0.3. Closed circles represent a sequence of random variables with mean zero, variance 0.3, and autocorrelation function $R(t_j - t_i) = \rho^{|t_j - t_i|}, \rho > 0$. The positive autocorrelation is reflected in the fact that runs of positive residuals alternate with runs of negative residuals. In other words, if an observation at time $t$ is above (below) average, it is very likely that an observation in the immediate past was also above (below) average. Positive autocorrelation in time series or spatial data are much more common than negative autocorrelation. The latter is often an indication of an improperly specified mean function, e.g., the process exhibits deterministic periodicity which has not been properly accounted for.

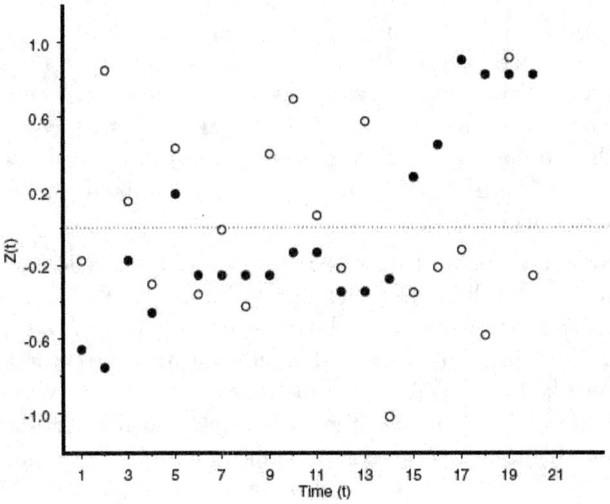


Figure 1.11 *A sequence of independent observations (open circles) and of correlated data (AR(1) process, closed circles),* $\sigma^2 = 0.3$.

### *1.4.2 Autocorrelation Functions in Space—Covariance and Semivariogram*

The autocovariance or autocorrelation function of a time series reflects the notion that values in a time series are not unrelated. If the serial correlation is positive, a high value at time $t$ is likely to be surrounded by high values at times $t-1$ and $t+1$. In the spatial case, autocorrelation is reflected by the fact that values at locations $\mathbf{s}_i$ and $\mathbf{s}_j$ are stochastically dependent. If this correlation is positive, we expect high (low) values to be surrounded by high (low) values. If it is negative, high (low) values should be surrounded by low (high) values. Tobler's *first law of geography* states that "everything is related to everything else, but near things are more related than distant things" (Tobler, 1970). It reflects an additional fact common to many statistical models for spatial data: correlations decrease with increasing spatial separation.

The **covariance function** of a spatial process is defined as

$$C(\mathbf{s},\mathbf{h}) = \mathrm{Cov}[Z(\mathbf{s}), Z(\mathbf{s}+\mathbf{h})] = \mathrm{E}[\{Z(\mathbf{s}) - \mu(\mathbf{s})\}\{(Z(\mathbf{s}+\mathbf{h}) - \mu(\mathbf{s}+\mathbf{h})\}]$$

and the autocorrelation function (or correlogram) is

$$R(\mathbf{s},\mathbf{h}) = \frac{C(\mathbf{h})}{\sqrt{\mathrm{Var}[Z(\mathbf{s})]\mathrm{Var}[Z(\mathbf{s}+\mathbf{h})]}}.$$

These definitions are obvious extensions of the covariance and correlation function for time series (equations (1.18) & (1.19)). The differences between temporal and spatial data are worth noting, however, because they are in part responsible for the facts that (i) many statistical methods for spatial data analysis have developed independently of developments in time series analysis, (ii) that spatial adaptations of time series models face particular challenges. First, time is directed; it flows from the past through the present into the future. Space is not directed in the same sense. There is no equivalent to past, present, or future. Second, edge effects are important for time series models, but much more so for spatial models. A time series has a starting and an end point. A configuration in a spatial domain has a bounding volume with potentially many points located close to the boundary. In time series applications it is customary to focus on serial correlations between the past and current values. In spatial applications we have to be concerned with correlations in all directions of the domain. In addition, the spatial dependencies may develop differently in various directions. This issue of **anisotropy** of a spatial process has no parallel in time series analysis.

Like time series analysis, however, many statistical methods of spatial data analysis make certain assumptions about the behavior of the covariance function $C(\mathbf{h})$, known as stationarity assumptions. We will visit various types of stationarity and their implications for the spatial process in detail in §2.2 and the reasons for the importance of stationarity are enunciated in §2.1. For now, it is sufficient to note that a commonly assumed type of stationarity, **second-order stationarity**, implies that the spatial process $Z(\mathbf{s})$ has a constant mean $\mu$ and a covariance function which is a function of the spatial separation between points only,

$$\mathrm{Cov}[Z(\mathbf{s}), Z(\mathbf{s}+\mathbf{h})] = C(\mathbf{h}).$$

Second-order stationarity implies the lack of importance of absolute coordinates. In the time series context it means that the covariance between two values separated by two days is the same, whether the first day is a Wednesday, or a Saturday. Under second-order stationarity, it follows directly that $\mathrm{Cov}[Z(\mathbf{s}), Z(\mathbf{s}) + \mathbf{0}] = C(\mathbf{0})$ does not depend on the spatial location. Consequently, $\mathrm{Var}[Z(\mathbf{s})] = \sigma^2$ is not a function of spatial location. The variability of the spatial process is the same everywhere. While we need to be careful to understand what random mechanism the expectation and variance of $Z(\mathbf{s})$ refer to (see §2.1), it emerges that second-order stationarity is the spatial equivalent of the *iid* random sample assumption in classical statistic.

Under second-order stationarity, the covariance and correlation function of the spatial process simplify, they no longer depend on $\mathbf{s}$:

$$\begin{aligned} \mathrm{Cov}[Z(\mathbf{s}), Z(\mathbf{s}+\mathbf{h})] &= C(\mathbf{h}) \\ R(\mathbf{h}) &= \frac{C(\mathbf{h})}{\sigma^2}. \end{aligned}$$

Expressing the second-order structure of a spatial process in terms of the covariance or the correlation function seems to be equivalent. This is not necessarily so; qualitatively different processes can have the same correlation functions (see Chapter problems).

Statisticians are accustomed to describe stochastic dependence between random variables through covariances or correlations. Originating in geostatistical applications that developed outside the mainstream of statistics, a common tool to capture the second-moment structure in spatial data is the **semivariogram**. The reasons to prefer the semivariogram over the covariance function in statistical applications are explored in §4.2. At this point we focus on the definition of the semivariogram, namely

$$\begin{aligned}\gamma(\mathbf{s},\mathbf{h}) &= \frac{1}{2}\mathrm{Var}[Z(\mathbf{s}) - Z(\mathbf{s}+\mathbf{h})] \qquad (1.20)\\ &= \frac{1}{2}\left\{\mathrm{Var}[Z(\mathbf{s})] + \mathrm{Var}[Z(\mathbf{s}+\mathbf{h})] - 2\mathrm{Cov}[Z(\mathbf{s}), Z(\mathbf{s}+\mathbf{h})]\right\}.\end{aligned}$$

If the process is second-order stationary, then it follows immediately that

$$\gamma(\mathbf{h}) = \frac{1}{2}\left\{2\sigma^2 - 2C(\mathbf{h})\right\} = C(\mathbf{0}) - C(\mathbf{h}). \qquad (1.21)$$

On these grounds the semivariogram and the covariance function contain the same information, and can be used interchangeably. For second-order stationary spatial processes with positive spatial correlation, the semivariogram has a characteristic shape. Since $\gamma(\mathbf{0}) = C(\mathbf{0}) - C(\mathbf{0}) = 0$, the semivariogram passes through the origin and must attain $\gamma(\mathbf{h}^*) = \sigma^2$ if the lag exceeds the distance $\mathbf{h}^*$ for which data points are correlated. Under Tobler's first law of geography, we expect $C(\mathbf{h})$ to decrease as $\mathbf{h}$ increases. The semivariogram then transitions smoothly from the origin to the **sill** $\sigma^2$ (Figure 1.12). The manner in which this transition takes place is important for the smoothness and continuity properties of the spatial process (see §2.3). The lag distance $h^* = ||\mathbf{h}^*||$, for which $C(h^*) = 0$, is termed the **range** of the spatial process. If the semivariogram achieves the sill only asymptotically, then the **practical range** is defined as the lag distance at which the semivariogram achieves 95% of the sill.

Geostatisticians are also concerned with semivariograms that do not pass through the origin. The magnitude of the discontinuity of the origin is termed the **nugget** effect after one of the possible explanations of the phenomenon: small nuggets of ore are distributed throughout a larger body of rock. This microscale variation creates a spatial process with sill $\theta_0$. The spatial structure of the microscale process cannot be observed unless the spacing of the sample observations is made smaller. Whereas nugget effects are commonly observed in sample semivariograms calculated from data, its very existence must be called into question if the spatial process is believed to be mean square continuous (§2.3). The second source for a possible nugget effect in a sample semivariogram is measurement error.

The semivariogram is not only a descriptive tool, it is also a structural tool through which the second-order properties of a spatial process can be studied. Because of its importance in spatial statistics, in particular for methods of spatial prediction, Chapter 4 is dedicated in its entirety to semivariogram analysis and estimation.

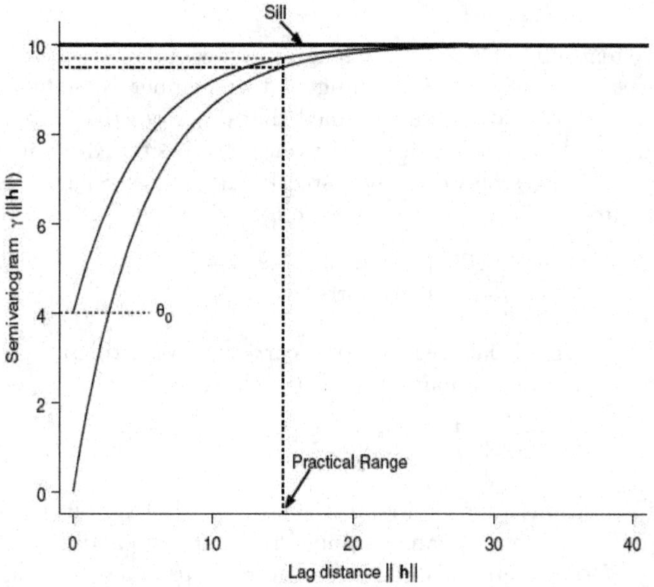


Figure 1.12 *Exponential semivariogram with sill* 10 *and practical range* 15. *Semivariogram not passing through the origin has a nugget effect of* $\theta_0 = 4$.

### *1.4.3 From Mantel's Statistic to the Semivariogram*

The measures of spatial association in §1.3.2 define spatial similarity (contiguity) through the weight matrix $\mathbf{W}$ and result in a single statistic that describes the extent to which data are spatially autocorrelated for the entire domain. The notion of spatial closeness comes into play through the $w_{ij}$, but the actual distances between sites are not directly used. First, this requires that each lattice site is identified with a representative point $\mathbf{s}_i$ in $\mathbb{R}^d$. Second, there is a finite set of distances for which the degree of spatial dependence can be investigated if the data are on a lattice. Consider a regular, rectangular lattice. We could define more than one set of neighborhood weights. For example, let $w_{ij}^{(1)}$ be first-order weights based on a queen definition as in Figure 1.9c) and let $w_{ij}^{(2)}$ be the second-order weights based on a queen's move. Then $w_{ij}^{(2)} = 1$ if site $j$ can be reached from site $i$ by a queen's move that passes over at least one additional tile on the chess board. The two statistics

$$M_1^{(1)} = \sum_{i=1}^{n-1} \sum_{j=i+1}^{n} w_{ij}^{(1)} U_{ij}$$

$$M_1^{(2)} = \sum_{i=1}^{n-1} \sum_{j=i+1}^{n} w_{ij}^{(2)} U_{ij}$$

measure the extent of the first-order-neighbor and second-order-neighbor spatial dependence, respectively. Continuing in this manner by extending the neighborhood, a set of statistics is obtained that describes the degree of spatial dependence with increasing distance from a given site. Alternatively, we can involve the distances between sites directly and define weights that yield a measure of autocorrelation at lag distance $h$:

$$w_{ij}(h) = \begin{cases} 1 & \text{if } ||\mathbf{s}_i - \mathbf{s}_j|| = h \\ 0 & \text{otherwise.} \end{cases}$$

For irregular (or small) lattices we can incorporate a lag tolerance $\epsilon > 0$ to ensure a sufficient number of pairs in each lag class:

$$w_{ij}(h, \epsilon) = \begin{cases} 1 & \text{if } h - \epsilon \leq ||\mathbf{s}_i - \mathbf{s}_j|| < h + \epsilon \\ 0 & \text{otherwise.} \end{cases}$$

These weight definitions are meaningful whenever an observation is associated with a point in $\mathbb{R}^d$, hence they can be applied to geostatistical data. Two common choices for $U_{ij}$ are squared differences and cross-products. The similarity statistics can be re-written as

$$M_1^* = \sum_{N(h)} (Z(\mathbf{s}_i) - Z(\mathbf{s}_j))^2$$

$$M_1^{**} = \sum_{N(h)} (Z(\mathbf{s}_i) - \overline{Z})(Z(\mathbf{s}_j) - \overline{Z}),$$

where $N(h)$ is the set of site pairs that are lag distance $h$ (or $h \pm \epsilon$) apart. The cardinality of this set is $|N(h)| = \sum_{i=1}^{n-1} \sum_{j=i+1}^{n} w_{ij}(h)$ (or $\sum\sum w_{ij}(h, \epsilon)$). If the random field has constant mean $\mu$, then $(Z(\mathbf{s}_i) - Z(\mathbf{s}_j))^2$ is an unbiased estimator of the variogram $2\gamma(\mathbf{s}_i, \mathbf{s}_j - \mathbf{s}_i)$. If the random field has furthermore constant variance $\sigma^2$, then $(Z(\mathbf{s}_i) - Z(\mathbf{s}_j))^2$ estimates $2(\sigma^2 - C(\mathbf{s}_i - \mathbf{s}_j))$. This suggests to estimate the semivariogram at lag vector $\mathbf{h}$ as

$$\widehat{\gamma}(\mathbf{h}) = \frac{1}{2|N(\mathbf{h})|} \sum_{N(\mathbf{h})} \left\{Z(\mathbf{s}_i) - Z(\mathbf{s}_j)\right\}^2.$$

This estimator, due to Matheron (1962, 1963), is termed the classical semivariogram estimator and a plot of $\widehat{\gamma}(\mathbf{h})$ versus $||\mathbf{h}||$ is known as the **empirical semivariogram**. Similarly,

$$\widehat{C}(\mathbf{h}) = \frac{1}{|N(\mathbf{h})|} \sum_{N(\mathbf{h})} (Z(\mathbf{s}_i) - \overline{Z})(Z(\mathbf{s}_j) - \overline{Z})$$

is known as the empirical covariance estimator. The properties of these estimators, their respective merits and demerits, and their utility for describing and modeling geostatistical data are explored in detail in Chapter 4.

## 1.5 The Effects of Autocorrelation on Statistical Inference

Since much of classical applied statistics rests on the assumption of *iid* observations, you might be tempted to consider correlation in the data as a nuisance at first. It turns out, however, (i) that ignoring autocorrelation has serious implications for the ensuing statistical inference, and (ii) that correlation in the data can be employed to your benefit. To demonstrate these points, we consider the following simple problem.

Assume that observations $Y_1, \cdots, Y_n$ are Gaussian distributed with mean $\mu_y$, variance $\sigma^2$ and covariances $\text{Cov}[Y_i, Y_j] = \sigma^2\rho \; (i \neq j)$. A second sample of size $n$ for variable $X$ has similar properties. $X_i \sim G(\mu_x, \sigma^2)$, $(i = 1, \cdots, n)$, $\text{Cov}[X_i, X_j] = \sigma^2\rho \, (i \neq j)$. The samples for $Y$ and $X$ are independent, $\text{Cov}[Y_i, X_j] = 0 \; (\forall i, j)$. Ignoring the fact that the $Y_i$ are correlated, one might consider the sample mean $n^{-1}\sum_{i=1}^{n} Y_i$ as the "natural" estimator of $\mu$. Some straightforward manipulations yield

$$\begin{aligned}
\text{Var}\left[\overline{Y}\right] &= \frac{1}{n^2}\sum_{i=1}^{n}\sum_{j=1}^{n}\text{Cov}[Y_i, Y_j] \\
&= \frac{1}{n^2}\left\{n\sigma^2 + n(n-1)\sigma^2\rho\right\} \\
&= \frac{\sigma^2}{n}\{1 + (n-1)\rho\}.
\end{aligned}$$

Assume that $\rho > 0$, so that $\text{Var}[\overline{Y}] > \sigma^2/n$; the sample mean is more dispersed than in a random sample. More importantly, we note that $\text{E}[\overline{Y}] = \mu_y$, regardless of the correlations, but that

$$\lim_n \text{Var}[\overline{Y}] = \sigma^2\rho.$$

The sample mean is *not* a consistent estimator of the population mean $\mu$. That is bad news.

A test of the hypothesis $H_0\colon \mu_x = \mu_y$ that proceeds as if the data were uncorrelated would use test statistic

$$Z^*_{obs} = \frac{\overline{Y} - \overline{X}}{\sigma\sqrt{2/n}},$$

whereas the correct test statistic would be

$$Z_{obs} = \frac{\overline{Y} - \overline{X}}{\sigma\sqrt{2\{1 + (n-1)\rho\}/n}} < Z^*_{obs}.$$

The test statistic $Z^*_{obs}$ does not account for the autocorrelation and is too large, $p$-values are too small, the evidence in the data against the null hypothesis is overstated. The test rejects more often than it should.

The effect of positive autocorrelation is that $n$ correlated observations do not provide the same amount of information than $n$ uncorrelated observations. Cressie (1993, p. 15) approaches this problem by asking "How many samples of the uncorrelated kind provide the same precision as a sample of correlated observations?" If $n$ denotes the number of correlated samples and $n'$ the number of uncorrelated samples, the **effective sample size** is calculated as

*Effective Sample Size*

$$n' = \frac{n}{1 + (n-1)\rho}. \tag{1.22}$$

To draw the conclusion from this demonstration that autocorrelation of data is detrimental would be incorrect. What the exercise conveys is that *ignoring* correlations and relying on the statistics known to perform well in *iid* samples is detrimental. How can the apparent loss of power in testing $H_0{:}\, \mu_x = \mu_y$ be recovered? For one, $\overline{X}$ and $\overline{Y}$ are not the most efficient estimators of $\mu_x$ and $\mu_y$ in this problem. The generalized least squares estimator

$$\widehat{\mu}_y = (\mathbf{1}'\mathbf{\Sigma}^{-1}\mathbf{Y})/(\mathbf{1}'\mathbf{\Sigma}^{-1}\mathbf{1}),$$

where $\mathbf{\Sigma} = \sigma^2\{(1-\rho)\mathbf{I} + \rho\mathbf{J}\}$, should be used instead of $\overline{Y}$. Test statistics should be derived based on $\widehat{\mu}_y - \widehat{\mu}_x$.

### 1.5.1 *Effects on Prediction*

Autocorrelations must be accounted for to achieve viable inferences. In other situations, the very presence of autocorrelations strengthens statistical abilities. An important case in point is the prediction of random variables. Consider again the simple model

$$\mathbf{Y} = \mathbf{1}\mu + \mathbf{e},$$

with $\mathbf{Y} = [Y_1, \cdots, Y_n]'$, $\mathbf{1} = [1, \cdots, 1]'$, $\mathbf{e} = [e_1, \cdots, e_n]'$, $\mathrm{E}[\mathbf{e}] = \mathbf{0}$, and

*Compound Symmetry, Exchangeable Correlations*

$$\mathrm{Var}[\mathbf{e}] = \mathbf{\Sigma} = \sigma^2 \begin{bmatrix} 1 & \rho & \rho & \cdots & \rho \\ \rho & 1 & \rho & \cdots & \rho \\ \rho & \rho & 1 & \cdots & \rho \\ \vdots & \cdots & \cdots & \cdots & \vdots \\ \rho & \cdots & \cdots & \ddots & 1 \end{bmatrix} = \sigma^2\{(1-\rho)\mathbf{I} + \rho\mathbf{J}\}.$$

This structure is termed the equicorrelation, compound symmetry, or exchangeable correlation structure. It arises naturally in situations with hierarchical random effects, e.g., models for split-plot designs or experiments involving sub-sampling. The compound symmetry structure is not commonly used in spatial statistics to model autocorrelations. It is not a reasonable correlation model for most spatial data since it does not take into account the spatial configuration. We select it here because the simple form of $\mathbf{\Sigma}$ enables us to carry out the manipulations that follow in closed form; $\mathbf{\Sigma}^{-1}$ is easily obtained.

Imagine that the prediction of a new observation $Y_0$ is of interest. Since the observed data are correlated, it is reasonable to assume that the new observation is also correlated with $\mathbf{Y}$; $\mathrm{Cov}[\mathbf{Y}, Y_0] = \mathbf{c} = \sigma^2\rho\mathbf{1}$. To find a suitable predictor $p(Y_0)$, certain restrictions are imposed. We want the predictor to be linear in the observed data, i.e., $p(Y_0) = \boldsymbol{\lambda}'\mathbf{Y}$. Furthermore, the predictor should be unbiased in the sense that $\mathrm{E}[p(Y_0)] = \mathrm{E}[Y_0]$. This constraint implies that $\boldsymbol{\lambda}'\mathbf{1} = 1$. If the measure of prediction loss is squared-error, we are led to the minimization of

$$\mathrm{E}[(p(Y_0) - Y_0)^2] \quad \text{subject to } \boldsymbol{\lambda}'\mathbf{1} = 1 \text{ and } p(Y_0) = \boldsymbol{\lambda}'\mathbf{Y}.$$

This can be rewritten as an unconstrained minimization problem using the Lagrange multiplier $m$,

$$\begin{aligned} & \min_{\lambda} && \mathrm{E}[(p(Y_0) - Y_0)^2] \text{ subject to } \boldsymbol{\lambda}'\mathbf{1} = 1 \\ \Leftrightarrow\; & \min_{\lambda,m} && \mathrm{Var}[p(Y_0)] + \mathrm{Var}[Y_0] - 2\mathrm{Cov}[p(Y_0), Y_0] - 2m(\boldsymbol{\lambda}'\mathbf{1} - 1) \\ \Leftrightarrow\; & \min_{\lambda,m} && \boldsymbol{\lambda}'\boldsymbol{\Sigma}\boldsymbol{\lambda} + \sigma^2 - 2\boldsymbol{\lambda}'\mathbf{c} - 2m(\boldsymbol{\lambda}'\mathbf{1} - 1). \end{aligned}$$

Taking derivatives with respect to $\boldsymbol{\lambda}$ and $m$ and setting these to zero leads to the system of equations

$$\begin{aligned} \mathbf{0} &= \boldsymbol{\Sigma}\boldsymbol{\lambda} - \mathbf{c} - m\mathbf{1} \\ 0 &= \boldsymbol{\lambda}'\mathbf{1} - 1. \end{aligned}$$

The solutions turn out to be

$$m = (1 - \mathbf{1}'\boldsymbol{\Sigma}^{-1}\mathbf{c})(\mathbf{1}'\boldsymbol{\Sigma}^{-1}\mathbf{1})^{-1} \tag{1.23}$$

$$\boldsymbol{\lambda}' = \left[\mathbf{c} + \mathbf{1}\frac{1 - \mathbf{1}'\boldsymbol{\Sigma}^{-1}\mathbf{c}}{\mathbf{1}'\boldsymbol{\Sigma}^{-1}\mathbf{1}}\right]' \boldsymbol{\Sigma}^{-1}. \tag{1.24}$$

After some algebraic manipulations, the best linear unbiased predictor (BLUP) can be expressed as $p(Y_0) = \boldsymbol{\lambda}'\mathbf{Y}$

$$p(Y_0) = \widehat{\mu} + \mathbf{c}'\boldsymbol{\Sigma}^{-1}(\mathbf{Y} - \mathbf{1}\widehat{\mu}), \tag{1.25}$$

where $\widehat{\mu}$ is the generalized least squares estimator: $\widehat{\mu} = (\mathbf{1}'\boldsymbol{\Sigma}^{-1}\mathbf{1})^{-1}\mathbf{1}'\boldsymbol{\Sigma}^{-1}\mathbf{Y}$. The minimized mean-squared error, the prediction variance, is

$$\sigma^2_{pred} = \sigma^2 - \mathbf{c}'\boldsymbol{\Sigma}^{-1}\mathbf{c} + (1 - \mathbf{1}'\boldsymbol{\Sigma}^{-1}\mathbf{c})^2(\mathbf{1}'\boldsymbol{\Sigma}^{-1}\mathbf{1})^{-1}. \tag{1.26}$$

We note in passing that expression (1.25) is known in the geostatistical vernacular as the ordinary kriging predictor and (1.26) is called the ordinary kriging variance. Spatial prediction and kriging are covered in detail in Chapters 5 and 6. In the current section we continue to explore the effects of autocorrelation on statistical inference without defining any one particular method.

If the data were uncorrelated ($\rho = 0$), the BLUP for $Y_0$ would be $p^*(Y_0) = \overline{Y}$ and the prediction variance would be $\sigma^2_{pred} = \sigma^2(1 + 1/n)$, a familiar result. If we were to repeat the exercise with the goal to estimate $\mu$, rather than to predict $Y_0$, subject to the same linearity and unbiasedness constraints, we would find $p(\mu) = (\mathbf{1}'\boldsymbol{\Sigma}^{-1}\mathbf{1})^{-1}\mathbf{1}'\boldsymbol{\Sigma}^{-1}\mathbf{Y} = \widehat{\mu}$ as the best predictor for $\mu$. In the case of uncorrelated data, the best predictor of the random variable $Y_0$ and the best estimator of its mean $\mathrm{E}[Y_0]$ are identical. When the correlations are taken into account, the predictor of $Y_0$ and the estimator of $\mathrm{E}[Y_0]$ differ,

$$p(Y_0) = p(\mu) + \mathbf{c}'\boldsymbol{\Sigma}^{-1}(Y - \mathbf{1}p(\mu)).$$

To investigate how the precision of the predicted value is affected by incorporating the correlations, we derive a scalar expression for $\sigma^2_{pred}$ which can be compared against $\sigma^2(1+1/n)$, the prediction error when data are not correlated. The inverse of $\mathbf{\Sigma}$ can be found by applying Theorem 8.3.4 in Graybill (1983, p. 190).

**Theorem 1.1** *If the $k \times k$ matrix $\boldsymbol{C}$ can be written as $\boldsymbol{C} = (a-b)\boldsymbol{I} + b\boldsymbol{J}$, then $\boldsymbol{C}^{-1}$ exists if and only if $a \neq b$ and $a \neq -(k-1)b$, and is given by*

$$\boldsymbol{C}^{-1} = \frac{1}{a-b}\left[\boldsymbol{I} - \frac{b}{a+(k-1)b}\boldsymbol{J}\right].$$

The proof is given on p. 191 of Graybill (1983).

The theorem can be applied to our situation since $\sigma^{-2}\mathbf{\Sigma} = (1-\rho)\mathbf{I}+\rho\mathbf{J}$. The condition that $\rho \neq -1/(n-1)$ is met. In fact, from $\text{Var}[Y_i] = \sigma^2 > 0$ it follows that $\text{Var}[\sum_{i=1}^n Y_i] = n\sigma^2(1+(n-1)\rho) > 0$ which implies $\rho > -1/(n-1)$. That the correlation coefficient is bounded from below is a simple consequence of equicorrelation. Applying Theorem 1.1 leads to $\mathbf{1}'\mathbf{\Sigma}^{-1}\mathbf{1} = n\sigma^{-2}/[1+(n-1)\rho]$. Finally, after some (tedious) algebra we obtain the prediction variance in the compound symmetry model:

$$\sigma^2_{pred} = \sigma^2\left[1 + \frac{1}{n}\frac{(\rho n)^2 + (1-\rho)^2}{1+(n-1)\rho}\right].$$

In order for the term $((\rho n)^2 + (1-\rho)^2)/(1+(n-1)\rho)$ to be less than one, we must have (provided $\rho > 0$)

$$\rho < \frac{n+1}{n^2+1}. \tag{1.27}$$

If the condition (1.27) is met, predictions in the compound symmetry model will be more precise than in the independence model. As the strength of the correlation increases predictions in the compound symmetry model can be less precise, however, because the effective sample size shrinks quickly.

### *1.5.2 Effects on Precision of Estimators*

The effect of ignoring autocorrelation in the data and proceeding with inference as if the data points were uncorrelated was discussed in §1.5. The effective sample size formula (1.22) allows a comparison of the precision of the arithmetic sample mean for the compound symmetry model and the case of uncorrelated data. The intuitive consequence of this expression is that positive autocorrelation results in a "loss of information." A sample of independent observations of size $n$ contains more information as a sample of autocorrelated observations of the same size. As noted, the arithmetic sample mean is not the appropriate estimator of the population mean in the case of correlated data. To further our understanding of the consequences of positive autocorrelation and the idea of "information loss," consider the following setting.

$$\begin{aligned} Y_i &= \mu + e_i \\ \mathrm{E}[e_i] &= 0 \\ \mathrm{Cov}[Y_i, Y_j] &= \sigma^2 \rho^{|i-j|} \\ i &= 1, \cdots, n. \end{aligned} \tag{1.28}$$

This model is known as the autoregressive model of first order, or simply the AR(1) model. Let $\mathbf{Y} = [Y_1, \cdots, Y_n]'$ and $\mathrm{Var}[\mathbf{Y}] = \mathbf{\Sigma}$. As for the compound symmetry model, an expression for $\mathbf{\Sigma}^{-1}$ is readily available. Graybill (1983, pp. 198–201) establishes that $\mathbf{\Sigma}^{-1}$ is a diagonal matrix of type 2, that is, $\sigma_{ij} \neq 0$ if $|i-j| \leq 1$, $\sigma_{ij} = 0$ if $|i-j| > 1$ and that

$$\mathbf{\Sigma}^{-1} = \frac{1}{\sigma^2(1-\rho^2)} \begin{bmatrix} 1 & -\rho & 0 & 0 & \cdots & 0 \\ -\rho & 1+\rho^2 & -\rho & 0 & \cdots & 0 \\ 0 & -\rho & 1+\rho^2 & -\rho & \cdots & 0 \\ 0 & 0 & -\rho & 1+\rho^2 & \cdots & \vdots \\ \vdots & \vdots & \vdots & \cdots & \ddots & -\rho \\ 0 & 0 & \cdots & 0 & -\rho & 1 \end{bmatrix}$$

The generalized least squares estimator of $\mu$ is

$$\widehat{\mu} = (\mathbf{1}'\mathbf{\Sigma}^{-1}\mathbf{1})^{-1}(\mathbf{1}'\mathbf{\Sigma}^{-1}\mathbf{Y}) \tag{1.29}$$

and some algebra leads to (see Chapter problems)

$$\mathrm{Var}[\widehat{\mu}] = \sigma^2 \frac{1+\rho}{(n-2)(1-\rho)+2}. \tag{1.30}$$

Two special cases are of particular interest. If $\rho = 0$, the variance is equal to $\sigma^2/n$ as it should, since $\widehat{\mu}$ is then the arithmetic mean. If the data points are perfectly correlated ($\rho = 1$) then $\mathrm{Var}[\widehat{\mu}] = \sigma^2$, the variance of a single observation. The variance of the estimator then does not depend on sample size. Having observed one observation, no additional information is accrued by sampling additional values; if $\rho = 1$, all further values would be identical to the first.

When you estimate $\mu$ in the autoregressive model by (1.29), the correlations are not ignored, the best possible linear estimator *is* being used. Yet, compared to a set of independent data of the same sample size, the precision of the estimator is reduced. The values in the body of Table 1.2 express how many times more variable $\widehat{\mu}$ is compared to $\sigma^2/n$, that is,

$$\left[\frac{1-\rho}{1+\rho} + \frac{2}{n}\frac{\rho}{1+\rho}\right]^{-1}.$$

This is not a relative efficiency in the usual sense. It does not set into relation the mean square errors of two competing estimators for the same set of data. It is a comparison of mean-squared errors for two estimators under two different data scenarios. We can think of the values in Table 1.2 as the relative excess variability (REV) incurred by correlation in the data.

Table 1.2 *Relative excess variability of the generalized least squares estimator (1.29) in AR(1) model.*

| Autocorrelation Parameter | | | | | | | | | | Sample Size |
|---|---|---|---|---|---|---|---|---|---|---|
| 0.0 | 0.1 | 0.2 | 0.3 | 0.4 | 0.5 | 0.6 | 0.7 | 0.8 | 0.9 | $n$ |
| 1.00 | 1.00 | 1.00 | 1.00 | 1.00 | 1.00 | 1.00 | 1.00 | 1.00 | 1.00 | 1 |
| 1.00 | 1.14 | 1.29 | 1.44 | 1.62 | 1.80 | 2.00 | 2.22 | 2.45 | 2.71 | 3 |
| 1.00 | 1.17 | 1.36 | 1.59 | 1.84 | 2.14 | 2.50 | 2.93 | 3.46 | 4.13 | 5 |
| 1.00 | 1.18 | 1.40 | 1.65 | 1.96 | 2.33 | 2.80 | 3.40 | 4.20 | 5.32 | 7 |
| 1.00 | 1.19 | 1.42 | 1.70 | 2.03 | 2.45 | 3.00 | 3.73 | 4.76 | 6.33 | 9 |
| 1.00 | 1.20 | 1.43 | 1.72 | 2.08 | 2.54 | 3.14 | 3.98 | 5.21 | 7.21 | 11 |
| 1.00 | 1.20 | 1.44 | 1.74 | 2.12 | 2.60 | 3.25 | 4.17 | 5.57 | 7.97 | 13 |
| 1.00 | 1.20 | 1.45 | 1.76 | 2.14 | 2.65 | 3.33 | 4.32 | 5.87 | 8.64 | 15 |
| 1.00 | 1.21 | 1.46 | 1.77 | 2.16 | 2.68 | 3.40 | 4.45 | 6.12 | 9.23 | 17 |
| 1.00 | 1.21 | 1.46 | 1.78 | 2.18 | 2.71 | 3.45 | 4.55 | 6.33 | 9.76 | 19 |
| 1.00 | 1.21 | 1.47 | 1.78 | 2.19 | 2.74 | 3.50 | 4.64 | 6.52 | 10.2 | 21 |
| 1.00 | 1.21 | 1.47 | 1.79 | 2.21 | 2.76 | 3.54 | 4.71 | 6.68 | 10.7 | 23 |
| 1.00 | 1.21 | 1.47 | 1.80 | 2.22 | 2.78 | 3.57 | 4.78 | 6.82 | 11.0 | 25 |
| 1.00 | 1.21 | 1.47 | 1.80 | 2.22 | 2.79 | 3.60 | 4.83 | 6.94 | 11.4 | 27 |
| 1.00 | 1.21 | 1.47 | 1.80 | 2.23 | 2.81 | 3.63 | 4.88 | 7.05 | 11.7 | 29 |
| 1.00 | 1.21 | 1.48 | 1.81 | 2.24 | 2.82 | 3.65 | 4.93 | 7.15 | 12.0 | 31 |
| 1.00 | 1.21 | 1.48 | 1.81 | 2.24 | 2.83 | 3.67 | 4.96 | 7.24 | 12.3 | 33 |
| 1.00 | 1.21 | 1.48 | 1.81 | 2.25 | 2.84 | 3.68 | 5.00 | 7.33 | 12.5 | 35 |
| 1.00 | 1.21 | 1.48 | 1.82 | 2.25 | 2.85 | 3.70 | 5.03 | 7.40 | 12.8 | 37 |
| 1.00 | 1.22 | 1.48 | 1.82 | 2.26 | 2.85 | 3.71 | 5.06 | 7.47 | 13.0 | 39 |
| 1.00 | 1.22 | 1.48 | 1.82 | 2.26 | 2.86 | 3.73 | 5.09 | 7.53 | 13.2 | 41 |
| 1.00 | 1.22 | 1.48 | 1.82 | 2.26 | 2.87 | 3.74 | 5.11 | 7.59 | 13.4 | 43 |
| 1.00 | 1.22 | 1.48 | 1.82 | 2.27 | 2.87 | 3.75 | 5.13 | 7.64 | 13.6 | 45 |
| 1.00 | 1.22 | 1.48 | 1.82 | 2.27 | 2.88 | 3.76 | 5.15 | 7.69 | 13.7 | 47 |
| 1.00 | 1.22 | 1.48 | 1.83 | 2.27 | 2.88 | 3.77 | 5.17 | 7.74 | 13.9 | 49 |

For any given sample size $n > 1$ the REV of (1.29) increases with $\rho$. No loss is incurred if only a single observation is collected, since then $\widehat{\mu} = Y_1$. The REV increases with sample size and this effect is more pronounced for $\rho$ large than for $\rho$ small. As is seen from (1.30) and the fact that $\mathrm{E}[\widehat{\mu}] = \mu$, the GLS estimator is consistent for $\mu$. Its precision increases with sample size for any given value $\rho < 1$. An important message that you can glean from these computations is that the most efficient estimator when data are (positively) correlated can be (much) more variable than the most efficient estimator for independent data. In designing simulation studies with correlated data these effects are often overlooked. The number of simulation runs sufficient to achieve small simulation variability for independent data is less than the needed number of runs for correlated data.

## 1.6 Chapter Problems

**Problem 1.1** Categorize the following examples of spatial data as to their data type:

(i) Elevations in the foothills of the Allegheny mountains;

(ii) Highest elevation within each state in the United States;

(iii) Concentration of a mineral in soil;

(iv) Plot yields in an uniformity trial;

(v) Crime statistics giving names of subdivisions where break-ins occurred in the previous year and property loss values;

(vi) Same as (v), but instead of the subdivision, the individual dwelling is identified;

(vii) Distribution of oaks and pines in a forest stand;

(viii) Number of squirrel nests in the pines of the stand in (vii).

**Problem 1.2** Consider $Y_1, \cdots, Y_n$, Gaussian random variables with mean $\mu$, variance $\sigma^2$, and $\text{Cov}[Y_i, Y_j] = \sigma^2\rho, \forall i \neq j$. Is $S^2 = (n-1)^{-1}\sum_{i=1}^{n}(Y_i - \overline{Y})^2$ an unbiased estimator of $\sigma^2$?

**Problem 1.3** Derive the mean and variance of the *BB* join count statistic (1.7) under the assumption of binomial sampling. Notice that $\text{E}[Z(\mathbf{s}_i)^k] = \pi, \forall k$, because $Z(\mathbf{s}_i)$ is an indicator variable. Also, under the null hypothesis of no autocorrelation, $\text{Var}[Z(\mathbf{s}_i)Z(\mathbf{s}_j)] = \pi^2 - \pi^4$.

**Problem 1.4** Is the variance of the *BB* join count statistic larger under binomial sampling or under hypergeometric sampling? Imagine that you are studying the retail volume of grocery stores in a municipality. The data are coded such that $Z(\mathbf{s}_i) = 1$ if the retail volume of the store at site $\mathbf{s}_i$ exceeds 20 million dollars per year, $Z(\mathbf{s}_i) = 0$ otherwise. The *BB* join count statistic with suitably chosen weights is used to test for spatial autocorrelation in the sale volumes. Discuss a situation when you would rely on the assumption of binomial sampling and one where hypergeometric sampling is appropriate.

**Problem 1.5** Let $\mathbf{W} = [w_{ij}]$ be a spatial contiguity matrix and let $u_i = Z(\mathbf{s}_i) - \overline{Z}$. Collect the $u_i$ into a vector $\mathbf{u} = [u_1, \cdots, u_n]'$ and standardize the weights such that $\sum_{i,j} w_{ij} = n$. Let $\mathbf{Y} = \mathbf{W}\mathbf{u}$ and consider the regression through the origin $\mathbf{Y} = \beta\mathbf{u} + \mathbf{e}$, $\mathbf{e} \sim (\mathbf{0}, \sigma^2\mathbf{I})$. What is measured by the slope $\beta$?

**Problem 1.6** Given a lattice of sites, assume that $Z(\mathbf{s}_i)$ is a binary variable. Compare the Black-White join count statistic to Moran's $I$ in this case (using the same contiguity definition). Are they the same? If not, are they very different? What advantages would one have in using the $BB$ (or $BW$) statistic in this problem that $I$ does not offer?

**Problem 1.7** Show that Moran's $I$ is a scale-free statistic, i.e., $Z(\mathbf{s})$ and $\lambda Z(\mathbf{s})$ yield the same $I$ statistic.

**Problem 1.8** Consider the simple $2 \times 3$ lattice with observations $Z(\mathbf{s}_1) = 5, Z(\mathbf{s}_2) = -3, Z(\mathbf{s}_3) = -6$, etc. For problems (i)–(iii) that follow, assume the rook definition of spatial connectivity.

| | Column 1 | Column 2 | Column 3 |
|---|---|---|---|
| Row 1 | 5 | −3 | −6 |
| Row 2 | 2 | 4 | −2 |

(i) Derive the mean and variance of Moran's $I$ under randomization empirically by enumerating the 6! permutations of the lattice. Compare your answer to the formulas for E[$I$] and Var[$I$].

(ii) Calculate the empirical $p$-value for the hypothesis of no spatial autocorrelation. Compare it against the $p$-value based on the Gaussian approximation under randomization. For this problem you need to know the variance of $I$ under randomization. It can be obtained from the following expression, given in Cliff and Ord (1981, Ch. 2):

$$\begin{aligned}
\mathrm{E}_r[I^2] &= \frac{n[n^2-3n+3]S_1 - nS_2 + 3w_{..}^2 - b[(n^2-n)S_1 - 2nS_2 + 6w_{..}^2}{(n-3)(n-2)(n-1)w_{..}^2} \\
S_1 &= \frac{1}{2}\sum_{i=1}^{n}\sum_{j=1}^{n}(w_{ij}+w_{ji})^2 \\
S_2 &= \sum_{i=1}^{n}\left\{\sum_{j=1}^{n} w_{ij} + \sum_{j=1}^{n} w_{ji}\right\}^2 \\
b &= n\frac{\sum_{i=1}^{n}(Z(\mathbf{s}_i)-\overline{Z})^4}{\left\{\sum_{i=1}^{n}(Z(\mathbf{s}_i)-\overline{Z})^2\right\}^2}
\end{aligned}$$

(iii) Prepare a histogram of the 6! $I$ values. Do they appear Gaussian?

**Problem 1.9** Consider $Y_1, \cdots, Y_n$, Gaussian random variables with mean $\mu$, variance $\sigma^2$, and $\mathrm{Cov}[Y_i, Y_j] = \sigma^2\rho(|i-j|=1)$. Assume that $\rho > 0$ and find the power function of the test for $H_0 : \mu = \mu_0$ versus $H_1 : \mu > \mu_0$, where the test statistic is given by

$$Z_{obs} = \frac{\overline{Y} - \mu_0}{\sqrt{\mathrm{Var}[\overline{Y}]}}.$$

Compare this power function to the power function that is obtained for $\rho = 0$.

**Problem 1.10** Using the same setup as in the previous problem, find the best linear unbiased predictor $p(Y_0)$ for a new observation based on $Y_1, \cdots, Y_n$. Compare its prediction variance $\sigma^2_{pred}$ to that of the predictor $\overline{Y}$.

**Problem 1.11** Show that (1.23) and (1.24) are the solutions to the constrained minimization problem that yields the best (under squared-error loss) linear unbiased predictor in §1.5.1. Establish that the solution is indeed a minimum.

**Problem 1.12** Establish algebraically the equivalence of $p(Y_0) = \boldsymbol{\lambda}'\mathbf{Z}(\mathbf{s})$ and (1.25), where $\boldsymbol{\lambda}$ is given by (1.24).

**Problem 1.13** Consider a random field $Z(\mathbf{s}) \sim G(\mu, \sigma^2)$ with covariance function $\text{Cov}[Z(\mathbf{s}), Z(\mathbf{s}+\mathbf{h})] = C(\mathbf{h})$ and let $\mathbf{s}_0$ be an unobserved location. If $[Z(\mathbf{s}_0), \mathbf{Z}(\mathbf{s})]'$ are jointly Gaussian distributed, find $\text{E}[Z(\mathbf{s}_0)|\mathbf{Z}(\mathbf{s})]$ and compare your finding to (1.25).

**Problem 1.14** For second-order stationary processes it is common—particularly in time series analysis—to describe the second-order properties through the auto-correlation function (the correlogram) $R(\mathbf{h}) = C(\mathbf{h})/C(\mathbf{0})$. The advantage of $R(\mathbf{h})$ is ease of interpretation since covariances are scale-dependent. It turns out, however, that the correlogram is an incomplete description of the second-order properties. Qualitatively different processes can have the same correlogram. Consider the autoregressive time series

$$\begin{aligned} Z_1(t) &= \rho Z(t-1) + e(t) \\ Z_2(t) &= \rho Z(t-1) + U(t), \end{aligned}$$

where $\text{E}[Z_1(t)] = \text{E}[Z_2(t)] = 0$, $e(t) \sim iid\ G(0, \sigma^2)$, and

$$U(t) \sim iid \begin{cases} 0 & \text{with probability } p \\ G(0, \sigma^2) & \text{with probability } 1-p \end{cases}.$$

(i) Find the covariance function, the correlogram, and the semivariogram for the processes $Z_1(t)$ and $Z_2(t)$.

(ii) Set $\rho = 0.5$ and simulate the two processes with $p = 0.5, 0.8, 0.9$. For each of the simulations generate a sequence for $t = 0, 1, \cdots, 200$ and graph the realization against $t$. Choose $\sigma^2 = 0.2$ throughout.

**Problem 1.15** Using the expression for $\boldsymbol{\Sigma}^{-1}$ in §1.5.2, derive the variance of the generalized least squares estimator in (1.30).

**Problem 1.16** Table 1.2 gives relative excess variabilities (REV) for the GLS estimator with variance (1.30) for several values of $\rho \geq 0$. Derive a table akin to Table 1.2 and discuss the REV if $-1 < \rho \leq 0$.

CHAPTER 2

# Some Theory on Random Fields

## 2.1 Stochastic Processes and Samples of Size One

A stochastic process is a family or collection of random variables, the members of which can be identified or located (indexed) according to some metric. For example, a time series $Y(t)$, $t = t_1, \cdots, t_n$, is indexed by the time points $t_1, \cdots, t_n$ at which the series is observed. Similarly, a spatial process is a collection of random variables that are indexed by some set $D \subset \mathbb{R}^d$ containing spatial coordinates $\mathbf{s} = [s_1, s_2, \cdots, s_d]'$. For a process in the plane, $d = 2$, and the longitude and latitude coordinates are often identified as $\mathbf{s} = [x, y]'$. If the dimension $d$ of the index set of the stochastic process is greater than one, the stochastic process is often referred to as a **random field**. In this text we are mostly concerned with spatial processes in $\mathbb{R}^2$, although higher-dimensional spaces are implicit in many derivations; spatio-temporal processes are addressed in Chapter 9. The name random "field" should not connote a two-dimensional plane or even an agricultural application. It is much more general.

In classical, applied statistics, stochastic process formulations of random experiments are uncommon. Its basis is steeped in the notion of random sampling, i.e., *iid* observations. To view the time series $Y(t)$ or the spatial data $Z(\mathbf{s})$ as a stochastic process is not only important because the observations might be correlated. It is the random mechanism that generates the data which is viewed differently from what you might be used to. To be more precise, think of $Z(\mathbf{s})$, the value of the attribute $Z$ at location $\mathbf{s}$, as the outcome of a random experiment $\omega$. Extending the notation slightly for the purpose of this discussion, we put $Z(\mathbf{s}, \omega)$ to make the dependency on the random experiment explicit. A particular realization $\omega$ produces a surface $Z(\cdot, \omega)$. Because the surface from which the samples are drawn is the result of this random experiment, $Z(\mathbf{s})$ is also referred to as a **random function**. As a consequence, the collection of $n$ georeferenced observations that make up the spatial data set do not represent a sample of size $n$. They represent the incomplete observation of a single realization of a random experiment; a sample of size one from an $n$-dimensional distribution. This raises another, important question: if we put statements such as $\mathrm{E}[Z(\mathbf{s})] = \mu(\mathbf{s})$, with respect to what distribution is the expectation being taken? The expectation represents the long-run average of the attribute at location $\mathbf{s}$ over the distribution of the possible realizations

$\omega$ of the random experiment,

$$\mathrm{E}[Z(\mathbf{s})] = \mathrm{E}_{\Omega}[Z(\mathbf{s}, \omega)]. \tag{2.1}$$

Imagine pouring sand from a bucket onto a surface. The sand distributes on the surface according to the laws of physics. We could—given enough resources—develop a model that predicts with certainty how the grains come to lie on the surface. By the same token, we can develop a deterministic model that predicts with certainty whether a coin will land on heads or tails, taking into account the angle and force at which the coin is released, the conditions of the air through which it travels, the conditions of the surface on which it lands, and so on. It is accepted, however, to consider the result of the coin-flip as the outcome of a random experiment. This probabilistic model is more parsimonious and economic than the deterministic model, and enables us to address important questions; e.g., whether the coin is fair. Considering the precise placement of the sand on the surface as the result of a random experiment is appropriate by similar reasoning. At issue is *not* that we consider the placement of the sand a random event. What *is* at issue is that the sand was poured only once; no matter at how many locations the depth of the sand is measured. If we are interested in the long-run average depth of the sand at a particular location $\mathbf{s}_0$, the expectation (2.1) tells us that we need to repeat the process of pouring the sand over and over again and consider the expected value with respect to the probability distribution of all surfaces so generated.

The implications are formidable. How are we to learn anything about the variability of a random process if only a single realization is available? In practical applications there is usually no replication in spatial data in the sense of observing several, independent realizations of the process. Are inferences about the long-run average really that important then? Are we then not more interested to model and predict the realized surface rather than some average surface? How are we to make progress with statistical inference based on a sample of size one? Fortunately, we can, provided that the random process has certain stationarity properties. The assumption of stationarity in random fields is often criticized, and sometimes justifiably so. Analyzing observations from a stochastic process as if the process were stationary—when it is not—can lead to erroneous inferences and conclusions. Without a good understanding of stationarity (and isotropy) issues, little progress can be made in the study of non-stationary processes. And in the words of Whittle (1954),

> The processes we mentioned can only as a first approximation be regarded as stationary, if they can be so regarded at all. However, the approximation is satisfactory sufficiently often to make the study of the stationary type of process worth while.

## 2.2 Stationarity, Isotropy, and Heterogeneity

A random field

$$\{Z(\mathbf{s}) : \mathbf{s} \in D \subset \mathbb{R}^d\} \tag{2.2}$$

is called a **strict** (or strong) stationary field if the spatial distribution is invariant under translation of the coordinates, i.e.,

*Spatial Distribution*

$$\Pr(Z(\mathbf{s}_1) < z_1, Z(\mathbf{s}_2) < z_2, \cdots, Z(\mathbf{s}_k) < z_k) =$$
$$\Pr(Z(\mathbf{s}_1 + \mathbf{h}) < z_1, Z(\mathbf{s}_2 + \mathbf{h}) < z_2, Z(\mathbf{s}_k + \mathbf{h}) < z_k),$$

for all $k$ and $\mathbf{h}$. A strictly stationary random field repeats itself throughout the domain.

As the name suggests, strict stationarity is a stringent condition; most statistical methods for spatial data analysis are satisfied with stationary conditions based on the moments of the spatial distribution rather than the distribution itself. This is akin to statistical estimation proceeding on the basis of the mean and the variance of a random variable only; least squares methods, for example, do not require knowledge of the data's joint distribution. **Second-order** (or weak) stationarity of a random field implies that $\mathrm{E}[Z(\mathbf{s})] = \mu$ and $\mathrm{Cov}[Z(\mathbf{s}), Z(\mathbf{s}+\mathbf{h})] = C(\mathbf{h})$. The mean of a second-order stationary random field is constant and the covariance between attributes at different locations is only a function of their spatial separation (the **lag**-vector) $\mathbf{h}$. Stationarity reflects the lack of importance of absolute coordinates. The covariance of observations spaced two days apart in a second-order stationary time series will be the same, whether the first day is a Monday or a Friday. The function $C(\mathbf{h})$ is called the **covariance function** of the spatial process and plays an important role in statistical modeling of spatial data. Strict stationarity implies second-order stationarity but the reverse is not true by the same token by which we cannot infer the distribution of a random variable from its mean and variance alone.

The existence of the covariance function $C(\mathbf{h})$ in a second-order stationary random field has important consequences. Since $C(\mathbf{h})$ does not depend on absolute coordinates and $\mathrm{Cov}[Z(\mathbf{s}), Z(\mathbf{s}+\mathbf{0})] = \mathrm{Var}[Z(\mathbf{s})] = C(\mathbf{0})$, it follows that the variability of a second-order stationary random field is the same everywhere. A second-order stationary spatial process has a constant mean, constant variance, and a covariance function that does not depend on absolute coordinates. Such a process is the spatial equivalent of a random sample in classical statistic in which observations have the same mean and dispersion (but are not correlated).

The covariance function $C(\mathbf{h})$ of a second-order stationary random field has several other properties. In particular,

*Properties of the Covariance Function $C(\mathbf{h})$*

(i) $C(\mathbf{0}) \geq 0$;

(ii) $C(\mathbf{h}) = C(-\mathbf{h})$, i.e., $C$ is an even function;

(iii) $C(\mathbf{0}) \geq |C(\mathbf{h})|$;

(iv) $C(\mathbf{h}) = \mathrm{Cov}[Z(\mathbf{s}), Z(\mathbf{s}+\mathbf{h})] = \mathrm{Cov}[Z(\mathbf{0}), Z(\mathbf{h})]$;

(v) If $C_j(\mathbf{h})$ are valid covariance functions, $j = 1, \cdots, k$, then $\sum_{j=1}^{k} b_j\, C_j(\mathbf{h})$ is a valid covariance function, if $b_j \geq 0\, \forall j$;

(vi) If $C_j(\mathbf{h})$ are valid covariance functions, $j = 1, \cdots, k$, then $\prod_{j=1}^{k} C_j(\mathbf{h})$ is a valid covariance function;

(vii) If $C(\mathbf{h})$ is a valid covariance function in $\mathbb{R}^d$, then it is also a valid covariance function in $\mathbb{R}^p$, $p < d$.

Properties (i) and (ii) are immediate, since $C(\mathbf{h}) = \text{Cov}[Z(\mathbf{s}), Z(\mathbf{s}+\mathbf{h})]$. At lag $\mathbf{h} = \mathbf{0}$ this yields the variance of the process. Since $C(\mathbf{h})$ does not depend on spatial location $\mathbf{s}$—otherwise the process would not be second-order stationary—we have $C(\mathbf{h}) = \text{Cov}[Z(\mathbf{s}), Z(\mathbf{s}+\mathbf{h})] = \text{Cov}[Z(\mathbf{t}-\mathbf{h}), Z(\mathbf{t})] = C(-\mathbf{h})$ for $\mathbf{t} = \mathbf{s}+\mathbf{h}$. Since $R(\mathbf{h}) = C(\mathbf{h})/C(\mathbf{0})$ is the autocorrelation function and is bounded $-1 \le R(\mathbf{h}) \le 1$, (iii) follows from the Cauchy-Schwarz inequality. The lack of importance of absolute coordinates that is characteristic for a stationary random field, is the reason behind (iv). This particular property will be helpful later to construct covariance functions for spatio-temporal data. Properties (i) and (iii) together suggest that the covariance function has a true maximum at the origin. In §2.3 it is shown formally that this is indeed the case.

Property (v) is useful to construct covariance models as linear combinations of basic covariance models. It is also an important mechanism in nonparametric modeling of covariances. The proof of this property is simple, once we have established what *valid* means. For a covariance function $C(\mathbf{s}_i - \mathbf{s}_j)$ of a second-order stationary spatial random field in $\mathbb{R}^d$ to be valid, $C$ must satisfy the positive-definiteness condition

*Positive Definiteness*

$$\sum_{i=1}^{k} \sum_{j=1}^{k} a_i a_j C(\mathbf{s}_i - \mathbf{s}_j) \ge 0, \tag{2.3}$$

for any set of locations and real numbers. This is an obvious requirement, since (2.3) is the variance of the linear combination $\mathbf{a}'[Z(\mathbf{s}_1), \cdots, Z(\mathbf{s}_k)]$.

In time series analysis, stationarity is just as important as with spatial data. A frequent device employed there to turn a non-stationary series into a stationary one is differencing of the series. Let $Y(t)$ denote an observation in the series at time $t$ and consider the random walk

$$Y(t) = Y(t-1) + e(t),$$

where the $e(t)$ are independent random variables with mean 0 and variance $\sigma^2$. Although we have $\text{E}[Y(t)] = \text{E}[Y(t-k)]$, the variance is not constant,

$$\text{Var}[Y(t)] = t\sigma^2,$$

and the covariance does depend on the origin, $\text{Cov}[Y(t), Y(t-k)] = (t-k)\sigma^2$. The random walk is not second-order stationary. However, the first differences $Y(t) - Y(t-1)$ are second-order stationary (see Chapter problems). A similar device is used in spatial statistics; even if $Z(\mathbf{s})$ is not second-order stationary, the increments $Z(\mathbf{s}) - Z(\mathbf{s}+\mathbf{h})$ might be. A process that has this characteristic is said to have **intrinsic** stationarity. It is often defined as follows: the process

$\{Z(\mathbf{s}) : \mathbf{s} \in D \subset \mathbb{R}^d\}$ is said to be intrinsically stationary if $\mathrm{E}[Z(\mathbf{s})] = \mu$ and

$$\frac{1}{2}\mathrm{Var}[Z(\mathbf{s}) - Z(\mathbf{s}+\mathbf{h})] = \gamma(\mathbf{h}).$$

The function $\gamma(\mathbf{h})$ is called the **semivariogram** of the spatial process. It can be shown that the class of intrinsic stationary processes is larger than the class of second-order stationary processes (Cressie, 1993, Ch. 2.5.2). To see that a second-order stationary process is also intrinsically stationary, it is sufficient to examine

*Relationship between Semivariogram and Covariance Function*

$$\begin{aligned}\mathrm{Var}[Z(\mathbf{s}) - Z(\mathbf{s}+\mathbf{h})] &= \mathrm{Var}[Z(\mathbf{s})] + \mathrm{Var}[Z(\mathbf{s}+\mathbf{h})] - 2\mathrm{Cov}[Z(\mathbf{s}), Z(\mathbf{s}+\mathbf{h})] \\ &= 2\{\mathrm{Var}[Z(\mathbf{s})] - 2C(\mathbf{h})\} \\ &= 2\{C(\mathbf{0}) - C(\mathbf{h})\} = 2\gamma(\mathbf{h}).\end{aligned}$$

On the other hand, intrinsic stationarity does not imply second-order stationarity.

Because of the relationship $\gamma(\mathbf{h}) = C(\mathbf{0}) - C(\mathbf{h})$, statistical methods for second-order stationary random fields can be cast in terms of the semivariogram or the covariance function. Whereas statisticians are more familiar with variances and covariances, many geostatisticians prefer the semivariogram. Working with $\gamma(\mathbf{h})$ compared to $C(\mathbf{h})$ has definite advantages, in particular when estimating these functions based on observed data (see §4.2). As parameters of the stochastic process under study, they can be viewed as re-parameterizations of the second-order structure of the process and thus as "equivalent." If the process is intrinsic but not second-order stationary, however, $C(\mathbf{h})$ is a non-existing parameter. One should then work with the semivariogram $\gamma(\mathbf{h})$.

The second moment structure of a weakly stationary random field is a function of the spatial separation $\mathbf{h}$, but the covariance function can depend on the direction. In the absence of this direction-dependence, that is, when the covariance function or the semivariogram depends only on the absolute distance between points, the function is termed **isotropic**. If the random field is second-order stationary with an isotropic covariance function, then $C(\mathbf{h}) = C^*(||\mathbf{h}||)$, where $||\mathbf{h}||$ is the Euclidean norm of the lag vector,

$$||(\mathbf{s}+\mathbf{h}) - \mathbf{s}|| = ||\mathbf{h}|| = \sqrt{h_1^2 + h_2^2}.$$

Similarly, if the semivariogram of an intrinsic (or weakly) stationary process is isotropic, then $\gamma(\mathbf{h}) = \gamma^*(||\mathbf{h}||)$. Note that $C$ and $C^*$ are two different functions (as are $\gamma$ and $\gamma^*$). In the sequel we will not use the "star" notation, however. It will be obvious from the context whether a covariance function or semivariogram is isotropic or anisotropic.

**Example 2.1** Realizations of an isotropic and anisotropic random field are shown in Figure 2.1. The random field in the left-hand panel of the figure exhibits geometric anisotropy, a particular kind of direction dependence of the covariance structure (see §4.3.7). Under geometric anisotropy the variance of the process is the same in all directions, but the strength of the spatial autocorrelation is not. The realization in the left-hand panel was generated with autocorrelations that are stronger in the East-West direction than in the North-South direction. The realization on the right-hand side of the panel has the same covariance structure in all directions as the anisotropic model in the East-West direction.

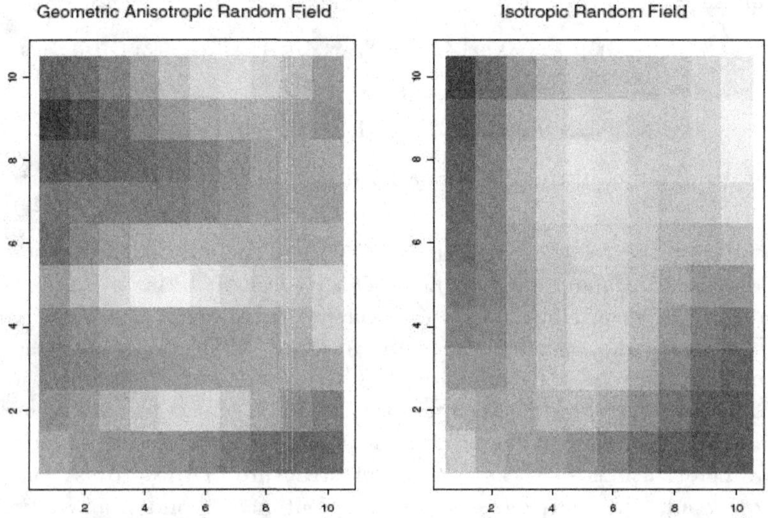


Figure 2.1 *Anisotropic (left) and isotropic (right) second-order stationary random fields. Adapted from Schabenberger and Pierce (2002, Figure 9.11).*

The lesser correlations in the North-South direction of the anisotropic field create a realization that is less smooth in this direction. The values of the random field change more slowly in the East-West direction. In the isotropic field, the same smoothness of the process is achieved in all directions of the domain. The concept of the smoothness of a spatial process is discussed in greater detail in §2.3. ☐

We close this section with an interesting and important question: "What is a Gaussian random field?" The assumption of Gaussian (i.e., *Normal*) distributed data is made commonly in the analysis of random samples in classical statistics. First, we prefer the name Gaussian over *Normal*, since there is really nothing normal about this distribution. Few data really follow this distribution law. The central position of the Gaussian distribution for statistical modeling and inference derives from such powerful results as the Central Limit Theorem and its mathematical simplicity and elegance. But what does it mean if the Gaussian label is attached to a spatial random field?

**Example 2.2** A famous data set used in many spatial statistics texts is the uniformity trial of Mercer and Hall (1911). On an area consisting of $20 \times 25$ experimental units a uniform wheat variety was planted and the grain yield was recorded for each of the units. These are lattice data since a field plot is a discrete spatial unit. We can identify a particular unit with a unique spatial location, however, e.g., the center of the field plot. From the histogram and the $Q$-$Q$ plot of the data one might conclude that the data came from a Gaussian distribution (Figure 2.2). That statement would ignore the spatial context of the data and the random mechanism. From a random field perspective, the 500 observations represent a single observation from a 500-dimensional spatial distribution.

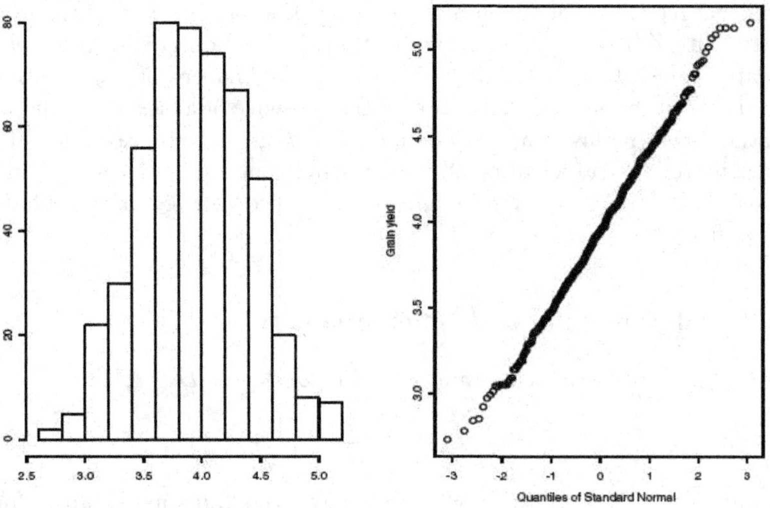


Figure 2.2 *Histogram of Mercer and Hall grain yields and normal QQ-plot.*

*Gaussian Random Field*

A random field $\{Z(\mathbf{s}) : \mathbf{s} \in D \subset \mathbb{R}^d\}$ is a Gaussian random field (GRF), if the cumulative distribution function

$$\Pr\left(Z(\mathbf{s}_1) < z_1, Z(\mathbf{s}_2) < z_2, \cdots, Z(\mathbf{s}_k) < z_k\right)$$

is that of a $k$-variate Gaussian random variable for all $k$. By the properties of the multivariate Gaussian distribution this implies that each $Z(\mathbf{s}_i)$ is a univariate Gaussian random variable. The reverse, unfortunately, is not true. Even if $Z(\mathbf{s}_i) \sim G(\mu(\mathbf{s}_i), \sigma^2(\mathbf{s}_i))$, this does not imply that the joint distribution of $Z(\mathbf{s}_1), \cdots, Z(\mathbf{s}_n)$ is multivariate Gaussian. This leap of faith is all too often made, however.

Just as the univariate Gaussian distribution is the magic ingredient of many classical statistical methods, spatial analysis for Gaussian random fields is more straightforward than for other cases. For example, the best linear unbiased predictor for the attribute $Z(\mathbf{s}_0)$ at an unobserved locations $\mathbf{s}_0$ in general is only *best* in this restricted class of predictors. If the random field is a GRF, then these linear predictors turn out to be *the* best predictors (under squared error loss) among all possible functions of the data (more on this in §5.2). Second-order stationarity does not imply strict stationarity of a random field. In a Gaussian random field, this implication holds.

The Gaussian distribution is often the default population model for continuous random variables in classical statistics. If the data are clearly non-Gaussian, practitioners tend to go at great length to invoke transformations that make the data look more Gaussian-like. In spatial statistics we need to make the important distinction between the type of spatial data—connected to the characteristics of the domain $D$—and the distributional properties of the attribute $Z$ being studied. The fact that the domain $D$ is continuous, i.e., the data are geostatistical, has no bearing on the nature of the attribute as discrete or continuous. One can observe the presence or absence of a disease in a spatially continuous domain. The fact that $D$ is discrete, does not impede the attribute $Z(\mathbf{s})$ at location $\mathbf{s}$ from possibly following the Gaussian law. Nor should continuity of the domain be construed as suggesting a Gaussian random field.

## 2.3 Spatial Continuity and Differentiability

The partial derivatives of the random field $\{Z(\mathbf{s}) : \mathbf{s} \in D \subset \mathbb{R}^d\}$,

$$\dot{Z}(\mathbf{s}) = \frac{\partial Z(\mathbf{s})}{\partial s_j},$$

are random variables whose stochastic behavior contains important information about the nature of the process; in particular its continuity. The more continuous a spatial process, the smoother and the more spatially structured are its realizations. Figure 2.3 shows realizations of four processes in $\mathbb{R}^1$. The processes have the same variance but increase in the degree of continuity from top to bottom. For a given lag $h$, the correlation function $R(h)$ of a highly continuous process will be larger than the correlation function of a less continuous process. As a consequence, neighboring values will change more slowly (Figure 2.4).

The modeler of spatial data needs to understand the differences in continuity between autocorrelation models and the implications for statistical inference. Some correlation models, such as the gaussian correlation model in Figure 2.4d, are more smooth than can be supported by physical or biological processes. With increasing smoothness of the spatial correlation model, statistical inference tends to be more sensitive to model mis-specification. Practitioners might argue whether to model a particular realization with an exponential (Figure 2.4b) or a spherical (Figure 2.4c) correlation model. The

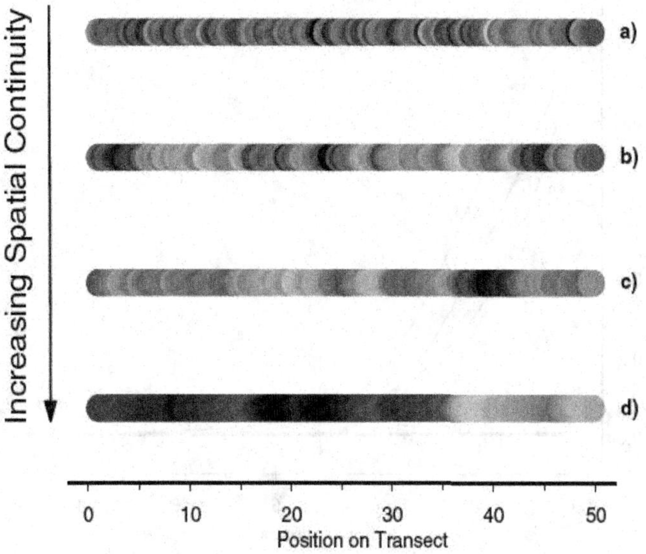


Figure 2.3 *Realizations of four processes on a transect of length* 50 *that differ in the degree of spatial continuity. More continuous processes have smoother realizations, their successive values are more similar. This is indicative of higher spatial autocorrelation for the same distance lag. Realization a) is that of a white noise process (uncorrelated data), b)–d) are those of Gaussian processes with exponential, spherical, and gaussian covariance function, respectively. Adapted from Schabenberger and Pierce (2002).*

estimates of the parameters governing the two models will differ for a particular set of data. The fitted correlation models may imply the same degree of continuity.

We have focused on inferring the degree of continuity from the behavior of the correlation (or covariance) model near the origin. This is intuitive, since the near-origin behavior governs the lag interval for which correlations are high. The theoretical reason behind this focus lies in mean square continuity and differentiability of the random field. Consider a sequence of random variables $\{X_n\}$. We say that $\{X_n\}$ is **mean square continuous** if there exists a random variable $X$ with $\mathrm{E}[X^2] < \infty$, such that $\mathrm{E}[(X_n - X)^2] \to 0$. For a spatial random field $\{Z(\mathbf{s}) : \mathbf{s} \in D \subset \mathbb{R}^d\}$ with constant mean and constant variance, mean-square continuity at $\mathbf{s}$ implies that

*Mean Square Continuity*

$$\lim_{\mathbf{h}\to\mathbf{0}} \mathrm{E}\left[(Z(\mathbf{s}) - Z(\mathbf{s}+\mathbf{h}))^2\right] = 0.$$

Since $\mathrm{E}[(Z(\mathbf{s}) - Z(\mathbf{s}+\mathbf{h}))^2] = 2\mathrm{Var}[Z(\mathbf{s})] - 2C(\mathbf{h}) = 2(C(\mathbf{0}) - C(\mathbf{h})) = 2\gamma(\mathbf{h})$,

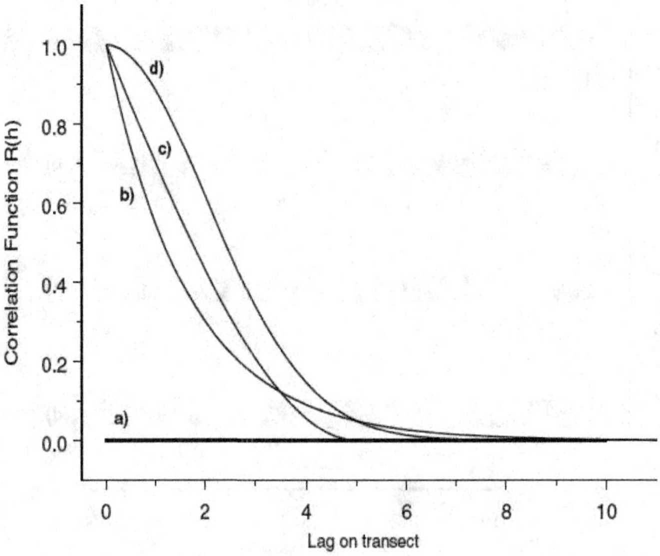


Figure 2.4 *Correlation functions of the spatial processes shown in Figure 2.3. The more sharply the correlation function decreases from the origin, the less continuous is the process.*

we conclude from

$$\lim_{\mathbf{h}\to\mathbf{0}} \mathrm{E}\left[(Z(\mathbf{s}) - Z(\mathbf{s}+\mathbf{h}))^2\right] = \lim_{\mathbf{h}\to\mathbf{0}} 2(C(\mathbf{0}) - C(\mathbf{h})),$$

that unless $C(\mathbf{h}) \to C(\mathbf{0})$ as $\mathbf{h} \to \mathbf{0}$, the random field cannot be continuous at $\mathbf{s}$. The random field will be mean square continuous, if and only if it is continuous at the origin. Mean square continuity can be verified through the behavior of the covariance function near $\mathbf{0}$.

*Nugget Effect*

As we will see in §4.2.2, some processes appear to have a semivariogram for which $\gamma(\mathbf{h}) \to c \equiv$ const., as $\mathbf{h} \to \mathbf{0}$. While empirical data may suggest that the semivariogram does not pass through the origin, explaining this phenomenon, known as the **nugget effect**, to the purists' satisfaction is an entirely different matter. A process that exhibits a discontinuity at the origin cannot be mean square continuous. Mean square continuity is important because some methods for spatial data analysis require the covariance function to be continuous. Examples are the spectral representation (§2.5) and some "nonparametric" methods of covariance function estimation (§4.6). It is argued on occasion, that functions should not be considered as covariance models for stochastic processes, unless they are continuous. This excludes models with nugget effect from consideration and reflects the sentiment that only the study of mean square continuous processes is worthwhile.

Mean square continuity by itself does not convey much about the smoothness of the process and how it is related to the covariance function. The smoothness concept is brought into focus by studying the partial derivatives of the random field. First, consider the special case of a weakly stationary spatial process on a transect, $\{Z(s) : s \in D \subset \mathbb{R}\}$, with mean $\mu$ and variance $\sigma^2$. Furthermore assume that data are collected at equally spaced intervals $\delta$. The gradient between successive observations is then

$$\dot{Z} = \frac{Z(s+\delta) - Z(s)}{\delta},$$

with $\mathrm{E}[\dot{Z}] = 0$ and variance

$$\begin{aligned} \mathrm{Var}[\dot{Z}] &= \delta^{-2}\{\mathrm{Var}[Z(s+\delta)] + \mathrm{Var}[Z(s)] - 2\mathrm{Cov}[Z(s+\delta), Z(s)]\} \\ &= 2\delta^{-2}\left\{\sigma^2 - C(\delta)\right\} \equiv \dot{\sigma}^2. \end{aligned} \quad (2.4)$$

For a second-order stationary random field, we know that $C(\mathbf{0}) = \sigma^2$ and hence $[dC(\delta)/d\delta]_{h=0} = 0$. Additional details can be garnered from (2.4), because

$$C(\delta) = \sigma^2 - \frac{\delta^2}{2}\dot{\sigma}^2 \quad (2.5)$$

can be the limiting form (as $\delta \to 0$) of $C(\delta)$ only if $\dot{\sigma}^2$ is finite. As a consequence of (2.5), the negative of the second derivative of $C(\delta)$ is the mean square derivative $\dot{\sigma}^2$; the covariance function has a true maximum at the origin.

Notice that (2.4) can be written as $\dot{\sigma}^2 = 2\delta^{-2}\{C(0) - C(\delta)\} = 2\delta^{-2}\gamma(\delta)$, where $\gamma(\delta)$ is the semivariogram of the $Z$ process. For the mean square derivative to be finite, the semivariogram cannot rise more quickly in $\delta$ than $\delta^2$. This condition is known as the **intrinsic hypothesis**. It is, in fact, slightly stronger than $2\gamma(\delta)/\delta^2 \to$ const., as $\delta \to \infty$. A valid semivariogram must satisfy $2\gamma(\delta)/\delta^2 \to 0$ as $\delta \to \infty$. For example, the power semivariogram model

*Intrinsic Hypothesis*

$$\gamma(\mathbf{h}; \boldsymbol{\theta}) = \begin{cases} 0 & \mathbf{h} = \mathbf{0} \\ \theta_1 + \theta_2 ||\mathbf{h}||^{\theta_3} & \mathbf{h} \neq \mathbf{0} \end{cases}$$

is a valid semivariogram for an intrinsically stationary process only if $0 \leq \theta_3 < 2$.

For a general process $Z(s)$ on $\mathbb{R}$ with covariance function $C$, define

$$\dot{Z}_h = \frac{Z(s+h) - Z(s)}{h}.$$

Stein (1999, Ch. 2.6) proves that $Z(s)$ is mean square differentiable, if and only if the second derivative of $C(h)$ evaluated at $h = 0$ exists and is finite. In general, $Z(s)$ is $m$-times mean square differentiable if and only if

$$\left[\frac{d^{2m}C(h)}{dh^{2m}}\right]_0$$

exists and is finite. The covariance function of $(d^m Z(s))/(ds^m)$ is

$$(-1)^m \frac{d^{2m}C(h)}{dh^{2m}}.$$

The smoothness of a spatial random field increases with the number of times it is mean square differentiable. The gaussian covariance model

$$C(\mathbf{s}_i - \mathbf{s}_j) = \sigma^2 \exp\left\{-3\frac{||\mathbf{s}_i - \mathbf{s}_j||^2}{\alpha^2}\right\}, \tag{2.6}$$

for example, is infinitely differentiable. A spatial random field with covariance (2.6) is infinitely smooth. Stein (1999, p. 30) argues that such smoothness is unrealistic for physical processes under normal circumstances.

## 2.4 Random Fields in the Spatial Domain

The representation

$$\{ Z(\mathbf{s}) : \mathbf{s} \in D \subset \mathbb{R}^d \} \tag{2.7}$$

is very general and reveals little about the structure of the random field under study. To be applicable, the formulation must be cast within a framework through which (i) statistical methods of analysis and inference can be derived, and (ii) the properties of statistical estimators as well as the properties of the random field itself can be studied. For second-order stationary random fields, the core components of any formulation are the mean function $\mathrm{E}[Z(\mathbf{s})] = \mu(\mathbf{s})$, the covariance function $C(\mathbf{h}) = \mathrm{Cov}[Z(\mathbf{s}), Z(\mathbf{s}+\mathbf{h})]$, and the properties of the index set $D$ (fixed continuous, fixed discrete, or random). Of the many possible formulations that add structure to (2.7), we present two that structure the random field in the spatial domain (§2.4.1 and §2.4.2), and the spectral representation in the frequency domain (§2.5). The distinction between spatial and spectral representation is coarsely whether $Z(\mathbf{s})$ is expressed in terms of functions of the observed coordinates $\mathbf{s}$, or in terms of a random field $X(\omega)$ that *lives* in a space consisting of frequencies.

Readers accustomed to traditional statistical modeling techniques such as linear, nonlinear, and generalized linear models will find the model representation in §2.4.1 most illustrative. Readers trained in the analysis of time series data in the spectral domain might prefer the representation in §2.5. The following discussion enunciates the relationships and correspondence between the three formulations. They have specific advantages and disadvantages. The model formulation will be the central representation for most of the remainder of this text. We invoke the spectral representation when it is mathematically more convenient to address an issue in the frequency domain, compared to the spatial domain.

### *2.4.1 Model Representation*

A statistical model is the mathematical representation of a data-generating mechanism. It is an abstraction of the physical, biological, chemical, etc. processes that generated the data; emphasizing those aspects of the process that matter for the analysis, and ignoring (or down-weighing) the inconsequential aspects. The most generic statistical models are a decomposition of a random response variable into a mathematical structure describing the mean and an additive stochastic structure describing variation and covariation among the responses. This simple decomposition is often expressed symbolically as

$$\text{Data} = \text{Structure} + \text{Error}.$$

The decomposition is immediately applicable to random fields, in particular, where we are concerned with their first- and second-moment structure. To motivate, recall that intrinsic and weak stationarity require constancy of the mean, $\mathrm{E}[Z(\mathbf{s})] = \mu$. If the mean of the random field changes with location, then $\mu(\mathbf{s})$ is called the large-scale trend of the random field. It is, of course, common to observe large-scale structure in data. By definition, the random field will be non-stationary and much of our preceding discussion seems to be called into question. What is the point of assuming stationarity, if its first requirement—constancy of the mean—is typically not met? The idea is then to not associate stationarity properties with the attribute $Z(\mathbf{s})$, but with its de-trended version. We can put

$$\mathbf{Z}(\mathbf{s}) = \mathbf{f}(\mathbf{X}, \mathbf{s}, \boldsymbol{\beta}) + \mathbf{e}(\mathbf{s}), \tag{2.8}$$

where $\mathbf{Z}(\mathbf{s}) = [Z(\mathbf{s}_1), \cdots, Z(\mathbf{s}_n)]'$, $\mathbf{X}$ is an $(n \times p)$ matrix of covariates, $\boldsymbol{\beta}$ is a vector of parameters, and $\mathbf{e}(\mathbf{s})$ is a random vector with mean $\mathbf{0}$ and variance $\mathrm{Var}[\mathbf{e}(\mathbf{s})] = \boldsymbol{\Sigma}(\boldsymbol{\theta})$. The function $f$ may be nonlinear, hence we need to define what the vector $\mathbf{f}$ represents. The elements of $\mathbf{f}$ are stacked as follows:

$$\mathbf{f}(\mathbf{X}, \mathbf{s}, \boldsymbol{\beta}) = \begin{bmatrix} f(\mathbf{x}_1, \mathbf{s}_1, \boldsymbol{\beta}) \\ f(\mathbf{x}_2, \mathbf{s}_2, \boldsymbol{\beta}) \\ \vdots \\ f(\mathbf{x}_n, \mathbf{s}_n, \boldsymbol{\beta}) \end{bmatrix}.$$

It follows from (2.8), that $\mathrm{E}[\mathbf{Z}(\mathbf{s})] = \mathbf{f}(\mathbf{X}, \mathbf{s}, \boldsymbol{\beta})$ represents the large-scale trend, the mean structure, of the spatial model. The variation and covariation of $\mathbf{Z}(\mathbf{s})$ is represented through the stochastic properties of $\mathbf{e}(\mathbf{s})$. The stationarity assumption is made for the error terms $\mathbf{e}(\mathbf{s})$ of the model, not for the attribute $\mathbf{Z}(\mathbf{s})$. The zero mean assumption of the model errors is a reflection of our belief that the model is correct *on average*. When modeling spatial data it is important to recognize the random mechanism this averaging process appeals to (see §2.1). The stationarity properties of the random field are reflected by the structure of $\mathrm{Var}[\mathbf{e}(\mathbf{s})] = \boldsymbol{\Sigma}(\boldsymbol{\theta})$. The entries of this covariance matrix can be built from the covariance function $C(\mathbf{h})$ of a second-order stationary process. The dependence of $\mathrm{Var}[\mathbf{e}(\mathbf{s})]$ on the vector $\boldsymbol{\theta}$ is added because in many applications the analyst explicitly parameterizes $C(\mathbf{h}) \equiv C(\mathbf{h}, \boldsymbol{\theta})$.

We make various simplifications and alterations to the basic structure (2.8) along the way. The large-scale structure will often be expressed as a linear function of the spatial coordinates, $\mathrm{E}[\mathbf{Z}(\mathbf{s})] = \mathbf{X}(\mathbf{s})\boldsymbol{\beta}$. The design or regressor matrix $\mathbf{X}$ is then a response surface model or other polynomial in the coordinates, hence the dependence of $\mathbf{X}$ on $\mathbf{s}$. The matrix $\mathbf{X}(\mathbf{s})$ can involve other variables apart from location information as is the case in spatial regression models (Chapter 6). The function $f$ is often a monotonic, invertible function, which enables us to extend generalized linear models theory to the spatial context. In that case we may choose to model

$$\begin{aligned} Z(\mathbf{s}) &= f(\mathbf{x}'(\mathbf{s})\boldsymbol{\beta}) + e(\mathbf{s}) \\ f^{-1}(\mathrm{E}[Z(\mathbf{s})]) &= \mathbf{x}'(\mathbf{s})\boldsymbol{\beta}. \end{aligned} \tag{2.9}$$

The model formulation of a spatial process is useful, because it is reminiscent of traditional linear or nonlinear statistical structures. It belies, however, which parameter component is most important to the modeler. Rarely is equal weight given to the vector $\boldsymbol{\beta}$ of mean parameters and the vector $\boldsymbol{\theta}$ of covariance parameters. In spatial regression or analysis of variance models, more emphasis is placed on inferences about the mean function and $\boldsymbol{\theta}$ is often considered a nuisance parameter. In applications of spatial prediction, the covariance structure and the parameter values $\boldsymbol{\theta}$ are critical, they "drive" the mean square prediction error.

The model formulation (2.8) is sometimes termed a direct formulation because the covariance function of the random field is explicitly defined through $\boldsymbol{\Sigma}(\boldsymbol{\theta})$. In linear mixed model theory for clustered correlated data, for example, you can distinguish correlation structures that are induced through hierarchical random effects from correlation structures that are parameterized explicitly (Schabenberger and Pierce, 2002, Ch. 7). For random field models, the distinction between a direct specification of the covariance structure in (2.8) and formulations in which the covariance function is incidental (induced) can also be made. Induced covariance functions are typical in statistical models for lattice data, in hierarchical models, and in some bivariate smoothing techniques. Direct modeling of the covariance function is a frequent device for geostatistical data.

#### *2.4.1.1 Modeling Covariation in Geostatistical Data Directly*

Consider a statistical model for the random field $Z(\mathbf{s})$ with additive error structure,

$$Z(\mathbf{s}) = \mu(\mathbf{s}) + e(\mathbf{s}),$$

where the errors have covariance function $\mathrm{Cov}[e(\mathbf{s}), e(\mathbf{s}+\mathbf{h})] = C_e(\mathbf{h})$. As with other statistical models, the errors can contain more than a single component. It is thus helpful to consider the following decomposition of the process (Cressie, 1993, Ch. 3.1):

*Scales of Variation*

$$Z(\mathbf{s}) = \mu(\mathbf{s}) + W(\mathbf{s}) + \eta(\mathbf{s}) + \epsilon(\mathbf{s}). \tag{2.10}$$

Although akin to a decomposition into sources of variability in an analysis of variance model, (2.10) is largely operational. The individual components may not be identifiable and/or separable. The mean function $\mu(\mathbf{s})$ is the **large-scale** trend of the random field. In terms of (2.8), we parameterize $\mu(\mathbf{s}) = f(\mathbf{x}, \mathbf{s}, \boldsymbol{\beta})$. The remaining components on the right-hand side of (2.10) are random processes. $W(\mathbf{s})$ is termed **smooth-scale** variation; it is a stationary process with covariance function $C_W(\mathbf{h})$ or semivariogram $\gamma_W(\mathbf{h})$. The range $r_W$ of this process, i.e., the lag distance beyond which points on the surface are (practically) uncorrelated, is larger than the smallest lag distance observed in the data. That is, $r_W > \min\{||\mathbf{s}_i - \mathbf{s}_j||\}, \forall i \neq j$. The spatial autocorrelation structure of the $W(\mathbf{s})$ process can be modeled and estimated from the data.

The second process, $\eta(\mathbf{s})$, is termed **micro-scale** variation by Cressie (1993, p. 112). It is also a stationary process, but its range is less than $\min\{||\mathbf{s}_i - \mathbf{s}_j||\}$. The spatial structure of the process cannot be estimated from the observed data. We can only hope to estimate $\text{Var}[\eta(\mathbf{s})] = \sigma^2_\eta$, but even this is not without difficulties. The final random component, $\epsilon(\mathbf{s})$, denotes white noise measurement error with variance $\text{Var}[\epsilon(\mathbf{s})] = \sigma^2_\epsilon$. Unless there are true replications in the data—which is usually not the case— $\sigma^2_\eta$ and $\sigma^2_\epsilon$ cannot be estimated separately. Many modelers thus consider spatial models of the form

$$Z(\mathbf{s}) = \mu(\mathbf{s}) + W(\mathbf{s}) + e^*(\mathbf{s}),$$

where $e^*(\mathbf{s}) = \eta(\mathbf{s}) + \epsilon(\mathbf{s})$. Since the three random components in (2.10) are typically independent, the variance-covariance matrix of $\mathbf{Z}(\mathbf{s})$ decomposes as

$$\text{Var}[\mathbf{Z}(\mathbf{s})] = \boldsymbol{\Sigma}(\boldsymbol{\theta}) = \boldsymbol{\Sigma}_W(\boldsymbol{\theta}_W) + \boldsymbol{\Sigma}_\eta(\boldsymbol{\theta}_\eta) + \boldsymbol{\Sigma}_\epsilon(\boldsymbol{\theta}_\epsilon).$$

With the decomposition (2.10) in place, we now define two types of models.

1. **Signal Model.** Let $S(\mathbf{s}) = \mu(\mathbf{s}) + W(\mathbf{s}) + \eta(\mathbf{s})$ denote the **signal** of the process. It contains all components which are spatially structured, either through deterministic or stochastic sources. The decomposition $Z(\mathbf{s}) = S(\mathbf{s}) + \epsilon(\mathbf{s})$ plays an important role in applications of spatial prediction. There, it is the signal $S(\mathbf{s})$ that is of interest to the modeler, not the noisy version $Z(\mathbf{s})$.

2. **Mean Model.** Let $e(\mathbf{s}) = W(\mathbf{s}) + \eta(\mathbf{s}) + \epsilon(\mathbf{s})$ denote the error process of the model and consider $Z(\mathbf{s}) = \mu(\mathbf{s}) + e(\mathbf{s})$. If the modeler focuses on the mean function but needs to account for autocorrelation structure in the data, the mean model is often the entry point for analysis. It is noteworthy that $e(\mathbf{s})$ contains different spatial error processes, some more structured than others. The idea of $W(\mathbf{s})$ and $\eta(\mathbf{s})$ is to describe small- and microscale stochastic fluctuations of the process. If one allows the mean function $\mu(\mathbf{s})$ to be flexible, then a locally varying mean function can absorb some of this random variation. In other words, "one modeler's mean function is another modeler's covariance structure." The early approaches to cope with spatial autocorrelation in field experiments, such as trend surface models and nearest-neighbor adjustments, attempted to model the mean structure

by adding local deterministic variation to the treatment structure to justify a model with uncorrelated errors (see §6.1).

Cliff and Ord (1981, Ch. 6) distinguish between **reaction** and **interaction** models. In the former, sites react to outside influences, e.g., plants react to the availability of nutrients in the root zone. Since this availability varies spatially, plant size or biomass will exhibit a regression-like dependence on nutrient availability. By this reasoning, nutrient-related covariates are included as regressors in the mean function $f(\mathbf{x}, \mathbf{s}, \boldsymbol{\beta})$. In an interaction model, sites react not to outside influences, but react with each other. Neighboring plants, for example, compete with each other for resources. Schabenberger and Pierce (2002, p. 601) conclude that "when the dominant spatial effects are caused by sites reacting to external forces, these effects should be part of the mean function $[f(\mathbf{x}, \mathbf{s}, \boldsymbol{\beta})]$. Interactive effects [...] call for modeling spatial variability through the autocorrelation structure of the error process."

The distinction between reactive and interactive models is not cut-and-dried. Significant autocorrelation does not imply an interactive model over a reactive one. Spatial autocorrelation can be spurious—if caused by large-scale trends—or real—if caused by cumulative small-scale, spatially varying components.

### *2.4.1.2 Modeling Covariation in Lattice Data Indirectly*

When the spatial domain is discrete, the decomposition (2.10) is not directly applicable, since the random processes $W(\mathbf{s})$ and $\eta(\mathbf{s})$ are now defined on a fixed, discrete domain. They no longer represent continuous spatial variation. As before, reactive effects can be modeled directly through the mean function $\mu(\mathbf{s})$. To incorporate interactive effects, the covariance structure of the model must be modified, however. One such modification gives rise to the simultaneous spatial autoregressive (SSAR) model. Let $\mu(\mathbf{s}_i)$ denote the mean of the random field at location $\mathbf{s}_i$. Then $Z(\mathbf{s}_i)$ is thought to consist of a mean contribution, contributions of *neighboring* sites, and random noise:

$$\begin{aligned} Z(\mathbf{s}_i) &= \mu(\mathbf{s}_i) + e(\mathbf{s}_i) \\ &= \mu(\mathbf{s}_i) + \sum_{j=1}^{n} b_{ij}\{Z(\mathbf{s}_i) - \mu(\mathbf{s}_i)\} + \epsilon(\mathbf{s}_i). \end{aligned} \qquad (2.11)$$

The coefficients $b_{ij}$ in (2.11) describe the spatial **connectivity** of the sites. If all $b_{ij} = 0$, the model reduces to a standard model with mean $\mu(\mathbf{s}_i)$ and uncorrelated errors $\epsilon(\mathbf{s}_i)$. The coefficients govern the spatial autocorrelation structure, but not directly. The responses at sites $\mathbf{s}_i$ and $\mathbf{s}_j$ can be correlated, even if $b_{ij} = 0$. To see this, consider a linear mean function $\mu(\mathbf{s}_i) = \mathbf{x}(\mathbf{s}_i)'\boldsymbol{\beta}$, collect the coefficients into a matrix $\mathbf{B} = [b_{ij}]$, and write the model as

*Simultaneous Spatial Auto-regression*

$$\begin{aligned} \mathbf{Z}(\mathbf{s}) &= \mathbf{X}(\mathbf{s})\boldsymbol{\beta} + \mathbf{B}(\mathbf{Z}(\mathbf{s}) - \mathbf{X}(\mathbf{s})\boldsymbol{\beta}) + \boldsymbol{\epsilon}(\mathbf{s}) \\ \boldsymbol{\epsilon}(\mathbf{s}) &= (\mathbf{I} - \mathbf{B})(\mathbf{Z}(\mathbf{s}) - \mathbf{X}(\mathbf{s})\boldsymbol{\beta}). \end{aligned}$$

It follows that $\text{Var}[\mathbf{Z}(\mathbf{s})] = (\mathbf{I} - \mathbf{B})^{-1}\text{Var}[\boldsymbol{\epsilon}(\mathbf{s})](\mathbf{I} - \mathbf{B}')^{-1}$. If $\text{Var}[\boldsymbol{\epsilon}(\mathbf{s})] = \sigma^2\mathbf{I}$, then $\text{Var}[\mathbf{Z}(\mathbf{s})] = \sigma^2(\mathbf{I} - \mathbf{B})^{-1}(\mathbf{I} - \mathbf{B}')^{-1}$. Although we may have $b_{ij} = 0$, we can have $\text{Cov}[Z(\mathbf{s}_i), Z(\mathbf{s}_j)] \neq 0$. Since $\mathbf{B}$ is a parameter of this model and contains many unknowns, the modeler typically parameterizes the neighborhood structure of the lattice model to facilitate estimation. A common choice is to put $\mathbf{B} = \rho\mathbf{W}$, where $\mathbf{W}$ is a user-defined spatial connectivity matrix and $\rho$ is a parameter to be estimated.

The spatial covariance structure of the lattice model is induced by the choice and parameterization of the $\mathbf{B}$ matrix. It is thus modeled indirectly. A random field representation that can be applied to discrete and continuous spatial domains and also induces covariances is based on convolutions of random noise with kernel functions.

### *2.4.2 Convolution Representation*

The most important characteristic of a second-order stationary process to be studied is its covariance structure. The mean, by definition of second-order stationarity, is constant and the magnitude of the mean may be of interest to the analyst. Beyond the simple problem to estimate the mean of the process, the second-order properties of the process are of primary interest. For geostatistical data, these properties can be described by the covariance function, the correlation function, or the semivariogram. For lattice data, the second-order structure is modeled through a neighborhood connectivity matrix and a parameterization of conditional or joint distributions of the data (see the previous subsection for introductory remarks and §6.2.2 for details). For point patterns, the first- and second-order properties of the process are described by the first- and second-order intensity functions (see §3.4).

The statistical model representation (2.8) is useful for geostatistical and lattice data. The second-order properties of the random field are explicit in the variance-covariance matrix $\boldsymbol{\Sigma}(\boldsymbol{\theta})$ of the model errors in geostatistical models or implied by the connectivity matrix $\mathbf{B}$ in lattice models. The model representation is convenient, because it has a familiar structure. It acknowledges the presence of spatial autocorrelation, but not how the autocorrelation originated.

Autocorrelation is the result of small-scale, stochastically dependent random innovations. Whereas random innovations at different locations are independent, the attribute being finally observed is the result of a *mixing* process that combines these innovations. This is the general idea behind the convolution representation of a stochastic process. It essentially relies on the idea that correlated data can be expressed as linear combinations of uncorrelated data. Consider *iid* Bernoulli($\pi$) random variables $X_1, \cdots, X_n$. Then $U = \sum_{i=1}^{k} X_i$ is a Binomial($k$,$\pi$) random variable and $V = \sum_{i=1}^{k+m} X_i$ is a Binomial($k+m$,$\pi$) random variable. Obviously, $U$ and $V$ are correlated because they share $k$ observations, $\text{Cov}[U, V] = \min(k, k+m)\pi(1-\pi)$. This idea can be generalized. Let $X_i$, $(i = 1, \cdots n)$, denote a sequence of independent random variables with common mean $\mu$ and common variance $\sigma^2$. Define a weight function $K(i,j)$. For simplicity we choose a weight function such that $\sum_{i=1}^n K(i,j) = 1$, but that is not a requirement. Correlated random variables can be created by considering the weighted averages $Y_j = \sum_{i=1}^n K(i,j)X_i$ and $Y_k = \sum_{i=1}^n K(i,k)X_i$. Then

$$\begin{aligned} \mathrm{Cov}[Y_j, Y_k] &= \mathrm{Cov}\left[\sum_{i=1}^n K(i,j)X_i, \sum_{i=1}^n K(i,k)X_i\right] \\ &= \sigma^2 \sum_{i=1}^n \sum_{m=1}^n K(i,j)K(m,k). \end{aligned} \tag{2.12}$$

The covariance between two weighted averages is governed by the weight functions.

To generate or represent a spatial random field $\{Z(\mathbf{s}) : \mathbf{s} \in D \subset \mathbb{R}^d\}$, we consider a white noise process $X(\mathbf{s})$ such that $\mathrm{E}[X(\mathbf{s})] = \mu_x$, $\mathrm{Var}[X(\mathbf{s})] = \sigma_x^2$, and $\mathrm{Cov}[X(\mathbf{s}), X(\mathbf{s}+\mathbf{h})] = C_x(\mathbf{h}) = 0, \mathbf{h} \neq \mathbf{0}$. The random field $X(\mathbf{s})$ is referred to as the **excitation field**. For geostatistical data with continuous $D$, the random field $Z(\mathbf{s})$ can be written in terms of the excitation field as

*Convolution Representation*

$$Z(\mathbf{s}) = \int_{\text{all } \mathbf{u}} K(\mathbf{s}-\mathbf{u})X(\mathbf{u})\, d\mathbf{u} = \int_{\text{all } \mathbf{v}} K(\mathbf{v})X(\mathbf{s}+\mathbf{v})\, d\mathbf{v}. \tag{2.13}$$

For a lattice process, integration is replaced with summation:

$$Z(\mathbf{s}) = \sum_{\mathbf{u}} K(\mathbf{s}-\mathbf{u})X(\mathbf{u}) = \sum_{\mathbf{v}} K(\mathbf{v})X(\mathbf{s}+\mathbf{v}). \tag{2.14}$$

Formulas (2.13) and (2.14) resemble kernel smoothers in nonparametric statistics. To evoke this parallel consider uncorrelated data $Y_1, \cdots, Y_n$, observed at design points $x_1 < x_2 < \cdots < x_n$. The data are generated according to some model

$$Y_i = f(x_i) + e_i,$$

where the $e_i$ are uncorrelated errors with mean zero and variance $\sigma^2$. A prediction of the mean of $Y$ at a particular point $x_0$ can be obtained as a weighted average of the $Y_i$. Since the mean of $Y$ changes with $x$ it is reasonable to give those observations at design points close to $x_0$ more weight than observations at design points for which $|x_0 - x_i|$ is large. Hence, a weight function $w(x_i, x_0, h)$ is considered, for example,

$$w(x_i, x_0, h) = \exp\left\{-(x_i - x_0)^2/h^2\right\}.$$

The weights are often standardized to sum to one, i.e.,

$$K(x_i, x_0, h) = \frac{w(x_i, x_0, h)}{\sum_{i=1}^n w(x_i, x_0, h)},$$

and the predicted value at point $x_0$ is

$$\widehat{f}(x_0) = \sum_{i=1}^{n} K(x_i, x_0, h) Y_i. \tag{2.15}$$

The predictor (2.15) is known as the Nadaraya-Watson estimator (Nadaraya, 1964; Watson, 1964). It is a convolution between a kernel function $K(\cdot)$ and the data. The parameter $h$, known as the bandwidth, governs the smoothness of the process. For larger $h$, the weights are distributed more evenly among the observations, and the prediction function $\hat{f}(x)$ will be smoother than for small $h$. Comparing (2.15) with (2.13) and (2.14), the function $K(\mathbf{s}-\mathbf{u})$ can be viewed as the kernel smoothing the white noise process $X(\mathbf{s})$. Several results follow immediately from the **convolution** representation (2.13).

**Lemma 2.1** *If $X(s)$ in (2.13) is a white noise process with mean $\mu_x$ and variance $\sigma_x^2$, then under some mild regularity conditions,*

*(i)* $\mathrm{E}[Z(s)] = \mu_x \int_{\mathbf{u}} K(u)\, du$;

*(ii)* $\mathrm{Cov}[Z(s), Z(s+h)] = \sigma_x^2 \int_{\mathbf{u}} K(u)K(u+h)\, du$;

*(iii) $Z(s)$ is a weakly stationary random field.*

*Proof.* The proof is straightforward and requires only standard calculus but it is dependent on being able to exchange the order of integration (the regularity conditions that permit application of Fubini's theorem). To show (iii), it is sufficient to establish that $\mathrm{E}[Z(\mathbf{s})]$ and $\mathrm{Cov}[Z(\mathbf{s}), Z(\mathbf{s}+\mathbf{h})]$ do not depend on $\mathbf{s}$. Provided the order of integration can be exchanged, tackling (i) yields

$$\begin{aligned}
\mathrm{E}[Z(\mathbf{s})] &= \int_X \int_{\mathbf{v}} K(\mathbf{v}) X(\mathbf{s}-\mathbf{v})\, d\mathbf{v} F(dx) \\
&= \int_{\mathbf{v}} K(\mathbf{v}) \int_X X(\mathbf{s}-\mathbf{v}) F(dx) d\mathbf{v} \\
&= \mu_x \int_{\mathbf{v}} K(\mathbf{v}) d\mathbf{v}.
\end{aligned}$$

To show (ii) assume that $\mu_x = 0$. The result holds in general for other values of $\mu_x$. Then,

$$\begin{aligned}
\mathrm{Cov}[Z(\mathbf{s}), Z(\mathbf{s}+\mathbf{h})] &= \int_{\mathbf{v}} \int_{\mathbf{t}} K(\mathbf{v})K(\mathbf{t})\, \mathrm{E}[X(\mathbf{s}-\mathbf{v})X(\mathbf{s}+\mathbf{h}-\mathbf{t})]\, d\mathbf{v}d\mathbf{t} \\
&= \int_{\mathbf{v}} \int_{\mathbf{t}} K(\mathbf{v})K(\mathbf{t})\, C_x(\mathbf{h}+\mathbf{v}-\mathbf{t})\, d\mathbf{v}d\mathbf{t}.
\end{aligned}$$

Since $X(\mathbf{s})$ is a white noise random field, only those terms for which $\mathbf{h}+\mathbf{v}-\mathbf{t} = \mathbf{0}$ need to be considered and the double integral reduces to

$$\mathrm{Cov}[Z(\mathbf{s}), Z(\mathbf{s}+\mathbf{h})] = \sigma_x^2 \int_{\mathbf{v}} K(\mathbf{v})K(\mathbf{h}+\mathbf{v})\, d\mathbf{v}. \tag{2.16}$$

Since the mean and covariance function of $Z(\mathbf{s})$ do not depend on spatial location, (iii) follows. This completes the proof of the lemma. Similar results, replacing integration with summation, can be established in the case of a discrete domain (2.14). □

### Example 2.3

To demonstrate the effect of convolving white noise we consider an excitation process $X(s)$ on the line and two kernel functions. The gaussian kernel function where the bandwidth $h$ corresponds to the standard deviation of the kernel and a uniform kernel function. The width of the uniform distribution was chosen so that its standard deviation also equals $h$. Figure 2.5 shows the realization of the excitation field and the realizations of the convolution (2.13) for bandwidths $h = 0.1; 0.05; 0.025$. For a given bandwidth, convolving with the uniform kernel produces realizations less smooth than those with the gaussian kernel; the uniform kernel distributes weights evenly. For a particular kernel function, the smoothness of the process decreases with the bandwidth.

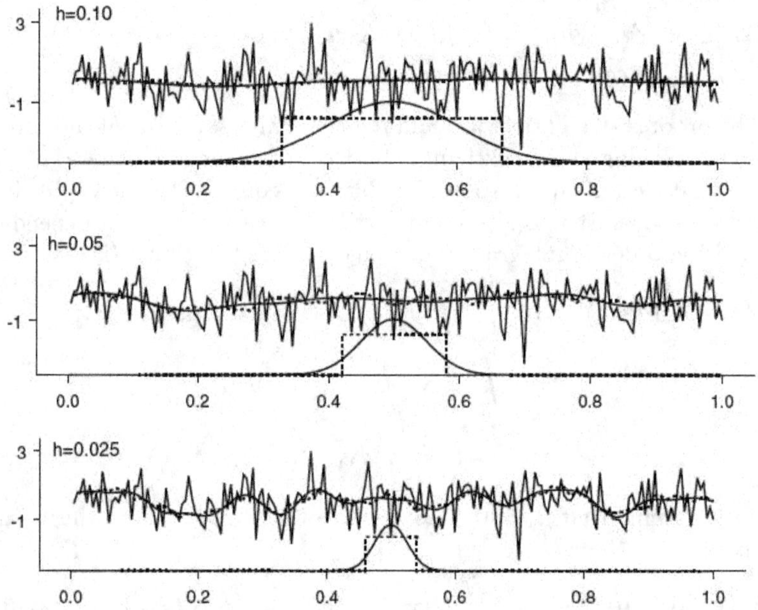


Figure 2.5 *Convolutions of Gaussian white noise with gaussian (solid line) and uniform (dotted line) kernel functions. The bandwidth h corresponds to the standard deviation of the gaussian kernel. The width of the uniform kernel was chosen to have equal standard deviation than the gaussian kernel. The jagged line represents the realization of the excitation field.*

The two kernel functions lead to distinctly different autocorrelation functions (Figure 2.6). For the uniform kernel autocorrelations decrease linearly with the lag. Once the lag exceeds one half of the window width, the correlation function is exactly zero. The gaussian kernel leads to a correlation function known as the gaussian model,

$$C(h) = \exp\left\{-h^2/\alpha^2\right\},$$

which is infinitely smooth (see §2.3). The slow decline of the correlation function with increasing lag—as compared to the correlation function under the uniform kernel—leads to smooth realizations (see Figure 2.5).

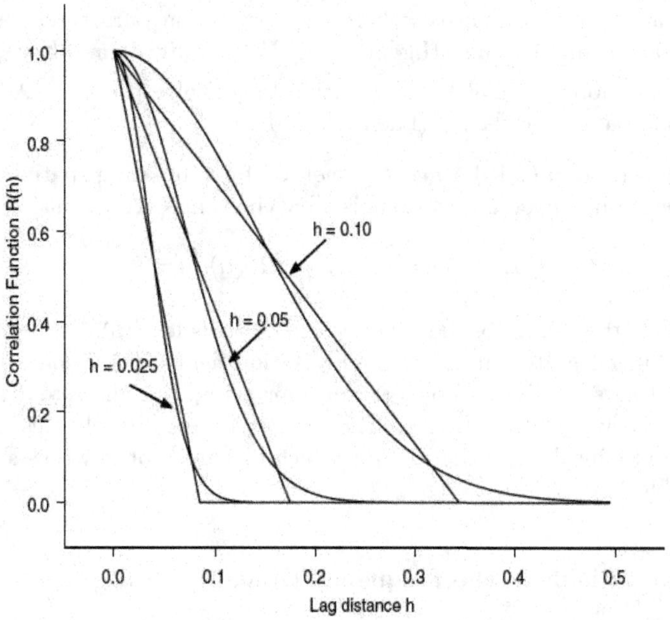


Figure 2.6 *Correlation functions determined according to (2.16) for the convolutions shown in Figure 2.5. The bandwidth h corresponds to the standard deviation of the gaussian kernel. The width of the uniform kernel was chosen to have standard deviation equal to that of the gaussian kernel.*

□

The convolution representation of a spatial process is useful in several regards.

- It describes a correlated process in terms of the excitation of a latent white noise process and defines the covariance structure of the process indirectly. Instead of asking what parametric model $\boldsymbol{\Sigma}(\boldsymbol{\theta})$ describes the spatial dependency structure, one can reformulate the question and ask which convolution kernel $K(\mathbf{s}-\mathbf{u},\gamma)$ could have given rise to the data. The estimation of the covariance parameter $\boldsymbol{\theta}$ has been supplanted by the estimation of the kernel parameters $\gamma$.
- By choosing the convolution kernel, a wide range of covariance structures can be modeled.

- Convolution techniques can be used to derive classes of covariance functions and semivariogram models based on moving averages that provide more flexibility than parametric models (see §4.6.2).
- Convolution techniques are appealing to model non-stationary processes (see §8.3.2). Two basic approaches are to
  - let the convolution kernel $K(\mathbf{u})$ depend on spatial location, i.e., $K(\mathbf{u}) = K_{\mathbf{s}}(\mathbf{u})$. For example, the kernel function for a process in $\mathbb{R}^2$ may be the bivariate Gaussian density, where the correlation parameter and variances are a function of $\mathbf{s}$ (Higdon, 1998; Higdon, Swall, and Kern, 1999).
  - use a location-invariant kernel function to convolve processes $X(\mathbf{s})$ that exhibit spatial correlation (Fuentes, 2001).
- The representation (2.13) suggests a method for simulating spatial random fields based on convolving white noise. By choosing kernels with

$$\int_{\mathbf{u}} K(\mathbf{u})d\mathbf{u} = 1, \qquad \int_{\mathbf{u}} \mathbf{u}K(\mathbf{u})d\mathbf{u} = 0,$$

  the mean and variance of the generated random field $Z(\mathbf{s})$ can be directed from the mean and variance of the excitation field $X(\mathbf{s})$. Such a method of simulating spatial data holds promise for generating discrete attributes $Z$ for which valid correlation models are not necessarily obvious, and for which valid joint distributions—from which to sample otherwise—are often intractable (see §7.4).

## 2.5 Random Fields in the Frequency Domain

Although common in the study of time series, frequency domain methods are not (yet) widely used in spatial statistics. One obstacle is probably the superficially more formidable mathematical treatment, another is unfamiliarity with the interpretation of key quantities such as the spectral density and the periodogram. Hopefully, the exposition that follows will allay these perceptions. The text in the following two subsections draws heavily on the excellent discussion in Priestley (1981, Ch. 4) and can only give the very basics of spectral analysis. Other texts that provide most readable introductions into spectral analysis (with focus on time series) are Bloomfield (1976) and Chatfield (1996). Both frequency-domain and space-domain representations of random fields are discussed in detail in the text by Vanmarcke (1983). Our exposition is an amalgam of these resources.

### *2.5.1 Spectral Representation of Deterministic Functions*

Representing functions through their frequency content has a long history in physical and engineering sciences. To see how the ideas connect to our study of random processes in the plane, we need to step back for a moment and consider deterministic functions. A (deterministic) function $f(s)$ is periodic if $f(s) = f(s + ip)$ for any $s$ and $i = \cdots, -2, -1, 0, 1, 2, \cdots$. If no $p$ exists for which this relationship holds for all $s$, then $f(s)$ is said to be non-periodic. Otherwise, the smallest $p$ for which $f(s) = f(s + ip), \forall s$ is called the period of the function. For example, $\cos(x)$ is periodic with period $2\pi$. Provided that the periodic function $f(s)$ with period $2p$ is absolutely integrable over $[-p, p]$, $f(s)$ can be written as a Fourier series

*Fourier Series*

$$f(s) = \frac{1}{2}a_0 + \sum_{n=1}^{\infty} \{a_n \cos(\pi ns/p) + b_n \sin(\pi ns/p)\}. \tag{2.17}$$

The condition of absolute integrability guarantees that the Fourier coefficients $a_0, a_1, \cdots, b_0, b_1, \cdots$ are finite. These coefficients are given by (Priestley, 1981, p. 194)

$$a_n = \frac{1}{p}\int_{-p}^{p} f(s)\cos(\pi ns/p)\, ds$$
$$b_n = \frac{1}{p}\int_{-p}^{p} f(s)\sin(\pi ns/p)\, ds.$$

In physical applications the **energy** dissipated by a process in a particular time interval is of importance. For the deterministic, periodic function with period $2p$, the total energy dissipated over the (time) interval $(-p, p)$ is a function of the Fourier coefficients. Following Priestley (1981, Ch. 4.4) define $c_0 = a_0/2$ and $c_i = (0.5(a_i^2 + b_i^2))^{1/2}, i = 1, 2, \cdots$. Then the total energy over $(-p, p)$ is $2p\sum_{i=0}^{\infty} c_i^2$. The **power** is defined as the energy per unit time, $\sum_{i=0}^{\infty} c_i^2$. The importance of the spectral representation for such functions lies in the fact that the coefficients $c_i^2$ yield the contribution to the power from the term in the Fourier series at frequency $i/(2p)$. The power spectrum is obtained by graphing $c_i^2$ against $i/(2p)$. For periodic functions this spectrum is discrete.

Most (deterministic) functions are not periodic, and adjustments must be made. We are still interested in the energy and power distribution, but the power spectrum will no longer be discrete; there will be power at all frequencies. In a sense, moving from Fourier analysis of periodic to non-periodic functions is somewhat akin to the changes incurred in switching the study of probability from discrete to continuous random variables. The switch is made by representing the function not as a Fourier series, but a Fourier integral, provided the function satisfies some additional regularity conditions. If $g(s)$ is a non-periodic, deterministic function, then, provided $g(s)$ decays to zero as $s \to \infty$ and $s \to -\infty$, and

$$\int_{-\infty}^{\infty} |g(s)|\, ds < \infty,$$

it can be represented as (Priestley, 1981, Ch. 4.5)

*Fourier Integral*

$$g(s) = \int_{-\infty}^{\infty} \{p(f)\cos(2\pi fs) + q(f)\sin(2\pi fs)\}\, df, \tag{2.18}$$

where $f$ denotes frequencies. It is convenient to change to angular frequencies $\omega = 2\pi f$ and to express (2.18) as a function of $\omega$. Notice that there is no consistency in the literature in this regard. Some authors prefer frequencies, others prefer angular frequencies. In the expressions that follow, the latter is more appealing in our opinion, since the terms in complex exponentials such as $\mathrm{i}\omega s$ are easier to read than $\mathrm{i}2\pi f s$.

The functions $p(f)$ and $q(f)$ in (2.18) are defined, unfortunately, through the inverse transformations,

$$p(\omega) = \int_{-\infty}^{\infty} g(s)\cos(\omega s)\,ds \quad q(\omega) = \int_{-\infty}^{\infty} g(s)\sin(\omega s)\,ds.$$

A more convenient formulation of these relationships is through a Fourier pair using complex exponentials ($\mathrm{i} \equiv \sqrt{-1}$):

$$g(s) = \frac{1}{\sqrt{2\pi}}\int_{-\infty}^{\infty} G(\omega)\exp\{\mathrm{i}\omega s\}\,d\omega \tag{2.19}$$

$$G(\omega) = \frac{1}{\sqrt{2\pi}}\int_{-\infty}^{\infty} g(s)\exp\{-\mathrm{i}\omega s\}\,ds. \tag{2.20}$$

An important identity can now be established, namely

$$\int_{-\infty}^{\infty} g^2(s)\,ds = \int_{-\infty}^{\infty} |G(\omega)|^2\,d\omega. \tag{2.21}$$

This shows that $|G(\omega)|^2$ represents the density of energy contributed by components of $g(s)$ in the vicinity of $\omega$. This interpretation is akin to the coefficients $c_i^2$ in the case of periodic functions, but now we are concerned with a continuous distribution of energy over frequencies. Again, this makes the point that studying frequency properties of non-periodic functions vs. periodic functions bears resemblance to the comparison of studying probability mass and density functions for discrete and continuous random variables.

Once the evolution from deterministic-periodic to deterministic-nonperiodic functions to realizations of stochastic processes is completed, the function $G(\omega)$ will be a random function itself and its average will be related to the density of power across frequencies.

One important special case of (2.19) occurs when the function $g(s)$ is real-valued and even, $g(s) = g(-s), \forall s$, since then the complex exponential can be replaced with a cosine function and does not require complex mathematics. The even deterministic function of greatest importance in this text is the covariance function $C(\mathbf{h})$ of a spatial process. But this mathematical convenience is not the only reason for our interest in (2.19) when $g(s)$ is a covariance function. It is the function that takes the place of $G(\omega)$ when we consider random processes rather than deterministic functions that is of such importance then, the spectral density function.

### *2.5.2 Spectral Representation of Random Processes*

At first glance, the extension of Fourier theory from deterministic functions to random processes appears to be immediate. After all, it was argued in §2.1 that the random processes of concern generate random **functions**. A single realization is then simply a function of temporal and/or spatial coordinates to which Fourier methods can be applied. This function—because it is the outcome of a random experiment—is, of course, most likely not periodic and one would have to consider Fourier integrals rather than Fourier series. The conditions of the previous subsection that permit the representation (2.19) are not necessarily met, however. First, it is required that the random process be second-order stationary. Consider a process $Z(s)$ on the line ($s \in D \subset \mathbb{R}^1$). The notion of a "self-replicating" process implied by stationarity is clearly at odds with the requirement that $Z(s) \to 0$ as $s \to \infty$ and $s \to -\infty$. How can the process decay to zero for large and small $s$ while maintaining the same behavior in-between? The second problem is that we observe only a single realization of the stochastic process. While we can treat this as a single function in the usual sense, our interest lies in describing the behavior of the process, not only that of the single realization. The transition from non-periodic deterministic functions to realizations of stationary random processes thus requires two important changes: (i) to recast the notion of energy and power spectra, (ii) to consider expectation operations in order to move from properties of a single realization to properties of the underlying process.

The solution to the first dilemma is to truncate the realization $Z(s)$ and to consider

$$\tilde{Z}(s) = \begin{cases} Z(s) & -S \leq s \leq S \\ 0 & s > S \text{ or } s < -S \end{cases}$$

instead. Now $\tilde{Z}(s)$ "decays" to zero as $|s| \to \infty$, and a Fourier integral can be applied, provided the condition of absolute integrability is also met:

$$\tilde{Z}(s) = \frac{1}{\sqrt{2\pi}} \int_{-\infty}^{\infty} \tilde{G}(\omega) \exp\{i\omega s\}\, d\omega \qquad (2.22)$$

$$\tilde{G}(\omega) = \frac{1}{\sqrt{2\pi}} \int_{-S}^{S} Z(s) \exp\{-i\omega s\}\, ds. \qquad (2.23)$$

Unfortunately, $|\tilde{G}(\omega)|^2$ cannot be viewed as the energy density of $Z(s)$, since (i) the truncation points $\pm S$ were arbitrary and (ii) consideration of the limit of $|\tilde{G}(\omega)|^2$ as $S \to \infty$ is not helpful. In that case we could have started with a Fourier pair based on $Z(s)$ in the first place, which is a problem because the stationary process does not decay to 0. Although the process has infinite energy on $(-\infty, \infty)$ it is possible that the power (energy per unit length) limit

$$\lim_{S \to \infty} \frac{1}{2S} |\tilde{G}(\omega)|^2 \qquad (2.24)$$

is finite ($\forall \omega$). The problem of a non-decaying realization is consequently tackled by focusing on the power of the function, rather than its energy. The conditions under which the limit (2.24) is indeed finite are surprisingly related to the rate at which the covariance function $C(h) = \mathrm{Cov}[Z(s), Z(s+h)]$ decays with increasing $h$; the continuity of the process. The problem of inferring properties of the process from a single realization is tackled by considering the expectation of (2.24). If it exists, the function

*Spectral Density*

$$s(\omega) = \lim_{S\to\infty} \mathrm{E}\left[\frac{1}{2S}|\tilde{G}(\omega)|^2\right] \tag{2.25}$$

is called the (power) **spectral density function** of the random process. We now establish the relationship between spectral density and covariance function for a process on the line because it yields a more accessible formulation than (2.25). The discussion will then extend to processes in $\mathbb{R}^d$.

### 2.5.3 *Covariance and Spectral Density Function*

Two basic approaches that connect the covariance function

$$C(h) = \mathrm{Cov}[Z(s), Z(s+h)]$$

to the spectral density function $s(\omega)$ are as follows.

(i) The realization $Z(s)$ is represented as a linear combination of sinusoids with random amplitudes and random phase angles. The covariance $C(h)$ at lag $h$ can then be expressed as a linear combination of variances at discrete frequencies and cosine terms. The variances are approximated as a rectangle, the width of which corresponds to a frequency interval, the height corresponds to the spectral mass. Upon taking the limit the spectral density function emerges.

(ii) The energy density $|\tilde{G}(\omega)|^2$ is expressed in terms of the convolution

$$\int_{-\infty}^{\infty} \tilde{Z}(s)\tilde{Z}(s-h)\,ds$$

which leads to $(2S)^{-1}|\tilde{G}(\omega)|^2$ as the Fourier transform of the sample covariance function. Taking expected values and the limit in (2.25) establishes the relationship between $s(\omega)$ and $C(h)$. This is the approach considered in Priestley (1981, pp. 212–213).

The development that follows considers approach (i) and is adapted from the excellent discussion in Vanmarcke (1983, Ch. 3.2–3.4). We commence by focusing on random processes in $\mathbb{R}^1$. The extensions to processes in $\mathbb{R}^d$ are immediate, the algebra more tedious. The extensions are provided at the end of this subsection.

#### 2.5.3.1 *Spectral Representation for Processes in* $\mathbb{R}^1$

Consider the second-order, real-valued stationary process $\{Z(s) : s \in D \subset \mathbb{R}^1\}$ with $\mathrm{E}[Z(s)] = \mu$, $\mathrm{Var}[Z(s)] = \sigma^2$. The random function $Z(s)$ can be expressed as a sum of $2K$ sinusoids

$$Z(s) = \mu + \sum_{j=-K}^{K} Y_j(s) = \mu + \sum_{j=-K}^{K} A_j \cos(\omega_j s + \phi_j). \tag{2.26}$$

The $A_j$ are random amplitudes and the $\phi_j$ are random phase angles distributed uniformly on $(0, 2\pi)$; $j = 1, \cdots, K$. All $A_j$ and $\phi_j$ are mutually independent. The $Y_j(s)$ are thus zero mean random variables because

$$(2\pi)^{-1} \int_0^{2\pi} \cos(a + \phi)\, d\phi = 0$$

and $\phi_j \perp A_j$. The frequencies are defined as $\omega_j = \pm[\Delta_\omega(2j-1)/2]$.

To determine the covariance function of the $Z$ process from (2.26) we use independence, simple geometry, and make use of property (iv) of the covariance function in §2.2: $\mathrm{Cov}[Z(s), Z(s+h)] = \mathrm{Cov}[Z(0), Z(h)]$. This leads to

$$\begin{aligned} C(h) &= \sum_{j=-K}^{K} \mathrm{E}[A_j \cos(\phi_j)\, A_j \cos(\omega_j h + \phi_j)] \\ &= \sum_{j=-K}^{K} \frac{1}{2}\mathrm{E}[A_j^2] \cos(\omega_j h) = \sum_{j=-K}^{K} \sigma_j^2 \cos(\omega_j h). \end{aligned} \tag{2.27}$$

It follows from (2.27) that $\mathrm{Var}[Z(s)] = C(0) = \sum_{j=-K}^{K} \sigma_j^2$. The variance $\sigma_j^2$ at each discrete frequency equals one-half of the average squared amplitude and the variability of the $Z$ process is being distributed over the discrete frequencies $\omega_j, (j = -K, \cdots, -1, 1, \cdots, K)$. The variance $\sigma_j^2$ can be thought of as the area of a rectangle centered at $\omega_j$, having width $\Delta_\omega$ and height $s(w_j)$:

$$\sigma_j^2 = s(\omega_j)\Delta_\omega.$$

We think of $s(\omega_j)$ as the spectral mass at frequency $w_j$. In the limit, as $\Delta\omega \to 0$ and $K \to \infty$, (2.27) becomes

$$C(h) = \int_{-\infty}^{\infty} \cos(\omega h) s(\omega)\, d\omega.$$

Using the Euler relationship $\exp\{i\omega h\} = \cos(\omega h) + i\sin(\omega h)$, the covariance can also be written as

*Fourier Pair: $C(h)$ and $s(\omega)$*

$$C(h) = \int_{-\infty}^{\infty} \exp\{i\omega h\}\, s(\omega)\, d\omega, \tag{2.28}$$

and the spectral density emerges as the Fourier transformation of $C(h)$:

$$s(\omega) = \frac{1}{2\pi} \int_{-\infty}^{\infty} C(h) \exp\{-i\omega h\}\, dh. \tag{2.29}$$

The relationship between spectral density and covariance function as a Fourier pair is significant on a number of levels.

- In §2.5.2 we introduced $s(\omega)$ as a limit of a scaled and averaged energy density and claimed that this is the appropriate function through which to study the power/energy properties of a random process in the frequency domain. Provided $s(\omega)$ exists! Since $C(h)$ is a non-periodic deterministic function, $s(\omega)$ exists when $C(h)$ satisfies the conditions in §2.5.1 for (2.20): $C(h)$ must decay to zero as $h \to \infty$ and must be absolutely integrable, $\int_{-\infty}^{\infty} |C(h)|\, dh < \infty$.

- Bochner's theorem states that for every **continuous** nonnegative function $C(h)$ with finite $C(0)$ there corresponds a nondecreasing function $dS(\omega)$ such that (2.28) holds. If $dS(\omega)$ is absolutely continuous then $dS(\omega) = s(\omega)d\omega$ (see §2.5.4 for the implications if $S(\omega)$ is a step-function). It is thus necessary that the covariance function is continuous which disallows a discontinuity at the origin (see §2.3). Bochner's theorem further tells us that $C(h)$ is positive-definite if and only if it has representation (2.28). This provides a method for constructing valid covariance functions for stochastic processes. If, for example, a function $g(h)$ is a candidate for describing the covariance in a stochastic process, then, if $g(h)$ can be expressed as (2.28) it is a valid model. If $g(h)$ does not have this representation it should not be considered. The construction of valid covariance functions from the spectral representation is discussed (for the isotropic case) in §4.3.1 and for spatio-temporal models in §9.3.
- Since the spectral density function is the Fourier transform of the covariance function, this suggests a simple method for estimating $s(\omega)$ from data. Calculate the sample covariance function at a set of lags and perform a Fourier transform. This is indeed one method to calculate an estimate of $s(\omega)$ known as the **periodogram** (see §4.7.1).

### *2.5.3.2 Extensions to $\mathbb{R}^d$*

The extensions of spectral representations to random fields of higher dimension are now given. To ease the transition, one additional representation of a stochastic process is introduced: in terms of a linear combination of a complex-valued white noise random field in the frequency domain. Let $\{Z(s) : \mathbf{s} \in D \subset \mathbb{R}^1\}$ be a (real-valued) second-order stationary random field with variance $\text{Var}[Z(s)] = \sigma^2$. Without loss of generality the mean of $Z(s)$ is assumed to be zero.

By the spectral representation theorem a real-valued random process $Z(s)$ with mean 0 can be represented as

$$Z(s) = \int_{\Omega} \exp\{i\omega s\} dX(\omega).$$

Where is the connection of this representation to (2.26)? The components $\exp\{i\omega s\} = \cos(\omega s) + i \sin(\omega s)$ play the same role as the cosine terms in (2.26), except now we include sine components. These components are associated with the infinitesimal frequency interval $d\omega$ and are amplified by a complex random amplitude $dX(\omega)$. These random components have very special properties. Their mean is zero and they are uncorrelated, representing a white noise process in the frequency domain. Because of their uncorrelatedness the covariance function can now be written as ($\overline{X}$ means complex conjugate)

$$\begin{aligned} C(h) &= \mathrm{Cov}[Z(0), Z(h)] = \mathrm{E}\left[\int_{\Omega_1} \overline{dX(\omega_1)} \int_{\Omega_2} dX(\omega_2)\right] \\ &= \int_{\Omega} \exp\{i\omega s\} \mathrm{E}\left[|dX(\omega)|^2\right]. \end{aligned}$$

Because of (2.28) it follows that $\mathrm{E}[|dX(\omega)|^2] = s(\omega)d\omega$.

This setup allows a direct extension to processes in $\mathbb{R}^d$. Let $\{Z(\mathbf{s}) : \mathbf{s} \in D \subset \mathbb{R}^d\}$ and let $X(\omega)$ be an orthogonal process—a process with zero mean and independent increments, possibly complex-valued—with $\mathrm{E}[|dX(\omega)|^2] = s(\omega)d\omega$. Then the process $Z(\mathbf{s})$ can be expressed in terms of $X(\omega)$ as

*Spectral Representation of $Z(s)$*

$$Z(\mathbf{s}) = \int_{-\infty}^{\infty} \cdots \int_{-\infty}^{\infty} \exp\{i\omega'\mathbf{s}\} dX(\omega) \tag{2.30}$$

and has covariance function

$$\begin{aligned} \mathrm{Cov}[Z(\mathbf{s}), Z(\mathbf{s}+\mathbf{h})] &= \mathrm{Cov}[Z(\mathbf{0}), Z(\mathbf{h})] = \mathrm{E}[Z(\mathbf{0})Z(\mathbf{h})] \\ &= \mathrm{E}\left[\int_{-\infty}^{\infty} \exp\{i\omega'\mathbf{0}\} dX(\omega) \int_{-\infty}^{\infty} \exp\{i\omega'\mathbf{h}\} dX(\omega)\right] \\ &= \mathrm{E}\left[\int_{-\infty}^{\infty} dX(\omega) \int_{-\infty}^{\infty} \exp\{i\omega'\mathbf{h}\} dX(\omega)\right] \\ &= \int_{-\infty}^{\infty} \exp\{i\omega'\mathbf{h}\} \mathrm{E}[|dX(\omega)|^2] \\ &= \int_{-\infty}^{\infty} \exp\{i\omega'\mathbf{h}\} s(\omega) d\omega, \end{aligned}$$

where the integrals are understood to be $d$-dimensional. Since $C(\mathbf{h})$ is an even function,$C(\mathbf{h}) = C(-\mathbf{h})$, if $Z(\mathbf{s})$ is real-valued we can also write

$$C(\mathbf{h}) = \int_{-\infty}^{\infty} \cdots \int_{-\infty}^{\infty} \cos(\omega'\mathbf{h}) s(\omega)\, d\omega.$$

The covariance function $C(\mathbf{h})$ and the spectral density function $s(\omega)$ form a Fourier transform pair,

$$C(\mathbf{h}) = \int_{-\infty}^{\infty} \cdots \int_{-\infty}^{\infty} \exp\{i\omega'\mathbf{h}\} s(\omega)\, d\omega \tag{2.31}$$

$$s(\omega) = \frac{1}{(2\pi)^d} \int_{-\infty}^{\infty} \cdots \int_{-\infty}^{\infty} \exp\{-i\omega'\mathbf{h}\} C(\mathbf{h})\, d\mathbf{h}, \tag{2.32}$$

Notice that this is an application of the inversion formula for characteristic functions. $C(\mathbf{h})$ is the characteristic function of a $d$-dimensional random variable whose cumulative distribution function is $S(\omega)$. The continuity of the characteristic function disallows spectral representations of stochastic processes whose covariance functions are discontinuous at the origin, whether the process is in $\mathbb{R}^1$ or in $\mathbb{R}^d$.

Also notice the similarity of (2.30) to the convolution representation (2.13). Both integrate over a stochastic process of independent increments. The convolution representation operates in the spatial domain, (2.30) operates in the frequency domain. This correspondence between convolution and spectral representation can be made more precise through linear filtering techniques (§2.5.6). But first we consider further properties of spectral density functions.

Up to this point we have tacitly assumed that the domain $D$ is continuous. Many stochastic processes have a discrete domain and lags are restricted to an enumerable set. For example, consider a process on a rectangular $r \times c$ row-column lattice. The elements of the lag vectors $\mathbf{h} = [h_1, h_2]'$ consist of the set $(h_1, h_2) : h_1, h_2 = 0, \pm 1, \pm 2, \cdots$. The first modification to the previous formulas is that integration in the expression for $s(\boldsymbol{\omega})$ is replaced by summation. Let $\boldsymbol{\omega} = [\omega_1, \omega_2]'$. Then,

$$s(\boldsymbol{\omega}) = \frac{1}{(2\pi)^2} \sum_{h_1=-\infty}^{\infty} \sum_{h_2=-\infty}^{\infty} C(\mathbf{h}) \cos(\omega_1 h_1 + \omega_2 h_2).$$

The second change is the restriction of the frequency domain to $[-\pi, \pi]$, hence

$$C(\mathbf{h}) = \int_{-\pi}^{\pi} \int_{-\pi}^{\pi} \cos(\boldsymbol{\omega}'\mathbf{h}) s(\boldsymbol{\omega})\, d\boldsymbol{\omega}.$$

Note that we still assume that the spectral density is continuous, even if the spatial domain is discrete. Continuity of the domain and continuity of the spectral distribution function $dS(\boldsymbol{\omega})$ are different concepts.

### 2.5.4 Properties of Spectral Distribution Functions

Assume that the stochastic process of concern is a real-valued, second-order stationary spatial process in $\mathbb{R}^2$ with continuous domain. Both lags and frequencies range over the real line and the relevant Fourier pair is

$$\begin{aligned} C(\mathbf{h}) &= \int_{-\infty}^{\infty} \int_{-\infty}^{\infty} \cos(\boldsymbol{\omega}'\mathbf{h}) s(\boldsymbol{\omega})\, d\boldsymbol{\omega} \\ s(\boldsymbol{\omega}) &= \frac{1}{(2\pi)^2} \int_{-\infty}^{\infty} \int_{-\infty}^{\infty} \cos(\boldsymbol{\omega}'\mathbf{h}) C(\mathbf{h})\, d\mathbf{h}. \end{aligned}$$

*Integrated Spectrum*

If $s(\boldsymbol{\omega})$ exists for all $\boldsymbol{\omega}$ we can introduce the **integrated spectrum**

$$S(\boldsymbol{\omega}) = S(\omega_1, \omega_2) = \int_{-\infty}^{\omega_1} \int_{-\infty}^{\omega_2} s(\vartheta_1, \vartheta_2)\, d\vartheta_1 d\vartheta_2. \qquad (2.33)$$

Several interesting properties of the integrated spectrum can be derived

- $S(\infty, \infty) = \text{Var}[Z(\mathbf{s})]$. This follows directly because $C(\mathbf{0}) = \text{Var}[Z(\mathbf{s})] \equiv \sigma^2$. The variance of the process thus represents the total power contributed by all frequency components and $s(\boldsymbol{\omega})d\boldsymbol{\omega}$ is interpreted as the contribution to the total power (variance) of the process from components in $Z(\mathbf{s})$ with frequency in $\omega_1 + d\omega_1 \times \omega_2 + d\omega_2$.

- $S(-\infty, -\infty) = 0$.
- $S(\boldsymbol{\omega})$ is a non-decreasing function of $\boldsymbol{\omega}$. That is, if $\omega_2 > \omega_1$ and $\vartheta_2 > \vartheta_1$, then

$$S(\omega_2, \vartheta_2) - S(\omega_2, \vartheta_1) - S(\omega_1, \vartheta_2) + S(\omega_1, \vartheta_1) \geq 0.$$

In the one-dimensional case this simply states that if $\omega_2 > \omega_1$ then $S(\omega_2) \geq S(\omega_1)$. This property follows from the fact that $s(\omega_1, \omega_2) \geq 0 \,\forall \omega_1, \omega_2$, which in turn is easily seen from (2.25).

The integrated spectrum behaves somewhat like a bivariate distribution function, except it does not integrate to one. This is easily remedied by introducing the **normalized integrated spectrum** $F(\boldsymbol{\omega}) = \sigma^{-2}S(\boldsymbol{\omega})$. This function shares several important properties with the cumulative distribution function of a bivariate random vector:

*Normalized Integrated Spectrum*

- $0 \leq F(\boldsymbol{\omega}) \leq 1$.
- $F(-\infty, -\infty) = 0$ and $F(\infty, \infty) = 1$.
- $F(\boldsymbol{\omega})$ is non-decreasing (in the sense established above).

Because of these properties, $F(\boldsymbol{\omega})$ is also called the spectral distribution function and the label *spectral density function* is sometimes attached to

*Normalized Spectral Density Function*

$$f(\omega_1, \omega_2) = \frac{\partial^2}{\partial \omega_1 \partial \omega_2} F(\omega_1, \omega_2),$$

if $f(\omega_1, \omega_2)$ exists. Since we have already termed $s(\boldsymbol{\omega})$ the spectral density function (sdf) and $f(\boldsymbol{\omega}) = s(\boldsymbol{\omega})/\sigma^2$, we call $f(\boldsymbol{\omega})$ the normalized sdf (nsdf) to avoid confusion. The sdf and covariance function form a Fourier pair; similarly, the nsdf and the correlation function $R(\mathbf{h}) = C(\mathbf{h})/C(\mathbf{0}) = \text{Corr}[Z(\mathbf{s}), Z(\mathbf{s}+\mathbf{h})]$ form a Fourier pair:

$$R(\mathbf{h}) = \int_{-\infty}^{\infty} \int_{-\infty}^{\infty} \cos(\boldsymbol{\omega}'\mathbf{h}) f(\boldsymbol{\omega})\, d\boldsymbol{\omega} \tag{2.34}$$

$$f(\boldsymbol{\omega}) = \frac{1}{(2\pi)^2} \int_{-\infty}^{\infty} \int_{-\infty}^{\infty} \cos(\boldsymbol{\omega}'\mathbf{h}) R(\mathbf{h})\, d\mathbf{h}. \tag{2.35}$$

The study of spectral distributions is convenient because (i), the integrated functions $S(\boldsymbol{\omega})$ and $F(\boldsymbol{\omega})$ have properties akin to cumulative distribution functions, and (ii), properties of the spectral functions teach us about properties of the covariance (correlation) function and vice versa. Concerning (i), consider integrating $F(\boldsymbol{\omega})$ over individual coordinates in the frequency domain, this will marginalize the spectral distribution function. In $\mathbb{R}^2$, for example, $F_1(\omega) = \int_{-\infty}^{\infty} F(\omega, \vartheta_2) d\vartheta_2$ is the spectral distribution function of the random field $Z(s_1, s_2)$ for a fixed coordinate $s_2$. The variance of the process in the other coordinate can be obtained from the marginal spectral density as $F_1(\infty) = \int_{-\infty}^{\infty} F(\infty, \vartheta_2) d\vartheta_2$. Because

$$f(\boldsymbol{\omega}) = \frac{\partial^d F(\omega_1, \cdots, \omega_d)}{\partial\omega_1 \cdots \partial\omega_d},$$

the spectral densities can also be marginalized,

$$f_1(\omega) = \int_{-\infty}^{\infty} \cdots \int_{-\infty}^{\infty} f(\omega, \omega_2, \cdots, \omega_d)\, d\omega_2, \cdots, d\omega_d.$$

Concerning (ii), it is easy to establish that

- for a real valued process $C(\mathbf{h})$ is an even function, and as a result, the sdf is also an even function, $s(\boldsymbol{\omega}) = s(-\boldsymbol{\omega})$;
- if $C(\mathbf{h})$ is reflection symmetric so that $C(h_1, h_2) = C(-h_1, h_2)$, the spectral density function is also reflection symmetric, $s(\omega_1, \omega_2) = s(-\omega_1, \omega_2)$. Conversely, reflection symmetry of $s(\boldsymbol{\omega})$ implies reflection symmetry of $C(\mathbf{h})$. The proof is straightforward and relies on the evenness of $C(\mathbf{h})$, $s(\boldsymbol{\omega})$, and $\cos(x)$;
- a random field in $\mathbb{R}^2$ is said to have a separable covariance function $C(\mathbf{h})$, if $C(\mathbf{h}) = C(h_1, h_2) \equiv C_1(h_1)C_2(h_2)$ for all $h_1, h_2$, where $C_1(\cdot)$ and $C_2(\cdot)$ are valid covariance functions. In this case the sdf also factors, $s(\omega_1, \omega_2) = s_1(\omega_1)s_2(\omega_2)$. Similarly, if the spectral density function can be factored as $s(\omega_1, \omega_2) = g(\omega_1)h(\omega_2)$, then the covariance function is separable.

### *2.5.5 Continuous and Discrete Spectra*

The relationship between $F(\boldsymbol{\omega})$ and $f(\boldsymbol{\omega})$ is reminiscent of the relationship between a cumulative distribution function and a probability density function (pdf) for a continuous random variable. In the study of random variables, $F(y) = \Pr(Y \leq y)$ always exists, whereas the existence of the pdf requires absolute continuity of $F(y)$. Otherwise we are led to a probability mass function $p(y) = \Pr(Y = y)$ of a discrete random variable (unless the random variable has a mixed distribution). A similar dividing line presents itself in the study of spectral properties. If $F(\boldsymbol{\omega})$ is absolutely continuous, then $f(\boldsymbol{\omega})$ exists. This in turn implies that the covariance function is absolute integrable, i.e., $C(\mathbf{h})$ must decrease quickly enough to 0 as the elements of $\mathbf{h}$ grow to $|\infty|$. If the covariance function does not diminish quickly enough, $F(\boldsymbol{\omega})$ exists but $f(\boldsymbol{\omega})$ may not. The apparent difficulty this presents in the Fourier expressions can be overcome by representing the transforms as Fourier-Stieltjes integrals, made possible by the celebrated Wiener-Khintchine theorem,

$$R(\mathbf{h}) = \int_{-\infty}^{\infty} \cdots \int_{-\infty}^{\infty} \exp\{i\boldsymbol{\omega}'\mathbf{h}\} dF(\boldsymbol{\omega}).$$

Priestley (1981, pp. 219–222) outlines the proof of the theorem and establishes the connection to Bochner's theorem.

Although this representation accommodates both continuous and non-continuous normalized integrated spectra, it does not convey readily the implications of these cases. Consider a process in $\mathbb{R}^1$. If $F(\omega)$ is a step function with jumps $p_i$ at an enumerable set of frequencies $\Omega = \{\omega_1, \omega_2, \cdots\}$, where $\sum_i p_i = 1$, then the spectrum is said to be discrete. In this case

$$\frac{dF(\omega)}{d\omega} = \begin{cases} 0 & \omega \in \Omega \\ \infty & \omega \notin \Omega \end{cases}$$

We can still conceive of a normalized spectral "density" function by using a Dirac delta function:

$$f(\omega) = \sum_{i=1}^{\infty} p_i \delta(\omega - \omega_i),$$

where

$$\delta(x) = \begin{cases} \infty & x = 0 \\ 0 & x \neq 0. \end{cases}$$

The spectral density of a process with discrete spectrum has infinite peaks at the frequencies $\omega_1, \omega_2, \cdots$, and is zero elsewhere. Processes with discrete spectrum have strictly periodic components and the spikes of infinite magnitude correspond to the frequencies of periodicity.

If $dF(\omega)/d\omega = f(\omega)$ exists for all $\omega$, then $F(\omega)$ is continuous and so is the nsdf. This type of spectrum is encountered for non-periodic functions. If a continuous spectrum exhibits a large—but finite—spike at some frequency $\omega^*$, it is indicated that the process has a near-periodic behavior at that frequency.

When examining continuous spectral densities it is helpful to remember that when a function $g(x)$ is relatively flat, its Fourier transform will be highly compressed and vice versa. If the autocorrelation function drops off sharply with increasing lag, the spectral density function will be rather flat. If high autocorrelations persist over a long range of lags, the spectral density function will be narrow and compressed near the zero frequency.

Consider a process in $\mathbb{R}^1$ with correlation function

$$\text{Corr}[Z(s), Z(s+h)] = \exp\{-3|h|/\alpha\} \qquad \alpha > 0.$$

This correlation function is known as the exponential correlation model, its genesis is more fully discussed in Chapter 4. Here we note that the parameter $\alpha$ represents the **practical range** of the process, i.e., the lag distance at which the correlations have diminished to (approximately) 0.05. The larger $\alpha$, the higher the correlations at a particular lag distance (Figure 2.7a). Substituting into (2.35) the nsdf can be derived as

$$\begin{aligned} f(\omega) &= \frac{1}{2\pi}\int_{-\infty}^{\infty} \cos(\omega h)\exp\{-3h/\alpha\}\, dh \\ &= \frac{1}{\pi}\int_{0}^{\infty} \cos(\omega h)\exp\{-3h/\alpha\}\, dh \\ &= \frac{3}{\pi}\frac{\alpha}{9+\omega^2\alpha^2}. \end{aligned}$$

The spectral density function for a process with $\alpha$ small is flatter than the sdf for a process with $\alpha$ large (Figure 2.7b).

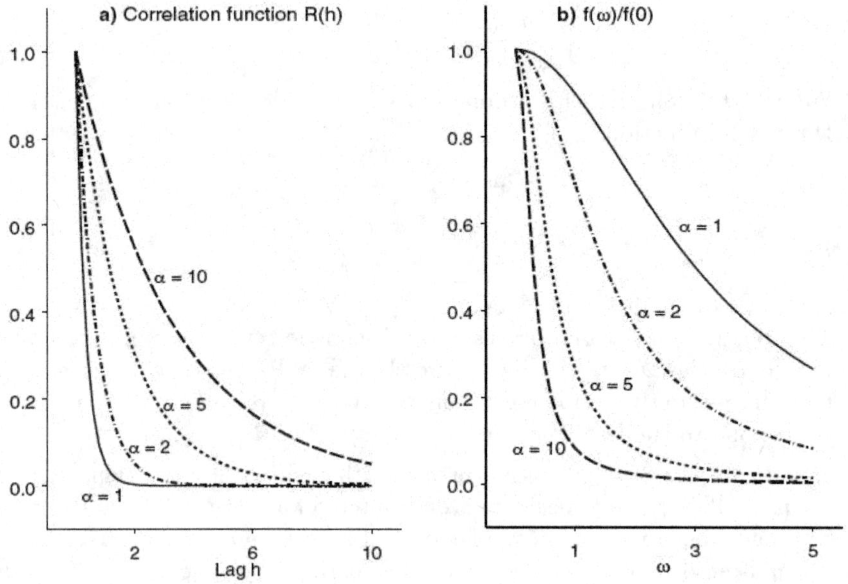


Figure 2.7 *Autocorrelation function and spectral density functions for processes with exponential correlation structure and different ranges. Panel b displays $f(\omega)/f(0)$ to amplify differences in shape, rather than scale.*

### *2.5.6 Linear Location-Invariant Filters*

At the beginning of §2.5 we noted the similarity between the spectral representation (2.30) of the field $Z(\mathbf{s})$ and the convolution representation (2.13). The correspondence can be made more precise by considering linear filtering techniques. Wonderful expositions of linear filtering and its applications to spectral analysis of random fields can be found in Thiébaux and Pedder (1987, Ch. 5.3) and in Percival and Walden (1993, Ch.5). Our comments are based on these works.

A **linear filter** is defined as a linear, location-invariant transformation of one variable to another. The **input** variable is the one to which the linear transformation is applied, the resulting variable is called the **output** of the filter. We are interested in applying filtering techniques to random fields, and denote the linear filtering operation as

$$Z(\mathbf{s}) = \mathcal{L}\{Y(\mathbf{s})\}, \tag{2.36}$$

to depict that $Z(\mathbf{s})$ is the output when the linear filter $\mathcal{L}$ is applied to the process $Y(\mathbf{s})$. A linear filter is defined through the properties

*Properties of a Linear Filter*

(i) For any constant $a$, $\mathcal{L}\{aY(\mathbf{s})\} = a\mathcal{L}\{Y(\mathbf{s})\}$;

(ii) If $Y_1(\mathbf{s})$ and $Y_2(\mathbf{s})$ are any two processes, then $\mathcal{L}\{Y_1(\mathbf{s}) + Y_2(\mathbf{s})\} = \mathcal{L}\{Y_1(\mathbf{s})\} + \mathcal{L}\{Y_2(\mathbf{s})\}$;

(iii) If $Z(\mathbf{s}) = \mathcal{L}\{Y(\mathbf{s})\}$ for all $\mathbf{s}$, then $Z(\mathbf{s}+\mathbf{h}) = \mathcal{L}\{Y(\mathbf{s}+\mathbf{h})\}$.

The linear filter is a scale-preserving, linear operator [(i) and (ii)] and is not affected by shifts in the origin (iii). It is easy to establish that the convolution (2.13) is a linear filter with input $X(\mathbf{s})$ (Chapter problems). It would be more appropriate to write the linear filter as $\mathcal{L}\{Y(\cdot)\} = Z(\cdot)$ since it associates a function with another function. The notation we choose here suggests that the filter associates a point with a point and should be understood to imply that the filter maps the function defined on a point-by-point basis by $Y(\mathbf{s})$ to the function defined on a point-by-point basis by $Z(\mathbf{s})$.

The excitation field of a convolution has a spectral representation, provided it is mean square continuous. Hence,

$$X(\mathbf{s}) = \int_{d\omega} \exp\{i\omega'\mathbf{s}\}\, U(d\omega), \tag{2.37}$$

where $U(\omega)$ is an orthogonal process in the frequency domain and $d\omega$ is an infinitesimal region in that space. Combining (2.37) with (2.13), we obtain

$$\begin{aligned} Z(\mathbf{s}) &= \int_{\mathbf{v}} K(\mathbf{v})X(\mathbf{s}-\mathbf{v})\, d\mathbf{v} = \int_{\mathbf{v}} K(\mathbf{v})d\mathbf{v} \int_{d\omega} \exp\{i\omega'(\mathbf{s}-\mathbf{v})\}\, U(d\omega) \\ &= \int_{\mathbf{u}} K(\mathbf{u}) \exp\{i\omega'\mathbf{u}\}\, d\mathbf{u} \int_{d\omega} \exp\{i\omega'\mathbf{s}\}\, U(d\omega) \\ &= \int_{d\omega} H(\omega) \exp\{i\omega'\mathbf{s}\}\, U(d\omega) = \mathcal{L}\{X(\mathbf{s})\}. \end{aligned} \tag{2.38}$$

The complex function $H(\omega)$ is called the **frequency response function** or the **transfer function** of the linear system. As the covariance function and the spectral density function, the kernel function $K(\mathbf{s})$ and the transfer function form a Fourier transform pair:

*Transfer Function*

$$\begin{aligned} H(\omega) &= \int_{\mathbf{u}} K(\mathbf{u}) \exp\{i\omega'\mathbf{u}\}\, d\mathbf{u} \\ K(\mathbf{s}) &= \frac{1}{(2\pi)^d} \int_{d\omega} H(\omega) \exp\{-i\omega'\mathbf{s}\}\, d\omega. \end{aligned}$$

Since $X(\mathbf{s})$ has a spectral representation, so does $Z(\mathbf{s})$. Given an orthogonal process $P(\omega)$ for which

$$Z(\mathbf{s}) = \int_{\omega} \exp\{i\omega'\mathbf{s}\}\, P(d\omega),$$

we see from (2.38) that the two frequency domain processes must be related,

$$P(d\omega) = H(\omega)U(d\omega).$$

As a consequence, the spectral densities of the $Z$ and $X$ processes are also related. Since $\mathrm{E}[|P(d\omega)|^2] = s_z(\omega)d\omega$ and $\mathrm{E}[|U(d\omega)|^2] = s_x(\omega)d\omega$, we have

$$s_z(\omega) = |H(\omega)|^2 s_x(\omega). \tag{2.39}$$

The spectral density function of the $Z$ process is the product of the squared modulus of the transfer function and the spectral density function of the input field $X(\omega)$. This result is important because it enables us to construct one spectral density from another, provided the transfer function and the spectral density function of either $Z$ or $X$ are known, for example

$$s_x(\omega) = |H(\omega)|^{-2} s_z(\omega).$$

*Filtering Complex Exponentials*

The transfer function relates the spectra of the input to a linear filter in a simple fashion. It is noteworthy that the relationship is independent of location. We can learn more about the transfer function $H(\omega)$. Consider the one-dimensional case for the moment. Since spectral representations involve complex exponentials, consider the function $\varepsilon_\omega(s) \equiv \exp\{i\omega s\}$ for a particular frequency $\omega$. When $\varepsilon_\omega(s)$ is processed by a linear location-invariant filter, we can write the output $y_\omega(s+h)$ as

$$\begin{aligned} y_\omega(s+h) &= \mathcal{L}\{\varepsilon_\omega(s+h)\} = \mathcal{L}\{\exp\{i\omega h\}\varepsilon_\omega(s)\} \\ &= \exp\{i\omega h\}\mathcal{L}\{\varepsilon_\omega(s)\} = \exp\{i\omega h\}y_\omega(s), \end{aligned}$$

for any shift $h$. We have made use here of the scale preservation and the location invariance properties of the linear filter. Since the result holds for any $s$, it also holds for $s = 0$ and we can write for the output of the filter

$$y_\omega(h) = \exp\{i\omega h\}y_\omega(0).$$

Since the shift $h$ can take on any value we also have

$$y_\omega(s) = \exp\{i\omega s\}y_\omega(0) = \varepsilon_\omega(s)H(\omega).$$

Is it justified to call the coefficient of $\exp\{i\omega s\}$ the transfer function? Let a realization of $X(s)$ be given by

$$x(s) = \int_{-\infty}^{\infty} \exp\{i\omega s\}\, u(d\omega).$$

Then $\mathcal{L}\{x(s)\} = \int_{-\infty}^{\infty} \exp\{i\omega s\}H(\omega)u(d\omega)$ and

$$\mathcal{L}\{X(s)\} = \int_{-\infty}^{\infty} \exp\{i\omega s\}\, H(\omega)U(d\omega),$$

which is (2.38). This result is important, because it allows us to find the transfer function of a filter by inputting the sequence defined by $\exp\{i\omega s\}$.

**Example 2.4** Consider a discrete autoregressive time-series of first order, the AR(1) process,

$$X_t = \alpha X_{t-1} + e_t,$$

where the $e_t$ are independent innovations with mean 0 and variance $\sigma^2$, and $\alpha$ is the autoregressive parameter, a constant. Define the linear filter $\mathcal{L}\{X_t\} = e_t$, so that $\mathcal{L}\{u_t\} = u_t - \alpha u_{t-1}$ for a sequence $\{u_t\}$. To find the transfer function of the filter, input the sequence $\{\exp\{i\omega t\}\}$:

$$\mathcal{L}\{\exp\{i\omega t\}\} = \exp\{i\omega t\} - \alpha\exp\{i\omega(t-1)\} = \exp\{i\omega t\}(1 - \alpha\exp\{-i\omega\}).$$

The transfer function of the filter is $H(\omega) = 1 - \alpha\exp\{-i\omega\}$ since this is the coefficient of $\exp\{i\omega t\}$ on the filter output. But the output of the filter was $e_t$, a sequence of independent, homoscedastic random innovations. The spectral densities $s_e(\omega)$ and $s_X(\omega)$ are thus related by (2.39) as

$$s_e(\omega) = |H(\omega)|^2 s_X(\omega),$$

and thus

$$s_X(\omega) = \frac{\sigma^2}{|1 - \alpha\exp\{-i\omega\}|^2}.$$

Linear filtering and the specific results for filtering complex exponentials made it easy to find the spectral density function of a stochastic process, here, the AR(1) process. □

### *2.5.7 Importance of Spectral Analysis*

The spectral representation of random fields may appear cumbersome at first; mathematics in the frequency domain require operations with complex-valued (random) variables and the interpretation of the spectrum is not (yet) entirely clear. That the spectral density function distributes the variability of the process over the frequency domain is appealing, but "so what?" The following are some reasons for the increasing importance of spectral methods in spatial data analysis.

- Mathematical proofs and derivations are often simpler in the frequency domain. The skillful statistician working with stochastic processes switches back and forth between the spatial and the frequency domain, depending on which "space" holds greater promise for simplicity of argument and derivation.
- The spectral density function and the covariance function of a stationary stochastic process are closely related; they form a Fourier transform pair. On this ground, studying the second-order properties of a random field via the covariance function or the spectral density can be viewed as equivalent. However,
  - the spectral density and the covariance are two different but complementary representations of the second-order properties of a stochastic

process. The covariance function emphasizes spatial dependency as a function of coordinate separation. The spectral density function emphasizes the association of components of variability with frequencies. From the covariance function we glean the degree of continuity and the decay of spatial autocorrelation with increasing point separation. From the spectral density we glean periodicity in the process;

- it is often difficult in practice to recognize the implications of different covariance structures for statistical inference from their mathematical form or from a graph of the functions alone. Processes that differ substantially in their stochastic properties can have covariance functions that appear rather similar when graphed. The spectral density function can amplify and highlight subtle differences in the second-order structure more so than the covariance functions.

- The spectral density function—as the covariance function—can be estimated from data via the periodogram (see §4.7.1). Computationally this does not provide any particular challenges beyond computing the sample covariances, at least if data are observed on a grid. Summary statistics calculated from data in the spatial domain are usually correlated. This correlation stems either from the fact that the same data point $Z(\mathbf{s}_i)$ is repeatedly used in multiple summaries, and/or from the spatial autocorrelation. The ordinates of the periodogram, the data-based estimate of the spectral density function, are—at least asymptotically—independent and have simple distributional properties. This enables you to construct test statistics with standard properties.
- The derivation of the spectral representation and its ensuing results requires mean-square continuous, second-order stationary random fields. Studying second-order properties of random fields in the spatial domain often requires, in addition, isotropy of the process. An example is the study of spatial dependence in point patterns (see §3.4). The $K$-function due to Ripley (1976) is a useful device to study stochastic dependence between random events in space. Many arguments favor the $K$-function approach, probably most of all, interpretability. It does, however, require isotropy. Establishing whether a point pattern is isotropic or anisotropic in the spatial domain is tricky. A spectral analysis requires *only* second-order stationarity and the stochastic dependence among events can be gleaned from an analysis of the pattern's spectra. In addition, the spectral analysis allows a simple test for anisotropy.

## 2.6 Chapter Problems

**Problem 2.1** Let $Y(t) = Y(t-1) + e(t)$ be a random walk with independent innovations $E(t) \sim (0, \sigma^2)$. Show that the process is not second-order stationary, but that the differences $D(t) = Y(t) - Y(t-1)$ are second-order stationary.

**Problem 2.2** For $\gamma(\mathbf{s}_i - \mathbf{s}_j)$ to be a valid semivariogram for a spatial process $Z(\mathbf{s})$, it must be conditionally negative-definite, i.e.,

$$\sum_{i=1}^{m}\sum_{j=1}^{m} a_i a_j \gamma(\mathbf{s}_i - \mathbf{s}_j) \leq 0,$$

for any number of sites $\mathbf{s}_1, \cdots, \mathbf{s}_m$ and real numbers $a_1, \cdots, a_m$ with $\sum_{i=1}^{m} a_i = 0$. Why is that?

**Problem 2.3** A multivariate Gamma random field can be constructed as follows. Let $X_i$, $i = 1, \cdots, n$, be independent Gamma$(\alpha_i, \beta)$ random variables so that $\mathrm{E}[X_i] = \alpha_i\beta$ and $\mathrm{Var}[X_i] = \alpha_i\beta^2$. Imagine a regularly spaced transect consisting of $Z(s_i) = X_0 + X_i$, $i = 1, \cdots, n$. Find the covariance function $C(i,j) = \mathrm{Cov}[Z(s_i), Z(s_j)]$ and the correlation function. Is this a stationary process? What is the distribution of the $Z(s_i)$?

**Problem 2.4** Consider the convolution representation (2.13) of a white noise excitation field for a random process on the line. If the convolution kernel is the gaussian kernel

$$K(s-u,h) = \frac{1}{\sqrt{2\pi h^2}} \exp\left\{ -\frac{1}{2}\frac{(s-u)^2}{h^2} \right\},$$

where $h$ denotes the bandwidth of the kernel, find an expression for the autocorrelation function.

**Problem 2.5** The convolution representation (2.13) of a random field can be used to simulate random fields. Assume you want to generate a random field $Z(\mathbf{s})$ in which the data points are autocorrelated, $\mathrm{Cov}[Z(\mathbf{s}), Z(\mathbf{s}+\mathbf{h})] = C(\mathbf{h})$, but marginally, the $Z(\mathbf{s})$ should be "Poisson-like." That is, you want $\mathrm{E}[Z(\mathbf{s})] = \lambda$, $\mathrm{Var}[Z(\mathbf{s})] = \lambda$, and the $Z(\mathbf{s})$ should be non-negative. Find a random field $X(\mathbf{s})$ that can be used as the excitation field. *Hint*: Consider $X$ such that $\mathrm{Var}[X] = \mathrm{E}[X]\theta$, for example, a negative binomial random variable.

**Problem 2.6** For a non-periodic, deterministic function $g(s)$, prove (2.21).

**Problem 2.7** Consider a random field in the plane with covariance function $C(\mathbf{h}) = C(h_1, h_2)$, where $h_1$ and $h_2$ are the coordinate shifts in the $x$- and $y$-directions. The covariance function is called **separable** if $C(h_1, h_2) = C_1(h_1)C_2(h_2)$. Here, $C_1$ is the covariance function of the random process $Z(x)$ along lines parallel to the $x$-axis. Show that separability of the covariance function implies separability of the spectral density.

**Problem 2.8** Establish that the convolution (2.13) is a linear, location-invariant filter.

**Problem 2.9** Establish that the derivative $\mathcal{L}\{Z(s)\} = dZ(s)/ds = W(s)$ is a linear filter. Use this result to find the spectral density of $W$ in terms of the spectral density of $Z$.

CHAPTER 3

# Mapped Point Patterns

## 3.1 Random, Aggregated, and Regular Patterns

The realization of a point process $\{Z(\mathbf{s}) : \mathbf{s} \in D \subset \mathbb{R}^2\}$ consists of an arrangement (pattern) of points in the random set $D$. These points are termed the events of the point process. If the events are only partially observed, the recorded pattern is called a **sampled** point pattern. When all events of a realization are recorded, the point pattern is said to be **mapped**. In this chapter we consider mapped point patterns in $\mathbb{R}^1$ (line processes) and $\mathbb{R}^2$ (spatial processes).

Since $D$ is a random set, the experiment that generates a particular realization can be viewed as a random draw of locations in $D$ at which events are observed. From this vantage point, all mapped point patterns are realizations of random experiments and the coarse distinction of patterns into (completely) random, clustered (spatially aggregated), and regular ones should not lead to the false impression that the latter two types of patterns are void of a random mechanism. A point pattern is called a **completely random pattern** if the following criteria are met. The average number of events per unit area—the intensity $\lambda(\mathbf{s})$—is homogeneous throughout $D$, the number of events in two non-overlapping subregions (Borel sets) $A_1$ and $A_2$ are independent, and the number of events in any subregion is Poisson distributed. Thus, events distribute uniformly and independently throughout the domain. The mathematical manifestation of complete spatial randomness is the homogeneous Poisson process (§3.2.2). It is a process void of any spatial structure and serves as the null hypothesis for many statistical investigations into point patterns. Observed point patterns are tested initially against the hypothesis of a complete spatial random (CSR) pattern. If the CSR hypothesis is rejected, then the investigator often follows up with more specific analyses that shed additional light on the nature of the spatial point pattern.

*Complete Spatial Randomness*

Diggle (1983) calls the homogeneous Poisson process an "unattainable standard." Most processes deviate from complete spatial randomness in some fashion. Events may be independent in non-overlapping subregions, but the intensity $\lambda(\mathbf{s})$ with which they occur is not homogeneous throughout $D$. More events will then be located in regions where the intensity is large, fewer events will be located in regions where $\lambda(\mathbf{s})$ is small. Events may occur with a constant (average) intensity $\lambda(\mathbf{s}) \equiv \lambda$ but exhibit some form of interaction. The presence of an event can attract or repel other events nearby. Accordingly, deviations from the CSR pattern are coarsely distinguished as aggregated (clustered) or regular patterns (Figure 3.1). Since the reason for this deviation may be deterministic spatial variation of the intensity function $\lambda(\mathbf{s})$ and/or the result of a stochastic element, we do not associate any mechanism with clustering or regularity yet. Different point process models achieve a certain deviation from the CSR pattern in different ways (see §3.7). At this point we distinguish random, regular, and clustered patterns simply on the basis of their relative patterns. In a clustered pattern, the average distance between an event $\mathbf{s}_i$ and its nearest-neighbor event is smaller than the same average distance in a CSR pattern. Similarly, in a regular pattern the average distance between an event and its nearest neighbor is larger than expected under complete spatial randomness.

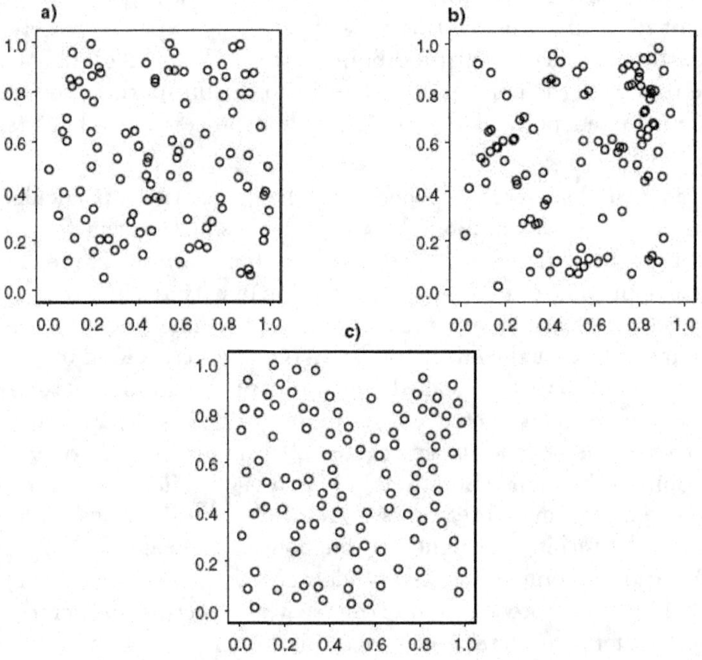


Figure 3.1 *Realizations of a completely random pattern (a), a Poisson cluster process (b), and a process with regularity (sequential inhibition, c). All patterns have* $n = 100$ *events on the unit square.*

Most observed patterns are snapshot realizations of a process that evolves in time. The temporal evolution affects the degree to which events are aggregated or clustered. A naturally regenerating oak stand, for example, will exhibit a highly clustered arrangement as newly established trees are close to the old trees from which acorns fell. Over time, differences in micro-site conditions lead to a (random) thinning of the trees. The spatial distribution appears less clustered. As trees grow older, competition leads to more regular patterns when trees have established the needed growing space.

## 3.2 Binomial and Poisson Processes

The homogeneous Poisson process (HPP) is the stochastic representation of complete spatial randomness, the point process equivalence of signal-free white noise. Statistical tests for comparing an observed point pattern against that expected from a HPP thus play a central role in point pattern analysis. In practical implementations of these tests, observed patterns are usually compared against a Binomial process. The connection between the Poisson and the Binomial process lies in the conditioning on the number of events. In simulation studies the number of events in simulated patterns is typically held fixed to equal the number of events in the observed pattern.

### *3.2.1 Bernoulli and Binomial Processes*

Let $\nu_d(D)$ denote Lebesgue measure of $D \in \mathbb{R}^d$. If $d = 1$, then $\nu_d(D)$ is the length of an interval, in $\mathbb{R}^2$ it measures the area of $D$ and in $\mathbb{R}^3$ the volume. We will refer to $\nu(A)$ simply as the volume of the Borel set $A$. If a single event $\mathbf{s}$ is distributed in $D$ such that $\Pr(\mathbf{s} \in A) = \nu(A)/\nu(D)$ for all sets $A \subset D$, this process containing a single point is termed a Bernoulli process. It is a rather uninteresting process but if $n$ Bernoulli processes are superposed to form a process of $n$ events in $D$ the resulting process is much more interesting and is termed a **Binomial point process.** Notice that we are following the same logic as in classical statistics, where the Binomial experiment is defined as $n$ independent and identical Bernoulli experiments with common success probability.

Point processes can be studied through either the stochastic properties of the event locations, or through a counting measure. The latter is often more intuitive, but the former is frequently the representation from which methods of simulating realizations of a point process model can be devised. If $Z(\mathbf{s})$ is a Binomial point process, then

$$\Pr(\mathbf{s}_1 \in A_1, \cdots, \mathbf{s}_n \in A_n) = \frac{\nu(A_1) \cdot \ldots \cdot \nu(A_n)}{\nu(D)^n}$$

for subregions $A_1, \cdots, A_n$ in $D$. In terms of a counting measure, let $N(A)$ denote the number of events in the (Borel) set $A \subset D$. In a Binomial process $N(A)$ is a Binomial random variable with sample size $n = N(D)$ and success probability $\pi(A) = \nu(A)/\nu(D)$.

The (first-order) **intensity** $\lambda(\mathbf{s})$ of a spatial point process measures the average number of events per unit area (volume). The intensity is defined as a limit since it is considered a function of points in $D$ on an area basis. Let $d\mathbf{s}$ denote an infinitesimal area (disc) in $\mathbb{R}^d$ centered at $\mathbf{s}$. Then the limit *Intensity*

$$\lambda(\mathbf{s}) = \lim_{\nu(d\mathbf{s}) \to 0} \frac{\mathrm{E}[N(d\mathbf{s})]}{\nu(d\mathbf{s})}$$

is the first-order intensity of the point process $Z(\mathbf{s})$. With the Binomial process, the average number of events in region $A$ is simply $n\pi(A)$, and for any Borel subset $A$ of $D$

$$\lambda(\mathbf{s}) = \lim_{\nu(d\mathbf{s})\to 0} \frac{n\pi(d\mathbf{s})}{\nu(d\mathbf{s})} = \lim_{\nu(d\mathbf{s})\to 0} \frac{n\nu(d\mathbf{s})/\nu(D)}{\nu(d\mathbf{s})} = \frac{n}{\nu(D)} \equiv \lambda.$$

*Homogeneity*

Since the first-order intensity does not change with spatial location, the Binomial process is a **homogeneous** (or **uniform**) process.

Points in non-overlapping subregions are not independent, however. Since the total number of events in $D$ is fixed, $m$ events in $A$ necessarily implies $n-m$ events in $D\backslash A$. Because of the correlation between the number of events in disjoint subregions, a Binomial process is not a completely spatial random process. It is a very important point process, however, for testing observed patterns against the CSR hypothesis. Whereas a CSR pattern is the result of a homogeneous Poisson process, in Monte Carlo tests of the CSR hypothesis one usually conditions the simulations to have the same number of events as the observed pattern. Conditioning a homogeneous Poisson process on the number of events yields a Binomial process.

### 3.2.2 Poisson Processes

There are many types of Poisson processes with relevance to spatial statistics. Among them are the homogeneous Poisson process, the inhomogeneous Poisson process, the Poisson cluster process, and the compound Poisson process. A process is referred to as *the* Poisson process if it has the following two properties:

*Homogeneous Poisson Process*

(i) If $N(A)$ denotes the number of events in subregion $A \subset D$, then $N(A) \sim \text{Poisson}(\lambda\nu(A))$, where $0 < \lambda < \infty$ denotes the constant intensity function of the process;

(ii) If $A_1$ and $A_2$ are two disjoint subregions of $D$, then $N(A_1)$ and $N(A_2)$ are independent.

Stoyan, Kendall, and Mecke (1995, p. 33) call (ii) the "completely random" property. It is noteworthy that property (ii) follows from (i) but that the reverse is not true. The number of events in $A$ can be distributed as a Poisson variable with a spatially varying intensity, but events can remain independent in disjoint subsets. We consider the combination of (i) and (ii) as the definition of complete spatial randomness. A point process that satisfies properties (i) and (ii) is called a homogeneous Poisson (or CSR) process.

If the intensity function $\lambda(\mathbf{s})$ varies spatially, property (i) is not met, but (ii) may still hold. A process of this kind is the **inhomogeneous** Poisson process (IPP). It is characterized by the following properties.

*Inhomogeneous Poisson Process*

(i) If $N(A)$ denotes the number of events in subregion $A \subset D$, then $N(A) \sim \text{Poisson}(\lambda(A))$, where $0 < \lambda(\mathbf{s}) < \infty$ is the intensity at location $\mathbf{s}$ and $\lambda(A) = \int_A \lambda(\mathbf{s})d\mathbf{s}$;

(ii) If $A_1$ and $A_2$ are two disjoint subregions of $D$, then $N(A_1)$ and $N(A_2)$ are independent.

The HPP is obviously a special case of the IPP where the intensity is constant. Stoyan et al. (1995) refer to the HPP as the stationary Poisson process and label the IPP the general Poisson process. Stationarity of point processes is explored in greater detail in §3.4. We note here that stationarity implies (at least) that the first-order intensity of the process is translation invariant which requires that $\lambda(\mathbf{s}) \equiv \lambda$. The inhomogeneous Poisson process is a non-stationary point process.

### *3.2.3 Process Equivalence*

The first-order intensity $\lambda(\mathbf{s})$ and the yet to be introduced second-order intensity $\lambda_2(\mathbf{s}_i, \mathbf{s}_j)$ (§3.4) capture the mean and dependence structure in a spatial point pattern. As the mean and covariance of two random variables $X$ and $Y$ provide an incomplete description of the bivariate distribution, these two intensity measures describe a point process incompletely. Quite different processes can have the same intensity measures $\lambda(\mathbf{s})$ and $\lambda_2(\mathbf{s}_i, \mathbf{s}_j)$ (for an example, see Baddeley and Silverman, 1984). In order to establish the equivalence of two point processes, their distributional properties must be studied. This investigation can focus on the distribution of the $n$-tuple $\{\mathbf{s}_1, \cdots, \mathbf{s}_n\}$ by considering the process as random sets of discrete points, or through distributions defined for random measures counting the number of points. We focus on the second approach. Let $N(A)$ denote the number of events in region (Borel set) $A$ with volume $\nu(A)$. The *finite-dimensional distributions* are probabilities of the form

*Finite-dimensional Distribution*

$$\Pr(N(A_1) = n_1, \cdots, N(A_k) = n_k),$$

where $n_1, \cdots, n_k \geq 0$ and $A_1, \cdots, A_k$ are Borel sets. The distribution of the counting measure is determined by the system of these probabilities for $k = 1, 2, \cdots$. It is convenient to focus on regions $A_1, \cdots, A_k$ that are mutually disjoint (non-overlapping). A straightforward system of probabilities that determines the distribution of a simple point process consists of the *zero-probability functionals*

*Void Probabilities*

$$P_N^0(A) = \Pr(N(A) = 0)$$

for Borel sets $A$. Stoyan et al. (1995, Ch. 4.1) refer to these functionals as *void-probabilities* since they give the probability that region $A$ is void of events. Notice that using zero-probability functionals for point process identification requires simple processes; no two events can occur at the same location.

Cressie (1993, p. 625) sketches the proof of the **equivalence theorem**, which states that two simple point processes with counting measures $N_1$ and $N_2$ are identically distributed if and only if their finite-dimensional distributions coincide for all integers $k$ and sets $A_1, \cdots, A_k$ and if and only if their void-probabilities are the same: $P_{N_1}^0(A) = P_{N_2}^0(A)\ \forall A$.

**Example 3.1** The equivalence theorem can be applied to establish the equivalence of a Binomial process and a homogeneous Poisson process on $D$ that is conditioned on the number of events. First note that for the Binomial process we have $N(A) \sim \text{Binomial}(n, \pi(A))$, where $\pi(A) = \nu(A)/\nu(D)$. Hence,

$$P_N^0(A) = \{1 - \pi(A)\}^n = \left[\frac{\nu(D) - \nu(A)}{\nu(D)}\right]^n \tag{3.1}$$

$$\Pr(N(A_1) = n_1, \cdots, N(A_k) = n_k) = \frac{n!}{n_1! \cdot \ldots \cdot n_k!} \times \frac{\nu(A_1)^{n_1} \cdot \ldots \cdot \nu(A_k)^{n_k}}{\nu(D)^n}, \tag{3.2}$$

for $A_1, \cdots, A_k$ disjoint regions such that $A_1 \cup \cdots \cup A_k = D$, $n_1 + \cdots + n_k = n$. Let $M(A)$ denote the counting measure in a homogeneous Poisson process with intensity $\lambda$. The void-probability in region $A$ is then given by

$$P_M^0(A) = \exp\{-\lambda\nu(A)\}.$$

Conditioning on the number of events $M(D) = n$, the void-probability of the conditioned process becomes

$$\begin{aligned} P^0_{M|M(D)=n}(A) &= \Pr(M(A) = 0|M(D) = n) \\ &= \frac{\Pr(M(A) = 0)\Pr(M(D \setminus A) = n)}{\Pr(M(D) = n)} \\ &= \frac{e^{-\lambda\nu(A)}[\lambda\nu(D \setminus A)]^n e^{-\lambda\nu(D\setminus A)}}{[\lambda\nu(D)]^n e^{-\lambda\nu(D)}} \\ &= \frac{\nu(D \setminus A)^n}{\nu(D)^n} = \left[\frac{\nu(D) - \nu(A)}{\nu(D)}\right]^n, \end{aligned}$$

which is (3.1). To establish that the Poisson process $M(A)$, given $M(D) = n$, is a Binomial process through the finite-dimensional distributions is the topic of Chapter problem 3.1. □

## 3.3 Testing for Complete Spatial Randomness

A test for complete spatial randomness addresses whether or not the observed point pattern could possibly be the realization of a homogeneous Poisson process (or a Binomial process for fixed $n$). Just as the stochastic properties of a point process can be described through random sets of points or counting measures, statistical tests of the CSR hypothesis can be based on counts of events in regions (so-called quadrats), or distance-based measures using the event locations. Accordingly, we distinguish between quadrat count methods (§3.3.3) and distance-based methods (§3.3.4).

With a homogeneous Poisson process, the number of events in region $A$ is a Poisson variate and counts in non-overlapping regions are independent. The distributional properties of quadrat counts are thus easy to establish, in particular for point patterns on rectangles. The distribution of test statistics based on quadrat counts is known at least asymptotically and allows closed-form tests. For irregularly shaped spatial domains, when considering edge effects, and for rare events (small quadrat counts), these approximations may not perform well. The sampling distribution of statistics based on distances between events or distances between sampling locations and events are much less understood, even in the case of a Poisson process. Although nearest-neighbor distributions can be derived for many processes, edge-effects and irregularly shaped domains are difficult to account for.

When sampling distributions are intractable or asymptotic results not reliable, one may rely on simulation methods. For point pattern analysis simulation methods are very common, if not the norm. Two of the basic tools are the Monte Carlo test and the examination of simulation envelopes.

### *3.3.1 Monte Carlo Tests*

A Monte Carlo test for CSR is a special case of a simulation test. The hypothesis is that an observed pattern $Z(\mathbf{s})$ could be the realization of a point process model $\Psi$. A test statistic $Q$ is chosen which can be evaluated for the observed pattern and for any realization simulated under the model $\Psi$. Let $q_0$ denote the realized value of the test statistic for the observed pattern. Then generate $g$ realizations of $\Psi$ and calculate their respective test statistics: $q_1 = q(\psi_1), \cdots, q_g = q(\psi_g)$. The statistic $q_0$ is combined with these and the set of $g+1$ values is ordered (ranked). Depending on the hypothesis and the choice of $Q$, either small or large values of $Q$ will be inconsistent with the model $\Psi$. For example, if $Q$ is the average distance between events and their nearest neighbors, then under aggregation one would expect $q_0$ to be small when $\Psi$ is a homogeneous Poisson process. Under regularity, $q_0$ should be large. If $\Psi$ is rejected as a data-generating mechanism for the observed pattern when $q_0 \le q_{(k)}$ or $q_0 \ge q_{(g+1-k)}$, where $q_{(k)}$ denotes the $k$th smallest value, this is a two-sided test with significance level $\alpha = 2k/(g+1)$.

Monte Carlo tests have numerous advantages. The $p$-values of the tests are exact in the sense that no approximation of the distribution of the test statistic is required. The $p$-values are inexact in the sense that the number of possible realizations under $\Psi$ is typically infinite. At least the number of realizations will be so large that enumeration is not possible. The number $g$ of simulations must be chosen sufficiently large. For a 5% level test $g = 99$ and for a 1% level test $g = 999$ have been recommended. As long as the model $\Psi$ can be simulated, the observed pattern can be compared against complex point processes by essentially the same procedure. Simulation tests thus provide great flexibility.

A disadvantage of simulation tests is that several critical choices are left to the user, for example, the number of simulations and the test statistic. Diggle (1983) cautions of "data dredging," the selection of non-sensible test statistics for the sake of rejecting a particular hypothesis. Even if sensible test statistics are chosen, the results of simulation tests may not agree. The power of this procedure is also difficult to establish, in particular, when applied to tests for point patterns. The alternative hypothesis for which the power is to be determined is not at all clear.

### *3.3.2 Simulation Envelopes*

A Monte Carlo test calculates a single test statistic for the observed pattern and each of the simulated patterns. Often, it is illustrative to examine not point statistics but functions of the point pattern. For example, let $h_i$ denote the distance from event $\mathbf{s}_i$ to the nearest other event and let $I(h_i \leq h)$ denote the indicator function which returns 1 if $h_i \leq h$. Then

$$\widehat{G}(h) = \frac{1}{n}\sum_{i=1}^{n} I(h_i \leq h)$$

is an estimate of the distribution function of nearest-neighbor event distances and can be calculated for any value of $h$. With a clustered pattern, we expect an excess number of short nearest-neighbor distances (compared to a CSR pattern). The method for obtaining simulation envelopes is similar to that used for a Monte Carlo test, but instead of evaluating a single test statistic for each simulation, a function such as $\widehat{G}(h)$ is computed. Let $\widehat{G}_0(h)$ denote the empirical distribution function based on the observed point pattern. Calculate $\widehat{G}_1(h), \cdots, \widehat{G}_g(h)$ from $g$ point patterns simulated under CSR (or any other hypothesis of interest). Calculate the percentiles of the investigated function from the $g$ simulations. For example, upper and lower 100% simulation envelopes are given by

$$\widehat{G}_l(h) = \min_{i=1,\cdots,g}\{\widehat{G}_i(h)\} \quad \text{and} \quad \widehat{G}_u(h) = \max_{i=1,\cdots,g}\{\widehat{G}_i(h)\}.$$

Finally, a graph is produced which plots $\widehat{G}_0(h)$, $\widehat{G}_l(h)$, and $\widehat{G}_u(h)$ against the theoretical distribution function $G(h)$, or, if $G(h)$ is not attainable, against the average empirical distribution function from the simulation,

$$\overline{G}(h) = \frac{1}{n}\sum_{i=1}^{g} \widehat{G}_i(h).$$

**Example 1.5 (Lightning strikes. Continued)** Recall the lightning data from §1.2.3 (p. 11). The pattern comprises 2,927 lightning flashes recorded by the National Lightning Detection Network within approximately 200 miles of the East coast of the United States during a span of four days in April 2003. Figure 3.2 displays the observed pattern and two bounding domains, the bounding box and the convex hull. Obviously, the pattern appears clustered, and this agrees with our intuition. Lightning strikes do not occur completely at random, they are associated with storms and changes in the electric charges of the atmosphere.

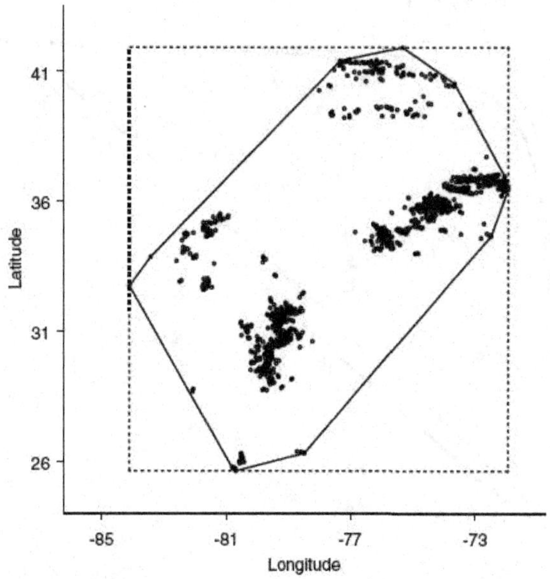


Figure 3.2 *Locations of lightning strikes, bounding rectangle, and convex hull.*

In order to perform a Monte Carlo-based analysis of these data we need to define the applicable domain. State boundaries are not adequate because the strikes were observed within a certain distance of the East coast (see Figure 1.7 on p. 13). Two common approaches are to choose the bounding rectangle and the convex hull of the observed pattern. Because the data were collected within a distance of approximately 200 miles within the coast line, the bounding box is not a very good representation of the domain. It adds too much "white space" and thus enhances the degree of clustering.

Figure 3.3 displays $\widehat{G}_0(h)$ based on the bounding box and the convex hull along with the simulation envelopes for $s = 500$. The extent of clustering is evident; the estimated $G$ functions step outside of the simulation envelopes immediately. The exaggerated impression of clustering one obtains by using the bounding rectangle for these data is also evident.

□

Simulation envelopes can be used for confirmatory inference in different ways.

- While generating the $g$ simulations one can calculate a test statistic along with the function of interest and thereby obtain all necessary ingredients for a Monte Carlo test. For example, one may use $\overline{h}$, the average nearest-neighbor distance.

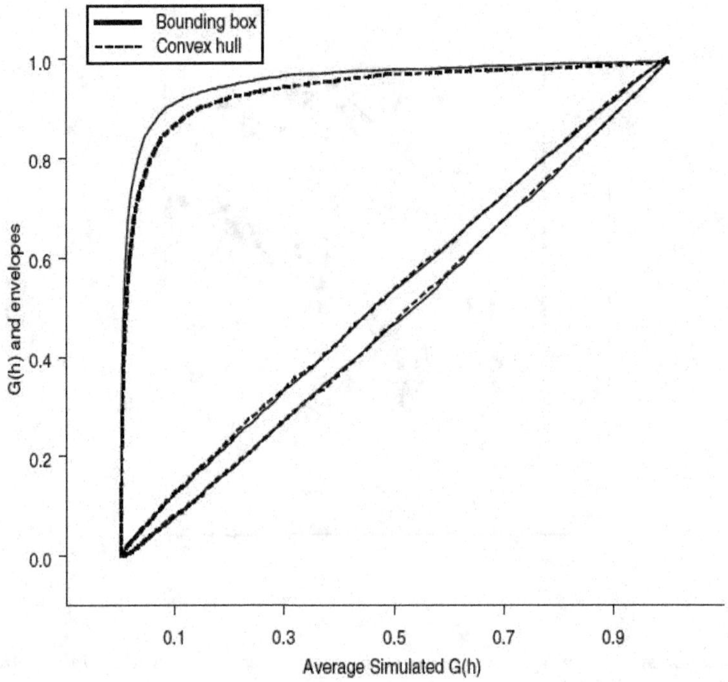


Figure 3.3 *G-function and simulation envelopes from 500 simulations on bounding box and convex hull.*

- If the null hypothesis is reasonable, the observed function $G(h)$ should fall within the simulation envelopes. When $G(h)$ and its envelopes are graphed against $h$ and a 95% upper simulation envelope is exceeded at a given small distance $h_0$, a Monte Carlo test with test statistic $G(h_0)$ would have rejected the null hypothesis in a one-sided test at the 5% level. It is thus common to calculate 95% simulation envelopes and examine whether $\widehat{G}(h)$ crosses the envelopes. It must be noted, however, that simulation envelopes are typically plotted against the theoretical $G(h)$ or $\overline{G}(h)$, not distance. Furthermore, unless the value of $h_0$ is set in advance, the Type-I error of this method is not protected.

### *3.3.3 Tests Based on Quadrat Counts*

#### *3.3.3.1 Goodness-of-Fit Test*

The most elementary test of CSR based on counting events in regions is based on dividing the domain $D$ into non-overlapping regions (quadrats) $A_1, \cdots, A_k$ of equal size such that $A_1 \cup \cdots \cup A_k = D$. Typically, the domain is assumed to be bounded by a rectangle and partitioned into $r$ rows and $c$ columns. If $n_{ij}$ is the number of events in quadrat $ij$, and $\overline{n} = n/(rc)$ is the expected number of events in any quadrat under CSR, then the standard Pearson Chi-square statistic is

*Index of Dispersion*

$$X^2 = \sum_{i=1}^{r}\sum_{j=1}^{c}\frac{(n_{ij}-\overline{n})^2}{\overline{n}}. \tag{3.3}$$

This test statistic is that of a Chi-square goodness-of-fit test of the hypothesis that the $n$ points are distributed uniformly *and* independently in $D$, or, in other words, whether the quadrat counts are independent Poisson variates with common mean. Fisher, Thornton, and MacKenzie (1922) used (3.3) in the latter sense to test whether bacterial density on plates can be described by the Poisson distribution. Note that the reference distribution for (3.3) is $\chi^2_{rc-1}$. Although (3.3) is written as a Chi-square statistic for a contingency table, the double summation is used to emphasize the row-column partition of the domain. Furthermore, no additional degree of freedom is lost to the estimation of $\overline{n}$, since $n$ is known in a mapped point pattern. An alternative expression for (3.3) is $X^2 = (rc-1)s^2/\overline{n}$, where $s^2$ is the sample variance of the $r \times c$ quadrat counts. If the pattern is CSR, then the ratio of sample variance and sample mean should be approximately 1. $X^2$ is thus also referred to as the **index of dispersion**. Note that Diggle (1983, p. 33) terms $I = X^2/(rc-1)$ as the index of dispersion.

The goodness-of-fit test based on quadrat counts is simple and the Chi-square approximation performs well provided that the expected number of events per quadrat exceeds 1 and $rc > 6$ (Diggle, 1983, p. 33). It is, however, very much influenced by the choice of the quadrat size.

**Example 3.2** For the three point patterns on the unit square shown in Figure 3.1, quadrat counts were calculated on a $r = 5 \times c = 5$ grid. Since each pattern contains $n = 100$ points, $\overline{n} = 4$ is common to the three realizations. The quadrat counts are shown in Table 3.1.

With a CSR process, the counts distribute evenly across the quadrats, whereas in the clustered pattern counts concentrate in certain areas of the domain. Consequently, the variability of quadrat counts, if events aggregate, exceeds the variability of the Poisson process. Clustered processes exhibit large values of $X^2$. The reverse holds for regular processes whose counts are under-dispersed relative to the homogeneous Poisson process. The CSR hypothesis based on the index of dispersion is thus rejected in the right tail against the clustered alternative and the left tail against the regular alternative.

The sensitivity of the goodness-of-fit test to the choice of quadrat size is evident when the processes are divided into $r = 3 \times c = 3$ quadrats. The left tail $X^2$ probabilities for 9 quadrats are 0.08, 0.99, and 0.37 for the CSR, clustered, and regular process, respectively. The scale on which the point pattern appears random, clustered, or regular, depends on the scale on which

Table 3.1 *Quadrat Counts in Simulated Point Patterns of Figure 3.1*

| | CSR Process, $c=$ | | | | | Cluster Process, $c=$ | | | | | Regular Process, $c=$ | | | | |
|---|---|---|---|---|---|---|---|---|---|---|---|---|---|---|---|
| $r=$ | 1 | 2 | 3 | 4 | 5 | 1 | 2 | 3 | 4 | 5 | 1 | 2 | 3 | 4 | 5 |
| 5 | 6 | 3 | 7 | 4 | 5 | 2 | 1 | 7 | 5 | 10 | 7 | 5 | 3 | 2 | 6 |
| 4 | 5 | 3 | 1 | 3 | 4 | 3 | 6 | 2 | 3 | 10 | 3 | 2 | 4 | 5 | 7 |
| 3 | 2 | 3 | 8 | 3 | 5 | 8 | 4 | 1 | 6 | 5 | 4 | 5 | 3 | 5 | 3 |
| 2 | 5 | 5 | 5 | 5 | 4 | 1 | 6 | 0 | 2 | 1 | 2 | 4 | 7 | 3 | 3 |
| 1 | 1 | 4 | 2 | 3 | 4 | 1 | 2 | 8 | 3 | 3 | 4 | 4 | 4 | 2 | 3 |
| | $X^2=17.0$ | | | | | $X^2=45.9$ | | | | | $X^2=14.5$ | | | | |
| | $\Pr(\chi^2_{24} \le X^2)=0.15$ | | | | | $\Pr(\chi^2_{24} \le X^2)=0.99$ | | | | | $\Pr(\chi^2_{24} \le X^2)=0.06$ | | | | |

the counts are aggregated. This is a special case of what is known in spatial data analysis as the **change of support** problem (see §5.7). □

#### *3.3.3.2 Analysis of Contiguous Quadrats*

The fact that quadrat counts can indicate randomness, regularity, and clustering depending on the scale of aggregation was the idea behind the method of contiguous quadrat aggregation proposed in an influential paper by Greig-Smith (1952). Whereas the use of randomly placed quadrats of differing size had been common a the time, Greig-Smith (1952) proposed a method of aggregating events at different scales by counting events in successively larger quadrats which form a grid in the domain. Initially, the domain is divided into a grid of $2^q \times 2^q$ quadrats. Common choices for $q$ are 4 or 5 leading to a basic aggregation into 256 or 1024 quadrats. Then the quadrats are successively combined into blocks consisting of $2\times 1$, $2\times 2$, $2\times 4$, $4\times 4$ quadrats and so forth. Consider a division of the domain into $16\times 16 = 256$ quadrats as shown in Figure 3.4.

Depending on whether the rectangular blocks that occur at every second level of aggregation are oriented horizontally or vertically, two modes of aggregation are distinguished. The entire pattern contains two blocks of 128 quadrats. Each of these contains two blocks of 64 quadrats. Each of these contains two blocks of 32 quadrats, and so forth. Let $N_{r,i}$ denote the number of events in the $i$th block of size $r$. The sum of squares between blocks of size $r$ is given by

$$SS_r = 2\sum_{i=1}^{m} N_{r,i}^2 - \sum_{j=1}^{m/2} N_{2r,j}^2.$$

The term sum of squares reveals the connection of the method with a (nested) analysis of variance. Blocks of size 1 are nested within blocks of size 2, these

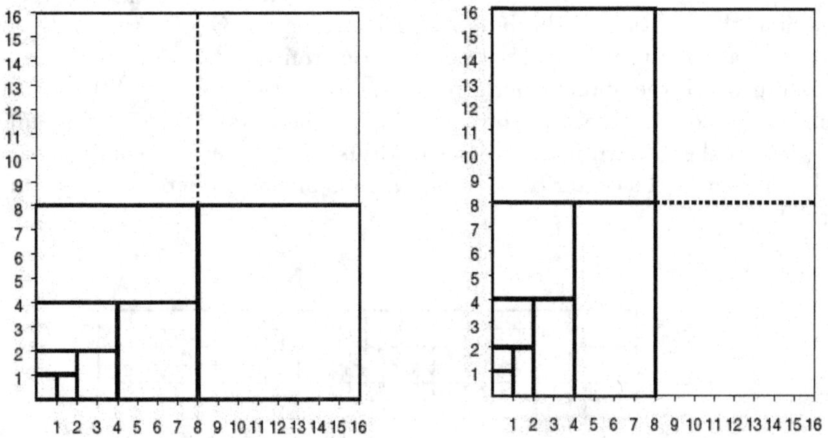


Figure 3.4 *Aggregation of quadrat counts into blocks of successively larger size for analysis of contiguous quadrats according to Greig-Smith (1952). Horizontal aggregation is shown in the left-hand panel, vertical aggregation in the right-hand panel. Dashed line shows the division of one block of size 128 into two blocks of size 64.*

are nested within blocks of size 4, and so forth. The mean square associated with blocks of size $r$ is then simply $MS_r = SS_r/2^{2q}$. The original Greig-Smith analysis consisted of plotting $MS_r$ against the block size $r$. Peaks or troughs in this graph are interpreted as indicative of clustered or regular patch sizes. Since $\text{Var}[MS_r]$ increases with the quadrat area, care must be exercised not to over-interpret the fluctuations in $MS_r$, in particular for larger block sizes. The Greig-Smith analysis thus provides a good application where simulation envelopes should be considered. The peaks and troughs in the $MS_r$ plot can then be interpreted relative to the variation that should be expected at that block size. To calculate simulation envelopes, the quadrat counts for the finest gridding are randomly permuted $s$ times among the grid locations.

**Example 3.3 Red cockaded woodpecker.** The red-cockaded Woodpecker is a federally endangered species sensitive to disruptions of its habitat because of its breeding habits. The species is known as a cooperative breeder where young male offspring remain with a family as nonbreeding helpers, sometimes for several years. The species builds its nest cavities in live pine trees. Because of the resistance exerted by live trees, building a cavity can take a long time and loss of nesting trees is more damaging than for species that can inhabit natural cavities or cavities vacated by other animals. Decline of the bird's population has been attributed to the reduction of natural habitat, the lack of old-growth, and the suppression of seasonal fires (Walters, 1990). Thirty clusters (=families) of birds were followed in the Fort Bragg area of North Carolina over a 23-month period (December 1994 to October 1996). For seventeen of the twenty-three months, one bird from each cluster was selected and observed for an entire day. At eight-minute intervals during that observation period, the location of the bird was geo-referenced. One of the goals of the study was to obtain an estimate of the homerange of the cluster, that is, the area in which the animals perform normal activities (Burt, 1943). Figure 3.5 shows the counts obtained from partitioning the 4,706 ft × 4,706 ft bounding square of the pattern into 32 × 32 quadrats of equal size. A concentration of the birds near the center of the study area is obvious, the data appear highly clustered.

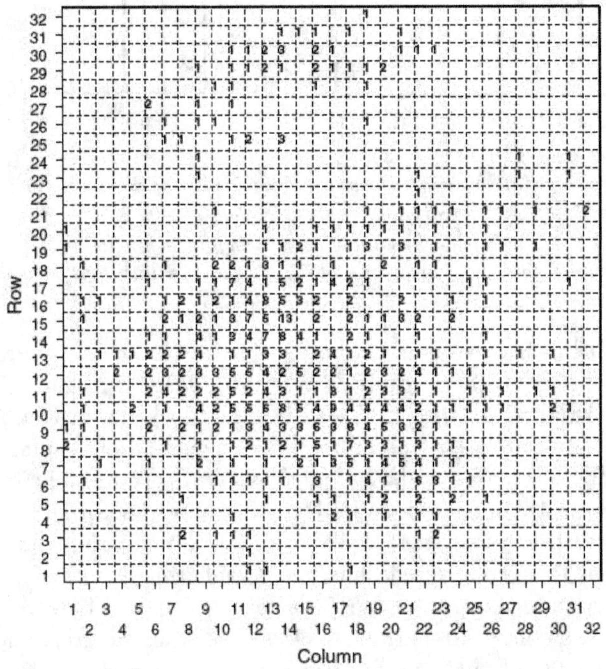


Figure 3.5 *Quadrat counts for woodpecker data. The bounding square was divided into 32 × 32 square quadrats of equal size. The width of a quadrat is approximately 147 ft. A total of 675 locations were recorded. Data kindly provided by Professor Jeff Walters, Department of Biology, Virginia Tech.*

The nested analysis of variance shows the dependence of $MS_r$ on the direction of aggregation. In particular for large block sizes, the differences for vertical and horizontal aggregation are considerable. Let $MS_r^{(v)}$ denote the mean square at block size $r$ from a vertical aggregation pattern and $MS_r^{(h)}$ the corresponding mean square from a horizontal pattern (see Figure 3.4). We recommend to eliminate the effects of the orientation of aggregation by considering the average $\overline{MS}_r = 0.5(MS_r^{(v)} + MS_r^{(h)})$. Next, $s = 200$ ran-

Table 3.2 *Nested analysis of variance for vertical and horizontal aggregation into contiguous quadrats*

| Block Size | No. of Blocks | $SS_r$ Vert. | $SS_r$ Horiz. | $MS_r$ Vert. | $MS_r$ Horiz. | $\overline{MS}_r$ | $\overline{MS}_r$ Env. Min | $\overline{MS}_r$ Env. Max |
|---|---|---|---|---|---|---|---|---|
| 1 | 1024 | 2299 | 2299 | 0.72 | 0.70 | 0.71 | 1.64 | 1.94 |
| 2 | 512 | 3869 | 3877 | 0.69 | 0.71 | 0.70 | 1.62 | 1.99 |
| 4 | 256 | 7025 | 7025 | 1.32 | 1.67 | 1.49 | 1.41 | 2.32 |
| 8 | 128 | 12701 | 12343 | 2.54 | 1.84 | 2.19 | 1.18 | 2.41 |
| 16 | 64 | 22803 | 22803 | 5.69 | 2.95 | 4.32 | 0.89 | 2.83 |
| 32 | 32 | 39779 | 42581 | 3.84 | 9.31 | 6.57 | 0.83 | 3.34 |
| 64 | 16 | 75631 | 75631 | 35.39 | 46.91 | 41.15 | 0.28 | 5.50 |
| 128 | 8 | 115013 | 103231 | 73.87 | 50.86 | 62.36 | 0.29 | 6.00 |
| 256 | 4 | 154383 | 154383 | 72.59 | 8.92 | 40.76 | 0.02 | 8.53 |
| 512 | 2 | 234425 | 299633 | 12.92 | 140.27 | 76.59 | 0.00 | 7.85 |
| 1024 | 1 | 455625 | 455625 | . | . | . | | |

dom permutations of the quadrat counts in Figure 3.5 were generated and the nested analysis of variance was repeated for each permutation. For the smallest block sizes the woodpecker distribution exhibits some regularity, but the effects of strong clustering are apparent for the larger block sizes (Table 3.2). Since $\text{Var}[MS_r]$ increases with block size, peaks and troughs must be compared to the simulation envelopes to avoid over-emphasizing spikes in the $MS_r$. Figure 3.6 shows significant clustering for block sizes of 32 and more quadrats. The spike of $\overline{MS}_r$ at $r = 128$ represents the mean square for blocks of $16 \times 8$ quadrats. This corresponds to a "patch size" of 2,352 ft $\times$ 1,176 ft.

The Woodpecker quadrat count data are vertically aggregated. Observations collected over time were accumulated into discrete spatial units. The analysis shows that both the size of the unit of aggregation as well as the spatial configuration (orientation) has an effect on the analysis. This is an example of a particular aspect of the change of support problem, the **modifiable areal unit problem** (MAUP, see §5.7). □

The graph of the mean squares by block size in the Greig-Smith analysis is an easily interpretable, exploratory tool. The analysis can attain a confirmatory character if it is combined with simulation envelopes or Monte Carlo tests. Closed-form significance tests have been suggested based on sums of squares and mean squares and the Chi-square and $F$-distributions (see, for example, Thompson, 1955, 1958; Zahl, 1977). Upton and Fingleton (1985, p. 53) conclude that "virtually all the significance tests are suspect." The analysis of contiguous quadrats also conveys information only about "scales of pattern" that coincide with blocks of size $2^k$, $k = 0, \cdots 2q - 1$. A peak or trough in the $MS_r$ plot for a particular value of $r$ could be induced by a patch

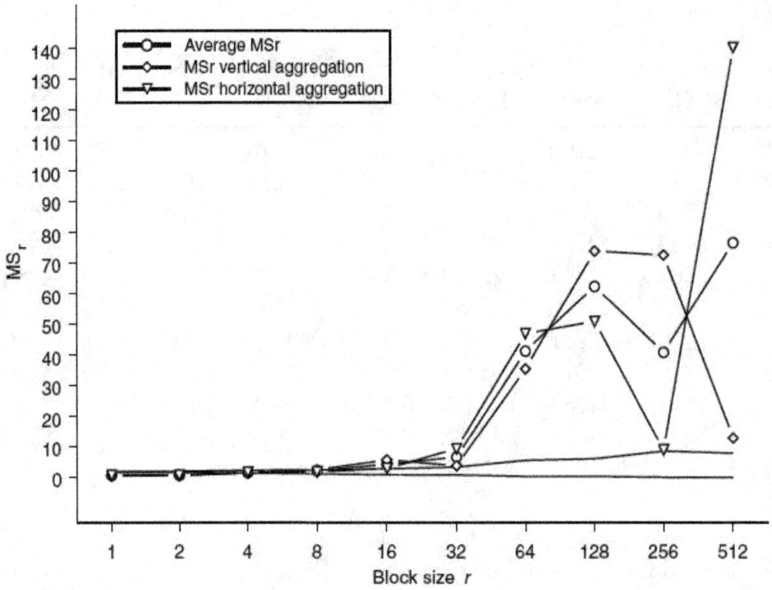


Figure 3.6 *Mean squares for nested, contiguous quadrat counts as a function of block size. Solid lines without symbols denote 100% simulation envelopes for* $\overline{MS}_r = 0.5(MS_r^{(v)} + MS_r^{(h)})$.

size for which the mean squares cannot be calculated. The effects of orientation of the grid on the analysis are formidable. Despite of these problems, the Greig-Smith analysis remains a popular tool.

#### *3.3.3.3 Application of Other Methods to Quadrat Counts*

The grouping of events into quadrats creates a lattice structure to which other statistical methods for spatial data can be applied. Since the counts are independent under CSR, finding significant spatial autocorrelation among the counts leads to a rejection of the CSR hypothesis. One approach is thus to compute Moran's $I$ or Geary's $c$ statistic and to apply the tests in §1.3.2.

**Example 3.3 (Red cockaded woodpecker. Continued)** Grouping the woodpecker events into a lattice of $10 \times 10$ quadrats leads to Table 3.3. The clustering of events is clearly evident. Both statistics provide strong evidence of clustering, based on a "queen" definition of the lattice neighborhood (Table 3.4). In Monte Carlo tests based on 200 permutations of the quadrat counts, no arrangement produced an $I$ or $c$ statistic more extreme than the observed value.

□

Table 3.3 *Quadrature counts for woodpecker data based on* $10 \times 10$ *square quadrats*

| Row | Column 1 | 2 | 3 | 4 | 5 | 6 | 7 | 8 | 9 | 10 |
|---|---|---|---|---|---|---|---|---|---|---|
| 10 | 0 | 0 | 0 | 5 | 9 | 4 | 5 | 0 | 0 | 0 |
| 9 | 0 | 2 | 2 | 6 | 3 | 4 | 1 | 0 | 0 | 0 |
| 8 | 0 | 0 | 6 | 3 | 3 | 0 | 1 | 0 | 2 | 2 |
| 7 | 1 | 0 | 2 | 1 | 1 | 3 | 5 | 4 | 2 | 2 |
| 6 | 2 | 1 | 4 | 19 | 15 | 14 | 6 | 4 | 3 | 1 |
| 5 | 3 | 4 | 17 | 42 | 41 | 8 | 9 | 5 | 1 | 0 |
| 4 | 2 | 14 | 21 | 34 | 31 | 31 | 22 | 7 | 4 | 4 |
| 3 | 5 | 5 | 11 | 25 | 31 | 45 | 33 | 3 | 0 | 1 |
| 2 | 2 | 0 | 3 | 5 | 6 | 10 | 19 | 8 | 0 | 0 |
| 1 | 0 | 0 | 3 | 5 | 0 | 3 | 2 | 2 | 0 | 0 |

Table 3.4 *Results for Moran's I and Geary's c analysis based on quadrat counts in Table 3.3*

| Method | Statistic | Observed Value | Expected Value | Standard Error | P-Value |
|---|---|---|---|---|---|
| Normality | $I$ | 0.7012 | −0.0101 | 0.0519 | $< .0001$ |
| Randomization | $I$ | 0.7012 | −0.0101 | 0.0509 | $< .0001$ |
| Normality | $c$ | 0.4143 | 1 | 0.0611 | $< .0001$ |
| Randomization | $c$ | 0.4143 | 1 | 0.0764 | $< .0001$ |

### *3.3.4 Tests Based on Distances*

The choice of shape and number of quadrats in CSR tests based on areal counts is a subjective element that can influence the outcome. Test statistics that are based on distances between events or between sample points and events eliminate this subjectiveness, but are more computationally involved. In this subsection tests based on distances between events are considered. Let $h_{ij}$ denote the inter-event distance between events at locations $\mathbf{s}_i$ and $\mathbf{s}_j$, $h_{ij} = ||\mathbf{s}_i - \mathbf{s}_j||$. The distance between event $\mathbf{s}_i$ and the nearest other event is called the nearest-neighbor distance and denoted $h_i$.

Sampling distributions of test statistics based on inter-event or nearest-neighbor distances are elusive, even under the CSR assumption. Ripley and Silverman (1978) described a closed-form quick test that is based on the first ordered inter-event distances. For example, if $t_1 = \min\{h_{ij}\}$ then $t_1^2$ has an exponential distribution under CSR. The consequences of recording locations inaccurately looms large for such a test, and Ripley and Silverman recommend to use tests based on the third-smallest inter-event distance.

To circumvent the problem of determining the sampling distribution and to account for irregularly shaped spatial domains, simulation based methods are common in CSR tests based on distances. Whereas there are $n(n-1)/2$ inter-event distances, there are only $n$ nearest-neighbor distances and using a tesselation or triangulation the nearest-neighbor distances can be quickly determined. Enumerating all inter-event distances and finding the smallest for each point in order to determine nearest-neighbor distances is only acceptable for small problems. Tests based on inter-event distances are thus computationally more intensive. The accuracy with which the point locations are determined intuitively appears to be more important when only the distances between points and their nearest neighbors are considered.

Reasonable test statistics for Monte Carlo tests are $\overline{h}$, the average nearest neighbor distance, $\overline{t}$, the average inter-event distance,

$$\widehat{G}(y_0) = \frac{\#(y_i \leq y_0)}{n},$$

the empirical estimate of the probability that the nearest-neighbor distance is at most $y_0$,

$$\widehat{H}(t_0) = 2\frac{\#(h_{ij} \leq h_0)}{n(n-1)},$$

the empirical estimate of the probability that the inter-event distance is at most $h_0$, and so forth. There are many other sensible choices provided that the test statistic is interpretable in the context of testing the CSR hypothesis. In a clustered pattern, for example, $\overline{h}$ tends to be smaller than in a CSR pattern and tends to be larger in a regular pattern (Figure 3.7). If $y_0$ is chosen small, $\widehat{G}(y_0)$ will be larger than expected under CSR in a clustered pattern and smaller in a regular pattern.

**Example 3.3 (Red cockaded woodpecker. Continued)** The average nearest-neighbor distance among the red cockaded woodpecker event locations is $\overline{h} = 0.0148$. In 200 simulations of CSR processes with the same number of events as the observed pattern, none of the average nearest-neighbor distances was less than 0.0148. The CSR hypothesis is rejected against the clustered alternative with $p$-value 0.00498. Figure 3.8 shows the empirical distribution function of nearest-neighbor distance and the upper and lower simulation envelopes. The upper envelope is exceeded for very short distances, evidence of strong clustering in the data.

□

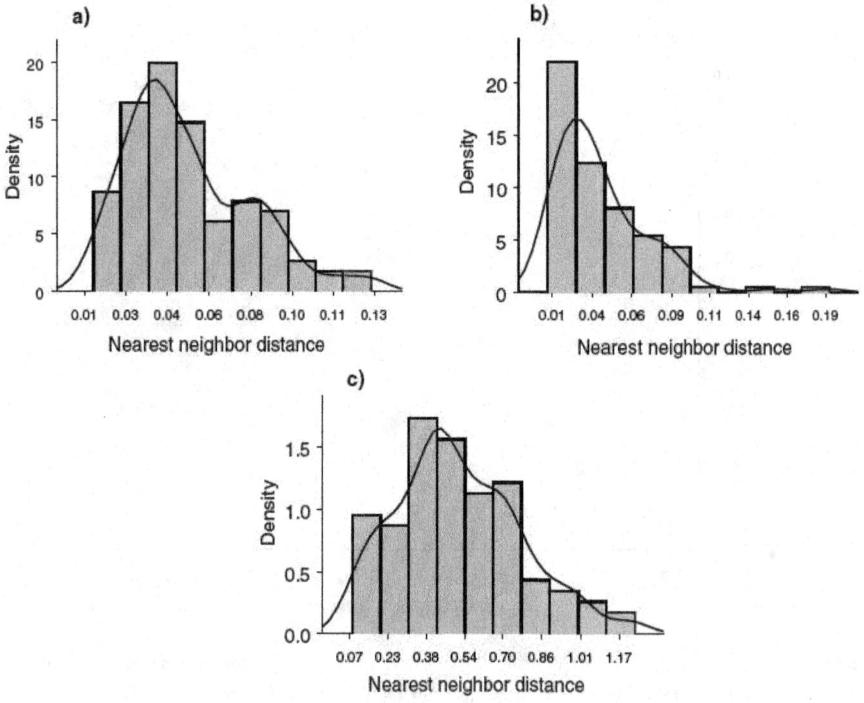


Figure 3.7 *Distribution of nearest-neighbor distances in the three point patterns of Figure 3.1. Completely random pattern (a), clustered pattern (b), and regular pattern (c).*

## 3.4 Second-Order Properties of Point Patterns

If $\lambda(\mathbf{s})$ plays a role in point pattern analysis akin to the mean function, what function of event locations expresses dependency of events? In order to capture spatial interaction, more than one event needs to be considered. The second-order intensity function sets into relationship the expected number of the cross-product of event counts in infinitesimal disks and the volumes of the disks as the disks are shrunk.

*Second-order Intensity*

$$\lambda_2(\mathbf{s}_i, \mathbf{s}_j) = \lim_{|d\mathbf{s}_i| \to 0, |d\mathbf{s}_j| \to 0} \frac{\mathrm{E}[N(d\mathbf{s}_i)N(d\mathbf{s}_j)]}{|d\mathbf{s}_i||d\mathbf{s}_j|}. \tag{3.4}$$

Stoyan et al. (1995, p. 112) refer to (3.4) as the second-order product density since it is the density of the second-order factorial moment measure.

A point process is homogeneous (uniform) if $\lambda(\mathbf{s}) = \lambda$. A process is stationary, if the second-order intensity depends only on event location differences, $\lambda_2(\mathbf{s}_i, \mathbf{s}_j) = \lambda_2^*(\mathbf{s}_i - \mathbf{s}_j)$. If the process is furthermore isotropic, the second-order intensity depends only on distance, $\lambda_2(\mathbf{s}_i, \mathbf{s}_j) = \lambda_2^*(||\mathbf{s}_i - \mathbf{s}_j||) = \lambda_2^*(h)$.

*Stationarity*

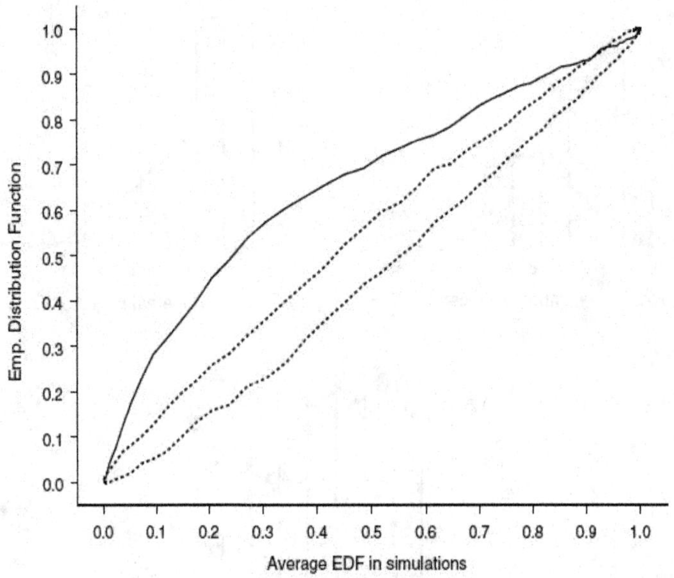


Figure 3.8 *EDF of nearest-neighbor distance (solid line) for woodpecker data and simulation envelopes (dashed lines) based on 200 simulations.*

In other words, the distribution of a stationary point process is translation invariant; the distribution of an isotropic pattern is rotation invariant. As previously, the star notation is dropped in what follows.

The covariance between event counts in regions $A$ and $B$ can be expressed in terms of the two intensities. Assuming a stationary, process,

$$\mathrm{Cov}[N(A), N(B)] = \int_A \int_B \lambda_2(\mathbf{s}_i - \mathbf{s}_j)\, d\mathbf{s}_i d\mathbf{s}_j + \lambda|A \cap B| - \lambda^2|A||B|. \quad (3.5)$$

The definition of $\lambda_2(\mathbf{s}_i, \mathbf{s}_j)$ resembles a cross-product expectation. If

$$\mathrm{Cov}[X, Y] = \mathrm{E}[XY] - \mathrm{E}[X]\mathrm{E}[Y],$$

then if $\mathrm{E}[XY] = \mathrm{E}[X]\mathrm{E}[Y]$, the random variables are uncorrelated. The covariance density function of the point process is similarly defined as

*Covariance Density Function*

$$C(\mathbf{s}_i - \mathbf{s}_j) = \lambda_2(\mathbf{s}_i - \mathbf{s}_j) - \lambda(\mathbf{s}_i)\lambda(\mathbf{s}_j). \quad (3.6)$$

If $\lambda_2(\mathbf{s}_i - \mathbf{s}_j) = \lambda(\mathbf{s}_i)\lambda(\mathbf{s}_j)$, then $C(\mathbf{s}_i - \mathbf{s}_j) = 0$ and $\mathrm{Cov}[N(A), N(B)] = 0$. From (3.5) or (3.6) it is clear that a stationary, isotropic process does not exhibit spatial dependency if $\lambda_2(h) = \lambda^2$. Beyond these simple relationships interpretation of the second-order intensity is difficult. The study of the dependence among events in a point pattern typically rests on functions of $\lambda_2$ that have a more accessible interpretation.

### 3.4.1 The Reduced Second Moment Measure—The K-Function

The $K$-function

*Ripley's K-Function*

$$K(h) = \frac{2\pi}{\lambda^2} \int_o^h x\lambda_2(x)\, dx \tag{3.7}$$

of Ripley (1976) is a function of $\lambda_2$ for stationary and isotropic processes. It is also known as the reduced second moment measure (Cressie, 1993), as the second reduced moment function (Stoyan et al., 1995), and as the second order reduced moment measure (Møller and Waagepetersen, 2003). Studying the second-order properties of a point pattern via the $K$-function is popular because the function has very appealing properties and interpretation.

- If the process is simple, $\lambda K(h)$ represents the expected number of extra events within distance $h$ from an arbitrary event. In the HPP with intensity $\lambda$ this expected number is $\lambda\pi h^2$ and the $K$-function for the HPP is simply $K(h) = \pi h^2$.
- If $K(h)$ is known for a particular point process, the second-order intensity is easily derived from (3.7),
$$\lambda_2(h) = \frac{\lambda^2}{2\pi h}\frac{\mathrm{d}K(t)}{\mathrm{d}h}.$$
- The definition for simple processes suggests a method of estimating $K(h)$ from an observed pattern as a function of the average number of events less than distance $h$ apart.
- In a clustered pattern an event is likely to be surrounded by events from the same cluster. The number of extra events within small distances will be large. In regular patterns the number of extra events for short distances will be small.
- $K(h)$ is not affected by events that are missing completely at random (MCAR). If not all events have been recorded—the pattern is not mapped—and the missing data process is MCAR, the observed pattern is a subset of the complete process whose events are retained or deleted in a sequence of *iid* Bernoulli trials. Such random thinning, also called $p$-thinning, reduces the intensity and the number of extra events by the same factor. The original process and the pattern which results from $p$-thinning have the same $K$-function (see §3.7.1).
- Other functions of $\lambda_2$ used in the study of dependence in point patterns are easily related to $K(h)$. For example, the pair-correlation function
$$R(h) = \frac{1}{2h\pi}\frac{\mathrm{d}K(h)}{\mathrm{d}h},$$
the radial distribution function
$$F(h) = \lambda\frac{\mathrm{d}K(h)}{\mathrm{d}h},$$
and the $L$-function
$$L(h) = \sqrt{K(h)/\pi}.$$

### 3.4.2 Estimation of K- and L-Functions

The first-order intensity of a homogeneous process does not depend on spatial location, $\lambda(\mathbf{s}) = \lambda$, and the natural estimator of the intensity within region A is

*Global Intensity Estimator*

$$\widehat{\lambda} = \frac{N(A)}{\nu(A)}. \qquad (3.8)$$

Recall that the $K$-function (3.7) is defined for stationary, isotropic point patterns and that $\lambda K(h) \equiv E(h)$ is the expected number of extra events within distance $h$. If $h_{ij}$ is the distance between events $\mathbf{s}_i$ and $\mathbf{s}_j$, a naïve moment estimator of $E(h)$ is

$$\tilde{E}(h) = \frac{1}{n}\sum_{i=1}^{n}\sum_{j\neq i}^{n} I(h_{ij} \leq h).$$

The inner sum yields the number of observed extra events within distance $h$ of event $\mathbf{s}_i$. The outer sum accumulates these counts. Since the process is stationary, the intensity is estimated with (3.8) and $\tilde{K}(h) = \hat{\lambda}^{-1}\tilde{E}(h)$.

Because events outside the study region are not observed, this estimator is negatively biased. If one calculates the extra events for an event near the boundary of the region, counts will be low because events outside the region are not taken into account. To adjust for these edge effects, various corrections have been applied. If one considers only those events for the computation of $K(h)$ whose distance $d_i$ from the nearest boundary exceeds $h$, one obtains

$$\begin{aligned} E^*(h) &= \sum_{i=1}^{n} \frac{\sum_{j=1\neq i}^{n} I(h_{ij} \leq h \text{ and } d_j > h)}{\sum_{j=1}^{n} I(d_j > h)} \\ \widehat{E}_d(h) &= \begin{cases} E^* & \sum_{j=1}^{n} I(d_j > h) > 0 \\ 0 & \text{otherwise.} \end{cases} \end{aligned}$$

Ripley's estimator (Ripley, 1976) applies weights $w(\mathbf{s}_i, \mathbf{s}_j)$ to each pair of observations that correspond to the proportion of the circumference of a circle that is within the study region, centered at $\mathbf{s}_i$, and with radius $h_{ij} = ||\mathbf{s}_i - \mathbf{s}_j||$. The estimator for $E(h)$ applying this edge correction is

$$\widehat{E}(h) = \frac{1}{n}\sum_{i=1}^{n}\sum_{j\neq i}^{n} w(\mathbf{s}_i, \mathbf{s}_j)^{-1} I(h_{ij} \leq h).$$

In either case,

$$\widehat{K}(h) = \widehat{\lambda}^{-1}\widehat{E}(h) \quad \text{or} \quad \widehat{K}(h) = \widehat{\lambda}^{-1}\widehat{E}_d(h).$$

Cressie (1993, p. 616) discusses related estimators of $K(h)$.

In statistical analyses one commonly computes $K(h)$ for a set of distances and compares the estimate against the $K$-function of the CSR process ($\pi h^2$). Unfortunately, important deviations between empirical and theoretical second-order behavior are often difficult to determine when $\widehat{K}(h)$ and $K(h)$ are overlayed in a plot. In addition, the variance of the estimated $K$-function increases quickly with $h$ and for large distances the behavior can appear erratic. Using a plug-in estimate, the estimated $L$-function

$$\widehat{L}(h) = \sqrt{\widehat{K}(h)/\pi}$$

has better statistical properties. For graphical comparisons of empirical and theoretical second-order behavior under CSR we recommend a plot of $\widehat{L}(h) - h$ versus $h$. The CSR model is the horizontal reference line at 0. Clustering of events manifests itself as positive values at short distances. Significance is assessed through Monte Carlo testing as described in §3.3.1 and in practice, we consider a plot of $\widehat{L}(h) - h$ versus $h$ together with the corresponding simulation envelopes computed under CSR as described in §3.3.2.

**Example 1.5 (Lightning strikes. Continued)** Based on the empirical distribution function of nearest neighbor distances we concluded earlier that the lightning data are highly clustered. If clustering is not the result of an inhomogeneous lightning intensity, but due to dependence of the events, a second-order analysis with $K$- or $L$-functions is appropriate. Figure 3.9 shows the observed $L$-functions for these data and simulation envelopes based on $s = 200$. The quick rise of the $L$-functions above the reference line for small distances is evidence of clustering. Whereas the simulation envelopes do not differ between an analysis on the bounding box and the convex hull, the empirical $L$-function in the former case overstates the degree of clustering because the bounding rectangle adds too much empty white space. □

### *3.4.3 Assessing the Relationship between Two Patterns*

The $K$-function considers only the location of events; it ignores any attribute values (marks) associated with the events. However, many point patterns include some other information about the events and this information is often binary in nature, e.g., which of two competing species of trees occurred at a particular location, whether or not an individual with a certain disease at a particular location is male or female, or whether or not a plant at a location was diseased. Diggle and Chetwynd (1991) refer to such processes as **labeled**. In cases such as these, we may wonder whether the nature of the spatial pattern is different for the two types of events. We discuss marked point patterns and multivariate spatial point processes in more generality in §3.6. In this section, we focus on the simple, yet common, case of a bivariate process with binary marks.

One generalization of $K(h)$ to a bivariate spatial point process is (Ripley, 1981; Diggle, 1983, p. 91)

$$\begin{aligned} K_{ij}(h) &= \lambda^{-1}\mathrm{E}[\#\text{of type } j \text{ events within distance } h \\ &\qquad\qquad \text{of a randomly chosen type } i \text{ event}]. \end{aligned}$$

Suppose the type $i$ events in $A$ are observed with intensity $\lambda_i$ at locations

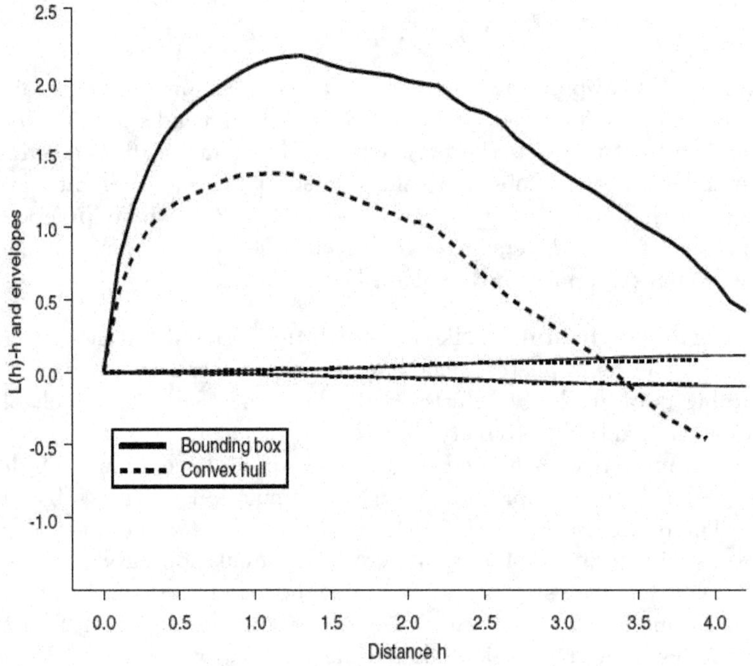


Figure 3.9 *L-functions and simulation envelopes from 200 simulations on bounding box and convex hull for lightning data. Edge correction is based on* $\widehat{E}_d(h)$.

referenced by $\mathbf{s}$, and the type $j$ events in $A$ are observed with intensity $\lambda_j$ at locations referenced by $\mathbf{u}$. Then, an edge-corrected estimator of $K_{ij}(h)$ is (Ripley, 1981)

$$\widehat{K}_{ij}(h) = [\widehat{\lambda}_i\widehat{\lambda}_j\nu(A)]^{-1}\sum_k\sum_l w(\mathbf{s}_k, \mathbf{u}_l)^{-1}I(h_{kl} \le h), \tag{3.9}$$

where $h_{kl} = ||\mathbf{s}_k - \mathbf{u}_l||$, and $w(\mathbf{s}_k, \mathbf{u}_l)$ is the proportion of the circumference of a circle centered at location $\mathbf{s}_k$ with radius $h_{kl}$ that lies inside $A$.

If the bivariate spatial process is stationary, the cross-$K$-functions are symmetric, i.e., $K_{12} = K_{21}$. However, $\widehat{K}_{12} \neq \widehat{K}_{21}$, so Lotwick and Silverman (1982) suggest using the more efficient estimator

$$K^*_{ij}(h) = \widehat{\lambda}_j\widehat{K}_{ij}(h) + \widehat{\lambda}_i\widehat{K}_{ji}(h)/\widehat{\lambda}_j + \widehat{\lambda}_i.$$

Under a null hypothesis of independence between the two spatial point processes, $K_{ij}(h) = \pi h^2$, regardless of the nature of the pattern of either type of event. Thus, again we work with the corresponding $L$-function, $L^*_{ij}(h) = (K^*_{ij}(h)/\pi)^{1/2}$, and under independence $L^*_{ij}(h) = h$. Values of $L^*_{ij}(h) - h > 0$

indicate attraction between the two processes at distance $h$. Values of $L^*_{ij}(h) - h < 0$ indicate repulsion. Unfortunately, hypothesis tests are more difficult in this situation since a complete bivariate model must be specified.

Diggle (1983) provides an alternative philosophy that is not based on CSR. Another way to define the null hypothesis of "no association" between the two processes is that each event is equally likely to be a type $i$ (or type $j$) event. This is known as the **random labeling hypothesis**. This hypothesis is subtly different than the independence hypothesis. The two scenarios arise from different random mechanisms. Under independence, the locations and associated marks are determined simultaneously. Under the random labeling hypothesis, locations arise from a univariate spatial point process and a second random mechanism determines the marks. Thus, the marks are determined independently of the locations. Diggle notes that the random labeling hypothesis neither implies nor is implied by the stricter notion of statistical independence between the two spatial point processes, and confusing the two scenarios can lead to "the analysis of data by methods which are largely irrelevant to the problem in hand" (Diggle, 1983, p. 93). The random labeling hypothesis always conditions on the set of locations of all observed events, and, under this hypothesis

*Random Labeling Hypothesis*

$$K_{11} = K_{22} = K_{12} \tag{3.10}$$

(Diggle and Chetwynd, 1991). In contrast, the independence approach conditions on the marginal structure of each process. Thus, the two approaches lead to different expected values for $K_{12}(h)$, to different tests, and to different interpretation.

Diggle and Chetwynd use the relationships in (3.10) to construct a test based on the difference of the $K$-functions

$$D(h) = K_{ii}(h) - K_{jj}(h).$$

They suggest estimating $D(h)$ by plugging in estimates of $K_{ii}$ and $K_{jj}$ obtained from (3.9) (adjusted so that $D(h)$ is unbiased, see Diggle and Chetwynd, 1991). Under the random labeling hypothesis, the expected value of $D(h)$ is zero for any distance $h$. Positive values of $D(h)$ suggest spatial clustering of type $i$ events over and above any clustering observed in the type $j$ events.

Diggle and Chetwynd (1991) derive the variance-covariance structure of $\widehat{D}(h)$ and give an approximate test based on the standard Gaussian distribution. However, Monte Carlo simulation is much easier. To test the random labeling hypothesis, we condition on the set of all $n_1 + n_2$ locations, draw a sample of $n_1$ from these (if we want to test for clustering in type $i$ events), assign the locations not selected to be of type $j$, and then compute $\widehat{D}(h)$ for each sampling. Under the random labeling hypothesis, the $n_1$ sampled locations reflect a random "thinning" (see §3.7.1) of the set of all locations. Also, Monte Carlo simulation enables us to consider different statistics for the comparison of the patterns, for example $D(h) = L_{ii}(h) - L_{jj}(h)$.

**Example 1.5 (Lightning strikes. Continued)** There are two types of lightning flashes between the earth surface (or objects on it) and a storm cloud. Most frequent are flashes where a channel of air with negative charge travels from the bottom of the cloud. Flashes with positive polarity commence from the upper regions of the cloud, occur less frequently, are more scattered, but also more destructive. We extracted from the lightning strikes data set those flashes that occurred off the coast of Florida, Georgia, and South Carolina. Of these 820 flashes, $n_p = 67$ were positive flashes.

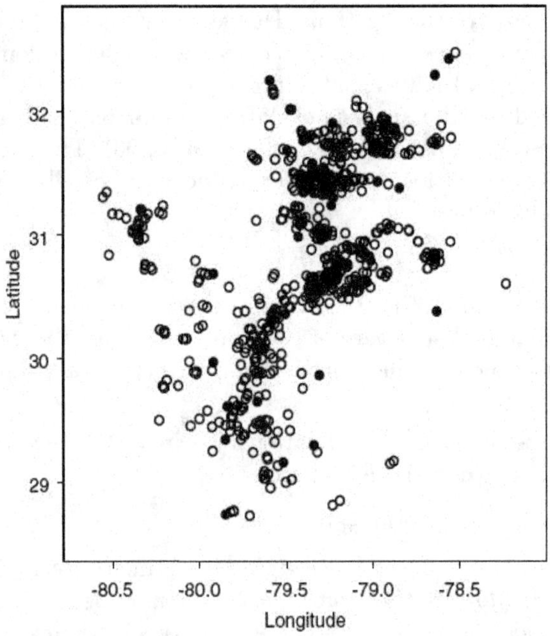


Figure 3.10 *Lightning flashes off the coast of Florida, Georgia, and South Carolina. Empty circles depict flashes with negative charge* ($n_n = 753$)*, closed circles depict flashes with positive charge* ($n_p = 76$)*.*

Figure 3.11 displays the $L$-functions for the two types of events, their difference, 5%, and 95% simulation envelopes for $D(h) = L_p(h) - L_n(h)$ based on 200 random assignment of labels to event. There is no evidence that lightning strikes of different polarity differ in their degree of clustering. In carrying out the simulations, the same bounding shape is assumed for all patterns, that based on the data for both event types combined.

□

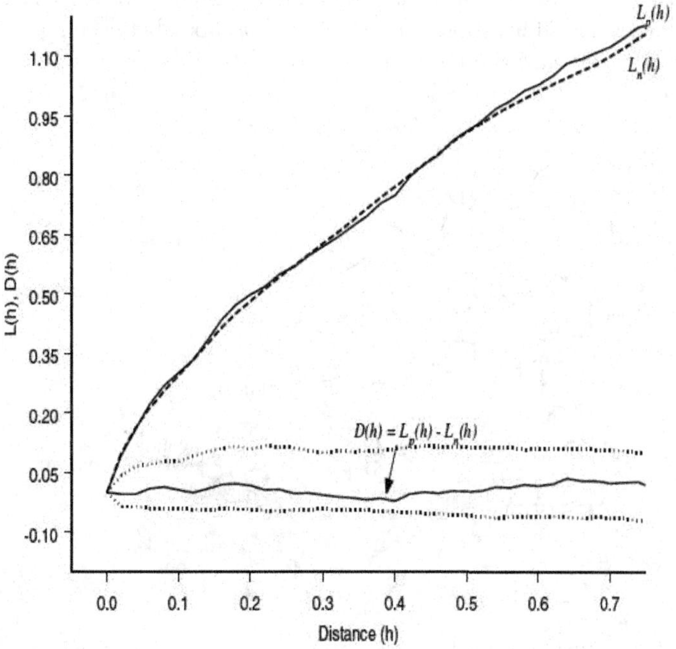


Figure 3.11 *Observed L-functions for flashes with positive and negative charge in pattern of Figure 3.10 and their difference. Dotted lines depict 5 and 95 percentile envelopes from 200 random labelings of polarity.*

## 3.5 The Inhomogeneous Poisson Process

Most processes deviate from complete spatial randomness in some fashion. For example, events may be independent in non-overlapping subregions, but the intensity $\lambda(\mathbf{s})$ with which they occur is not homogeneous throughout $D$. More events will then be located in regions where the intensity is large, and fewer events will be located in regions where $\lambda(\mathbf{s})$ is small. Thus, the resulting point pattern often appear clustered. In fact, this is typical for geographical processes based on human populations. Residences where people live are clustered into cities, towns, school districts, and neighborhoods. In analyzing events that relate to people, we have to adjust for this geographical variation in our spatial analyses.

**Example 3.4 GHCD 9 infant birth weights.** As an example, consider the case-control study of Rogers, Thompson, Addy, McKeown, Cowen, and DeCoulfé (2000) for which the study area comprised 25 contiguous counties in southeastern Georgia, collectively referred to as Georgia Health Care District

9 (GHCD9) (Figure 3.12). One of the purposes of the study was to examine geographic risk factors associated with the risk of having a very low birth weight (VLBW) baby, one weighing less than 1,500 grams at birth. Cases were identified from all live-born, singleton infants born between April 1, 1986 and March 30, 1988 and the locations of the mothers' addresses are shown in Figure 3.13.

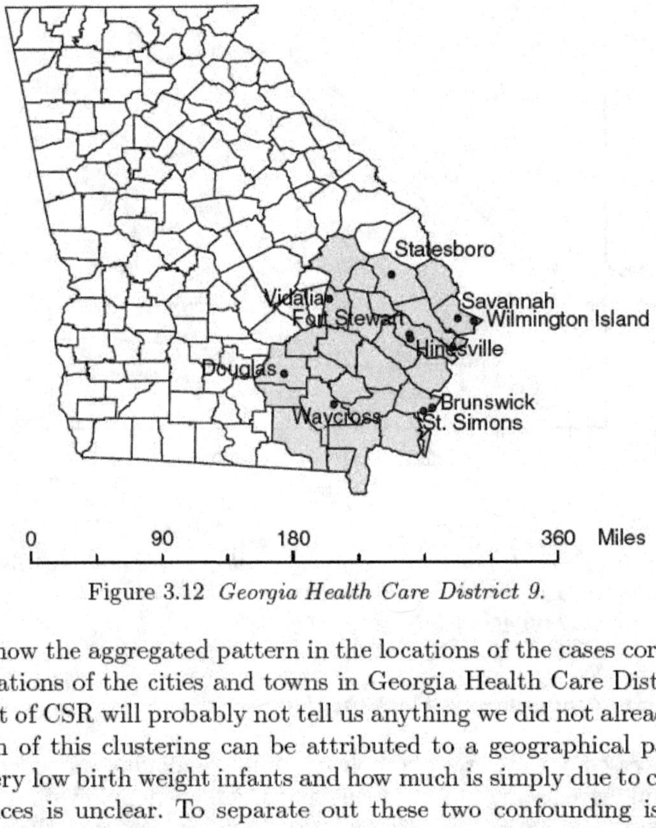


Figure 3.12 *Georgia Health Care District 9.*

Notice how the aggregated pattern in the locations of the cases corresponds to the locations of the cities and towns in Georgia Health Care District 9. A formal test of CSR will probably not tell us anything we did not already know. How much of this clustering can be attributed to a geographical pattern in cases of very low birth weight infants and how much is simply due to clustering in residences is unclear. To separate out these two confounding issues, we need to compare the geographic pattern in the cases to that based on a set of controls that represent the geographic pattern in infants who were born with normal birth weights. Controls were selected for this study by drawing a 3% random sample of all live-born infants weighing more than 2,499 grams at birth. This sampling was constrained so that the controls met the same residency and time frame requirements as the case subjects. Their geographic distribution is shown in Figure 3.14. Notice that the locations of both the cases and the controls appear to be clustered. We can use the controls to quantify a background geographical variation in infant birth weights and then assess whether there are differences in the observed spatial pattern for babies born with very low birth weights. In order to do this, we need a different null hypothesis than the one provided by CSR. One that is often used is called the **constant risk hypothesis** (Waller and Gotway, 2004). Under the

*Constant Risk Hypothesis*

Figure 3.13 *Cases of very low birth weight babies in Georgia Health Care District 9 from Rogers et al. (2000). The locations have been randomly relocated to protect confidentiality.*

constant risk model, the probability of being an event is the same, regardless of location (e.g., each baby born to a mother residing in Georgia Health Care District 9 has the same risk of being born with a very low birth weight). Under the constant risk hypothesis, we expect more events in areas with more individuals. Clusters of cases in high population areas could violate CSR but would not necessarily violate the constant risk hypothesis. Thus, choosing the constant risk hypothesis as a null model allows us to refine the question of interest from "are the cases clustered?" (the answer to which we already know is probably "yes") to the question "are the cases more clustered than we would expect under the constant risk hypothesis?" Answering this latter question allows adjustment for any patterns that might occur among all the individuals within a domain of interest. □

Statistical methods that allow us to use the constant risk hypothesis as a null model are often based on an inhomogeneous intensity function. When point processes are studied through counting measures, the number of events in region $A$, $N(A)$, is a random variable and the usual expectations can be constructed from the mass function of $N(A)$. Since $N(A)$ is an aggregation of events in region $A$, there is a function $\lambda(\mathbf{s})$ whose integration over $A$ yields its expected value. Earlier, this function was termed the (first-order) intensity and defined as a limit,

*First-order Intensity*

$$\lambda(\mathbf{s}) = \lim_{|d\mathbf{s}| \to 0} \frac{\mathrm{E}[N(d\mathbf{s})]}{|d\mathbf{s}|}. \tag{3.11}$$

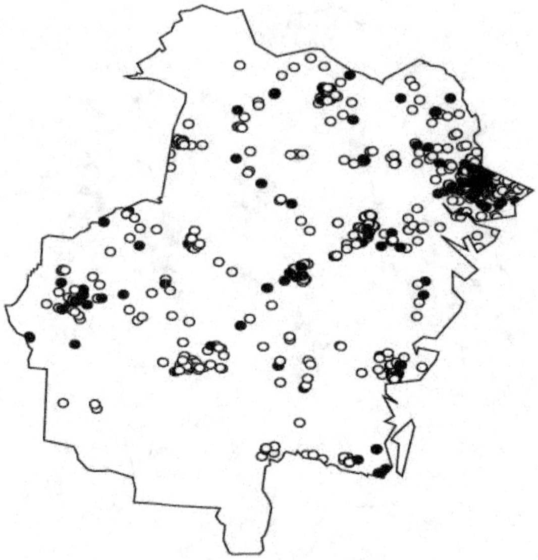

Figure 3.14 *Cases of very low birth weight babies in Georgia Health Care District 9 and Controls from Rogers et al. (2000). The locations have been randomly relocated to protect confidentiality.*

The connection with the average number of events in region $A$ is simply

$$\mathrm{E}[N(A)] = \mu(A) = \int_A \lambda(\mathbf{s})\, d\mathbf{s}.$$

Studying point patterns through $\lambda(\mathbf{s})$ rather than through $\mathrm{E}[N(A)]$ is often mathematically advantageous because it eliminates the dependency on the size (and shape) of the area A. In practical applications, when an estimate of the intensity function is sought, an area context is required.

### *3.5.1 Estimation of the Intensity Function*

Even for homogeneous processes it is useful to study the intensity of events more locally, for example, to determine whether to proceed with an analysis of the second-order behavior. In practice, spatially variable estimates $\widehat{\lambda}(\mathbf{s})$ of the intensity at location s are obtained by nonparametric smoothing of quadrat counts or by methods of density estimation.

To see the close relationship between density estimation and intensity estimation consider a random sample $y_1, \cdots, y_n$ from the distribution of random variable $Y$. An estimate of the density function $f(y)$ at $y_0$ can be found from the number of sample realizations within a certain distance $h$ from $y_0$,

$$\hat{f}(y_0) = \frac{1}{nh} \sum_{i=1}^{n} k\left(\frac{y_i - y_0}{h}\right), \qquad (3.12)$$

where $k(t)$ is the uniform density on $-1 \le t \le 1$,

$$k(t) = \begin{cases} 0 & |y_i - y_0| > h \\ 1 & \text{otherwise.} \end{cases}$$

If the neighborhood $h$ is small, $\hat{f}(y_0)$ is a nearly unbiased estimate of $f(y_0)$ but suffers from large variability. With increasing window width $h$, the estimate becomes smoother, less variable, and more biased. A mean-squared-error-based procedure such as cross-validation can be used to determine an appropriate value for the parameter $h$.

Rather than choosing a uniform kernel function that gives equal weight to all points within the window $y_0 \pm h$, we use modal kernel functions. Popular kernel functions are the Gaussian kernel

$$k(t) = \frac{1}{\sqrt{2\pi}} \exp\left\{-t^2/2\right\},$$

the quadratic kernel

$$k(t) = \begin{cases} 0.75(1-t^2) & |t| \le 1 \\ 0 & \text{otherwise} \end{cases}$$

or the minimum variance kernel

$$k(t) = \begin{cases} \frac{3}{8}(3-5t^2) & |t| \le 1 \\ 0 & \text{otherwise.} \end{cases}$$

The choice of the kernel function is less important in practice than the choice of the bandwidth. A function $k(u)$ can serve as a kernel provided that $\int k(u)du = 1$ and $\int uk(u) = 0$.

For a spatial point pattern, density estimation produces an estimate of the probability of observing an event at location $\mathbf{s}$ and integrates to one over the domain $A$. The relationship between the density $f_A(\mathbf{s})$ on $A$ and the intensity $\lambda(\mathbf{s})$ is

$$\lambda(\mathbf{s}) = f_A(\mathbf{s}) \int_A \lambda(\mathbf{u})d\mathbf{u} = f_A(\mathbf{s})\mu(A).$$

The intensity and density are proportional. For a process in $\mathbb{R}^1$, we modify the density estimator (3.12) as

$$\widehat{\lambda}(s_0) = \frac{1}{\nu(A)h} \sum_{i=1}^{n} k\left(\frac{s_i - s_0}{h}\right), \tag{3.13}$$

to obtain a kernel estimator of the first-order intensity.

In the two-dimensional case, the univariate kernel function needs to be replaced by a function that can accommodate two coordinates. A product-kernel function is obtained by multiplying two univariate kernel functions. This choice is frequently made for convenience, it implies the absence of interaction between the coordinates. The bandwidths can be chosen differently for the two dimensions, but kernels with spherical contours are common. If the $x_i$ and $y_i$ denote the coordinates of location $\mathbf{s}_i$, then the product-kernel approach leads to the intensity estimator

*Kernel Intensity Estimator*

$$\widehat{\lambda}(\mathbf{s}_0) = \frac{1}{\nu(A)h_x h_y} \sum_{i=1}^{n} k\left(\frac{x_i - x_0}{h_x}\right) k\left(\frac{y_i - y_0}{h_y}\right), \qquad (3.14)$$

where $h_x$ and $h_y$ are the bandwidths in the respective directions of the coordinate system. The independence of the coordinates can be overcome with bivariate kernel functions. For example, elliptical contours can be achieved with a bivariate Gaussian kernel function with unequal variances. A non-zero covariance of the coordinates introduces a rotation.

The expressions above do not account for edge effects, which can be substantial. Diggle (1985) suggested an edge-corrected kernel intensity estimator with a single bandwidth

*Kernel Intensity Estimator*

$$\widehat{\lambda}(\mathbf{s}) = \frac{1}{p_h(\mathbf{s})} \sum_{i=1}^{n} \frac{1}{h^2} k\left(\frac{\mathbf{s} - \mathbf{s}_i}{h}\right).$$

The denominator $p_h(\mathbf{s}) = \int_A h^{-2}k((\mathbf{s}-\mathbf{u})/h)d\mathbf{u}$ serves as the edge correction.

### 3.5.2 Estimating the Ratio of Intensity Functions

In many applications, the goal of the analysis is the comparison of spatial patterns between two groups (e.g., between males and females, between cases and controls). Suppose we have $n_1$ events of one type and $n_2$ events of another type and let $\lambda_1(\mathbf{s})$ and $\lambda_2(\mathbf{s})$ be their corresponding intensity functions. It seems natural to estimate the ratio $\lambda_1(\mathbf{s})/\lambda_2(\mathbf{s})$ by the ratio of the corresponding kernel density estimates $\widehat{\lambda}_1(\mathbf{s})/\widehat{\lambda}_2(\mathbf{s})$, where $\widehat{\lambda}_i(\mathbf{s})$ is given in (3.14). Since the intensity function is proportional to the density function, Kelsall and Diggle (1995) suggest inference (conditional on $n_1$ and $n_2$) based on

$$\widehat{r}(\mathbf{s}) = \log\{\widehat{f}_1(\mathbf{s})/\widehat{f}_2(\mathbf{s})\},$$

where $f_1$ and $f_2$ are the densities of the two processes and $\widehat{f}_1$ and $\widehat{f}_2$ are their corresponding kernel density estimators. Mapping $\widehat{r}(\mathbf{s})$ provides a spatial picture of the (logarithm of) the probability of observing an event of one type rather than an event the other type at location $\mathbf{s}$ in $D$.

**Example 3.4 (Low birth weights. Continued)** Applying this procedure to the case-control data considered in the previous section we obtain the surface shown in Figure 3.15. This shows the relative risk of a VLBW birth at every location within Georgia Health Care District 9. Naturally, the eye is drawn to areas with the highest risk, but care must be taken in interpreting the results. First, in drawing such a map, we implicitly assume that $r(\mathbf{s})$ is a continuous function of location $\mathbf{s}$, which is somewhat unappealing. There are probably many locations where it is impossible for people to live and for which $r(\mathbf{s})$ is

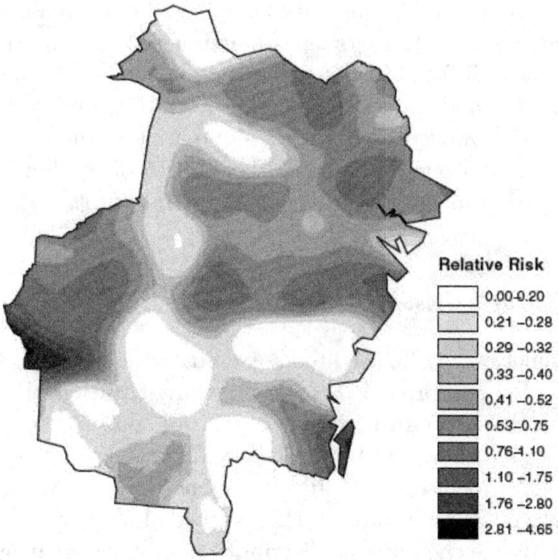


Figure 3.15 *Relative risk of very low birth weight babies in Georgia Health Care District 9. Conclusions are not epidemiologically valid since locations and case/control status were altered to preserve confidentiality.*

inherently zero. Second, zero estimates for $f_2(\mathbf{s})$ are clearly problematic for computation of $\widehat{r}(\mathbf{s})$, but certainly can occur and do have meaning as part of the estimation of $f_2$.

An advantage to choosing a kernel with infinite tails such as the bivariate Gaussian kernel, as we have done here, is that the estimate of $f_2(\mathbf{s})$ is non-zero for all locations. Third, the choice for the bandwidths is critical and different choices can have a dramatic effect on the resulting surface. In constructing the map in Figure 3.15 we experimented with a variety of choices for the bandwidth and the underlying grid for which the kernel density estimates are obtained. The combination of the two (bandwidth and grid spacing) reflects the tradeoff between resolution and stability. A fine grid with a small bandwidth will allow map detail, but the resulting estimates may be unstable. A coarse grid with a large bandwidth will produce more stable estimates, but much of the spatial variation in the data will be smoothed away. Also, there are large regions within Georgia Health Care District 9 without controls. This leads to unstable estimates for certain bandwidths. We began by choosing the bandwidths according to automatic selection criteria (e.g., cross validation, Wand and Jones, 1995), but found the results were visually uninteresting; the resulting map appeared far too smooth. Because of the large gaps where there are no controls, we took the same bandwidth for the cases as for the controls and then increased it systematically until we began to lose stability in the estimates. We may have actually crossed the threshold here: the area with a high relative risk on the western edge of the domain may be artificially high, reflecting estimate instability and edge effects, rather than a high relative risk. This illustrates the importance of careful estimation and interpretation of the results, particularly if formal inference (e.g., hypothesis tests, see Kelsall and Diggle, 1995 and Waller and Gotway, 2004) will be conducted using the resulting estimates. However, even with a few potential anomalies, and the odd contours that result from the kernel, Figure 3.15 does allow us to visualize the spatial variation in the risk of very low birth weight. □

### *3.5.3 Clustering and Cluster Detection*

While the $K$-function can be used to assess clustering in events that arise from a homogeneous Poisson process, the assumption of stationarity upon which it is based precludes its use for inhomogeneous Poisson processes. Thus, Cuzick and Edwards (1990) adapted methods based on nearest neighbor distances (described in §3.3) for use with inhomogeneous Poisson processes. Instead of assuming events occur uniformly in the absence of clustering, a group of controls is used to define the baseline distribution and nearest neighbor statistics are based on whether the nearest neighbor to each case is another case or a control. The null hypothesis of no clustering is that each event is equally likely to have been a case or a control, i.e., the random labeling hypothesis.

Let $\{\mathbf{s}_1, \ldots \mathbf{s}_n\}$ denote the locations of all events and assume $n_1$ of these are cases and $n_2$ are controls. Let

$$\delta_i = \begin{cases} 1 & \text{if } \mathbf{s}_i \text{ is a case} \\ 0 & \text{if } \mathbf{s}_i \text{ is a control.} \end{cases}$$

and

$$d_i = \begin{cases} 1 & \text{if the nearest neighbor to } \mathbf{s}_i \text{ is a case} \\ 0 & \text{if the nearest neighbor to } \mathbf{s}_i \text{ is a control.} \end{cases}$$

The test statistic represents the number of the $q$ nearest neighbors of cases that are also cases,

$$T_q = \sum_{i=1}^{n} \delta_i d_i^k,$$

where $q$ is specified by the user. For inference, Cuzick and Edwards (1990) derive an asymptotic test based on the Gaussian distribution. A Monte Carlo test based on the random labeling hypothesis is also applicable.

**Example 3.4 (Low birth weights. Continued)** We use Cuzick and Edward's NN test to assess whether there is clustering in locations of babies born with very low birth weights in Georgia Health Care District 9. This test is not entirely applicable to this situation in that it assumes each event location must be either a case or a control. However, because people live in apartment buildings, there can be multiple cases and/or controls at any location; we cannot usually measure a person's location so specifically. This situation is common when addresses in urban areas have been geocoded. In Georgia Health Care District 9, there were 7 instances of duplicate locations, each containing from 2–3 locations. Thus, in order to use Cuzick and Edward's NN test, we randomly selected one record from each of these groups.

In order to make the test we need to specify $q$, the number of nearest neighbors. Small choices for $q$ tend to make the test focus more locally, while larger values of $q$ allow a more regional assessment of clustering. Thus, to some degree, different values of $q$ indicate the scale of any observed clustering. With this in mind, we chose $q = 1, 5, 10, 20$ and the results are summarized in Table 3.5.

Table 3.5 *Results from Cuzick and Edward's NN Test Applied to Case/Control Data in Georgia Health Care District 9. The p-values were obtained from Monte Carlo simulation. Conclusions are not epidemiologically valid since locations and case/control status were altered to preserve confidentiality.*

| $q$ | $T_q$ | p-value |
|---|---|---|
| 1 | 81 | 0.0170 |
| 5 | 388 | 0.0910 |
| 10 | 759 | 0.2190 |
| 20 | 1464 | 0.3740 |

The results in Table 3.5 seem to indicate that there is some clustering among the cases at very local levels. As $q$ is increased, the test statistics are not significant, indicating that, when considering Georgia Health Care District 9 overall, there is no strong evidence for clustering among locations of babies born with very low birth weights. Note, however, that $T_{q_2}$ is correlated with $T_{q_1}$ for $q_1 < q_2$ since the $q_2$ nearest neighbors include the $q_1$ nearest neighbors. Ord (1990) suggests using contrasts between statistics (e.g., $T_{q_2} - T_{q_1}$) since they exhibit considerably less correlation and can be interpreted as excess cases between the $q_1$ and the $q_2$ nearest neighbors of cases. □

As Stroup (1990) notes, if we obtain a significant result from Cuzick and Edwards' test, the first response should be "Where's the cluster?" However, answering this question is not what this, or other methods for detecting spatial clustering, are designed to do. Besag and Newell (1991) stress the need to clearly distinguish between the concepts of detecting *clustering* and detecting *clusters*. Cluster*ing* is a *global* tendency for events to occur near other events. A *cluster* is a *local* collection of events that is inconsistent with the hypothesis of no clustering, either CSR or constant risk (Waller and Gotway, 2004). Cuzick and Edwards' test, and tests based on the $K$-function are tests of clustering. We need different methods to detect clusters.

Interest in cluster detection has grown in recent years because of the public health and policy issues that surround their interpretation. For example, a cluster of people with a rare disease could indicate a common, local environmental contaminant. A cluster of burglarized residences can alert police to "hot spots" of crime that warrant increased surveillance. Thus, what is needed in such situations is a test that will: 1) detect clusters; 2) determine whether they contain a significantly higher or lower number of events than we would expect; and 3) identify the location and extent of the cluster. There are several methods for cluster detection (see, e.g., Waller and Gotway, 2004, for a comprehensive discussion and illustration), but the most popular is the spatial scan statistic developed by Kulldorff and Nagarwalla (1995) and Kulldorff (1997) and popularized by the SatScan software (Kulldorff and International Management Services, Inc., 2003).

Scan statistics use moving windows to compare a value (e.g., a count of events or a proportion) within the window to the value outside of the window. Kulldorff (1997) uses circular windows with variable radii ranging from the smallest inter-event distance to a user-defined upper bound (usually one half the width of the study area). The spatial scan statistic may be applied to circles centered at specified grid locations or centered on the set of observed event locations.

The spatial scan statistic developed by Kulldorff (1997) considers local likelihood ratio statistics that compare the likelihood under the the constant risk hypothesis to various alternatives where the proportion of cases within the window is greater than that outside the window. Let $C$ denote the total number of cases and let $c$ be the total number of cases within a window. Under the assumption that the number of cases follows a Poisson distribution, the likelihood function for a given window is proportional to

$$\left(\frac{c}{n}\right)^c \left(\frac{C-c}{C-n}\right)^{C-c} I(\cdot),$$

where $n$ is the number expected assuming constant risk assumption over the study domain. $I(\cdot)$ denotes the indicator function which, when high proportions are of interest, is equal to 1 when the window has more cases than expected.

The likelihood function is maximized over all windows and the window with the maximum likelihood function is called "the most likely cluster." Significance is determined by using Monte Carlo simulation. Using random labeling, cases are randomly assigned to event locations, the likelihood function is computed for each window, and the maximum value of this function is determined. In this way the distribution of the *maximum* likelihood function is simulated (Turnbull, Iwano, Burnett, Howe, and Clark, 1990). As a result, the spatial scan statistic provides a single $p$-value for the study area, and avoids the multiple testing problem that plagues many other approaches.

**Example 3.4 (Low birth weights. Continued)** We use the spatial scan statistic developed by Kulldorff (1997) and Kulldorff and International Management Services, Inc. (2003) to find the most likely cluster among locations of babies born with very low birth weights in Georgia Health Care District 9. We note that this "most likely" cluster may not be at all "likely," and thus we rely on the $p$-value from Monte Carlo testing to determine its significance. We assumed a Poisson model and allowed the circle radii to vary from the smallest inter-event distance to one half of the largest inter-event distance. The results from the scan give the location and radius of the circular window that constitutes the most likely cluster and a $p$-value from Monte Carlo testing. The results are shown in Figure 3.16.

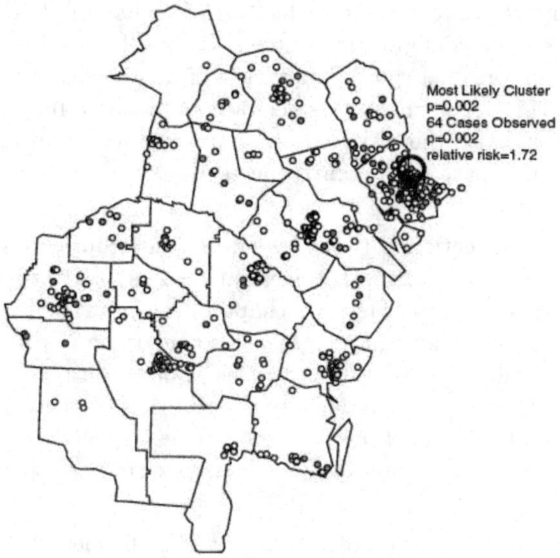


Figure 3.16 *Results from the spatial scan statistic. Conclusions are not epidemiologically valid since locations and case/control status were altered to preserve confidentiality.*

The most likely cluster is approximately 12 km in diameter and is located just north of the city of Savannah. Under the constant risk hypothesis, 37.1 cases were expected here, but 64 were observed, leading to a relative risk of 1.72. The $p$-value of 0.002 indicates that the proportion of cases observed in this window is significantly higher than we would expect under the constant risk hypothesis. Note that these results are somewhat consistent with those from Cuzick and Edwards' test (Table 3.5) that indicated local clustering among the cases. Note, however, that the power of spatial statistics can also vary spatially. We have greater power to detect abnormalities in regions where more events occurred. Since such regions are often urban, the results of the spatial scan should be interpreted in this context. □

## 3.6 Marked and Multivariate Point Patterns

### *3.6.1 Extensions*

Up to this point we have focused on the random distribution of events throughout a spatial domain. Little attention was paid to whether some additional attribute is observable at the event location or whether the events were all of the same type. Consider, for example, the distribution of trees throughout a forest stand. Most foresters would not be satisfied with knowing where the trees are. Tree attributes such as breast height diameter, age, height, and species are important, as well as understanding whether these attributes are related to the spatial configuration of trees. It is well established, for example, that in managed forests a tree's diameter is highly influenced by its available horizontal growing area, whereas its height is primarily a function of soil quality. This suggests that the distribution of tree diameters in a forest stand is related to the intensity of a point process that governs the distribution of trees.

To make the connection between events and attributes observed at event locations more precise, we recall the notation $Z(\mathbf{s})$ for the attribute $Z$ observed at location $\mathbf{s}$. In the first two chapters the notation $Z(\mathbf{s})$ was present throughout and it appears that we lost it somehow in discussing point patterns. It never really left us, but up to this "point" patterns were just that: points. The focus was on studying the distribution of the events itself. The "unmarked" point pattern of previous sections is a special case of the marked pattern, where the distribution of $Z$ is degenerate (a mark space with a single value).

In the vernacular of point process theory, $Z$ is termed the mark variable. It is a random variable, its support is called the mark space. The mark space can be continuous or discrete; the diameter or height of a tree growing at $\mathbf{s}$, the depth of the lunar crater with center $\mathbf{s}$, the value of goods stolen during a burglary, are examples of marked processes with continuous mark variable. The number of eggs in a birds nest at $\mathbf{s}$ or the grass species growing at $\mathbf{s}$ are cases of discrete mark variables. Figure 3.17 is an example of a point process with a binary mark variable. The spatial events represent tree locations in a forest in Lansing, MI. The mark variable associated with each location indicates whether the tree is a hickory or a maple.

So why are we treating marked point processes separately from, say, geostatistical data? Well, we are and we are not. The "big" difference between the two types of data is the randomness of the spatial domain, of course. In geostatistical data, the domain is continuous and observations are collected at a finite number of points. The sample locations can be determined by a random mechanism, such as stratified sampling, or be chosen by a deterministic method. In either case, the samples represent an incomplete observation of a random function and the random choosing of the sample locations does not enter into a geostatistical analyses as a source of randomness. A mapped,

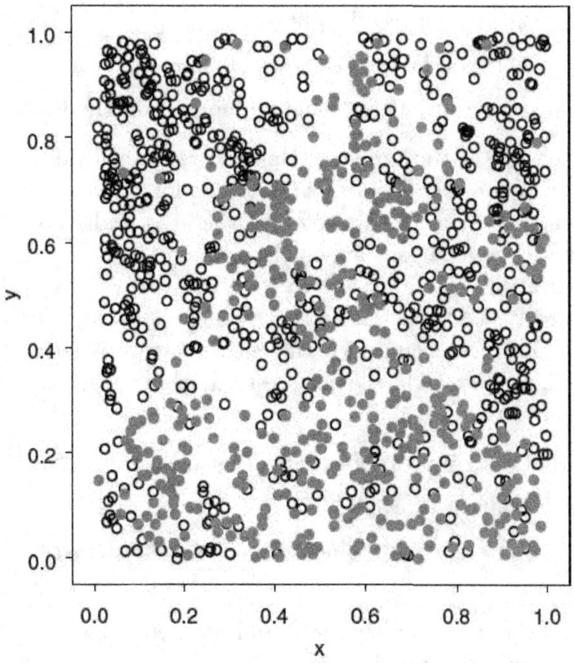


Figure 3.17 *Marked point pattern. Distribution of hickories (open circles) and maples (closed circles) in a forest in Lansing, MI (Gerrard, 1969; Diggle, 1983). The mark variable is discrete with two levels.*

marked point pattern, on the other hand, represents the complete observation of all event locations. There are no other locations at which the attribute $Z$ could have been observed. Consequently, the notion of a continuous random surface expanding over the domain does not arise and predicting the value of $Z$ at an unobserved location appears objectionable. Why would one want to predict the diameter of trees that do not exist? For planning purposes, for example. In order to do so, one views the marked point process conditional and treats the event locations as if they were non-stochastic.

Because there are potentially two sources of randomness in a marked point pattern, the randomness of the mark variable at $\mathbf{s}$ given that an event occurred at that location and the distribution of events according to a stochastic process, one can choose to study either conditional on the other, or to study them jointly. If the mark variable is not stochastic as in §3.1–3.4, one can still ask questions about the distribution of events. When the mark variable is stochastic, we are also interested in studying the distributional properties of $Z$. In the tree example we may inquire about

- the degree of randomness, clustering, or regularity in the distribution of trees;
- the mean and variance of tree heights;
- the correlation between heights of trees at different locations;
- whether the distribution of tree heights depends on the location of trees.

A second extension of the unmarked point processes leads to multivariate point patterns, which are collections of patterns for events of different types. Møller and Waagepetersen (2003) refer to them as multitype patterns. Let $\mathbf{s}_1^m, \cdots, \mathbf{s}_{n_1}^m$ denote the locations at which events of type $m = 1, \cdots, M$ occur and assume that a multivariate process generates the events in $D \in \mathbb{R}^2$. The counting measure $N_m(A)$ represents the number of events of type $m$ in the Borel set $A$ and the event counts for the entire pattern is the $(M \times 1)$ vector $\mathbf{N}(A) = [N_1(A), \cdots, N_M(A)]$. Basic questions that arise with multivariate point patterns concern

- the multinomial distribution of events of each type;
- the spatial distribution of events;
- whether the proportions with which events of different types occur depend on location.

The connection between multivariate and marked patterns is transparent. Rather than counting the number of events in each of the $M$ patterns one could combine the patterns into a single pattern of $\sum_{m=1}^M n_m$ events and associate with each event location a mark variable that indicates the pattern type. A particularly important case is that of a two-level mark variable, the bivariate point process (Figure 3.18).

A unified treatment of marked, unmarked, and multivariate point processes is possible by viewing the process metric as the product space between the space of the mark variable and the spatial domain. For more details along these lines the reader is referred to the texts by Stoyan, Kendall, and Mecke (1995) and by Møller and Waagepetersen (2003).

### 3.6.2 Intensities and Moment Measures for Multivariate Point Patterns

Recall that for a univariate point pattern the first- and second-order intensities are defined as

$$\begin{aligned} \lambda(\mathbf{s}) &= \lim_{|d\mathbf{s}| \to 0} \frac{\mathrm{E}[N(d\mathbf{s})]}{|d\mathbf{s}|} \\ \lambda_2(\mathbf{s}_i, \mathbf{s}_j) &= \lim_{|d\mathbf{s}_i| \to 0, |d\mathbf{s}_j| \to 0} \frac{\mathrm{E}[N(d\mathbf{s}_i)N(d\mathbf{s}_j)]}{|d\mathbf{s}_i||d\mathbf{s}_j|}. \end{aligned}$$

For a multivariate pattern $N_m(A)$ is the count of events of type $m$ in the region (Borel set) $A$ of the $m$th pattern. The intensities for the component

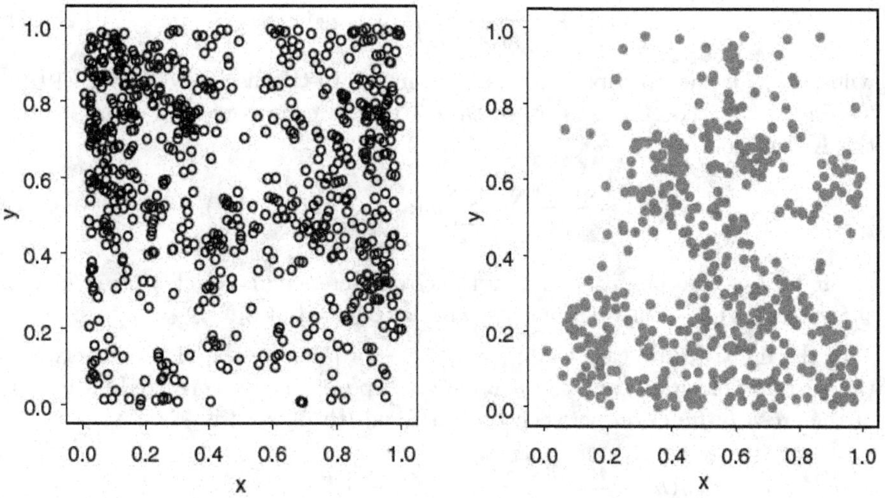


Figure 3.18 *Lansing tree data as a bivariate point pattern. The bivariate pattern is the collection of the two univariate patterns. Their superposition is a marked process (Figure 3.17).*

patterns are similarly defined as

$$\lambda_m(\mathbf{s}) = \lim_{|d\mathbf{s}|\to 0} \frac{\mathrm{E}[N_m(d\mathbf{s})]}{|d\mathbf{s}|}$$

$$\lambda_{m,2}(\mathbf{s}_i,\mathbf{s}_j) = \lim_{|d\mathbf{s}_i|\to 0, |d\mathbf{s}_j|\to 0} \frac{\mathrm{E}[N_m(d\mathbf{s}_i)N_m(d\mathbf{s}_j)]}{|d\mathbf{s}_i||d\mathbf{s}_j|}.$$

Summary statistics and estimates of the first- and second-order properties can be computed as in the univariate case for each of the $k$ pattern types. In addition, one can now draw on measures that relate properties between pattern types. For example, the (second-order) cross-intensities

*Cross-pattern Intensity*

$$\lambda_{ml,2}(\mathbf{s}_i,\mathbf{s}_j) = \lim_{|d\mathbf{s}_i^m|\to 0, |d\mathbf{s}_j^l|\to 0} \frac{\mathrm{E}[N_m(d\mathbf{s}_i{}^m)N_l(d\mathbf{s}_j{}^l)]}{|d\mathbf{s}_i{}^m||d\mathbf{s}_j{}^l|}.$$

is an obvious extension of $\lambda_{m,2}$.

As in the univariate case, the second-order intensities are difficult to interpret and reduced moment measures are used instead. The $K$-functions for the individual component patterns are obtained as in the univariate case under the usual assumptions of stationarity and isotropy. To study the interaction between pattern types, recall that for a univariate pattern $\lambda K(h)$ represents the expected number of additional events within distance $h$ of an arbitrary event. Considering in a multivariate case the expected number of events of type $m$ within distance $h$ of an event of type $l$ under stationarity and isotropy of $\mathbf{N}$ in $\mathbb{R}^2$ leads to

*Cross K-Functions*

$$K_{ml}(h) = \frac{1}{\lambda_m \lambda_l} 2\pi \int_0^h x\lambda_{ml,2}(x)\, dx, \tag{3.15}$$

where $\lambda_{ml,2}$ is the isotropic cross-pattern intensity (Hanisch and Stoyan, 1979; Cressie, 1993). For the univariate case, Ripley's edge corrected estimator of the $K$-function is

$$\widehat{K}(h) = \frac{\nu(A)}{n^2} \sum_{i=1}^{n} \sum_{j \neq i}^{n} w(\mathbf{s}_i, \mathbf{s}_j)^{-1} I(h_{ij} \leq h),$$

where $w(\mathbf{s}_i, \mathbf{s}_j)$ is the proportion of the circumference of a circle that is within the study region, passes through $\mathbf{s}_j$ and is centered at $\mathbf{s}_i$.

In the multivariate case let $h_{ij}^{ml} = ||\mathbf{s}_i{}^m - \mathbf{s}_j{}^l||$ denote the distance between the $i$th point of type $m$ and the $j$th point of type $l$. An edge corrected estimator of the cross $K$-function between the $m$th and $l$th pattern is

$$\widehat{K}_{ml}(h) = \frac{\nu(A)}{n_m n_l} \sum_{i=1}^{n_m} \sum_{j=1}^{n_l} w(\mathbf{s}_i{}^m - \mathbf{s}_j{}^l)^{-1} I(h_{ij}^{ml} \leq h),$$

where $w(\mathbf{s}_i{}^m - \mathbf{s}_j{}^l)$ is the proportion of the circumference of a circle within $A$ that passes through $\mathbf{s}_j{}^l$ and is centered at $\mathbf{s}_i{}^m$ (Hanisch and Stoyan, 1979; Cressie, 1993, p. 698).

## 3.7 Point Process Models

The homogeneous Poisson process provides the natural starting point for a statistical investigation of an observed point pattern. Rejection of the CSR hypothesis does not come as a great surprise in many applications and you are naturally confronted with the question "What kind of pattern is it?" If the CSR test suggests a clustered pattern, one may want to compare, for example, the observed $K$-function to simulated $K$-functions from a cluster process.

We can only skim the surface of point process models in this chapter. A large number of models have been developed and described for clustered and regular alternatives, details can be found in, e.g., Diggle (1983), Cressie (1993), Stoyan, Kendall, and Mecke (1995), and Møller and Waagepetersen (2004). The remainder of this chapter draws on these sources as well as on Appendix A9.9.11 in Schabenberger and Pierce (2002). The models were chosen for their representativeness for a particular data-generating mechanism, and because of their importance in theoretical and applied statistics. When you analyze an observed spatial point pattern, keep in mind that based on a single realization of the process unambiguous identification of the event-generating point process model may not be possible. For example, an inhomogeneous

Poisson process and a Cox Process (see below) lead to clustering of events. The mechanisms are entirely different, however. In case of the IPP, events in non-overlapping regions are independent and clustering arises because the intensity function varies spatially. In the Cox process, clustering occurs because events are dependent, the (average) intensity may be homogeneous. Certain Poisson cluster processes, where one point process generates *parent* events and a second process places *offspring* events around the locations of the parent events, can be made equivalent to a Poisson process with a randomly varying intensity.

Processes that are indistinguishable based on a single realization, can have generating mechanism that suggest very different biological and physical interpretation. It behooves the analyst to consider process models whose genesis are congruent with the subject-matter theory. Understanding the genesis of the process models also holds important clues about how to simulate realizations from the model.

### *3.7.1 Thinning and Clustering*

One method of deriving a point process model is to apply a defined operation to an existing process. Among the basic operations discussed by Stoyan et al. (1995, Ch. 5) are superpositioning, thinning, and clustering. If $Z_1(\mathbf{s})$, $Z_2(\mathbf{s}), \cdots, Z_k(\mathbf{s})$ are point processes, then their superposition

$$Z(\mathbf{s}) = \bigcup_{i=1}^{k} Z_i(\mathbf{s})$$

is also a point process. If the $Z_i(\mathbf{s})$ are mutually independent homogeneous Poisson processes with intensities $\lambda_1, \cdots, \lambda_k$, then $Z(\mathbf{s})$ is a homogeneous Poisson process with intensity $\sum_{i=1}^{k} \lambda_i$.

More important than the combining of processes is the operation by which events in one process are eliminated based on some probability $p$; thinning. Stoyan et al. (1995) distinguish the following types of thinning

- **$p$-thinning.** Each point in the pattern is retained with probability $p$ and eliminated with probability $1-p$. The retention decisions can be represented as $N(A)$ independent Bernoulli trials with common success probability $p$.
- **$p(\mathbf{s})$-thinning.** The retention probabilities are given by the deterministic function $0 \leq p(\mathbf{s}) \leq 1$.
- **$\pi$-thinning.** The thinning function is stochastic, a random field. A thinning is obtained by drawing a realization $p(\mathbf{s})$ of the random function $\pi(\mathbf{s})$ and applying $p(\mathbf{s})$-thinning.

These types of thinning are obvious generalizations, with $p$-thinning being the most special case. They are important operations, because the properties of the resultant process can relate quite easily to the properties of the original process to which thinning is applied. For this reason, one often applies thinning operations to a homogeneous Poisson process, because its properties are well understood.

**Example 3.5** An inhomogeneous Poisson process can be constructed as the $p(\mathbf{s})$-thinning of a homogeneous Poisson process. The basic result is the following. If a Poisson process $Z(\mathbf{s})$ with intensity $\lambda$ is subject to $p(\mathbf{s})$-thinning, then the resulting process $Z^*(\mathbf{s})$ is also a Poisson process with intensity $\lambda p(\mathbf{s})$. To generate an inhomogeneous Poisson process on $A$ with intensity $\alpha(\mathbf{s})$, commence by generating a homogeneous Poisson process on $A$ with intensity $\lambda = \max_A\{\alpha(\mathbf{s})\}$. The thinning rule is to retain points from the original pattern with probability $p(\mathbf{s}) = \alpha(\mathbf{s})/\lambda$. This is the idea behind the Lewis-Shedler algorithm for simulating Poisson processes with heterogeneous intensity (Lewis and Shedler, 1979). □

Further properties of thinned processes are as follows.

1. If $\mu(A) = \int \lambda(\mathbf{s})ds$ is the intensity measure of the original process and $\mu^*(A)$ is the measure of the thinned process, then under $p(\mathbf{s})$-thinning
$$\mu^*(A) = \int p(\mathbf{s})\mu(ds).$$
2. If the original process has intensity $\lambda$, then the thinned process has intensity $\lambda^*(\mathbf{s}) = \lambda p(\mathbf{s})$. For $\pi$-thinning of a process with intensity $\lambda$, the resulting intensity is $\lambda\mathrm{E}[\pi(\mathbf{s})]$.
3. If $Z(\mathbf{s})$ is a Poisson process subject to $p(\mathbf{s})$-thinning, then the thinned process $Z^*$ and the process $Z \setminus Z^*$ of the removed points are independent Poisson processes with intensities $\lambda p(\mathbf{s})$ and $\lambda(1-p(\mathbf{s}))$, respectively (Møller and Waagepetersen, 2003, p. 23).
4. The $p$-thinning of a stationary process yields a stationary process. $p(\mathbf{s})$-thinning does not retain stationarity. The $\pi$-thinning of a stationary process is a stationary process, provided that the random field $\pi(\mathbf{s})$ is stationary.
5. The $K$-function of a point process is not affected by $p$-thinning. The intensity and the expected number of extra events within distance $h$ from an arbitrary event are reduced by the same factor.
6. The $K$-function of a $\pi$-thinned process can be constructed from the $K$-function of the original process and the mean and covariance of the $\pi(\mathbf{s})$ process. If the random field $\pi(\mathbf{s})$ has mean $\xi$ and covariance function $C(||\mathbf{h}||) = \mathrm{E}[\pi(\mathbf{0})\pi(\mathbf{h})] - \xi^2$, then
$$K^*(h) = \int_0^h \frac{1}{\xi^2}(C(u) + \xi^2)dK(u).$$
7. The $\pi$-thinning of a Poisson process yields a Cox process (see below).

In an inhomogeneous Poisson process regions where the intensity is higher receive more events per unit area than regions in which the intensity is low. The result is a clustered appearance of the point pattern. Thinning with a location-dependent probability function, whether it is deterministic or stochastic, thus leads to clustered patterns. Areas with high retention probability have greater density of events. Although you can achieve aggregation of events by thinning, clustering as a point process operation refers to a different technique: the event at $\mathbf{s}_i$ is replaced with the realization of a separate point process that has $n_i$ events. Each realization of the second process is referred to as a cluster. The final process consists of the union (superposition) of the events in the clusters. A convenient framework in which to envision clustering operations is that of "parent" and "offspring" processes. First, a point process generates $k$ events, call these the parent process and parent events. At each parent event, $n_i$ offspring events are generated according to a bivariate distribution function which determines the coordinates of the offspring. A bivariate density with small dispersion groups offsprings close to the parents, forming distinct clusters. The Poisson cluster process (see below) arises when the parent process is a Poisson process.

### 3.7.2 *Clustered Processes*

#### 3.7.2.1 *Cox Process*

An inhomogeneous Poisson process creates aggregated patterns. Regions where $\lambda(\mathbf{s})$ is high receive a greater density of points compared to regions with low intensity. If the intensity function is itself the realization of a stochastic process, the resulting point process model is known as a doubly stochastic process, or Cox process. The random intensity is denoted $\Lambda(\mathbf{s})$ and $\lambda(\mathbf{s})$ is a particular realization. Since conditional on $\Lambda(\mathbf{s}) = \lambda(\mathbf{s})$ we obtain an inhomogeneous Poisson process, the (conditional) realizations of a Cox process are non-stationary, yet the process may still be stationary, since properties of the process are reckoned also with respect to the distribution of the random intensity measure.

Writing $\mu(A) = \int_A \Lambda(\mathbf{s})d\mathbf{s}$, then

$$\Pr(N(A) = n) = \mathrm{E}_\Lambda \left[ \frac{1}{n!} \mu(A)^n \exp\{-\mu(A)\} \right],$$

where $\mu(A)$ is a random variable. On the contrary, in the inhomogeneous Poisson process we have $\Pr(N(A) = n) = \frac{1}{n!}\mu(A)^n \exp\{-\mu(A)\}$ with $\mu(A)$ a constant. Similarly, the first- and second-order intensities of a Cox process are determined as expected values:

$$\begin{aligned} \lambda &= \mathrm{E}[\Lambda(\mathbf{s})] \\ \lambda_2(\mathbf{s}_1, \mathbf{s}_2) &= \mathrm{E}[\Lambda(\mathbf{s}_1)\Lambda(\mathbf{s}_2)]. \end{aligned}$$

If the random intensity measure $\Lambda(\mathbf{s})$ is stationary, so is the Cox process and similarly for isotropy. The intensity function $\Lambda(\mathbf{s}) = W\lambda$, where $W$ is a non-negative random variable, for example, enables one to model clustered data without violating first-order stationarity, whereas the inhomogeneous Poisson process is a non-stationary process. Furthermore, events in disjoint regions in a Cox process are generally not independent.

An interesting case is where $\Lambda(\mathbf{s}) = \Lambda$ is a Gamma$(\alpha, \beta)$ random variable with probability density function, mean, and variance

$$\begin{aligned} f(\Lambda) &= \frac{1}{\beta^\alpha \Gamma(\alpha)} \Lambda^{\alpha-1} \exp\{-\Lambda/\beta\} \qquad \Lambda > 0 \\ \mathrm{E}[\Lambda] &= \alpha\beta \\ \mathrm{Var}[\Lambda] &= \alpha\beta^2. \end{aligned}$$

The number of events in region $A$ is then a negative binomial random variable. Also, since the variance of a negative Binomial random variable exceeds that of a Poisson variable with the same mean, it is seen how the additional variability in the process that stems from the uncertainty in $\Lambda$ increases the dispersion of the data. As Stoyan et al. (1995, p. 156) put it, Cox processes are "super-Poissonian." The variance of the counts exceeds those of a Poisson process with the same intensity, they are overdispersed. Also, since clustered processes exhibit more variation in the location of events than the CSR process, the Cox process is a point process model that implies clustering.

A "brute-force" method to simulation of a Cox process is a two-step procedure that first generates a realization of the intensity process $\Lambda(\mathbf{s})$. In the second step, an inhomogeneous Poisson process is generated based on the realization $\lambda(\mathbf{s})$.

#### *3.7.2.2 Poisson Cluster Process*

The Poisson cluster process derives by applying a clustering operation to the events in a Poisson process. Because of this generating mechanism, it is an appealing point process model for investigations that involve "parents" and "offspring." The original process and the clustering are motivated as follows:

(i) The parent process is the realization of an inhomogeneous Poisson process with intensity $\lambda(\mathbf{s})$.

(ii) Each parent produces a random number $N$ of offspring.

(iii) The positions of the offspring relative to their parents are distributed according to a bi-variate distribution function $f(\mathbf{s})$.

(iv) The final process consists of the locations of the offspring only.

The Neyman-Scott process (Neyman and Scott, 1972) is a special case of the Poisson cluster process. It is characterized by the following simplifications of (ii) and (iii):

(ii*) The number of offspring are realized independently and identically for each parent according to a discrete probability mass function $\Pr(N = k) = p_n$.

(iii*) The positions of the offspring relative to their parents are distributed independently and identically.

We note that some authors consider as Neyman-Scott processes those Poisson cluster processes in which postulate (i) is also modified to allow only stationary parent processes, see, e.g., Stoyan et al. (1995, p. 157) for this view, and Cressie (1993, p. 662) for the more general case. When the parent process is stationary with intensity $\rho$, further simplifications arise. For example, if additionally, $f(\mathbf{s})$ is radially symmetric, the Neyman-Scott process is also stationary. The intensity of the resulting cluster process is $\lambda = \rho\mathrm{E}[N]$.

If the parent process is a homogeneous Poisson process with intensity $\rho$ and $f(\mathbf{s})$ is radially symmetric, it can be shown that the second-order intensity is

$$\lambda_2(h) = \rho^2\mathrm{E}\,[N]^2 + \rho\mathrm{E}\,[N(N-1)]\,f(h).$$

In this expression $h$ denotes the Euclidean distance between two events in the same cluster (offspring from the same parent). Invoking the relationship between the second-order intensity and the $K$-function, the latter can be derived as

$$K(h) = \pi h^2 + \frac{1}{\rho}\frac{\mathrm{E}\,[N(N-1)]}{\mathrm{E}[N]^2}F(h),$$

where $F(h)$ is the cumulative distribution function of $h$.

Suppose, for example, that the parent process is a homogeneous Poisson process with intensity $\rho$, $f(\mathbf{s})$ is a bivariate Gaussian density with mean 0 and variance $\sigma^2\mathbf{I}$, and the number of offspring per parent has a Poisson distribution with mean $\mathrm{E}[N] = \mu$. You can then show (see Chapter problems) that

$$K(h) = \pi h^2 + \frac{1}{\rho}\left(1 - \exp\left\{-\frac{h^2}{4\sigma^2}\right\}\right). \tag{3.16}$$

Notice that the $K$-function does not depend on the mean number of offspring per parent ($\mu$). The explicit formula for the $K$-function of a Neyman-Scott process allows fitting a theoretical $K$-function to an observed $K$-function (Cressie, 1993, p. 666; Diggle, 1983, p. 74).

The mechanisms of aggregating events by way of a random intensity function (Cox process) and by way of a clustering operation (Poisson cluster process) seem very different, yet for a particular realization of a point pattern they may not be distinguishable. In fact, the theoretical results by Bartlett (1964) show the equivalence of Neyman-Scott processes to certain Cox processes (the equivalence of the more general Poisson cluster processes to Cox processes has not been established (Cressie, 1993, p. 664)). Mathematically, the equivalence exists between Cox processes whose random intensity function is of the form

$$\Lambda(\mathbf{s}) \propto \sum_{i=1}^{\infty} k(\mathbf{s} - \mathbf{s}_i), \tag{3.17}$$

and Neyman-Scott processes with inhomogeneous Poisson parent process and $N \sim \text{Poisson}(\mu)$. Think of the intensity function (3.17) as placing bivariate densities at locations $\mathbf{s}_1, \mathbf{s}_2, \cdots$, for example, bivariate densities of independent Gaussian variables. A realization of $\Lambda(\mathbf{s})$ determines where the densities are centered (the parent locations in the Neyman-Scott process). The densities themselves determine the intensity of events near that center (the offspring in the Neyman-Scott process). The realization of this Cox process is no different from the Neyman-Scott process that places the parents at the centers of the densities and generates offspring with the same density about them.

### *3.7.3 Regular Processes*

In order to generate events with greater regularity than the homogeneous Poisson process, we can invoke a very simple requirement: no two events can be closer than some minimum permissible distance $\delta$. For example, start with a homogeneous Poisson process and apply a thinning that retains points with the probability that there are no points within distance $\delta$. Such processes are referred to as hard-core processes. The Matérn models I and II and Diggle's simple sequential inhibition process are variations of this theme.

Matérn (1960) constructed regular point processes of two types which are termed the Matérn models I and II, respectively (see also Matérn 1986). Model I starts with a homogeneous Poisson process $Z_0$ with intensity $\rho$. Then all pairs of events that are separated by a distance of less than $\delta$ are thinned. The remaining events form the more regular spatial point process $Z_1$. If $\mathbf{s}_i$ is an event of $Z_0$, the probability of its retention in $Z_1$ is $\Pr(\text{no other point within distance } \delta \text{ of } \mathbf{s}_i)$. Since the process is CSR, this leads to

$$\Pr(\mathbf{s}_i \text{ is retained}) = \Pr(N(\pi\delta^2) = 0) = \exp\left\{-\rho\pi\delta^2\right\}.$$

and the intensity of the resulting process is

$$\lambda = \rho \exp\left\{-\rho\pi\delta^2\right\} < \rho.$$

The second-order intensity is given by $\lambda_2(h) = \rho^2 p(h)$, where $p(h)$ is the probability that two events distance $h$ apart are retained. The function $p(h)$ is given by (Diggle, 1983, p. 61; Cressie, 1993, p. 670)

$$p(h) = \begin{cases} 0 & h < \delta \\ \exp\{-\rho U_\delta(h)\} & h \geq \delta, \end{cases}$$

where $U_\delta(h)$ is the area of the union of two circles with radius $\delta$ and distance $h$ apart. The $K$-function can be obtained by integration,

$$K(h) = \frac{2\pi}{\lambda^2} \int_0^h x\lambda_2(x)dx.$$

Matérn's second model also commences with a homogeneous Poisson process $Z_0$ with intensity $\rho$ and marks each event $\mathbf{s}$ independently with a random variable $M(\mathbf{s})$ from a continuous distribution function. Often $M(\mathbf{s})$ is taken to be uniform on $(0,1)$. The event $\mathbf{s}$ is deleted, if another event $\mathbf{u}$ is closer than the minimum permissible distance $\delta$ *and* if its mark is less than the mark at $\mathbf{u}$. Put differently, the event $\mathbf{s}$ is retained, if there is no other point within distance $\delta$ with a mark less than $M(\mathbf{s})$. Diggle (1983, p. 61) refers to the mark variable $M(\mathbf{s})$ as the "time of birth" of the event $\mathbf{s}$. An event of $Z_0$ is then removed, if it lies within a distance $\delta$ of an older event. You keep the "oldest" events.

The intensity of the resulting process, which comprises all points not thinned, is

$$\lambda = \frac{1}{\pi\delta^2}\left(1 - \exp\left\{-\rho\pi\delta^2\right\}\right).$$

Cressie (1993, p. 670) gives an expression for the second-order intensity.

Diggle et al. (1976) consider the following procedure that leads to a regular process termed simple sequential inhibition. Place a disk of radius $\delta$ at random in the region $A$. Determine the remaining points in $A$ for which you could place a disk of radius $\delta$ that would not overlap with the first disk. Select the center point of the next disk at random from a uniform distribution of these points. Continue in this fashion, choosing at each stage the disk center at random from the points at which the next disk would not overlap with any of the previous disks. The process stops when a pre-specified number of disks has been placed or no additional disk can be placed without overlapping previously placed disks. This model is appealing for regular patterns where events have an inhibition distance such as cell nuclei that are surrounded by cell mass. The simple sequential inhibition process is a Matérn model II conditioned on the total number of points (Ripley, 1977).

## 3.8 Chapter Problems

**Problem 3.1** Let $N(A)$ denote the counting measure in region $A \subset D$ of a homogeneous Poisson process. The finite-dimensional distribution of the process for non-overlapping intervals $A_1, \cdots, A_k$ is given by

$$\begin{aligned}\Pr(N(A_1) = n_1, \cdots, N(A_k) = n_k) &= \frac{\lambda^n \nu(A_1)^{n_1} \cdot \ldots \cdot \nu(A_k)^{n_k}}{n_1! \cdot \ldots \cdot n_k!} \\ &\times \exp\left\{-\sum_{i=1}^{k} \lambda\nu(A_i)\right\}.\end{aligned}$$

Show that by conditioning on $N(D) = n_1 + \cdots + n_k \equiv n$, the finite-dimensional distribution equals that of a Binomial process (i.e., (3.2)).

**Problem 3.2** For the homogeneous Poisson process, find unbiased estimators of $\lambda$ and $\lambda^2$.

**Problem 3.3** Data are sampled independently from two respective Gaussian populations with variance $\sigma^2 = 1.5^2$. Five observations are drawn from each population. The realized values are

| | Sampled Values | | | | |
|---|---|---|---|---|---|
| Population | $y_{i1}$ | $y_{i2}$ | $y_{i3}$ | $y_{i4}$ | $y_{i5}$ |
| $i = 1$ | 9.7 | 10.2 | 10.9 | 8.6 | 10.3 |
| $i = 2$ | 8.2 | 6.1 | 9.6 | 8.2 | 8.9 |

Calculate the $p$-value for $H_0{:}\,\mu_1 = \mu_2$ against the two-sided alternative by using the appropriate standard test. Then, perform Monte Carlo tests with $s = 19, 49, 99$, and $s = 999$ repetitions. Compare the simulated $p$-values against the $p$-value from the standard test.

**Problem 3.4** The random intensity function of the Cox process induces spatial dependency between the events. You can think of this mechanism as creating stochastic dependency through shared random effects. To better understand and appreciate this important mechanism consider the following two statistical scenarios

1. A one-way random effects model can be written as $Y_{ij} = \mu + \alpha_i + \epsilon_{ij}$ where the $\epsilon_{ij}$ are *iid* random variables with zero mean and variance $\sigma^2_\epsilon$ and the $\alpha_i$ are independent random variables with mean 0 and variance $\sigma^2_\alpha$. Also, $\text{Cov}[\alpha_i, \epsilon_{ij}] = 0$. Determine $\text{Cov}[Y_{ij}, Y_{kl}]$. How does "sharing" of random effects relate to the covariance of observations for $i = k$ and $i \neq k$?
2. Let $Y_1|\lambda, \cdots, Y_n|\lambda$ be a random sample from a Poisson distribution with mean $\lambda$. If $\lambda \sim \text{Gamma}(\alpha, \beta)$, find the mean and variance of $Y_i$ as well as $\text{Cov}[Y_i, Y_j]$.

**Problem 3.5** (Stein, 1999, Ch. 1.4) Consider a Poisson process on the line with intensity $\lambda$. Let $A$ denote an interval on the line with length $|A|$, and $N(A)$ the number of events in $A$. Define $Z(t) = N((t-1, t+1])$. Find the mean and covariance function of $Z(t)$.

**Problem 3.6** Let $\{Z(\mathbf{s}) : \mathbf{s} \in D \subset \mathbb{R}^2\}$ be a Cox process with random intensity function $\Lambda(\mathbf{s}) = \Lambda$. The intensity is distributed as a $\text{Gamma}(\alpha, \beta)$ random variable,

$$f(\lambda) = \frac{1}{\beta^\alpha \Gamma(\alpha)} \lambda^{\alpha - 1} \exp\{-\lambda/\beta\}.$$

Find the distribution of $N(A)$, the number of events in region $A$. How does $\text{Var}[N(A)]$ relate to the variance of $N(A)$ in a homogeneous Poisson process with intensity $\lambda = \alpha\beta$?

**Problem 3.7** Generate an inhomogeneous Poisson process on the unit square with intensity function $\lambda(\mathbf{s}) = \lambda(x, y) = \exp\{5x + 2y\}$. Divide the unit square into a $4 \times 4$ grid and perform a Chi-square goodness-of-fit test for CSR based on quadrat counts. If you want to generate this process conditional on having exactly $n = 100$ events, how must the generating algorithm be adjusted?

**Problem 3.8** For an inhomogeneous Poisson process with intensity function $\lambda(\mathbf{s})$, show that the density function of $\{\mathbf{s}_1, \cdots, \mathbf{s}_n\}$ is given by

$$f(\{\mathbf{s}_1, \cdots, \mathbf{s}_n\}) = \begin{cases} \exp\{-\mu(A)\} & n = 0 \\ \exp\{-\mu(A)\} \prod_{i=1}^{n} \lambda(\mathbf{s}_i)/n! & n \geq 1. \end{cases}$$

**Problem 3.9** A special case of the Neyman-Scott process is as follows. A homogeneous Poisson process with intensity $\lambda$ generates parent events. Each parent produces a random number $N$ of offspring, $N \sim \text{Poisson}(\mu)$, and the positions of the offspring relative to their parents are *iid* with a radial symmetric, standard bivariate Gaussian density. That is, the offspring event $\mathbf{s} = [x, y]'$ has distribution $G_2(\mathbf{0}, \sigma^2\mathbf{I})$. The $K$-function of this process in $\mathbb{R}^2$ is given by

$$K(h) = \pi h^2 + \frac{\mathrm{E}[N(N-1)]}{\lambda m^2} F(h),$$

where $F(h)$ is the cumulative distribution function of the distance

$$\left\{(x_1 - x_2)^2 + (y_1 - y_2)^2\right\}^{1/2}$$

between two events in the same cluster (Cressie, 1993, p. 665).

(i) Show that the mean of the squared distance to an offspring from its parent is $2\sigma^2$.

(ii) Show that the density of $H$, the distance between two events in the same cluster is $f(h) = h/(2\sigma^2) \exp\left\{-h^2/(4\sigma^2)\right\}$.

(iii) Show that the $K$ function is $K(h) = \pi h^2 + \lambda^{-1}(1 - \exp\{-h^2/(4\sigma^2)\})$.

(iv) Explain why the $K$-function does not depend on the mean number of offspring per parent ($\mu$).

(v) Under what condition do this process and the homogeneous Poisson process have the same second-order properties?

CHAPTER 4

# Semivariogram and Covariance Function Analysis and Estimation

## 4.1 Introduction

Two important features of a random field are its mean and covariance structure. The former represents the large-scale changes of $Z(\mathbf{s})$, the latter the variability due to small- and micro-scale stochastic sources. In Chapter 2, we gave several different representations of the stochastic dependence (second-order structure) between spatial observations. Direct and indirect specifications based on model representations (§2.4.1), representations based on convolutions (§2.4.2) and spectral decompositions (§2.5). In the case of a spatial point process, the second-order structure is represented by the second-order intensity, and by the $K$-function in the isotropic case (Chapter 3). If a spatial random field has model representation $\mathbf{Z}(\mathbf{s}) = \boldsymbol{\mu}(\mathbf{s}) + \mathbf{e}(\mathbf{s})$, where $\mathbf{e}(\mathbf{s}) \sim (\mathbf{0}, \boldsymbol{\Sigma})$, the spatial dependence structure is expressed through the variance-covariance matrix $\boldsymbol{\Sigma}$. The semivariogram and covariance function of a spatial process with fixed, continuous domain were introduced in §2.2, since these parameters require that certain stationarity conditions be met. The variance-covariance matrix of $\mathbf{e}(\mathbf{s})$ is not bound by any stationarity requirements, it simply captures the variances and covariances of the process. In addition, the model representation does not confine $\mathbf{e}(\mathbf{s})$ to geostatistical applications, the domain may be a lattice, for example. In practical applications, $\boldsymbol{\Sigma}$ is unknown and must be estimated from the data. Unstructured variance-covariance matrices that are common in multivariate statistical methods are uncommon in spatial statistics. There is typically structure to the spatial covariances, for example, they may decrease with increasing lag. And without true replications, there is no hope to estimate the entries in an unspecified variance-covariance matrix. Parametric forms are thus assumed so that $\boldsymbol{\Sigma} \equiv \boldsymbol{\Sigma}(\boldsymbol{\theta})$ and $\boldsymbol{\theta}$ is estimated from the data. The techniques employed to parameterize $\boldsymbol{\Sigma}$ vary with circumstances. In a lattice model $\boldsymbol{\Sigma}$ is defined indirectly by the choice of a neighborhood matrix and an autoregressive structure. For geostatistical data, $\boldsymbol{\Sigma}$ is constructed directly from a model for the continuous spatial autocorrelation among observations. The importance of choosing the correct model for $\boldsymbol{\Sigma}(\boldsymbol{\theta})$ also depends on the application. Consider a spatial model

$$\mathbf{Z}(\mathbf{s}) = \mathbf{X}\boldsymbol{\beta} + \mathbf{e}(\mathbf{s}), \qquad \mathbf{e}(\mathbf{s}) \sim (\mathbf{0}, \boldsymbol{\Sigma}(\boldsymbol{\theta})),$$

where the primary interest is inference about $\boldsymbol{\beta}$, for example, confidence intervals and hypothesis tests about the mean. When $\boldsymbol{\theta}$ is estimated from data,

*Estimated GLS Estimator*

the estimated generalized least squares estimator is given by

$$\widehat{\beta}_{egls} = (\mathbf{X}'\mathbf{\Sigma}(\widehat{\boldsymbol{\theta}})^{-1}\mathbf{X})^{-1}\mathbf{X}'\mathbf{\Sigma}(\widehat{\boldsymbol{\theta}})^{-1}\mathbf{Z}(\mathbf{s}).$$

Provided that $\widehat{\boldsymbol{\theta}}$ is a consistent estimator and that $\mathbf{\Sigma}(\widehat{\boldsymbol{\theta}})$ satisfies some regularity conditions as $n \to \infty$, the EGLS estimator of $\widehat{\beta}$ is consistent (see §6.2). A judicious choice of the covariance model for $\mathbf{\Sigma}(\boldsymbol{\theta})$ will suffice for this analysis. Contrariwise, assume that our interest is in predicting the value of a new observation $Z(\mathbf{s}_0)$ at the unobserved location $\mathbf{s}_0$, and that the underlying spatial model is

$$\mathbf{Z}(\mathbf{s}) = \mu\mathbf{1} + \mathbf{e}(\mathbf{s}), \qquad \mathbf{e}(\mathbf{s}) \sim (\mathbf{0}, \mathbf{\Sigma}(\boldsymbol{\theta})).$$

The best linear unbiased predictor (BLUP) under squared-error loss for this problem is derived in §5.2 as

$$p_{ok}(\mathbf{Z}; \mathbf{s}_0) = \widehat{\mu} + \mathbf{c}(\boldsymbol{\theta})'\mathbf{\Sigma}(\boldsymbol{\theta})^{-1}(\mathbf{Z}(\mathbf{s}) - \mathbf{1}\widehat{\mu}).$$

Here, $\widehat{\mu}$ is the generalized least squares estimator of the (spatially constant) mean $\mu$ and $\mathbf{c}(\boldsymbol{\theta}) = \mathrm{Cov}[Z(\mathbf{s}_0), \mathbf{Z}(\mathbf{s})]'$. The correctness of the covariance model is important in this situation, since it "drives" the spatial predictor through $\mathbf{c}(\boldsymbol{\theta})$ and $\mathbf{\Sigma}(\boldsymbol{\theta})$. Much more attention needs to be paid to the selection of the covariance model compared to situations where the parameters of the mean function are of primary interest.

The predictor $p_{ok}(\mathbf{Z}; \mathbf{s}_0)$ cannot be computed unless $\boldsymbol{\theta}$ is known, which typically it is not. The optimal predictor $p_{ok}(\mathbf{Z}; \mathbf{s}_0)$ is thus inaccessible just as the generalized least squares estimator

$$\widehat{\beta}_{gls} = (\mathbf{X}'\mathbf{\Sigma}(\boldsymbol{\theta})^{-1}\mathbf{X})^{-1}\mathbf{X}'\mathbf{\Sigma}(\boldsymbol{\theta})^{-1}\mathbf{Z}(\mathbf{s})$$

is inaccessible. Besides issues concerning the selection of appropriate models for $\mathbf{\Sigma}$, the analyst is concerned with finding a good estimator of $\boldsymbol{\theta}$. In this chapter we discuss various estimation strategies for $\boldsymbol{\theta}$ under the assumption of stationarity of the $Z(\mathbf{s})$ process. In Chapter 5 we extend the problem of estimating $\boldsymbol{\theta}$ to the situation where $\mathrm{E}[Z(\mathbf{s})]$ is not constant, that is, the random field exhibits a large-scale trend. Furthermore, we assess there the ramifications of working with estimated, rather than known, quantities (using $\widehat{\boldsymbol{\theta}}$ instead of $\boldsymbol{\theta}$).

Predicting the random field at unobserved locations is of particular importance for geostatistical data. By definition, a complete sampling of the random field surface is impossible and the user may be interested in predicting the amount $Z$ at an arbitrary location $\mathbf{s}_0$. For lattice data, complete observations of the domain is common, since sites are associated with discrete spatial units. In this chapter we discuss methods for representing spatial dependence in geostatistical data (§4.2), common models for parameterizing this dependence (§4.3), and statistical approaches to estimation (§4.4 and §4.5). Because modeling the spatial dependence is so important for spatial prediction, these issues have received much attention and have given rise to a considerable array of models and estimation methods. The primary tool that we investigate for estimating the spatial dependence is the semivariogram. In what follows, the reader is reminded that stationarity assumptions are implicit and much damage can be done by applying semivariogram estimators and semivariogram models to data from non-stationary spatial processes.

## 4.2 Semivariogram and Covariogram

### *4.2.1 Definition and Empirical Counterparts*

Let $\{Z(\mathbf{s}) : \mathbf{s} \in D \subset \mathbb{R}^d\}$ be a spatial process and define

$$\begin{aligned}\gamma^*(\mathbf{s}_i, \mathbf{s}_j) &= \frac{1}{2}\mathrm{Var}[Z(\mathbf{s}_i) - Z(\mathbf{s}_j)] \\ &= \frac{1}{2}\{\mathrm{Var}[Z(\mathbf{s}_i)] + \mathrm{Var}[Z(\mathbf{s}_j)] - 2\mathrm{Cov}[Z(\mathbf{s}_i), Z(\mathbf{s}_j)]\}. \quad (4.1)\end{aligned}$$

If $\gamma^*(\mathbf{s}_i, \mathbf{s}_j) \equiv \gamma(\mathbf{s}_i - \mathbf{s}_j)$, a function of the coordinate difference $\mathbf{s}_i - \mathbf{s}_j$ only, then we call $\gamma(\mathbf{s}_i - \mathbf{s}_j)$ the semivariogram of the spatial process. If $Z(\mathbf{s})$ is intrinsically stationary (§2.2), then $\gamma(\mathbf{s}_i - \mathbf{s}_j)$ is a parameter of the stochastic process. In the absence of stationarity, $\gamma^*$ remains a valid function from which the variance-covariance matrix $\mathrm{Var}[\mathbf{Z}(\mathbf{s})] = \mathbf{\Sigma}$ can be constructed, but it should not be referred to as the semivariogram. The function $2\gamma(\mathbf{s}_i - \mathbf{s}_j)$ is referred to as the variogram although the literature is not consistent in this regard. Some authors define $\gamma(\mathbf{s}_i - \mathbf{s}_j)$ through (4.1) and refer to it as the variogram. Chilès and Delfiner (1999, p. 31), for example, define $\gamma(\mathbf{h}) = \frac{1}{2}\mathrm{Var}[Z(\mathbf{s}) - Z(\mathbf{s}+\mathbf{h})]$, term it the variogram because it "tends to become established for its simplicity" and acknowledge that $\gamma(\mathbf{h})$ is *also* called the semivariogram. There is nothing "established" about being off the mark by factor 2. For clarity, we refer throughout to $\gamma$ as the semivariogram and to $2\gamma$ as the variogram. The savings in ink are disproportionate to the confusion created when "semi" is dropped.

The name variogram is most often associated with the work by Matheron (1962, 1963). Jowett (1955a,b) used the term variogram sparingly and called the equivalent of $\gamma(\mathbf{h})$ in the time series context the *serial variation function.* (We wish he would have used the term semivariogram sparingly.) Jowett (1955c) termed what is now known as the empirical semivariogram (§4.4) the *serial variation curve.* Statistical computing packages are also notorious for calculating a semivariogram but labeling it the variogram. In the S+SpatialStats® manual, for example, Kaluzny et al. (1998, p. 68) define the semivariogram as in (4.1), but refer to it as the variogram for "conciseness." The VARIOGRAM procedure in SAS/STAT® computes the semivariogram and uses the variogram label in the output data set.

If the spatial process is not only intrinsic, but second-order stationary, the semivariogram can be expressed in terms of the covariance function $C(\mathbf{s}_i - \mathbf{s}_j) = \mathrm{Cov}[Z(\mathbf{s}_i), Z(\mathbf{s}_j)]$ as

$$\gamma(\mathbf{s}_i - \mathbf{s}_j) = C(\mathbf{0}) - C(\mathbf{s}_i - \mathbf{s}_j), \quad (4.2)$$

making use of the fact that $\text{Var}[Z(\mathbf{s})] = C(\mathbf{0})$ under second-order stationarity. Note that the name semivariogram is used both for the function $\gamma(\mathbf{s}_i - \mathbf{s}_j)$ as well as the graph of $\gamma(\mathbf{h})$ against $\mathbf{h}$. When working with covariances, $C(\mathbf{s}_i - \mathbf{s}_j)$ is the covariance function, the graph of $C(\mathbf{h})$ against $\mathbf{h}$ is referred to as the **covariogram**. Similarly, a graph of the correlation function $R(\mathbf{h})$ against $\mathbf{h}$ is termed the **correlogram**. In the spirit of parallel language, we sometimes will use the term *covariogram* even if technically the term *covariance function* may be more appropriate.

Because of the simple relationship between semivariogram and covariance function, it seems immaterial which function is used to study the spatial dependence of a process. Since the class of intrinsic stationary processes contains the class of second-order stationary processes, the semivariogram of a second-order stationary process can be constructed from the covariance function by (4.2). If the process is intrinsic but not second-order stationary, the covariance function is not a parameter of the process. Our preference to work with the semivariogram $\gamma$ and not the variogram $2\gamma$ is partly due to the fact that $\gamma(\mathbf{s}_i - \mathbf{s}_j) \to C(\mathbf{0})$ provided $C(\mathbf{s}_i - \mathbf{s}_j) \to 0$. An unbiased estimate of the semivariogram for lag distances at which data are (practically) uncorrelated, is an unbiased estimate of the variance of the process.

In geostatistical applications, it is common to work with the semivariogram, rather than the covariance function. Statisticians, on the other hand, are trained in expressing dependency between random variables in terms of covariances. The reasons are not just convenience and training, and a nice interpretation of the semivariogram sill in a second-order stationary process. When you are estimating the spatial dependence from data, the ambivalence between covariance function and semivariogram gives way to differences in statistical properties of the *empirical* estimators. Details on empirical semivariogram estimators are given in §4.4. Here we address briefly the issue of bias when working with (semi-)variograms and with covariances. Let $Z(\mathbf{s}_1), \cdots, Z(\mathbf{s}_n)$ denote the observations from a spatial process with constant but unknown mean. Since then

$$\gamma(\mathbf{s}_i - \mathbf{s}_j) = \frac{1}{2}\,\mathrm{E}[(Z(\mathbf{s}_i) - Z(\mathbf{s}_j))^2],$$

*Matheron Estimator*

a simple, moment-based estimator due to Matheron (1962, 1963) is

$$\widehat{\gamma}(\mathbf{s}_i - \mathbf{s}_j) = \frac{1}{2|N(\mathbf{s}_i - \mathbf{s}_j)|} \sum_{N(\mathbf{s}_i - \mathbf{s}_j)} \{Z(\mathbf{s}_i) - Z(\mathbf{s}_j)\}^2,$$

where $N(\mathbf{s}_i - \mathbf{s}_j)$ is the set of location pairs with coordinate difference $\mathbf{s}_i - \mathbf{s}_j$ and $|N(\mathbf{s}_i - \mathbf{s}_j)|$ is the number of distinct pairs in this set. The corresponding *natural* estimator of the covariance function $C(\mathbf{h})$ is

*Empirical Covariogram*

$$\widehat{C}(\mathbf{s}_i - \mathbf{s}_j) = \frac{1}{|N(\mathbf{s}_i - \mathbf{s}_j)|} \sum_{N(\mathbf{s}_i - \mathbf{s}_j)} (Z(\mathbf{s}_i) - \overline{Z})(Z(\mathbf{s}_j) - \overline{Z}), \qquad (4.3)$$

where $\overline{Z} = n^{-1}\sum_{i=1}^{n} Z(\mathbf{s}_i)$. Let $\mathbf{s}_i - \mathbf{s}_j = \mathbf{h}$. The estimator $\widehat{\gamma}(\mathbf{h})$ is unbiased for $\gamma(\mathbf{h})$ if $Z(\mathbf{s})$ is intrinsically stationary. If the mean is estimated from the data, then $\widehat{C}(\mathbf{h})$ is a biased estimator of the covariance function at lag $\mathbf{h}$. Furthermore,

$$\begin{aligned}\widehat{\gamma}(\mathbf{h}) &= \frac{1}{2|N(\mathbf{h})|}\sum_{N(\mathbf{h})}\left\{Z(\mathbf{s}_i)-\overline{Z}-Z(\mathbf{s}_j)+\overline{Z}\right\}^2\\ &= \frac{1}{|N(\mathbf{h})|}\sum_{N(\mathbf{h})}\left\{Z(\mathbf{s}_i)-\overline{Z}\right\}^2-\widehat{C}(\mathbf{h}),\end{aligned}$$

but $\widehat{C}(\mathbf{0}) = n^{-1}\sum_{i=1}^n \left\{Z(\mathbf{s}_i)-\overline{Z}\right\}^2$. As a consequence, $\widehat{C}(\mathbf{0})-\widehat{C}(\mathbf{h}) \neq \widehat{\gamma}(\mathbf{h})$ and a semivariogram estimate constructed from (4.3) will also be biased. As $|N(\mathbf{h})|/n \to 1$, the bias disappears.

The semivariogram estimator $\widehat{\gamma}(\mathbf{h})$ has other appealing properties. For example, $\widehat{\gamma}(\mathbf{0}) = 0 = \gamma(\mathbf{0})$ and $\widehat{\gamma}(\mathbf{h}) = \widehat{\gamma}(-\mathbf{h})$, sharing properties of the semivariogram $\gamma(\mathbf{h})$. If the data contain a linear large-scale trend,

$$\mathbf{Z}(\mathbf{s}) = \mathbf{X}(\mathbf{s})\boldsymbol{\beta} + \mathbf{e}(\mathbf{s}),$$

then the spatial dependency in the model errors $\mathbf{e}(\mathbf{s})$ is often estimated from the least squares residuals. Since $\text{Var}[\mathbf{e}(\mathbf{s})] = \boldsymbol{\Sigma}$ is unknown—otherwise there is no need for semivariogram or covariance function estimation—the ordinary least squares residuals

$$\widehat{\mathbf{e}}(\mathbf{s}) = \left(\mathbf{I}-\mathbf{X}(\mathbf{s})(\mathbf{X}(\mathbf{s})'\mathbf{X}(\mathbf{s}))^{-1}\mathbf{X}(\mathbf{s})'\right)\mathbf{Z}(\mathbf{s})$$

are often used. Although the semivariogram estimated from $\widehat{\mathbf{e}}(\mathbf{s})$ is a biased estimator for the semivariogram of $\mathbf{e}(\mathbf{s})$, this bias is less than the bias of the covariance function estimator based on $\widehat{\mathbf{e}}(\mathbf{s})$ for $C(\mathbf{h})$ (Cressie and Grondona, 1992; Cressie, 1993, p. 71 and §3.4.3).

The advantages of the classical semivariogram estimator over the covariance function estimator (4.3) stem from the fact that the unknown—but constant—mean is not important for estimation of $\gamma(\mathbf{h})$. The semivariogram filters the mean. This must not be interpreted as robustness of variography to arbitrary mean. First, the semivariogram is a parameter of a spatial process only under intrinsic or second-order stationary, both of which require $\text{E}[Z(\mathbf{s})] = \mu$. Second, the semivariogram reacts rather poorly to changes in the mean with spatial locations. Let $Z(\mathbf{s}) = \mu(\mathbf{s}) + e(\mathbf{s})$, where $\text{E}[e(\mathbf{s})] = \mathbf{0}$, $\gamma_e(\mathbf{h}) = \gamma_z(\mathbf{h})$. It is easy to show (Chapter problem 4.1) that

$$\text{E}[\widehat{\gamma}_z(\mathbf{h})] = \gamma_e(\mathbf{h}) + \frac{1}{2|N(\mathbf{h})|}\sum_{N(\mathbf{h})}\left\{\mu(\mathbf{s}_i)-\mu(\mathbf{s}_j)\right\}^2, \tag{4.4}$$

the empirical semivariogram is positively biased.

It is important to keep separate the question of whether a statistical method is expressed in terms of the semivariogram or the covariance function and the question of whether one should prefer the empirical semivariogram or the empirical covariogram. Because the empirical covariogram is a biased estimator and the empirical semivariogram is unbiased, is not justification to favor statistical techniques that express dependence in terms of the semivariogram. The bias would only be of concern, if the method of estimating the parameters in $\boldsymbol{\theta}$ actually drew on the empirical covariogram. For example, (restricted) maximum likelihood techniques (see §4.5.2) express the likelihood in terms of covariances, but the expression (4.3) is never formed. Finally, the empirical semivariogram or covariogram is hardly ever the end result of a statistical analysis. In a confirmatory analysis you need to estimate the parameter vector $\boldsymbol{\theta}$ so that large-scale trends can be estimated efficiently, and for spatial predictions. If you fit a theoretical semivariogram model $\gamma(\boldsymbol{\theta})$ to the empirical semivariogram by nonlinear least squares, for example, the least squares estimates of $\boldsymbol{\theta}$ will be biased; most nonlinear least squares estimates are.

### *4.2.2 Interpretation as Structural Tools*

The behavior of the covariance function near the origin and its differentiability were studied in §2.3 to learn about the continuity and smoothness of a second-order stationary random field. Recall that a mean square continuous random field must be continuous everywhere, and that a random field cannot be mean square continuous unless it is continuous at the origin. Hence, $C(\mathbf{h}) \to C(\mathbf{0})$ as $\mathbf{h} \to \mathbf{0}$ which implies that $\gamma(\mathbf{h}) \to 0$ as $\mathbf{h} \to \mathbf{0}$. Furthermore, we must have $\gamma(\mathbf{0}) = 0$, of course. Mean square continuity of a random field implies that the semivariogram is continuous at the origin. The notion of smoothness of a random field was then brought into focus in §2.3 by studying the partial derivatives of the process. The more often a random field is mean square differentiable, the higher its degree of smoothness.

The semivariogram is not only a device to derive the spatial dependency structure in a random field and to build the variance-covariance matrix of $\mathbf{Z}(\mathbf{s})$, which is needed for model-based statistical inferences. It is a structural tool which in itself conveys much information about the behavior of a random field. For example, semivariograms that increase slowly from the origin and/or exhibit quadratic behavior near the origin, imply processes more smooth than those whose semivariogram behaves linear near the origin.

For a second-order stationary random field, the (isotropic) semivariogram $\gamma(||\mathbf{h}||) \equiv \gamma(h)$ has a very typical form (Figure 1.12, page 29). It rises from the origin and if $C(h)$ decreases monotonically with increasing $h$, then $\gamma(h)$ will approach $\text{Var}[Z(\mathbf{s})] = \sigma^2$ either asymptotically or exactly at a particular lag $h^*$. The asymptote itself is termed the **sill** of the semivariogram and the lag $h^*$ at which the sill is reached is called its **range**. Observations $Z(\mathbf{s}_i)$ and $Z(\mathbf{s}_j)$ for which $||\mathbf{s}_i - \mathbf{s}_j|| \geq h^*$ are uncorrelated. If the semivariogram reaches the sill asymptotically, the **practical range** is defined as the lag $h^*$ at which $\gamma(h) = 0.95 \times \sigma^2$. Semivariograms that do not reach a sill occur frequently. This could be due to

*Sill and Range*

- Non-stationarity of the process, e.g., the mean of $Z(\mathbf{s})$ is not constant across the domain;
- An intrinsically stationary process. The intrinsic hypothesis states that a variogram must satisfy

$$2\frac{\gamma(\mathbf{h})}{||\mathbf{h}||^2} \to 0 \qquad \text{as } ||\mathbf{h}|| \to 0.$$

- The process is second-order stationary, but the largest lag for which the semivariogram can be estimated is shorter than the range of the process. The lag distance at which the semivariogram would flatten has not been observed.

In practice, empirical semivariograms $\widehat{\gamma}(\mathbf{h})$ calculated from a set of data often suggest that the semivariogram does not pass through the origin. This intercept of the semivariogram has been termed the **nugget** effect $c_0$, $c_0 = \lim_{h\to 0}\gamma(h) \neq 0$. If the random field under study is mean square continuous, such a discontinuity at the origin must not exist. Following Matérn (1986, Ch. 2.2), define $Q_d$ as the class of all functions that are valid covariance functions in $\mathbb{R}^d$, $Q'_d$ as the subclass of functions which are continuous everywhere except possibly at the origin, and $Q''_d$ as the subclass of covariance functions continuous everywhere. Matérn shows that if $C(\mathbf{h}) \in Q'_d$ it can be written as $C(h) = aC_0(\mathbf{h}) + bC_1(\mathbf{h})$, where $a, b \geq 0$, $C_1(\mathbf{h}) \in Q''_d$, and

*Nugget Effect*

$$C_0(\mathbf{h}) = \begin{cases} 1 & \text{if } \mathbf{h} = \mathbf{0} \\ 0 & \text{otherwise.} \end{cases} \tag{4.5}$$

It follows that if $Z(\mathbf{s})$ has a covariance function in $Q'_d$, it can be decomposed as $Z(\mathbf{s}) = U(\mathbf{s}) + \nu(\mathbf{s})$, where $U(\mathbf{s})$ is a process with covariance function in $Q''_d$ and $\nu(\mathbf{s})$ has covariance function (4.5). Matérn (1986, p. 12) calls $U(\mathbf{s})$ the **continuous component** and $\nu(\mathbf{s})$ the **chaotic component** in the decomposition. The variance of the latter component is the nugget effect of the semivariogram. The chaotic component is not necessarily completely spatially unstructured; it can be further decomposed. Recall the decomposition $Z(\mathbf{s}) = \mu(\mathbf{s}) + W(\mathbf{s}) + \eta(\mathbf{s}) + \epsilon(\mathbf{s})$ from §2.4, where $W(\mathbf{s})$ depicts smooth-scale spatial variation, $\eta(\mathbf{s})$ micro-scale variation, and $\epsilon(\mathbf{s})$ is pure measurement error. The micro-scale process $\eta(\mathbf{s})$ is a stationary spatial process whose semivariogram has sill $\text{Var}[\eta(\mathbf{s})] = \sigma^2_\eta$. It represents spatial structure but cannot be observed unless data points are collected at lag distances smaller than the range of the $\eta(\mathbf{s})$ process. The measurement error component has variance $\text{Var}[\epsilon(\mathbf{s})] = \sigma^2_\epsilon$ and the nugget effect of a semivariogram is

$$c_0 = \sigma^2_\eta + \sigma^2_\epsilon.$$

The name was coined by Matheron (1962) in reference to small nuggets of ore distributed throughout a larger body of rock. The small nuggets constitute a microscale process with spatial structure. Matheron's definition thus appeals to the micro-scale process and Matérn's definition to the measurement error process. In practice, $\eta(\mathbf{s})$ and $\epsilon(\mathbf{s})$ cannot be distinguished unless there are replicate observations at the same spatial locations. The modeler who encounters a nugget effect in a semivariogram thus needs to determine on non-statistical grounds whether the effect is due to micro-scale variation or measurement error. The choice matters for deriving best spatial predictors and measures of their precision (§5). Software packages are not consistent in this regard.

In the presence of a nugget effect, the variance of a second-order stationary process is $\text{Var}[Z(\mathbf{s})] = c_0 + \sigma_0^2$, where $\sigma_0^2$ is the **partial sill**. Since the nugget reduces the smoothness of the process, a common measure for the degree of spatial structure is the **relative structured variability**

*Relative Structured Variability*

$$RSV = \left( \frac{\sigma_0^2}{\sigma_0^2 + c_0} \right) \times 100\%. \tag{4.6}$$

This is a rather crude measure for the degree of structure (or smoothness) of a random field. Besides the relative magnitude of the discontinuity at the origin of the semivariogram it does not incorporate other features of the process that represents the continuous component, e.g., its mean square differentiability.

The range of the semivariogram is often considered an important parameter. In ecological applications it has been related to the size of patches that form after human intervention. It is not clear why the distance at which observations are no longer spatially correlated should be equal to the diameter of patches. Consider observations $Z(\mathbf{s}_1)$, $Z(\mathbf{s}_2)$, and $Z(\mathbf{s}_3)$. If $||\mathbf{s}_1 - \mathbf{s}_2|| < h^*$, where $h^*$ is the range, but $||\mathbf{s}_1 - \mathbf{s}_3|| > h^*$, $||\mathbf{s}_2 - \mathbf{s}_3|| < h^*$ then $\text{Cov}[Z(\mathbf{s}_1), Z(\mathbf{s}_2)] \neq 0$, $\text{Cov}[Z(\mathbf{s}_1), Z(\mathbf{s}_3)] = 0$, but $Z(\mathbf{s}_3)$ and $Z(\mathbf{s}_2)$ are correlated. In spatial prediction $Z(\mathbf{s}_3)$ can impact $Z(\mathbf{s}_1)$ through its correlation with $Z(\mathbf{s}_2)$. Chilès and Delfiner (1999, p. 205) call this the **relay effect** of spatial autocorrelation. The relative structured variability measures that component of spatial continuity that is reflected by the nugget effect. Other measures, which incorporate the shape of the semivariogram (and the range) have been proposed. Russo and Bresler (1981) and Russo and Jury (1987) consider **integral scales**. If $R(h) = C(h)/C(0)$ is the autocorrelation function of an isotropic process, then the scales for processes in $\mathbb{R}^1$ and $\mathbb{R}^2$ are

*Integral Scales*

$$I_1 = \int_0^\infty R(h)\, dh \qquad I_2 = \left\{ 2 \int_0^\infty R(h) h\, dh \right\}^{1/2}.$$

The idea of an integral scale is to consider distances over which observations are highly correlated, rather than the distance at which observations are no longer correlated. Processes with greater continuity have larger integral scales, correlations wear off more slowly. Solie, Raun, and Stone (1999) argue that integral scales provide objective measures for the distance at which soil and plant variables are highly correlated and are useful when this distance cannot be determined based on subject matter alone.

## 4.3 Covariance and Semivariogram Models

### *4.3.1 Model Validity*

In §4.3.1–4.3.5 we consider isotropic models for the covariance function and the semivariogram of a spatial process (accommodating anisotropy is discussed in §4.3.7). We start from models for covariance functions because valid semivariograms for second-order stationary processes can be constructed from valid covariance functions. For example, if $C(h)$ is the covariance function of an isotropic process with variance $\sigma^2$ and no nugget effect, then

$$\gamma(h) = \begin{cases} 0 & h = 0 \\ \sigma^2(1 - C(h))) & h > 0. \end{cases}$$

Not every mathematical function can serve as a model for the spatial dependency in a random field, however. Let $C(h)$ be the isotropic covariance function of a second-order stationary field and $\gamma(h)$ the isotropic semivariogram of a second-order or intrinsically stationary field. Then the following hold:

- If $C(h)$ is valid in $\mathbb{R}^d$, then it is also valid in $\mathbb{R}^s$, $s < d$ (Matérn, 1986, Ch. 2.3). If $\gamma(h)$ is valid in $\mathbb{R}^d$, it is also valid in $\mathbb{R}^s$, $s < d$.
- If $C_1(h)$ and $C_2(h)$ are valid covariance functions, then $aC_1(h) + bC_2(h)$, $a, b \geq 0$, is a valid covariance function.
- If $\gamma_1(h)$ and $\gamma_2(h)$ are valid semivariograms, then $a\gamma_1(h) + b\gamma_2(h)$, $a, b \geq 0$, is a valid semivariogram.
- A valid covariance function $C(\mathbf{h})$ is a positive-definite function, that is,

$$\sum_{i=1}^{k}\sum_{j=1}^{k} a_i a_j C(\mathbf{s}_i - \mathbf{s}_j) \geq 0,$$

for any set of real numbers $a_1, \cdots, a_k$ and sites. By Bochner's theorem this implies that $C(\mathbf{h})$ has spectral representation (§2.5)

$$C(\mathbf{h}) = \int_{-\infty}^{\infty} \cdots \int_{-\infty}^{\infty} \exp\{i\boldsymbol{\omega}'\mathbf{h}\} dS(\boldsymbol{\omega}).$$

In the isotropic case, the spectral representation of the covariance function in $\mathbb{R}^d$ becomes (Matérn, 1986, p. 14; Yaglom, 1987, p. 106; Cressie, 1993, p. 85; Stein, 1999, Ch. 2.10)

$$C(h) = \int_0^{\infty} \Omega_d(h\omega) dH(\omega) \tag{4.7}$$

$$\Omega_d(t) = \left(\frac{2}{t}\right)^v \Gamma(d/2) J_v(t), \tag{4.8}$$

where $v = d/2 - 1$, $J_v$ is the Bessel function of the first kind of order $v$ (§4.9.1), and $H$ is a non-decreasing function on $[0, \infty)$ with $\int_0^\infty dH(\omega) < \infty$. This is known as the Hankel transform of $H()$ of order $v$. The function

$H$ is related to the spectral distribution function of the process through $H(u) = \int_{||\omega||<u} dS(\omega)$ (Stein, 1999, p. 43).

*Basis Functions*

We call $\Omega_d$ the **basis function** of the covariance model in $\mathbb{R}^d$. For processes in $\mathbb{R}^d$, $d \leq 3$, $\Omega_1(t) = \cos(t)$, $\Omega_2(t) = J_0(t)$, $\Omega_3(t) = \sin(t)/t$. Also, for $d \to \infty$, $\Omega_d(t) \to \exp\{-t^2\}$ (Figure 4.1).

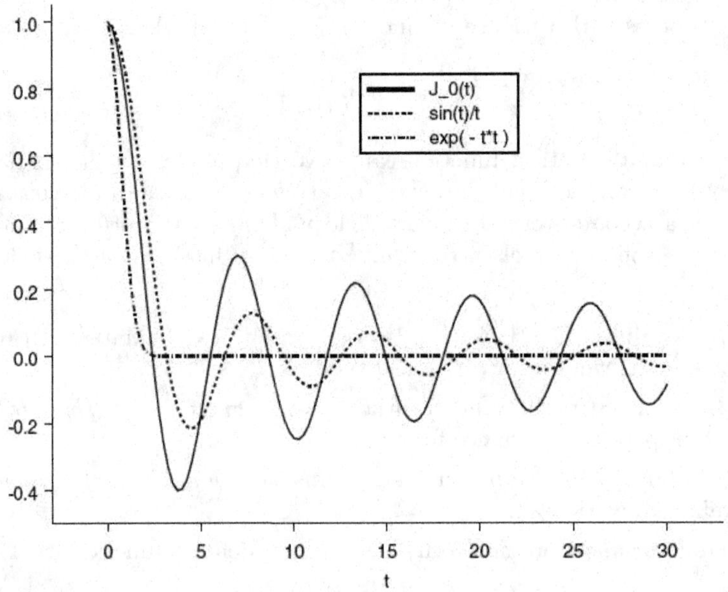


Figure 4.1 *Basis functions for processes in $\mathbb{R}^2$, $\mathbb{R}^3$, and $\mathbb{R}^\infty$. With increasing dimension, the basis functions have fewer sign changes. Covariance functions are non-increasing unless the basis function permits at least one sign change. A process in $\mathbb{R}^\infty$ does not permit negative autocorrelation at any lag distance.*

- A valid semivariogram $\gamma(\mathbf{h})$ is conditionally negative definite, that is,

$$2\sum_{i=1}^{m}\sum_{j=1}^{m} a_i a_j \gamma(\mathbf{s}_i - \mathbf{s}_j) \leq 0,$$

  for any real numbers $a_1, \cdots, a_m$ such that $\sum_{i=1}^{m} a_i = 0$ and any finite number of sites. A valid (isotropic) semivariogram also has a spectral representation, namely

$$\gamma(h) = \frac{1}{2}\int_0^\infty \omega^{-2}(1 - \Omega_d(\omega h))dH(\omega),$$

  with $\int_0^\infty (1+\omega^2)^{-1} dH(\omega) < \infty$.
- A necessary condition for $\gamma(\mathbf{h})$ to be a valid semivariogram is that $2\gamma(\mathbf{h})$

grows more slowly than $||\mathbf{h}||^2$. This is often referred to as the *intrinsic hypothesis.*

### 4.3.2 The Matérn Class of Covariance Functions

Based on the spectral representation (4.7)–(4.8) of isotropic covariance functions, Matérn (1986) constructed a flexible class of covariance functions,

*Matérn Class*

$$C(h) = \sigma^2 \frac{1}{\Gamma(\nu)} \left(\frac{\theta h}{2}\right)^{\nu} 2K_\nu(\theta h) \qquad \nu > 0, \theta > 0, \tag{4.9}$$

where $K_\nu$ is the modified Bessel function of the second kind of order $\nu > 0$. The parameter $\theta$ governs the range of the spatial dependence, the smoothness of the process increases with $\nu$. Properties of the Bessel functions are given in §4.9.2; it is seen there that for fixed $\nu$ and $t \to 0$

$$K_\nu(t) \approx \frac{\Gamma(\nu)}{2} \left(\frac{t}{2}\right)^{-\nu}.$$

Hence $\sigma^2$ is the variance of the process. Expression (4.9) is only one of several possible parameterizations of this family of covariance functions. Others are given in §4.7.2.

We commence the list of isotropic models with the Matérn class because of its generality. $C(h)$ given by (4.9) is valid in $\mathbb{R}^d$ and its smoothness increases with $\nu$. Although $\theta$ is related to the (practical) range of the process, the range is itself a function of $\nu$. For particular values of $\nu$, the range is easily determined however, as (4.9) takes on simple forms. As $\nu \to \infty$ the limiting covariance model is known as the **gaussian** model

*"G"aussian Model*

$$C(h) = \sigma^2 \exp\left\{-\theta h^2\right\} = \sigma^2 \exp\left\{-3\frac{h^2}{\alpha^2}\right\}. \tag{4.10}$$

The second parameterization is common in geostatistical applications where $\alpha$ is the practical range, the distance at which the correlations have decreased to $\approx 0.05$ or less ($\exp\{-3\} = 0.04978$, to be more exact). Other important cases in the Matérn class of covariance functions are obtained for $\nu = 1/2$ and $\nu = 1$. In the former case, the resulting model is known as the **exponential** model. Using the following results regarding Bessel functions (§4.9.2),

$$\begin{aligned} K_\nu(t) &= \frac{\pi}{2} \frac{I_{-\nu}(t) - I_\nu(t)}{\sin(\pi\nu)} \\ I_{-1/2}(t) &= \sqrt{\frac{2}{\pi t}} \cosh(t) \qquad I_{1/2}(t) = \sqrt{\frac{2}{\pi t}} \sinh(t) \\ \sinh(t) &= \frac{1}{2}\left(e^t - e^{-t}\right) \qquad \cosh(t) = e^{-t} + \sinh(t), \end{aligned}$$

one obtains

$$K_{1/2}(t) = \sqrt{\frac{\pi}{2t}}\, e^{-t}$$

*Exponential Model*

and substitution into (4.9) yields (recall that $\Gamma(1/2) = \sqrt{\pi}$)

$$C(h) = \sigma^2 \exp\{-\theta h\} = \sigma^2 \exp\left\{-3\frac{h}{\alpha}\right\}. \quad (4.11)$$

The second parameterization is again common in geostatistical applications where $\alpha$ denotes the practical range. The exponential model is the continuous-time analog of the first-order autoregressive time series covariance structure. It enjoys popularity not only in spatial applications, but also in modeling longitudinal and repeated measures data (see, e.g., Jones, 1993; Schabenberger and Pierce, 2002, Ch. 7). The model for $\nu = 1$,

*Whittle Model*

$$C(h) = \sigma^2 \theta h K_1(\theta h), \quad (4.12)$$

was suggested by Whittle (1954). He considered the exponential model as the "elementary" covariance function in $\mathbb{R}^1$ and (4.12) as the "elementary" model in $\mathbb{R}^2$. A process $Z(t)$ in $\mathbb{R}^1$ with exponential correlation can be represented by the stochastic differential equation

$$\left(\frac{d}{dt} + \theta\right) Z(t) = \epsilon(t),$$

where $\epsilon(t)$ is a white noise process. Whittle (1954) and Jones and Zhang (1997) consider this the elementary stochastic differential equation in $\mathbb{R}^1$. In $\mathbb{R}^2$, with coordinates $x$ and $y$, Whittle awards this distinction to the stochastic Laplace equation

$$\left(\frac{\partial^2}{\partial x^2} + \frac{\partial^2}{\partial y^2} - \theta^2\right) Z(x, y) = \epsilon(x, y).$$

A process represented by this equation has correlation function

$$R(h) = \theta h K_1(\theta h),$$

a Whittle model. Whittle (1954) concludes that "the exponential function has no divine right in two dimensions" and calls processes in $\mathbb{R}^2$ with exponential covariance function "artificial"; finding it "difficult to visualize a physical mechanism" that has covariance function (4.11).

We strongly feel that the exponential model has earned its place among the isotropic covariance models for modeling spatial data. In fitting these models to data, the exponential model has a definite advantage over Whittle's model. It does not require evaluation of infinite series (§4.9.2).

If there is an "artificial" model for the spatial dependence, it is the gaussian model (4.10). Because it is the limiting model for $\nu \to \infty$ it is infinitely differentiable. Physical and biological processes with this type of smoothness are truly artificial. The name is unfortunately somewhat misleading. Covariance model (4.10) is called the "gaussian" model because of the functional similarity of the spectral density of a process with that covariance function to the Gaussian probability density function (§4.7.2). It does *not* command the same respect as the Gaussian distribution. We choose lowercase notation to distinguish the covariance model (4.10) from the Gaussian distribution.

Figure 4.2 shows semivariograms derived from several different covariance functions in the Matérn class (4.9). For the same value of $\theta$, the semivariogram rises more quickly from the origin as $\nu$ decreases. The Whittle model with $\nu = 1$ is slightly quadratic near the origin but much less so than the gaussian model ($\nu \rightarrow \infty$). All models in this class are for second-order stationary processes with positive spatial autocorrelation that decreases with distance.

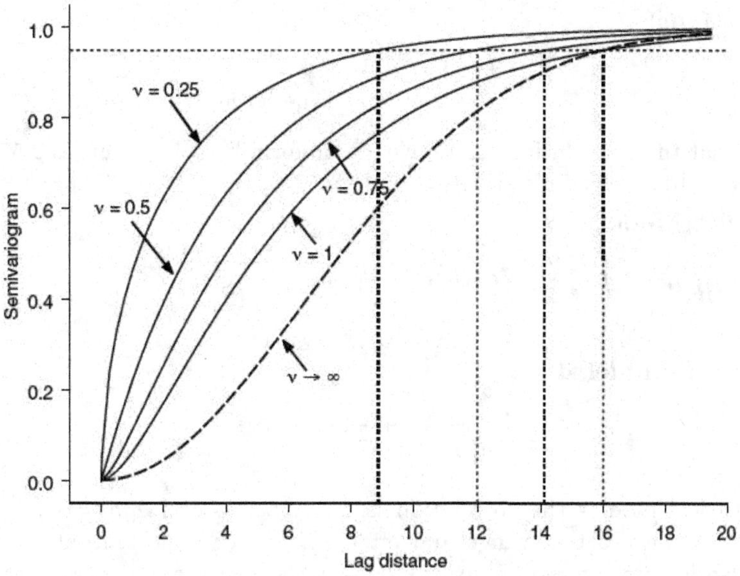


Figure 4.2 *Semivariograms constructed from covariance functions in the Matérn class for different values of the smoothness parameter* $\nu$, $\theta = 0.25$, *and* $\sigma^2 = 1$. *The model for* $\nu \rightarrow \infty$ *is the gaussian model and was chosen to have the same (practical) range as the Whittle model with* $\nu = 1$. *Vertical lines indicate practical ranges and the horizontal line denotes 95% of the sill.*

### *4.3.3 The Spherical Family of Covariance Functions*

A second-order stationary random field can be represented by the convolution of a kernel function and a white noise random field (§2.4.2). The covariance of the resulting random field is then simply the convolution of the kernels,

$$\text{Cov}[Z(\mathbf{s}), Z(\mathbf{s}+\mathbf{h})] = \sigma_x^2 \int_{\mathbf{u}} K(\mathbf{u})K(\mathbf{u}+\mathbf{h})\, d\mathbf{u}.$$

Chilès and Delfiner (1999, p. 81) present an interesting family of isotropic covariance functions by choosing as kernel function the indicator function of the sphere of $\mathbb{R}^d$ with diameter $\alpha$,

$$K_d(u) = \begin{cases} 1 & u \leq \alpha/2 \\ 0 & \text{otherwise.} \end{cases}$$

The family of isotropic covariance functions so constructed can be written as

$$C_d(h) \propto \begin{cases} \int_{h/\alpha}^{1} \left(1 - u^2\right)^{(d-1)/2} du & h \leq \alpha \\ 0 & \text{otherwise.} \end{cases}$$

The autocorrelation functions for $d = 1, 2, 3$ are

- **Tent Model**, $d = 1$:

$$R_1(h) = \begin{cases} 1 - \frac{h}{\alpha} & h \leq \alpha \\ 0 & \text{otherwise.} \end{cases}$$

  Note that the convolutions in $\mathbb{R}^1$ with a uniform kernel in Figures 2.5 and 2.6 (page 61) yielded a tent correlation function.

- **Circular Model**, $d = 2$:

$$R_2(h) = \begin{cases} \frac{2}{\pi}\left(\arccos\{h/\alpha\} - \frac{h}{\alpha}\sqrt{1 - h^2/\alpha^2}\right) & h \leq \alpha \\ 0 & \text{otherwise.} \end{cases}$$

- **Spherical Model**, $d = 3$:

$$R_1(h) = \begin{cases} 1 - \frac{3}{2}\frac{h}{\alpha} + \frac{1}{2}\left(\frac{h}{\alpha}\right)^3 & h \leq \alpha \\ 0 & \text{otherwise.} \end{cases} \tag{4.13}$$

In spherical models the correlation is exactly zero at lag $h = \alpha$, hence these models have a true range and often exhibit a visible kink at $h = \alpha$. The near-origin behavior of semivariograms in the spherical family is linear or close-to-linear (Figure 4.3). Because (4.13) is valid in $\mathbb{R}^3$, it is often considered *the* spherical model. The popularity of the spherical covariance function and its semivariogram

*"The" Spherical Model*

$$C(h) = \sigma^2\left(1 - \frac{3}{2}\frac{h}{\alpha} + \frac{1}{2}\left(\frac{h}{\alpha}\right)^3\right) \tag{4.14}$$

$$\gamma(h) = \sigma^2\left(\frac{3}{2}\frac{h}{\alpha} - \frac{1}{2}\left(\frac{h}{\alpha}\right)^3\right) \tag{4.15}$$

are a mystery to Stein (1999, p. 52), who argues that perhaps "there is a mistaken belief that there is some statistical advantage in having the autocorrelation function being exactly zero beyond some finite distance."

### *4.3.4 Isotropic Models Allowing Negative Correlations*

The second-order stationary models discussed so far permit only positive autocorrelation, the semivariogram is a non-decreasing function (the covariance

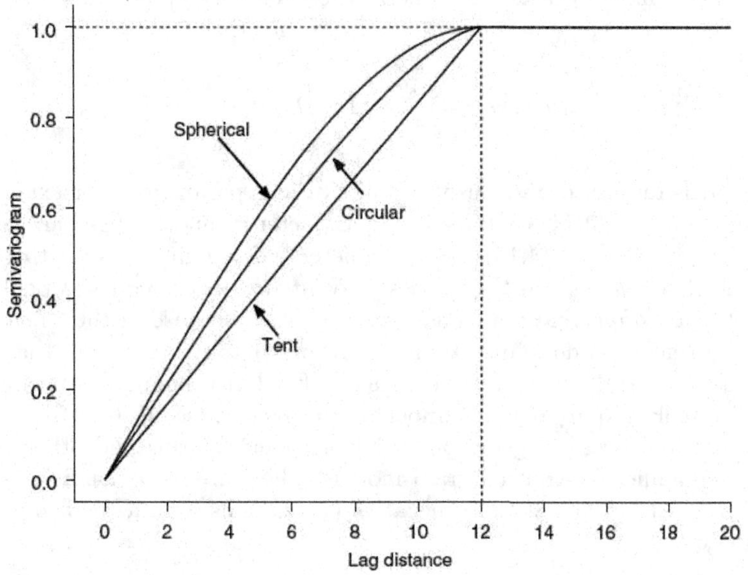


Figure 4.3 *Semivariograms constructed from covariance functions in the spherical family with* $\alpha = 12$ *and sill* $\sigma^2 = 1$.

function is non-increasing). Negative spatial association can be built into models for the spatial dependency but positive and negative autocorrelations can not change arbitrarily in a second-order stationary process.

### *4.3.4.1 The Nonparametric Approach*

The spectral representation (4.7) of the isotropic covariance function suggests a method to construct covariance functions that allow for negative and positive spatial association. The basis function $\Omega_d(t)$ is an oscillating function unless $d = \infty$. In $\mathbb{R}^2$, for example, $\Omega_d(t) = J_0(t)$, the Bessel function of the first kind of zero order (Figure 4.1 and §4.9.1). Choose $dH(\omega) = f(\omega)d\omega$ and

$$f(\omega) = \begin{cases} \sigma^2 & \omega = 1 \\ 0 & \text{otherwise.} \end{cases}$$

Then $C(h) = \sigma^2\Omega(h)$ and it is seen that $J_0(h)$ is a valid correlation function in $\mathbb{R}^2$. Since linear combinations of valid covariance function are valid—provided the coefficients in the linear combination are positive—a flexible class of covariance models for second-order stationary processes can be constructed as

$$C(h, \mathbf{t}, \mathbf{w}) = \sum_{i=1}^{p} w_i \Omega_d(ht_i), \tag{4.16}$$

where $\mathbf{t} = [t_1, \cdots, t_p]'$ is a set of nodes and $\mathbf{w} = [w_1, \cdots, w_p]'$ is an associated set of weights. This is a discrete representation of (4.7) where the non-decreasing function $F(\omega)$ has been replaced with a step function that has positive mass $w_1, \cdots, w_p$ at nodes $t_1, \cdots, t_p$. Notice that $C(0) = \sum_{i=1}^{p} w_i$ and

$$\gamma(h, \mathbf{t}, \mathbf{w}) = \sum_{i=1}^{p} w_i \left(1 - \Omega_d(ht_i)\right).$$

The models discussed previously contain various parameters, for example, the spherical model (4.14) contains a sill parameter $\sigma^2$ and a range parameter $\alpha$. In the representation (4.16) the "parameters" of the model consist of the placement of the nodes and the masses associated with the nodes. Models of form (4.16) are often referred to as nonparametric because of their flexibility. Their formulation does involve unknown constants (parameters!) that are estimated from the data. To achieve sufficient flexibility, nonparametric models may actually require a fair number of parameters (see §4.6). Parameter proliferation and overfitting are potential problems for this and other nonparametric families of semivariogram models. There are some candidates for parametric models that allow for negative correlations with a small number of parameters.

#### *4.3.4.2 Hole Models*

Any isotropic covariance model in $\mathbb{R}^d$ has representation (4.7) and a parametric model can be constructed from (4.8) as

$$C(h) = 2^v \, \Gamma(d/2) \left(\frac{h}{\alpha}\right)^{1-v} J_v(h/\alpha), \tag{4.17}$$

where $\alpha > 0$ and $v = d/2 - 1$. Matérn (1986, p. 16) shows that permissible correlations in isotropic are bounded from below. If $R(h)$ denotes the autocorrelation function, then $R(h) > -0.403$ in $\mathbb{R}^2$ and $R(h) > -0.218$ in $\mathbb{R}^3$. Models that allow negative correlations in excess of these values are thus not permissible as isotropic covariance models. Since

$$J_{1/2}(t) = \sqrt{\frac{2}{\pi t}} \sin\{t\}, \tag{4.18}$$

evaluating (4.17) for $v = 0.5$ using (4.18) yields a model valid in $\mathbb{R}^3$ known as the **cardinal-sine** model:

*Cardinal-sine Model*

$$C(h) = \left(\frac{\alpha}{h}\right) \sin\{h/\alpha\}. \tag{4.19}$$

This model is also known as the **hole-effect** or **wave** model. The corresponding semivariogram model with sill $\sigma^2$ is

$$\gamma(h) = \sigma^2 \left(1 - \left(\frac{\alpha}{h}\right) \sin\{h/\alpha\}\right). \tag{4.20}$$

The "practical" range for this model is defined as the lag distance at which the first peak is no greater than $1.05\sigma^2$ or the first valley is no less than $0.95\sigma^2$. It is approximately $6.5 \times \pi\alpha$ (Figure 4.4b).

### *4.3.5 Basic Models Not Second-Order Stationary*

The two basic isotropic models for processes that are not second-order stationary are the linear and the power model. The former is a special case of the latter. The **power** model is given in terms of the semivariogram *Power Models*

$$\gamma(h) = \theta h^{\lambda}, \tag{4.21}$$

with $\theta \geq 0$ and $0 \leq \lambda < 2$. If $\lambda \geq 2$, the model violates the intrinsic hypothesis. For $\lambda = 1$, the linear semivariogram model results. As mentioned earlier, a linear semivariogram can be indicative of an intrinsically, but not second-order stationary process. It could also be an indication of a second-order stationary process whose semivariogram behaves linearly near the origin (spherical model, for example), but whose sill has not been reached across the observed lag distances (Figure 4.4a).

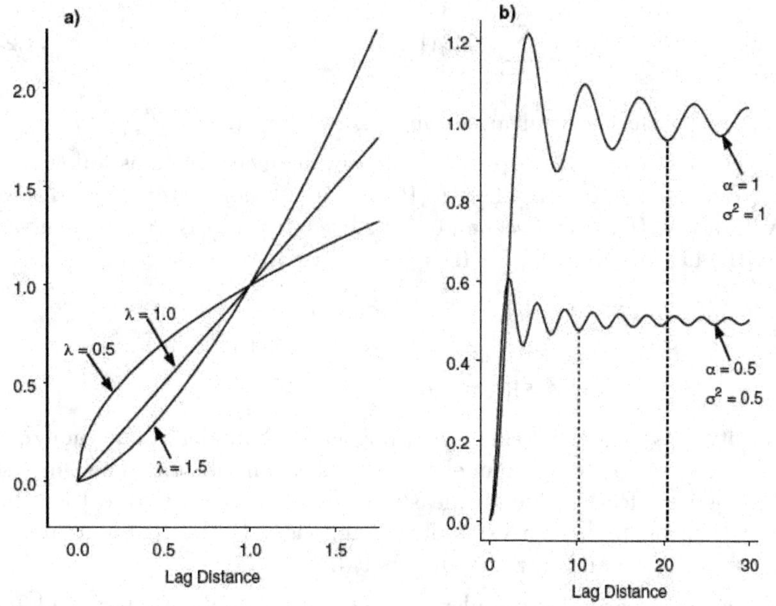


Figure 4.4 *Power semivariogram (a) and hole (cardinal-sine) models (b).*

### 4.3.6 Models with Nugget Effects and Nested Models

The basic second-order stationary parametric covariance functions and semivariograms of the previous subsections are often considered too restrictive to model the complexities of spatial dependence in geostatistical data. None of the models presented there incorporates a nugget effect, for example. One device to introduce such an effect into the semivariogram is through nesting of models. Recall from §4.3.1 that $aC_1(h)+bC_2(h)$ is a valid covariance function for a second-order stationary process if $a, b \geq 0$ and $C_1(h)$ and $C_2(h)$ are valid covariance functions.

Assume that the random field $Z(\mathbf{s})$ with $\mathrm{E}[Z(\mathbf{s})] = \mu$ consists of orthogonal, zero-mean components $U_1(\mathbf{s}), \cdots, U_p(\mathbf{s})$ and can be decomposed as

$$Z(\mathbf{s}) = \mu + \sum_{j=1}^{p} a_j U_j(\mathbf{s}), \tag{4.22}$$

In the geostatistical literature this decomposition is termed the **linear model of regionalization**. Let $C_z(\mathbf{h})$ denote the covariance function of the $Z(\mathbf{s})$ process. Then, because $\mathrm{Cov}[U_j(\mathbf{s}), U_k(\mathbf{s})] = 0 \, \forall j \neq k$,

$$\begin{aligned} C_z(\mathbf{h}) &= \mathrm{Cov}[Z(\mathbf{s}), Z(\mathbf{s}+\mathbf{h})] \\ &= \sum_{j=1}^{p}\sum_{k=1}^{p} a_j a_k \mathrm{Cov}[U_j(\mathbf{s}), U_k(\mathbf{s}+\mathbf{h})] \\ &= \sum_{j=1}^{p} a_j^2 C_j(\mathbf{h}). \end{aligned} \tag{4.23}$$

Similarly, we obtain the semivariogram of $Z(\mathbf{s})$ as $\gamma_z(\mathbf{h}) = \sum_{j=1}^{p} a_j^2 \gamma_j(\mathbf{h})$.

A nugget effect can be incorporated into any semivariogram as follows. Let $Z(\mathbf{s}) = \sqrt{c_0} U_1(\mathbf{s}) + \sigma_0 U_2(\mathbf{s})$, where $U_1(\mathbf{s})$ is a white noise process with mean 0 and variance 1. $U_2(\mathbf{s})$ is a second-order stationary process whose semivariogram $\gamma_2(\mathbf{h})$ has sill 1. Then, $C_2(\mathbf{h}) = 1 - \gamma_2(\mathbf{h})$, and

$$\begin{aligned} \gamma_z(\mathbf{h}) &= c_0 + \sigma_0^2 \gamma_2(\mathbf{h}) \\ &= c_0 + \sigma_0^2 (1 - C_2(\mathbf{h})) \\ \mathrm{Var}[Z(\mathbf{s})] &= c_0 + \sigma_0^2. \end{aligned}$$

The quantity $c_0$ is the nugget effect. Any nugget effect model can be thought of as a nested model where one model component is white noise. This suggests that the nugget effect is due to measurement error. If it is due to microscale variation, then the corresponding component $U_1(\mathbf{h})$ has a no-nugget semivariogram $\gamma_1(\mathbf{h})$ and $c_0$ represents its sill.

Nesting semivariograms is popular in geostatistical applications to add flexibility to models. When the process is believed to consist of several components that operate on different spatial scales, then nesting models with different ranges is attractive to estimate the scales of the respective processes.

For example, spatial variation in a soil nutrient may be driven by micro-environmental conditions on a small scale, land-use and soil-type on a medium spatial scale, and geology on a large scale. Nesting three semivariograms to estimate the respective ranges of the three processes has appeal. However, it is usually difficult to justify why these processes should be orthogonal (independent). Without it, (4.23) does not hold. The orthogonality assumption is more tenable if a white noise measurement error process is nested with one other model to create a nugget effect.

### *4.3.7 Accommodating Anisotropy*

If the covariance function of a second-order stationary process is anisotropic, the spatial structure is direction dependent. Whereas in the isotropic case, iso-correlation contours are spherical, a particular case of anisotropy gives rise to elliptical contours (Figure 4.5). This case is known as **geometric anisotropy** and can be corrected by a linear transformation of the coordinate system. Following Matérn (1986, p. 19), let $Z_1(\mathbf{s})$ be a stationary process in $\mathbb{R}^d$ with covariance function $C_1(\mathbf{h})$, mean $\mu$, and variance $\sigma^2$. Let $\mathbf{B}_{(d\times d)}$ be a real matrix and consider the stochastic process $Z(\mathbf{s}) = Z_1(\mathbf{Bs})$. Because $Z_1(\mathbf{s})$ is stationary, we have $\mathrm{E}[Z(\mathbf{s})] = \mu$ and $\mathrm{Var}[Z(\mathbf{s})] = \sigma^2$. Furthermore,

$$\begin{aligned}\mathrm{Cov}[Z(\mathbf{s}), Z(\mathbf{s}+\mathbf{h})] = C(\mathbf{h}) &= \mathrm{Cov}[Z_1(\mathbf{Bs}), Z_1(\mathbf{B}(\mathbf{s}+\mathbf{h}))] \\ &= C_1(\mathbf{Bs} - \mathbf{B}(\mathbf{s}+\mathbf{h})) = C_1(-\mathbf{Bh}) \\ &= C_1(\mathbf{Bh}).\end{aligned}$$

Hence, if $C_1(\mathbf{h})$ is isotropic, then $C(\mathbf{h}) = C_1(||\mathbf{Bh}||)$ is a geometrically anisotropic covariance function.

To correct for geometric anisotropy this transformation of the coordinate system can be reversed. If $\mathbf{s} = [x, y]'$ is a coordinate in $\mathbb{R}^2$ such that the process $Z(\mathbf{s})$ is geometrically anisotropic, then $Z(\mathbf{s}^*) = Z(\mathbf{As})$ has isotropic covariance function if $\mathbf{A} = \mathbf{B}^{-1}$. A linear transformation $\mathbf{s}^* = \mathbf{As}$ of Euclidean space provides the appropriate space to express the covariance. The geometric anisotropy shown in Figure 4.5 is corrected by (i) a rotation of the coordinate system to align the major and minor axes of the elliptical contours and (ii), a compression of the major axis to make contours spherical. Hence,

$$\mathbf{A} = \begin{bmatrix} 1 & 0 \\ 0 & \lambda \end{bmatrix} \begin{bmatrix} \cos\theta & -\sin\theta \\ \sin\theta & \cos\theta \end{bmatrix},$$

where $\lambda$ is the anisotropy ratio. A geometric anisotropy manifests itself in semivariograms that have the same shape and sill in the direction of the major and minor axes, but different ranges. The parameter $\lambda$ equals the ratio of the ranges in these two directions. Geometric anisotropy is common for processes that evolve along particular directions. For example, airborne pollution will likely exhibit anisotropy in the prevailing wind direction and perpendicular to it.

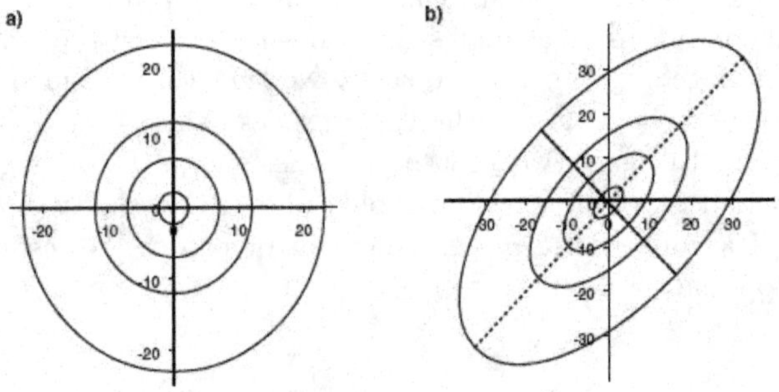


Figure 4.5 *Contours of iso-correlation (0.8, 0.5, 0.3, 0.1) for two processes with exponential correlation function. The isotropic model (a) has spherical correlation contours. The elliptic contours in panel b) correspond to a 45 degree rotation and a ratio of* $\lambda = 0.5$ *between the two major axes. The axes depict lag distances in the* $(x, y)$ *(a) and* $(x^*, y^*)$ *coordinate systems (b).*

In a process exhibiting **zonal anisotropy** the covariance function depends on only some components of the lag vector. A case of zonal anisotropy exists if the semivariogram sills vary with direction. Typically, the range in the direction with the shorter range also has the smaller sill. Zonal anisotropy can then be modeled by nesting an isotropic model and a model which depends only on the lag-distance in the direction $\theta$ of the greater sill (Goovaerts, 1997, p. 93),

$$\gamma(\mathbf{h}) = \gamma_1(||\mathbf{h}||) + \gamma_2(h_\theta).$$

Chilès and Delfiner (1999, p. 96) warn about zonal models that partition the coordinates, because certain linear combinations can have zero variance. In $\mathbb{R}^2$ let $Z(\mathbf{s}) = Z_1(x) + Z_2(y)$. If the components $Z_1(x)$ and $Z_2(y)$ are orthogonal, then, by §4.3.6, $\gamma_z(\mathbf{h}) = \gamma_1(h_x) + \gamma_2(h_y)$. Let $\mathbf{h}_u = [u, 0]'$ and $\mathbf{h}_v = [0, v]'$ be two vectors shifting the coordinates. Then

$$\text{Var}[Z(\mathbf{s}) - Z(\mathbf{s} + \mathbf{h}_v) - Z(\mathbf{s} + \mathbf{h}_u) + Z(\mathbf{s} + \mathbf{h}_u + \mathbf{h}_v)] = \text{Var}[X(\mathbf{s})] = 0,$$

because

$$\begin{aligned} X(\mathbf{s}) &= Z(\mathbf{s}) - Z(\mathbf{s} + \mathbf{h}_v) - Z(\mathbf{s} + \mathbf{h}_u) + Z(\mathbf{s} + \mathbf{h}_u + \mathbf{h}_v) \\ &= Z_1(x) + Z_2(y) - Z_1(x) - Z_2(y + v) - Z_1(x + u) - Z_2(y) \\ &\quad + Z_1(x + u) + Z_2(y + v) = 0. \end{aligned}$$

## 4.4 Estimating the Semivariogram

### *4.4.1 Matheron's Estimator*

To learn about the semivariogram from a set of observed data, $Z(\mathbf{s}_1), \cdots, Z(\mathbf{s}_n)$, one could plot the squared differences $\{Z(\mathbf{s}_i) - Z(\mathbf{s}_j)\}^2$ against the lag distance $\mathbf{h}$ (or $||\mathbf{h}||$). Such a graph is appropriately termed the empirical semivariogram cloud because it is usually not very informative and "clouds" the big picture. The number of pairwise differences can be very large, lag distances may be unique for irregularly spaced data, and extreme observations cause many "outliers" in the cloud. Since $\{Z(\mathbf{s}_i) - Z(\mathbf{s}_j)\}^2$ estimates unbiasedly the variogram at lag $\mathbf{s}_i - \mathbf{s}_j$, provided the mean of the random field is constant, a more useful estimator is obtained by summarizing the squared differences. The semivariogram estimator that averages the squared differences of points that are distance $\mathbf{s}_i - \mathbf{s}_j = \mathbf{h}$ apart is known commonly as the **classical** or Matheron estimator since it was proposed by Matheron (1962):

*Matheron Estimator*

$$\widehat{\gamma}(\mathbf{h}) = \frac{1}{2|N(\mathbf{h})|} \sum_{N(\mathbf{h})} \{Z(\mathbf{s}_i) - Z(\mathbf{s}_j)\}^2 . \tag{4.24}$$

The set $N(\mathbf{h})$ consists of location pairs $(\mathbf{s}_i, \mathbf{s}_j)$ such that $\mathbf{s}_i - \mathbf{s}_j = \mathbf{h}$ and $|N(\mathbf{h})|$ denotes the number of distinct pairs in $N(\mathbf{h})$. When data are sparse or irregularly shaped, the number of distinct pairs in $N(\mathbf{h})$ may not be sufficient to obtain a stable estimate at lag $\mathbf{h}$. Typical recommendations are that at least 30 (better 50) pairs of locations should be available at each lag. If the number of pairs is smaller, lags are grouped into lag classes so that $\widehat{\gamma}(\mathbf{h})$ is the average squared difference of site pairs that satisfy $\mathbf{s}_i - \mathbf{s}_j = \mathbf{h} \pm \epsilon$. The choice of the tolerance $\epsilon$ is left to the user. A graph of $\widehat{\gamma}(\mathbf{h})$ against $||\mathbf{h}||$ is called the Matheron semivariogram or the **empirical** semivariogram

Among the appealing properties of the Matheron estimator—which are partly responsible for its widespread use—are simple computation, unbiasedness, evenness, and attaining zero at zero lag: $\mathrm{E}[\widehat{\gamma}(\mathbf{h})] = \gamma(\mathbf{h})$, $\widehat{\gamma}(\mathbf{h}) = \widehat{\gamma}(-\mathbf{h})$, $\widehat{\gamma}(\mathbf{0}) = 0$. It is difficult in general to determine distributional properties and moments of semivariogram estimators without further assumptions. The estimators at two different lag values are usually correlated because (i) observations at that lag class are spatially correlated, and (ii) the same points are used in estimating the semivariogram at the two lags. Because the Matheron estimator is based on squared differences, more progress has been made in establishing (approximate) moments and distributions than for some of its competitors.

Consider $Z(\mathbf{s})$ to be a Gaussian random field so that $(2\gamma(\mathbf{h}))^{-1}\{Z(\mathbf{s}) - Z(\mathbf{s}+\mathbf{h})\}^2 \sim \chi^2_1$ and

$$\mathrm{Var}[\{Z(\mathbf{s}) - Z(\mathbf{s}+\mathbf{h})\}^2] = 2 \times 4\gamma(\mathbf{h})^2,$$

Cressie (1985) shows that the variance of (4.24) at lag $\mathbf{h}_i$ can be approximated as

$$\text{Var}[\widehat{\gamma}(\mathbf{h}_i)] \approx 2\frac{\gamma(\mathbf{h}_i)^2}{|N(\mathbf{h}_i)|}. \tag{4.25}$$

The approximation ignores the correlations between $Z(\mathbf{s}_i)-Z(\mathbf{s}_j)$ and $Z(\mathbf{s}_k)-Z(\mathbf{s}_l)$ (see §4.5.1). If it holds, consistency of the Matheron estimator is easily ascertained from (4.25), since $\widehat{\gamma}(\mathbf{h})$ is unbiased. The expression (4.25) also tells us what to expect for large lag values. In practice empirical semivariograms are common that appear ill-behaved and erratic for large lags. Since the semivariogram $\gamma(\mathbf{h}_i)$ of a second-order stationary process rises until it reaches the sill, the numerator of (4.25) increases sharply in $\mathbf{h}_i$ as long as the semivariogram has not reached the sill. Even then, the variance of the Matheron estimator does not remain constant. The number of pairs from which $\widehat{\gamma}(\mathbf{h})$ can be computed decreases sharply with $\mathbf{h}$.

**Example 4.1** Table 4.1 demonstrates these effects for data on a $10 \times 10$ lattice with an exponential semivariogram (practical range 7 and sill 10). The lag distances were grouped in unit lag classes with a lag tolerance of 0.5 units. The number of distance pairs increases until the fourth lag class and decreases afterwards. The approximate variance of the Matheron estimator increases slowly for small lags because the increase in $\gamma(\mathbf{h})$ is almost offset by an increase in the number of pairs. For distances exceeding the practical range, $\gamma(\mathbf{h})$ increases slowly but the variance of $\widehat{\gamma}(\mathbf{h})$ increases sharply due to lack of observation pairs. The recommendation to use $\widehat{\gamma}(\mathbf{h}_i)$ only if at least 30 (or 50) pairs are available in lag class $i$ is based on the idea to keep the variation of $\widehat{\gamma}(\mathbf{h})$ at bay. It does not lead to a homoscedastic set of empirical semivariogram values. □

One could, however, group the pairs into lag classes such that each class contains a number of observations that makes the empirical semivariogram values homoscedastic under a particular semivariogram model. This leads to a concentration of lag classes at small lags and sparsity at large lags. The overall shape of the semivariogram may be difficult to determine. Since one would have to know the form and parameters of the true semivariogram, this is an impractical proposition in any case. But even choosing lag classes that have the same number of points may inappropriately group the lag distances.

**Example 4.2 C/N ratios.** Figure 4.6 displays the 195 locations on an agricultural field at which the total soil carbon and total soil nitrogen percentages were measured. These data were kindly provided by Dr. Thomas G. Mueller, Department of Agronomy, University of Kentucky, and represent a subset of the data used in Chapter 9 of Schabenberger and Pierce (2002). The field had been in no-tillage management for more than ten years when strips of the field were chisel-plowed. The data considered here correspond to these plowed parts of the field.

The data are geostatistical and irregularly spaced. The Euclidean distances

Table 4.1 *Approximate variance of Matheron estimator (4.24) on a* $10 \times 10$ *lattice for one unit lag classes with a tolerance of 0.5 units. Exponential semivariogram with practical range 7 and sill 10.*

| Lag Class Value | Number of Distances in Class | Average Lag Distance | Semi-variogram Value | Variance according to (4.25) |
|---|---|---|---|---|
| 1 | 342 | 1.196 | 4.011 | 0.094 |
| 2 | 448 | 2.152 | 6.023 | 0.162 |
| 3 | 520 | 3.036 | 7.278 | 0.203 |
| 4 | 850 | 4.062 | 8.246 | 0.160 |
| 5 | 608 | 5.131 | 8.891 | 0.260 |
| 6 | 684 | 6.078 | 9.261 | 0.251 |
| 7 | 522 | 7.049 | 9.512 | 0.347 |
| 8 | 444 | 7.984 | 9.673 | 0.422 |
| 9 | 368 | 8.996 | 9.788 | 0.521 |
| 10 | 94 | 10.005 | 9.862 | 2.069 |
| 11 | 60 | 10.925 | 9.907 | 3.272 |
| 12 | 8 | 12.042 | 9.942 | 24.714 |
| 13 | 2 | 12.728 | 9.957 | 99.147 |

between observations range between 5 and 565.8 feet. In order to obtain lag classes with at least 50 observations per class, we decided on 35 lag classes of width 6 feet. The resulting Matheron estimator of the empirical semivariogram is shown in Figure 4.7.

It is customary not to compute the empirical semivariogram up to the largest possible lag class. The number of available pairs shrinks quickly for larger lags and the variability of the empirical semivariogram increases. A common recommendation is to compute the empirical semivariogram up to about one half of the maximum separation distance in the data, although this is only a general guideline. It is important to extend the empirical semivariogram far enough so that the important features of the spatial dependency structure can be discerned but not so far as to hinder model selection and interpretation due to lack of reliability. The empirical semivariogram of the C/N ratios appears quite "well-behaved." It rises from what appears to be the origin up to a distance of 100 feet and has a sill between 0.25 and 0.30. A spherical or exponential model may fit this empirical semivariogram well. We will return to these data throughout the chapter.

The question of possible anisotropy can be investigated by computing the empirical semivariogram surface or by constructing directional empirical semivariograms. To compute the semivariogram surface you divide the domain into non-overlapping regions of equal size, typically rectangles or squares. If $\delta_x$ and $\delta_y$ are the dimensions of the rectangles in the two main directions, we compute a point on the surface at location $\mathbf{h}$ by averaging the pairs separated

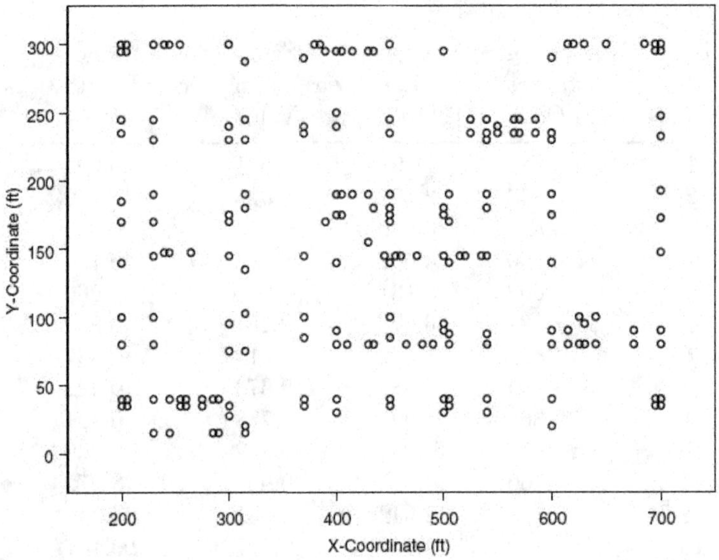


Figure 4.6 *Sample locations in chisel-plowed components of an agricultural field. At each location a measurement of soil carbon (%) and soil nitrogen (%) was obtained.*

by $\mathbf{h} \pm [\delta_x, \delta_y]'$. The surface is then typically smoothed or contoured and the display is centered at $\mathbf{h} = [0, 0]'$. For the C/N ratio data, Figure 4.8 shows the result of smoothing the semivariance surface.

There does not appear to be a discernible pattern in any direction that would differ substantively in another direction. Based on this surface we consider isotropic covariance models for the C/N ratio henceforth.

One problem with the semivariogram surface is that the division of the domain needs to be fine enough to convey the semivariogram detail in all directions while at the same time to provide sufficient pairs of points to obtain a reliable estimator of the semivariogram. The second approach of exploring anisotropy, the construction of directional semivariograms, suffers in general from the same problem. Because the number of directions for which the empirical semivariogram is computed is relatively small, the issue of data sparsity is less pronounced. Figure 4.9 shows empirical semivariograms for the C/N ratio data computed at 0, 45, 90, and 135 degrees (from N) with angle tolerances of $\pm 22.5$ degrees. The angle tolerance was chosen so that segments do not overlap. For example, the first semivariogram collects all pairs in a ray between $-22.5$ and 22.5 degrees from N. The conclusion for these data based on Figure 4.9 is the same as with the semivariance surface. There does not

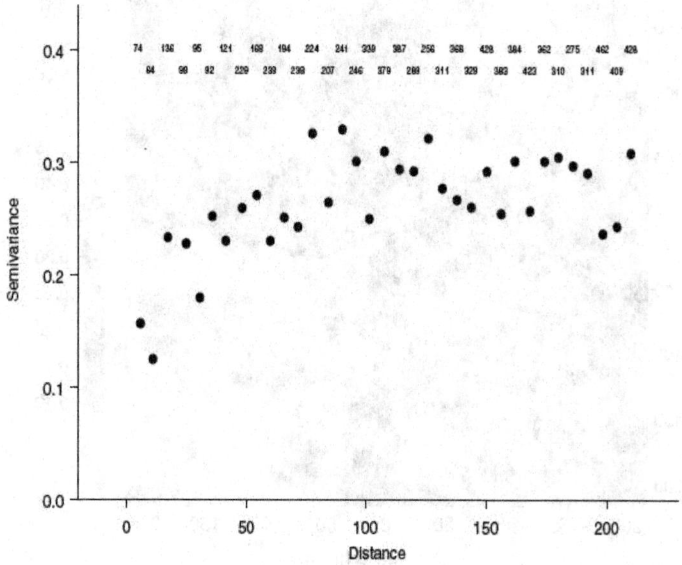


Figure 4.7 *Empirical semivariogram based on the Matheron (classical) estimator for C/N ratios. Numbers across the top denote the number of pairs within the lag class.*

appear to be direction dependence in the spatial covariance structure of the C/N ratios.

□

Considerations of heteroscedasticity arise with all semivariogram estimators, they are not unique to the Matheron estimator. Because of the simple form of the estimator's approximate variance, the concepts are highlighted most clearly for this estimator. A problem that is typical for the Matheron estimator, however, is its sensitivity to outlying observations. In developing the other estimators in this section, robustness or resistance of the estimator was an important consideration. Even a single extreme observation can negatively affect the empirical semivariogram estimate, because the squared differences $\{Z(\mathbf{s}_i) - Z(\mathbf{s}_j)\}^2$ magnify the deviation between the outlier $Z(\mathbf{s}_j)$ and other values. In addition, an observation contributes to $\widehat{\gamma}(\mathbf{h}_i)$ at several lags and outliers "spread" contamination.

**Example 4.3 Four point semivariogram.** Consider the simplified example from Schabenberger and Pierce (2002, p. 588) of a spatial data set with five locations. Let $Z([x, y])$ denote the observed value at coordinate $\mathbf{s} = [x, y]$.

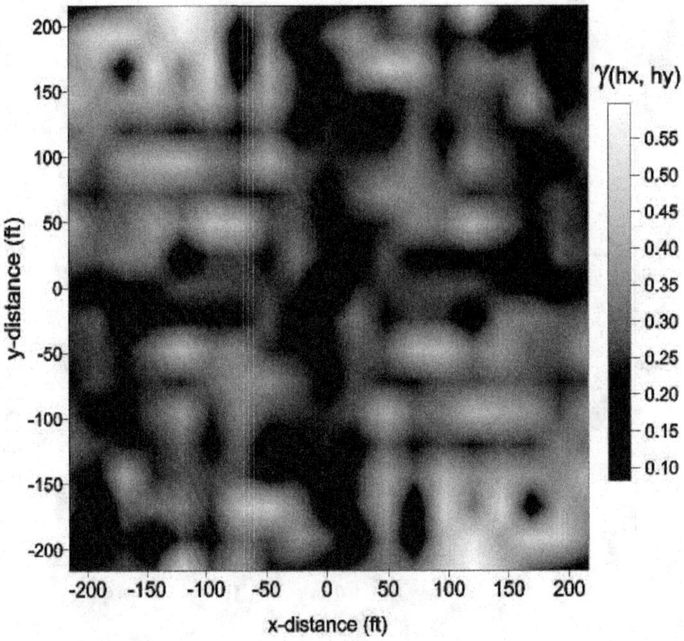


Figure 4.8 *Smoothed semivariance surface for C/N ratio data.*

The data are

$$\begin{aligned} Z([1,1]) &= 1 \qquad Z([1,4]) = 4 \\ Z([2,2]) &= 2 \qquad Z([3,1]) = 3 \\ Z([3,4]) &= 20 \end{aligned}$$

To see the impact of $Z([3,4])$ on the computation of the Matheron estimator notice that there are two pairs for each of five lag classes. The respective estimates are

$$\begin{aligned} \widehat{\gamma}(\sqrt{2}) &= \frac{1}{4}\left\{(1-2)^2 + (2-3)^2)\right\} = 1/2 \\ \widehat{\gamma}(2) &= \frac{1}{4}\left\{(1-3)^2 + (4-20)^2)\right\} = 65 \\ \widehat{\gamma}(\sqrt{5}) &= \frac{1}{4}\left\{(4-2)^2 + (20-2)^2)\right\} = 82 \\ \widehat{\gamma}(3) &= \frac{1}{4}\left\{(4-1)^2 + (20-3)^2)\right\} = 74.5 \\ \widehat{\gamma}(\sqrt{13}) &= \frac{1}{4}\left\{(3-4)^2 + (20-1)^2)\right\} = 90.5 \end{aligned}$$

The extreme observation $Z([3,4])$ contributes to estimates at four of the five lag values. In each case it produces the largest squared difference. If the

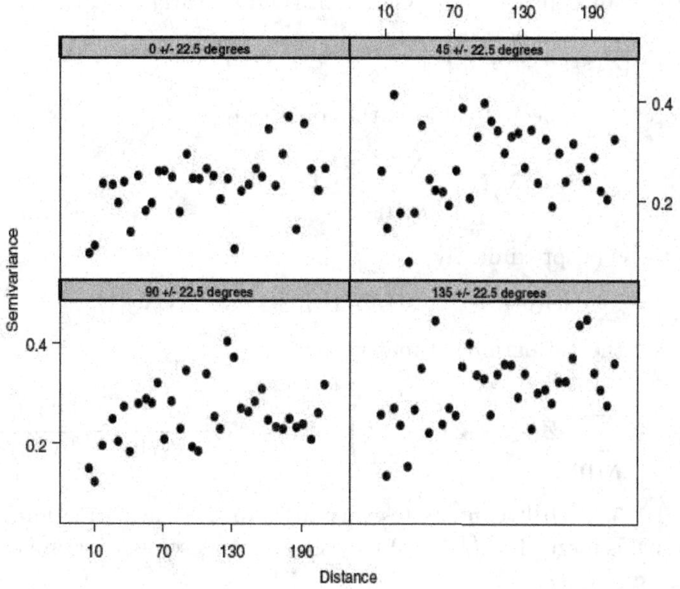


Figure 4.9 *Directional empirical semivariograms for C/N ratio.*

observation is removed from the data, the estimates are $\widehat{\gamma}(2) = 2$, $\widehat{\gamma}(\sqrt{5}) = 2$, $\widehat{\gamma}(3) = 4.5$, $\widehat{\gamma}(\sqrt{13}) = 1/2$. □

Unless a measurement has been obtained in error, removing extreme observations is not the correct course of action. This also reduces the number of pairs available at the lag classes. In order to retain observations but reduce their negative influence, one can downweigh the observation or choose a statistic that is less affected by its value.

### *4.4.2 The Cressie-Hawkins Robust Estimator*

Cressie and Hawkins (1980) suggested an estimator that alleviates the negative impact of outlying observations by eliminating squared differences from the calculation. It is often referred to as the robust semivariogram estimator; we refer to it as the Cressie-Hawkins (CH) estimator. Its genesis is as follows.

In a Gaussian random field all bivariate distributions of $[Z(\mathbf{s}_i), Z(\mathbf{s}_j)]$ are Gaussian and

$$\frac{Z(\mathbf{s}_i) - Z(\mathbf{s}_j)}{\sqrt{2\gamma(\mathbf{s}_i - \mathbf{s}_j)}} \sim G(0,1),$$

$$\frac{(Z(\mathbf{s}_i) - Z(\mathbf{s}_j))^2}{2\gamma(\mathbf{s}_i - \mathbf{s}_j)} \sim \chi^2_1.$$

Cressie and Hawkins (1980) note that the fourth root transformation of $(Z(\mathbf{s}_i)-Z(\mathbf{s}_j))^2$ yields an approximately Gaussian random variable with mean

$$\mathrm{E}\left[|Z(\mathbf{s}_i)-Z(\mathbf{s}_j)|^{1/2}\right] \approx \frac{1}{2}\pi^{-1/2}\Gamma(0.75)\times\gamma(\mathbf{s}_i-\mathbf{s}_j)^{1/4}.$$

Furthermore, the expected value of the fourth power of

$$\frac{1}{|N(\mathbf{h})|}\sum_{|N(\mathbf{h})|}|Z(\mathbf{s}_i)-Z(\mathbf{s}_j)|^{1/2}$$

turns out to be (approximately)

$$2\gamma(\mathbf{h})(0.457+0.494/|N(\mathbf{h})|+0.045/|N(\mathbf{h})|^2).$$

This suggests the variogram estimator

$$\left(\frac{1}{|N(\mathbf{h})|}\sum_{N(\mathbf{h})}|Z(\mathbf{s}_i)-Z(\mathbf{s}_j)|^{1/2}\right)^4 \Big/ \left(0.457+\frac{0.494}{|N(\mathbf{h})|}+\frac{0.045}{|N(\mathbf{h})|^2}\right).$$

The term $0.045/|N(\mathbf{h})|^2$ contributes very little to the bias correction, particularly if $|N(\mathbf{h})|$ is large. The (robust) Cressie-Hawkins semivariogram estimator is finally given by

*Cressie-Hawkins Estimator*

$$\tilde{\gamma}(\mathbf{h})=\frac{1}{2}\left\{\frac{1}{|N(\mathbf{h})|}\sum_{N(\mathbf{h})}|Z(\mathbf{s}_i)-Z(\mathbf{s}_j)|^{1/2}\right\}^4 \Big/ \left(0.457+\frac{0.494}{|N(\mathbf{h})|}\right). \quad (4.26)$$

Because the square root differences are averaged first and the resulting average is then raised to the fourth power, the first term in (4.26) is much less affected by extreme values than the average of the squared differences in the Matheron estimator. The robust estimator is not unbiased, but the term in the denominator serves to achieve approximate unbiasedness.

The attribute *robust* of the CH estimator refers to small amounts of contamination in a Gaussian process. It is under this premise that Hawkins and Cressie (1984) investigated the robustness of (4.26): white noise $\epsilon(\mathbf{s})$ was added to an intrinsically stationary process such that $\epsilon(\mathbf{s})$ is $G(0,\sigma_0^2)$ with probability $1-p$ and $G(0,k\sigma_0^2)$ with probability $p$. To be more specific, let

$$Z(\mathbf{s})=\mu+S(\mathbf{s})+\epsilon(\mathbf{s}),$$

where $S(\mathbf{s})$ is a Gaussian random field with semivariogram $\gamma_S(\mathbf{h})$. For some value $\mathbf{h}$ it is assumed that $\gamma_S(\mathbf{h}) = m\sigma_0^2$. One could thus think of $S(\mathbf{s})$ as second-order stationary with sill $m\sigma_0^2$. The particular model investigated was

$$\epsilon(\mathbf{s}) \sim \begin{array}{ll} G(0,\sigma_0^2) & \text{with probability } 0.95 \\ G(0,9\sigma_0^2) & \text{with probability } 0.05. \end{array}$$

Under this Gaussian contamination model $\widehat{\gamma}(\mathbf{h})$ is no longer an unbiased estimator of the semivariogram of the uncontaminated process (the process with $p=0$). It has positive bias, its values are too large. From Hawkins and

Cressie (1984) and Cressie (1993, p. 82, Table 2.2) it is seen that $\widetilde{\gamma}(\mathbf{h})$ is less biased than $\widehat{\gamma}(\mathbf{h})$ if the relative nugget effect is small. Similarly, if the nugget $\sigma_0^2$ is small relative to the semivariogram of the intrinsically stationary process, then the variability of $\widetilde{\gamma}(\mathbf{h})$ is less than that of the Matheron estimator. The CH estimator will typically show less variation at small lags and also result in generally smaller values than (4.24).

However, at $m = 1$ the variability of $\widehat{\gamma}(\mathbf{h})$ and $\widetilde{\gamma}(\mathbf{h})$ are approximately the same and the robust estimator is more variable for $m > 1$. In that case the contamination of the data plays a minor role compared to the stochastic variation in $S(\mathbf{s})$. As Hawkins and Cressie (1984) put it: "The loss of efficiency as $m \to \infty$ may be thought of as a premium paid by the robust estimators on normal data to insure against the effects of possible outliers."

As shown by Hawkins (1981), the $|Z(\mathbf{s}_i) - Z(\mathbf{s}_j)|^{0.5}$ are less correlated than the squared differences $(Z(\mathbf{s}_i) - Z(\mathbf{s}_j))^2$. This is a reason to prefer the CH estimator over the Matheron estimator when fitting a semivariogram model by weighted (instead of generalized) least squares to the empirical semivariogram (see §4.5).

**Example 4.3 (Four point semivariogram. Continued)** For the four lag distances in this simple example the estimates according to equation (4.26) are

$$\begin{aligned}
\widetilde{\gamma}(\sqrt{2}) &= \frac{1}{2}\left\{\frac{1}{2}\left(\sqrt{|1-2|}+\sqrt{|2-3|}\right)\right\}^4 \Big/ 0.704 = 0.71 \\
\widetilde{\gamma}(2) &= \frac{1}{2}\left\{\frac{1}{2}\left(\sqrt{|1-3|}+\sqrt{|4-20|}\right)\right\}^4 \Big/ 0.704 = 38.14 \\
\widetilde{\gamma}(\sqrt{5}) &= \frac{1}{2}\left\{\frac{1}{2}\left(\sqrt{|4-2|}+\sqrt{|20-2|}\right)\right\}^4 \Big/ 0.704 = 45.5 \\
\widetilde{\gamma}(3) &= \frac{1}{2}\left\{\frac{1}{2}\left(\sqrt{|4-1|}+\sqrt{|20-3|}\right)\right\}^4 \Big/ 0.704 = 52.2 \\
\widetilde{\gamma}(\sqrt{13}) &= \frac{1}{2}\left\{\frac{1}{2}\left(\sqrt{|3-4|}+\sqrt{|20-1|}\right)\right\}^4 \Big/ 0.704 = 36.6
\end{aligned}$$

The influence of the extreme observation $Z[(3,4])$ is clearly suppressed. □

### *4.4.3 Estimators Based on Order Statistics and Quantiles*

The robustness attribute of the CH estimator refers to small amounts of contamination in a Gaussian process. It is not a resistant estimator, because it is not stable under gross contamination of the data. Furthermore, the CH and the Matheron estimators have unbounded influence functions and a breakdown point of 0%. The influence function of an estimator measures the effect of infinitesimal contamination of the data on the statistical properties of the estimator (Hampel et al., 1986). The breakdown point is the percentage of data that can be replaced by arbitrary values without explosion (implosion) of the estimator.

The median absolute deviation (MAD), for example, is an estimator of scale with a 50% breakdown point and a smooth influence function. For a set of numbers $x_1, \cdots, x_n$, the MAD is

$$\text{MAD} = b\,\text{median}_i\{|x_i - \text{median}_j(x_j)|\}, \tag{4.27}$$

where $\text{median}_i(x_i)$ denotes the median of the $x_i$. The factor $b$ is chosen to yield approximate unbiasedness and consistency. If $x_1, \cdots, x_n$ are independent realizations from a $G(\mu, \sigma^2)$, for example, the MAD will be consistent for $\sigma$ for $b = 1.4826$.

Rousseeuw and Croux (1993) suggested a robust estimator of scale which also has a 50% breakdown point but a smooth influence function. Their $Q_n$ estimator is given by the $k$th order statistic of the $n(n-1)/2$ inter-point distances. Let $h = \lfloor n/2 \rfloor + 1$ and $k = \binom{h}{2}$. Then,

$$Q_n = c\{|x_i - x_j|; i < j\}_{(k)}. \tag{4.28}$$

For Gaussian data, the multiplicative factor that gives consistency for the standard deviation is $c = 2.2191$. The $Q_n$ estimator has positive small-sample bias (see Table 1 for $n \leq 40$ in their paper) which can be corrected (Croux and Rousseeuw, 1992).

Genton (1998a, 2001) considers the modification that leads from (4.24) to (4.26) not sufficient to impart robustness and develops a robust estimator of the semivariogram based on $Q_n$. If spatial data $Z(\mathbf{s}_1), \cdots, Z(\mathbf{s}_n)$ are observed, let $N(\mathbf{h})$ denote the set of pairwise differences $T_i = Z(\mathbf{s}_i) - Z(\mathbf{s}_i + \mathbf{h})$, $i = 1, \cdots, n(n-1)/2$. Next, calculate $Q_{|N(\mathbf{h})|}$ for the $T_i$ and return as the semivariogram estimator at lag $\mathbf{h}$

*Genton Estimator*

$$\overline{\gamma}(\mathbf{h}) = \frac{1}{2} Q^2_{|N(\mathbf{h})|}. \tag{4.29}$$

Since $Q_n$ has a 50% breakdown point, $\overline{\gamma}(\mathbf{h})$ has a 50% breakdown point in terms of the process of differences $T_i$, but not necessarily in terms of the $Z(\mathbf{s}_i)$. Genton (2001) establishes through simulation that (4.29) will be resistant to roughly 30% of outliers among the $Z(\mathbf{s}_i)$.

Another approach of "robustifying" the empirical semivariogram estimator is to consider quantiles of the distribution of $\{Z(\mathbf{s}_i) - Z(\mathbf{s}_j)\}^2$ or $|Z(\mathbf{s}_i) - Z(\mathbf{s}_j)|$ instead of arithmetic averages (as in (4.24) and (4.26)). If $[Z(\mathbf{s}), Z(\mathbf{s}+\mathbf{h})]'$ are bivariate Gaussian with common mean, then

$$\begin{aligned}
\frac{1}{2}\{Z(\mathbf{s}) - Z(\mathbf{s}+\mathbf{h})\}^2 &\sim \gamma(\mathbf{h})\chi^2_1 \\
\frac{1}{2}|Z(\mathbf{s}) - Z(\mathbf{s}+\mathbf{h})| &\sim \sqrt{\frac{1}{2}\gamma(\mathbf{h})}|U| \qquad U \sim G(0,1).
\end{aligned}$$

Let $q^{(p)}_{|N(\mathbf{h})|}$ denote the $p$th quantile. Then

$$\widehat{\gamma}_p(\mathbf{h}) = q^{(p)}_{|N(\mathbf{h})|}\left\{\frac{1}{2}[Z(\mathbf{s}) - Z(\mathbf{s}+\mathbf{h})]^2\right\}$$

estimates $\gamma(\mathbf{h}) \times \chi^2_{p,1}$. A median-based estimator ($p = 0.5$) would be

$$\begin{aligned}\widehat{\gamma}_p(\mathbf{h}) &= \frac{1}{2}\,\text{median}_{|N(\mathbf{h})|}\{[Z(\mathbf{s}) - Z(\mathbf{s}+\mathbf{h})]^2\}/0.455 \\ &= \frac{1}{2}\left(\text{median}_{|N(\mathbf{h})|}\{|Z(\mathbf{s}) - Z(\mathbf{s}+\mathbf{h})|^{1/2}\}\right)^4/0.455.\end{aligned}$$

The latter expression is (2.4.13) in Cressie (1993, p. 75).

## 4.5 Parametric Modeling

The empirical semivariogram $\widehat{\gamma}(\mathbf{h})$ is an unbiased estimator of $\gamma(\mathbf{h})$, but it provides estimates only at a finite set of lags or lag classes. In order to obtain estimates of $\gamma(\mathbf{h})$ at any arbitrary lag the empirical semivariogram must be smoothed. A nonparametric kernel smoother will not suffice since it is not guaranteed that the resulting fit is a conditionally negative-definite function. The common approach is to fit one of the parametric semivariogram models of §4.3 or to apply the "nonparametric" semivariogram representation (§4.6). Although fitting a parametric semivariogram model to the empirical semivariogram by a least squares method is by far the most common approach, it is not the only parametric technique.

Modeling techniques that fit a parametric model to the observed data $Z(\mathbf{s}_1), \cdots, Z(\mathbf{s}_n)$ are distinguished from those approaches that fit a model to pseudo-data. In a pseudo-data approach the response being modeled is derived from $Z(\mathbf{s}_1), \cdots, Z(\mathbf{s}_n)$ and the construction of the pseudo-data often involves subjective choices; for example, the semivariogram cloud consists of pseudo-data $T_{ij} = Z(\mathbf{s}_i) - Z(\mathbf{s}_j)$. The Matheron and the Cressie-Hawkins estimators

$$\widehat{\gamma}(\mathbf{h}) = \text{average}(T^2_{ij}) \qquad \widetilde{\gamma}(\mathbf{h}) = (\text{average}(|T_{ij}|^{1/2})^4,$$

are functions of the semivariogram cloud values that depend on the number and width of lag classes, the maximum lag for which the empirical semivariogram is calculated, the minimum number of pairs per lag class, and so forth. Although the subjectivity inherent in the empirical semivariogram estimators is allayed if the $T_{ij}$ are not averaged, the user must decide whether to model $(Z(\mathbf{s}_i) - Z(\mathbf{s}_j))^2$, $Z(\mathbf{s}_i)Z(\mathbf{s}_j)$, or some other form of pseudo-response.

The least squares methods fit a semivariogram model to $\widehat{\gamma}(\mathbf{h})$ or $\widetilde{\gamma}(\mathbf{h})$. Maximum likelihood (ML) and restricted (residual) maximum likelihood (REML) estimation use the observed data directly, usually assuming a Gaussian random field. Other estimating-function-based methods such as generalized estimating equations (GEE) and composite likelihood (CL) also utilize pseudo-data. No single method can claim uniform superiority. In the following subsections we discuss the various approaches and their respective merits and demerits. To distinguish the empirical semivariogram $\gamma(\mathbf{h})$ and its estimate $\widehat{\gamma}(\mathbf{h})$ from the semivariogram model being fit, we introduce the notation $\gamma(\mathbf{h}, \boldsymbol{\theta})$ for the latter. The vector $\boldsymbol{\theta}$ contains all unknown parameters to be estimated from the data. The model may be a single, isotropic semivariogram function as in §4.3.2–4.3.5, a model with nugget effect, an anisotropic, or a nested model.

### 4.5.1 Least Squares and the Semivariogram

The geometric least squares principle enables us to estimate the parameters in a model describing the mean of a random vector, taking into account the variation and covariation of the vector elements. To apply least squares estimation to semivariogram modeling, the mean of the "response" being modeled must be (a function of) the semivariogram. Hence, the empirical semivariogram estimators of §4.4 serve as the data for this process. Consider an empirical semivariogram estimator at $k$ lags. For example, a semivariogram model $\gamma(\mathbf{h}, \boldsymbol{\theta})$ can be fit to the pseudo-data

$$\widehat{\gamma}(\mathbf{h}) = [\widehat{\gamma}(\mathbf{h}_1), \cdots, \widehat{\gamma}(\mathbf{h}_k)]',$$

or

$$\widetilde{\gamma}(\mathbf{h}) = [\widetilde{\gamma}(\mathbf{h}_1), \cdots, \widetilde{\gamma}(\mathbf{h}_k)]',$$

or another empirical estimator. We concentrate in this section on the Matheron estimator. The necessary steps in the derivation can be repeated for the other estimators.

Least squares methods do not make distributional assumptions about $\widehat{\gamma}(\mathbf{h})$ apart from the first two moments. They consider a statistical model of the form

$$\widehat{\gamma}(\mathbf{h}) = \gamma(\mathbf{h}, \boldsymbol{\theta}) + \mathbf{e}(\mathbf{h}), \tag{4.30}$$

where $\gamma(\mathbf{h}, \boldsymbol{\theta}) = [\gamma(\mathbf{h}_1, \boldsymbol{\theta}), \cdots, \gamma(\mathbf{h}_k, \boldsymbol{\theta})]'$. It is assumed that the $(k \times 1)$ vector of errors in this model has mean $\mathbf{0}$. The variance-covariance matrix of the errors, $\text{Var}[\mathbf{e}(\mathbf{h})] = \mathbf{R}$, typically depends on $\boldsymbol{\theta}$ also. We shall write $\mathbf{R}(\boldsymbol{\theta})$ if it is necessary to make this dependence explicit. The appropriate course of action is then to minimize the generalized sum of squares

$$(\widehat{\gamma}(\mathbf{h}) - \gamma(\mathbf{h}, \boldsymbol{\theta}))' \, \mathbf{R}(\boldsymbol{\theta})^{-1} \, (\widehat{\gamma}(\mathbf{h}) - \gamma(\mathbf{h}, \boldsymbol{\theta})) \, . \tag{4.31}$$

If $\mathbf{R}$ does not depend on $\boldsymbol{\theta}$, this is a standard nonlinear generalized least squares problem; it is solved iteratively. Otherwise, an iterative re-weighting scheme is employed since updates to $\widehat{\boldsymbol{\theta}}$ should be followed by updates to $\mathbf{R}(\widehat{\boldsymbol{\theta}})$. The difficulty of minimizing the generalized sum of squares does not lie with the presence of a weight matrix. It lies in obtaining $\mathbf{R}$. Following Cressie (1985, 1993), the basic ingredients are derived as follows.

To shorten notation let $T_{ij} = Z(\mathbf{s}_i) - Z(\mathbf{s}_j)$, $\mathbf{h}_{ij} = \mathbf{s}_i - \mathbf{s}_j$, and assume that $T_{ij} \sim G(0, 2\gamma(\mathbf{h}_{ij}, \boldsymbol{\theta}))$. Hence, $\text{E}[T_{ij}^2] = 2\gamma(\mathbf{h}_{ij}, \boldsymbol{\theta})$ and $\text{Var}[T_{ij}^2] = 8\gamma(\mathbf{h}_{ij}, \boldsymbol{\theta})^2$. To find $\text{Cov}[T_{ij}^2, T_{kl}^2]$, it is helpful to rely on the following result (Chapter problem 4.3): if $X \sim G(0,1)$ and $Y \sim G(0,1)$ with $\mathrm{Corr}[X,Y] = \rho$, then $\mathrm{Corr}[X^2,Y^2] = \rho^2$. Hence,

$$\mathrm{Cov}[T_{ij}^2, T_{kl}^2] = \sqrt{\mathrm{Var}[T_{ij}^2]\mathrm{Var}[T_{kl}^2]}\mathrm{Corr}[T_{ij}, T_{kl}]^2.$$

Since

$$\begin{aligned}\mathrm{Cov}[T_{ij}, T_{kl}] &= \mathrm{E}[T_{ij}T_{kl}] \\ &= \mathrm{E}[Z(\mathbf{s}_i)Z(\mathbf{s}_k) - Z(\mathbf{s}_i)Z(\mathbf{s}_l) - Z(\mathbf{s}_j)Z(\mathbf{s}_k) + Z(\mathbf{s}_j)Z(\mathbf{s}_l)]\end{aligned}$$

and $\mathrm{E}[Z(\mathbf{s}_i)Z(\mathbf{s}_j)] = C(\mathbf{0}) - \gamma(\mathbf{h}_{ij}, \boldsymbol{\theta}) + \mu^2$, we have

$$\mathrm{Corr}[T_{ij}, T_{kl}]^2 = \frac{\{\gamma(\mathbf{h}_{ij}, \boldsymbol{\theta}) + \gamma(\mathbf{h}_{jk}, \boldsymbol{\theta}) - \gamma(\mathbf{h}_{jl}, \boldsymbol{\theta}) - \gamma(\mathbf{h}_{ik}, \boldsymbol{\theta})\}^2}{4\gamma(\mathbf{h}_{ij})\gamma(\mathbf{h}_{kl})},$$

which is (2.6.10) in Cressie (1993, p. 96). Finally,

$$\mathrm{Cov}[T_{ij}^2, T_{kl}^2] = 2\left\{\gamma(\mathbf{h}_{ij}, \boldsymbol{\theta}) + \gamma(\mathbf{h}_{jk}, \boldsymbol{\theta}) - \gamma(\mathbf{h}_{jl}, \boldsymbol{\theta}) - \gamma(\mathbf{h}_{ik}, \boldsymbol{\theta})\right\}^2. \quad (4.32)$$

If $i = k$ and $j = l$, (4.32) reduces to $8\gamma(\mathbf{h}_{ij}, \boldsymbol{\theta})^2$, of course. The variance of the Matheron estimator at lag $\mathbf{h}_m$ is now obtained as

$$\mathrm{Var}[2\widehat{\gamma}(\mathbf{h}_m)] = \frac{1}{|N(\mathbf{h}_m)|^2}\mathrm{Var}\left[\sum_{N(\mathbf{h}_m)} T_{ij}^2\right] = \frac{1}{|N(\mathbf{h}_m)|^2}\sum_{i,j}\sum_{k,l}\mathrm{Cov}[T_{ij}^2, T_{kl}^2].$$

Cressie (1985) suggests approximating the diagonal entries of $\mathbf{R}(\boldsymbol{\theta})$ as

$$\mathrm{Var}[\widehat{\gamma}(\mathbf{h}_m)] \approx 2\frac{\gamma(\mathbf{h}_m, \boldsymbol{\theta})^2}{|N(\mathbf{h}_m)|}. \quad (4.33)$$

This is the appropriate variance formula if the $T_{ij}^2$ are uncorrelated, and if the Gaussian assumption holds. The weighted least squares (WLS) approach to semivariogram fitting replaces $\mathbf{R}(\boldsymbol{\theta})$ by the diagonal matrix $\mathbf{W}(\boldsymbol{\theta})$ whose entries are given by (4.33). Instead of the generalized sum of squares (4.31), this approach minimizes the weighted sum of squares

$$\begin{aligned}&(\widehat{\gamma}(\mathbf{h}) - \gamma(\mathbf{h}, \boldsymbol{\theta}))'\,\mathbf{W}(\boldsymbol{\theta})^{-1}\,(\widehat{\gamma}(\mathbf{h}) - \gamma(\mathbf{h}, \boldsymbol{\theta})) \\ &= \sum_{m=1}^{k}\frac{|N(\mathbf{h}_m)|}{2\gamma(\mathbf{h}_m, \boldsymbol{\theta})^2}\{\widehat{\gamma}(\mathbf{h}_m) - \gamma(\mathbf{h}_m, \boldsymbol{\theta})\}^2. \quad (4.34)\end{aligned}$$

Because the off-diagonal entries of $\mathbf{R}(\boldsymbol{\theta})$ are appreciable, the WLS criterion is a poor approximation of (4.31). Since (4.34) can be written as a weighted sum of squares over the $k$ lag classes, it is a simple matter to fit a semivariogram model with a nonlinear statistics package, provided it can accommodate weights. A further "simplification" is possible if one assumes that $\mathbf{R} = \phi\mathbf{I}$. This ordinary least squares (OLS) approach ignores the correlation and the unequal dispersion among the $\widehat{\gamma}(\mathbf{h}_m)$. Zimmerman and Zimmerman (1991) found that the ordinary least squares and weighted least squares estimators of the semivariogram performed more or less equally well. One does not lose much by assuming that the $\widehat{\gamma}(\mathbf{h}_m)$ have equal variance. The greatest loss of efficiency is not incurred by employing OLS over WLS, but by not incorporating the correlations among the $\widehat{\gamma}(\mathbf{h}_m)$.

The covariance and correlation structure of $2\widehat{\gamma}(\mathbf{h})$ has been studied by Genton (1998b) under the assumption that $Z(\mathbf{s})$ is Gaussian and by Genton (2000) for elliptically contoured distributions (see also Genton, He, and Liu, 2001). The derivations rest on writing the Matheron estimator as

$$2\widehat{\gamma}(\mathbf{h}) = Z(\mathbf{s})'\mathbf{A}(\mathbf{h})Z(\mathbf{s}),$$

where $\mathbf{A}(\mathbf{h})$ is a spatial design matrix of the data at lag $\mathbf{h}$. Applying known results for quadratic forms in Gaussian random variables, $Z(\mathbf{s}) \sim G(\boldsymbol{\mu}, \boldsymbol{\Sigma}(\boldsymbol{\theta}))$, yields

$$\begin{aligned}
\mathrm{E}[2\widehat{\gamma}(\mathbf{h})] &= \mathrm{tr}[\mathbf{A}(\mathbf{h})\boldsymbol{\Sigma}(\boldsymbol{\theta})] \\
\mathrm{Var}[2\widehat{\gamma}(\mathbf{h})] &= 2\mathrm{tr}[\mathbf{A}(\mathbf{h})\boldsymbol{\Sigma}(\boldsymbol{\theta})\mathbf{A}(\mathbf{h})\boldsymbol{\Sigma}(\boldsymbol{\theta})] \\
\mathrm{Cov}[2\widehat{\gamma}(\mathbf{h}_i), 2\widehat{\gamma}(\mathbf{h}_j)] &= 2\mathrm{tr}[\mathbf{A}(\mathbf{h}_i)\boldsymbol{\Sigma}(\boldsymbol{\theta})\mathbf{A}(\mathbf{h}_j)\boldsymbol{\Sigma}(\boldsymbol{\theta})],
\end{aligned}$$

where tr is the trace operator.

As is the case in (4.33), these expressions depend on the unknown parameters. Genton (1998b) assumes that the data are only "slightly correlated" and puts $\boldsymbol{\Sigma}(\boldsymbol{\theta}) \propto \mathbf{I}$. It seems rather strange to assume that the data are uncorrelated in order to model the parameters of the data dependence. Genton (2000) shows that if the distribution of the data is elliptically contoured, $\boldsymbol{\Sigma} = \phi\mathbf{I} + \mathbf{1a}' + \mathbf{a1}'$, $\phi \in \mathbb{R}$, and $\boldsymbol{\Sigma}$ is positive definite, the correlation structure of the Matheron estimator is

$$\mathrm{Corr}[2\widehat{\gamma}(\mathbf{h}_i), 2\widehat{\gamma}(\mathbf{h}_j)] = \frac{\mathrm{tr}[\mathbf{A}(\mathbf{h}_i)\mathbf{A}(\mathbf{h}_j)]}{\mathrm{tr}[\mathbf{A}^2(\mathbf{h}_i)]\mathrm{tr}[\mathbf{A}^2(\mathbf{h}_j)]}.$$

### *4.5.2 Maximum and Restricted Maximum Likelihood*

Estimating the parameters of a spatial random field by likelihood methods requires that the spatial distribution (§2.2) be known and is only developed for the case of the Gaussian random field. We consider here the case of a constant mean, $\mathrm{E}[Z(\mathbf{s})] = \mu$, congruent with a second-order or intrinsic stationarity assumption. Likelihood estimation does not impose this restriction, however, and we will relax the constant mean assumption in §5.5.2 in the context of spatial prediction with spatially dependent mean function and unknown covariance function. In the meantime, let $\mathbf{Z} = [Z(\mathbf{s}_1), \cdots, Z(\mathbf{s}_n)]'$ denote the vector of observations and assume $\mathbf{Z}(\mathbf{s}) \sim G(\mu\mathbf{1}, \boldsymbol{\Sigma}(\boldsymbol{\theta}))$. The variance-covariance matrix of $\mathbf{Z}(\mathbf{s})$ has been parameterized so that for any estimate $\widehat{\boldsymbol{\theta}}$ the variances and covariances can be estimated as $\widehat{\mathrm{Var}}[\mathbf{Z}(\mathbf{s})] = \boldsymbol{\Sigma}(\widehat{\boldsymbol{\theta}})$. The negative of twice the Gaussian log likelihood is

*Minus 2 log Likelihood*

$$\begin{aligned}
\varphi(\mu; \boldsymbol{\theta}; \mathbf{Z}(\mathbf{s})) = {} & \ln\{|\boldsymbol{\Sigma}(\boldsymbol{\theta})|\} + n\ln\{2\pi\} \\
& + (\mathbf{Z}(\mathbf{s}) - \mathbf{1}\mu)'\boldsymbol{\Sigma}(\boldsymbol{\theta})^{-1}(\mathbf{Z}(\mathbf{s}) - \mathbf{1}\mu), \qquad (4.35)
\end{aligned}$$

and is minimized with respect to $\mu$ and $\boldsymbol{\theta}$. If $\boldsymbol{\theta}$ is known, the minimum of (4.35) can be expressed in closed form,

$$\widetilde{\mu} = \left(\mathbf{1}'\boldsymbol{\Sigma}(\boldsymbol{\theta})^{-1}\mathbf{1}\right)^{-1}\mathbf{1}'\boldsymbol{\Sigma}(\boldsymbol{\theta})^{-1}\mathbf{Z}(\mathbf{s}), \tag{4.36}$$

the generalized least squares estimator.

**Example 4.4** Notice that the first term in (4.36) is a scalar, the inverse of the sum of the elements of the inverse variance-covariance matrix. In the special case where $\boldsymbol{\Sigma}(\boldsymbol{\theta}) = \theta\mathbf{I}$, we obtain $\mathbf{1}'\boldsymbol{\Sigma}(\boldsymbol{\theta})^{-1}\mathbf{1} = n/\theta$ and the generalized least squares estimator is simply the sample mean,

$$\widetilde{\mu} = \frac{\theta}{n}\mathbf{1}'\boldsymbol{\Sigma}(\boldsymbol{\theta})^{-1}\mathbf{Z}(\mathbf{s}) = \frac{\theta}{n}\frac{1}{\theta}\sum_{i=1}^{n} Z(\mathbf{s}_i) = \overline{Z}.$$

□

Substituting (4.36) into (4.35) yields a (negative) log likelihood function that depends on $\boldsymbol{\theta}$ only. We say that $\mu$ has been profiled from the objective function and term the resulting function as (twice the negative) profile log-likelihood. This function is then minimized, typically by numerical, iterative methods because the log likelihood is usually a nonlinear function of the covariance parameters. Once the maximum likelihood estimates $\widehat{\boldsymbol{\theta}}_{ml}$ have been obtained, their value is substituted into the expressions for the profiled parameters to obtain their likelihood estimates. The maximum likelihood estimate (MLE) of $\mu$ is simply

$$\widehat{\mu}_{ml} = \left(\mathbf{1}'\boldsymbol{\Sigma}(\widehat{\boldsymbol{\theta}}_{ml})^{-1}\mathbf{1}\right)^{-1}\mathbf{1}'\boldsymbol{\Sigma}(\widehat{\boldsymbol{\theta}}_{ml})^{-1}\mathbf{Z}(\mathbf{s}). \tag{4.37}$$

This is known as an estimated generalized least squares estimator (EGLSE). The case of independent and homoscedastic observations again surfaces immediately as a special case. Then we can write $\boldsymbol{\Sigma}(\boldsymbol{\theta}) = \theta\mathbf{I}$ and the EGLS estimator does not depend on $\widehat{\theta}$. Furthermore, there is a closed-form expression for $\theta$ in this case too,

$$\widehat{\theta}_{ml} = \frac{1}{n}\sum_{i=1}^{n}\left(Z(\mathbf{s}_i) - \overline{Z}\right)^2. \tag{4.38}$$

Maximum likelihood estimators have appealing statistical properties. For example, under standard regularity conditions they are asymptotically Gaussian and efficient. In finite samples they are often biased, however, and the bias of covariance and variance parameters is typically negative. To illustrate, consider again the case of independent observations with equal variance $\theta$ and $\mu$ unknown. The MLE of $\theta$ is given by (4.38) and has bias $\mathrm{E}[\widehat{\theta}_{ml} - \theta] = -\theta/n$. If $\mu$ were known, the MLE would be $\widehat{\theta}_{ml} = n^{-1}\sum_{i=1}^{n}(Z(\mathbf{s}_i) - \mu)^2$, which is an unbiased estimator of $\theta$. The bias of MLEs for variances and covariances is due to the fact that the method makes no allowance for the loss of degrees of freedom that is incurred when mean parameters are estimated. Restricted maximum likelihood (also known as residual maximum likelihood; REML) estimation mitigates the bias in MLEs. In certain balanced situations, the bias is removed completely (Patterson and Thompson, 1971).

The idea of REML estimation is to estimate variance and covariance parameters by maximizing the likelihood of $\mathbf{KZ}(\mathbf{s})$ instead of maximizing the likelihood of $\mathbf{Z}(\mathbf{s})$, where the matrix $\mathbf{K}$ is chosen so that $\mathrm{E}[\mathbf{KZ}(\mathbf{s})] = \mathbf{0}$. Because of this property, $\mathbf{K}$ is called a matrix of error contrasts and its function to "take out the mean" explains the name *residual* maximum likelihood.

REML estimation is developed only for the case of a linear mean function, otherwise it is not clear how to construct the matrix of error contrasts $\mathbf{K}$. The more general case of a linear regression structure is discussed in §5.5.3. The matrix $\mathbf{K}$ is not unique, but fortunately, this does not matter for parameter estimation and inference (Harville, 1974). In the case considered in this chapter, namely $\mathrm{E}[Z(\mathbf{s})] = \mu$, a simple choice is the $(n-1) \times n$ matrix

$$\mathbf{K} = \begin{bmatrix} 1-\frac{1}{n} & -\frac{1}{n} & \cdots & -\frac{1}{n} \\ -\frac{1}{n} & 1-\frac{1}{n} & \cdots & -\frac{1}{n} \\ \vdots & \vdots & \ddots & \vdots \\ -\frac{1}{n} & -\frac{1}{n} & \cdots 1-\frac{1}{n} & \end{bmatrix}.$$

Then $\mathbf{KZ}(\mathbf{s})$ is the $(n-1) \times 1$ vector of differences from the sample mean. Notice that differences are taken for all but one observation, this is consistent with the fact that the estimation of $\mu$ by $\overline{Z}$ amounts to the loss of a single degree of freedom.

*Minus 2 Restricted log Likelihood*

Now, $\mathbf{KZ}(\mathbf{s}) \sim G(\mathbf{0}, \mathbf{K}\boldsymbol{\Sigma}(\boldsymbol{\theta})\mathbf{K}')$ and the restricted maximum likelihood estimates are those values of $\boldsymbol{\theta}$ that minimize

$$\begin{aligned} \varphi_R(\boldsymbol{\theta}; \mathbf{KZ}(\mathbf{s})) &= \ln\{|\mathbf{K}\boldsymbol{\Sigma}(\boldsymbol{\theta})\mathbf{K}'|\} + (n-1)\ln\{2\pi\} + \\ & \quad \mathbf{Z}(\mathbf{s})'\mathbf{K}'\left(\mathbf{K}\boldsymbol{\Sigma}(\boldsymbol{\theta})\mathbf{K}'\right)^{-1}\mathbf{KZ}(\mathbf{s}). \end{aligned} \quad (4.39)$$

Akin to the profiled maximum likelihood, (4.39) does not contain information about the mean $\mu$. In ML estimation, $\boldsymbol{\mu}$ was profiled out of the log likelihood, in REML estimation the likelihood being maximized is that of a different set of data, $\mathbf{KZ}(\mathbf{s})$ instead of $\mathbf{Z}(\mathbf{s})$. Hence, there is no restricted maximum likelihood estimator of $\mu$. Instead, the estimator obtained by evaluating (4.36) at the REML estimates $\widehat{\boldsymbol{\theta}}_{reml}$ is an estimated generalized least squares estimator:

$$\widehat{\mu}_{reml} = \left(\mathbf{1}'\boldsymbol{\Sigma}(\widehat{\boldsymbol{\theta}}_{reml})^{-1}\mathbf{1}\right)^{-1} \mathbf{1}'\boldsymbol{\Sigma}(\widehat{\boldsymbol{\theta}}_{reml})^{-1}\mathbf{Z}(\mathbf{s}). \quad (4.40)$$

This distinction between ML and REML estimation is important, because the log likelihood (4.35) can be used to test hypotheses about $\mu$ and $\boldsymbol{\theta}$ with a likelihood-ratio test. Likelihood ratio comparisons based on (4.39) are meaningful only when they relate to the covariance parameters in $\boldsymbol{\theta}$ and the models have the same mean structure. This issue gains importance when the mean is modeled with a more general regression structure, for example, $\mu = \mathbf{x}'(\mathbf{s})\boldsymbol{\beta}$, see §5.3.3 and Chapter 6.

### *4.5.3 Composite Likelihood and Generalized Estimating Equations*

#### *4.5.3.1 Generalized Estimating Equations*

The idea of using generalized estimating equations (GEE) for the estimation of parameters in statistical models was made popular by Liang and Zeger (1986) and Zeger and Liang (1986) in the context of longitudinal data analysis. The technique is an application of estimating function theory and quasi-likelihood. Let $\mathbf{T}$ denote a random vector whose mean depends on some parameter vector $\boldsymbol{\theta}$, $\mathrm{E}[\mathbf{T}] = f(\boldsymbol{\theta})$. Furthermore, denote as $\mathbf{D}$ the matrix of first derivatives of the mean function with respect to the elements of $\boldsymbol{\theta}$. If $\mathrm{Var}[\mathbf{T}] = \boldsymbol{\Sigma}$, then

$$U(\boldsymbol{\theta}; \mathbf{T}) = \mathbf{D}'\boldsymbol{\Sigma}^{-1}(\mathbf{T} - f(\boldsymbol{\theta})),$$

is an unbiased estimating function for $\boldsymbol{\theta}$ in the sense that $\mathrm{E}[U(\boldsymbol{\theta}; \mathbf{T})] = \mathbf{0}$ and an estimate $\widehat{\boldsymbol{\theta}}$ can be obtained by solving $U(\boldsymbol{\theta}; \mathbf{t}) = \mathbf{0}$ (Heyde, 1997). The optimal estimating function in the sense of Godambe (1960) is the (likelihood) score function. In estimating problems where the score is unaccessible or intractable—as is often the case when data are correlated—$U(\boldsymbol{\theta}; \mathbf{T})$ nevertheless implies a consistent estimator of $\boldsymbol{\theta}$. The efficiency of this estimator increases with *closeness* of $U$ to the score function. For correlated data, where $\boldsymbol{\Sigma}$ is unknown or contains unknown parameters, Liang and Zeger (1986) and Zeger and Liang (1986) proposed to substitute a "working" variance-covariance matrix $\mathbf{W}(\boldsymbol{\alpha})$ for $\boldsymbol{\Sigma}$ and to solve instead the estimating equation

$$U^*(\boldsymbol{\theta}; \mathbf{T}) = \mathbf{D}'\mathbf{W}(\boldsymbol{\alpha})^{-1}(\mathbf{T} - f(\boldsymbol{\theta})) \equiv \mathbf{0}.$$

If the parameter vector $\boldsymbol{\alpha}$ can be estimated and $\widehat{\boldsymbol{\alpha}}$ is a consistent estimator, then for any particular value of $\widehat{\boldsymbol{\alpha}}$

$$U_{gee}(\boldsymbol{\theta}; \mathbf{T}) = \mathbf{D}'\mathbf{W}(\widehat{\boldsymbol{\alpha}})^{-1}(\mathbf{T} - f(\boldsymbol{\theta})) \equiv \mathbf{0} \qquad (4.41)$$

is an unbiased estimating equation. The root is a consistent estimator of $\boldsymbol{\theta}$, provided that $\mathbf{W}(\widehat{\boldsymbol{\alpha}})$ satisfies certain properties; for example, if $\mathbf{W}$ is block-diagonal, or has specific mixing properties (see Fuller and Battese, 1973; Zeger, 1988).

Initially, the GEE methodology was applied to the estimation of parameters that model the mean of the observed responses. Later, it was extended to the estimation of association parameters, variances, and covariances (Prentice, 1988; Zhao and Prentice, 1990). This process commences with the construction of a vector of pseudo-data. For example, let $T_{ij} = (Y_i - \mu_i)(Y_j - \mu_j)$, then $\mathrm{E}[T_{ij}] = \mathrm{Cov}[Y_i, Y_j]$ and after parameterizing the covariances, the GEE methodology can be applied. Now assume that the data comprise the incomplete sampling of a geostatistical process, $\mathbf{Z}(\mathbf{s}) = [Z(\mathbf{s}_1), \cdots, Z(\mathbf{s}_n)]'$ and consider the pseudo-data

$$\begin{aligned} T_{ij}^{(1)} &= (Z(\mathbf{s}_i) - \mu(\mathbf{s}_i))(Z(\mathbf{s}_j) - \mu(\mathbf{s}_j)) \\ T_{ij}^{(2)} &= Z(\mathbf{s}_i) - Z(\mathbf{s}_j) \end{aligned}$$

$$T_{ij}^{(3)} = (Z(\mathbf{s}_i) - Z(\mathbf{s}_j))^2.$$

In a stationary process $\mathrm{E}[T_{ij}^{(1)}] = C(\mathbf{s}_i - \mathbf{s}_j)$, $\mathrm{Var}[T_{ij}^{(2)}] = \mathrm{E}[T_{ij}^{(3)}] = 2\gamma(\mathbf{s}_i - \mathbf{s}_j)$. Focus on the $n(n-1)/2$ squared differences $T_{ij}^{(3)}$ of the unique pairs for the moment and let $\mathbf{h}_{ij} = \mathbf{s}_i - \mathbf{s}_j$. Parameterizing the mean as $\mathrm{E}[T_{ij}^{(3)}] = 2\gamma(\mathbf{h}_{ij}, \boldsymbol{\theta})$, a generalized estimating equation for $\boldsymbol{\theta}$ is

$$U_{gee}(\boldsymbol{\theta}; \mathbf{T}^{(3)}) = 2\frac{\partial \gamma(\mathbf{h}, \boldsymbol{\theta})}{\partial \boldsymbol{\theta}'}\mathbf{W}^{-1}(\mathbf{T}^{(3)} - 2\gamma(\mathbf{h}, \boldsymbol{\theta})) \equiv \mathbf{0}, \tag{4.42}$$

where the $\mathbf{h}_{ij}$ were collected into vector $\mathbf{h}$ and

$$\gamma(\mathbf{h}, \boldsymbol{\theta}) = [\gamma(\mathbf{h}_{11}, \boldsymbol{\theta}), \cdots, \gamma(\mathbf{h}_{n-1,n}, \boldsymbol{\theta})]'.$$

In cases where $\mathbf{T}$ is pseudo-data and $\boldsymbol{\theta}$ models covariances or correlations, it is common to consider the identity matrix as the working structure $\mathbf{W}$ (Prentice, 1988; Zhao and Prentice, 1990; McShane, Albert, and Palmatier, 1997). Otherwise, third and fourth moments of the distribution of $Z(\mathbf{s})$ are necessary. For this choice of a working variance-covariance matrix, the estimation problem has the structure of a nonlinear least squares problem. The generalized estimating equations can be written as

*GEE "Score"*

$$U_{gee}(\boldsymbol{\theta}; \mathbf{T}^{(3)}) = 2\sum_{i=1}^{n-1}\sum_{j=i+1}^{n} \frac{\partial \gamma(\mathbf{h}_{ij}, \boldsymbol{\theta})}{\partial \boldsymbol{\theta}'}\left(T_{ij}^{(3)} - 2\gamma(\mathbf{h}_{ij}, \boldsymbol{\theta})\right) \equiv \mathbf{0}. \tag{4.43}$$

GEE estimates can thus be calculated as the ordinary (nonlinear) least squares estimates in the model

$$T_{ij}^{(3)} = 2\gamma(\mathbf{h}_{ij}, \boldsymbol{\theta}) + \delta_{ij} \qquad \delta_{ij} \sim iid\ (0, \phi)$$

with a Gauss-Newton algorithm. Notice that this is the same as fitting the semivariogram model by OLS to the semivariogram cloud consisting of $\{Z(\mathbf{s}_i) - Z(\mathbf{s}_j)\}^2$, instead of fitting the model to the Matheron semivariogram estimator.

#### *4.5.3.2 Composite Likelihood*

In the GEE approach only the model for the mean of the data (or pseudo-data) is required, the variance-covariance matrix is supplanted by a "working" structure. The efficiency of GEE estimators increases with the closeness of the working structure to $\mathrm{Var}[\mathbf{T}]$. The essential problems that lead to the consideration of generalized estimating equations are the intractability of the likelihood function and the difficulties in modeling $\mathrm{Var}[\mathbf{T}]$. No likelihood is used at any stage of the estimation problem. A different—but as we will show, related—strategy is to consider the likelihood of components of $\mathbf{T}$, rather than the likelihood of $\mathbf{T}$. This is an application of the composite likelihood (CL) idea (Lindsay, 1988; Lele, 1997; Heagerty and Lele, 1998).

Let $Y_i$, $(i = 1, \cdots, n)$ denote random variables with known (marginal) distribution and let $\ell(\boldsymbol{\theta}; y_i)$ denote the log likelihood function for $Y_i$. Then $\partial \ell(\boldsymbol{\theta}; y_i)/\partial \boldsymbol{\theta}'$ is a true score function and

*Component "Score"*

$$S(\boldsymbol{\theta}; y_i) = \frac{\partial \ell(\boldsymbol{\theta}; y_i)}{\boldsymbol{\theta}'} = 0$$

is an unbiased estimating function for $\boldsymbol{\theta}$. Unless the $Y_i$ are mutually independent, the sum of the component scores $S(\boldsymbol{\theta}; y_i)$ is not the score function for the entire data. Nevertheless,

$$U_{cl}(\boldsymbol{\theta}; \mathbf{Y}) = \sum_{i=1}^{n} S(\boldsymbol{\theta}; \mathbf{Y})$$

remains an unbiased estimating function. Setting $U_{cl}(\boldsymbol{\theta}; \mathbf{y}) \equiv \mathbf{0}$ and solving for $\boldsymbol{\theta}$ leads to the composite likelihood estimate (CLE) $\widehat{\boldsymbol{\theta}}_{cl}$.

Several approaches exist to apply the composite likelihood idea to the estimation of semivariogram parameters for spatial data. We focus here on the pseudo-data $T_{ij}^{(2)} = Z(\mathbf{s}_i) - Z(\mathbf{s}_j)$ from the previous section (notice that the GEEs were developed in terms of $T_{ij}^{(3)}$). Other setups are considered in Lele (1997), Curriero and Lele (1999), and the Chapter problems. To perform CL inference, a distribution must be stipulated. Assuming that the $T_{ij}^{(2)}$ are Gaussian and the random field is (at least intrinsically) stationary, we obtain the component score function

$$\begin{aligned} S(\boldsymbol{\theta}; T_{ij}^{(2)}) &= \frac{\partial \ell(\boldsymbol{\theta}; T_{ij}^{(2)})}{\partial \boldsymbol{\theta}'} \\ &= \frac{\partial \gamma(\mathbf{h}_{ij}, \boldsymbol{\theta})}{\partial \boldsymbol{\theta}'} \frac{1}{4\gamma(\mathbf{h}_{ij}, \boldsymbol{\theta})^2} \left( T_{ij}^{(3)} - 2\gamma(\mathbf{h}_{ij}, \boldsymbol{\theta}) \right). \end{aligned}$$

The composite likelihood *score* function is then

*Composite Likelihood "Score"*

$$CS(\boldsymbol{\theta}; \mathbf{T}^{(2)}) = 2 \sum_{i=1}^{n-1} \sum_{j=i+1}^{n} \frac{\partial \gamma(\mathbf{h}_{ij}, \boldsymbol{\theta})}{\partial \boldsymbol{\theta}'} \frac{1}{8\gamma(\mathbf{h}_{ij}, \boldsymbol{\theta})^2} \left( T_{ij}^{(3)} - 2\gamma(\mathbf{h}_{ij}, \boldsymbol{\theta}) \right). \quad (4.44)$$

Comparing (4.44) to (4.43), it is seen that the respective estimating equations differ only by the factor $1/(8\gamma(\mathbf{h}_{ij}, \boldsymbol{\theta})^2)$. This factor is easily explained. Under the distributional assumption made here for the $T_{ij}^{(2)}$ it follows that $T_{ij}^{(3)}/2\gamma(\mathbf{h}_{ij}, \boldsymbol{\theta}) \sim \chi_1^2$ and thus

$$\mathrm{Var}\left[T_{ij}^{(3)}\right] = 8\gamma(\mathbf{h}_{ij}, \boldsymbol{\theta})^2.$$

The composite likelihood estimating equation (4.44) is a variance-weighted version of the generalized estimating equation (4.43). Put differently, GEE estimates for the semivariogram based on pseudo-data $T_{ij}^{(3)}$ are identical to composite likelihood estimate if the GEE working structure takes into account the unequal dispersion of the pseudo-data. Composite likelihood estimates can thus be calculated by (nonlinear) **weighted** least squares in the model

$$T_{ij}^{(3)} = 2\gamma(\mathbf{h}_{ij}, \boldsymbol{\theta}) + \delta_{ij} \qquad \delta_{ij} \sim (0, 8\gamma(\mathbf{h}_{ij}, \boldsymbol{\theta})^2)$$

with a Gauss-Newton algorithm.

The derivation of the composite likelihood estimates commenced with the pseudo-data $T_{ij}^{(2)}$. One could have also started by defining the pseudo-data for CL estimation as $T_{ij}^{(3)}$ as in the case of the GEEs. It is left as an exercise (Chapter problem 4.6) to show that $CS(\boldsymbol{\theta}; \mathbf{t}^{(2)}) = CS(\boldsymbol{\theta}; \mathbf{t}^{(3)})$.

### *4.5.4 Comparisons*

#### *4.5.4.1 Semivariogram or Semivariogram Cloud. What Are the Data?*

The composite likelihood and generalized estimating equation approaches can be viewed as generalizations of the least squares methods in §4.5.1, where the data consist of the empirical semivariogram cloud, rather than the empirical semivariogram. The process of averaging the semivariogram cloud into lag classes has important consequences for the statistical properties of the estimators, as well as for the practical implementation. Let the number of lag classes be denoted by $K$. Fitting a semivariogram model to the empirical semivariogram can be viewed as the case of $K$ fixed, even if $n \to \infty$. If the sample size grows, so does the number of "lag classes" when the empirical semivariogram cloud are the least-squares data. Hence, if the data for fitting consist of the $T_{ij}^{(3)}$, $K \to \infty$ as $n \to \infty$.

Important results on the consistency and asymptotic efficiency of least-squares estimators can be found in Lahiri, Lee, and Cressie (2002); see also Cressie (1985). If certain regularity conditions on the semivariogram model are met, and if $K$ is fixed, then OLS, WLS, and GLS estimators are consistent and asymptotically Gaussian distributed under an increasing-domain asymptotic model. If the asymptotic model is what Lahiri et al. (2002) term a *mixed-increasing-domain* asymptotic structure, the least-squares estimators remain consistent and asymptotic Gaussian, but their rate of convergence is slower than in the pure increasing-domain asymptotic model. Under the mixed structure, an increasing domain is simultaneously filled in with additional observations. The resultant retention of small lag distances with high spatial dependence reduces the rate of convergence. Nevertheless, the consistency and asymptotic normality are achieved under either asymptotic structure. A further interesting result in Lahiri et al. (2002), is the asymptotic efficiency of the OLS, WLS, and GLS estimators when the number of lag classes equals the number of semivariogram parameters.

A very different picture emerges when the data for semivariogram estimation by least-squares methods are given by the empirical semivariogram cloud $(T_{ij}^{(3)}, ||\mathbf{s}_i - \mathbf{s}_j||)$. Each lag class then contains only a single observation, and the asymptotic structure assumes that $K \to \infty$ as the number of sampling sites increases. At first glance, one might conjecture in this case that composite likelihood estimators are more efficient than their GEE counterparts, because the GEE approach with working independence structure does not take into account the unequal dispersion of the pseudo-data. Recall that CL estimation as described above entails WLS fitting of the semivariogram model to the empirical semivariogram cloud, while GEE estimation with working independence structure is OLS estimation. Write the two non-linear models as

$$\begin{aligned} T_{ij}^{(3)} &= 2\gamma(\mathbf{h}_{ij}, \boldsymbol{\theta}) + \epsilon_{ij}, \qquad \epsilon_{ij} \text{ iid } (0, \phi) \\ T_{ij}^{(3)} &= 2\gamma(\mathbf{h}_{ij}, \boldsymbol{\theta}) + \epsilon_{ij}, \qquad \epsilon_{ij} \text{ iid } (0, 8\gamma(\mathbf{h}_{ij}, \boldsymbol{\theta})). \end{aligned}$$

The mean and variance are functionally dependent in the CL approach. As a consequence, the WLS weights depend on the parameter vector which biases the estimating function. Fedorov (1974) established in the case of uncorrelated data, that WLS estimators are inconsistent if the weights depend on the model parameters. Fedorov's results were applied to semivariogram estimation by Müller (1999). To demonstrate the consequence of parameter-dependent weights in the asymptotic structure where $n \to \infty$, $K \to \infty$, we apply the derivations in Müller (1999) to our CL estimator of the semivariogram. Recall that this estimator of $\boldsymbol{\theta}$ is

$$\widehat{\boldsymbol{\theta}}_{cl} = \arg\min_{\boldsymbol{\theta}} \sum_{i=1}^{n-1} \sum_{j=i+1}^{n} \frac{1}{8\gamma(\mathbf{h}_{ij}, \boldsymbol{\theta})} \left( T_{ij}^{(3)} - 2\gamma(\mathbf{h}_{ij}, \boldsymbol{\theta}) \right).$$

Now replace site indices with lag classes and let $K \to \infty$. If $\boldsymbol{\theta}^*$ denotes the true parameter vector, then the limit of the WLS criterion is

$$\sum_{i=1}^{\infty} \frac{1}{8\gamma(h_k, \boldsymbol{\theta})} \left( 2\gamma(h_k, \boldsymbol{\theta}^*) - 2\gamma(h_k \boldsymbol{\theta}) + \epsilon_k \right)^2.$$

Applying the law of large numbers and rearranging terms, we find that this is equivalent to the minimization of

$$\frac{3}{2} \sum_{i=1}^{\infty} \left( \frac{\gamma(h_k, \boldsymbol{\theta}^*)}{\gamma(h_k, \boldsymbol{\theta})} \right)^2 + \frac{1}{3} - \frac{2}{3} \frac{\gamma(h_k, \boldsymbol{\theta}^*)}{\gamma(h_k, \boldsymbol{\theta})}$$

which, in turn, is equivalent to the minimization of

$$\sum_{i=1}^{\infty} \left( \frac{\gamma(h_k, \boldsymbol{\theta}^*)}{\gamma(h_k, \boldsymbol{\theta})} - \frac{1}{3} \right)^2.$$

The particular relationship between the mean and variance of $T_{ij}^{(3)}$ and the dependence of the variance on model parameters has created a situation where the semivariogram evaluated at the CL estimator is not consistent for $\gamma(h_k, \boldsymbol{\theta}^*)$. Instead, it consistently estimates $3\gamma(h_k, \boldsymbol{\theta}^*)$. The "bias correction" for the CL estimator is remarkably simple.

#### *4.5.4.2 Ordinary, Weighted, or Generalized Least Squares*

If correlated data are fit by least squares methods, we would like to take into account the variation and covariation of the observations. In general, GLS estimation is more efficient than WLS estimation, and it, in turn, is more efficient than OLS estimation. We are tacitly implying here that the covariance matrix of the data is known for GLS estimation, and that the weights are known for WLS estimation. The preceding discussion shows, however, that if one works with the semivariogram cloud, one may be better off fitting the semivariogram model by OLS, than by weighted least squares without bias correction, because the weights depend on the semivariogram parameters. Since the bias correction is so simple (multiply by 1/3), it is difficult to argue in favor of OLS. Müller (1999) also proposed an iterative scheme to obtain consistent estimates of the semivariogram based on WLS. In the iteratively re-weighted algorithm the weights are computed for current estimates of $\boldsymbol{\theta}$ and held fixed. Based on the non-linear WLS estimates of $\boldsymbol{\theta}$, the weights are re-computed and the semivariogram model is fit again. The process continues until changes in the parameter estimates from two consecutive fits are sufficiently small. The performance of the iteratively re-weighted estimators was nearly as good as that of the iterated GLS estimator in the simulation study of Müller (1999).

The difficulty of a pure or iterated GLS approach lies in the determination of the full covariance structure of the pseudo-data and the possible size of the covariance matrix. For Gaussian random fields, variances and covariances of the $T_{ij}^{(3)}$ are easy to ascertain and can be expressed as a function of semivariogram values. For the empirical semivariogram in a Gaussian random field, the covariance matrix of the Matheron estimator $\widehat{\gamma}(\mathbf{h})$ is derived in Genton (1998b). In either case, the covariance matrix depends on $\boldsymbol{\theta}$ and an iterative approach seems prudent. Based on a starting value $\boldsymbol{\theta}_0$, compute $\text{Var}[\mathbf{T}^{(3)}]$ or $\text{Var}[\widehat{\gamma}(\mathbf{h})]$ and estimate the first update $\beta_1$ by (estimated) generalized least squares. Recompute the variance-covariance matrix and repeat the GLS step. This process continues until changes in subsequent estimates of $\boldsymbol{\theta}$ are minor. The difficulty with applying GLS estimation to the semivariogram cloud is the size of the data vector. Since the set of pseudo-data contains up to $n(n-1)/2$ points, compared to $K \ll n$ if you work with the empirical semivariogram, building and inverting $\text{Var}[\mathbf{T}^{(3)}]$ quickly becomes computationally prohibitive as $n$ grows.

If the distribution of the data is elliptically contoured, the simplification described in Genton (2000) can be put in place, provided the covariance matrix of the data is of the form described there (see also §4.5.1 in this text). This eliminates the covariance parameters $\boldsymbol{\theta}$ from the GLS weight matrix.

#### *4.5.4.3 Binning Versus Not Binning*

Unless the data are regularly spaced, least-squares fitting of semivariogram models invariably entails the grouping of squared differences (or functions thereof) into lag classes. Even for regularly spaced data, binning is often necessary to achieve a recommended number of pairs in each lag class. The process of binning itself is not without controversy. The choice (number and spacing) of lag classes affects the resulting semivariogram cloud. The choice of the largest lag class for which to calculate the empirical semivariogram can eliminate values with large variability. The user who has a particular semivariogram model in mind may be tempted to change the width and number of lag classes so that the empirical semivariogram resembles the theoretical model. Combined with trimming values for larger lags, the process is slanted towards creating a set of data to which a model fits well. The process of fitting a statistical model entails the development of a model that supports the data; not the development of a set of data that supports a model.

The CL and GEE estimation methods are based on the semivariogram cloud and avoid the binning process. As discussed above, the choice of parameter-dependent weights can negatively affect the consistency of the estimates, however, and a correction may be necessary. Estimation based on the semivariogram cloud can also "trim" values at large lags, akin to the determination of a largest lag class in least-squares fitting. Specifically, let $w_{ij}$ denote a weight associated with $\{Z(\mathbf{s}_i) - Z(\mathbf{s}_j)\}^2$. For example, take

$$w_{ij} = \begin{cases} 1 & \text{if } ||\mathbf{s}_i - \mathbf{s}_j|| \leq c \\ 0 & \text{if } ||\mathbf{s}_i - \mathbf{s}_j|| > c, \end{cases}$$

and modify the composite likelihood score equation as

$$CS(\boldsymbol{\theta}; \mathbf{T}^{(2)}) = 2\sum_{i=1}^{n-1} \sum_{j=i+1}^{n} w_{ij} \frac{\partial \gamma(\mathbf{h}_{ij}, \boldsymbol{\theta})}{\partial \boldsymbol{\theta}'} \frac{1}{8\gamma(\mathbf{h}_{ij}, \boldsymbol{\theta})^2} \left( T_{ij}^{(3)} - 2\gamma(\mathbf{h}_{ij}, \boldsymbol{\theta}) \right),$$

to exclude pairs from estimation whose distance exceeds $c$. One might also use weights that depend on the magnitude of the residual $(t_{ij}^{(3)} - 2\gamma(\mathbf{h}_{ij}, \boldsymbol{\theta}))$ to "robustify" the estimator.

The ML and REML estimators also avoid the binning process altogether. In fact, squared differences between observed values do not play a role in likelihood estimation. The objective functions (4.35) and (4.39) involve generalized sums of squares between observations and their means, not squared differences between the $Z(\mathbf{s}_i)$. It is thus not correct to cite as a disadvantage of likelihood methods that one cannot eliminate from estimation pairs at large lags. The lag structure figures into the structure of the covariance matrix $\boldsymbol{\Sigma}(\boldsymbol{\theta})$, which reflects the variation and covariation of the data. Likelihood methods use an objective function built on the data, not an objective function built on pseudo-data that was crafted by the analyst based on the spatial configuration.

Comparing CL with ML (or REML) estimation, it is obvious that composite likelihood methods are less efficient, since $CS(\boldsymbol{\theta}; \mathbf{T}^{(2)})$ is not the score function of $\mathbf{Z}(\mathbf{s})$. Computationally, CL and GEE estimation are more efficient, however. Minimizing the maximum or restricted maximum likelihood score function is an iterative process which involves inversion of the $(n \times n)$ matrix $\boldsymbol{\Sigma}(\boldsymbol{\theta})$ at every iteration. For $n$ large, this is a time-consuming process. In CL/GEE estimation the task of inverting an $(n \times n)$ matrix once at each iteration is replaced with the manipulation of small $(q \times q)$ matrices, where $q$ is the number of elements in $\boldsymbol{\theta}$. On the other hand, CL/GEE estimation manipulates $(n-1)/2$ times as many "data" points.

**Example 4.2 (C/N ratios. Continued)** For the C/N ratio data we applied the previously discussed estimation methods assuming an exponential covariance structure or semivariogram. Table 4.2 displays the parameter estimates for five estimation methods with and without a nugget effect and Figure 4.10 displays the fitted semivariograms. The least-squares fits used the classical empirical semivariogram (Matheron estimator).

It is noteworthy in Table 4.2 that the inclusion of a nugget effect tends to raise the estimate of the range, a common phenomenon. In other words, the decrease in spatial continuity due to measurement error is compensated to some degree by an increase in the range which counteracts the decline in the spatial autocorrelations on short distances. Unfortunately, the OLS, WLS, CL, and GEE estimation methods do not produce reliable standard errors for the parameter estimates and the necessity for inclusion of a nugget effect must be determined on non-statistical grounds. These methods estimate the semivariogram parameters from pseudo-data and do not account properly for the covariation among the data points.

Table 4.2 *Estimated parameters for C/N ratios with exponential covariance structure; see also Figure 4.10.*

| Method | Nugget | Estimates of (Partial) Sill | Practical Range |
|---|---|---|---|
| OLS | 0. | 0.279 | 45.0 |
| OLS | 0.111 | 0.174 | 85.3 |
| WLS | 0. | 0.278 | 31.4 |
| WLS | 0.117 | 0.166 | 79.4 |
| CL | 0. | 0.278 | 32.6 |
| CL | 0.107 | 0.176 | 75.1 |
| GEE | 0. | 0.281 | 45.9 |
| GEE | 0.110 | 0.173 | 77.2 |
| REML | 0. | 0.318 | 41.6 |
| REML | 0.118 | 0.215 | 171.1 |

Only likelihood-based methods produce a reliable basis for model comparisons on statistical grounds. The negative of two times the restricted log likelihoods for the nugget and no-nugget models are 264.83 and 277.79, respectively. The likelihood ratio statistic to test whether the presence of the nugget effect significantly improves the model fit is $277.79 - 264.83 = 12.96$ and is significant; $\Pr(\chi_1^2 > 12.96) < 0.00032$. Based on this test, the model should contain a nugget effect. On the other hand, the REML method produces by far the largest estimate of the variance of the process (0.318 and $0.215 + 0.118 = 0.333$), and the REML estimate of the range in the no-nugget model (171.1) appears large compared to other methods. Recall that OLS, WLS, CL, and GEE estimates are obtained from a data set in which the largest lag does not coincide with the largest distance in the data. Pairs at large lags are often excluded from the analysis. In our case, only data pairs with lags less than $6 \times 35 = 210$ feet were used in the OLS/WLS/CL/GEE analyses. The ML and REML methods cannot curtail the data.

The consequences of using the empirical semivariogram cloud (GEE/CL) versus the empirical semivariogram (OLS/WLS) are minor for these data. The OLS and GEE estimates are quite close, as are the WLS and CL estimates. This is further amplified in a graph of the fitted semivariograms (Figure 4.10). The CL and WLS fits are nearly indistinguishable. The same holds for the OLS and GEE fits.

Performing a weighted analysis does, however, affect the estimate of the (practical) range in the no-nugget models. Maybe surprisingly, the CL estimates do not exhibit the large bias that was mentioned earlier. Based on the previous derivations, one would have expected the CL estimator of the practical range to be much larger on average than the consistent WLS estimator. Recall that the lack of consistency of the CL estimator—which is a weighted version of the GEE estimator—was established based on an asymptotic model in which the number of observations as well as the number of lag classes grows to infinity. In this application, we curtailed the max lag class (to $6 \times 35 = 210$ feet), a practice we generally recommend for composite likelihood and generalized estimating equation estimation. An asymptotic model under which the domain and the number of observation increases is not meaningful in this application. The experimental field cannot be arbitrarily increased. We accept the CL estimators without "bias" correction.

When choosing between pseudo-data based estimates and ML/REML estimates, the fitted semivariograms are sometimes displayed together with the empirical semivariograms. The CL/GEE and ML/REML estimates in general will fair visually less favorably compared to OLS/WLS estimates. CL/GEE and ML/REML estimates do not minimize a (weighted) sum of squares between the model and the empirical semivariogram. The least squares estimates obviously fit the empirical semivariogram best; that is their job. This does not imply that least squares yields the best estimates from which to reconstruct the second-order structure of the spatial process. □

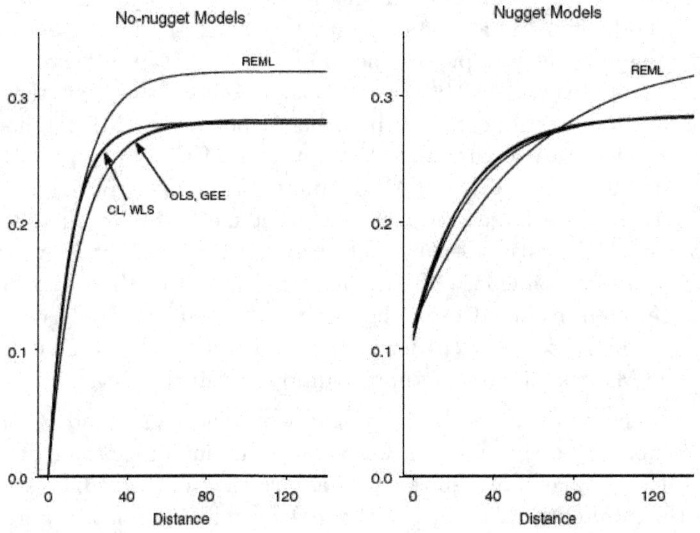


Figure 4.10 *Fitted semivariograms for C/N data with and without nugget effect constructed from parameter estimates in Table 4.2.*

## 4.6 Nonparametric Estimation and Modeling

Choosing a valid parametric semivariogram or covariogram model and fitting it to the semivariogram cloud or the empirical semivariogram ensures that the predicted variogram or covariogram has the needed properties: conditional negative definiteness of the semivariogram and positive definiteness of the covariance function. On the downside, however, one is restricted to a relatively small number of semivariogram models. Few empirical semivariograms exhibit in practical applications the *textbook* behavior that makes choice of a semivariogram model a simple matter. Often, the empirical semivariogram appears erratic or wavy. The large sample variance of $\widehat{\gamma}(\mathbf{h})$ and related estimators for large lags and small number of pairs makes separating noise from structure sometimes difficult (see §4.4).

To achieve greater flexibility, the modeler can resort to more complicated semivariogram models that accommodate waviness and rely on nesting of models (§4.3.6). Few parametric models incorporate positive and negative autocorrelation and nesting of semivariograms is not without controversy. Stein (1999, p. 14), for example, cautions about the common practice of nesting spherical semivariograms. Since linear combinations of valid covariance functions yield valid covariance functions, we should not give up on nested models too quickly. The trick, presumably, is in taking linear combinations of the right models.

What is termed the "nonparametric" approach to semivariogram modeling consists of choosing a family of semivariogram models that is sufficiently flexible to accommodate a wider range of shapes than the models described in §4.3. The moniker ought not connote a rank-based approach. The resulting models are parametric in the sense that they depend on a fixed number of unknown quantities that are estimated from the data. The three approaches we describe in this section have in common that the fitting process can be viewed as a weighted nonlinear regression problem. The attribute "nonparametric" is deserved, because, as in the case of nonparametric regression models, certain model parameters govern the smoothness of the resulting semivariogram estimate and need to be chosen by some mechanism.

Whereas in the development of nonparametric (local) methods much energy has been spent on the problem of determining the appropriate smoothness of a fit based on data, nonparametric semivariogram modeling has not reached this stage yet. Not having solved all issues related to the determination of the degree of smoothness should not deter from exploring the intriguing ideas underpinning these models. A word of caution is nevertheless in order, because the nonparametric approach might enable a blackbox approach to geostatistical analysis. It is not difficult to envision a scenario where a nonparametric semivariogram is derived by completely data driven methods with little or no interference from the analyst. The fitted semivariogram is then used to construct and solve the kriging equations to produce spatial predictions. The result of this blackbox analysis is a map of observed and predicted value which forms the basis of decisions. If the estimated semivariogram drives the results of spatial prediction, one should not develop this important determinant in a black box. It is OK to look.

Nonparametric semivariogram fitting makes use of the linear combination property of covariance functions. The "linear combination" often resolves to integration and the basic model components being integrated or combined are typically more elementary than the parametric models in §4.3. We distinguish two basic approaches in this section; one based on the spectral representation of a spatial random field, the other on a moving average representation of the semivariogram itself.

### *4.6.1 The Spectral Approach*

In §2.5.2 it was shown that the class of valid covariance functions in $\mathbb{R}^d$ can be expressed as

$$C(\mathbf{h}) = \int_{-\infty}^{\infty} \cdots \int_{-\infty}^{\infty} \cos(\boldsymbol{\omega}'\mathbf{h}) S(d\boldsymbol{\omega}).$$

and in the isotropic case we have

$$C(h) = \int_0^{\infty} \Omega_d(h\omega) F(d\omega), \qquad (4.45)$$

where $\Omega_d(t)$ is defined through (4.8) in §4.3.1.

From (4.45) it is seen that this representation is a special case of a nested model, where a linear combination is taken of the basic covariance functions $\Omega_d(h\omega)$.

#### 4.6.1.1 Using Step Functions

A convenient representation of the covariogram can be achieved by considering step functions for $F(\omega)$ in (4.45). If $F$ has positive mass at nodes $t_1, \cdots, t_p$ with respective masses $w_1, \cdots, w_p$, then the covariance function of the process can be represented as

$$C(h, \mathbf{t}, \mathbf{w}) = \sum_{i=1}^{p} w_i \Omega_d(ht_i). \tag{4.46}$$

Notice that $C(0) = \text{Var}[Z(\mathbf{s})] = \int_0^\infty F(d\omega)$. In the discrete representation (4.46), the corresponding result is $C(0) = \sum_{i=1}^{p} w_i$, so that the semivariogram is given by

*Nonparametric Semivariogram (Spectral)*

$$\gamma(h, \mathbf{t}, \mathbf{w}) = \sum_{i=1}^{p} w_i \left(1 - \Omega_d(ht_i)\right). \tag{4.47}$$

In general, an estimate of $\gamma(h)$ can be obtained by fitting $\gamma(h, \mathbf{t}, \mathbf{w})$ to the semivariogram cloud or the empirical semivariogram by a constrained, iterative (reweighted) least squares approach. Details are given by Shapiro and Botha (1991) and Cherry, Banfield, and Quimby (1996). Gorsich and Genton (2000) use a least-squares objective function to find an estimator for the derivative of the semivariogram.

As part of this procedure the user must select the placement of the nodes $t_i$ and the estimation is constrained because only nodes with $w_i \geq 0$ are permissible. Ecker and Gelfand (1997) use a Bayesian approach where either the nodes $t_i$ or the weights $w_i$ are random which is appealing because the number of nodes can be substantially reduced.

The number of nodes varies widely between authors, from five to twenty, to several hundred. Shapiro and Botha (1991) choose $p$ to be one less than the number of points in the empirical semivariogram cloud and spread the nodes out evenly, the spacing being determined on an ad-hoc basis. Because of the large number of nodes relative to the number of points in the empirical semivariogram, these authors impose additional constraints to prevent overfitting and noisy behavior. For example, by bounding the slope of the fitted function, or by ensuring monotonicity or convexity.

Gorsich and Genton (2000) point out the importance of choosing the smallest and largest node wisely to capture the low frequency components of the semivariogram and to avoid unnecessary oscillations. Genton and Gorsich (2002) argue that the number of nodes should be smaller than the number of unique lags and the nodes should coincide with zeros of the Bessel functions. If $k$ denotes the number of lags, these authors recommend to choose $t_1, \cdots, t_p$ such that $p < k$ and $J_v(t_i) = 0$.

#### *4.6.1.2 Using Parametric Kernel Functions*

As with other nonparametric methods it is of concern to determine the appropriate degree of smoothing that should be applied to the data. The selection of number and placement of nodes and the addition of constraints on the derivative of the fitted semivariogram affect the smoothness of the resulting fit. The smoothness of the basis function $\Omega_d$ also plays an important role. The nodes $t_1, \cdots, t_p$, together with the largest node $t_p$ and the lag distances for which the covariogram is to be estimated determine the number of sign changes of the basis function. The more sign changes are permitted, the less smooth the resulting covariogram. The nonparametric approach discussed so far rests on replacing $F(\omega)$ with a step function and estimating the height of the steps given the nodes. This reduces integration to summation. An alternative approach that circumvents the step function idea uses parametric kernel functions and can be implemented if $dF(\omega) = f(\omega)d\omega$.

Let $f(\boldsymbol{\theta}, \omega)$ denote a function chosen by the user. For a particular value of $\boldsymbol{\theta}$ the covariance of two observations spaced $||\mathbf{h}|| = h$ distance units apart is

$$C(\boldsymbol{\theta}, h) = \int_0^\infty \Omega_d(h\omega) f(\boldsymbol{\theta}, \omega)\, d\omega. \tag{4.48}$$

If $\boldsymbol{\theta}$ can be estimated from the data, an estimate of the covariogram at lag $h$ can be obtained as

$$\widehat{C}(\boldsymbol{\theta}, h) = C(\widehat{\boldsymbol{\theta}}, h) = \int_0^\infty \Omega_d(h\omega) f(\widehat{\boldsymbol{\theta}}, \omega)\, d\omega.$$

The main distinction between this and the previous approach lies in the solution of the integration problem. In the step function approach $F(\omega)$ is discretized to change integration to summation. The "unknowns" in the step function approach are the number of nodes, their placement, and the weights. Some of these unknowns are fixed a priori and the remaining unknowns are estimated from the data. In the kernel approach we fix a priori only the class of functions $f(\boldsymbol{\theta}, \omega)$. Unknowns in the estimation phase are the parameters that index the kernel function $f(\boldsymbol{\theta}, \omega)$. If the integral in (4.48) cannot be solved in closed form, we can resort to a quadrature or trapezoidal rule to calculate $C(\widehat{\boldsymbol{\theta}}, h)$ numerically. The parametric kernel function approach for semivariogram estimation is appealing because of its flexibility and parsimony. Valid semivariograms can be constructed with a surprisingly small number of parameters. The process of fitting the semivariogram can typically be carried out by nonlinear least squares, often without constraints.

The need for $F$ to be non-decreasing and $\int_0^\infty f(\boldsymbol{\theta}, \omega) d\omega < \infty$ suggests to draw on probability density functions in the construction of $f(\boldsymbol{\theta}, \omega)$. A particularly simple, but powerful, choice is as follows. Suppose $G(\boldsymbol{\theta})$ is the cumulative distribution function (cdf) of a $U(\theta_l, \theta_u)$ random variable so that $\boldsymbol{\theta} = [\theta_l, \theta_u]'$.

Then define

$$F(\boldsymbol{\theta}, \omega) = \begin{cases} 0 & \omega < 0 \\ G(\boldsymbol{\theta}) & 0 \leq \omega \leq 1 \\ 1 & \omega > 1. \end{cases} \tag{4.49}$$

The kernel $f(\boldsymbol{\theta}, \omega)$ is positive only for values of $\omega$ between 0 and 1. As a consequence, the largest value for which the basis function $\Omega_d(h\omega)$ is evaluated, corresponds to the largest semivariogram lag. This largest lag may imply too many or too few sign changes of the basis function (see Figure 4.1 on page 142). In the latter case, you can increase the bounds on $\omega$ in (4.49) and model

$$F(\boldsymbol{\theta}, \omega) = \begin{cases} 0 & \omega < 0 \\ G(\boldsymbol{\theta}) & 0 \leq \omega \leq b \\ 1 & \omega > b. \end{cases}$$

The weighting of the basis functions is controlled by the shape of the kernel between 0 and $b$. Suppose that $\theta_l = 0$ and $G(\boldsymbol{\theta})$ is a uniform cdf. Then all values of $\Omega(h\omega)$ receive equal weight $1/\theta_u$ for $0 \leq \omega \leq \theta_u$, and weight 0 everywhere else. By shifting the lower and upper bound of the uniform cdf, different parts of the basis functions are weighted. For $\theta_l$ small, the product $h\omega$ is small for values where $f(\boldsymbol{\theta}, \omega) \neq 0$ and the basis functions will have few sign changes on the interval $(0, h\omega)$ (Figure 4.11).

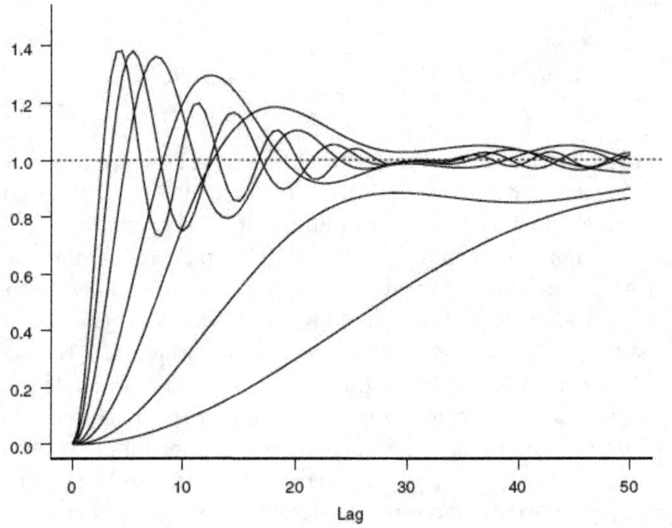


Figure 4.11 *Semivariograms with sill 1.0 constructed from covariance functions (4.48) for processes in* $\mathbb{R}^2$ *with uniform kernels of width* $\theta_u - \theta_l = 0.2$ *for various values of* $\theta_l = -0.1, 0, 0.1, 0.2, 0.4, 0.6.0.8$. *As* $\theta_l$ *increases the semivariogram becomes more and more wavy. The basis function is* $\Omega_d(t) = J_0(t)$.

To adjust for the variance of the process, a sill parameter is added and we model

*Parametric Kernel Estimator*

$$C(\boldsymbol{\theta}, h) = \sigma^2 \int_0^b \Omega(h\omega) f(\boldsymbol{\theta}, \omega)\, d\omega. \tag{4.50}$$

The fitting process starts with the calculation of an empirical semivariogram at $k$ lag classes. Any of the estimators in §4.4 can be used. It is important, however, to ensure that a sufficient number of pairs are available at all lag classes. Nonparametric semivariogram estimators allow for positive and negative autocorrelations and hence can have a wavy appearance. If the fit follows a wavy empirical semivariogram estimator, you want to be confident that this behavior is not spurious, caused by large variability in $\widehat{\gamma}(\mathbf{h}_k)$ due to an insufficient number of pairs. Our fitting criterion is a nonlinear, ordinary least squares criterion,

$$Q(\boldsymbol{\theta}) = \sum_{i=1}^{k} \left(\widehat{\gamma}(h_i) - (\sigma^2 - C(\boldsymbol{\theta}, h_i))\right)^2. \tag{4.51}$$

where $C(\boldsymbol{\theta}, h)$ is given in (4.50). The parameter $\sigma^2$ is akin to the sill of classical semivariogram models in the sense that the nonparametric semivariogram oscillates about $\sigma^2$ and will approach it asymptotically. In practical implementation, the integral in (4.50) can often be approximated with satisfactory accuracy by a sum, applying a trapezoidal or quadrature rule. This is helpful if the fitting procedure allows array processing, such as the NLIN or NLMIXED procedures of SAS/STAT®. An example of fitting a semivariogram with the parametric kernel approach is presented at the conclusion of §4.6.3.

The uniform kernel is simple to work with but you may not want to weigh the basis functions equally for values of $\omega$ where $f(\boldsymbol{\theta}, \omega)$ is nonzero. Kernel functions with unequal weighing can be constructed easily by drawing on other probability densities. For example,

$$f(\mu, \xi, \omega) = \begin{cases} 0 & \omega < 0 \\ \exp\{-0.5(\omega - \mu)^2/\xi\} & 0 \le \omega \le b \\ 1 & \omega > b. \end{cases} \tag{4.52}$$

is a two-parameter kernel derived from the Gaussian density. The kernel can be scaled $f(\boldsymbol{\theta}, \omega), 0 \le \omega \le b$ so that it integrates to one, for example,

$$f(\mu, \xi, \omega) = \frac{\exp\{-0.5(\omega - \mu)^2/\xi\}}{\int_0^b \exp\{-0.5(\omega - \mu)^2/\xi\}}.$$

### *4.6.2 The Moving-Average Approach*

The family of semivariogram (or covariogram) models in the previous subsection is derived by starting from a spectral representation of $C(\mathbf{h})$. The search for flexible models can also commence from a representation in the spatial domain, the convolution representation. Recall from page 58 that a random field $Z(\mathbf{s})$ can be expressed in terms of a kernel function $K(\mathbf{u})$ and a white noise excitation field $X(\mathbf{s})$ as

$$Z(\mathbf{s}) = \int_{\mathbf{u}} K(\mathbf{s}-\mathbf{u})X(\mathbf{u})\,d\mathbf{u}.$$

It was shown in §2.4.2 that

$$\mathrm{Cov}[Z(\mathbf{s}), Z(\mathbf{s}+\mathbf{h})] = \sigma_x^2 \int_{\mathbf{u}} K(\mathbf{u})K(\mathbf{h}+\mathbf{u})\,d\mathbf{u},$$

and thus

$$\mathrm{Var}[Z(\mathbf{s})] = \sigma_x^2 \int_{\mathbf{u}} K(\mathbf{u})^2\,d\mathbf{u}. \tag{4.53}$$

These results are put to use in finding the semivariogram of the convolved process:

$$\begin{aligned}
\gamma(\mathbf{h}) &= \mathrm{Var}[Z(\mathbf{s}) - Z(\mathbf{s}+\mathbf{h})] \\
&= \mathrm{Var}\left[\int_{\mathbf{u}} K(\mathbf{s}-\mathbf{u})X(\mathbf{u})\,d\mathbf{u} - \int_{\mathbf{u}} K(\mathbf{s}+\mathbf{h}-\mathbf{u})X(\mathbf{u})\,d\mathbf{u}\right] \\
&= \mathrm{Var}\left[\int_{\mathbf{u}} (K(\mathbf{s}-\mathbf{u}) - K(\mathbf{s}+\mathbf{h}-\mathbf{u}))\,X(\mathbf{u})\,d\mathbf{u}\right] \\
&= \mathrm{Var}\left[\int_{\mathbf{u}} P(\mathbf{s}-\mathbf{u},\mathbf{h})X(\mathbf{u})\,d\mathbf{u}\right]
\end{aligned}$$

The last expression is the variance of a random field $U(\mathbf{s})$ with kernel $P(\mathbf{u},\mathbf{h})$. Applying (4.53) one finds

$$\begin{aligned}
\gamma(\mathbf{h}) &= \mathrm{Var}[Z(\mathbf{s}) - Z(\mathbf{s}+\mathbf{h})] = \int_{\mathbf{u}} P(\mathbf{s}-\mathbf{u},\mathbf{h})^2\,d\mathbf{u} \\
&= \int_{\mathbf{u}} (K(\mathbf{s}-\mathbf{u}) - K(\mathbf{s}+\mathbf{h}-\mathbf{u}))^2\,d\mathbf{u}.
\end{aligned}$$

Because $\mathbf{s}$ is arbitrary and $C(\mathbf{h})$ is an even function, the expression simplifies to

*Nonparametric Semivariogram (Moving Average)*

$$\gamma(\mathbf{h}) = \int_{\mathbf{u}} (K(\mathbf{u}) - K(\mathbf{u}-\mathbf{h}))^2\,d\mathbf{u}, \tag{4.54}$$

the moving average formulation of the semivariogram.

**Example 4.5** Recall Example 2.3, where white noise was convolved with a uniform and a Gaussian kernel function. The resulting correlation functions in Figure 2.5 on page 60 show a linear decline of the correlation for the uniform kernel up to some range $r$. The correlation remains zero afterward. Obviously, this correlation function corresponds to a linear isotropic semivariogram model

$$\gamma(h) = \begin{cases} \theta|h| & |h| \le r \\ \theta r & |h| > r \end{cases}$$

with sill $\theta r (h, r > 0)$. This is easily verified with (4.54). For a process in $\mathbb{R}^1$ define $K(u) = \sqrt{\theta/2}$ if $0 \le u \le r$ and 0 elsewhere. Then $(K(u) - K(u-h))^2$ is

0 where the rectangles overlap and $\theta/2$ elsewhere. For example, if $0 \le h \le r$, we obtain

$$\begin{aligned}\gamma(h) &= \int_0^h \frac{1}{2}\theta\,du + \int_r^{h+r} \frac{1}{2}\theta\,du \\ &= \frac{1}{2}\theta h + \frac{1}{2}\theta h = \theta h,\end{aligned}$$

and $\gamma(h) = 2\int_0^r \theta/2\,du = \theta r$ for $h > r$ or $h < -r$. □

Barry and Ver Hoef (1996) have drawn on these ideas and extended the basic moving average procedure in more than one dimension, also allowing for anisotropy. Their families of variogram models are based on moving averages using piecewise linear components. From the previous discussion it is seen that any valid—square integrable—kernel function $K(\mathbf{u})$ can be used to construct nonparametric semivariograms from moving averages.

The approach of Barry and Ver Hoef (1996) uses linear structures which yields explicit expressions for the integral in (4.54). For a one-dimensional process you choose a range $c > 0$ and divide the interval $(0, c]$ into $k$ equal subintervals of width $w = c/k$. Let $f(u, \theta_i)$ denote the rectangular function with height $\theta_i$ on the $i$th interval,

$$f(u, \theta_i) = \begin{cases} \theta_i & (i-1) < u/w \le i \\ 0 & \text{otherwise.} \end{cases} \tag{4.55}$$

The moving average function $K(u)$ is a step function with steps $\theta_1, \cdots, \theta_k$,

$$K(u|c, k) = \sum_{i=1}^{k} f(u, \theta_i).$$

When the lag distance $h$ for which the semivariogram is to be calculated is an integer multiple of the width, $h = jw, j = 1, 2, \cdots$, the semivariogram becomes

$$\gamma(h) = w\left(\sum_{i=1}^{k} \theta_i^2 - \sum_{i=j+1}^{k} \theta_i\theta_{i-j}\right).$$

When the lag distance $|h|$ equals or exceeds the range $c$, that is, when $j \ge k$, the second sum vanishes and the semivariogram remains flat at $\sigma^2 = w\sum_{i=1}^{k}\theta_i^2$, the sill value. When $h$ is not an integer multiple of the width $w$, the semivariogram value is obtained by interpolation.

For processes in $\mathbb{R}^2$, the approach uses piecewise planar functions that are constant on the rectangles of a grid. The grid is formed by choosing ranges $c$ and $d$ in two directions and then dividing the $(0,0) \times (c, d)$ rectangle into $k \times l$ rectangles. Instead of the constant function on the line (4.55), the piecewise planar functions assign height $\theta_{i,j}$ whenever a point falls inside the $i$th, $j$th sub-rectangle (see Barry and Ver Hoef, 1996, for details). Note that the piecewise linear model in (4.55) is not valid for processes in $\mathbb{R}^2$.

The piecewise linear (planar) formulation of Barry and Ver Hoef combines the flexibility of other nonparametric techniques to provide flexible and valid forms with a true range beyond which the semivariogram does not change. Whereas in parametric modeling the unknown range parameter is estimated from the data, here it is chosen a-priori by the user (the constant $c$). The empirical semivariogram cloud can aid in determining that constant, or it can be chosen as the distance beyond which interpretation of the semivariogram is not desired (or needed), for example, one half of the largest point distance. Based on results from simulation studies, Barry and Ver Hoef (1996) recommend $k = 15$ subdivisions of $(0, c]$ in one dimension and use 20 sub-rectangles ($k = 5, l = 4$) for modeling the semivariogram of the Wolfcamp Aquifer data (Cressie, 1993, p. 214) in two dimensions. The nonparametric semivariogram is fit to the empirical semivariogram or semivariogram cloud by (weighted) nonlinear least squares to estimate the $\theta$ parameters.

### *4.6.3 Incorporating a Nugget Effect*

Any semivariogram model can be furnished with a nugget effect using the method in §4.3.6; nonparametric models are no exception. There is, however, a trade-off between estimating the parameters of a nonparametric model that govern the smoothness of the semivariogram and estimating the nugget effect. The sum of the weights $\sum_{i=1}^{p} w_i$ in the spectral approach and the sum of the squared step heights $\sum_{i=1}^{k} \theta_i^2$ in the moving average approach represent the partial sill $\sigma_0^2$ in the presence of a nugget effect $c_0$. The nonlinear least squares objective function is adjusted accordingly. When the nugget effect is large, the process contains a lot of background noise and the nonparametric semivariogram estimate tends to be not smooth, the nonparametric coefficients $w_i, \theta_i^2$ tend to be large. Since the sill $\sigma^2 = c_0 + \sigma_0^2$ is fixed, the nugget estimate will be underestimated. A large estimate of the nugget effect, on the other hand, leads to an artificially smooth nonparametric semivariogram that is not sufficiently flexible because of small weights.

Barry and Ver Hoef (1996) recommend estimating the nugget effect separately from the nonparametric coefficients, for example, by fitting a line to the first few lags of the empirical semivariogram cloud, obtaining the nugget estimate $\widehat{c}_0$ as the intercept. The data used in fitting the nonparametric semivariogram is then shifted by that amount provided $\widehat{c}_0 > 0$.

**Example 4.2 (C/N ratios. Continued)** We applied the parametric kernel approach introduced on page 181 to the C/N ratio data to fit a semivariogram with nugget effect. The empirical semivariogram (Matheron estimator) is shown in Figure 4.7 on page 157. Based on the empirical semivariogram and our previous analyses of these data, we decided to model a nugget effect in two ways. The nugget effect was first estimated based on fitting a linear model to the first five lag classes by weighted least squares. The resulting estimate was $\widehat{c}_0 = 0.1169$ and it was held fixed in the subsequent fitting of the semivariogram. In the second case, the nugget was estimated simultaneously with the parameters of the kernel function.

For the function $f(\boldsymbol{\theta}, \omega)$ in (4.50), we chose a very simple kernel,

$$f(\theta_u, \omega) = \begin{cases} 1/\theta_u & 0 \leq \omega \leq b \\ 0 & \text{otherwise.} \end{cases}$$

This is a uniform kernel with lower bound fixed at 0. The Bessel function of the first kind serves as the basis function and the covariance model is the two-parameter model

$$C(\sigma^2, \theta_u, h) = \sigma^2 \int_0^{\theta_u} J_0(h\omega) 1/\theta_u \, d\omega.$$

The nugget effect was held fixed at the same value as in the moving average approach. Figure 4.12 shows fitted semivariograms for $b = 1$ and $b = 2$ for the case of an externally estimated nugget effect ($\widehat{c}_0 = 0.1169$). The parameter estimates were $\widehat{\theta}_u = 0.1499$ and $\widehat{\sigma}^2 = 0.166$ for $b = 1$ and $\widehat{\theta}_u = 0.1378$ and $\widehat{\sigma}^2 = 0.166$ for $b = 2$. Increasing the limit of integration had the effect of reducing the upper limit of the kernel, but not proportionally to the increase. As a result, the Bessel functions are evaluated up to a larger abscissa, and the fitted semivariogram for $b = 2$ is less smooth.

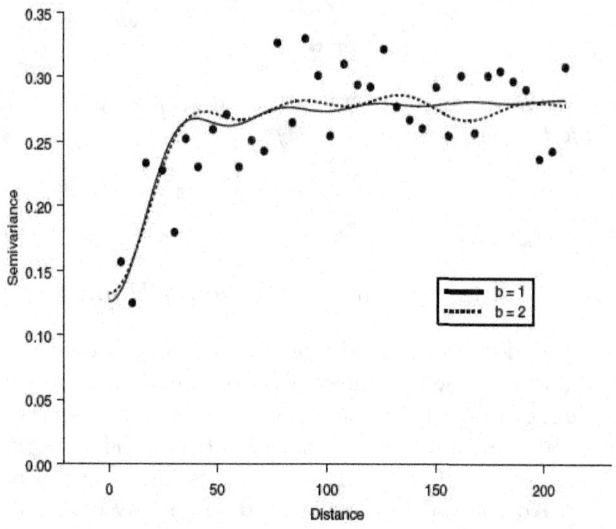


Figure 4.12 *Fitted semivariograms for C/N ratios with uniform kernel functions and $b = 1, 2$. Nugget effect estimated separately from kernel parameters.*

When the nugget effect is estimated simultaneously with the kernel parameter $\theta_u$, we observe a phenomenon similar to that reported by Barry and Ver

Hoef (1996). The simultaneous estimate of the nugget effect is smaller than the externally obtained estimate. The trade-off between the nugget effect and the sill is resolved in the optimization by decreasing the smoothness of the fit, at the nugget effect's expense. The estimate of the nugget effect for the semivariograms in Figure 4.13 is 0.039 for $b = 1$ (0.056 for $b = 2$) and the estimate of the kernel parameter increased to 0.335 for $b = 1$ (0.291 for $b = 2$).

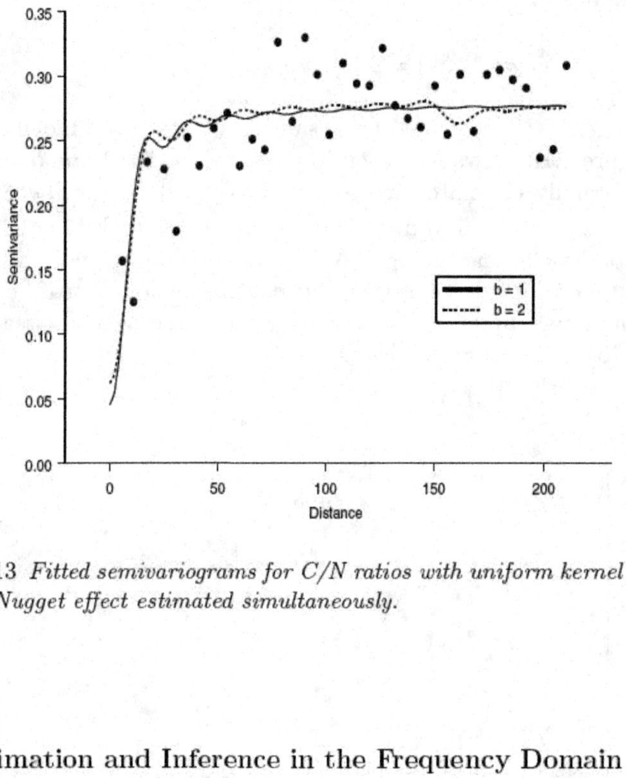


Figure 4.13 *Fitted semivariograms for C/N ratios with uniform kernel functions and* $b = 1, 2$. *Nugget effect estimated simultaneously.*

□

## 4.7 Estimation and Inference in the Frequency Domain

The study of the second-order properties of a stochastic process can be carried out in the spatial domain based on the semivariogram, covariance function, or correlation function, or in the frequency (spectral) domain based on the spectral density function. The empirical estimators of the semivariogram and/or the covariance function, such as (4.24) or (4.3), estimate the corresponding process parameter. An analysis in the spatial domain may proceed as follows

1. Compute the empirical semivariogram based on the Matheron estimator, $\widehat{\gamma}(\mathbf{h})$;
2. Select a theoretical semivariogram model $\gamma(\mathbf{h}; \boldsymbol{\theta})$;
3. Estimate $\boldsymbol{\theta}$ by least squares fitting of $\gamma(\mathbf{h}; \boldsymbol{\theta})$ to the empirical semivariogram;

4. Use the estimated model $\gamma(\mathbf{h}; \widehat{\boldsymbol{\theta}})$ in further calculations, for example, to solve a spatial prediction problem (Chapter 5).

In the frequency domain, similar steps are performed. Instead of the semivariogram, we work with the spectral density, however. The first step, then, is to compute an empirical estimator of $s(\boldsymbol{\omega})$, the **periodogram**. Having selected a theoretical model for the spatial dependency, the spectral density model $s(\boldsymbol{\omega}; \boldsymbol{\theta})$, the parameter vector $\boldsymbol{\theta}$ is estimated. Further calculations are then based on the estimated spectral density $s(\boldsymbol{\omega}; \widehat{\boldsymbol{\theta}})$. For example, one could construct the semivariances or covariances of the process from the estimated spectral density function in order to solve a prediction problem.

Recall that the covariance function and the spectral density function of a second-order stationary stochastic process form a Fourier transform pair (§2.5),

$$s(\boldsymbol{\omega}) = \frac{1}{(2\pi)^d} \int_{\mathbb{R}^d} C(\mathbf{u}) \exp\{-i\boldsymbol{\omega}'\mathbf{u}\}\, d\mathbf{u}.$$

Because you can switch between the spectral density and the covariance function by means of a (inverse) Fourier transform, it is sometimes noted that the two approaches are "equivalent." This is not correct in the sense that space-domain and frequency-domain methods for studying the second-order properties of a random field represent different aspects of the process. The space-domain analysis expresses spatial dependence as a function of separation in spatial coordinates. It is a second-order method because it informs us about the covariation of points in the process and its dependence on spatial separation. Spectral methods do not study covariation of points but the manner in which a function dissipates energy (or power, which is energy per unit interval) at certain frequencies. The second step in the process, the selection of an appropriate spectral density function, is thus arguably more difficult than in the spatial domain, where the shape of the empirical semivariogram suggests the model. By choosing from a sufficiently flexible family of processes—which leads to a flexible family of spectral densities—this issue can be somewhat defused. The Matérn class is particularly suited in this respect.

A further, important, difference between estimating the spectral density function from the periodogram compared to estimating the semivariogram from its empirical estimator, lies in the distributional properties. We discussed in §4.5.1 that the appropriate least-squares methodology in fitting a semivariogram model is generalized least squares, because the "data" $\widehat{\gamma}(h_i)$ are not independent. The dependency stems from the spatial autocorrelation and the sharing of data points. GLS estimation is nearly impractical, however, because of the difficulty of computing a dense variance matrix for the empirical semivariogram values. Weighted or ordinary least squares are used instead. The spectral approach has a considerable advantage in that the periodogram values are—at least asymptotically—independent. A weighted least squares approach based on the periodogram is more justifiable than a weighted least squares approach based on the empirical semivariogram.

### *4.7.1 The Periodogram on a Rectangular Lattice*

The spectral density and the covariance function are related to each other through a Fourier transform. Considering that the asymptotic properties of the empirical estimators are vastly different, it may come as a surprise that the estimates are related in the same fashion as the process quantities; the periodogram turns out to be the Fourier transform of the sample covariance function.

In what follows we focus on the case where the domain $D$ is discrete and $Z$ is real-valued. Specifically, we assume that the data are observed on a rectangular $r \times c$ row-column lattice. Letting $u$ and $v$ denote a row and column position, respectively, $Z(u, v)$ represents the attribute in row $u$ and column $v$. The covariance function can then be expressed as $\mathrm{Cov}[Z(u,v), Z(u+j, v+k)] = C(j,k)$ and the integral in the spectral density function can be replaced case by a double summation:

$$\begin{aligned} s(\omega_1, \omega_2) &= \frac{1}{(2\pi)^2} \sum_{j=-\infty}^{\infty} \sum_{k=-\infty}^{\infty} C(j,k) \exp\{-i(\omega_1 j + \omega_2 k)\} \\ &= \frac{1}{(2\pi)^2} \sum_{j=-\infty}^{\infty} \sum_{k=-\infty}^{\infty} C(j,k) \cos\{\omega_1 j + \omega_2 k\} \qquad (4.56) \\ -\pi &< \omega_1 < \pi \qquad -\pi < \omega_2 < \pi. \end{aligned}$$

The periodogram is the sample based estimate of $s(\omega_1, \omega_2)$. It is defined as

$$I(\omega_1, \omega_2) = \frac{1}{(2\pi)^2} \frac{1}{rc} \left| \sum_{u=1}^{r} \sum_{v=1}^{c} Z(u,v) \exp\{-i(\omega_1 u + \omega_2 v)\} \right|^2, \qquad (4.57)$$

*Fourier Frequencies*

and is computed for the set of frequencies

$$S = \left\{ (\omega_1, \omega_2) : \omega_1 = -\frac{2\pi}{r} \left\lfloor \frac{r-1}{2} \right\rfloor, \cdots, \frac{2\pi}{r} \left\lfloor \frac{r}{2} \right\rfloor \quad \omega_2 = -\frac{2\pi}{c} \left\lfloor \frac{c-1}{2} \right\rfloor, \cdots, \frac{2\pi}{c} \left\lfloor \frac{c}{2} \right\rfloor \right\},$$

where $\lfloor \cdot \rfloor$ is the greatest integer (floor) function. These frequencies, which are multiples of $2\pi/r$ and $2\pi/c$, are known as the Fourier frequencies. The connection between (4.57) and the spectral density as the Fourier transform of the covariance function (4.56) is not obvious in this formulation. We now establish this connection between the periodogram and the sample covariance function for the case of a one-dimensional process $Z(1), Z(2), \cdots, Z(r)$. The operations are similar for the two-dimensional case, the algebra more tedious, however (see Chapter problems).

#### *4.7.1.1 Periodogram and Sample Covariances in $\mathbb{R}^1$*

In the one-dimensional case, the Fourier frequencies can be written as $\omega_j = 2\pi j/r$, $j = \lfloor (j-1)/2 \rfloor, \cdots, \lfloor r/2 \rfloor$ and the periodogram becomes

$$
\begin{aligned}
(2\pi)I(\omega_j) &= \frac{1}{r}\left|\sum_{u=1}^{r} Z(u)\exp\{-\mathrm{i}\omega_j u\}\right|^2 \\
&= \frac{1}{r}\left(\sum_{u=1}^{r} Z(u)\exp\{-\mathrm{i}\omega_j u\}\right)\left(\overline{\sum_{u=1}^{r} Z(u)\exp\{\mathrm{i}\omega_j u\}}\right).
\end{aligned}
$$

Using the Euler relation $\exp\{\mathrm{i}x\} = \cos(x) + \mathrm{i}\sin(x)$, the periodogram can be expressed in terms of trigonometric functions:

$$
\begin{aligned}
2\pi I(\omega_j) &= \frac{1}{r}\sum_{u=1}^{r} Z(u)\{\cos(\omega_j u) - \mathrm{i}\sin(\omega_j u)\}\sum_{u=1}^{r} Z(u)\{\cos\omega_j u + \mathrm{i}\sin(\omega_j u)\} \\
&= \frac{1}{r}\sum_{u=1}^{r}\sum_{p=1}^{r} Z(u)Z(p)\cos(\omega_j u)\ \cos(\omega_j p) + \\
&\sum_{u=1}^{r}\sum_{p=1}^{r} Z(u)Z(p)\sin(\omega_j u)\ \sin(\omega_j p).
\end{aligned}
$$

At this point we make use of the fact that, by definition of the Fourier frequencies, $\sum_u \cos(\omega_j u) = 0$, and hence we can subtract any value from $Z$ without altering the sums. For example,

$$
\sum_{u=1}^{r} Z(u)\cos(\omega_j u) = \sum_{u=1}^{r}(Z(u) - \overline{Z})\cos(\omega_j u).
$$

Using the further fact that $\cos(a)\cos(b) + \sin(a)\sin(b) = \cos(a-b)$, we arrive at

$$
\begin{aligned}
2r\pi I(\omega_j) &= \sum_{u=1}^{r}\sum_{p=1}^{r}(Z(u) - \overline{Z})(Z(p) - \overline{Z})\cos(\omega_j u)\ \cos(\omega_j p) + \\
&\sum_{u=1}^{r}\sum_{p=1}^{r}(Z(u) - \overline{Z})(Z(p) - \overline{Z})\sin(\omega_j u)\ \sin(\omega_j p) \\
&= \sum_{u=1}^{r}\sum_{p=1}^{r}(Z(u) - \overline{Z})(Z(p) - \overline{Z})\cos(\omega_j(u-p)). \qquad (4.58)
\end{aligned}
$$

This expression involves cross-products of deviations from the sample mean and must be related to a covariance estimate. The sample autocovariance function at lag $k$ for this one-dimensional process can be written as

$$
\widehat{C}(k) = \begin{cases} r^{-1}\sum_{u=k+1}^{r}(Z(u-k) - \overline{Z})(Z(u) - \overline{Z}) & k \geq 0 \\ r^{-1}\sum_{u=-k+1}^{r}(Z(u+k) - \overline{Z})(Z(u) - \overline{Z}) & k < 0. \end{cases}
$$

Since $\cos(x) = \cos(-x)$, (4.58) can be written as

$$2\pi I(\omega_j) = \widehat{C}(0) + 2\sum_{k=1}^{r-1}\cos(\omega_j k)\widehat{C}(k) = \sum_{k=-r+1}^{r-1}\cos(\omega_j k)\widehat{C}(k).$$

#### 4.7.1.2 The Periodogram in $\mathbb{R}^2$

Similar operations as in the pervious paragraphs can be carried out for a two-dimensional lattice process. It can be established that for $\omega_1 \neq 0$ and $\omega_2 \neq 0$ the periodogram is the Fourier transform of the sample covariance function,

*Periodogram*

$$\begin{aligned} I(\omega_1, \omega_2) &= \frac{1}{(2\pi)^2}\frac{1}{rc}\left|\sum_{u=1}^{r}\sum_{v=1}^{c} Z(u,v)\exp\{-\mathrm{i}(\omega_1 u + \omega_2 v)\}\right|^2 \\ &= \frac{1}{(2\pi)^2}\sum_{j=-r+1}^{r-1}\sum_{k=-c+1}^{c-1}\widehat{C}(j,k)\exp\{-\mathrm{i}(\omega_1 j + \omega_2 k)\} \\ &= \frac{1}{(2\pi)^2}\sum_{j=-r+1}^{r-1}\sum_{k=-c+1}^{c-1}\widehat{C}(j,k)\cos\{\omega_1 j + \omega_2 k\}, \qquad (4.59) \end{aligned}$$

and $I(0,0) = 0$. The expression (4.59) suggests a simple method to obtain the periodogram. Compute the sample covariance function $\widehat{C}(j,k)$ for the combination of lags $j = -r+1, \cdots, r-1$, $k = -c+1, \cdots, c-1$. Once the sample covariance has been obtained for all relevant lags, cycle through the set of frequencies $S$ and compute (4.59).

The covariance function at lag $j$ in the row direction and lag $k$ in the column direction is estimated from sample moments as

*Sample Covariance Function*

$$\widehat{C}(j,k) = \frac{1}{rc}\sum_{l}^{L}\sum_{m}^{M}(Z(l,m) - \overline{Z})(Z(l+j, m+k) - \overline{Z}), \qquad (4.60)$$

where $l = \max\{1, 1-j\}$, $L = \min\{r, r-j\}$, $m = \max\{1, 1-k\}$, $M = \min\{c, c-k\}$ (Ripley, 1981, p. 79). The sample covariance function is an even function, $\widehat{C}(j,k) = \widehat{C}(-j,-k)$, a property of the covariance function $C(j,k)$. Computational savings are realized by utilizing the evenness property of the covariance function and its estimate. Evenness is also a property of the spectral density *and* the periodogram (4.59). It is thus sufficient to compute the periodogram only for the set of frequencies $S_2$, which removes from $S$ the points with $\omega_2 < 0$ and the set $\{(\omega_1, \omega_2) : \omega_1 < 0, \omega_2 = 0\}$. Since periodogram analyses do not require isotropy of the process, no such assumption is made in computing (4.60). In fact, the periodogram can be used to test for isotropy ($C(j,k) = C(k,j)\forall(j,k)$) and other forms of second-order symmetry, for example, reflection symmetry ($C(j,k) = C(-j,k)\,\forall(j,k)$).

#### *4.7.1.3 Interpretation*

To illustrate the interpretation of the periodogram, we consider by way of example three situations. Recall the simulated data displayed in panels a and d of Figure 1.1 (page 6). Data were assigned independently to the positions of a $10 \times 10$ lattice in Figure 1.1a and were rearranged to achieve a high degree of spatial autocorrelation in Figure 1.1d. The data and their row/column positions in the latter panel are shown in Table 4.3.

Table 4.3 *Data for* $10 \times 10$ *lattice in Figure 1.1d.*

| | Column | | | | | | | | | |
|---|---|---|---|---|---|---|---|---|---|---|
| Row | 1 | 2 | 3 | 4 | 5 | 6 | 7 | 8 | 9 | 10 |
| 1 | 3.55 | 3.52 | 4.41 | 4.04 | 4.43 | 5.01 | 4.81 | 4.97 | 5.02 | 4.63 |
| 2 | 3.56 | 3.91 | 3.39 | 3.22 | 4.29 | 5.77 | 4.61 | 5.02 | 4.14 | 4.24 |
| 3 | 4.26 | 4.26 | 4.32 | 3.99 | 5.40 | 5.83 | 4.92 | 3.55 | 2.68 | 3.22 |
| 4 | 4.29 | 5.84 | 5.24 | 5.30 | 5.27 | 5.48 | 5.02 | 2.98 | 3.15 | 2.71 |
| 5 | 5.75 | 4.80 | 5.50 | 4.51 | 4.71 | 4.62 | 5.26 | 4.15 | 3.10 | 3.97 |
| 6 | 6.20 | 5.81 | 4.45 | 5.14 | 4.62 | 5.38 | 5.31 | 4.50 | 4.84 | 3.94 |
| 7 | 6.87 | 6.23 | 6.18 | 5.61 | 6.00 | 5.75 | 5.67 | 5.32 | 5.03 | 4.68 |
| 8 | 6.73 | 7.24 | 6.53 | 5.20 | 4.65 | 5.19 | 4.83 | 5.11 | 5.35 | 5.82 |
| 9 | 6.42 | 7.27 | 6.14 | 5.95 | 5.52 | 4.78 | 5.40 | 5.32 | 5.36 | 6.17 |
| 10 | 7.52 | 6.49 | 6.18 | 5.46 | 5.12 | 5.55 | 5.09 | 5.69 | 5.99 | 5.91 |

Figure 4.14 and 4.15 show the sample covariance functions (4.60) for the spatially uncorrelated data in Figure 1.1a and the highly correlated data in Figure 1.1d. The covariances are close to zero everywhere, except for $(j,k) = (0,0)$, where the sample covariance function estimates the variance of the process. In the correlated case, covariances are substantial for small $j$ and $k$. Notice the evenness of the sample covariance function, $\widehat{C}(j,k) = \widehat{C}(-j,-k)$, and the absence of reflection symmetry, $\widehat{C}(j,k) \neq \widehat{C}(-j,k)$. The periodograms are displayed in Figures 4.16 and 4.17. For spatially uncorrelated data, the periodogram is more or less evenly distributed. High and low ordinates occur for large and small frequencies (note that $I(0,0) = 0$). Strong positive spatial association among the responses is reflected in large periodogram ordinates for small frequencies.

#### *4.7.1.4 Properties*

The important result from which we derive properties of the periodogram (4.59) and related quantities is as follows. Under an increasing domain asymptotic model, where $r \to \infty$, $c \to \infty$, and $r/c$ tends to a non-zero constant,

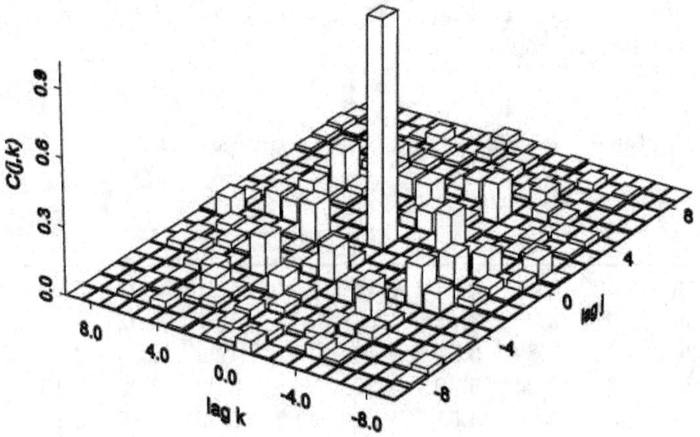


Figure 4.14 *Sample covariance function* $\widehat{C}(j,k)$ *for the spatially uncorrelated data of Figure 1.1a.*

then, at the non-zero Fourier frequencies,

$$2\frac{I(\omega_1,\omega_2)}{s(\omega_1,\omega_2)} \xrightarrow{d} \chi_2^2.$$

Hence, asymptotically, $\mathrm{E}[I(\omega_1,\omega_2)] = s(\omega_1,\omega_2)$, $\mathrm{Var}[I(\omega_1,\omega_2)] = s(\omega_1,\omega_2)^2$. The periodogram is an asymptotically unbiased, inconsistent estimator of the spectral density function. The inconsistency of the periodogram—the variance does not depend on $r$ or $c$—is not a problem when a spectral density is fit to the periodogram by least squares. Other techniques, for example, confirmatory inference about the spectral density, may require a consistent estimate of $s(\omega)$. In this case it is customary to smooth or average the periodogram ordinates to achieve consistency. This is the case in spectral analyses of point patterns (see §4.7.3).

A further property of the periodogram ordinates is their asymptotic independence. Under the asymptotic model stated above,

$$\mathrm{Cov}[I(\omega_k), I(\omega_l)] = \begin{cases} s(\omega)^2 & \omega_k = \omega_l \\ 0 & \omega_k \neq \omega_l. \end{cases}$$

It is this result that favors OLS/WLS estimation in the frequency domain over the spatial domain.

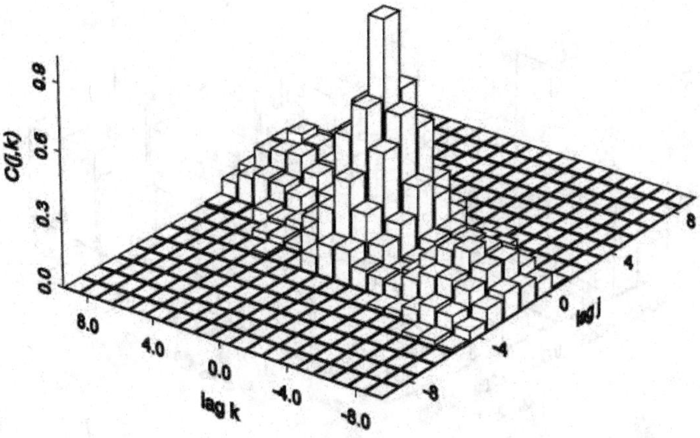


Figure 4.15 *Sample covariance function $\widehat{C}(j,k)$ for the highly spatially correlated data of Figure 1.1d and Table 4.3.*

The conditions under which these asymptotic results hold are detailed in Pagano (1971). Specifically, it is required that the random field is second-order stationary with finite variance and has a spectral density. Then we can write $Z(u,v)$ as the discrete convolution of a white noise excitation field $\{X(s,t) : (s,t) \in I \times I\}$,

$$Z(u,v) = \sum_{j=-\infty}^{\infty} \sum_{k=-\infty}^{\infty} a(j,k) X(u-j, v-k),$$

for a sequence of constants $a(\cdot,\cdot)$. It is important to note that it is not required that $Z(u,v)$ is a Gaussian random field (as is sometimes stated). Instead, it is only necessary that the excitation field $X(s,t)$ satisfies a central limit condition,

$$\frac{1}{\sqrt{rc}} \sum_{s=1}^{r} \sum_{t=1}^{c} X(s,t) \to G(0,1).$$

The asymptotic results are very appealing. For a finite sample size, however, the periodogram is a biased estimator of the spectral density function. Fuentes (2001) gives the following expression for the expected value of the periodogram

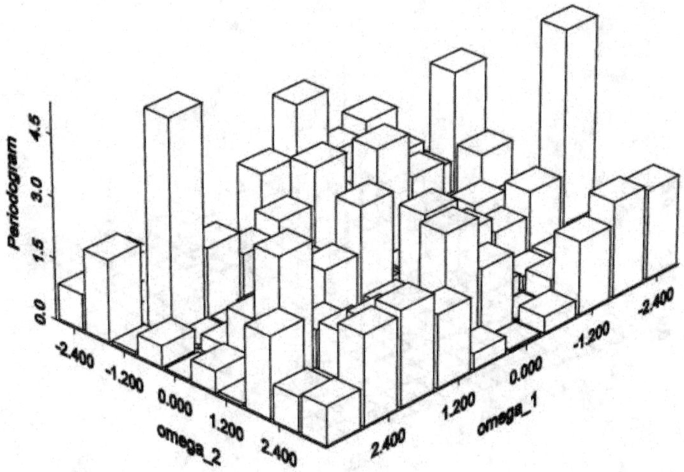


Figure 4.16 *Periodogram $I(\omega_1, \omega_2)$ for the spatially uncorrelated data of Figure 1.1a.*

at $\omega$ in $\mathbb{R}^2$:

$$I(\omega) = \frac{1}{rc(2\pi)^2} \int s(\alpha) W(\alpha - \omega)\, d\alpha, \tag{4.61}$$

*Fejér's Kernel*

with

$$W(\alpha) = \left( \frac{\sin^2(r\alpha_1/2)}{\sin^2(\alpha_1/2)} \right) \left( \frac{\sin^2(c\alpha_2/2)}{\sin^2(\alpha_2/2)} \right),$$

and $\alpha$ is a non-zero Fourier frequency. The bias comes about because the function $W$ has subsidiary peaks (sidelobes), large values away from $\omega$. This allows contributions to the integral from parts of $s(\cdot)$ far away from $\omega$; the phenomenon is termed **leakage**. The function $W(\cdot)$ operates as a kernel function in equation (4.61), it is known as Fejér's kernel. Leakage occurs when the kernel has substantive sidelobes. The kernel then transfers power from other regions of the spectral density to $\omega$. The bias can be substantial when the process has high dynamic range. This quantity is defined by Percival and Walden (1993, p. 201) as

*Dynamic Range*

$$10 \log_{10} \left( \frac{\max\{s(\omega)\}}{\min\{s(\omega)\}} \right). \tag{4.62}$$

And, the bias of $I(\omega)$ in processes with high dynamic range affects particularly those frequencies where the spectral density is small.

The two common methods to combat leakage in periodogram estimation

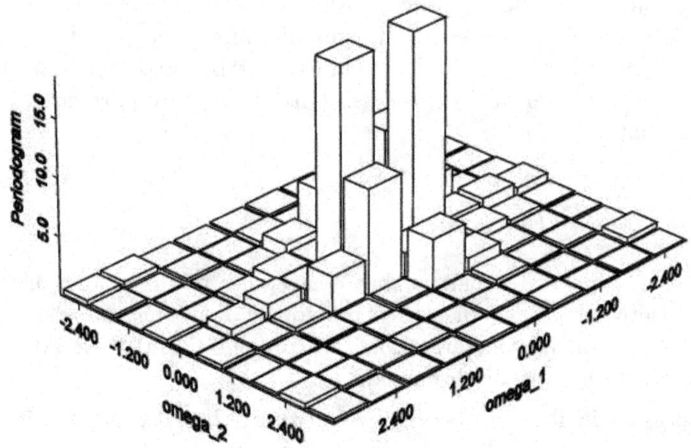


Figure 4.17 *Periodogram $I(\omega_1, \omega_2)$ for the highly spatially correlated data of Figure 1.1d and Table 4.3.*

are **tapering** and **pre-whitening** of the data. We are mentioning these techniques only in passing, they are not without controversy. The interested reader is referred to the monograph by Percival and Walden (1993) for the theory as well as the various arguments for and against tapering and pre-whitening. Data tapering replaces $Z(\mathbf{s})$ with $h(\mathbf{s})Z(\mathbf{s})$, where the function $h(\mathbf{s})$ is termed a taper function or data taper. The tapering controversy is ignited by the fact that the operation in the spatial domain, creating the product $h(\mathbf{s})Z(\mathbf{s})$, is a weighing of the observations by $h(\mathbf{s})$. Data tapers typically give smaller weights to observations near the boundary of the domain, so it can be viewed as a method of adjusting for edge effects. The weighing is not one that gives more weight to observations with small variance, as is customary in statistics. Negative sentiments range from "losing information" and "throwing away data" to likening tapering and tampering. The real effect of tapering is best seen in the frequency domain. Its upshot is to replace in (4.61) the Fejér kernel with a kernel function that has smaller sidelobes, hence reducing leakage. Tapering applies weights to data in order to change the resulting kernel function in the frequency domain.

Pre-whitening is a filtering technique where the data are processed with a linear filter. The underlying idea is that (i) the spectral densities of the original and the filtered data are related (see §2.5.6), and that (ii) the filtered process has a smaller dynamic range. The spectral density of the transformed process (the filter output) is the product of the spectral density of the filter input and the filter transfer function $H(\boldsymbol{\omega})$, see equation (2.39) on page 76. Then, the periodogram is estimated from the filtered data and this is used to construct the periodogram of the original process. Ideally, filtering would create a white noise process, since it has the smallest dynamic range (zero). But in order to do this we need to know either the variance matrix $\text{Var}[Z(\mathbf{s})] = \boldsymbol{\Sigma}$, or the spectral density function of the process. This is the very parameter we are trying to estimate.

### 4.7.2 Spectral Density Functions

We now give the spectral densities that correspond to the second-order stationary and isotropic models discussed previously. For some models, e.g., the Matérn class, several parameterizations are presented and their advantages and disadvantages are briefly discussed.

Since processes in $\mathbb{R}^1$ are necessarily isotropic, the spectral density for a process with continuous domain can be constructed via

$$s(\omega) = \frac{1}{2\pi} \int_{-\infty}^{\infty} \cos(\omega h) C(h)\, dh.$$

In $\mathbb{R}^d$, the "brute-force" method to derive the spectral density, when the covariance function is isotropic, is to compute

$$s(\boldsymbol{\omega}) = \frac{1}{(2\pi)^d} \int_{\mathbb{R}^d} \cos(\boldsymbol{\omega}'\mathbf{h}) C(||\mathbf{h}||)\, d\mathbf{h}.$$

Another device by which to represent the spectral density as a function of a scalar frequency argument is to consider the one-dimensional Fourier transform of the covariance function on the line. Let $\mathbf{h} = [h_1, h_2, \cdots, h_d]'$ and $h = ||\mathbf{h}||$. By isotropy

$$C(\mathbf{h}) = C(h_1, h_2, \cdots, h_d) = C(||\mathbf{h}||, 0, \cdots, 0) = C_r(h).$$

Vanmarcke (1983, p. 103) calls $C_r(h)$ the radial covariance function. The spectral density function is then obtained as

$$s(\omega) = \frac{1}{2\pi} \int_{-\infty}^{\infty} \cos(\omega h) C_r(h)\, dh.$$

A related, but not identical approach, is to derive the $d$-dimensional spectral density $s(\omega_1, \omega_2, \cdots, \omega_d)$ and to reduce it to a radial function,

$$s(\omega_1, \omega_2, \cdots, \omega_d) \equiv s_r \left( \left\{ \omega_1^2 + \omega_2^2 + \cdots + \omega_d^2 \right\}^{1/2} \right).$$

For $d = 3$ the line and radial spectra are related through

$$s_r(\omega) = -\omega \frac{ds(\omega)}{d\omega}.$$

Table 4.4 gives expressions for the spectral densities in $\mathbb{R}^1$ for some common covariance models in different parameterizations.

Table 4.4 *Covariance and spectral density functions for second-order stationary processes in $\mathbb{R}^1$. The parameter $\alpha$ is a function of the range and $\sigma^2$ is the variance of the process.*

| Model | $C(h)$ | $s(\omega)$ |
|---|---|---|
| Tent model | $\sigma^2\{1-\frac{h}{\alpha}\}I(h\le\alpha)$ | $\dfrac{\sigma^2\alpha}{\pi}\dfrac{1-\cos\omega\alpha}{\omega^2\alpha^2}$ |
| Exponential | $\sigma^2\exp\{-h/\alpha\}$ | $\dfrac{\sigma^2\alpha}{\pi(1+\omega^2\alpha^2)}$ |
| Exponential | $\sigma^2\exp\{-3h/\alpha\}$ | $\dfrac{3\sigma^2\alpha}{\pi(9+\omega^2\alpha^2)}$ |
| "G"aussian | $\sigma^2\exp\{-h^2/\alpha^2\}$ | $\dfrac{\alpha\sigma^2}{2\sqrt{\pi}}\,e^{-(\alpha\omega)^2/4}$ |
| "G"aussian | $\sigma^2\exp\{-3h^2/\alpha^2\}$ | $\dfrac{\sigma^2\alpha/\sqrt{3}}{2\sqrt{\pi}}\,e^{-(\alpha\omega)^2/12}$ |
| Spherical | $\sigma^2\{1-\frac{3h}{2\alpha}+\frac{h^3}{2\alpha^3}\}I(h\le\alpha)$ | $\dfrac{3\sigma^2}{\pi\omega^2\alpha}\left\{\dfrac{1}{2}+\dfrac{1-\cos\omega\alpha}{(\omega\alpha)^2}-\dfrac{\sin\omega\alpha}{\omega\alpha}\right\}$ |
| Matérn class | $\frac{\sigma^2}{\Gamma(\nu)}\left(\frac{\alpha h}{2}\right)^\nu 2K_\nu(\alpha h)$ | $\sigma^2\dfrac{\alpha^{2\nu}\Gamma(\nu+\frac{1}{2})}{\Gamma(\nu)\Gamma(\frac{1}{2})}(\alpha^2+\omega^2)^{-(\nu+\frac{1}{2})}$ |
| Matérn class | $\frac{\pi^{1/2}\phi}{\Gamma(\nu+1/2)\alpha^{2\nu}}\left(\frac{\alpha h}{2}\right)^\nu 2K_\nu(\alpha h)$ | $\phi(\alpha^2+\omega^2)^{-(\nu+\frac{1}{2})}$ |
| Matérn class | $\frac{\sigma^2}{\Gamma(\nu)}\left(\frac{h\sqrt{\nu}}{\rho}\right)^\nu 2K_\nu\left(\frac{2h\sqrt{\nu}}{\rho}\right)$ | $\dfrac{\sigma^2 g(\rho,\nu)}{\left\{1+(\frac{\rho\omega}{2\sqrt{\nu}})^2\right\}^{\nu+1/2}}$ |

Notice the functional similarity of $s(\omega)$ for the gaussian models and the Gaussian probability density function from which these models derive their name.

The second parameterization of the Matérn class in Table 4.4 is given by Stein (1999, p. 31). In $\mathbb{R}^1$, it is related to the first parameterization through

$$\phi = \sigma^2\frac{\theta^{2\nu}\Gamma(\nu+\frac{1}{2})}{\Gamma(\nu)\Gamma(\frac{1}{2})}.$$

In general, for a second-order stationary process in $\mathbb{R}^d$, the covariance function in the Matérn class and the associated spectral density based on the Stein parameterization are

$$C(\mathbf{h}) = \frac{\pi^{d/2}\phi}{2^{\nu-1}\Gamma(\nu+d/2)\alpha^{2\nu}}(\alpha||\mathbf{h}||)^{\nu}K_{\nu}(\alpha||\mathbf{h}||) \quad (4.63)$$

$$s(\boldsymbol{\omega}) = \phi(\alpha^2+||\boldsymbol{\omega}||^2)^{-(\nu+d/2)}. \quad (4.64)$$

Whereas this parameterization leads to a simple expression for the spectral density function, the covariance function unfortunately depends on the dimension of the domain ($d$). Furthermore, the range of the process is a function of $\alpha^{-1}$ but depends strongly on $\nu$ (Stein, 1999, p. 50). Fuentes (2001) argues—in the context of spatial prediction (kriging, Chapter 5)—that if the degree of smoothness, $\nu$, has been correctly determined, the kriging predictions are asymptotically optimal, provided $\nu$ and $\phi$ are not spatially varying. The low-frequency values then have little effect on the predicted values. As a consequence, she focuses on the high frequency values and uses (in $\mathbb{R}^d$)

$$s(\boldsymbol{\omega}) = \phi\left(||\boldsymbol{\omega}||^2\right)^{\nu-d/2},$$

an approximation to (4.64) as $||\boldsymbol{\omega}|| \to \infty$.

The third parameterization of the Matérn class was used in Handcock and Stein (1993) and Handcock and Wallis (1994). The range parameter $\rho$ is less dependent on the smoothness parameter $\nu$ than in the other two parameterizations. For the exponential model ($\nu = 1/2$) the practical range in the Handcock-Stein-Wallis parameterization is $3\rho/\sqrt{2}$. In $\mathbb{R}^d$, the (isotropic) spectral density is

$$s(\omega) = \frac{\sigma^2 g(\rho,\nu,d)}{\left\{1+(\frac{\rho\omega}{2\sqrt{\nu}})^2\right\}^{\nu+d/2}}$$

with

$$g(\rho,\nu,d) = \frac{\rho^d\Gamma(\nu_d/2)(4\nu)^{\nu}}{\Gamma(\nu)\pi^{d/2}}.$$

The function $g(\rho,\nu)$ in Table 4.4 is $g(\rho,\nu,1)$.

### *4.7.3 Analysis of Point Patterns*

Among the advantages of spectral methods for spatial data is the requirement that the process be second-order stationary, but it does not have to be isotropic (§2.5.7). This is particularly important for point pattern analysis, because the tool most commonly used for second-order analysis in the spatial domain, the $K$-function, requires stationarity *and* isotropy. The previous discussion of periodogram analysis focused on the case of equally spaced, gridded data. This is not a requirement of spectral analysis but offers computational advantages in making available the Fast Fourier Transform (FFT). Spatial locations in a point pattern are irregularly spaced by the very nature of the process and $Z$ is a degenerate attribute unless the process is marked. The spectral approach to point pattern analysis proceeds as follows. An estimate of the spectral density of the observed point process is compared against the spectral density function of a known second-order stationary point process and tested for agreement. To obtain the spectral density function, the covariance function of the process is required, however. The covariance functions discussed so far are not directly applicable to point processes since the domain is random and not fixed as for geostatistical or lattice data.

#### 4.7.3.1 Bartlett's Covariance Function

Bartlett (1964) approached the development of a covariance function in a point process by first assuming that the process is orderly, that is,

$$\lim_{|d\mathbf{s}|\to 0} \Pr(N(d\mathbf{s}) > 1) = 0.$$

This eliminates the possibility of multiple events at location $\mathbf{s}$. Thus, in the limit, $\mathrm{E}[N(d\mathbf{s})^2] = \mathrm{E}[N(d\mathbf{s})]$, and

$$\lambda = \lim_{|d\mathbf{s}|\to 0} \frac{\mathrm{E}[N(d\mathbf{s})]}{|d\mathbf{s}|} = \lim_{|d\mathbf{s}|\to 0} \frac{\mathrm{E}[N(d\mathbf{s})^2]}{|d\mathbf{s}|}. \tag{4.65}$$

*Complete Covariance Density*

Bartlett (1964) then defined the complete covariance density function as

$$\lim_{|d\mathbf{s}_i|,|d\mathbf{s}_j|\to 0} \frac{\mathrm{Cov}[N(d\mathbf{s}_i), N(d\mathbf{s}_j)]}{|d\mathbf{s}_i||d\mathbf{s}_j|} = C(\mathbf{s}_i - \mathbf{s}_j) + \delta(\mathbf{s}_i - \mathbf{s}_j), \tag{4.66}$$

where $C(\mathbf{s}_i - \mathbf{s}_j) = \lambda_2(\mathbf{s}_i - \mathbf{s}_j) - \lambda^2$ is the autocovariance density function (§3.3.4) and $\delta(\mathbf{u})$ is the Dirac delta function

$$\delta(\mathbf{u}) = \begin{cases} \infty & \mathbf{u} = \mathbf{0} \\ 0 & \mathbf{u} \neq \mathbf{0}. \end{cases}$$

Substituting into (4.66) and using (4.65) yields

$$\lim_{|d\mathbf{s}_i|,|d\mathbf{s}_j|\to 0} \frac{\mathrm{Cov}[N(d\mathbf{s}_i), N(d\mathbf{s}_j)]}{|d\mathbf{s}_i||d\mathbf{s}_j|} = \begin{cases} C(\mathbf{s}_i - \mathbf{s}_j) & \mathbf{s}_i \neq \mathbf{s}_j \\ \lim_{|d\mathbf{s}_i|\to 0} \frac{\lambda}{|d\mathbf{s}_i|} - \lambda^2 & \mathbf{s}_i = \mathbf{s}_j. \end{cases}$$

The Dirac delta function is involved in (4.66) because the variance of $N(d\mathbf{s})/|d\mathbf{s}|$ should go to infinity in the limit as $|d\mathbf{s}|$ is shrunk.

#### 4.7.3.2 Spectral Density and Periodogram

We now can take the Fourier transform of (4.66) to obtain the spectral density function

$$\begin{aligned} s(\boldsymbol{\omega}) &= \frac{1}{(2\pi)^2} \left\{ \int_{\mathbf{u}} e^{-\mathrm{i}\boldsymbol{\omega}'\mathbf{u}} \{v(\mathbf{u}) + \lambda\delta(\mathbf{u})\}\, d\mathbf{u} \right\} \\ &= \frac{1}{(2\pi)^2} \left\{ \lambda + \int_{\mathbf{u}} e^{-\mathrm{i}\boldsymbol{\omega}'\mathbf{u}} v(\mathbf{u}) d\mathbf{u} \right\}. \end{aligned}$$

As before, $s(\omega)$ is estimated through the periodogram $I(\omega) = J(\omega)\overline{J}(\omega)$. Following Renshaw and Ford (1983), we put

$$J(\omega) = \frac{1}{2\pi}\sum_{k=1}^{n} Z(\mathbf{s}_j)\exp\{i\omega'\mathbf{L}^{-1}\mathbf{s}_k\}, \tag{4.67}$$

where $\overline{J}(\omega)$ is the complex conjugate of $J(\omega_{pq})$, and $\{\omega\} = \{[2\pi p, 2\pi q]'\}$ for $p = 0, 1, 2, \cdots$ and $q = 0, \pm 1, \pm 2, \cdots$. The matrix $\mathbf{L}$ is diagonal with entries $L_x$ and $L_y$ such that $D = L_x L_y$. It is thus assumed that the bounding shape of the point process is a rectangle. The term $\mathbf{L}^{-1}\mathbf{s}$ scales the process to the unit square and the intensity $\lambda$ is estimated by the number of events $n$. The periodogram is then given by

$$\begin{aligned} I(\omega) &= J(\omega)\overline{J}(\omega) \\ &= \frac{1}{(2\pi)^2}\sum_{j=1}^{n}\sum_{k=1}^{n}\exp\left\{-i\omega'\mathbf{L}^{-1}(\mathbf{s}_j - \mathbf{s}_k)\right\}. \end{aligned} \tag{4.68}$$

Mugglestone and Renshaw (1996a) recommend that the periodogram be computed for frequencies constructed from $p = 0, \cdots, 16$ and $q = -15, \cdots, 16$ if $n < 100$.

Renshaw and Ford (1983) give a polar representation of $I(\omega)$. Note that $I(\omega)$ is calculated for the frequencies $\omega_p = 2\pi p$ and $\omega_q = 2\pi q$ for some integer values of $p = 0, 1, \cdots$ and $q = -c, -c+1, \cdots, c, c+1$. Let $\rho = \sqrt{p^2 + q^2}$, the Euclidean distance on the $p$–$q$ lattice, and $\theta = \tan^{-1}(q/p)$. If $P(\alpha_{\rho\theta})$ is the periodogram in polar coordinates that corresponds to $I(\omega_p, \omega_q)$, then Renshaw and Ford (1983) define the $R$- and $\Theta$- spectra by averaging periodogram ordinates with similar values of $\rho$ or $\theta$:

*Polar Spectra*

$$\begin{aligned} S_R(\rho) &= \frac{1}{n_\rho}\sum_{\tau}\sum_{\theta} P(\alpha_{\tau\theta}) \\ S_\Theta(\theta) &= \frac{1}{n_\theta}\sum_{\rho}\sum_{\tau} P(\alpha_{\rho\tau}). \end{aligned}$$

The numbers $n_\rho$ and $n_\theta$ denote the number of ordinates for which $\tau$ falls within a specified tolerance. The result is that ordinates are averaged in arcs around the origin in the $R$-spectrum and in rays emanating from the origin in the $\Theta$-spectrum (Figure 4.18).

The $R$-spectrum gives insight about clustering or regularity of events, the $\Theta$-spectrum about the isotropy of the process. The $R$-spectrum is interpreted along the same lines as the $K$-function. If $S_R(\rho)$ takes on large values for small $\rho$, the process is clustered. Small values of $S_R(\rho)$ for small $\rho$ implies regularity.

#### 4.7.3.3 Inference Based on the Periodogram

Mugglestone and Renshaw (1996b) establish that (4.68) is an unbiased estimator of $s(\omega)$ as $n \to \infty$. These authors also provide an extension of a proof

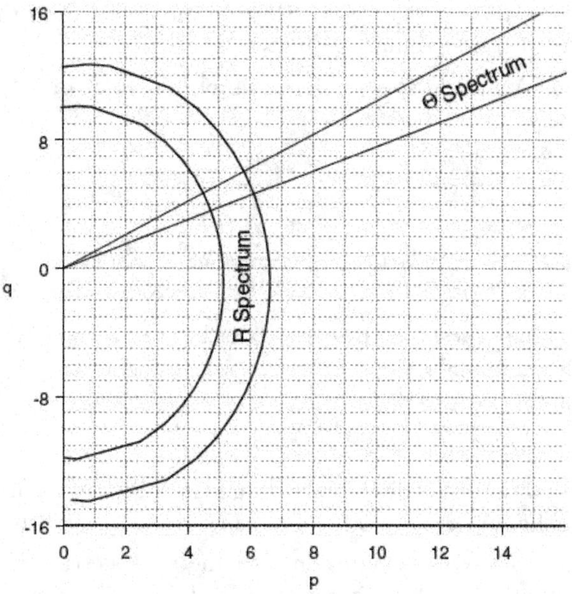


Figure 4.18 *Polar spectra according to Renshaw and Ford (1983).*

by Brillinger (1972) for point processes in $\mathbb{R}^1$ to the two-dimensional case (Mugglestone and Renshaw, 1996a). If the process is second-order stationary and the spectral density is continuous, then

$$2\frac{I(\omega)}{s(\omega)} \xrightarrow{d} \chi_2^2 \text{ if } \omega \neq \mathbf{0} \tag{4.69}$$

$$2\frac{I(\omega) - (2\pi)^{-2}\lambda}{s(\mathbf{0})} \xrightarrow{d} \chi_1^2 \text{ if } \omega = \mathbf{0} \tag{4.70}$$

$$\text{Cov}[I(\omega_1), I(\omega_2)] \to 0, \text{ if } \omega_1 \neq \omega_2. \tag{4.71}$$

Furthermore, the CSR process has spectral density $s(\omega) = \lambda/(2\pi)^2$. Combining this result with (4.69) and (4.71) enables us to derive test statistics for the CSR hypothesis for a test without simulations. In addition, we can make use of the polar spectra and the fact that the spectral analysis does not require isotropy of the process, to develop a test for isotropy (and other forms of dependence symmetry).

First, asymptotically, any sum of periodogram ordinates is a sum of independent scaled Chi-square random variables. For example, under the CSR hypothesis,

$$\frac{8\pi^2}{\lambda} \sum_{p,q \neq 0} I(\omega_p, \omega_q) \sim \chi_{2m}^2,$$

where $m$ is the number of periodogram ordinates in the sum. Bartlett (1978) constructs a goodness-of-fit test of the CSR hypothesis based on this result. The frequency plane is gridded, and ordinates are totaled in each cell.

Under CSR, the $R$- and $\Theta$-spectra have simple (asymptotic) Chi-square distributions, namely

$$\begin{aligned}\frac{(2\pi)^2}{\lambda}S_R(\rho) &\sim \frac{1}{2n_\rho}\chi^2_{2n_\rho}\\ \frac{(2\pi)^2}{\lambda}S_\Theta(\theta) &\sim \frac{1}{2n_\theta}\chi^2_{2n_\theta}.\end{aligned}$$

A test of the CSR hypothesis can be carried out by comparing $(2\pi)^2 S_R(\rho)/\widehat{\lambda}$ against the confidence bounds

$$\frac{1}{2n_\rho}\times\left(\chi^2_{\alpha/2,2n_\rho},\chi^2_{1-\alpha/2,2n_\rho}\right).$$

Values of the $R$-spectrum outside of these bounds suggest a departure from CSR (clustering if the upper, regularity if the lower boundary is crossed).

By construction of the $\Theta$-spectrum, an asymmetric spectrum is a sign of anisotropy in the spatial point process. An $F$-test is easily constructed, by comparing periodogram ordinates reflected about the origin. In an isotropic pattern we expect $S_\Theta(\theta) = S_\Theta(-\theta)$. A test for isotropy in the direction $\theta$ uses test statistic

$$F = \frac{S_\Theta(\theta)}{S_\Theta(-\theta)}$$

and null distribution $F_{2n_\theta,2n_{-\theta}}$.

## 4.8 On the Use of Non-Euclidean Distances in Geostatistics

In §2.2 we defined isotropic semivariograms and covariance functions in terms of the Euclidean norm, $||\mathbf{s}_i - \mathbf{s}_j|| = \sqrt{(x_i - x_j)^2 + (y_i - y_j)^2}$, for $\mathbf{s}_i = [x_i, y_i]'$. The use of this norm was implicit in all isotropic models described in §4.3. Even more implicit was the use of this norm in proving that these models are "valid," i.e., the isotropic semivariogram models are conditionally negative definite and the isotropic covariance models are positive definite. However, in many practical applications this distance measure may not be realistic. For example, mountains, irregularly-shaped domains, water bodies (e.g., lakes, bays, estuaries) and partitions in a building can present barriers to movement. Although two points on either side of a barrier may be physically close, it may be unrealistic to assume that they are related. For example, Little et al. (1997) note that for measurements made in estuarine streams, distances should be measured "as the fish swims" and compute **water distance** (the shortest path between two sites that may be traversed entirely over water) using a GIS. In some contexts, water distance may be calculated along an irregular one-dimensional transect (Cressie and Majure, 1997; Rathbun, 1998), but branching estuaries or large barriers prohibit this approach. For more complex applications such as these, Kern and Higdon (2000) define an algorithm to compute **polygonal distance** that compensates for irregularities in the spatial domain. Krivoruchko and Gribov (2004) solve similar problems using **cost weighted distance**, a common raster function in GIS. Unfortunately, as we illustrate below, not all isotropic covariance function and semivariogram models remain valid when based on non-Euclidean distances.

### *4.8.1 Distance Metrics and Isotropic Covariance Functions*

A **metric space** is a nonempty set $\mathcal{S}$ together with a real valued function $d : \mathcal{S} \times \mathcal{S} \to [0, \infty)$ which satisfies the following conditions:

1. $d(\mathbf{s}_i, \mathbf{s}_j) \geq 0$, with equality holding if and only if $\mathbf{s}_i = \mathbf{s}_j$.
2. $d(\mathbf{s}_i, \mathbf{s}_j) = d(\mathbf{s}_j, \mathbf{s}_i)$ (symmetry).
3. $d(\mathbf{s}_i, \mathbf{s}_j) \leq d(\mathbf{s}_i, \mathbf{s}_k) + d(\mathbf{s}_k, \mathbf{s}_j)$ (triangle inequality).

The function $d$ is called a metric on $\mathcal{S}$. In our context, $\mathcal{S}$ is the collection of spatial locations $\mathbf{s} \in D \subset \mathbb{R}^n$, and $d(\mathbf{s}_i, \mathbf{s}_j)$ is considered to be the distance from $\mathbf{s}_i$ to $\mathbf{s}_j$. Common examples in $\mathcal{S} = \mathbb{R}^2$ include:

$$d(\mathbf{s}_i, \mathbf{s}_j) = |x_i - x_j| + |y_i - y_j| \quad \text{Rectangular or City Block distance;}$$
$$d(\mathbf{s}_i, \mathbf{s}_j) = \sqrt{(x_i - x_j)^2 + (y_i - y_j)^2} \quad \text{Euclidean distance;}$$
$$d(\mathbf{s}_i, \mathbf{s}_j) = \max\{|x_i - x_j|, |y_i - y_j|\}.$$

Curriero (1996, 2004) gives a simple example that clearly demonstrates the problem with using non-Euclidean distances with isotropic models for covariance functions. He considers a regular two dimensional lattice of four points with unit spacing. The matrix of the distances among all four points based on the city block distance definition is

$$\begin{bmatrix} 0 & 1 & 1 & 2 \\ 1 & 0 & 2 & 1 \\ 1 & 2 & 0 & 1 \\ 2 & 1 & 1 & 0 \end{bmatrix}.$$

Using these distances in an isotropic gaussian model for the covariance function (4.10) with $\sigma^2 = 20$ and $\alpha = 2\sqrt{3}$, gives the following covariance matrix

$$\begin{bmatrix} 20.00 & 15.58 & 15.58 & 7.36 \\ 15.58 & 20.00 & 7.36 & 15.58 \\ 15.58 & 7.36 & 20.00 & 15.58 \\ 7.36 & 15.58 & 15.58 & 20.00 \end{bmatrix}.$$

Unfortunately, one of the characteristic roots of this matrix is negative. Consequently, this matrix is not positive definite and so the gaussian model based on city block distances cannot be a valid covariance function in $\mathbb{R}^2$. However, the exponential model remains valid when used with the city block metric (Curriero, 2004).

It is fairly straightforward to show that a particular model based on a specified distance metric is *not* a valid covariance function. However, proving that a particular model based on a specified distance metric *is* valid is much more difficult. It is not sufficient to show that the corresponding covariance matrix is positive definite. While this may suffice for a particular problem with specified locations, the validity may change when based on a different spatial configuration, including that which may be considered for spatial prediction.

Other than checking the validity of a particular model with respect to a chosen distance matrix as described above (and then hoping for the best), the most common way of ensuring valid models for covariance functions and semivariograms based on non-Euclidean distance metrics uses transformations of the spatial coordinates. This is called **isometric embedding** (Curriero, 1996, 2004). The metric space $(\mathcal{S}, d)$ is isometrically embedded in a Euclidean space of dimension $k$, if there exists points $\mathbf{s}_i^*, \mathbf{s}_j^* \in \mathbb{R}^k$ and a function $\phi : \mathcal{S} \rightarrow \mathbb{R}^k$ such that

$$d(\mathbf{s}_i, \mathbf{s}_j) = ||\mathbf{s}_i^* - \mathbf{s}_j^*||,$$

where $\phi(\mathbf{s}_i) = \mathbf{s}_i^*$ and $||\mathbf{s}_i^* - \mathbf{s}_j^*||$ is the Euclidean distance metric in $\mathbb{R}^k$. Exact isometric embedding can be difficult to accomplish theoretically. However, it can be done approximately, using a technique known as multidimensional scaling.

### *4.8.2 Multidimensional Scaling*

Consider a $(n \times n)$ symmetric matrix $\mathbf{D}$ with typical element $d_{ij} \geq 0$ and $d_{ii} = 0$. Then $\mathbf{D}$ is said to be a matrix of distances or dissimilarities. Note that the $d_{ij}$ are not necessarily Euclidean distances. They can represent some other measure of how "far" two items are apart. In the space deformation approach of Sampson and Guttorp (1992) for modeling non-stationary spatial covariances, for example, the $d_{ij}^2$ are estimates of $\widehat{\text{Var}}[Z(\mathbf{s}_i) - Z(\mathbf{s}_j)]$; see §8.2.2.

The technique of multidimensional scaling (MDS; Mardia, Kent, and Bibby, 1979, Ch. 14) is applied when the matrix $\mathbf{D}$ is known, but the point locations that produced the distances are unknown. The objective is to find points $\mathbf{s}_1, \cdots, \mathbf{s}_n$ in $\mathbb{R}^k$ such that

$$\widehat{d}_{ij} = ||\mathbf{s}_i - \mathbf{s}_j|| \approx d_{ij}.$$

In other words, find a configuration of points in $k$ dimensional space such that distances among these points are close to the measures of dissimilarity which were given. One aspect of MDS is to determine the dimension $k$. For interpretability of the solution you want $k$ to be small. When it is known that $\mathbf{D}$ is constructed from points in $\mathbb{R}^d$, then $k \leq d$.

A feature of solutions to the MDS problem is their invariance to rotation and translation. If $\mathbf{D}$ was constructed from known coordinates, a solution can be transformed so that its coordinates are comparable to those of the original configuration (see §4.8.2.2). An application is the translation of non-Euclidean distances (e.g., city block distances) into Euclidean distances.

#### *4.8.2.1 The Classical, Metric Solution*

Mardia, Kent, and Bibby (1979, Ch. 14) distinguish the classical solution—also called a metric solution—from the non-metric solution that is based on ranks of distances and iterative optimization. A metric solution determines the point configuration directly and can serve as the solution of the MDS problem or as the starting configuration of a non-metric technique. The classical solution consists of choosing as point configuration the $k$ (scaled) eigenvectors that correspond to the $k$ largest (positive) eigenvalues of the matrix

$$\mathbf{B} = -\frac{1}{2}(\mathbf{I} - \mathbf{J}/n)'\mathbf{D}^{[2]}(\mathbf{I} - \mathbf{J}/n).$$

The notation $\mathbf{A}^{[p]}$ stands for a matrix whose elements are $a_{ij}^p$. The eigenvectors are scaled so that if $\mathbf{s}_i{}^*$ is the $i$th eigenvector of $\mathbf{B}$ and $\lambda_i$ is the $i$th (positive) eigenvalue, then $\mathbf{s}_i{}^{*\prime}\mathbf{s}_i{}^* = \lambda_i$.

Denote the solution so obtained as $\mathbf{S}$ and let $\widehat{d}_{ij} = ||\mathbf{s}_i - \mathbf{s}_j||$. The discrepancy between $\mathbf{D}$ and the fit can be measured by

$$\psi = \text{trace}\{(\mathbf{B} - \widehat{\mathbf{B}})^2\}.$$

The classical solution minimizes this trace among all configurations that have distance matrix $\mathbf{D}$ for a given value of $k$. If $\boldsymbol{\lambda} = [\lambda_{(1)}, \cdots, \lambda_{(n)}]'$ denotes the vector of ordered eigenvalues of $\mathbf{B}$, then

$$r = \sum_{i=1}^{k} \lambda_{(i)} \Big/ \sum_{i=1}^{n} |\lambda_i|$$

is a measure of the agreement between the metric solution and the distance matrix $\mathbf{D}$.

**Example 4.6 Classical MDS.** Assume the following six point locations in $\mathbb{R}^2$: $\mathbf{s}_1 = [0, 65], \mathbf{s}_2 = [115, 50], \mathbf{s}_3 = [225, 120], \mathbf{s}_4 = [175, 65]$, $\mathbf{s}_5 = [115, 132.5]$, $\mathbf{s}_6 = [30, 105]$. The matrix of Euclidean distances is

$$\mathbf{D}_1 = \begin{bmatrix} 0 & 115.9 & 231.6 & 175 & 133.3 & 50 \\ 115.9 & 0 & 130.4 & 61.8 & 82.5 & 101.2 \\ 231.6 & 130.3 & 0 & 74.3 & 110.7 & 195.6 \\ 175 & 61.8 & 74.3 & 0 & 90.3 & 150.4 \\ 133.3 & 82.5 & 110.7 & 90.3 & 0 & 89.3 \\ 50 & 101.2 & 195.6 & 150.4 & 89.3 & 0 \end{bmatrix}$$

and the matrix of city-block distances is

$$\mathbf{D}_2 = \begin{bmatrix} 0.0 & 130.0 & 280.0 & 175.0 & 182.5 & 70.0 \\ 130.0 & 0.0 & 180.0 & 75.0 & 82.5 & 140.0 \\ 280.0 & 180.0 & 0.0 & 105.0 & 122.5 & 210.0 \\ 175.0 & 75.0 & 105.0 & 0.0 & 127.5 & 185.0 \\ 182.5 & 82.5 & 122.5 & 127.5 & 0.0 & 112.5 \\ 70.0 & 140.0 & 210.0 & 185.0 & 112.5 & 0.0 \end{bmatrix}.$$

The (ordered) eigenvalues of $\mathbf{B}_1$ are $\boldsymbol{\lambda}_1 = [36375.07, 5405.14, 0, 0, 0, 0]'$ and those of $\mathbf{B}_2$ are $\boldsymbol{\lambda}_2 = [47754.86, 14646.25, 0, -2196.03, -5775.36]'$. For $k = 2$, $r_1 = 1$ and $r_2 = 0.818$. A classical solution with $k = 2$ will reproduce perfectly the distance matrix $\mathbf{D}_1$, but not the distance matrix $\mathbf{D}_2$. The fitted distances are

$$\widehat{\mathbf{D}}_1 = \mathbf{D}_1 = \begin{bmatrix} 0.0 & 115.9 & 231.6 & 175 & 133.3 & 50.0 \\ 115.9 & 0.0 & 130.4 & 61.8 & 82.5 & 101.2 \\ 231.6 & 130.3 & 0.0 & 74.3 & 110.7 & 195.6 \\ 175.0 & 61.8 & 74.3 & 0.0 & 90.3 & 150.4 \\ 133.3 & 82.5 & 110.7 & 90.3 & 0.0 & 89.3 \\ 50.0 & 101.2 & 195.6 & 150.4 & 89.3 & 0.0 \end{bmatrix}$$

and

$$\widehat{\mathbf{D}}_2 = \begin{bmatrix} 0.0 & 125.8 & 280.9 & 192.6 & 177.1 & 107.7 \\ 125.8 & 0.0 & 169.9 & 67.2 & 99.5 & 131.7 \\ 280.9 & 169.9 & 0.0 & 130.7 & 113.7 & 221.7 \\ 192.6 & 67.2 & 130.7 & 0.0 & 114.8 & 187.2 \\ 177.1 & 99.5 & 113.7 & 114.8 & 0.0 & 108.1 \\ 107.7 & 131.7 & 221.7 & 187.2 & 108.1 & 0.0 \end{bmatrix}.$$

The MDS solutions do not "match" the original coordinates in the sense that they are centered at $\mathbf{0}$ and are not yet rotated and scaled (Table 4.5). This can be accomplished with a Procrustes rotation (see below).

Table 4.5 Classical Solutions to MDS for $D_1$ and $D_2$

| x | y | MDS for $\mathbf{D}_1$ | | MDS for $\mathbf{D}_2$ | |
|---|---|---|---|---|---|
| 0 | 65.0 | −112.04 | −12.33 | −134.88 | 22.99 |
| 115 | 50.0 | 0.61 | −39.89 | −11.41 | 46.98 |
| 225 | 120.0 | 117.65 | 17.57 | 141.48 | −27.03 |
| 175 | 65.0 | 61.90 | −31.59 | 52.22 | 68.49 |
| 115 | 132.5 | 9.69 | 42.10 | 29.04 | −43.94 |
| 30 | 105.0 | −77.82 | 24.13 | −76.45 | −67.51 |

□

The non-metric solution to the MDS problem is a rank-based method in which the "stress" criterion

$$\sum_{i<j} \frac{(f(d_{ij}) - \widehat{d}_{ij})^2}{\sum_{i<j} \widehat{d}_{ij}^2}$$

is minimized subject to a monotonicity requirement,

$$d_{ij} < d_{kl} \Rightarrow f(d_{ij}) \leq f(d_{kl}) \, \forall \, i < j, k < l.$$

The process is iterative, starting from an initial configuration which can be the geographical locations (if known, as in Sampson and Guttorp, 1992), or the result of a classical MDS solution. The algorithm is known as the Shepard-Kruskal algorithm.

*4.8.2.2 Procrustes Rotation With Scaling*

Consider two point configurations $\mathbf{S}_1$ and $\mathbf{S}_2$. For example, $\mathbf{S}_1$ is the original point configuration from which $\mathbf{D}$ was computed based on city block distances and $\mathbf{S}_2$ is the classical MDS solution. Or, $\mathbf{S}_1$ and $\mathbf{S}_2$ correspond to the classical and non-metric MDS solutions. We would like to translate the configuration $\mathbf{S}_2$ so that its coordinates match that of $\mathbf{S}_1$ and compute a measure of closeness between the two configurations. The technique that accomplishes these two goals is termed a Procrustes rotation of $\mathbf{S}_2$ relative to $\mathbf{S}_1$ (Mardia, Kent, and Bibby, 1979, Ch. 14.7).

Denote the points in $\mathbf{S}_i$ as $\mathbf{s}_1^{(i)}, \cdots, \mathbf{s}_n^{(i)}$. The Procrustes rotation (with scaling) of $\mathbf{S}_2$ relative to $\mathbf{S}_1$ is obtained by minimizing

$$SS(\mathbf{A}, \mathbf{b}) = \sum_{i=1}^{n} (\mathbf{s}_i^{(1)} - \mathbf{s}_i^{(2)*})'(\mathbf{s}_i^{(1)} - \mathbf{s}_i^{(2)*})$$

subject to $\mathbf{s}_i^{(2)*} = c\mathbf{A}'\mathbf{s}_i^{(2)} + \mathbf{b}$. The solutions to this least squares problem are

$$\begin{aligned} \mathbf{b} &= \overline{\mathbf{s}}^{(1)} - \mathbf{A}'\overline{\mathbf{s}}^{(2)} \\ \mathbf{A} &= \mathbf{U}\mathbf{V}' \\ c &= \mathrm{tr}\{\boldsymbol{\Delta}\}/\mathrm{tr}\{\mathbf{S}_2\mathbf{S}_2'\} \end{aligned}$$

and $\mathbf{U}\boldsymbol{\Delta}\mathbf{V}'$ is the singular value decomposition of $\mathbf{S}_2'\mathbf{S}_1$. A Procrustes rotation without scaling has $c = 1.0$.

**Example 4.6 (Classical MDS. Continued)** Applying a Procrustes rotation of the MDS solutions in Table 4.5 with respect to the original configuration yields the coordinates in Table 4.6. As expected, the solution for $\mathbf{D}_1$ maps perfectly in the observed coordinates; recall that $r_1 = 1$. The coordinates for the $\mathbf{D}_2$ solution do not agree with the original point location because $\mathbf{D}_2$ is based on city block distances. The configuration $\mathbf{S}_2$ is the arrangement of points that will yield a matrix of Euclidean distances closest to $\mathbf{D}_2$.

□

Table 4.6 *Procrustes rotations for MDS solutions*

| x | y | $\mathbf{S}_1^*$ | | $\mathbf{S}_2^*$ | |
|---|---|---|---|---|---|
| 0 | 65.0 | 0 | 65.0 | 9.11 | 47.72 |
| 115 | 50.0 | 115 | 50.0 | 109.47 | 50.99 |
| 225 | 120.0 | 225 | 120.0 | 215.30 | 135.75 |
| 175 | 65.0 | 175 | 65.0 | 162.81 | 45.56 |
| 115 | 132.5 | 115 | 132.5 | 124.79 | 128.94 |
| 30 | 105.0 | 30 | 105.0 | 38.51 | 128.54 |

## 4.9 Supplement: Bessel Functions

### *4.9.1 Bessel Function of the First Kind*

The Bessel function of the first kind of order $\nu$ is defined by the series

$$J_\nu(t) = \left(\frac{t}{2}\right)^\nu \sum_{i=0}^{\infty} \frac{(-\frac{1}{4}t^2)^i}{i!\,\Gamma(\nu+i+1)}. \tag{4.72}$$

We use the notation $J_n(t)$ if the Bessel function has integer order. A special case is $J_0(t)$, the Bessel function of the first kind of order 0. It appears as the basis function in spectral representations of isotropic covariance functions in $\mathbb{R}^2$ (§4.3.1). Bessel functions of the first kind of integer order satisfy (Abramowitz and Stegun, 1964)

$$\begin{aligned}
J_{n+1}(t) &= \frac{2n}{t}J_n(t) - J_{n-1}(t) \\
J_n'(t) &= \frac{1}{2}\left(J_{n-1}(t) - J_{n+1}(t)\right) \\
&= J_{n-1}(t) - \frac{n}{t}J_n(t) \\
J_{-n}(t) &= (-1)^n J_n(t)
\end{aligned}$$

### *4.9.2 Modified Bessel Functions of the First and Second Kind*

There are two types of modified Bessel functions. Of particular importance for spatial modeling are the modified Bessel functions of the second kind $K_\nu(t)$ of (real) order $\nu$. They appear as components of the Matérn class of covariance functions for second-order stationary processes (see §4.3.2):

$$K_\nu(t) = \frac{\pi}{2}\frac{I_{-\nu}(t) - I_\nu(t)}{\sin(\pi\nu)}. \tag{4.73}$$

The function $I_\nu(t)$ in (4.73) is the modified Bessel function of the first kind, defined by

$$I_\nu(t) = \left(\frac{t}{2}\right)^\nu \sum_{i=0}^{\infty} \frac{(\frac{1}{4}t^2)^i}{i!\,\Gamma(\nu+i+1)} \left(\frac{t}{2}\right)^{2k}.$$

Since computation of these functions can be numerically expensive, approximations can be used for $t \to 0$:

$$K_0(t) \approx -\ln\{t\}; \qquad K_\nu(t) \approx \frac{\Gamma(\nu)}{2}\left(\frac{t}{2}\right)^{-\nu} \quad \text{for } \nu > 0.$$

Other important results regarding modified Bessel functions (Abramowitz and Stegun, 1964; Whittaker and Watson, 1927) are ($n$ denoting integer and $\nu$ denoting real order)

$$\begin{aligned}
K_\nu(t) &= K_{-\nu}(t) \\
K_{n+1}(t) &= K_{n-1}(t) + \frac{2n}{t}K_n(t) \\
K_n'(t) &= \frac{n}{t}K_n(t) - K_{n+1}(t) \Rightarrow K_0'(t) = -K_1(t) \\
K_\nu(t) &= \frac{(2t)^\nu\Gamma(t+1/2)}{\sqrt{\pi}}\cos\{\nu\pi\}\int_0^\infty (u^2+t^2)^{-\nu-1/2}\cos\{u\}\,du \\
I_\nu(0) &= \begin{cases} 1 & \nu = 0 \\ 0 & \nu > 0 \end{cases} \\
I_n(t) &= \frac{1}{\pi}\int_0^\pi e^{z\cos\theta}\cos\{n\theta\}\,d\theta \\
I_n(t) &= I_{-n}(t) \\
I_{1/2}(t) &= \sqrt{\frac{2}{\pi t}}\sinh\{t\} \qquad I_{-1/2}(t) = \sqrt{\frac{2}{\pi t}}\cosh\{t\} \\
I_n'(t) &= \frac{n}{t}I_n(t) + I_{n+1}(t) \Rightarrow I_0'(t) = I_1(t)
\end{aligned}$$

Some of these properties have been used in §4.3.2 to establish that the Matèrn model for $\nu = 1/2$ reduces to the exponential covariance function.

A Fortran program (`rkbesl`) to calculate $K_{n+\alpha}(t)$ for non-negative $t$ and non-negative order $n+\alpha$ is distributed as part of the `SPECFUN` package (Cody, 1987). It is available at `www.netlib.org`.

## 4.10 Chapter Problems

**Problem 4.1** Verify

(i) that the empirical semivariogram (use the Matheron estimator) is a biased estimator of $\gamma(\mathbf{h})$ under trend contamination of the random field;

(ii) that the mean of $\widehat{\gamma}(\mathbf{h})$ is given by (4.4).

Simulate data from a simple linear regression $Y_i = \alpha + \beta x_i + e_i$, where the errors are independently and identically distributed with mean 0 and variance

$\sigma^2$. Let $x_i = i$ and consider $x_i$ as the location on a transect at which $Y_i$ was observed. Calculate the empirical semivariogram estimator for different values of $\alpha$, $\beta$, and $\sigma^2$. How does the choice of $\alpha$ affect the semivariogram?

**Problem 4.2** (Schabenberger and Pierce, 2002, pp. 431–433) Let $Y_i = \beta + e_i$, $(i = 1, \cdots, n)$, where $e_i \sim iid\ G(0, \sigma^2)$.

(i) Find the maximum likelihood estimators of $\beta$ and $\sigma^2$.

(ii) Define a random vector

$$\mathbf{U}_{(n-1\times 1)} = \begin{bmatrix} Y_1 - \overline{Y} \\ Y_2 - \overline{Y} \\ \vdots \\ Y_{n-1} - \overline{Y} \end{bmatrix}$$

and find the maximum likelihood estimator of $\sigma^2$ based on $\mathbf{U}$. Is it possible to estimate $\beta$ from the likelihood of $\mathbf{U}$?

(iii) Show that the estimator of $\sigma^2$ found in (ii) is the **restricted** maximum likelihood estimator for $\sigma^2$ based on the random vector $\mathbf{Y} = [Y_1, \cdots, Y_n]'$.

**Problem 4.3** Consider random variables $X \sim G(0,1)$ and $Y \sim G(0,1)$ with $\text{Corr}[X,Y] = \rho$. Show that $\text{Corr}[X^2, Y^2] = \rho^2$. *Hint: Consider* $X$, $U \sim G(0,1)$ *with* $X \perp U$ *and define* $Y = \rho X + \sqrt{1-\rho^2}U$.

**Problem 4.4** The Matheron estimator can be written as a quadratic form in the observed data,

$$\widehat{\gamma}(\mathbf{h}) = \frac{1}{2} Z(\mathbf{s})' \mathbf{A}(\mathbf{h}) Z(\mathbf{s}).$$

The matrix $\mathbf{A}$ is called a spatial design matrix (Genton, 1998b).

(i) Imagine $n$ regularly spaced data points in $\mathbb{R}^1$. Give the matrix $\mathbf{A}(\mathbf{h})$ for this case.

(i) Give the matrix $\mathbf{A}(\mathbf{h})$ when the data are on a $3 \times 3$ lattice.

**Problem 4.5** Derive the composite likelihood score equation (4.44) under the assumption that $Z(\mathbf{s}_i) - Z(\mathbf{s}_j)$ are zero mean Gaussian random variables.

**Problem 4.6** The derivation of the composite likelihood estimator of the semivariogram in §4.5.3 commenced by assuming that $T_{ij}^{(2)} = Z(\mathbf{s}_i) - Z(\mathbf{s}_j) \sim G(0, 2\gamma(\mathbf{h}_{ij}, \boldsymbol{\theta}))$ and then built a composite score function from the score contributions of the $T_{ij}^{(2)}$. With this distributional assumption for $T_{ij}^{(2)}$, derive the composite likelihood score function from the distribution of $T_{ij}^{(3)}$.

**Problem 4.7** Establish the connection between the periodogram (4.57) and the covariance function (4.60) for data on a rectangular $r \times c$ lattice. That is, show that the periodogram is the Fourier transform of the sample covariance function.

**Problem 4.8** Determine the dynamic range, see equation (4.61) on 196, for some of the spectral densities shown in Table 4.4. Which models are susceptible to bias in periodogram estimation due to leakage? What is the dynamic range of a white noise process?

**Problem 4.9** Show that the spectral density function $s(\boldsymbol{\omega})$ for a homogeneous Poisson process has complete covariance density function $\lambda(\mathbf{s}_i - \mathbf{s}_j)$ and spectral density $s(\boldsymbol{\omega}) = (2\pi)^{-2}\lambda$.

**Problem 4.10** One of the appealing features of the method of "nonparametric" semivariogram fitting that relies on kernel functions rather than step functions (see §4.6.1), is to allow estimation by unconstrained optimization. This is only true if the function $f(\boldsymbol{\theta}, \omega)$ can be parameterized without constraints. Take, for example, the uniform kernel model (4.49) on page 182. The parameters must satisfy $\theta > 0$, $0 \leq \theta_l < \theta_u$. Give a parameterization that satisfies these constraints. That is, give functions $g$, $g_l$, and $g_u$ such that $\theta = g(\psi)$, $\theta_l = g_l(\psi_l)$, and $\theta_u = g_u(\psi_u, \psi_l)$ where $[\psi, \psi_l, \psi_u] \in \mathbb{R}^3$.

CHAPTER 5

# Spatial Prediction and Kriging

## 5.1 Optimal Prediction in Random Fields

Consider the random field $\{Z(\mathbf{s}) : \mathbf{s} \in D \subset \mathbb{R}^d\}$ observed at locations $\mathbf{s}_1, \cdots, \mathbf{s}_n$, and the corresponding data vector $\mathbf{Z}(\mathbf{s}) = [Z(\mathbf{s}_1), \cdots, Z(\mathbf{s}_n)]'$. The domain $D$ is fixed and continuous, hence we are dealing with geostatistical data. Our sample is an incomplete observation of the surface $Z(\mathbf{s}, \omega)$ that is the outcome of a random experiment with realization $\omega$. One of the pervasive problems in spatial statistics is the prediction of $Z$ at some specified location $\mathbf{s}_0 \in D$. This can be a location that is part of the set of locations where $Z(\mathbf{s}, \omega)$ has been observed, or a new (= unobserved) location.

Before we elaborate on general details of the spatial prediction problem, a few comments about the distinction of *prediction* and *estimation* are in order. These terms are often used interchangeably. For example, in a simple linear regression model, $Y_i = \beta_0 + \beta_1 x_i + \epsilon_i$, where the errors are uncorrelated, the regression coefficients $\beta_0$ and $\beta_1$ are estimated as $\widehat{\beta}_0$ and $\widehat{\beta}_1$ and then used to calculate predicted values $\widehat{Y}_0 = \widehat{\beta}_0 + \widehat{\beta}_1 x_0$. It is not clear in this situation whether $\widehat{Y}_0$ is supposed to be a predictor of $Y_0$, the response at $x_0$, or an estimator of $\mathrm{E}[Y_0]$. The fuzziness of the distinction in standard situations with uncorrelated errors stems from the fact that $\widehat{Y}_0$ is the best (linear unbiased) estimator of the *fixed, non-random* quantity $\mathrm{E}[Y_0]$ and also the best (linear unbiased) predictor of the *random* variable $Y_0$. Although the distinction between estimating a fixed quantity and predicting a random quantity may seem overly pedantic, the importance of the distinction becomes clear when we consider the uncertainty associated with the two quantities. The prediction error associated with using $\widehat{Y}_0$ as a predictor of $Y_0$ is larger due to the variability incurred from predicting a new observation.

In the case of a spatial random field, we can also focus on either prediction or estimation. Assume that a random field for geostatistical data (§2.4.1) has model representation

$$\mathbf{Z}(\mathbf{s}) = \mathbf{X}(\mathbf{s})\boldsymbol{\beta} + \mathbf{e}(\mathbf{s}), \ \mathbf{e}(\mathbf{s}) \sim (\mathbf{0}, \boldsymbol{\Sigma}).$$

We may be interested in estimating $\mathrm{E}[\mathbf{Z}(\mathbf{s})] = \mathbf{X}(\mathbf{s})\boldsymbol{\beta}$ or in predicting $\mathbf{Z}(\mathbf{s})$. In geostatistical applications prediction is often more important than mean estimation. First, the expectation $\mathrm{E}[Z(\mathbf{s})]$ is reckoned with respect to the distribution of the possible realizations $\omega$ at location $\mathbf{s}$. Appealing to this distribution may not be helpful. Second, one is often interested in the *actual* amount $Z(\mathbf{s})$ that is there, not some conceptual average amount.

**Example 5.1** A public shooting range has been in operation for seven years in a national forest, operated by the U.S. Forest Service. The lead concentration on the range is observed at sampling locations $\mathbf{s}_1, \cdots, \mathbf{s}_n$ by collecting the soil on a $50 \times 50$ cm square at each location, sieving the non-lead materials and weighing the lead content. The investigators are interested in determining the lead concentration at all locations on the shooting range. Estimating the mean lead concentration at location $\mathbf{s}$ appeals to a universe of similar shooting ranges. There may be no other shooting ranges like the one under consideration. What matters to the investigators is not how much lead is at location $\mathbf{s}_0$ on average across many other—conceptual—shooting ranges. What matters is to determine the amount of lead on the shooting range that was sampled. □

**Example 5.2** An oil company plans to install an oil well in a particular field. Sample wells are installed at locations $\mathbf{s}_1, \cdots, \mathbf{s}_n$. From this sample it is to be determined where on the field the actual oil well is to be installed. The average yield on similar fields is not of importance. What matters is where on *this* field the well should be placed. □

Even if estimation of the mean, the average amount at location $\mathbf{s}$, is not of primary interest, the pursuit of spatial prediction cannot escape matters of estimation. Both the mean and covariance structure of a random field are typically unknown. Unless the mean is known or constant, it must be estimated from the data. In the model above, estimates for $\boldsymbol{\beta}$ must be found before a predictor can be computed. And estimates of the mean will in most cases depend on the covariance structure, which requires estimates of $\boldsymbol{\theta}$.

Geostatistical prediction methods are statistical tools for predicting $g(Z(\mathbf{s}_0))$ from a set of observed data. They are typically known as methods of **kriging**, a term coined by G. Matheron in honor of the South African mining engineer D.G. Krige, whose work on ore-grade estimation in the Witwatersrand gold mines laid preliminary groundwork for the field of geostatistics (Krige, 1951; Matheron, 1963). Much of the same methodology that forms the basis of the field of geostatistics was developed around the same time in meteorology by L. S. Gandin, where it is often referred to as *objective analysis* (Gandin, 1963). Chilès and Delfiner (1999, p. ix) note that "it is from Matheron that geostatistics emerged as a discipline in its own right," and the widespread adoption of the term "kriging" is undoubtedly a testament to his influence. An interesting discussion on the origins of kriging and the development of optimal spatial prediction can be found in Cressie (1990).

One of the earliest books on geostatistics is Journel and Huijbregts (1978) and this remains an excellent source for the theory of kriging. More introductory texts include Clark (1979), Isaaks and Srivastava (1989), Armstrong (1999), Olea (1999), and Webster and Oliver (2001). In particular, Olea (1999) and Webster and Oliver (2001) are fairly comprehensive with many practical examples. A more advanced treatment of geostatistics and spatial prediction methods is given in Deutsch and Journel (1992) and the extension thereof by Goovaerts (1997), and the books of Cressie (1993) and Chilès and Delfiner (1999). Our writing in this chapter borrows considerably from Cressie (1993) and Chilès and Delfiner (1999). In particular, we follow the notation set forth in Cressie (1993).

The function $g(Z(\mathbf{s}_0))$ of interest can be varied. For example, we may be interested in predicting the $Z$ process itself, $g(Z(\mathbf{s}_0)) = Z(\mathbf{s}_0)$. Or, we may be interested in the average amount in a particular region $B$ (block) of volume $|B|$,

$$g(Z(B)) = \frac{1}{|B|}\int_B Z(\mathbf{u})\,d\mathbf{u}.$$

Or, we may be interested in binary variables indicating whether $Z(\mathbf{s})$ exceeds a certain environmentally-safe threshold level $c$,

$$g(Z(\mathbf{s}_0)) = \begin{cases} 1 & Z(\mathbf{s}_0) > c \\ 0 & Z(\mathbf{s}_0) \leq c. \end{cases}$$

In what follows, the notation $p(\mathbf{Z}; g(Z(\mathbf{s}_0)))$ denotes the predictor of $g(Z(\mathbf{s}_0))$ at location $\mathbf{s}_0$ based on the observed data vector $\mathbf{Z}(\mathbf{s})$. If $g(Z(\mathbf{s}_0)) \equiv Z(\mathbf{s}_0)$, a case on which we shall focus now, the "shorthand" $p(\mathbf{Z}; \mathbf{s}_0)$ will be used.

The derivation of a predictor commences with the choice of a function $L(Z(\mathbf{s}_0), p(\mathbf{Z}; \mathbf{s}_0))$ that measures the **loss** incurred by using $p(\mathbf{Z}; \mathbf{s}_0)$ as a predictor of $Z(\mathbf{s}_0)$. The squared-error loss function

*Squared-error Loss*

$$L(Z(\mathbf{s}_0), p(\mathbf{Z}; \mathbf{s}_0)) = (Z(\mathbf{s}_0) - p(\mathbf{Z}; \mathbf{s}_0))^2$$

is most commonly used and it is the only loss function considered in this text. It is not necessarily the most reasonable loss function, for example, because it is symmetric. In some applications, under- and over-predictions need to be penalized to different degrees. See the work by Zellner (1986), for example, on estimation and prediction using asymmetric loss functions. Having said that, the compelling reasons for choosing squared-error loss outweigh competing functions in most instances. Among these reasons are the following.

1. Statistical properties of predictors are fairly easily examined under squared-error loss. It essentially requires knowing only the first two moments of the process.
2. The loss function $L$ depends on the data and is thus a random variable. This leads us to the use of statistical measures of risk, e.g., finding predictors that minimize the average loss $\mathrm{E}[L(Z(\mathbf{s}_0), p(\mathbf{Z}; \mathbf{s}_0))]$. Under squared-error loss, the average loss is simply the mean-squared prediction error (MSPE) of using $p(\mathbf{Z}; \mathbf{s}_0)$ as a predictor of $Z(\mathbf{s}_0)$:
$$\mathrm{E}\left[(Z(\mathbf{s}_0) - p(\mathbf{Z}; \mathbf{s}_0))^2\right] = MSE[p(\mathbf{Z}; \mathbf{s}_0); Z(\mathbf{s}_0)].$$
3. The predictor that minimizes the mean-squared error has a particularly

appealing form; it is the conditional expectation of $Z(\mathbf{s}_0)$ given the observed data (see Chapter problems),

$$p^0(\mathbf{Z}; \mathbf{s}_0) = \mathrm{E}[Z(\mathbf{s}_0)|\, \mathbf{Z}(\mathbf{s})]. \tag{5.1}$$

By the law of iterated expectations, $p^0(\mathbf{Z}; \mathbf{s}_0)$ is unbiased in the sense that $\mathrm{E}[p^0(\mathbf{Z}; \mathbf{s}_0)] = \mathrm{E}[Z(\mathbf{s}_0)]$.

4. Finally, in the Gaussian case, finding the conditional expectation function is simple. Consider a $(n \times 1)$ random vector $\mathbf{W}$. Partition $\mathbf{W} = [\mathbf{U}, \mathbf{V}]'$, where $\mathbf{U}$ is of dimension $(u \times 1)$ and $\mathbf{V}$ has dimension $(v = (n-u) \times 1)$. Partition the mean vector and variance-covariance matrix accordingly:

$$\mathrm{E}[\mathbf{W}] = \mathrm{E}\begin{bmatrix}\mathbf{U}\\ \mathbf{V}\end{bmatrix} = \begin{bmatrix}\boldsymbol{\mu}_u\\ \boldsymbol{\mu}_v\end{bmatrix} \qquad \mathrm{Var}[\mathbf{W}] = \begin{bmatrix}\boldsymbol{\Sigma}_u & \boldsymbol{\Sigma}_{uv}\\ \boldsymbol{\Sigma}'_{uv} & \boldsymbol{\Sigma}_v\end{bmatrix}.$$

If $\mathbf{W}$ is Gaussian, then

$$\begin{aligned} \mathrm{E}[\mathbf{U}|\,\mathbf{V}] &= \boldsymbol{\mu}_u + \boldsymbol{\Sigma}_{uv}\boldsymbol{\Sigma}_v^{-1}(\mathbf{V} - \boldsymbol{\mu}_v) \\ \mathrm{Var}[\mathbf{U}|\,\mathbf{V}] &= \boldsymbol{\Sigma}_u - \boldsymbol{\Sigma}_{uv}\boldsymbol{\Sigma}_v^{-1}\boldsymbol{\Sigma}'_{uv}. \end{aligned} \tag{5.2}$$

In terms of a Gaussian spatial random field, where $Z(\mathbf{s}_0)$ and $\mathbf{Z}(\mathbf{s})$ are jointly multivariate Gaussian with $\mathrm{E}[Z(\mathbf{s}_0)] = \mu(\mathbf{s}_0)$, $\mathrm{Cov}[\mathbf{Z}(\mathbf{s}), Z(\mathbf{s}_0)] = \boldsymbol{\sigma}$, $\mathrm{Var}[\mathbf{Z}(\mathbf{s})] = \boldsymbol{\Sigma}$, and $\mathrm{E}[\mathbf{Z}(\mathbf{s})] = \boldsymbol{\mu}(\mathbf{s})$, (5.2) becomes

*Conditional Mean in GRF*

$$\mathrm{E}[Z(\mathbf{s}_0)|\,\mathbf{Z}(\mathbf{s})] = \mu(\mathbf{s}_0) + \boldsymbol{\sigma}'\boldsymbol{\Sigma}^{-1}(\mathbf{Z}(\mathbf{s}) - \boldsymbol{\mu}(\mathbf{s})). \tag{5.3}$$

This conditional expectation is linear in the observed data and establishing its statistical properties is comparatively straightforward.

Under squared-error loss the conditional mean is the best predictor and the mean-squared prediction error can be written as

$$\mathrm{E}[(Z(\mathbf{s}_0) - p^0(\mathbf{Z}; \mathbf{s}_0))^2] = \mathrm{Var}[Z(\mathbf{s}_0)] - \mathrm{Var}[p^0(\mathbf{Z}; \mathbf{s}_0)]. \tag{5.4}$$

In the case of the Gaussian random field, where $p^0(\mathbf{Z}; \mathbf{s}_0)$ is given by (5.3), we obtain

*Conditional Variance in GRF*

$$\mathrm{E}[(Z(\mathbf{s}_0) - p^0(\mathbf{Z}; \mathbf{s}_0))^2] = \mathrm{Var}[Z(\mathbf{s}_0)] - \boldsymbol{\sigma}'\boldsymbol{\Sigma}^{-1}\boldsymbol{\sigma}. \tag{5.5}$$

The result (5.4) is rather stunning. The term on the left-hand side must be positive and variances are typically not subtracted. Somehow, the variation of the best predictor under squared-error loss must be guaranteed to be less than the variation of the random field itself (establishing (5.4) is a Chapter problem). More importantly, this relationship conveys the behavior one should expect from a predictor that performs well, that is, a predictor with small mean-squared prediction error. It is a predictor that varies a lot. At first, this seems contrary to the results learned in classical statistics where one searches for those estimators of unknown quantities that have small mean square error. In the search for UMVU estimators, this means finding the estimator with least dispersion. There is a heuristic explanation for the fact that variable predictors will perform well in this situation, however. Figure 5.1 shows a realization $Z(t)$ of a temporal process. In order to predict the value of the series at time $t = 20$, three prediction functions are shown. The sample mean $\overline{Z}$ and two kernel smoothers. As the smoothness of the prediction function decreases, the variability of the predictor increases. The prediction function that will be close on average to the value of the series is one that is allowed to vary a lot. Based on these arguments one would expect the "best" predictor to follow the data even more closely than the prediction functions in Figure 5.1. One would expect the best predictor to interpolate the time series at the observed points. Kriging predictors in mean square continuous random fields have this property, they honor the data.

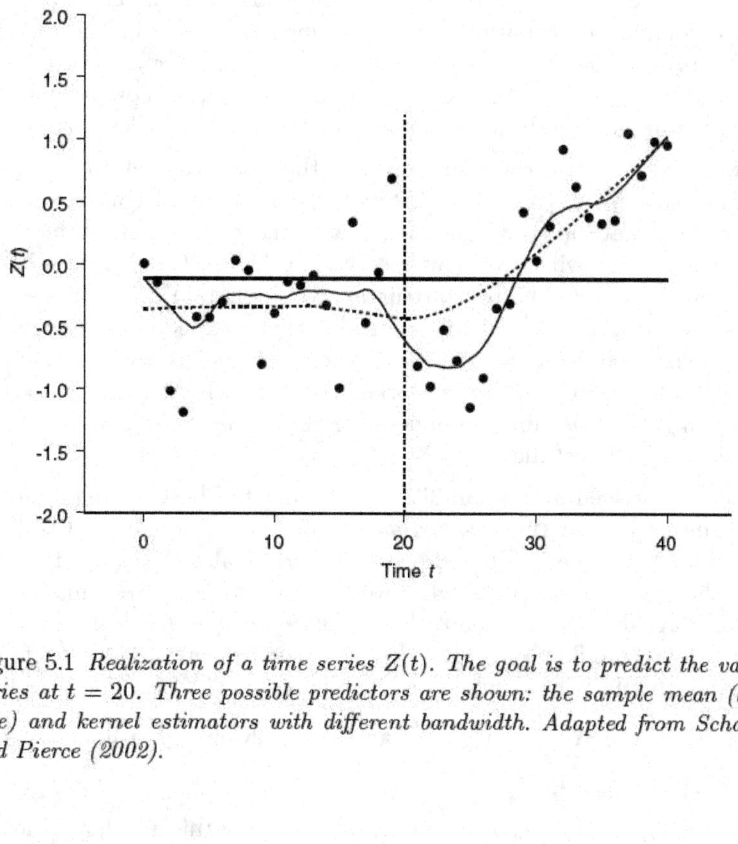


Figure 5.1 *Realization of a time series $Z(t)$. The goal is to predict the value of the series at $t = 20$. Three possible predictors are shown: the sample mean (horizontal line) and kernel estimators with different bandwidth. Adapted from Schabenberger and Pierce (2002).*

**Example 5.3 Simple linear regression.** We close this introductory section into spatial prediction by re-visiting a familiar (non-spatial) case, the simple linear regression model for uncorrelated, homoscedastic data. Denote the model as $Y_i = \alpha + \beta x_i + \epsilon_i$, $i = 1, \cdots, n$, $\text{Cov}[\epsilon_i, \epsilon_j] = 0$ if $i \neq j$, and $\text{Cov}[\epsilon_i, \epsilon_i] = \sigma^2$. In a first course in statistics, much ado is made about the difference between a $\xi \times 100\%$ *confidence* interval for $\text{E}[Y_0]$ and a $\xi \times 100\%$ *prediction* interval for $Y_0 = \alpha + \beta x_0 + \epsilon_0$, a new observation. The familiar formulas, if the errors are Gaussian, are

$$\begin{aligned}
\widehat{y}_0 &\pm t_{1-\xi/2,n-2}\widehat{\sigma}\sqrt{\frac{1}{n}+\frac{(x_0-\overline{x})^2}{S_{xx}}} \\
\widehat{y}_0 &\pm t_{1-\xi/2,n-2}\widehat{\sigma}\sqrt{1+\frac{1}{n}+\frac{(x_0-\overline{x})^2}{S_{xx}}} \\
\widehat{y}_0 &= \widehat{\alpha}+\widehat{\beta}x_0 \\
S_{xx} &= \sum_{i=1}^{n}(x_i-\overline{x})^2.
\end{aligned}$$

The "only" difference seems to be the additional "1+" under the square root, and a common explanation for the distinction is that in one case we consider $\text{Var}[\widehat{y}_0]$ and for the prediction interval we consider the variance of the difference $\text{Var}[\widehat{y}_0 - y_0]$. We can examine the distinction now in terms of the problem of finding the best predictor under squared-error loss.

First, we need to settle the issue whether the new observation is dependent on the data $\mathbf{Y} = [Y_1, \cdots, Y_n]'$. Since the fitted model assumes that the observed data are uncorrelated, there is no need to assume that a dependency would exist with any new observation; hence, $\text{Cov}[Y_i, Y_0] = 0, \forall i$. Under squared error loss, the best predictor of $Y_0$ is $\text{E}[Y_0|\mathbf{Y}]$; in our case the conditional expectation is equal to the unconditional expectation because of the independence. So, $\text{E}[Y_0] = \alpha + \beta x_0$ is the best predictor. Since $\alpha$ and $\beta$ are unknown, we turn to the Gauss-Markov theorem, which instructs us that $\widehat{y}_0 = \widehat{\alpha} + \widehat{\beta}x_0$ is the best linear unbiased predictor, where $\widehat{\alpha}$ and $\widehat{\beta}$ are the ordinary least squares estimators.

We have now arrived at the familiar result, that the best *predictor* of the random quantity $Y_0$ and the best *estimator* of the fixed quantity $\text{E}[Y_0]$ are the same. But are the mean-squared prediction errors also the same? In order to answer this question based on what we know up to now, we cannot draw on equation (5.4), because $\widehat{y}_0$ is not the conditional expectation. Instead, we draw on the following first principle: the mean-squared error $MSE[U; f(\mathbf{Y})]$ for estimating (predicting) $U$ based on some function $f(\mathbf{Y})$ is

$$\text{Var}[U - f(\mathbf{Y})] = \text{Var}[U] + \text{Var}[f(\mathbf{Y})] - 2\text{Cov}[U, f(\mathbf{Y})],$$

provided $\text{E}[U] = \text{E}[f(\mathbf{Y})]$.

In the simple linear regression example, we can apply this as follows, taking note that

$$\text{Var}[\widehat{Y}_0] = \sigma^2\left(\frac{1}{n}+\frac{(x_0-\overline{x})^2}{S_{xx}}\right).$$

Now, consider both the estimation problem and the prediction problem in this context.

(i) Estimation of $\text{E}[Y_0]$: $U = \text{E}[Y_0]$, $f(\mathbf{Y}) = \widehat{Y}_0$, $\text{Var}[U] = 0$, $\text{Cov}[U, f(\mathbf{Y})] =$

0. As a consequence,

$$MSE[U; f(\mathbf{Y})] = \text{Var}[U - f(\mathbf{Y})] = \text{Var}[f(\mathbf{Y})] = \sigma^2 \left( \frac{1}{n} + \frac{(x_0 - \overline{x})^2}{S_{xx}} \right).$$

(ii) Prediction of $Y_0$: $U = Y_0$, $f(\mathbf{Y}) = \widehat{Y}_0$, $\text{Var}[U] = \sigma^2$, $\text{Cov}[U, f(\mathbf{Y})] = 0$. As a consequence,

$$MSE[U; f(\mathbf{Y})] = \text{Var}[U - f(\mathbf{Y})] = \sigma^2 \left( 1 + \frac{1}{n} + \frac{(x_0 - \overline{x})^2}{S_{xx}} \right).$$

The additional factor "1+" represents $\text{Var}[Y_0]$. It is thus not quite correct to say that in the estimation (confidence interval) case we consider the variance of $\widehat{Y}_0$ and in the prediction case the variance of the difference $Y_0 - \widehat{Y}_0$. In both cases we consider the variance of a difference between the target $U$ and the prediction function. In the case of estimation, $U$ is a constant and for *that* reason we drop $\text{Var}[U]$ and the covariance between $U$ and $f(\mathbf{Y})$. In the case of prediction, $U$ is a random variable. The covariance term is eliminated *now* for a different reason, because of the independence assumption. □

As we apply prediction theory to spatial data in the sections that follow, keep in mind that we started the previous example by asking how the new observation relates to the observed data. If $\mathbf{Z}(\mathbf{s}) = [Z(\mathbf{s}_1), \cdots, Z(\mathbf{s}_n)]'$ are spatially correlated, then it is only reasonable to assume that a new observation $Z(\mathbf{s}_0)$ is part of the *same* process, and hence correlated with the observed data. In this case the best estimator of the mean $\text{E}[Z(\mathbf{s}_0)]$ and the best predictor of the random variable $Z(\mathbf{s}_0)$ will differ. In other applications of prediction for correlated data, you may need to revisit such assumptions.

**Example 5.4** Imagine an observational clinical study in which patients are repeatedly measured over time. Statistical models for such data often assume that data from a particular subject are serially correlated and that data from different patients are uncorrelated. To predict a future observation of a patient who participated in the study, you would assume that the future observation is correlated with the observed values for that patient and determine the correlation according to the covariance structure that was applied to the observed data. To predict the response of a patient who has not participated in the study, an assumption of dependence with any of the observed data points is not reasonable, if you assumed in the fitted model that patients' responses were independent. □

## 5.2 Linear Prediction—Simple and Ordinary Kriging

Under squared-error loss one should utilize the conditional mean function for predictions. Not only does $p^0(\mathbf{Z}; \mathbf{s}_0)$ minimize the Bayes risk *Bayes Risk*

$$\text{E}[L(Z(\mathbf{s}_0), p(\mathbf{Z}; \mathbf{s}_0))],$$

it is also "unbiased" in the sense that $\mathrm{E}[p^0(\mathbf{Z}; \mathbf{s}_0)] = \mathrm{E}[Z(\mathbf{s}_0)]$. This follows directly from the law of iterated expectations; $\mathrm{E}[\mathrm{E}[Y|X]] = \mathrm{E}[Y]$. If $\{Z(\mathbf{s}) : \mathbf{s} \in D \subset \mathbb{R}^d\}$ is a Gaussian random field, $p^0(\mathbf{Z}; \mathbf{s}_0)$ is also linear in $\mathbf{Z}(\mathbf{s})$, the observed data. In general, however, $p^0(\mathbf{Z}; \mathbf{s}_0)$ is not a linear function of the data and establishing the statistical properties of the best predictor under squared-error loss can be difficult; even intractable. Thus, in statistical practice the search for good estimators is restricted to particular classes of estimators. The properties of linearity in the observed data and unbiasedness are commonly imposed because of mathematical tractability and the mistaken impression that unbiasedness is an intrinsically "good" feature. Not surprisingly, similar constraints are imposed on prediction functions. In the Gaussian case no additional restrictions are called for, since $p^0(\mathbf{Z}; \mathbf{s}_0)$ already is linear and unbiased. For the general case, this new consideration of what constitutes a *best* predictor leads to what are called Best Linear Unbiased Predictors (BLUPs).

Consider random variables $X$ and $Y$ with joint density function $f(x, y)$ (or mass function $p(x, y)$). We are given the value of $X$ and wish to predict $Y$ from it, subject to the condition that the predictor $p(X)$ satisfies $\mathrm{E}[p(X)] = \mathrm{E}[Y] \equiv \mu_y$ and subject to a linearity condition, $p(X) = \alpha + \beta X$. Under squared-error loss this amounts to finding the function $p(X)$ that minimizes

$$
\begin{aligned}
MSE[p(X); Y] &= \mathrm{E}\left[(Y - p(X))^2\right] \text{ subject to} \\
p(X) &= \alpha + \beta X \qquad \beta = (\mu_x - \mu_y)/\alpha.
\end{aligned}
$$

The solutions to this minimization problem are (Chapter problem 5.4)

$$
\begin{aligned}
\alpha &= \mu_y - \frac{\sigma_{xy}}{\sigma_x^2}\mu_x \qquad \beta = \frac{\sigma_{xy}}{\sigma_x^2} \\
p(X) &= BLUP(Y|X) = \mu_y - \frac{\sigma_{xy}}{\sigma_x^2}(\mu_x - X), \qquad (5.6)
\end{aligned}
$$

where $\sigma_{xy} = \mathrm{Cov}[X, Y]$, $\sigma_x^2 = \mathrm{Var}[X]$.

**Example 5.3 (Simple linear regression. Continued)** How do these results relate to the simple linear regression (SLR) model? In order to make the connection between (5.6) and prediction in the SLR model, we assume that the conditional mean function is given by

$$
Y_i|x = \alpha + \beta x_i + \epsilon_i, \quad \epsilon_i \sim iid\ (0, \sigma^2),\ i = 1, \cdots, n.
$$

The ordinary least squares estimators of $\alpha$ and $\beta$ and the predictor of $Y|x$—based on the sample—are

$$
\begin{aligned}
\widehat{\alpha} &= \overline{Y} - \widehat{\beta}\overline{x} \qquad \widehat{\beta} = \frac{s_{xy}^2}{s_{xx}^2} \\
\widehat{Y} &= \overline{Y} - \widehat{\beta}(\overline{x} - x) = \widehat{\alpha} + \widehat{\beta}x, \qquad (5.7)
\end{aligned}
$$

where $s_{xy}^2 = (1/(n-1))\sum_{i=1}^n (x_i - \overline{x})Y_i$, $s_{xx}^2 = (1/(n-1))\sum_{i=1}^n (x_i - \overline{x})^2$.

These expressions are based on the sample estimators of the unknown quantities in (5.6). Based on the discussion that follows, you will be able to show that (5.7) is indeed the best linear unbiased predictor of $Y|x$ if the means, variance of $X$, and covariance are unknown. By the Gauss-Markov theorem we know that (5.7) is also the best linear unbiased estimator (BLUE) of $\mathrm{E}[Y|x]$. The differences in their mean-squared prediction errors were established previously. □

### 5.2.1 *The Mean Is Known—Simple Kriging*

Consider spatial data $\mathbf{Z}(\mathbf{s}) = [Z(\mathbf{s}_1), \cdots, Z(\mathbf{s}_n)]'$ and assume

$$\mathbf{Z}(\mathbf{s}) = \boldsymbol{\mu}(\mathbf{s}) + \mathbf{e}(\mathbf{s}), \qquad \mathbf{e}(\mathbf{s}) \sim (\mathbf{0}, \boldsymbol{\Sigma}). \tag{5.8}$$

Thus, $\mathrm{E}[\mathbf{Z}(\mathbf{s})] = \boldsymbol{\mu}(\mathbf{s})$ and $\mathrm{Var}[\mathbf{Z}(\mathbf{s})] = \boldsymbol{\Sigma}$, and we initially assume both $\boldsymbol{\mu}(\mathbf{s})$ and $\boldsymbol{\Sigma}$ are known. The goal is to find the predictor of $Z(\mathbf{s}_0)$, $p(\mathbf{Z};\mathbf{s}_0)$, that minimizes $\mathrm{E}[(p(\mathbf{Z};\mathbf{s}_0) - Z(\mathbf{s}_0))^2]$. This is a formidable problem, so we refine it by considering linear predictors of the form $p(\mathbf{Z};\mathbf{s}_0) = \lambda_0 + \boldsymbol{\lambda}'\mathbf{Z}(\mathbf{s})$, where $\lambda_0$ and the elements of $\boldsymbol{\lambda} = [\lambda_1, \cdots \lambda_n]'$ are unknown coefficients to be determined. Then $\mathrm{E}[(p(\mathbf{Z};\mathbf{s}_0) - Z(\mathbf{s}_0))^2]$ can be written as

$$\mathrm{E}[(p(\mathbf{Z};\mathbf{s}_0) - Z(\mathbf{s}_0))^2] = \mathrm{E}[(\lambda_0 + \boldsymbol{\lambda}'\mathbf{Z}(\mathbf{s}) - Z(\mathbf{s}_0))^2].$$

Adding and subtracting the term $E[p(\mathbf{Z};\mathbf{s}_0) - Z(\mathbf{s}_0)] \equiv \lambda_0 + \boldsymbol{\lambda}'\boldsymbol{\mu}(\mathbf{s}) - \mu(\mathbf{s}_0)$, where $\mu(\mathbf{s}_0) = \mathrm{E}[Z(\mathbf{s}_0)]$, we obtain

$$\begin{aligned} \mathrm{E}[(p(\mathbf{Z};\mathbf{s}_0) - Z(\mathbf{s}_0))^2] &= \mathrm{Var}[\boldsymbol{\lambda}'\mathbf{Z}(\mathbf{s}) - Z(\mathbf{s}_0)] \\ &\quad + (\lambda_0 + \boldsymbol{\lambda}'\boldsymbol{\mu}(\mathbf{s}) - \mu(\mathbf{s}_0))^2. \end{aligned}$$

Since both terms are nonnegative, the expected mean-squared prediction error will be minimized when each term is as small as possible. Clearly, the second term is minimized by taking $\lambda_0 = \mu(\mathbf{s}_0) - \boldsymbol{\lambda}'\boldsymbol{\mu}(\mathbf{s})$. If $\mathrm{Var}[Z(\mathbf{s}_0)] = \sigma^2$ and $\boldsymbol{\sigma} = \mathrm{Cov}[\mathbf{Z}(\mathbf{s}), Z(\mathbf{s}_0)]$, then the second term can be written as

$$\mathrm{Var}[\boldsymbol{\lambda}'\mathbf{Z}(\mathbf{s}) - Z(\mathbf{s}_0)] = \sigma^2 + \boldsymbol{\lambda}'\boldsymbol{\Sigma}\boldsymbol{\lambda} - 2\boldsymbol{\sigma}'\boldsymbol{\lambda}. \tag{5.9}$$

Calculus can be used to determine the minimum of this objective function. Differentiating with respect to $\boldsymbol{\lambda}$ and equating to zero gives $\boldsymbol{\lambda}'\boldsymbol{\Sigma} = \boldsymbol{\sigma}'$. From matrix algebra theory, these equations have a unique solution if $\boldsymbol{\Sigma}$ is nonsingular (so that duplicate measurements at the same location are excluded). Thus, the optimal choices for $\lambda_0$ and $\boldsymbol{\lambda}$ are

$$\begin{aligned} \lambda_0 &= \mu(\mathbf{s}_0) - \boldsymbol{\lambda}'\boldsymbol{\mu}(\mathbf{s}); \\ \boldsymbol{\lambda} &= \boldsymbol{\Sigma}^{-1}\boldsymbol{\sigma} \end{aligned}$$

and the optimal linear predictor is

$$p_{sk}(\mathbf{Z};\mathbf{s}_0) = \lambda_0 + \boldsymbol{\lambda}'\mathbf{Z}(\mathbf{s}) = \mu(\mathbf{s}_0) + \boldsymbol{\sigma}'\boldsymbol{\Sigma}^{-1}(\mathbf{Z}(\mathbf{s}) - \boldsymbol{\mu}(\mathbf{s})). \tag{5.10}$$

*Simple Kriging Predictor*

In geostatistical parlance, (5.10) is called the **simple kriging** predictor. It is the best linear predictor under squared-error loss. The predictor is equivalent to the conditional mean in a Gaussian random field, (5.3). It is thus *the* best predictor (linear or not) under squared-error loss *if* $\mathbf{Z}(\mathbf{s})$ is a GRF. If the joint distribution of the data is not Gaussian, then (5.10) is the *best* predictor among those that are linear in $\mathbf{Z}(\mathbf{s})$. It is also unbiased, but since we did not impose an unbiasedness constraint as part of the minimization, the simple kriging predictor is best in the class of all linear predictors. Kriging is often referred to as optimal spatial prediction, but optimality considerations are confined to this sub-class of predictors unless the random field is Gaussian.

Substitution of $\boldsymbol{\lambda}' = \boldsymbol{\sigma}'\boldsymbol{\Sigma}^{-1}$ into the expression for the mean-squared error in (5.9) yields the minimized mean-squared prediction error, also called the (simple) kriging variance

*Simple Kriging Variance*

$$\sigma^2_{sk}(\mathbf{s}_0) = \sigma^2 - \boldsymbol{\sigma}'\boldsymbol{\Sigma}^{-1}\boldsymbol{\sigma}, \tag{5.11}$$

which agrees with $\text{Var}[Z(\mathbf{s}_0)|\,\mathbf{Z}(\mathbf{s})]$ in (5.5) in the Gaussian case. The simple kriging variance depends on the prediction location through the vector $\boldsymbol{\sigma}$ of covariances between $Z(\mathbf{s}_0)$ and the observed data.

It was noted in §5.1 on heuristic grounds, that a good predictor should be variable in the sense that it follows the observed data closely. The simple kriging predictor has an interesting property that it shares with many other types of kriging predictors. Consider predicting at locations where data are actually observed. Thus, the predictor $p_{sk}(\mathbf{Z};\mathbf{s}_0)$ becomes $p_{sk}(\mathbf{Z};[\mathbf{s}_1,\cdots,\mathbf{s}_n]')$, and in (5.10) we replace $\text{Cov}[Z(\mathbf{s}_0),\mathbf{Z}(\mathbf{s})] = \boldsymbol{\sigma}'$ with $\text{Cov}[\mathbf{Z}(\mathbf{s}),\mathbf{Z}(\mathbf{s})] = \boldsymbol{\Sigma}$ and $\mu(\mathbf{s}_0)$ with $\boldsymbol{\mu}(\mathbf{s})$ to obtain

$$\begin{aligned} p_{sk}(\mathbf{Z};[\mathbf{s}_1,\cdots,\mathbf{s}_n]') &= \boldsymbol{\mu}(\mathbf{s}) + \boldsymbol{\Sigma}\boldsymbol{\Sigma}^{-1}(\mathbf{Z}(\mathbf{s}) - \boldsymbol{\mu}(\mathbf{s})) \\ &= \mathbf{Z}(\mathbf{s}). \end{aligned}$$

Thus, the simple kriging predictor interpolates the observed data. It is an "exact" interpolator or said to "honor the data." Historically, many disciplines have considered this to be a very desirable property; one that should be asked of any spatial predictor. However, in some situations, smoothing, as is typically done in most regression situations, may be more appealing. For example, when the semivariogram of the spatial process contains a nugget effect, it is not necessarily desirable to interpolate the data. If the nugget effect consists of micro-scale variability only, then a structured portion of the spatial variability has not been observed and honoring the observed data is reasonable. If the nugget effect contains a measurement error component, that is, $Z(\mathbf{s}) = S(\mathbf{s}) + \epsilon(\mathbf{s})$, where $\epsilon(\mathbf{s})$ is the measurement error at location $\mathbf{s}$, then we do not want the predictor to interpolate the data. We are then not interested in the amount that has been erroneously measured, but the amount $S(\mathbf{s})$ that is actually there. The predictor should be an interpolator of the signal $S(\mathbf{s})$, not the observed amount $Z(\mathbf{s})$. In §5.4.3 these issues regarding kriging with and without a measurement error will be revisited in more detail.

The term *simple kriging* is unfortunate on another ground. There is nothing *simple* or common about the requirement that the mean $\mu(\mathbf{s})$ of the random field be known. An exception is best linear unbiased prediction of residuals from a regression fit. If $\mathbf{Z}(\mathbf{s}) = \mathbf{X}(\mathbf{s})\boldsymbol{\beta} + \boldsymbol{\epsilon}(\mathbf{s})$, then the vector of ordinary least squares (OLS) residuals

$$\widehat{\boldsymbol{\epsilon}}(\mathbf{s}) = \mathbf{Z}(\mathbf{s}) - \mathbf{X}(\mathbf{s})(\mathbf{X}(\mathbf{s})'\mathbf{X}(\mathbf{s}))^{-1}\mathbf{X}(\mathbf{s})'\mathbf{Z}(\mathbf{s}) \tag{5.12}$$

has (known) mean $\mathbf{0}$—provided that the mean model $\mathbf{X}(\mathbf{s})\boldsymbol{\beta}$ has been specified correctly.* One application of simple kriging is thus the following method intended to cope with the problem of non-stationarity that arises from large-scale trends in the mean of $\mathbf{Z}(\mathbf{s})$. It is sometimes incorrectly labelled as "Universal Kriging," which it is not (see §5.3.3).

*"Universal Kriging," but not quite*

1. Specify a linear spatial model $\mathbf{Z}(\mathbf{s}) = \mathbf{X}(\mathbf{s})\boldsymbol{\beta} + \mathbf{e}(\mathbf{s})$.
2. Fit the model by OLS to obtain $\widehat{\boldsymbol{\beta}}_{ols} = (\mathbf{X}(\mathbf{s})'\mathbf{X}(\mathbf{s}))^{-1}\mathbf{X}(\mathbf{s})'\mathbf{Z}(\mathbf{s})$.
3. Perform simple kriging on the OLS residuals (5.12) to obtain $p_{sk}(\widehat{\boldsymbol{\epsilon}}; \widehat{\epsilon}(\mathbf{s}_0))$.
4. Obtain the *kriging* predictor of $Z(\mathbf{s}_0)$ as $\mathbf{x}(\mathbf{s}_0)'\widehat{\boldsymbol{\beta}} + p_{sk}(\widehat{\boldsymbol{\epsilon}}; \widehat{\epsilon}(\mathbf{s}_0))$.

Although it appears to be a reasonable thing to do, there are many problems with this approach:

- Since $\text{Var}[\mathbf{e}(\mathbf{s})] = \boldsymbol{\Sigma}$, ordinary least squares estimation is not most efficient. The large-scale trend parameters $\boldsymbol{\beta}$ should be estimated by generalized least squares (GLS):

$$\widehat{\boldsymbol{\beta}}_{gls} = (\mathbf{X}(\mathbf{s})'\boldsymbol{\Sigma}^{-1}\mathbf{X}(\mathbf{s}))^{-1}\mathbf{X}(\mathbf{s})\boldsymbol{\Sigma}^{-1}\mathbf{Z}(\mathbf{s}).$$

- GLS requires knowledge of $\boldsymbol{\Sigma}$. If $\boldsymbol{\Sigma}$ is unknown, one can resort to estimated GLS (EGLS), where $\boldsymbol{\Sigma}$ is replaced with an estimate

$$\widehat{\boldsymbol{\beta}}_{egls} = (\mathbf{X}(\mathbf{s})'\widehat{\boldsymbol{\Sigma}}^{-1}\mathbf{X}(\mathbf{s}))^{-1}\mathbf{X}(\mathbf{s})\widehat{\boldsymbol{\Sigma}}^{-1}\mathbf{Z}(\mathbf{s}).$$

  How to obtain this estimate is not clear. Using the OLS residuals (or even the GLS residuals) and estimating the covariance function or semivariogram of $\boldsymbol{\epsilon}(\mathbf{s})$ from $\widehat{\boldsymbol{\epsilon}}(\mathbf{s})$ leads to biased estimators. The fitted residuals comply to constraints that the unobservable model errors $\boldsymbol{\epsilon}(\mathbf{s})$ are not subject to, e.g., $\mathbf{1}'\widehat{\boldsymbol{\epsilon}}_{ols}(\mathbf{s}) = 0$ if $\mathbf{X}(\mathbf{s})$ contains an intercept. The fitted residuals will exhibit more negative correlations than $\mathbf{e}(\mathbf{s})$.

- $\mathbf{x}(\mathbf{s}_0)'\widehat{\boldsymbol{\beta}} + p_{sk}(\widehat{\boldsymbol{\epsilon}}(\mathbf{s}); \widehat{\epsilon}(\mathbf{s}_0))$ is an unbiased predictor of $Z(\mathbf{s}_0)$, provided the model was correctly specified, but it is not the BLUP. The practical ramification is that the simple kriging variance inferred from the residuals does not incorporate the variability associated with estimating $\boldsymbol{\beta}$, regardless of whether OLS or GLS is used. Thus, the uncertainty associated with predictions from this approach is estimated too low.

* Expressions involving inverses of functions of the regression matrix $\mathbf{X}(\mathbf{s})$ are written in terms of regular inverses in this and the following chapter. The $\mathbf{X}(\mathbf{s})$ matrix is an $(n \times p)$ matrix of rank $k$ and $k$ can be less than $p$. In this case it is safe to substitute generalized inverses in expressions such as $(\mathbf{X}(\mathbf{s})'\mathbf{X}(\mathbf{s}))^{-1}$ and $(\mathbf{X}(\mathbf{s})'\boldsymbol{\Sigma}^{-1}\mathbf{X}(\mathbf{s}))^{-1}$. However, we always assume that the variance matrix $\boldsymbol{\Sigma}$ is positive definite and hence non-singular.

### 5.2.2 *The Mean Is Unknown and Constant—Ordinary Kriging*

The simple kriging predictor is used when the mean $\boldsymbol{\mu}(\mathbf{s})$ in model (5.8) is known. With this model, the mean can change with spatial location. If $\mathrm{E}[\mathbf{Z}(\mathbf{s})]$ is unknown but constant across locations, $\mathrm{E}[\mathbf{Z}(\mathbf{s})] \equiv \mu\mathbf{1}$, best linear unbiased prediction under squared-error loss is known as **ordinary kriging**.

We need to find the predictor $p(\mathbf{Z}; \mathbf{s}_0)$ of $Z(\mathbf{s}_0)$ that minimizes $\mathrm{E}[(p(\mathbf{Z}; \mathbf{s}_0) - Z(\mathbf{s}_0))^2]$, when the data follow the model

$$\mathbf{Z}(\mathbf{s}) = \mu\mathbf{1} + \mathbf{e}(\mathbf{s}), \qquad \mathbf{e}(\mathbf{s}) \sim (\mathbf{0}, \boldsymbol{\Sigma}).$$

Thus, $\mathrm{E}[\mathbf{Z}(\mathbf{s})] = \mu\mathbf{1}$ and $\mathrm{Var}[\mathbf{Z}(\mathbf{s})] = \boldsymbol{\Sigma}$, where $\mu$ is an unknown constant and $\boldsymbol{\Sigma}$ is known.

As in the development of the simple kriging predictor, we consider linear predictors of the form $p(\mathbf{Z}; \mathbf{s}_0) = \lambda_0 + \boldsymbol{\lambda}'\mathbf{Z}(\mathbf{s})$, where $\lambda_0$ and the elements of the vector $\boldsymbol{\lambda} = [\lambda_1, \cdots \lambda_n]'$ are unknown coefficients to be determined. Repeating the development in §5.2.1 gives $\lambda_0 = \mu - \boldsymbol{\lambda}'\mu\mathbf{1}$. However, this does not determine the value of $\lambda_0$ since $\mu$ is unknown. When the mean in unknown, there is no best linear predictor in the class of all linear predictors. Thus, we refine the problem by further restricting the class of linear predictors to those that are also unbiased. Since the mean of $\mathbf{Z}(\mathbf{s})$ does not depend on $\mathbf{s}$, it is reasonable to posit also that $\mathrm{E}[Z(\mathbf{s}_0)] = \mu$. Then we require for unbiasedness that $\mathrm{E}[p(\mathbf{Z}; \mathbf{s}_0)] = \mathrm{E}[Z(\mathbf{s}_0)]$ or equivalently, $\mathrm{E}[\lambda_0 + \boldsymbol{\lambda}'\mathbf{Z}(\mathbf{s})] = \mathrm{E}[Z(\mathbf{s}_0)]$, which implies that $\lambda_0 + \mu(\boldsymbol{\lambda}'\mathbf{1} - 1) = 0$. Since this must hold for every $\mu$, it must hold for $\mu = 0$ and so the unbiasedness constraint requires that $\lambda_0 = 0$ and $\boldsymbol{\lambda}'\mathbf{1} = 1$.

Now our problem is to choose weights $\boldsymbol{\lambda} = [\lambda_1, \cdots, \lambda_n]'$ that minimize

$$\mathrm{E}[(\boldsymbol{\lambda}'\mathbf{Z}(\mathbf{s}) - Z(\mathbf{s}_0))^2] \qquad \text{subject to } \boldsymbol{\lambda}'\mathbf{1} = 1.$$

This can be accomplished as an unconstrained minimization problem introducing the Lagrange multiplier $m$:

$$\arg\min_{\boldsymbol{\lambda},m} Q = \arg\min_{\boldsymbol{\lambda},m} \mathrm{E}[(\boldsymbol{\lambda}'\mathbf{Z}(\mathbf{s}) - Z(\mathbf{s}_0))^2] - 2m(\boldsymbol{\lambda}'\mathbf{1} - 1). \tag{5.13}$$

#### 5.2.2.1 *Ordinary Kriging in Terms of the Covariance Function*

Expanding the expectation in (5.13), putting $\mathrm{Var}[Z(\mathbf{s}_0)] = \sigma^2 = C(\mathbf{0})$, and assuming $\mathbf{Z}(\mathbf{s})$ is second-order stationary, we obtain

$$Q = C(\mathbf{0}) + \boldsymbol{\lambda}'\boldsymbol{\Sigma}\boldsymbol{\lambda} - 2\boldsymbol{\lambda}'\boldsymbol{\sigma} - 2m(\boldsymbol{\lambda}'\mathbf{1} - 1).$$

Taking derivatives with respect to $\boldsymbol{\lambda}$ and $m$ and setting to zero yields the system of equations

$$\begin{aligned}
\frac{\partial Q}{\partial \boldsymbol{\lambda}} &= 2\boldsymbol{\Sigma}\boldsymbol{\lambda} - 2\boldsymbol{\sigma} - 2m\mathbf{1} \equiv \mathbf{0} \\
\frac{\partial Q}{\partial m} &= -2(\boldsymbol{\lambda}'\mathbf{1} - 1) \equiv 0.
\end{aligned}$$

The factor 2 in front of the Lagrange multiplier was chosen to allow cancellation. It is left as an exercise (Chapter problem 5.8) to show that the solutions to this problem are (Cressie, 1993, p. 123)

*Ordinary Kriging Weights*

$$\boldsymbol{\lambda}' = \left(\boldsymbol{\sigma} + \mathbf{1}\frac{1-\mathbf{1}'\boldsymbol{\Sigma}^{-1}\boldsymbol{\sigma}}{\mathbf{1}'\boldsymbol{\Sigma}^{-1}\mathbf{1}}\right)'\boldsymbol{\Sigma}^{-1} \quad (5.14)$$

$$m = \frac{1-\mathbf{1}'\boldsymbol{\Sigma}^{-1}\boldsymbol{\sigma}}{\mathbf{1}'\boldsymbol{\Sigma}^{-1}\mathbf{1}}, \quad (5.15)$$

and that the minimized mean-squared prediction error, the ordinary kriging variance, is (Cressie, 1993, p. 123)

*Ordinary Kriging Variance*

$$\begin{aligned}\sigma^2_{ok}(\mathbf{s}_0) &= C(\mathbf{0}) - \boldsymbol{\lambda}'\boldsymbol{\sigma} + m \\ &= C(\mathbf{0}) - \boldsymbol{\sigma}'\boldsymbol{\Sigma}^{-1}\boldsymbol{\sigma} + \frac{(1-\mathbf{1}'\boldsymbol{\Sigma}^{-1}\boldsymbol{\sigma})^2}{\mathbf{1}'\boldsymbol{\Sigma}^{-1}\mathbf{1}}. \quad (5.16)\end{aligned}$$

Notice that $\sigma^2_{sk} < \sigma^2_{ok}$, since the last term on the right-hand side of (5.16) is positive. Not knowing the mean of the random field increases the mean-squared prediction error. The expression $p_{ok}(\mathbf{Z};\mathbf{s}_0) = \boldsymbol{\lambda}'\mathbf{Z}(\mathbf{s})$, where $\boldsymbol{\lambda}$ is given by (5.14), hides the fact that the unknown mean of the random field is actually estimated implicitly. The formulation of the ordinary kriging predictor we prefer is

*GLS form of Ordinary Kriging*

$$p_{ok}(\mathbf{Z};\mathbf{s}_0) = \widehat{\mu} + \boldsymbol{\sigma}'\boldsymbol{\Sigma}^{-1}(\mathbf{Z}(\mathbf{s}) - \mathbf{1}\widehat{\mu}) \quad (5.17)$$

(Toutenburg, 1982, p. 141, Cressie, 1993, p. 173; Gotway and Cressie, 1993). This formulation shows the correspondence between ordinary and simple kriging, as well as the connection to the best predictor in the Gaussian random field more clearly. Comparing (5.17) and (5.10), it appears that "all" that is required to accommodate an unknown mean is to replace $\mu$ with an estimate $\widehat{\mu}$. It is important to note that not just any estimate will do. The algebraic manipulations leading from $\boldsymbol{\lambda}'\mathbf{Z}(\mathbf{s})$ to (5.17) reveal that $\mu$ must be estimated by its best linear unbiased estimator, which in this case is the generalized least squares estimator (Chapter problem 5.9; Goldberger, 1962):

$$\widehat{\mu} = (\mathbf{1}'\boldsymbol{\Sigma}^{-1}\mathbf{1})^{-1}\mathbf{1}'\boldsymbol{\Sigma}^{-1}\mathbf{Z}(\mathbf{s}). \quad (5.18)$$

In the derivation of $p_{ok}(\mathbf{Z};\mathbf{s}_0)$ presented above, the assumption of second-order stationarity of the process was implicit; e.g., by putting $\text{Var}[Z(\mathbf{s}_0)] = C(\mathbf{0})$. If the process is not second-order stationary, then $C(\mathbf{h})$ is not defined. The variance-covariance matrix of the model errors as well as the covariance vector between $Z(\mathbf{s}_0)$ and $\mathbf{Z}(\mathbf{s})$ can be constructed, however. In the second-order stationary case the $(i,j)^{\text{th}}$ element of $\boldsymbol{\Sigma}$ is given by $C(\mathbf{s}_i - \mathbf{s}_j)$, which is $C(||\mathbf{s}_i - \mathbf{s}_j||)$ if the process is isotropic. For a non-stationary process we construct $\boldsymbol{\Sigma} = [C(\mathbf{s}_i,\mathbf{s}_j)]$ and $\boldsymbol{\sigma} = \text{Cov}[\mathbf{Z}(\mathbf{s}), Z(\mathbf{s}_0)] = [C(\mathbf{s}_0,\mathbf{s}_1),\cdots,C(\mathbf{s}_0,\mathbf{s}_n)]'$ and calculate $\boldsymbol{\lambda}$ and $m$ in the usual way (equations (5.14) and (5.15)). The ordinary kriging variance for the case of a non-stationary covariance function becomes

$$\sigma^2_{ok}(\mathbf{s}_0) = C(\mathbf{s}_0,\mathbf{s}_0) - \boldsymbol{\lambda}'\boldsymbol{\sigma} + m.$$

Thus, it is not necessary to assume $C(\mathbf{s}_i, \mathbf{s}_j) = C(\mathbf{s}_i - \mathbf{s}_j)$ to solve the kriging problem. The actual process of best linear unbiased prediction does not involve any type of stationarity. But because modeling the covariance function or semivariogram in the absence of stationarity is typically difficult, issues of stationarity cannot be separated from best linear prediction. In practical applications $\text{Var}[\mathbf{Z}(\mathbf{s})]$ is not known.

### *5.2.2.2 Ordinary Kriging in Terms of the Semivariogram*

The kriging equations can also be expressed in terms of the semivariogram of the process. This is helpful for spatial prediction with processes that are intrinsically, but not second-order stationary. The derivation of the kriging weights $\boldsymbol{\lambda}$ in terms of $\gamma(\mathbf{h})$ follows along the same lines as before. The criterion (5.13) is now expanded in terms of the matrix $\boldsymbol{\Gamma} = [\gamma(\mathbf{s}_i - \mathbf{s}_j)]$ and the vector $\boldsymbol{\gamma}(\mathbf{s}_0) = [\gamma(\mathbf{s}_0 - \mathbf{s}_1), \cdots, \gamma(\mathbf{s}_0 - \mathbf{s}_n)]'$:

$$Q = -\boldsymbol{\lambda}'\boldsymbol{\Gamma}\boldsymbol{\lambda} + 2\boldsymbol{\lambda}'\boldsymbol{\gamma}(\mathbf{s}_0) - 2m(\boldsymbol{\lambda}'\mathbf{1} - 1).$$

Differentiating with respect to $\boldsymbol{\lambda}$ and $m$, setting to zero and solving the two simultaneous equations yields (Cressie, 1993, p. 122)

$$\begin{aligned}
\boldsymbol{\lambda}' &= \left(\boldsymbol{\gamma}(\mathbf{s}_0) + \mathbf{1}\frac{1 - \mathbf{1}'\boldsymbol{\Gamma}^{-1}\boldsymbol{\gamma}(\mathbf{s}_0)}{\mathbf{1}'\boldsymbol{\Gamma}^{-1}\mathbf{1}}\right)' \boldsymbol{\Gamma}^{-1} && (5.19)\\
m &= -\frac{1 - \mathbf{1}'\boldsymbol{\Gamma}^{-1}\boldsymbol{\gamma}(\mathbf{s}_0)}{\mathbf{1}'\boldsymbol{\Gamma}^{-1}\mathbf{1}} && (5.20)\\
p_{ok}(\mathbf{Z}; \mathbf{s}_0) &= \boldsymbol{\lambda}'\mathbf{Z}(\mathbf{s}) && (5.21)\\
\sigma^2_{ok}(\mathbf{s}_0) &= \boldsymbol{\lambda}'\boldsymbol{\gamma}(\mathbf{s}_0) + m = 2\boldsymbol{\lambda}'\boldsymbol{\gamma}(\mathbf{s}_0) - \boldsymbol{\lambda}'\boldsymbol{\Gamma}\boldsymbol{\lambda}. && (5.22)
\end{aligned}$$

### *5.2.3 Effects of Nugget, Sill, and Range*

The ordinary kriging predictor at location $\mathbf{s}_0$ is $p_{ok}(\mathbf{Z}; \mathbf{s}_0) = \boldsymbol{\lambda}'\mathbf{Z}(\mathbf{s})$, where the kriging weights $\boldsymbol{\lambda}$ are given by (5.14). The two important components, the "driving forces" behind the ordinary kriging weights, are the vector $\boldsymbol{\sigma} = \text{Cov}[Z(\mathbf{s}_0), \mathbf{Z}(\mathbf{s})]$ and the variance-covariance matrix $\boldsymbol{\Sigma} = \text{Var}[\mathbf{Z}(\mathbf{s})]$. To be more precise, it is the inverse covariance matrix and the vector of covariances between attributes at prediction locations and observed locations that drive the kriging weights. The important point we wish to make is that the spatial dependency structure has great impact on ordinary (and simple) kriging predictions.

We are not concerned with the issue that $\boldsymbol{\Sigma}$ is unknown, that its elements (parameters) need to be estimated, and that the variability of the estimates is not reflected in the standard expressions for the kriging error, such as equation (5.16). These issues are considered in §5.5. Of concern now is the fact that $\boldsymbol{\Sigma}$ is unknown, and you adopt a parametric model to capture the spatial dependence, for example, one of the semivariogram models of §4.3. Obviously, the choice of model has an impact on the kriging predictions. In general, however, the precise implications are difficult to describe, because different features of competing models can enhance or suppress a particular effect in model comparisons. For example, predictions based on highly continuous covariance models tend to be more smooth than predictions using a model of lesser continuity (see §2.3). A gaussian semivariogram model will generally produce smoother predicted surfaces than an exponential model (with the same practical range and sill). Intuitively, it is the near-origin behavior of the semivariogram or covariance function that has an important impact on kriging predictions. Since the nugget-only model is the model of least continuity, and a model with nugget effect can be thought of as a nested model, an exponential model without nugget effect can produce predictions that are more smooth than those from a gaussian model with nugget effect. Statements and conclusions about the effect of semivariogram parameters on ordinary kriging predictions thus need to be made with care and appropriate caveats. We consider here a simple example that draws on Isaaks and Srivastava (1989) to illustrate some important points, *ceteris paribus*.

**Example 5.5** Isaaks and Srivastava (1989, pp. 291, 301–307) use a small data set of seven observations and one prediction location to examine the effect of semivariogram parameters on ordinary kriging predictions. We use a data set of the same size, the observed data locations and their attribute values are as follows

| $i$ | $\mathbf{s}_i$ | $Z(\mathbf{s}_i)$ |
|---|---|---|
| 1 | [5,20] | 100 |
| 2 | [20,2] | 70 |
| 3 | [25,32] | 60 |
| 4 | [8,39] | 90 |
| 5 | [10,17] | 50 |
| 6 | [35,20] | 80 |
| 7 | [38,10] | 40 |

The prediction location is $\mathbf{s}_0 = [20, 20]$, and the sample mean of the observed data is $\overline{Z} = 70.0$. The observed locations surround the prediction location. Notice that, in contrast to Isaaks and Srivastava (1989), two locations are equidistant from the prediction location ($\mathbf{s}_1$ and $\mathbf{s}_6$, Figure 5.2).

The kriging weights (5.14), predictions, and kriging variance (5.16), are computed for the following series of semivariogram models.

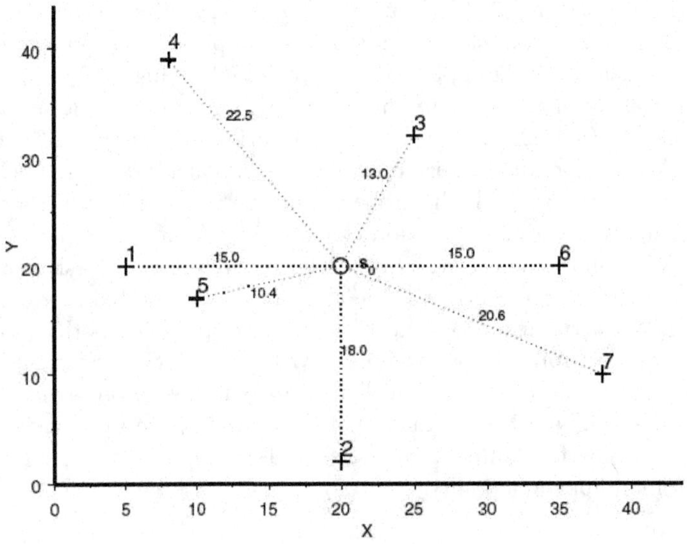


Figure 5.2 *Seven observed locations and target prediction location. The dotted rays show Euclidean distance between the observed location and the target location.*

| Model | Practical Range | Sill | Nugget | $\gamma(h)$ | Type |
|---|---|---|---|---|---|
| A | 20 | 10 | 0 | $10\left(1-e^{-3h/20}\right)$ | Exponential |
| B | 10 | 10 | 0 | $10\left(1-e^{-3h/10}\right)$ | Exponential |
| C | 20 | 10 | 5 | $5+5\left(1-e^{-3h/20}\right)$ | Exp. + nugget |
| D | – | – | 10 | 10 | Nugget only |
| E | 20 | 20 | 0 | $20\left(1-e^{-3h/20}\right)$ | Exponential |
| F | 20 | 10 | 0 | $10\left(1-e^{-3(h/20)^2}\right)$ | Gaussian |

Models A and B differ in the range, models A and C in the relative nugget effect. Model D is a nugget-only model in which data are not spatially correlated. A comparison of models A and E highlights the effect of the variability of the random field. The final model has the same variability and practical range as the exponential model A, but a much higher degree of spatial continuity. Model F exhibits large short-range correlations.

The kriging weights sum to 1.0 in all cases (within roundoff error) as needed (Table 5.1); recall that ordinary kriging weights are derived subject to the constraint $\boldsymbol{\lambda}'\mathbf{1} = 1$. Interestingly, as the degree of spatial continuity decreases, so does the "variation" among the kriging weights $\lambda_i$. Model D, the nugget-only model, assigns the same weight $\lambda_i = 1/n$ to each observation. The resulting

Table 5.1 *Kriging weights for predicting* $Z(s_0)$ *with semivariograms A–F.* $\lambda_i$ *denotes the kriging weight for the attribute* $Z(s_i)$. $s_\lambda$ *is the "standard deviation" of the seven kriging weights for a particular model.*

| Model | $p_{ok}$ | $\sigma^2_{ok}$ | $\lambda_1$ | $\lambda_2$ | $\lambda_3$ | $\lambda_4$ | $\lambda_5$ | $\lambda_6$ | $\lambda_7$ |
|---|---|---|---|---|---|---|---|---|---|
| A | 66.23 | 9.74 | 0.08 | 0.13 | 0.20 | 0.10 | 0.24 | 0.15 | 0.09 |
| B | 69.04 | 11.25 | 0.12 | 0.14 | 0.16 | 0.14 | 0.16 | 0.14 | 0.13 |
| C | 68.64 | 10.63 | 0.12 | 0.14 | 0.17 | 0.12 | 0.18 | 0.14 | 0.12 |
| D | 70.00 | 11.43 | 0.14 | 0.14 | 0.14 | 0.14 | 0.14 | 0.14 | 0.14 |
| E | 66.23 | 19.48 | 0.08 | 0.13 | 0.20 | 0.10 | 0.24 | 0.15 | 0.09 |
| F | 44.52 | 6.67 | -0.35 | 0.08 | 0.28 | 0.06 | 0.75 | 0.18 | 0.01 |
| | $\lvert\lvert \mathbf{s}_0 - \mathbf{s}_i \rvert\rvert$ | | 15.0 | 18.0 | 13.0 | 22.5 | 10.4 | 15.0 | 20.6 |

| Model | A | B | C | D | E | F |
|---|---|---|---|---|---|---|
| $s_\lambda$ | 0.06 | 0.01 | 0.03 | 0.00 | 0.06 | 0.33 |

predictor is the sample mean. Recall that models A and B are identical, except for the practical range. In the model with larger range (A), short-distance correlations are higher, creating greater heterogeneity in the weights.

The results in Table 5.1 also demonstrate that points that are separated by more than the range do not have zero kriging weights. Also, the kriging weight of these points is not $1/n$, unless all observations are uncorrelated (as in model D). For example, $\mathbf{s}_1$ is further from the prediction location than the (practical) range in model B, yet its kriging weight is 0.12. Points more distant than the range are spatially correlated with other points that are less distant from the target location than the range. This is called the **relay** effect (Chilès and Delfiner, 1999, p. 205).

Models A and E are identical, except for the sill of the semivariogram. The variability of the random field is twice as large for E, than for A. This has no effect on the kriging weights, and hence the kriging predictor is the same under the two models. The kriging variance, however, increases accordingly with the variability of the random field.

Finally, model F is highly continuous, with large short-distance correlations. Since the range and sill of model F are identical to those of model A, the gaussian model's short-distance correlations exceed those of the exponential model. The kriging weights show the most "variation" of the models in Table 5.1 and the value closest to the prediction location, $Z(\mathbf{s}_5)$, receives the most weight. It contributes 3/4 of its own value to $p_{ok}(\mathbf{Z}; \mathbf{s}_0)$, accounting for more than half of the predicted amount. Maybe surprisingly, this model yields a negative kriging weight for the observation at $\mathbf{s}_1$. A similar effect can be noted for model A, but it is less pronounced. The weight for $Z(\mathbf{s}_1)$ is considerably less than that for observation $Z(\mathbf{s}_5)$, although they occupy very similar points in the spatial configuration. It is exactly *because* they occupy similar positions that $Z(\mathbf{s}_1)$ receives small weight, and even a negative weight in model F. The effect of $Z(\mathbf{s}_1)$ is **screened** by the observation at $\mathbf{s}_5$, because it lies "behind" it relative to the prediction location.

In the derivation of the kriging weights only a "sum-to-one" constraint was imposed on the kriging weights, but not a positivity constraint. On first glance, a negative kriging weight may seem undesirable. If weights can be negative, so could possibly the predicted values. Spatial attributes are often positive, however, e.g., yields, concentrations, counts. When the weights are restricted to be positive, then all predicted values lie between the minimum and maximum observed value. Szidarovsky et al. (1987) derive a version of kriging with only positive weights. While this predictor has attractive advantages for obtaining predicted values of nonnegative processes, the extra constraint may lead to unacceptably large kriging standard errors (Cressie, 1993, p. 143). □

## 5.3 Linear Prediction with a Spatially Varying Mean

In section §5.2 we considered models of the form (5.8):

$$\mathbf{Z}(\mathbf{s}) = \boldsymbol{\mu}(\mathbf{s}) + \mathbf{e}(\mathbf{s}), \qquad \mathbf{e}(\mathbf{s}) \sim (\mathbf{0}, \boldsymbol{\Sigma}).$$

With simple kriging in §5.2.1, we assume that the mean $\boldsymbol{\mu}(\mathbf{s})$ is known. With ordinary kriging in §5.2.2, we assume $\boldsymbol{\mu}(\mathbf{s})$ is unknown but not spatially varying, $\boldsymbol{\mu}(\mathbf{s}) \equiv \mu\mathbf{1}$. Either assumption is not necessarily met, although there are cases where it is easy to verify a constant mean, for example, when we work with fitted residuals. Recall from §2.4.1 the operational decomposition of variability in a random field into large-scale trend $\mu(\mathbf{s})$, smooth, small-scale variation $W(\mathbf{s})$, micro-scale variation $\eta(\mathbf{s})$, and measurement error $\epsilon(\mathbf{s})$:

$$Z(\mathbf{s}) = \mu(\mathbf{s}) + e(\mathbf{s}) = \mu(\mathbf{s}) + W(\mathbf{s}) + \eta(\mathbf{s}) + \epsilon(\mathbf{s}).$$

The model underlying ordinary kriging predictions assumes that $\mu(\mathbf{s}) = \mu$ and that all variation in the data is associated with the spatial dependency structure $W(\mathbf{s}) + \eta(\mathbf{s})$ plus some white noise $\epsilon(\mathbf{s})$. On the other hand, the model

$$\mathbf{Z}(\mathbf{s}) = \mathbf{X}(\mathbf{s})\boldsymbol{\beta} + \boldsymbol{\epsilon}, \qquad \boldsymbol{\epsilon} \sim (\mathbf{0}, \sigma^2\mathbf{I})$$

assumes that apart from white noise all variability is associated with changes in the mean function. The fact that we consider such very different models for modeling (and predicting) spatial data is due to the adage that "one modeler's fixed effect (regressor variable) is another modeler's random effect (spatial dependency)." Historically, estimation and prediction in models for spatial data started at the two extremes: regression models with uncorrelated errors (statistics) and correlated errors with a constant mean (geostatistics). Both approaches provide considerable simplifications over the case that is probably most relevant: the combination of a spatially varying mean and spatial dependency, which gives rise to models of the form

$$\mathbf{Z}(\mathbf{s}) = \boldsymbol{\mu}(\mathbf{s}) + \mathbf{e}(\mathbf{s}), \qquad \mathbf{e}(\mathbf{s}) \sim (\mathbf{0}, \boldsymbol{\Sigma}).$$

In this section we make the assumption that $\boldsymbol{\mu}(\mathbf{s})$ is linear in some covariates, so that

$$\mathbf{Z}(\mathbf{s}) = \mathbf{X}(\mathbf{s})\boldsymbol{\beta} + \mathbf{e}(\mathbf{s}), \qquad \mathbf{e}(\mathbf{s}) = W(\mathbf{s}) + \eta(\mathbf{s}) + \epsilon(\mathbf{s}) \sim (\mathbf{0}, \boldsymbol{\Sigma}). \qquad (5.23)$$

Any one or several of the random components of $\mathbf{e}(\mathbf{s})$ may be zero (or assumed to be zero) at times. The common thread of models of form (5.23) is linearity of the mean function and a spatially correlated error process. The notation $\mathbf{X}(\mathbf{s})$ is used to emphasize that the $(n \times p)$ matrix $\mathbf{X}$ depends on spatial coordinates. The columns of this regressor matrix are usually comprised of spatial variables, although there are applications where the columns of $\mathbf{X}$ do not depend on spatial coordinates at all. Furthermore, $\mathbf{X}$ may contain dummy (design) variables and can be of less than full column rank. This situation arises when experimental data with design and treatment structure are modeled spatially. Occasionally, the dependence of $\mathbf{X}(\mathbf{s})$ on $\mathbf{s}$ will be omitted for brevity.

The random processes $W(\mathbf{s})$, $\eta(\mathbf{s})$, and $\epsilon(\mathbf{s})$ have mean zero and measurement errors are uncorrelated,

$$\mathrm{Cov}[\epsilon(\mathbf{s}_i), \epsilon(\mathbf{s}_j)] = \begin{cases} 0 & \mathbf{s}_i \neq \mathbf{s}_j \\ \sigma_\epsilon^2 & \mathbf{s}_i = \mathbf{s}_j. \end{cases}$$

As a result, the error process of (5.23) is a zero-mean stochastic process with

$$\mathrm{Var}[\mathbf{e}(\mathbf{s})] = \boldsymbol{\Sigma}_W + \boldsymbol{\Sigma}_\eta + \sigma_\epsilon^2 \mathbf{I} \equiv \boldsymbol{\Sigma}.$$

The spatial stochastic (second-order) structure is represented by the variance-covariance matrix components of $\boldsymbol{\Sigma}$. The models of interest in the analysis of spatial data can coarsely be classified into two categories. Models for uncorrelated errors assume that $W(\mathbf{s}) = \eta(\mathbf{s}) \equiv 0$ and hence only the measurement error process $\epsilon(\mathbf{s})$ remains. (One can generalize this process by allowing the variances to vary with spatial location, $\mathrm{Var}[\epsilon(\mathbf{s}_i)] = \sigma_{\epsilon,i}^2$.) Models with correlated errors assume that $\boldsymbol{\Sigma}$ is not diagonal. Based on the discussion in previous chapters it seems somewhat odd to analyze spatial data as if random disturbances were uncorrelated, since we know that spatial data at nearby locations usually exhibit similarities. There is (some) justification for this approach, however—beyond the desire for an analysis that is simple; such methods are the focus of §5.3.1 and §5.3.2. Statistical techniques for linear models with uncorrelated errors are simple, well-known, and commercial software packages to perform the numerical chores of estimation, prediction, and hypothesis testing are readily available. By comparison, statistical software for modeling correlated data is a much more recent development and the mathematical-statistical theory of correlated error models is more complicated and still evolving. The temptation to bring models for spatial data into the classical linear model (regression) framework is understandable.

If the process contains a smooth-scale spatial component, $W(\mathbf{s})$, then the smooth fluctuations in the spatial signal are handled in an uncorrelated error model by allowing the mean function $\mathbf{X}(\mathbf{s})\boldsymbol{\beta}$ to be sufficiently flexible. In other words, the mean function is parameterized to capture local behavior. With geostatistical data this can be accomplished parametrically by expressing the mean as a polynomial function of the spatial coordinates. As the local fluctuations of the spatial signal become more pronounced, higher-order terms must be included (§5.3.1). A non-parametric alternative is to model local behavior by applying $d$-dimensional smoothing or to localize estimation (§5.3.2). The degree of smoothness is then governed by a smoothing parameter (bandwidth).

The contemporary approach, however, is to assume that some spatial stochastic structure is present which conveys in the presence of $W(\mathbf{s})$ and $\eta(\mathbf{s})$, hence $\boldsymbol{\Sigma}$ will be a non-diagonal matrix. (The argument that $\text{Var}[\mathbf{W}(\mathbf{s})] = \sigma^2_W\mathbf{I}$, $\text{Var}[\boldsymbol{\eta}(\mathbf{s})] = \sigma^2_\eta\mathbf{I}$ and hence $\boldsymbol{\Sigma}$ is diagonal is vacuous. The individual random components would not be identifiable.) Modeling then comprises parameterization of the mean function, i.e., proper choice of the columns of $\mathbf{X}(\mathbf{s})$, as well as parameterization of $\text{Var}[\mathbf{Z}(\mathbf{s})]$. Unstructured variance-covariance matrices for spatial data are a non-sensible option because of the large number of parameters that would have to be estimated. We usually place some parametric structure on the variance-covariance matrix. To make the parametric nature of the covariance matrix more explicit, model (5.23) should be written as

*Spatial Linear Model*

$$\mathbf{Z}(\mathbf{s}) = \mathbf{X}(\mathbf{s})\boldsymbol{\beta} + \mathbf{e}(\mathbf{s}), \quad \mathbf{e}(\mathbf{s}) \sim (\mathbf{0}, \boldsymbol{\Sigma}(\boldsymbol{\theta})). \tag{5.24}$$

The $(q \times 1)$ vector $\boldsymbol{\theta}$ contains the unknown parameters of the error process. This is the model for which prediction theory is presented in §5.3.3.

### 5.3.1 Trend Surface Models

The idea of the trend surface approach is to model the mean function in (5.24) with a highly parameterized fixed effects structure, comprised of functions of the spatial coordinates, $\mathbf{s}_i = [x_i, y_i]'$. For example, a first-degree trend surface model is

$$Z(\mathbf{s}_i) = \beta_0 + \beta_1 x_i + \beta_2 y_i + \epsilon_i, \quad \epsilon_i \sim iid\,(0, \sigma^2).$$

If $\text{E}[Z(\mathbf{s}_i)] = \mu$, and the $Z(\mathbf{s}_i)$ are correlated, then this model is incorrect in several places: $\beta_0 + \beta_1 x_i + \beta_2 y_i$ is not the model for the mean and the errors are not *iid*. By over-parameterizing the mean, the model accounts for variability that is associated with the spatial random structure. The approach pretends that the models

$$\mathbf{Z}(\mathbf{s}) = \mathbf{1}\mu + \mathbf{e}(\mathbf{s}), \quad \mathbf{e}(\mathbf{s}) \sim (\mathbf{0}, \boldsymbol{\Sigma})$$

and

$$\mathbf{Z}(\mathbf{s}) = \mathbf{X}(\mathbf{s})\boldsymbol{\beta} + \boldsymbol{\epsilon}, \quad \boldsymbol{\epsilon} \sim (\mathbf{0}, \sigma^2\mathbf{I})$$

are *equivalent* representations of the spatial variability seen in the realization of a random field. If $\mathbf{e}(\mathbf{s})$ contains a smooth, small-scale component, then $\mathbf{X}(\mathbf{s})\boldsymbol{\beta}$ captures the local behavior of the random field, not its mean. If $\mathbf{e}(\mathbf{s})$ contains a nugget effect, its variance should equal $\sigma^2$ if the fixed effects structure $\mathbf{X}(\mathbf{s})\boldsymbol{\beta}$ removed all smooth, small-scale variation. In practice, it is impossible to separate out all the potential components of a spatial process from a single set of spatial data, and both versions of the model may fit the data equally well, although their interpretations will be dramatically different.

The name trend surface model stems from the parameterization of $\mathbf{X}(\mathbf{s})\boldsymbol{\beta}$ in terms of polynomial trend surfaces in $d$ coordinates of degree $p$. For a process in $\mathbb{R}^2$, we write

*Trend Surface Model*

$$
\begin{aligned}
Z(\mathbf{s}_i) &= \mathbf{x}(\mathbf{s}_i)\boldsymbol{\beta} + \epsilon_i \\
\mathbf{x}(\mathbf{s}_i)\boldsymbol{\beta} &= \sum_{k=0}^{p}\sum_{m=0}^{p} \beta_{km} x_i^k y_i^m \quad k+m \le p \qquad (5.25)\\
\epsilon_i &\sim (0, \sigma^2) \\
\mathrm{Cov}[\epsilon_i, \epsilon_j] &= 0 \ \forall i \ne j.
\end{aligned}
$$

Estimation and prediction theory is straightforward in models with uncorrelated (and homoscedastic) errors. Specifically, the regression coefficients are estimated by OLS as

$$\widehat{\boldsymbol{\beta}}_{ols} = (\mathbf{X}(\mathbf{s})'\mathbf{X}(\mathbf{s}))^{-1}\mathbf{X}(\mathbf{s})'\mathbf{Z}(\mathbf{s}),$$

and the estimate of the residual variance is

$$\widehat{\sigma}^2 = \frac{1}{n-k}\sum_{i=1}^{n}\left(Z(\mathbf{s}_i) - \widehat{\mathrm{E}}[Z(\mathbf{s}_i)]\right)^2.$$

The BLUE of $\mathrm{E}[Z(\mathbf{s}_i)]$ and the BLUP of $Z(\mathbf{s}_i)$ are the same, the best estimator and the best predictor are

$$\widehat{Z}(\mathbf{s}_0) = \widehat{\mathrm{E}}[Z(\mathbf{s}_0)] = \mathbf{x}'(\mathbf{s}_0)\widehat{\boldsymbol{\beta}}_{ols}.$$

The mean-squared prediction error for $Z(\mathbf{s}_0)$ based on $\widehat{Z}(\mathbf{s}_0)$ is

$$MSE[Z(\mathbf{s}_0); \widehat{Z}(\mathbf{s}_0)] = \sigma^2\left(1 + \mathbf{x}'(\mathbf{s}_0)(\mathbf{X}(\mathbf{s})'\mathbf{X}(\mathbf{s}))^{-1}\mathbf{x}(\mathbf{s}_0)\right),$$

where we assumed that the new data point $Z(\mathbf{s}_0)$ is uncorrelated with the observed data. This is consistent with the uncorrelated error assumption of model (5.25).

The number of regression coefficients in a trend surface model increases quickly with the degree of the polynomial, $\boldsymbol{\beta}$ is a vector of length $(p+1)(p+2)/2$. To examine whether the mean function has been made sufficiently flexible, the residuals from the fit can be examined for residual spatial autocorrelation.

In the model

$$\mathbf{Z}(\mathbf{s}) = \mathbf{X}(\mathbf{s})\boldsymbol{\beta} + \boldsymbol{\epsilon}$$

it is of interest to test $H_0$: $\text{Var}[\epsilon] \propto \mathbf{I}$. Since $\epsilon$ is unobservable, we draw instead on

$$\widehat{\epsilon} = \mathbf{Z}(\mathbf{s}) - \mathbf{X}(\mathbf{s})(\mathbf{X}(\mathbf{s})'\mathbf{X}(\mathbf{s}))^{-1}\mathbf{X}(\mathbf{s})'\mathbf{Z}(\mathbf{s}) = \mathbf{MZ}(\mathbf{s}).$$

If $\text{Var}[\epsilon] = \sigma^2\mathbf{I}$ then $\text{Var}[\widehat{\epsilon}] = \sigma^2\text{M}$. When the data are on a lattice, a test for autocorrelation can be carried out based on Moran's $I$ statistic for regression residuals ((1.16), p. 23 and §6.1.2). For geostatistical data, the empirical semivariogram of the ordinary least squares residuals can be examined, although this will yield a biased estimator of the semivariogram of $\epsilon$. The bias is typically negligible for small lags. More details on residual analysis in models with uncorrelated errors and models with correlated errors as well as the implications for semivariogram estimation are forthcoming in Chapter 6.

Judging fitted residuals based on autocorrelation statistics such as Moran's $I$, or by interpreting the empirical semivariogram can lead to very different conclusions about the degree of smoothness. Rook, queen, and bishop definitions of spatial connectivity only examine immediate neighbors on the lattice and thus focus on very small-scale, localized variation. The empirical semivariogram displays information on numerous lag classes and gives a picture of the small-, medium-, and long-range spatial variation. We examine the two approaches with a simulated data example.

**Example 5.6** Figure 5.3a displays data simulated on a $20 \times 20$ grid with exponential covariance function, range 9, nugget 0.25, and sill 0.75. The spatial autocorrelations in this process are not very pronounced due to the medium range, the presence of a nugget, and the comparatively low continuity of the exponential covariance function.

A test of autocorrelation using Moran's $I$ indicates spatial dependence in these data (Table 5.2, $p = 0$). The "beyond-independent-data" dependence in the residuals is gradually decreased as the order of the trend surface increases (Table 5.2). Although the value of $I_{res}$ decreases with increasing $p$, $\text{E}[I_{res}]$ depends on $\mathbf{X}(\mathbf{s})$ and $I_{res}$ has the familiar expectation of $-1/(n-1)$ only for $p = 0$. For $p = 14$, the residual autocorrelation due to the autocorrelation in the process has been substantially reduced. But note that this results in a regression model with 120 coefficients. Further increases in the surface degree lead to models that are hopelessly overfit and ill-conditioned. Figure 5.3b shows the predicted surface from the polynomial trend surface with $p = 14$.

A different perspective can be gained by choosing a goodness-of-fit criterion that penalizes for the number of parameters. The last column of Table 5.2 gives values of Akaike's AIC criterion for maximum likelihood estimation (Akaike, 1974). The smallest value is achieved for $p = 11$, yielding a fit less smooth than model selection based on $I_{res}$.

To determine an appropriate trend surface degree we can also examine the empirical semivariogram of the ordinary least squares residuals for different values of $p$. Since the empirical semivariogram estimator based on fitted residuals is biased, in particular for larger lags, we focus on the behavior near the

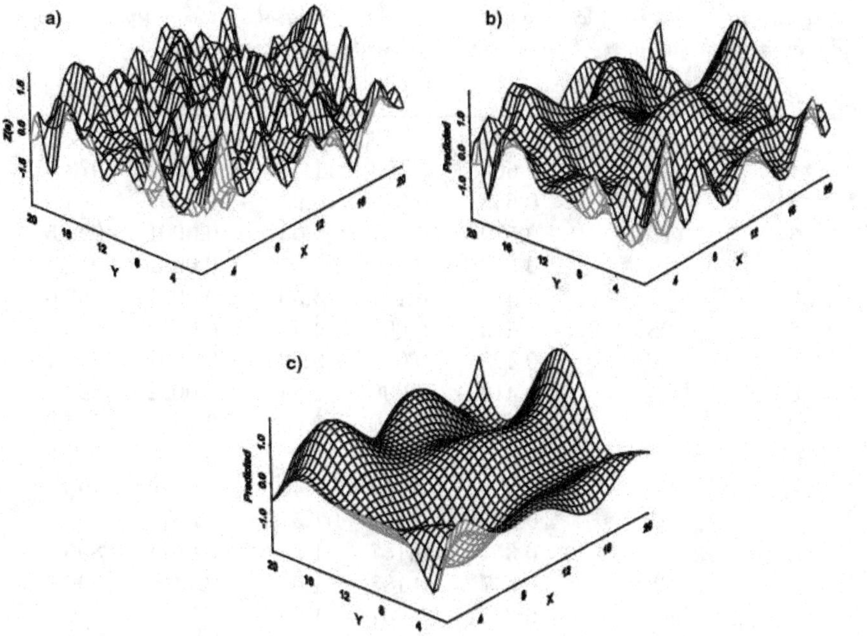


Figure 5.3 *Trend surface modeling. a: simulated spatial arrangements on a* $20 \times 20$ *grid with exponential covariance function, range 9, nugget 0.25, and sill 0.75. Panels b) and c) show predicted mean values from polynomial trend surface with degrees* $p = 14$ *and* $p = 8$.

origin. The semivariogram of the original data shows the spatial structure clearly as well as the nugget effect (Figure 5.4). A trend surface of fifth degree still exhibits some near-origin spatial structure in the semivariogram. For $p = 8$ this structure has disappeared and the empirical semivariogram appears to decline with increasing lag distance. This apparent decline is a combination of a biased estimator and a statistical model that overfits the mean function. Notice that the nugget effect remains present in all semivariograms; it is the non-spatially structured source of variability in the data.

Based on the results displayed in Figure 5.4, a trend surface model of degree $p = 8$ seems adequate. The predicted surface for $p = 8$ is shown in Figure 5.3c. It is considerably more smooth than the surface with $p = 14$. Note that with the rook definition of spatial neighborhood, the $I_{res}$ statistic focuses entirely on the relationships between residuals of nearest neighbors, emphasizing, and perhaps over-emphasizing, local behavior. A trend surface with $p = 11$, as selected based on the AIC criterion, appears to be a good compromise. Note

Table 5.2 *Fit statistics for trend surface models of different order fit to data shown in Figure 5.3a. $Z_{obs}$ and p-values refer to the test of residual spatial autocorrelation based on the $I^*$ statistic with a rook definition of spatial connectivity. The second column gives the number of regression coefficients (intercept included).*

| $p$ | $(p+1)(p+2)/2$ | $\widehat{\sigma}^2$ | $I_{res}$ | $Z_{obs}$ | $p$-value | AIC |
|---|---|---|---|---|---|---|
| 0 | 1 | 0.733 | 0.473 | 13.17 | $< 0.0001$ | 1012.1 |
| 1 | 3 | 0.662 | 0.403 | 11.46 | $< 0.0001$ | 973.4 |
| 2 | 6 | 0.643 | 0.375 | 10.92 | $< 0.0001$ | 964.3 |
| 3 | 10 | 0.579 | 0.297 | 9.03 | $< 0.0001$ | 926.5 |
| 4 | 15 | 0.552 | 0.266 | 8.51 | $< 0.0001$ | 912.8 |
| 5 | 21 | 0.489 | 0.181 | 6.50 | $< 0.0001$ | 869.6 |
| 6 | 28 | 0.443 | 0.095 | 4.46 | $< 0.0001$ | 837.3 |
| 7 | 36 | 0.428 | 0.052 | 3.69 | 0.0001 | 830.4 |
| 8 | 45 | 0.410 | 0.006 | 2.84 | 0.0022 | 821.3 |
| 9 | 55 | 0.398 | −0.019 | 2.64 | 0.0041 | 817.6 |
| 10 | 66 | 0.386 | −0.046 | 2.42 | 0.0077 | 814.3 |
| 11 | 78 | 0.360 | −0.098 | 1.49 | 0.0674 | 795.7 |
| 12 | 91 | 0.362 | −0.117 | 1.57 | 0.0577 | 808.9 |
| 13 | 105 | 0.351 | −0.147 | 1.36 | 0.0871 | 805.7 |
| 14 | 120 | 0.347 | −0.183 | 1.05 | 0.1466 | 809.6 |
| 15 | 136 | 0.343 | −0.207 | 1.08 | 0.1400 | 813.0 |

that $p = 11$ is the largest value in Table 5.2 for which the $I_{res}$ statistic is not significant at the 0.05 level. □

### 5.3.2 *Localized Estimation*

Trend surface models require a large number of regression coefficients to capture rather simple spatial variation. A high polynomial degree is needed in order for the predictions to be locally adequate everywhere. The flexibility of $\widehat{Z}(\mathbf{s}_0) = \mathbf{x}(\mathbf{s}_0)'\widehat{\boldsymbol{\beta}}$ is achieved through having many regressors in $\mathbf{x}$. As the prediction location is changed, the elements of $\mathbf{x}(\mathbf{s}_0)$ change considerably to produce good predictions although the same vector $\widehat{\boldsymbol{\beta}}$ is used for predictions. Could we not achieve the same (or greater) flexibility by keeping the order of the trend surface low, but allowing $\boldsymbol{\beta}$ to vary spatially? This naturally leads to the idea of fitting a model locally, instead of the trend surface model (5.25), which is a global model. A global model has one set of parameters that apply everywhere, regardless of spatial location.

If the order of the trend surface is reduced, for example to $p = 1$, we do not expect the resulting plane to be a good fit over the entire domain. We can expect, however, for the plane to fit well at a given point $\mathbf{s}_0$ and in its immediate neighborhood. The idea of localized estimation is to assign weights

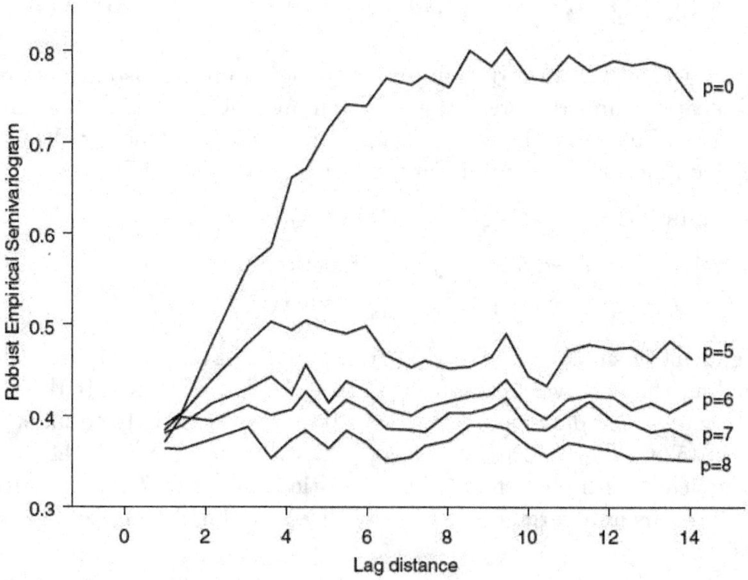


Figure 5.4 *Empirical Cressie-Hawkins semivariogram estimators for the original data ($p = 0$) and the ordinary least squares residuals for different trend surface degrees.*

to data $Z(\mathbf{s}_1), \cdots, Z(\mathbf{s}_n)$ that reflect the extent to which the low-rank model applied at $\mathbf{s}_0$ is expected to fit at other locations. The weights control the influence of observations on the estimation of the model at $\mathbf{s}_0$. The benefits of a parsimonious model $\mathbf{x}(\mathbf{s}_0)'\boldsymbol{\beta}$ for any one site is traded against estimation of local regression coefficients and the determination of the weight function.

Local estimation in this spirit is often referred to as non-parametric regression. Kernel estimation, for example, moves a window over the data and estimates the mean at a particular location as the weighted average of the data points within the window. This window may extend over the entire data range if the weight function reaches zero only asymptotically, or consist of a set of nearest neighbors. Local polynomial regression applies the same idea but fits a polynomial model at each prediction location rather than a constant mean. An equivalent representation of local estimation—which we prefer—is in terms of weighted linear regression models.

Assume that a prediction is desired at $\mathbf{s}_0$ and that the behavior of the realized surface near $\mathbf{s}_0$ can be expressed as a polynomial of first degree. An ordinary least squares fit at $\mathbf{s}_0$ is obtained by minimizing

$$Q(\mathbf{s}_0, \lambda) = \sum_{i=1}^{n} W(||\mathbf{s}_i - \mathbf{s}_0||, \lambda) \left(Z(\mathbf{s}_i) - \beta_{00} - \beta_{01}x_i - \beta_{02}y_i\right)^2 .$$

The (kernel) weight function $W(||\mathbf{s}_i - \mathbf{s}_0||, \lambda)$ depends on the distance between observed locations and the prediction location and the smoothing parameter (bandwidth) $\lambda$. Collecting the $W(||\mathbf{s}_i - \mathbf{s}_0||, \lambda)$ into a diagonal matrix $\mathbf{W}(\mathbf{s}_0, \lambda)$, the objective function can be written more clearly as

$$Q(\mathbf{s}_0, \lambda) = (\mathbf{Z}(\mathbf{s}) - \mathbf{X}(\mathbf{s})\boldsymbol{\beta}_0)'\mathbf{W}(\mathbf{s}_0, \lambda)(\mathbf{Z}(\mathbf{s}) - \mathbf{X}(\mathbf{s})\boldsymbol{\beta}_0).$$

This is a weighted least squares objective function in the model

$$\mathbf{Z}(\mathbf{s}) = \mathbf{X}(\mathbf{s})\boldsymbol{\beta}_0 + \mathbf{e}_0, \quad \mathbf{e}_0 \sim (\mathbf{0}, \mathbf{W}(\mathbf{s}_0, \lambda)^{-1}) \tag{5.26}$$

The assignment of small weight to $Z(\mathbf{s}_j)$ compared to $Z(\mathbf{s}_i)$, say, is equivalent to pretending that the variance of $Z(\mathbf{s}_j)$ exceeds that of $Z(\mathbf{s}_i)$. If $W(||\mathbf{s}_i - \mathbf{s}_0||, \lambda) = 0$, then the data point $Z(\mathbf{s}_i)$ can be removed entirely to allow the inversion in (5.26). This representation of local estimation as a weighted estimation problem is entirely general. Any statistical model that can be written in terms of means and variances can be localized by this device (see Chapter problems).

A special case of local polynomial estimation is LOESS regression (Cleveland, 1979), where the (tri-cube) weight function achieves exactly zero and the estimation uses robust re-weighing of residuals. In general, the choice of the weight function is less important than the choice of the bandwidth $\lambda$, a reasonable choice of $W$ will lead to reasonable results. Common choices in $\mathbb{R}^1$ are the Epanechnikov kernel

$$W_e(d, \lambda) = \begin{cases} \frac{3}{4}\lambda^{-1}(1 - (d/\lambda)^2) & -\lambda \le d \le \lambda \\ 0 & \text{otherwise,} \end{cases}$$

and the Gaussian kernel

$$W_g(d, \lambda) = \frac{1}{\lambda\sqrt{2\pi}} \exp\left\{-\frac{1}{2}\left(\frac{d}{\lambda}\right)^2\right\}.$$

To construct a kernel weight function in $\mathbb{R}^d$, one can either draw on a multivariate kernel or compute the kernel as the product of $d$ univariate kernel functions. The two approaches are not identical, of course. For example, a bivariate Gaussian kernel function can be constructed from the bivariate Gaussian distribution function allowing an elliptical contour of the weight function. The Gaussian product kernel

$$\begin{aligned} W_g(\mathbf{s}_i - \mathbf{s}_j, \lambda) &= \frac{1}{2\pi\lambda^2} \exp\left\{-\frac{1}{2}\left(\frac{x_i - x_j}{\lambda}\right)^2\right\} \exp\left\{-\frac{1}{2}\left(\frac{y_i - y_j}{\lambda}\right)^2\right\} \\ &= \frac{1}{2\pi\lambda^2} \exp\left\{-\frac{1}{2\lambda^2}||\mathbf{s}_i - \mathbf{s}_j||^2\right\} \end{aligned}$$

has a common bandwidth for the major axes of the coordinate system and spherical weight contours.

The two important choices made in local estimation are the degree of the local polynomial and the bandwidth. Locally constant means lead to estimates which suffer from edge bias. For spatial data this is an important consideration because many data points fall near the bounding box or the convex hull of a set of points.

### *5.3.3 Universal Kriging*

Suppose we have data $Z(\mathbf{s}_1), \cdots, Z(\mathbf{s}_n)$ at spatial locations $\mathbf{s}_1, \cdots, \mathbf{s}_n$, and we want to predict $Z(\mathbf{s}_0)$ at location $\mathbf{s}_0$ where we do not have an observation. Further suppose that the form of the general linear model holds for both the data and the unobservables:

$$\begin{aligned} \mathbf{Z}(\mathbf{s}) &= \mathbf{X}(\mathbf{s})\boldsymbol{\beta} + \mathbf{e}(\mathbf{s}), \\ Z(\mathbf{s}_0) &= \mathbf{x}(\mathbf{s}_0)'\boldsymbol{\beta} + e(\mathbf{s}_0), \end{aligned}$$

where $\mathbf{x}(\mathbf{s}_0)$ is the $(p \times 1)$ vector of explanatory values associated with location $\mathbf{s}_0$. As before, we assume a general variance-covariance matrix for the data, $\text{Var}[\mathbf{Z}(\mathbf{s})] = \boldsymbol{\Sigma}$, but we also assume the data and the unobservables are spatially correlated, so that $\text{Cov}[\mathbf{Z}(\mathbf{s}), Z(\mathbf{s}_0)] = \boldsymbol{\sigma}$, an $(n \times 1)$ vector, and $\text{Var}[Z(\mathbf{s}_0)] = \sigma_0$.

The goal is to find the optimal linear predictor, one that is unbiased and has minimum variance in the class of linear, unbiased predictors. Thus, we consider predictors of the form $\mathbf{a}'\mathbf{Z}(\mathbf{s})$, and find the vector $\mathbf{a}$ so that $\mathbf{a}'\mathbf{Z}(\mathbf{s})$ is the best linear unbiased predictor of $Z(\mathbf{s}_0)$. Statistically, this problem becomes: find the vector $\mathbf{a}$ that minimizes the mean-squared prediction error

$$\begin{aligned} \text{E}[(\mathbf{a}'\mathbf{Z}(\mathbf{s}) - Z(\mathbf{s}_0))^2] &= \text{Var}[\mathbf{a}'\mathbf{Z}(\mathbf{s})] + \text{Var}[Z(\mathbf{s}_0)] - 2\text{Cov}[\mathbf{a}'\mathbf{Z}(\mathbf{s}), Z(\mathbf{s}_0)] \\ &= \mathbf{a}'\boldsymbol{\Sigma}\mathbf{a} + \sigma_0 - 2\mathbf{a}'\boldsymbol{\sigma}, \qquad (5.27) \end{aligned}$$

subject to $\text{E}[\mathbf{a}'\mathbf{Z}(\mathbf{s})] = \text{E}[Z(\mathbf{s}_0)]$. This unbiasedness condition implies

$$\mathbf{a}'\mathbf{X}(\mathbf{s})\boldsymbol{\beta} = \mathbf{x}(\mathbf{s}_0)'\boldsymbol{\beta} \quad \forall \boldsymbol{\beta},$$

which gives $\mathbf{a}'\mathbf{X}(\mathbf{s}) = \mathbf{x}(\mathbf{s}_0)'$.

To minimize this function subject to the constraint, we use the method of Lagrange multipliers. The Lagrangian is

$$\mathcal{L} = \mathbf{a}'\boldsymbol{\Sigma}\mathbf{a} + \sigma_0 - 2\mathbf{a}'\boldsymbol{\sigma} + 2\mathbf{m}'(\mathbf{X}(\mathbf{s})'\mathbf{a} - \mathbf{x}(\mathbf{s}_0)),$$

where $\mathbf{m}$ is a $p \times 1$ vector of Lagrange multipliers. Differentiating with respect to $\mathbf{a}$ and $\mathbf{m}$ gives

$$\begin{aligned} \frac{\partial \mathcal{L}}{\partial \mathbf{a}} &= 2\boldsymbol{\Sigma}\mathbf{a} - 2\boldsymbol{\sigma} + 2\mathbf{X}(\mathbf{s})\mathbf{m} \\ \frac{\partial \mathcal{L}}{\partial \mathbf{m}} &= 2(\mathbf{X}(\mathbf{s})'\mathbf{a} - \mathbf{x}(\mathbf{s}_0)). \end{aligned}$$

Equating each term to zero and rearranging gives

$$\begin{aligned}\Sigma\mathrm{a} + \mathbf{X}(\mathrm{s})\mathrm{m} &= \boldsymbol{\sigma}\\ \mathbf{X}(\mathrm{s})'\mathrm{a} &= \mathrm{x}(\mathrm{s}_0).\end{aligned}$$

Solving these equations for a we obtain:

$$\begin{aligned}\mathrm{a} &= \Sigma^{-1}(\boldsymbol{\sigma} - \mathbf{X}(\mathrm{s})(\mathbf{X}(\mathrm{s})'\Sigma^{-1}\mathbf{X}(\mathrm{s}))^{-1}(\mathbf{X}(\mathrm{s})'\Sigma^{-1}\boldsymbol{\sigma} - \mathrm{x}(\mathrm{s}_0))) \quad (5.28)\\ &= \Sigma_X{}^{-}\boldsymbol{\sigma} + \Sigma^{-1}\mathbf{X}(\mathrm{s})(\mathbf{X}(\mathrm{s})'\Sigma^{-1}\mathbf{X}(\mathrm{s}))^{-1}\mathrm{x}(\mathrm{s}_0)\\ &\text{with } \Sigma_X^{-} = \Sigma^{-1} - \Sigma^{-1}\mathbf{X}(\mathrm{s})(\mathbf{X}(\mathrm{s})'\Sigma^{-1}\mathbf{X}(\mathrm{s}))^{-1}\mathbf{X}(\mathrm{s})'\Sigma^{-1}.\end{aligned}$$

Note that $(\mathbf{X}(\mathrm{s})'\Sigma^{-1}\mathbf{X}(\mathrm{s}))^{-1}\mathbf{X}'\Sigma^{-1}\mathbf{Z}(\mathrm{s}) = \widehat{\boldsymbol{\beta}}_{gls}$, and the best linear unbiased predictor of $Z(\mathrm{s}_0)$ can be written as

*Universal Kriging Predictor*

$$\mathrm{a}'\mathbf{Z}(\mathrm{s}) = p_{uk}(\mathbf{Z};\mathrm{s}_0) = \mathrm{x}(\mathrm{s}_0)'\widehat{\boldsymbol{\beta}}_{gls} + \boldsymbol{\sigma}'\Sigma^{-1}(\mathbf{Z}(\mathrm{s}) - \mathbf{X}(\mathrm{s})\widehat{\boldsymbol{\beta}}_{gls}). \quad (5.29)$$

We use the subscript "UK" to denote this as the *universal kriging* predictor, to distinguish it from the simple and ordinary kriging predictors of §5.2.1 and §5.2.2. Obviously, if we assume a constant mean $\mathrm{E}[\mathbf{Z}(\mathrm{s})] = \mathbf{1}\mu$, then the universal kriging predictor reduces to the ordinary kriging predictor (compare to equations (5.17) and (5.18))

$$p_{ok}(\mathbf{Z};\mathrm{s}_0) = \mathbf{1}\widehat{\boldsymbol{\beta}}_{gls} + \boldsymbol{\sigma}'\Sigma^{-1}(\mathbf{Z}(\mathrm{s}) - \mathbf{1}\widehat{\boldsymbol{\beta}}_{gls}),$$

with

$$\widehat{\boldsymbol{\beta}}_{gls} = (\mathbf{1}'\Sigma^{-1}\mathbf{1})^{-1}\mathbf{1}'\Sigma^{-1}\mathbf{Z}(\mathrm{s}).$$

The minimized mean-squared prediction error provides a measure of uncertainty associated with the universal kriging predictor. This is obtained by substituting the optimal weights, a, given in equation (5.28), into equation (5.27), yielding the *kriging variance*

$$\begin{aligned}\sigma^2_{uk}(\mathrm{s}_0) &= \mathrm{a}'\Sigma\mathrm{a} + \sigma_0 - 2\mathrm{a}'\boldsymbol{\sigma}\\ &= \sigma_0 - \boldsymbol{\sigma}'\Sigma^{-1}\boldsymbol{\sigma}\\ &\quad + (\mathrm{x}(\mathrm{s}_0)' - \boldsymbol{\sigma}'\Sigma^{-1}\mathbf{X}(\mathrm{s}))(\mathbf{X}(\mathrm{s})'\Sigma^{-1}\mathbf{X}(\mathrm{s}))^{-1}\\ &\quad \times (\mathrm{x}(\mathrm{s}_0)' - \boldsymbol{\sigma}'\Sigma^{-1}\mathbf{X}(\mathrm{s}))'. \quad (5.30)\end{aligned}$$

As with ordinary kriging, universal kriging can also be done in terms of the semivariogram (see Cressie, 1993, pp. 153–154).

To predict $r$ variables, $\mathbf{Z}(\mathrm{s}_0)$, simultaneously, we extend the model above to

$$\mathrm{E}\begin{bmatrix}\mathbf{Z}(\mathrm{s})\\ \mathbf{Z}(\mathrm{s}_0)\end{bmatrix} = \begin{bmatrix}\mathbf{X}(\mathrm{s})\boldsymbol{\beta}\\ \mathbf{X}(\mathrm{s}_0)\boldsymbol{\beta}\end{bmatrix} \quad (5.31)$$

$$\mathrm{Var}\begin{bmatrix}\mathbf{Z}(\mathrm{s})\\ \mathbf{Z}(\mathrm{s}_0)\end{bmatrix} = \begin{bmatrix}\Sigma_{ZZ} & \Sigma_{Z0}\\ \Sigma_{0Z} & \Sigma_{00}\end{bmatrix}. \quad (5.32)$$

Here, $\Sigma_{ZZ}$ is the $n \times n$ variance-covariance matrix of the data, $\Sigma_{00}$ is the $r \times r$ variance-covariance matrix among the unobservables, and $\Sigma_{0Z}$ is the $r \times n$

variance-covariance matrix between the data and the unobservables. With this model, the best linear unbiased predictor (BLUP) (Goldberger, 1962; Gotway and Cressie, 1993) is

$$\hat{\mathbf{Z}}(\mathbf{s}_0) = \mathbf{X}(\mathbf{s}_0)\widehat{\boldsymbol{\beta}}_{gls} + \boldsymbol{\Sigma}_{0Z}\boldsymbol{\Sigma}_{ZZ}^{-1}(\mathbf{Z}(\mathbf{s}) - \mathbf{X}(\mathbf{s})\widehat{\boldsymbol{\beta}}_{gls}), \tag{5.33}$$

and the associated mean-squared prediction error is given by

$$\begin{aligned} \mathrm{E}[(\hat{\mathbf{Z}}(\mathbf{s}_0) - \mathbf{Z}(\mathbf{s}_0))^2] &= \boldsymbol{\Sigma}_{00} - \boldsymbol{\Sigma}_{0Z}\boldsymbol{\Sigma}_{ZZ}^{-1}\boldsymbol{\Sigma}_{Z0} \\ &\quad + (\mathbf{X}(\mathbf{s}_0) - \boldsymbol{\Sigma}_{0Z}\boldsymbol{\Sigma}_{ZZ}^{-1}\mathbf{X}(\mathbf{s}))(\mathbf{X}(\mathbf{s})'\boldsymbol{\Sigma}_{ZZ}^{-1}\mathbf{X}(\mathbf{s}))^{-1} \\ &\quad \times (\mathbf{X}(\mathbf{s}_0) - \boldsymbol{\Sigma}_{0Z}\boldsymbol{\Sigma}_{ZZ}^{-1}\mathbf{X}(\mathbf{s}))'. \end{aligned} \tag{5.34}$$

Equations (5.33) and (5.34) are obvious extensions of (5.29) and (5.30) to the multi-predictor case.

## 5.4 Kriging in Practice

### *5.4.1 On the Uniqueness of the Decomposition*

For the purpose of predictions, we can model the spatial variation entirely through the covariates, entirely as small-scale variation characterized by the semivariogram or $\boldsymbol{\Sigma}(\boldsymbol{\theta})$, or through some combination of covariates and residual spatial autocorrelation. Thus, the decomposition of the data into covariates plus spatially correlated error as depicted through equation (5.24) is not unique. However, our choice impacts both the interpretation of our model and the magnitude of the prediction standard errors.

For example, suppose we accidentally left out an important spatially-varying covariate (say $x_{p+1}$) when we defined $\mathbf{X}(\mathbf{s})$. If we do a good job of fitting both models, the model omitting $x_{p+1}$ may fit as well as the model including $x_{p+1}$. So we could have two competing models defined by parameters $(\boldsymbol{\beta}_1, \mathbf{e}(\mathbf{s})_1)$ and $(\boldsymbol{\beta}_2, \mathbf{e}(\mathbf{s})_2)$ with comparable fit. If $\mathbf{X}(\mathbf{s})_1\boldsymbol{\beta}_1 \neq \mathbf{X}(\mathbf{s})_2\boldsymbol{\beta}_2$, then the interpretations in the two models could be *very* different, although both models are valid representations of the spatial variation in the data. The predicted surfaces based on these two models will be similar, but the standard errors and the interpretation of covariates effects will be substantially different (see e.g., Cressie, 1993, pp. 212–224, and Gotway and Hergert, 1997).

As you will see in §5.5, the question of how to estimate the unknown parameters of the spatial correlation structure—when the mean is spatially varying—is an important aspect of spatial prediction. If the mean is constant, then the techniques of §4.4 and §4.5 can be applied to obtain estimators of the covariance and/or semivariogram parameters. It is tempting from this vantage point to adopt an "ordinary-kriging-at-all-cost" attitude and to model spatial variation entirely through the small-scale variation. For example, because the semivariogram filters the (unknown but) constant mean, not knowing $\mu$ is of no consequence in semivariogram estimation. An incorrect assumption of a constant large-scale mean can be dangerous for your spatial analysis, however.

Schabenberger and Pierce (2002, p. 614) give the following example, where data are generated on a transect according to the deterministic functions

$$\begin{aligned} Z_1(t) &= 1 + 0.5t \\ Z_2(t) &= 1 + 0.22t + 0.022t^2 - 0.0013t^3. \end{aligned}$$

Note that "data" so generated is deterministic, there is no random variation. If one computes the Matheron estimator of the empirical semivariogram from these data, the graphs in Figure 5.5 result. A power semivariogram model was fit to the empirical semivariogram in the left-hand panel. A gaussian semivariogram fits the empirical semivariogram in the right-hand panel well. Not accounting for the large-scale structure may lead you to attribute deterministic spatial variation—because the large-scale trend is non-random—to random sources. The spatial "dependency" one is inclined to infer from Figure 5.5 is entirely spurious.

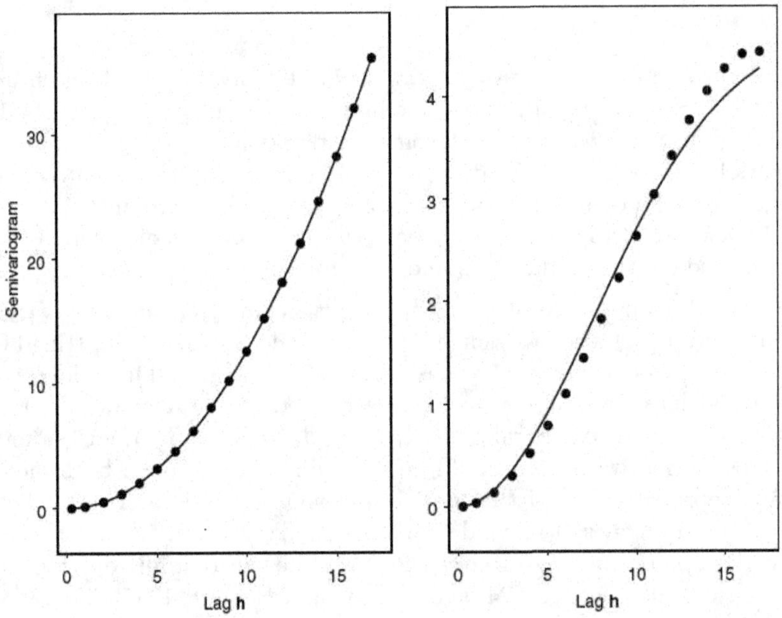


Figure 5.5 *Empirical semivariograms (dots) and fitted models for data from deterministic trend. Left panel is for $Z_1(t)$, right panel is for $Z_2(t)$. From Schabenberger and Pierce (2002, p. 614).*

### *5.4.2 Local Versus Global Kriging*

Kriging is a statistical method for interpolation, very similar to interpolating splines and inverse-distance squared interpolation. The most common application of kriging is the construction of smooth contour maps for visualization. In this type of application, variables are predicted on a regular grid—hence this process is often called *gridding*—and the values are then displayed using a surface or contour map. Thus, since many predictions are required, kriging is usually done using only some of the "neighboring" data in order to speed the computations. The search neighborhood is usually taken to be a circle centered on the grid node unless the study domain is elongated (e.g., elliptical) or the spatial process is anisotropic. If there are many points in the neighborhood, we may further restrict the calculations and use just the closest $m$ points. (In many computer programs, defaults for $m$ vary from 6 to 24). These parameters (search neighborhood size, shape, and number of points used for interpolation) can affect the nature of the interpolated surface.

The use of local search neighborhoods can result in great computational savings when working with large data sets. However, the search neighborhood should be selected with care as it will affect the characteristics of the predicted surface. Global kriging using all the data produces a relatively smooth surface. Small search neighborhoods with few data points for prediction will produce surfaces that show more detail, but this detail may be misleading if predictions are unstable. In general, at least 6–12 points should be used for each predicted value; kriging with more than 25 points is often unnecessary. If the data are sparse, or the relative nugget effect is large, distant points may have important information and the search neighborhood should be increased to include them. When the data are evenly distributed within the domain, a simple search that defines the neighborhood by the closest points is usually adequate. However, when observations are clustered, very irregularly spaced, or located on widely-spaced transects, *quadrant* or *octant* searches can be useful. Here, one divides the neighborhood around each target node into quadrants or octants and uses the nearest 2–3 points from each quadrant or octant in the interpolation. This ensures neighbors from several different directions, and not just the closest points, will be used for prediction. It is also possible to change the search strategy node by node, thus using an adaptive search technique.

Kriging with a local neighborhood as just described is not entirely a method of local estimation in the sense that the local models in §5.3.2 were estimating all parameters of a linear model locally. In the local kriging approach the parameters of the spatial dependency structure are not spatially varying; the semivariogram is not re-estimated within each local neighborhood. When working with covariates and trend surfaces, universal kriging is also often done globally, since the model assumes that the relationship between the data and the covariates holds across the entire domain. If the domain is very large (e.g., the entire United States) this assumption may not be reasonable. In such cases, the covariance function and any trend surface functions or covariates can be determined and estimated within local moving windows and then kriging (either ordinary or universal) can also be done locally within these moving windows (see e.g., Haas, 1990). This is one approach to modeling a non-stationary covariance function; see Chapter 8.

**Example 4.2 (C/N ratios. Continued)** For the C/N ratio data, first introduced in §4.4.1 on page 154, we modeled an exponential semivariogram parametrically in §4.5 and presented estimates for nugget and no-nugget models in Table 4.2 on page 176. Now we compute ordinary kriging predictions on a regular 10 feet × 10 feet grid of points for the REML and OLS parameter estimates. To solve the kriging problem for the 1560 prediction locations efficiently, we define a local quadrant search neighborhood by dividing a circle with radius equal to the range of the semivariogram into 4 sections and choose the 5 points nearest to the prediction location in each quadrant.

Figure 5.6 displays the predictions obtained using the OLS and REML estimates of the exponential semivariogram model with and without the nugget effect. Also shown (in the third panel from the top) are the predictions obtained using the REML estimates in the nugget model but using a search radius equal to that of the range estimated using OLS.

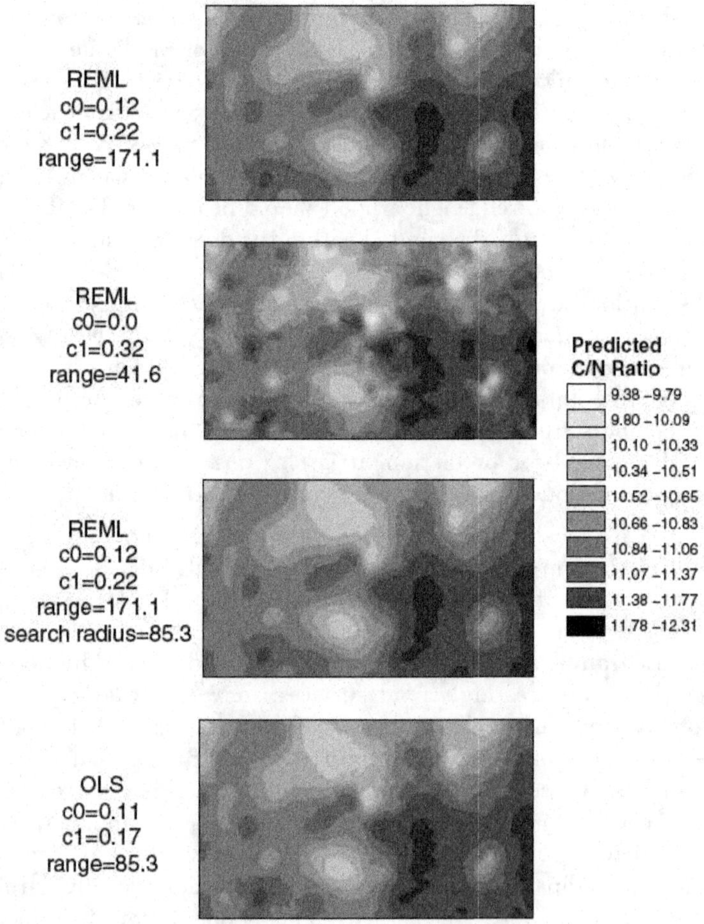


Figure 5.6 *Ordinary kriging predictions for C/N ratio data.*

The predictions from the nugget models are generally similar, showing an area of high ratios in the south-east portion of the field and various pockets of low C/N ratios. The predictions based on the OLS-fitted semivariogram model shown in the bottom panel are less smooth than those obtained using the corresponding REML-fitted semivariogram model. The estimates of the relative structured variability are about the same ($RSV_{ols} = 39\%$, $RSV_{reml} = 35\%$), but the smaller OLS estimate of the range (85.3 versus 171.1) creates a process with less continuity. You can glean this from the more "frizzled" contour edges in the lower panel. Overall, the predictions based on the two sets of estimates are close (Figure 5.7).

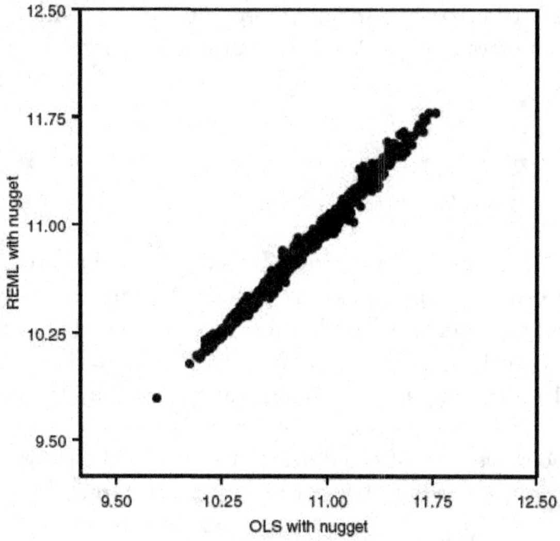


Figure 5.7 *Comparison of predictions in nugget models based on REML and OLS estimates.*

When using statistical software to perform these (any) calculations it is important to be familiar with the software features and defaults. For example, whether local or global kriging predictions are calculated, how kriging neighborhoods are determined, etc. When the kriging neighborhood depends on the semivariogram parameters, as is the case here, then changes in the parameter estimates can have spurious effects on the predicted surface and the prediction standard errors. The comparison between the top and bottom panels in Figure 5.6 is possibly not only affected by the large REML estimate of the range, but also by a local search radius that is twice as large. While this effect appears to be minor in this application (comparing the first and third panel in Figure 5.6), it can have a larger impact in others.

The least smooth predictions are those from the REML no-nugget model.

This may be surprising at first, since adding a nugget effect reduces the continuity of the process. However, the lack of a nugget effect is more than offset in this case by a small range. Recall that earlier we rejected the hypothesis of a zero nugget effect based on the REML analysis (see page 177). It would be incorrect to attribute the more erratic appearance of the predicted C/N ratios in the second panel to an analysis that reveals more "detail" about the C/N surface and so would be preferable on those grounds. Since statistical inference is conditional on the selected model, the predictions in the second panel must be dismissed if we accept the necessity of a nugget effect, regardless of how informative the resulting map appears to be.

Figure 5.8 displays contour maps of the standard errors corresponding to the kriging predictions in Figure 5.6. The standard errors are small near the location of the observed data (compare to Figure 4.6 on page 156). At the data locations the standard errors are exactly zero, since the predictions honor the data. The standard error maps basically trace the observed locations. □

### *5.4.3 Filtering and Smoothing*

The universal kriging predictor "honors" the data. The predicted surface passes through the data points, i.e., the predicted values at locations where data are measured are identical to the observed values. Thus, while kriging produces a smooth surface, it does not smooth the data like least squares or loess regression. In some applications such smoothing may be desirable, however. For example, when the data are measured with error, it would be better to predict a less noisy version of the data that removes the measurement error instead of requiring the prediction surface to pass through the noisy data.

**Example 5.7** Particle concentrations are measured daily at several monitoring sites throughout a state. Based on these measurements, a daily map of state-wide particle concentrations is produced. Which value should be displayed on the map for the monitoring sites? If the particle concentrations were measured without error, then the recorded values represent the concentrations actually present. If the measurements are contaminated with error, one would be interested not in the amounts that had been measured, but rather in the actual concentrations. □

Following the ideas in Cressie (1993), suppose we really want to make inferences about a spatial process, $S(\mathbf{s})$, but instead can only measure the process $Z(\mathbf{s})$, where

$$Z(\mathbf{s}) = S(\mathbf{s}) + \epsilon(\mathbf{s}), \quad \mathbf{s} \in D,$$

with $\mathrm{E}[\epsilon(\mathbf{s})] = 0$, $\mathrm{Var}[\epsilon(\mathbf{s})] = \sigma^2_\epsilon$, $\mathrm{Cov}[\epsilon(\mathbf{s}_i), \epsilon(\mathbf{s}_j)] = 0$, for all $i \neq j$ and $S(\mathbf{s})$ and $\epsilon(\mathbf{s})$ are independent. Further suppose that $S(\mathbf{s})$ can be described with

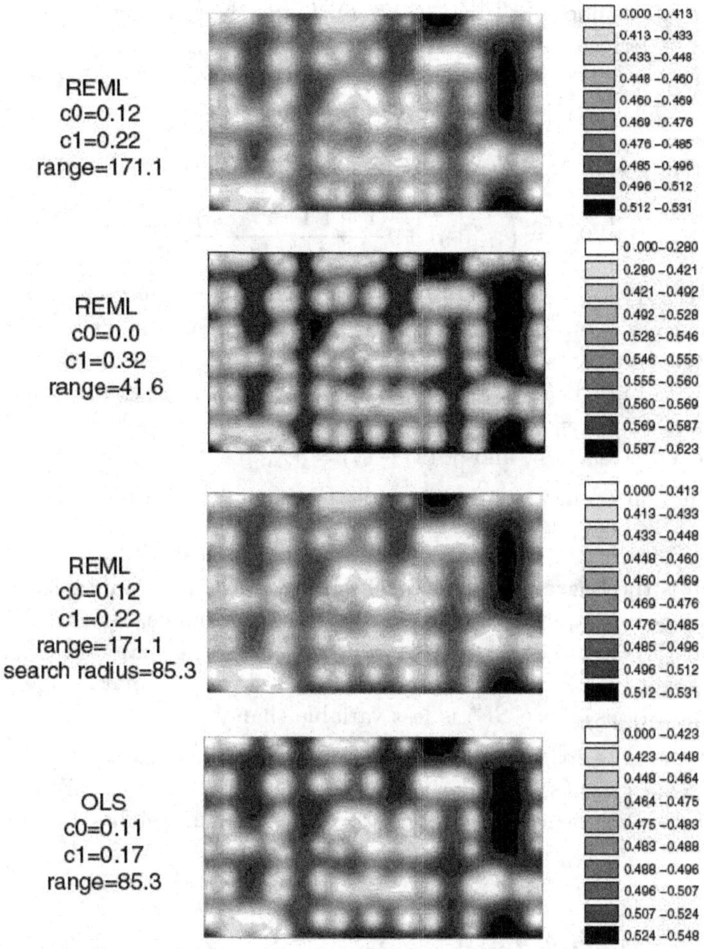


Figure 5.8 *Standard error maps corresponding to kriging predictions in Figure 5.6.*

the general linear model with autocorrelated errors

$$S(\mathbf{s}) = \mathbf{x}(\mathbf{s})'\boldsymbol{\beta} + \upsilon(\mathbf{s})$$

where $\mathrm{E}[\upsilon(\mathbf{s})] = 0$, and $\mathrm{Cov}[\upsilon(\mathbf{s}_i), \upsilon(\mathbf{s}_j)] = \mathrm{Cov}[S(\mathbf{s}_i), S(\mathbf{s}_j)] = C_S(\mathbf{s}_i, \mathbf{s}_j)$. Thus,

$$Z(\mathbf{s}) = \mathbf{x}(\mathbf{s})'\boldsymbol{\beta} + \upsilon(\mathbf{s}) + \epsilon(\mathbf{s}),$$

and $\mathrm{E}[Z(\mathbf{s})] = \mathbf{x}(\mathbf{s})'\boldsymbol{\beta}$. In terms of the operational decomposition given in equation (2.10) on page 54, this is a signal model and $S(\mathbf{s})$ corresponds to a random process comprised of $\mu(\mathbf{s}) + W(\mathbf{s}) + \eta(\mathbf{s})$. The best linear unbiased predictor of $S(\mathbf{s})$ remains of the form $\boldsymbol{\lambda}^{*\prime}\mathbf{Z}(\mathbf{s})$, where the weights, $\{\lambda_i^*\}$, are derived by minimizing $\mathrm{E}[(\boldsymbol{\lambda}^{*\prime}\mathbf{Z}(\mathbf{s}) - S(\mathbf{s}_0))^2]$ subject to the unbiasedness criterion $\mathrm{E}[\boldsymbol{\lambda}^{*\prime}\mathbf{Z}(\mathbf{s})] = \mathrm{E}[S(\mathbf{s}_0)]$.

If we first assume that $E[Z(\mathbf{s})] = \mathbf{x}(\mathbf{s})'\boldsymbol{\beta} \equiv \mu$, the ordinary kriging predictor of $S(\mathbf{s}_0)$ is derived in a manner analogous to that given in §5.2.2. Using the expressions in terms of the semivariogram, this predictor is

$$p_{ofk}(\mathbf{Z}; S(\mathbf{s}_0)) = \boldsymbol{\lambda}^{*\prime}\mathbf{Z}(\mathbf{s})$$

with optimal weights satisfying (Cressie, 1993, p. 128)

$$\boldsymbol{\lambda}^{*\prime} = \left(\boldsymbol{\gamma}^*(\mathbf{s}_0) + \mathbf{1}\frac{1 - \mathbf{1}'\boldsymbol{\Gamma}^{-1}\boldsymbol{\gamma}^*(\mathbf{s}_0)}{\mathbf{1}'\boldsymbol{\Gamma}^{-1}\mathbf{1}}\right)'\boldsymbol{\Gamma}^{-1}.$$

The matrix $\boldsymbol{\Gamma}$ is the same as in the ordinary kriging equations (5.19)–(5.22), with elements $\gamma_z(\mathbf{s}_i - \mathbf{s}_j)$. The vector $\boldsymbol{\gamma}^*(\mathbf{s}_0) = [\gamma^*(\mathbf{s}_0 - \mathbf{s}_1), \cdots, \gamma^*(\mathbf{s}_0 - \mathbf{s}_n)]'$ is slightly different since, in the minimization, the elements of this matrix are derived from $E[(Z(\mathbf{s}_0) - Z(\mathbf{s}_i))^2] = \gamma_s(\mathbf{s}_0 - \mathbf{s}_i) + \sigma^2_\epsilon \equiv \gamma^*(\mathbf{s}_0 - \mathbf{s}_i)$. At prediction locations, $\mathbf{s}_0 \neq \mathbf{s}_i$, and $\gamma^*(\mathbf{s}_0 - \mathbf{s}_i) = \gamma_z(\mathbf{s}_0 - \mathbf{s}_i)$, $i = 1, \cdots, n$. At data locations, $\mathbf{s}_0 = \mathbf{s}_i$, and $\gamma^*(\mathbf{s}_0 - \mathbf{s}_i) = \sigma^2_\epsilon$.

*Filtered Kriging Variance*

The minimized mean-squared prediction error is given by

$$\sigma^2_{ofk}(\mathbf{s}_0) = \boldsymbol{\lambda}^{*\prime}\boldsymbol{\gamma}^*(\mathbf{s}_0) + m - \sigma^2_\epsilon,$$

where $m$ is the Lagrange multiplier from the constrained minimization. Note that this is not equal to the ordinary kriging variance, $\sigma^2_{ok}(\mathbf{s}_0)$, defined in equation (5.22), unless $\sigma^2_\epsilon = 0$. Prediction standard errors associated with filtered kriging are smaller than those associated with ordinary kriging (except at data locations) since $S(\cdot)$ is less variable than $Z(\cdot)$.

If the mean is spatially varying, i.e., if we consider the more general case where $E[Z(\mathbf{s})] = \mathbf{x}(\mathbf{s})'\boldsymbol{\beta}$, a derivation analogous to that for universal kriging (in terms of the covariance matrix) gives the optimal weights as solutions to

$$\begin{aligned} \boldsymbol{\Sigma}\mathbf{a} + \mathbf{X}(\mathbf{s})\mathbf{m} &= \boldsymbol{\sigma}^* \\ \mathbf{X}(\mathbf{s})'\mathbf{a} &= \mathbf{x}(\mathbf{s}_0), \end{aligned}$$

where $\boldsymbol{\sigma}^* = \text{Cov}[\mathbf{Z}(\mathbf{s}), S(\mathbf{s}_0)]$ and thus has elements $C_S(\mathbf{s}_i, \mathbf{s}_j)$. However, we cannot estimate $C_S(\mathbf{s}_i, \mathbf{s}_j)$ directly; we obtain this quantity only through $C_Z(\mathbf{s}_i, \mathbf{s}_j)$ which is equal to

$$C_Z(\mathbf{s}_i, \mathbf{s}_j) = \text{Cov}[Z(\mathbf{s}_i), Z(\mathbf{s}_j)] = \begin{cases} C_S(\mathbf{s}_i, \mathbf{s}_j) + \sigma^2_\epsilon & \mathbf{s}_i = \mathbf{s}_j \\ C_S(\mathbf{s}_i, \mathbf{s}_j) & \mathbf{s}_i \neq \mathbf{s}_j. \end{cases}$$

Thus, at the prediction locations where $\mathbf{s}_i \neq \mathbf{s}_j$, $C_Z(\mathbf{s}_i, \mathbf{s}_j) = C_S(\mathbf{s}_i, \mathbf{s}_j)$, and $\boldsymbol{\sigma}^* = \boldsymbol{\sigma}$. However, at the data locations, $C_Z(\mathbf{s}_i, \mathbf{s}_j) = C_S(\mathbf{s}_i, \mathbf{s}_j) + \sigma^2_\epsilon$, so we use $\boldsymbol{\sigma}^* = C_Z(\mathbf{s}_i, \mathbf{s}_j) - \sigma^2_\epsilon$.

The modification to the kriging predictor comes into play when we want to predict at locations where we already have data. This is called **filtering**, since we are removing the error from our observed data through a prediction process. It is also **smoothing**, since the filtered kriging predictor smooths the data, with larger values of $\sigma^2_\epsilon$ resulting in more smoothing.

Historically, the different terms (filtering and smoothing) arise from the time series literature which identifies three distinct types of prediction problems.

The first is **forecasting**: prediction of future data. The second is **filtering**: prediction of data at the present time. The third is **smoothing**: prediction of past data. In spatial statistics, we cannot distinguish past, present, and future and so we only have two types of prediction problems: prediction of data at new locations where no measurements were made, and prediction of data at observed locations where measurements were recorded. Thus, in spatial statistics, filtering and smoothing refer to the same prediction problem.

**Example 5.8** Schabenberger and Pierce (2002, Ch. A9.9.5) give a small example to demonstrate the effects of measurement error on kriging predictions and their standard errors, which we adapt here. Figure 5.9 shows four observed locations on a grid, $\mathbf{s}_1 = [0,0]'$, $\mathbf{s}_2 = [4,0]'$, $\mathbf{s}_3 = [0,8]'$, and $\mathbf{s}_4 = [4,8]'$. The observed values are $Z(\mathbf{s}_1) = 5$, $Z(\mathbf{s}_2) = 10$, $Z(\mathbf{s}_3) = 15$, and $Z(\mathbf{s}_4) = 6$, with arithmetic average $\overline{Z} = 9$. The locations $\mathbf{s}_0{}^{(1)}$–$\mathbf{s}_0{}^{(4)}$ are prediction locations. Notice that one of the prediction locations is also an observed location, $\mathbf{s}_0{}^{(3)} = \mathbf{s}_2$.

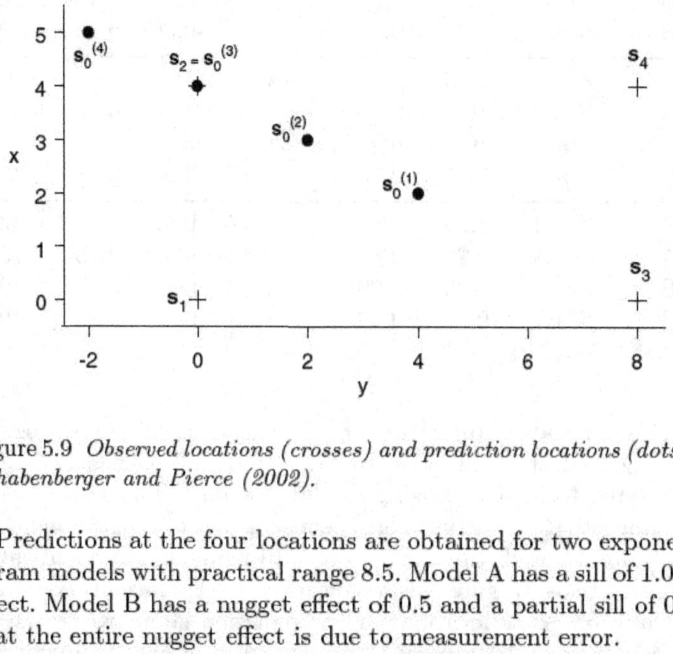


Figure 5.9 *Observed locations (crosses) and prediction locations (dots). Adapted from Schabenberger and Pierce (2002).*

Predictions at the four locations are obtained for two exponential semivariogram models with practical range 8.5. Model A has a sill of 1.0 and no nugget effect. Model B has a nugget effect of 0.5 and a partial sill of 0.5. We assume that the entire nugget effect is due to measurement error.

In the absence of a nugget effect the predictions of $Z(\mathbf{s}_0)$ and $S(\mathbf{s}_0)$ agree in value and precision (Table 5.3). In the presence of a nugget effect (model B), predictions of $Z(\mathbf{s}_0)$ and $S(\mathbf{s}_0)$ agree in value but predictions of the signal are more precise than those of the error-contaminated process. The difference between the two kriging variances is the magnitude of the nugget effect.

At the observed location $\mathbf{s}_2 = \mathbf{s}_0{}^{(3)}$, the kriging predictor under model A honors the data. Notice that the kriging weights are zero except for $\lambda_2 = 1$. Since the predicted value is identical to the observed value, the kriging variance is zero. In the presence of a nugget effect (model B) the predictor of the signal does not reproduce the observed value and the kriging variance is not zero. □

Table 5.3 *Kriging predictions of $Z(\mathbf{s}_0)$ and $S(\mathbf{s}_0)$ under semivariogram models A and B. Adapted from Schabenberger and Pierce (2002, Ch. 9.9.5).*

| | | | Prediction | | | | | |
|---|---|---|---|---|---|---|---|---|
| $\mathbf{s}_0$ | Model | Target | Value | Variance | $\lambda_1$ | $\lambda_2$ | $\lambda_3$ | $\lambda_4$ |
| [2, 4] | A | $Z(\mathbf{s}_0)$ | 9.0 | 0.92 | 0.25 | 0.25 | 0.25 | 0.25 |
| [2, 4] | A | $S(\mathbf{s}_0)$ | 9.0 | 0.92 | 0.25 | 0.25 | 0.25 | 0.25 |
| [2, 4] | B | $Z(\mathbf{s}_0)$ | 9.0 | 1.09 | 0.25 | 0.25 | 0.25 | 0.25 |
| [2, 4] | B | $S(\mathbf{s}_0)$ | 9.0 | 0.59 | 0.25 | 0.25 | 0.25 | 0.25 |
| [3, 2] | A | $Z(\mathbf{s}_0)$ | 8.77 | 0.78 | 0.25 | 0.48 | 0.12 | 0.15 |
| [3, 2] | A | $S(\mathbf{s}_0)$ | 8.77 | 0.78 | 0.25 | 0.48 | 0.12 | 0.15 |
| [3, 2] | B | $Z(\mathbf{s}_0)$ | 8.83 | 1.04 | 0.26 | 0.36 | 0.18 | 0.20 |
| [3, 2] | B | $S(\mathbf{s}_0)$ | 8.83 | 0.54 | 0.26 | 0.36 | 0.18 | 0.20 |
| [4, 0] | A | $Z(\mathbf{s}_0)$ | 10.0 | 0. | 0 | 1 | 0 | 0 |
| [4, 0] | A | $S(\mathbf{s}_0)$ | 10.0 | 0. | 0 | 1 | 0 | 0 |
| [4, 0] | B | $Z(\mathbf{s}_0)$ | 10.0 | 0. | 0 | 1 | 0 | 0 |
| [4, 0] | B | $S(\mathbf{s}_0)$ | 9.25 | 0.30 | 0.17 | 0.60 | 0.11 | 0.12 |
| [5, −2] | A | $Z(\mathbf{s}_0)$ | 9.26 | 0.88 | 0.17 | 0.57 | 0.13 | 0.13 |
| [5, −2] | A | $S(\mathbf{s}_0)$ | 9.26 | 0.88 | 0.17 | 0.57 | 0.13 | 0.13 |
| [5, −2] | B | $Z(\mathbf{s}_0)$ | 9.03 | 1.09 | 0.23 | 0.40 | 0.18 | 0.19 |
| [5, −2] | B | $S(\mathbf{s}_0)$ | 9.03 | 0.59 | 0.23 | 0.40 | 0.18 | 0.19 |

**Example 4.2 (C/N ratios. Continued)** For the exponential semivariogram model with a nugget effect fitted using REML (Table 4.2 on page 176), we computed the filtered (ordinary) kriging predictions on the same grid as in Figure 5.6, assuming that the entire nugget effect is due to measurement error. The map of filtered predictions in Figure 5.10 is practically identical to the unfiltered predictions (Figure 5.6 on page 246) because only very few grid points are also observed locations. The prediction standard errors are not bounded below by zero, because at the observed locations the prediction variance is nonzero. At unobserved locations, the prediction standard errors are smaller compared to those obtained without filtering of the $Z(\mathbf{s})$ process. A scatterplot of the standard errors of the filtered and ordinary kriging predictions emphasizes these findings (Figure 5.11). Ordinary kriging standard errors are zero at the data locations. At unobserved locations they are larger than the standard errors form filtered kriging. □

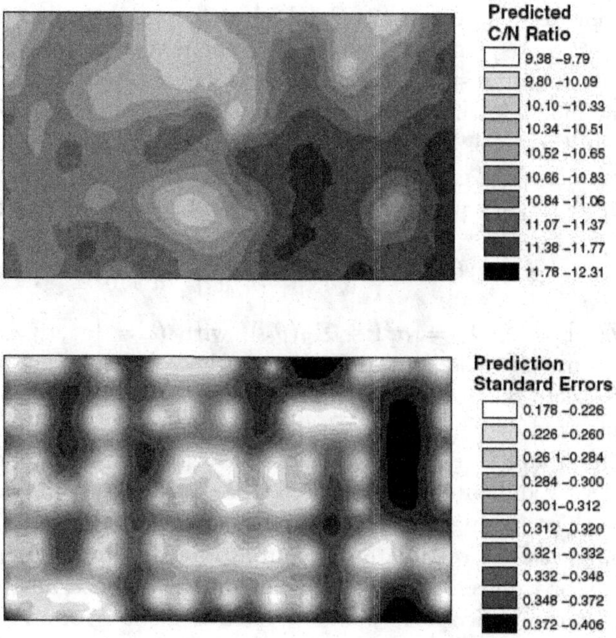


Figure 5.10 *Prediction and standard error maps for filtered kriging. Compare to top panels in Figures 5.6 and 5.8.*

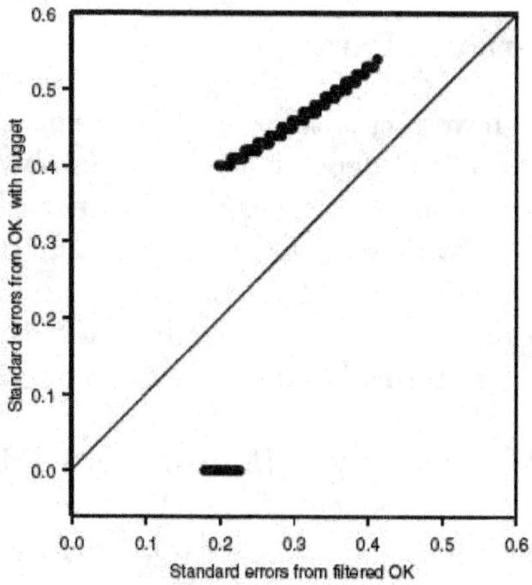


Figure 5.11 *Prediction standard errors for filtered and ordinary kriging.*

We may not always be concerned solely with spatial prediction. We may also want to estimate $\boldsymbol{\beta}$ and make inferences about the effect of the covariates on our outcome of interest. To use these methods, we often specify the form of $\text{Var}[\mathbf{Z}(\mathbf{s})]$ and model $\text{Var}[\mathbf{Z}(\mathbf{s})]$ parametrically in order to reduce the number of parameters. In this case, $\text{Var}[\mathbf{Z}(\mathbf{s})] = \sigma_\epsilon^2\mathbf{I} + \boldsymbol{\Sigma}_S$, thereby focusing modeling on the elements of $\boldsymbol{\Sigma}_S$. For example, suppose we assume that $S(\mathbf{s})$ is an isotropic second-order stationary process with exponential covariance function

$$C_S(h;\boldsymbol{\theta}_S) = \begin{cases} c_0 + c_e & h = 0, \\ c_e \exp\{-h/a_e\} & h > 0. \end{cases}$$

Then $\text{Var}[\mathbf{Z}(\mathbf{s})] = \boldsymbol{\Sigma}(\boldsymbol{\theta}) = \sigma_\epsilon^2\mathbf{I} + \boldsymbol{\Sigma}_S(\boldsymbol{\theta}_S)$, with $\boldsymbol{\theta} = [\sigma_\epsilon^2, c_0, c_e, a_e]'$. As a result, we assume $Z(\mathbf{s})$ is a spatial process whose behavior at the origin is due to both micro-scale variation characterized by $c_0$, and to measurement error characterized by $\sigma_\epsilon^2$, i.e., $\text{Var}[Z(\mathbf{s})] = \sigma_\epsilon^2 + C_S(0,\boldsymbol{\theta}_S) = \sigma_\epsilon^2 + c_0 + c_e$. We then estimate $\boldsymbol{\beta}$ and $\boldsymbol{\theta}$ using the methods described in §5.5.1–§5.5.2. It will probably be very difficult to separate all three variance components $\sigma_\epsilon^2, c_0$, and $c_e$, using just a single realization of $Z(\mathbf{s})$. Thus, we either need more information, such as an estimate of $\sigma_\epsilon^2$ from repeated measurements at the same location, or we need to make some assumptions concerning the amount of the overall nugget effect that is due to measurement error and the amount that is due to micro-scale variation. For example, if we assume the overall nugget effect, $c_0^Z$, is comprised of 75% measurement error and 25% micro-scale variation, then $\sigma_\epsilon^2 = 0.75c_0^Z$, and $c_0 = 0.25c_0^Z$. In this case, only one parameter needs to be estimated, $c_0^Z$.

## 5.5 Estimating Covariance Parameters

So far in this chapter we have tacitly assumed that quantities such as $\boldsymbol{\Sigma}$, $\boldsymbol{\Gamma}$, $\boldsymbol{\sigma}$, and $\boldsymbol{\gamma}$, that describe the spatial dependency between observed, and new and observed data, are known. But of course, they are not. The usual approach is to parameterize the semivariogram or covariance function and to estimate the parameters of this model. Then, the relevant expressions for the various predictors and their precision are evaluated by substituting the estimated parameters, a process aptly termed "plug-in" estimation.

**Example 5.9** Suppose you want to perform ordinary kriging in terms of covariances and you choose the exponential model $\gamma(h,\boldsymbol{\theta}) = \sigma^2(1-\exp\{-h/\alpha\})$ as the semivariogram model. Once you have obtained estimates $\widehat{\boldsymbol{\theta}} = [\widehat{\sigma}^2, \widehat{\alpha}]'$, you estimate the semivariogram as

$$\widehat{\gamma}(h) = \gamma(h,\widehat{\boldsymbol{\theta}}) = \widehat{\sigma}^2(1-\exp\{-3h/\widehat{\alpha}\})$$

by plugging-in the estimates into the expression for the model. In order to estimate covariances under this model, you can invoke the relationship $C(h) = C(0) - \gamma(h)$ and estimate

$$\widehat{C}(h) = \widehat{C}(0) - \widehat{\gamma}(h)$$

$$\begin{aligned} &= \gamma(\infty, \widehat{\boldsymbol{\theta}}) - \gamma(h, \widehat{\boldsymbol{\theta}}) \\ &= \widehat{\sigma}^2 - \widehat{\sigma}^2(1 - \exp\{-3h/\widehat{\alpha}\}) \\ &= \widehat{\sigma}^2 \exp\{-3h/\widehat{\alpha}\} = C(h, \widehat{\boldsymbol{\theta}}). \end{aligned}$$

□

Several important issues are connected with this practice:

- how to select a model for the covariance function or semivariogram;
- how to estimate the parameters in $\boldsymbol{\theta}$;
- how to adjust inference in light of the fact that $\boldsymbol{\theta}$ is a constant, yet $\widehat{\boldsymbol{\theta}}$ is a random variable.

Our efforts in this section focus on the last two points. We assume that the model has been chosen correctly. Also, we will generically refer to parameters of the spatial dependence as covariance parameters, regardless of whether modeling is based on the semivariogram or the covariance function.

Fortunately, when the mean of the random field is constant, the techniques in Chapter 4, in particular the parametric techniques in §4.5, can be used to estimate the covariance parameters. If the mean is spatially varying, things are considerably more complicated. One complication is that the GLS estimate of $\boldsymbol{\beta}$ also depends on the covariance parameters; $\boldsymbol{\beta} \equiv \boldsymbol{\beta}(\boldsymbol{\theta})$. It seems natural then to use plug-in estimation for $\boldsymbol{\beta}$; $\widehat{\boldsymbol{\beta}} = \boldsymbol{\beta}(\widehat{\boldsymbol{\theta}})$. If you consider estimation of $\boldsymbol{\theta}$ by least squares fitting of a semivariogram model based on the empirical semivariogram, for example, this requires an assumption of intrinsic stationarity (for semivariograms) and second-order stationarity (for covariance functions); see §4.5.1. These assumptions are of course not met in the general linear model where $\mu(\mathbf{s}) = \mathbf{x}(\mathbf{s})'\boldsymbol{\beta} \neq \mu$. To see this, consider estimating $\gamma(\cdot)$ using the classical semivariogram estimator given in equation (4.24). Under the model defined by equation (5.24),

$$\begin{aligned} \mathrm{E}[(Z(\mathbf{s}_i) - Z(\mathbf{s}_j))^2] &= \mathrm{Var}[Z(\mathbf{s}_i) - Z(\mathbf{s}_j)] + \{\mu(\mathbf{s}_i) - \mu(\mathbf{s}_j)\}^2 \\ &= 2\gamma(\mathbf{s}_i - \mathbf{s}_j) + \left\{\sum_{k=1}^{p} \beta_k(x_k(\mathbf{s}_i) - x_k(\mathbf{s}_j))\right\}^2 \quad (5.35) \end{aligned}$$

with $x_1(\mathbf{s}_i) \equiv 1$ for all $i$ (Cressie, 1993, p. 165). The empirical semivariogram no longer estimates the true, theoretical semivariogram. If the covariates are trend surface functions, the trend often manifests itself in practice by an empirical semivariogram that increases rapidly with $||\mathbf{h}||$ (often quadratically). Analogous problems occur when estimating the covariance function. In light of this problem, we need to re-examine the techniques from §4.5 for the case of a spatially varying mean. This is the topic of §5.5.1–§5.5.3.

The third issue mentioned above, regarding the variability of $\widehat{\boldsymbol{\theta}}$ is of importance whether the mean is constant or spatially varying. The implications here are whether a predictor that is best in some sense retains this property when it is evaluated at estimates of the covariance parameters and how to adjust the estimate of the prediction error. Take, for example, the ordinary kriging predictor $p_{ok}(\mathbf{Z}; \mathbf{s}_0)$ and its kriging variance, written in terms of $\boldsymbol{\theta}$ to emphasize the dependence on a particular model

$$\begin{aligned} p_{ok}(\mathbf{Z}; \mathbf{s}_0) &= \left(\boldsymbol{\sigma}(\boldsymbol{\theta}) + \mathbf{1}\frac{1 - \mathbf{1}'\boldsymbol{\Sigma}(\boldsymbol{\theta})^{-1}\boldsymbol{\sigma}(\boldsymbol{\theta})}{\mathbf{1}'\boldsymbol{\Sigma}(\boldsymbol{\theta})^{-1}\mathbf{1}}\right)\boldsymbol{\Sigma}(\boldsymbol{\theta})^{-1}\mathbf{Z}(\mathbf{s}), \\ \sigma^2_{ok}(\mathbf{s}_0) &= C(\mathbf{0}) - \boldsymbol{\sigma}(\boldsymbol{\theta})'\boldsymbol{\Sigma}(\boldsymbol{\theta})^{-1}\boldsymbol{\sigma}(\boldsymbol{\theta}) + \frac{(1 - \mathbf{1}'\boldsymbol{\Sigma}(\boldsymbol{\theta})^{-1}\boldsymbol{\sigma}(\boldsymbol{\theta}))^2}{\mathbf{1}'\boldsymbol{\Sigma}(\boldsymbol{\theta})^{-1}\mathbf{1}}. \end{aligned}$$

*Ordinary Kriging Plug-in Predictor*

The plug-in predictor

$$\widehat{p}_{ok}(\mathbf{Z}; \mathbf{s}_0) = \left(\boldsymbol{\sigma}(\widehat{\boldsymbol{\theta}}) + \mathbf{1}\frac{1 - \mathbf{1}'\boldsymbol{\Sigma}(\widehat{\boldsymbol{\theta}})^{-1}\boldsymbol{\sigma}(\widehat{\boldsymbol{\theta}})}{\mathbf{1}'\boldsymbol{\Sigma}(\widehat{\boldsymbol{\theta}})^{-1}\mathbf{1}}\right)\boldsymbol{\Sigma}(\widehat{\boldsymbol{\theta}})^{-1}\mathbf{Z}(\mathbf{s}) \qquad (5.36)$$

is no longer the best linear unbiased predictor of $Z(\mathbf{s}_0)$. It is an *estimate* of the BLUP, a so-called EBLUP. Also, this EBLUP will not be invariant to your choice of $\widehat{\boldsymbol{\theta}}$. Different estimation methods yield different estimates of the covariance parameters, which affects the predictions—unless you are predicting at observed locations without measurement error; all predictors honor the data, regardless of "how" you obtained $\widehat{\boldsymbol{\theta}}$. Not only is (5.36) no longer best, we do not know its prediction error. The common practice of evaluating $\sigma^2_{ok}(\mathbf{s}_0)$ at $\widehat{\boldsymbol{\theta}}$ does not yield an estimate of the prediction error of $\widehat{p}_{ok}(\mathbf{Z}; \mathbf{s}_0)$. It yields an estimate of the prediction error of $p_{ok}(\mathbf{Z}; \mathbf{s}_0)$. In other words, by substituting estimated covariance parameters into the expression for the predictor, we obtain an estimate of the predictor. By substituting into the expression for the prediction variance we get an estimate of the prediction error of a different predictor, not for the one we are using. How to determine, or at least approximate, the prediction error of plug-in predictors, is the topic of §5.5.4.

### *5.5.1 Least Squares Estimation*

Consider again the spatial linear model

$$\mathbf{Z}(\mathbf{s}) = \mathbf{X}(\mathbf{s})\boldsymbol{\beta} + \mathbf{e}(\mathbf{s}), \qquad \mathbf{e}(\mathbf{s}) \sim (\mathbf{0}, \boldsymbol{\Sigma}(\boldsymbol{\theta})).$$

The universal kriging predictor, written in the GLS form, is

$$p_{uk}(\mathbf{Z}; \mathbf{s}_0) = \mathbf{x}(\mathbf{s}_0)'\widehat{\boldsymbol{\beta}}_{gls} + \boldsymbol{\sigma}(\boldsymbol{\theta})'\boldsymbol{\Sigma}(\boldsymbol{\theta})^{-1}(\mathbf{Z}(\mathbf{s}) - \mathbf{X}(\mathbf{s})\widehat{\boldsymbol{\beta}}_{gls}),$$

*GLS Estimator*

where

$$\widehat{\boldsymbol{\beta}}_{gls} = \left(\mathbf{X}(\mathbf{s})'\boldsymbol{\Sigma}(\boldsymbol{\theta})^{-1}\mathbf{X}(\mathbf{s})\right)^{-1}\mathbf{X}(\mathbf{s})'\boldsymbol{\Sigma}(\boldsymbol{\theta})^{-1}\mathbf{Z}(\mathbf{s}) \qquad (5.37)$$

is the generalized least squares estimator of the fixed effects. How are we going to respond to the fact that $\boldsymbol{\theta}$ is unknown? In order to estimate $\boldsymbol{\theta}$ by least squares fitting of a semivariogram model, we cannot use the empirical semivariogram based on the observed data $\mathbf{Z}(\mathbf{s})$, because the mean of $\mathbf{Z}(\mathbf{s})$ is not constant. From equation (5.35) we see that the resulting semivariogram would be biased, and there is little hope to find reasonable estimates for $\boldsymbol{\theta}$ this way. What we need is the semivariogram of $\mathbf{e}(\mathbf{s})$, not that of $\mathbf{Z}(\mathbf{s})$. If $\boldsymbol{\beta}$ were known, then we could construct "data" $\mathbf{Z}(\mathbf{s}) - \mathbf{X}(\mathbf{s})\boldsymbol{\beta}$ and unbiasedly estimate the semivariogram, because the error process would be observable. But $\boldsymbol{\beta}$ is not known, otherwise we would not find ourselves in the situation to contemplate a linear model for a spatially varying mean. To compute the GLS estimate of $\boldsymbol{\beta}$, (5.37), requires knowledge of the covariance parameters. If instead, we use a plug-in estimator for $\boldsymbol{\beta}$, the estimated generalized least squares estimate (EGLS)

*EGLS Estimator*

$$\widehat{\boldsymbol{\beta}}_{egls} = \left(\mathbf{X}(\mathbf{s})'\boldsymbol{\Sigma}(\widehat{\boldsymbol{\theta}})^{-1}\mathbf{X}(\mathbf{s})\right)^{-1}\mathbf{X}(\mathbf{s})'\boldsymbol{\Sigma}(\widehat{\boldsymbol{\theta}})^{-1}\mathbf{Z}(\mathbf{s}), \tag{5.38}$$

we are still left with the problem of having to find a reasonable estimator for $\boldsymbol{\theta}$. And we just established that this is not likely by way of least squares techniques without knowing the mean. Schabenberger and Pierce (2002, pp. 613–615) refer to this circular argument as the "cat and mouse game of universal kriging." In order to estimate the mean you need to have an estimate of the covariance parameters, which you can not get by least squares without knowing the mean. Percival and Walden (1993, p. 219) refer to a similar problem in spectral analysis of time series—where in order to derive a filter for pre-whitening of a spectral density estimate one needs to know the spectral density of the series—as a "cart and horse" problem.

The approach that is taken to facilitate least squares estimation of covariance parameters in the case of a spatially varying mean is to compute initially an estimate of the mean that does not depend on $\boldsymbol{\theta}$. Then use this estimate to detrend the data and estimate the semivariogram by least squares based on the empirical semivariogram of the residuals. Since the estimate $\widehat{\boldsymbol{\theta}}$ depends on how we estimated $\boldsymbol{\beta}$ initially—we know it was not by GLS or EGLS—the process is often repeated (iterated). The resulting "simultaneous" estimation scheme for $\boldsymbol{\theta}$ and $\boldsymbol{\beta}$ is based on the methods described in §4.5.1 and is termed iteratively re-weighted generalized least squares (IRWGLS) :

*IRWGLS Algorithm*

1. Obtain a starting estimate of $\boldsymbol{\beta}$, say $\widehat{\boldsymbol{\beta}}$;
2. Compute residuals $\mathbf{r} = \mathbf{Z}(\mathbf{s}) - \mathbf{X}(\mathbf{s})\widehat{\boldsymbol{\beta}}$;
3. Estimate and model the semivariogram of the residuals using techniques described in §4.4 and §4.5, obtaining $\widehat{\boldsymbol{\theta}}$ by minimizing either equation (4.31) or equation (4.34);
4. Obtain a new estimate of $\boldsymbol{\beta}$ using equation (5.38);
5. Repeat steps 2–4 until the relative or absolute change in estimates of $\boldsymbol{\beta}$ and $\boldsymbol{\theta}$ are small.

The starting value in step 1 is almost always obtained by ordinary least squares. In the first go-around of the IRWGLS algorithm you are working with OLS residuals, once past step 4, you are working with (E)GLS residuals. The semivariogram estimator based on the residuals $\mathbf{r} = \mathbf{Z}(\mathbf{s}) - \mathbf{X}\widehat{\boldsymbol{\beta}}$ (step 3 above) is biased. It is important to understand that this bias is different from the bias mentioned earlier that is due to not accounting for the change in the mean, namely equation (5.35). The bias in the semivariogram based on OLS residuals stems from the fact that the residuals fail to share important properties of the model errors $\mathbf{e}(\mathbf{s})$. For example, their variance is not $\boldsymbol{\Sigma}$, they are rank-deficient and heteroscedastic. Frankly, the only thing $\mathbf{e}(\mathbf{s})$ and $\mathbf{Z}(\mathbf{s})-\mathbf{X}(\mathbf{s})\widehat{\boldsymbol{\beta}}_{ols}$ have in common is a zero mean. The semivariogram of $\mathbf{Z}(\mathbf{s})-\mathbf{X}(\mathbf{s})\widehat{\boldsymbol{\beta}}_{ols}$ is not the semivariogram of $\mathbf{e}(\mathbf{s})$. A detailed discussion of the properties of OLS and GLS residuals, and the diagnosis of the covariance model choice based on residuals is deferred until Chapter 6. At this point it suffices to note that the bias in the estimated semivariogram based on OLS residuals increases with the lag. Cressie (1993, p. 166) thus argues that the bias can be controlled by fitting the semivariogram model by weighted nonlinear least squares as described in §4.5.1. Because the weights are proportional to the approximate variance of the empirical semivariogram, empirical semivariogram values at large lags—where the bias is greater—are down-weighted.

By iterating the above steps, that is, by repeating steps 2–4 in the IRWGLS algorithm, the bias problem is not solved. The empirical semivariogram of the *generalized* least squares residuals is also a biased estimator of the semivariogram. Since both $\widehat{\boldsymbol{\beta}}_{ols}$ and $\widehat{\boldsymbol{\beta}}_{egls}$ are unbiased estimators of $\boldsymbol{\beta}$, most of the trend is removed in the very first step of the algorithm, and you may pick up comparably little additional structure in subsequent iterations. What matters for efficient estimation of the large-scale trend, and for spatial prediction, is that the estimator of $\boldsymbol{\theta}$ is efficient and has as little bias as possible. If you estimate the covariogram instead of the semivariogram in step 3, then a single OLS fit and OLS residuals may be preferable over an iterated algorithm.

The final IRWGLS estimator of $\boldsymbol{\beta}$ is an EGLS estimator. If we denote the estimator of $\boldsymbol{\theta}$ obtained from IRWGLS by $\widehat{\boldsymbol{\theta}}_{IRWGLS}$, the IRWGLS estimator of $\boldsymbol{\beta}$ is

*IRWGLS Estimator*

$$\widehat{\boldsymbol{\beta}}_{IRWGLS} = (\mathbf{X}(\mathbf{s})'\boldsymbol{\Sigma}(\widehat{\boldsymbol{\theta}}_{IRWGLS})^{-1}\mathbf{X}(\mathbf{s}))^{-1}\mathbf{X}(\mathbf{s})'\boldsymbol{\Sigma}(\widehat{\boldsymbol{\theta}}_{IRWGLS})^{-1}\mathbf{Z}(\mathbf{s}), \tag{5.39}$$

and its variance-covariance matrix is usually estimated as

$$\widehat{\text{Var}}[\widehat{\boldsymbol{\beta}}_{IRWGLS}] = (\mathbf{X}(\mathbf{s})'\boldsymbol{\Sigma}(\widehat{\boldsymbol{\theta}}_{IRWGLS})^{-1}\mathbf{X}(\mathbf{s}))^{-1}. \tag{5.40}$$

Since steps 2–4 are repeated, the process is iterative, but in contrast to a numerical optimization technique, such as the Newton-Raphson algorithm, it is difficult to study the overall behavior of the IRWGLS procedure. There is, for example, no guarantee that the process "converges" in the usual sense or that some "extremum" has been found when the process stops. When the algorithm comes to a halt, think of it as lack of progress, rather than convergence. Any continuance would lead to the same estimates of $\boldsymbol{\theta}$ and these would lead to the same estimates of $\boldsymbol{\beta}$. We recommend not to monitor the absolute change in parameter estimates alone, but to use a relative criterion in step 5. For example, if $\widehat{\boldsymbol{\beta}}^{(u)}$ and $\widehat{\boldsymbol{\beta}}^{(u+1)}$ are the estimates from two successive iterations, compute

$$\delta_j^{(\beta)} = \frac{|\widehat{\beta}_j^{(u)} - \widehat{\beta}_j^{(u+1)}|}{0.5\left(|\widehat{\beta}_j^{(u)}| + |\widehat{\beta}_j^{(u+1)}|\right)} \quad j = 1, \cdots, p,$$

$$\delta_k^{(\theta)} = \frac{|\widehat{\theta}_k^{(u)} - \widehat{\theta}_k^{(u+1)}|}{0.5\left(|\widehat{\theta}_k^{(u)}| + |\widehat{\theta}_k^{(u+1)}|\right)} \quad k = 1, \cdots, q,$$

and stop when $\max\{\delta_k^{(\theta)}, \delta_j^{(\beta)}\} < 10^{-6}$.

### 5.5.2 *Maximum Likelihood*

In the context of estimating parameters of the covariance function, we had briefly discussed maximum likelihood (ML) estimation in §4.5.2. In order to proceed with ML estimation, we must make a distributional assumption for $\mathbf{Z}(\mathbf{s})$. It is not sufficient to specify the first two moments only, as was previously the case. ML estimation for spatial models is developed only for the Gaussian case (Mardia and Marshall, 1984), we assume that $\mathbf{Z}(\mathbf{s}) \sim G(\mathbf{X}(\mathbf{s})\boldsymbol{\beta}, \boldsymbol{\Sigma}(\boldsymbol{\theta}))$.

In contrast to the IRWGLS approach, ML estimation is truly simultaneous estimation of mean and covariance parameters. This fact may be obstructed by profiling of $\boldsymbol{\beta}$ (see below), but the important point is that the ML estimates are the simultaneous solution to the problem of minimizing the negative of twice the Gaussian log likelihood

*Minus 2 log Likelihood*

$$\begin{aligned}\varphi(\boldsymbol{\beta};\boldsymbol{\theta};\mathbf{Z}(\mathbf{s})) &= \ln\{|\boldsymbol{\Sigma}(\boldsymbol{\theta})|\} + n\ln\{2\pi\} \\ &\quad + (\mathbf{Z}(\mathbf{s}) - \mathbf{X}(\mathbf{s})\boldsymbol{\beta})'\boldsymbol{\Sigma}(\boldsymbol{\theta})^{-1}(\mathbf{Z}(\mathbf{s}) - \mathbf{X}(\mathbf{s})\boldsymbol{\beta})). \end{aligned} \quad (5.41)$$

If $\mathbf{X}$ is an $(n \times p)$ matrix of rank $k$, this optimization problem involves $k + q$ parameters. Because the elements of $\boldsymbol{\Sigma}$ are usually nonlinear functions of the elements of $\boldsymbol{\theta}$, the process is typically iterative. From some starting values $[\boldsymbol{\theta}^{(0)}, \boldsymbol{\beta}^{(0)}]$, one computes successive updates according to a nonlinear optimization technique; for example, by way of the Newton-Raphson, Quasi-Newton, or some other suitable algorithm.

Fortunately, the size of the optimization problem can be substantially reduced. First, note that usually a scalar parameter can be factored from the variance-covariance matrix. We write $\boldsymbol{\Sigma}(\boldsymbol{\theta}) = \sigma^2\boldsymbol{\Sigma}(\boldsymbol{\theta}^*)$ where $\boldsymbol{\theta}^*$ is a $((q-1) \times 1)$ vector with its elements possibly adjusted to reflect the factoring of $\sigma^2$. For example, if the process has an exponential covariance structure, then $\boldsymbol{\Sigma}(\boldsymbol{\theta}^*)$ is the autocorrelation matrix with exponential correlation structure. Second, a closed form expression can be obtained for the parameters $\boldsymbol{\beta}$ and $\sigma^2$ given $\boldsymbol{\theta}$. This enables us to remove these parameters from the optimization, a process termed **profiling**. To profile $\boldsymbol{\beta}$, take derivatives of (5.41) with respect to $\boldsymbol{\beta}$ and solve. The result is the GLS estimator

$$\widehat{\boldsymbol{\beta}} = (\mathbf{X}(\mathbf{s})'\boldsymbol{\Sigma}(\boldsymbol{\theta})^{-1}\mathbf{X}(\mathbf{s}))^{-1}\mathbf{X}(\mathbf{s})'\boldsymbol{\Sigma}(\boldsymbol{\theta})^{-1}\mathbf{Z}(\mathbf{s}). \quad (5.42)$$

Substituting this expression into (5.41) yields an objective function for minimization profiled for $\beta$,

$$\varphi_\beta(\boldsymbol{\theta}; \mathbf{Z}(\mathbf{s})) = \ln\{|\sigma^2 \boldsymbol{\Sigma}(\boldsymbol{\theta}^*)|\} + n\ln\{2\pi\} + \sigma^{-2}\mathbf{r}'\boldsymbol{\Sigma}(\boldsymbol{\theta}^*)^{-1}\mathbf{r}, \qquad (5.43)$$

where

$$\mathbf{r} = \mathbf{Z}(\mathbf{s}) - (\mathbf{X}(\mathbf{s})'\boldsymbol{\Sigma}(\boldsymbol{\theta})^{-1}\mathbf{X}(\mathbf{s}))^{-1}\mathbf{X}(\mathbf{s})'\boldsymbol{\Sigma}(\boldsymbol{\theta})^{-1}\mathbf{Z}(\mathbf{s})$$

is the GLS residual. To profile $\sigma^2$ from the objective function, note that its MLE is

$$\widehat{\sigma}^2_{ml} = \frac{1}{n}\mathbf{r}'\boldsymbol{\Sigma}(\boldsymbol{\theta}^*)^{-1}\mathbf{r}.$$

*Minus 2 Profiled log Likelihood*

Substituting again yields the negative of twice the profiled log likelihood,

$$\varphi_{\beta,\sigma}(\boldsymbol{\theta}^*; \mathbf{Z}(\mathbf{s})) = \ln\{|\boldsymbol{\Sigma}(\boldsymbol{\theta}^*)|\} + n\ln\{\widehat{\sigma}^2\} + n(\ln\{2\pi\} - 1). \qquad (5.44)$$

*MLE of $\beta$*

Minimizing (5.44) is an optimization problem with only $q-1$ parameters. Upon convergence you obtain $\widehat{\boldsymbol{\theta}}_{ml}$ from $\widehat{\sigma}^2_{ml}$ and $\widehat{\boldsymbol{\theta}}^*_{ml}$, and $\widehat{\boldsymbol{\beta}}_{ml}$ by evaluating (5.42) at the maximum likelihood estimate $\widehat{\boldsymbol{\theta}}_{ml}$ of $\boldsymbol{\theta}$:

$$\widehat{\boldsymbol{\beta}}_{ml} = \left(\mathbf{X}(\mathbf{s})'\boldsymbol{\Sigma}(\widehat{\boldsymbol{\theta}}_{ml})^{-1}\mathbf{X}(\mathbf{s})\right)^{-1}\mathbf{X}(\mathbf{s})'\boldsymbol{\Sigma}(\widehat{\boldsymbol{\theta}}_{ml})^{-1}\mathbf{Z}(\mathbf{s}). \qquad (5.45)$$

The savings in computing time are substantial when parameters can be profiled from the optimization. For example, in a model ($\mathbb{R}^2$) where the large-scale trend (mean function) is modeled as a trend surface of second degree and $\boldsymbol{\Sigma}(\boldsymbol{\theta})$ has a spherical no-nugget covariance structure, there are eight parameters ($q = 2, k = 6$). The iterative optimization is carried out for a single parameter, the range of the spatial process. Given an estimate of the range parameter, the correlation matrix $\boldsymbol{\Sigma}(\boldsymbol{\theta}^*)$ can be computed which yields the MLE of $\sigma^2$ and that of $\boldsymbol{\beta}$. Note that (5.45) could also be evaluated at $\boldsymbol{\Sigma}(\widehat{\boldsymbol{\theta}}^*_{ml})$. The profiled parameter $\sigma^2$ does not affect the point estimate of the regression parameters. Additional computational savings can be gained by using other algorithms that avoid repeated inversion of large, unstructured matrices (Mardia and Marshall, 1984; Zimmerman, 1989).

One of the advantages of likelihood estimation is the ability to estimate the variance-covariance matrix of the parameter estimates based on the observed or expected information matrix. The observed information matrix equals $0.5\mathbf{H}$, where $\mathbf{H}$ is the Hessian (second derivative) matrix of (5.41). Standard errors of the maximum likelihood estimators are obtained as the diagonal elements of $2\mathbf{H}^{-1}$ or $2\mathrm{E}[\mathbf{H}]^{-1}$. Because

$$\mathrm{E}\left[\frac{\partial\varphi(\boldsymbol{\beta}; \boldsymbol{\theta}; \mathbf{Z}(\mathbf{s}))}{\partial\boldsymbol{\beta}\,\partial\boldsymbol{\theta}}\right] = \mathbf{0},$$

the information matrix is block-diagonal. Even if $\boldsymbol{\beta}$ is profiled from the optimization, the leading block of the inverse information matrix is

$$(\mathbf{X}(\mathbf{s})\boldsymbol{\Sigma}(\boldsymbol{\theta})^{-1}\mathbf{X}(\mathbf{s}))^{-1}$$

and the variance-covariance matrix of the regression effects is estimated as

$$\widehat{\mathrm{Var}}(\widehat{\beta}_{ml}) = (\mathbf{X}(\mathbf{s})'\mathbf{\Sigma}(\widehat{\boldsymbol{\theta}}_{ml})^{-1}\mathbf{X}(\mathbf{s}))^{-1}. \tag{5.46}$$

Thus, the variance-covariance matrix of $\widehat{\beta}_{ml}$ has the same form as that of equation (5.40), but with $\widehat{\theta}_{IRWGLS}$ replaced with $\widehat{\theta}_{ml}$.

For full ML estimation without profiling, the inverse of the information matrix for $\omega = [\beta', \theta']'$ can be written as (see Breusch, 1980 and Judge et al., 1985, p. 182)

*Inverse Fisher Information*

$$I(\omega)^{-1} = \begin{bmatrix} (\mathbf{X}(\mathbf{s})'\mathbf{\Sigma}(\boldsymbol{\theta})^{-1}\mathbf{X}(\mathbf{s})) & \mathbf{0} \\ \mathbf{0}' & \frac{1}{2}\Delta'(\mathbf{\Sigma}(\boldsymbol{\theta})^{-1} \otimes \mathbf{\Sigma}(\boldsymbol{\theta})^{-1})\Delta \end{bmatrix}^{-1}, \tag{5.47}$$

where $\mathbf{0}$ denotes the $(p \times q)$ zero matrix, and $\Delta = \Delta(\boldsymbol{\theta}) = \partial(\mathrm{vec}\ (\mathbf{\Sigma}(\boldsymbol{\theta})))/\partial\boldsymbol{\theta}'$ is a $(n^2 \times q)$ matrix that contains the partial derivatives of each element of $\mathbf{\Sigma}(\boldsymbol{\theta})$ with respect to each element in $\boldsymbol{\theta}$. The matrix operator vec$(\cdot)$ stacks the columns of a matrix into a single vector, so vec$(\mathbf{\Sigma}(\boldsymbol{\theta}))$ is a $(n^2 \times 1)$ vector. The symbol $\otimes$ denotes the matrix direct product multiplying each element in the first matrix by every element in the second matrix producing an $n^2 \times n^2$ matrix.

Detailed expressions for the Hessian matrix with respect to the covariance parameters for the $\boldsymbol{\theta}$ and $\boldsymbol{\theta}^*$ parameterizations are given in Wolfinger, Tobias, and Sall (1994) for ML and restricted maximum likelihood estimation. These authors also present expressions to convert the $(q-1) \times (q-1)$ Hessian in terms of $\boldsymbol{\theta}^*$ into the $q \times q$ Hessian for the $\boldsymbol{\theta}$ parameterization.

### 5.5.3 *Restricted Maximum Likelihood*

Restricted (or residual) maximum likelihood (REML) estimates are often preferred over MLEs because the latter exhibit greater negative bias for estimates of covariance parameters. The culprit of this bias—roughly—lies in the failure of ML estimation to account for the number of mean parameters in the estimation of the covariance parameters. The most famous—and simplest—example is that of an *iid* sample from a $G(\mu, \sigma^2)$ distribution, where $\mu$ is unknown. The MLE for $\sigma^2$ is

$$\widehat{\sigma}^2_{ml} = \frac{1}{n}\sum_{i=1}^{n}(Y_i - \overline{Y})^2,$$

which has bias $-\sigma^2/n$. The REML estimator of $\sigma^2$ is unbiased:

$$\widehat{\sigma}^2_{reml} = \frac{1}{n-1}\sum_{i=1}^{n}(Y_i - \overline{Y})^2.$$

Similarly, in a regression model with Gaussian, homoscedastic, uncorrelated errors, the ML and REML estimators for the residual variance are

$$\widehat{\sigma}^2_{ml} \quad = \quad \frac{1}{n}(\mathbf{Y} - \mathbf{X}\widehat{\beta})'(\mathbf{Y} - \mathbf{X}\widehat{\beta})$$

$$\widehat{\sigma}^2_{reml} = \frac{1}{n-k}(\mathbf{Y} - \mathbf{X}'\widehat{\beta})'(\mathbf{Y} - \mathbf{X}\widehat{\beta}),$$

respectively. The ML estimator is again a biased estimator.

For the spatial model $\mathbf{Z}(\mathbf{s}) \sim G(\mathbf{X}(\mathbf{s})\boldsymbol{\beta}, \boldsymbol{\Sigma}(\boldsymbol{\theta}))$, the REML adjustment consists of performing ML estimation not for $\mathbf{Z}(\mathbf{s})$, but for $\mathbf{KZ}(\mathbf{s})$, where the $((n-k) \times n)$ matrix $\mathbf{K}$ is chosen so that $\mathrm{E}[\mathbf{KZ}(\mathbf{s})] = \mathbf{0}$ and $\mathrm{rank}[\mathbf{K}] = n - k$. Because of these properties, the matrix $\mathbf{K}$ is called a matrix of error contrast, which supports the alternate name as *residual* maximum likelihood. Although REML estimation is well-established in statistical theory and applications, in the geostatistical arena it appeared first in work by P. Kitanidis and co-workers in the mid-1980's (Kitanidis, 1983; Kitanidis and Vomvoris, 1983; Kitanidis and Lane, 1985).

An important aspect of REML estimation lies in the handling of $\boldsymbol{\beta}$. In §5.5.2, the $\boldsymbol{\beta}$ vector was profiled from the log likelihood to reduce the size of the optimization problem. This led to the objective function $\varphi_{\beta}(\boldsymbol{\theta}; \mathbf{Z}(\mathbf{s}))$ in (5.43). When you consider ML estimation for the $n - k$ vector $\mathbf{KZ}(\mathbf{s})$, the fixed effects $\boldsymbol{\beta}$ have seemingly disappeared from the objective function. Minus twice the log likelihood of $\mathbf{KZ}(\mathbf{s})$ is

$$\begin{aligned}\varphi_R(\boldsymbol{\theta}; \mathbf{KZ}(\mathbf{s})) &= \ln\{|\mathbf{K}\boldsymbol{\Sigma}(\boldsymbol{\theta})\mathbf{K}'|\} + (n-k)\ln\{2\pi\} \\ &\quad + \mathbf{Z}(\mathbf{s})'\mathbf{K}'\left(\mathbf{K}\boldsymbol{\Sigma}(\boldsymbol{\theta})\mathbf{K}'\right)^{-1}\mathbf{KZ}(\mathbf{s}). \qquad (5.48)\end{aligned}$$

This is an objective function about $\boldsymbol{\theta}$ only. It is possible to write $\varphi_R(\boldsymbol{\theta}; \mathbf{KZ}(\mathbf{s}))$ in terms of an estimate $\widehat{\boldsymbol{\beta}}$ and we will do so shortly to more clearly show the relationship between the ML and REML objective functions. You need to keep in mind, however, that there is no "REML estimator of $\boldsymbol{\beta}$." Minimizing $\varphi_R(\boldsymbol{\theta}; \mathbf{KZ}(\mathbf{s}))$ yields $\widehat{\boldsymbol{\theta}}_{reml}$. What is termed $\widehat{\boldsymbol{\beta}}_{reml}$ and is computed as

*REML "Estimator" of $\boldsymbol{\beta}$*

$$\widehat{\boldsymbol{\beta}}_{reml} = \left(\mathbf{X}'\boldsymbol{\Sigma}(\widehat{\boldsymbol{\theta}}_{reml})^{-1}\mathbf{X}\right)^{-1}\mathbf{X}'\boldsymbol{\Sigma}(\widehat{\boldsymbol{\theta}}_{reml})^{-1}\mathbf{Z}(\mathbf{s})$$

is simply an EGLS estimator evaluated at $\widehat{\boldsymbol{\theta}}_{reml}$.

We now rewrite (5.48) to eliminate the matrix $\mathbf{K}$ from the expression. First notice that if $\mathrm{E}[\mathbf{KZ}(\mathbf{s})] = \mathbf{0}$, then $\mathbf{KX}(\mathbf{s}) = \mathbf{0}$. If $\boldsymbol{\Sigma}(\boldsymbol{\theta})$ is positive definite, Searle et al. (1992, pp. 451–452) shows that

$$\mathbf{K}'\left(\mathbf{K}\boldsymbol{\Sigma}(\boldsymbol{\theta})\mathbf{K}'\right)^{-1}\mathbf{K} = \boldsymbol{\Sigma}(\boldsymbol{\theta})^{-1} - \boldsymbol{\Sigma}(\boldsymbol{\theta})^{-1}\mathbf{X}(\mathbf{s})\boldsymbol{\Omega}(\boldsymbol{\theta})\mathbf{X}(\mathbf{s})'\boldsymbol{\Sigma}(\boldsymbol{\theta})^{-1},$$

where $\boldsymbol{\Omega}(\boldsymbol{\theta}) = (\mathbf{X}(\mathbf{s})'\boldsymbol{\Sigma}(\boldsymbol{\theta})^{-1}\mathbf{X}(\mathbf{s}))^{-1}$. This identity and $\boldsymbol{\Omega}\mathbf{X}(\mathbf{s})'\boldsymbol{\Sigma}(\boldsymbol{\theta})^{-1}\mathbf{Z}(\mathbf{s}) = \widehat{\boldsymbol{\beta}}$ yields

$$\mathbf{Z}(\mathbf{s})'\mathbf{K}'\left(\mathbf{K}\boldsymbol{\Sigma}(\boldsymbol{\theta})\mathbf{K}'\right)^{-1}\mathbf{KZ}(\mathbf{s}) = \mathbf{r}'\boldsymbol{\Sigma}(\boldsymbol{\theta})^{-1}\mathbf{r}.$$

In fundamental work on REML estimation, Harville (1974, 1977) established important further results. For example, he shows that if

$$\mathbf{K}'\mathbf{K} = \mathbf{I} - \mathbf{X}(\mathbf{s})(\mathbf{X}(\mathbf{s})'\mathbf{X}(\mathbf{s}))^{-1}\mathbf{X}(\mathbf{s})'$$

and $\mathbf{KK}' = \mathbf{I}$, then minus twice the log likelihood of $\mathbf{KZ}(\mathbf{s})$ can be written as

$$\varphi_R(\boldsymbol{\theta}; \mathbf{KZ}(\mathbf{s})) = \ln\{|\boldsymbol{\Sigma}(\boldsymbol{\theta})|\} + \ln\{|\mathbf{X}(\mathbf{s})'\boldsymbol{\Sigma}(\boldsymbol{\theta})^{-1}\mathbf{X}(\mathbf{s})|\}$$

$$- \ln\{|\mathbf{X}(\mathbf{s})'\mathbf{X}(\mathbf{s})|\} + \mathbf{r}'\boldsymbol{\Sigma}(\boldsymbol{\theta})^{-1}\mathbf{r}$$
$$+ (n-k)\ln\{2\pi\}.$$

Harville (1977) points out that $(n-k) \times n$ matrices whose rows are linearly independent rows of $\mathbf{I} - \mathbf{X}(\mathbf{s})(\mathbf{X}(\mathbf{s})'\mathbf{X}(\mathbf{s}))^{-1}\mathbf{X}(\mathbf{s})'$ will lead to REML objective functions that differ by a constant amount. The amount does not depend on $\boldsymbol{\theta}$ or $\boldsymbol{\beta}$. The obvious choice as a REML objective function for minimization is

*Minus two Restricted log Likelihood*

$$\begin{aligned} \varphi_R(\boldsymbol{\theta}; \mathbf{KZ}(\mathbf{s})) &= \ln\{|\boldsymbol{\Sigma}(\boldsymbol{\theta})|\} + \ln\{|\mathbf{X}(\mathbf{s})'\boldsymbol{\Sigma}(\boldsymbol{\theta})^{-1}\mathbf{X}(\mathbf{s})|\} \\ &\quad + \mathbf{r}'\boldsymbol{\Sigma}(\boldsymbol{\theta})^{-1}\mathbf{r} + (n-k)\ln\{2\pi\}. \end{aligned}$$

In this form, minus twice the REML log likelihood differs from (5.41) by the terms $\ln\{|\mathbf{X}(\mathbf{s})'\boldsymbol{\Sigma}(\boldsymbol{\theta})^{-1}\mathbf{X}(\mathbf{s})|\}$ and $k\ln\{2\pi\}$. As with ML estimation, a scale parameter can be profiled from $\boldsymbol{\Sigma}(\boldsymbol{\theta})$. The REML estimator of this parameter is

$$\widehat{\sigma}^2_{reml} = \frac{1}{n-k}\mathbf{r}'\boldsymbol{\Sigma}(\boldsymbol{\theta}^*)^{-1}\mathbf{r}$$

and upon substitution one obtains minus twice the profiled REML log likelihood

$$\begin{aligned} \varphi_{R,\sigma}(\boldsymbol{\theta}^*; \mathbf{KZ}(\mathbf{s})) &= \ln\{|\boldsymbol{\Sigma}(\boldsymbol{\theta}^*)|\} + \ln\{|\mathbf{X}(\mathbf{s})'\boldsymbol{\Sigma}(\boldsymbol{\theta}^*)^{-1}\mathbf{X}(\mathbf{s})|\} \\ &\quad + (n-k)\ln\{\widehat{\sigma}^2\} + (n-k)(\ln\{2\pi\} - 1). \quad (5.49) \end{aligned}$$

Wolfinger, Tobias, and Sall (1994) give expressions for the gradient and Hessian of the REML log likelihood with and without profiling of $\sigma^2$.

There is a large literature on the use of ML and REML for spatial modeling and this is an area of active research in statistics. Searle, Casella and McCulloch (1992) provide an introduction to REML estimation in linear models and Littell, Milliken, Stroup and Wolfinger (1996) adapt some of these results to the spatial case. Cressie and Lahiri (1996) provide the distributional properties of REML estimators in a spatial setting.

### *5.5.4 Prediction Errors When Covariance Parameters Are Estimated*

A common theme of statistical modeling when errors are correlated is the plug-in form of expressions involving the unknown covariance parameters. The "plugging" occurs when we compute predictions of new observations, estimates of the mean, and estimates of precision. This has important consequences for the statistical properties of the estimates (predictions) and our ability to truthfully report their uncertainty. From the introductory remarks of this section recall the case of the ordinary kriging predictor and its variance.

$$\begin{aligned} p_{ok}(\mathbf{Z}; \mathbf{s}_0) &= \left(\boldsymbol{\sigma}(\boldsymbol{\theta}) + \mathbf{1}\frac{1 - \mathbf{1}'\boldsymbol{\Sigma}(\boldsymbol{\theta})^{-1}\boldsymbol{\sigma}(\boldsymbol{\theta})}{\mathbf{1}'\boldsymbol{\Sigma}(\boldsymbol{\theta})^{-1}\mathbf{1}}\right)\boldsymbol{\Sigma}(\boldsymbol{\theta})^{-1}\mathbf{Z}(\mathbf{s}), \\ \sigma^2_{ok}(\mathbf{s}_0) &= C(\mathbf{0}) - \boldsymbol{\sigma}(\boldsymbol{\theta})'\boldsymbol{\Sigma}(\boldsymbol{\theta})^{-1}\boldsymbol{\sigma}(\boldsymbol{\theta}) + \frac{(1 - \mathbf{1}'\boldsymbol{\Sigma}(\boldsymbol{\theta})^{-1}\boldsymbol{\sigma}(\boldsymbol{\theta}))^2}{\mathbf{1}'\boldsymbol{\Sigma}(\boldsymbol{\theta})^{-1}\mathbf{1}}. \end{aligned}$$

Applying "plugging" to both yields

$$\begin{aligned}
\widehat{p}_{ok}(\mathbf{Z}; \mathbf{s}_0) &= \left(\boldsymbol{\sigma}(\widehat{\boldsymbol{\theta}}) + \mathbf{1}\frac{1 - \mathbf{1}'\boldsymbol{\Sigma}(\widehat{\boldsymbol{\theta}})^{-1}\boldsymbol{\sigma}(\widehat{\boldsymbol{\theta}})}{\mathbf{1}'\boldsymbol{\Sigma}(\widehat{\boldsymbol{\theta}})^{-1}\mathbf{1}}\right)\boldsymbol{\Sigma}(\widehat{\boldsymbol{\theta}})^{-1}\mathbf{Z}(\mathbf{s}) \\
\widehat{\sigma}^2_{ok}(\mathbf{s}_0) &= \widehat{C}(\mathbf{0}) - \boldsymbol{\sigma}(\widehat{\boldsymbol{\theta}})'\boldsymbol{\Sigma}(\widehat{\boldsymbol{\theta}})^{-1}\boldsymbol{\sigma}(\widehat{\boldsymbol{\theta}}) + \\
&\quad \frac{(1 - \mathbf{1}'\boldsymbol{\Sigma}(\widehat{\boldsymbol{\theta}})^{-1}\boldsymbol{\sigma}(\widehat{\boldsymbol{\theta}}))^2}{\mathbf{1}'\boldsymbol{\Sigma}(\widehat{\boldsymbol{\theta}})^{-1}\mathbf{1}}.
\end{aligned}$$

$\widehat{p}_{ok}(\mathbf{Z}; \mathbf{s}_0)$ is not the ordinary kriging predictor, it is an estimate thereof. Similarly, $\widehat{\sigma}^2_{ok}(\mathbf{s}_0)$ is not the variance of the ordinary kriging predictor, it is an estimate thereof; and because of the nonlinear involvement of $\widehat{\boldsymbol{\theta}}$, it is a biased estimate of $\sigma^2_{ok}(\mathbf{s}_0)$. More importantly, $\sigma^2_{ok}(\mathbf{s}_0)$ is not the prediction error of $\widehat{p}_{ok}(\mathbf{Z}; \mathbf{s}_0)$, but that of $p_{ok}(\mathbf{Z}; \mathbf{s}_0)$. Intuitively, one would expect the prediction error of $\widehat{p}_{ok}(\mathbf{Z}; \mathbf{s}_0)$ to exceed that of the ordinary kriging predictor, because not knowing $\boldsymbol{\theta}$ has introduced additional variability into the system; $\widehat{\boldsymbol{\theta}}$ is a random vector. So, if $\sigma^2_{ok}(\mathbf{s}_0)$ is not the prediction error we should compute if $\widehat{p}_{ok}(\mathbf{Z}; \mathbf{s}_0)$ is our predictor, are we not making things even worse by evaluating $\sigma^2_{ok}(\mathbf{s}_0)$ at $\widehat{\boldsymbol{\theta}}$? Is that not a biased estimate of the wrong quantity?

The consequences of plug-in estimation for estimating variability are not germane to spatial models. To address the issues in more generality, we consider in this section a more generic model and notation. The model at issue is

$$\mathbf{Z} = \mathbf{X}\boldsymbol{\beta} + \mathbf{e}, \qquad \mathbf{e} \sim (\mathbf{0}, \boldsymbol{\Sigma}(\boldsymbol{\theta})),$$

a basic correlated error model with a linear mean and parameterized variance matrix. In the expressions that follow, an overline, $\bar{\ }$, denotes a general estimator or predictor, $\tilde{\ }$ denotes a quantity evaluated if $\boldsymbol{\theta}$ is known, and $\hat{\ }$ is used when the quantity is evaluated at the estimate $\widehat{\boldsymbol{\theta}}$.

Following Harville and Jeske (1992), our interest is in predicting a quantity $\omega$ with the properties

$$\begin{aligned}
\mathrm{E}[\omega] &= \mathbf{x}_0'\boldsymbol{\beta} \\
\mathrm{Var}[\omega] &= \sigma^2(\boldsymbol{\theta}) \\
\mathrm{Cov}[\mathbf{Z}, \omega] &= \boldsymbol{\sigma}(\boldsymbol{\theta}),
\end{aligned}$$

where $\mathbf{x}_0'\boldsymbol{\beta}$ is estimable. If $\overline{\omega}$ is any predictor of $\omega$, then the quality of our prediction under squared-error loss is measured by the mean-squared error

$$\mathrm{mse}[\overline{\omega}, \omega] = \mathrm{E}[(\overline{\omega} - \omega)^2].$$

A special case is that of $\mathrm{Var}[\omega] = \sigma^2 = 0$. The problem is then one of estimating $\mathbf{x}_0'\boldsymbol{\beta}$ and

$$\mathrm{mse}[\overline{\omega}, \omega] = \mathrm{mse}[\mathbf{x}_0'\overline{\boldsymbol{\beta}}, \mathbf{x}_0'\boldsymbol{\beta}] = \mathbf{x}_0'\mathrm{E}[(\overline{\boldsymbol{\beta}} - \boldsymbol{\beta})(\overline{\boldsymbol{\beta}} - \boldsymbol{\beta})']\mathbf{x}_0.$$

If $\overline{\boldsymbol{\beta}}$ is unbiased for $\boldsymbol{\beta}$, then this mean-squared error equals the variance of $\mathbf{x}_0'\overline{\boldsymbol{\beta}}$.

Under the stated conditions, the GLS estimator of $\boldsymbol{\beta}$ and the BLUP of $\omega$ are

$$\begin{aligned}\widetilde{\boldsymbol{\beta}} &= (\mathbf{X}'\boldsymbol{\Sigma}(\boldsymbol{\theta})^{-1}\mathbf{X})^{-1}\mathbf{X}'\boldsymbol{\Sigma}(\boldsymbol{\theta})^{-1}\mathbf{Z} \\ \widetilde{\omega} &= \mathbf{x}_0'\widetilde{\boldsymbol{\beta}} + \boldsymbol{\sigma}(\boldsymbol{\theta})'\boldsymbol{\Sigma}(\boldsymbol{\theta})^{-1}(\mathbf{Z} - \mathbf{X}\widetilde{\boldsymbol{\beta}}).\end{aligned}$$

It is easy to establish that $\mathrm{E}[\widetilde{\boldsymbol{\beta}}] = \boldsymbol{\beta}$, and that $\widetilde{\omega}$ is unbiased in the sense that $\mathrm{E}[\widetilde{\omega}] = \mathrm{E}[\omega] = \mathbf{x}_0'\boldsymbol{\beta}$. The respective mean-squared errors are given by

$$\begin{aligned}\mathrm{mse}[\mathbf{x}_0'\widetilde{\boldsymbol{\beta}}, \mathbf{x}_0'\boldsymbol{\beta}] &= (\mathbf{X}'\boldsymbol{\Sigma}(\boldsymbol{\theta})^{-1}\mathbf{X})^{-1} = \boldsymbol{\Omega}(\boldsymbol{\theta}) && (5.50)\\ \mathrm{mse}[\widetilde{\omega}, \omega] &= \mathrm{E}[(\widetilde{\omega} - \omega)^2] = \mathbf{p}'\boldsymbol{\Sigma}(\boldsymbol{\theta})\mathbf{p} + \sigma^2(\boldsymbol{\theta}) - 2\mathbf{p}'\boldsymbol{\sigma}(\boldsymbol{\theta}) && (5.51)\\ \mathbf{p} &= (\mathbf{x}_0' - \boldsymbol{\sigma}(\boldsymbol{\theta})'\boldsymbol{\Sigma}(\boldsymbol{\theta})^{-1}\mathbf{X})(\mathbf{X}'\boldsymbol{\Sigma}(\boldsymbol{\theta})^{-1}\mathbf{X})^{-1}\boldsymbol{\Sigma}(\boldsymbol{\theta})^{-1} \\ &\quad + \boldsymbol{\sigma}(\boldsymbol{\theta})'\boldsymbol{\Sigma}(\boldsymbol{\theta})^{-1}\end{aligned}$$

If $\boldsymbol{\theta}$ is unknown, we plug in the estimate $\widehat{\boldsymbol{\theta}}$ and obtain the EGLS estimator and EBLUP

$$\begin{aligned}\widehat{\boldsymbol{\beta}} &= (\mathbf{X}'\boldsymbol{\Sigma}(\widehat{\boldsymbol{\theta}})^{-1}\mathbf{X})^{-1}\mathbf{X}'\boldsymbol{\Sigma}(\widehat{\boldsymbol{\theta}})^{-1}\mathbf{Z} = \boldsymbol{\Omega}(\widehat{\boldsymbol{\theta}})\mathbf{X}'\boldsymbol{\Sigma}(\widehat{\boldsymbol{\theta}})^{-1}\mathbf{Z} \\ \widehat{\omega} &= \mathbf{x}_0'\widehat{\boldsymbol{\beta}} + \boldsymbol{\sigma}(\widehat{\boldsymbol{\theta}})'\boldsymbol{\Sigma}(\widehat{\boldsymbol{\theta}})^{-1}(\mathbf{Z} - \mathbf{X}\widehat{\boldsymbol{\beta}}).\end{aligned}$$

The estimator $\widehat{\boldsymbol{\beta}}$ and the predictor $\widehat{\omega}$ remain unbiased. The question, then, is how to compute the variance of $\widehat{\boldsymbol{\beta}}$ and the mean square error $\mathrm{mse}[\widehat{\omega}, \omega]$? Before looking into the details, let us consider the plug-in quantity

$$\boldsymbol{\Omega}(\widehat{\boldsymbol{\theta}}) = (\mathbf{X}'\boldsymbol{\Sigma}(\widehat{\boldsymbol{\theta}})^{-1}\mathbf{X})^{-1},$$

which is commonly used as an estimate of $\mathrm{Var}[\widehat{\boldsymbol{\beta}}]$. There are two major problems. First, it is not unbiased for $\boldsymbol{\Omega}(\boldsymbol{\theta})$; $\mathrm{E}[\boldsymbol{\Omega}(\widehat{\boldsymbol{\theta}})] \neq \boldsymbol{\Omega}(\boldsymbol{\theta})$. We could use it as a biased estimator of (5.50), however. Second, even if it was unbiased, $\boldsymbol{\Omega}(\boldsymbol{\theta})$ is not the variance of $\widehat{\boldsymbol{\beta}}$. We need

- an estimator of the mean-squared error that takes into account the fact that $\boldsymbol{\theta}$ was estimated and hence that $\widehat{\boldsymbol{\theta}}$ is a random variable;
- a computationally feasible method for evaluating the mean-squared error.

Now let us return to the more general case of predicting $\omega$ with $\widehat{\omega}$, keeping in mind that finding the variance of $\widehat{\boldsymbol{\beta}}$ is a special case of determining $\mathrm{mse}[\widehat{\omega}, \omega]$. Progress can be made by considering only those estimators $\overline{\boldsymbol{\theta}}$ that have certain properties. For example, Kackar and Harville (1984) and Harville and Jeske (1992) consider even, translation invariant estimators. Christensen (1991, Ch. VI.5) considers "residual-type" statistics; see also Eaton (1985). Suffice it to say that ML and REML estimators have the needed properties. Kackar and Harville (1984) decompose the prediction error into

$$\widehat{\omega} - \omega \;=\; \widetilde{\omega} - \omega \;+\; \widehat{\omega} - \widetilde{\omega} = e_1(\boldsymbol{\theta}) + e_2(\boldsymbol{\theta}).$$

If $\widehat{\boldsymbol{\theta}}$ is translation invariant, then $e_1(\boldsymbol{\theta})$ and $e_2(\boldsymbol{\theta})$ are distributed independently, and

$$\mathrm{mse}[\widehat{\omega}, \omega] = \mathrm{mse}[\widetilde{\omega}, \omega] + \mathrm{Var}[\widehat{\omega} - \widetilde{\omega}].$$

By choosing to ignore the fact that $\boldsymbol{\theta}$ was estimated, the mean-squared prediction error is underestimated by the amount $\text{Var}[\widehat{\omega} - \widetilde{\omega}]$. Hence, $\text{mse}[\widehat{\omega}, \omega] \geq \text{mse}[\widetilde{\omega}, \omega]$. If you follow the practice of plugging $\widehat{\boldsymbol{\theta}}$ into expressions that apply if $\boldsymbol{\theta}$ were known, you would estimate $\text{mse}[\widehat{\omega}, \omega]$ by evaluating (5.51) at $\widehat{\boldsymbol{\theta}}$. This yields the (estimated) mean-squared error of the wrong quantity, and can be substantially biased.

Kackar and Harville (1984) propose a correction term, and Harville and Jeske (1992) provide details about estimation. First, using a Taylor series, $\text{Var}[\widehat{\omega} - \widetilde{\omega}]$ is approximated as $\text{tr}\{\mathbf{A}(\boldsymbol{\theta})\mathbf{B}(\boldsymbol{\theta})\}$, where

$$\begin{aligned} \mathbf{A} &= \text{Var}\left[\frac{\partial \widetilde{\omega}}{\partial \boldsymbol{\theta}}\right] \\ \mathbf{B} &= \text{mse}[\widehat{\boldsymbol{\theta}}, \boldsymbol{\theta}] = \text{E}[(\widehat{\boldsymbol{\theta}} - \boldsymbol{\theta})(\widehat{\boldsymbol{\theta}} - \boldsymbol{\theta})']. \end{aligned}$$

Although we now have an expression—at least an approximate one—for the mean-squared prediction error based on a translation-invariant estimator $\widehat{\boldsymbol{\theta}}$, we do not seem to have helped the cause very much. We now have the improved mean-squared error expression

$$\text{mse}[\widehat{\omega}, \omega] \doteq \text{mse}[\widetilde{\omega}, \omega] + \text{tr}\{\mathbf{A}(\boldsymbol{\theta})\mathbf{B}(\boldsymbol{\theta})\}, \tag{5.52}$$

but $\text{mse}[\widetilde{\omega}, \omega]$, $\mathbf{A}(\boldsymbol{\theta})$ and $\mathbf{B}(\boldsymbol{\theta})$ depend on $\boldsymbol{\theta}$. To evaluate the matrices, we are going to plug in $\widehat{\boldsymbol{\theta}}$ again and, no surprise here, incur yet another source of bias. This does appear like a vicious cycle. Our initial estimate of prediction error was not reliable, because $\widehat{\boldsymbol{\theta}}$ was estimated and we simply plugged it into the expression for the unknown $\boldsymbol{\theta}$. Which prompts us to derive an approximation of the actual prediction error, (5.52). Unfortunately, it also depends on $\boldsymbol{\theta}$, and to evaluate it we plug in $\widehat{\boldsymbol{\theta}}$ again, which biases the result. Fortunately, we are already working with the correct prediction error, and the bias can be managed.

Now we introduce $\boldsymbol{\theta}$ into the symbolic mean squared error representation to underline the fact that the mse[ , ] expressions depend on $\boldsymbol{\theta}$. First, Prasad and Rao (1990) show that the plug-in estimator $\text{tr}\{\mathbf{A}(\widehat{\boldsymbol{\theta}})\mathbf{B}(\widehat{\boldsymbol{\theta}})\}$ is approximately unbiased for $\text{tr}\{\mathbf{A}(\boldsymbol{\theta})\mathbf{B}(\boldsymbol{\theta})\}$. Then, Harville and Jeske (1992) derive that

$$\text{E}[\text{mse}[\widetilde{\omega}, \omega, \widehat{\boldsymbol{\theta}}] - \text{tr}\{\mathbf{A}(\widehat{\boldsymbol{\theta}})\mathbf{B}(\widehat{\boldsymbol{\theta}})\}] \doteq \text{mse}[\widetilde{\omega}, \omega, \boldsymbol{\theta}].$$

Evaluating the improved estimator (5.52) at $\widehat{\boldsymbol{\theta}}$, we still fall short of the prediction error on average, but things have improved. Half of the bias is gone. An approximately unbiased estimator of the mean-squared prediction error $\text{mse}[\widehat{\omega}, \omega]$ is then

*Prasad-Rao MSE Estimator*

$$\text{mse}[\widetilde{\omega}, \omega, \widehat{\boldsymbol{\theta}}] + 2\text{tr}\{\mathbf{A}(\widehat{\boldsymbol{\theta}})\mathbf{B}(\widehat{\boldsymbol{\theta}})\}. \tag{5.53}$$

The bias of the plug-in mean squared error estimator can be substantive, in particular when sample size is small. However, as Zimmerman and Cressie (1992) note, the magnitude of this bias and the improvement offered by (5.53) depends on the validity of the Gaussian assumption, whether or not $\widehat{\boldsymbol{\theta}}$ is unbiased, the nature of the covariance model used for $\boldsymbol{\Sigma}(\widehat{\boldsymbol{\theta}})$, the spatial configuration of the data, and the strength of spatial autocorrelation. Based on examples in Zimmerman and Zimmerman (1991) and Zimmerman and Cressie (1992), Zimmerman and Cressie (1992) offer the following general guidelines. The performance of the plug-in mean-squared error estimator ($\text{mse}[\widetilde{\omega}, \omega, \widehat{\boldsymbol{\theta}}]$, the estimated kriging variance) as an estimator of the true prediction mean-squared error can often be improved upon when the spatial correlation is weak, but it is often adequate and sometimes superior to the alternative estimators such as (5.53) when the spatial correlation is strong. Zimmerman and Cressie (1992) suggest that corrections of the type used in (5.53) should only be used when $\widehat{\boldsymbol{\theta}}$ is unbiased, or $\boldsymbol{\Sigma}(\widehat{\boldsymbol{\theta}})$ is negatively biased, and the spatial correlation is weak. In other words, the use of a plug-in estimator of the kriging variance is fine for most spatial problems with moderate to strong spatial autocorrelation.

## 5.6 Nonlinear Prediction

The development of the simple and ordinary kriging predictors requires no distributional assumptions other than those pertaining to the first two moments of the random field. Thus, simple kriging is always the best linear predictor and ordinary kriging is always the best linear unbiased predictor, regardless of the underlying distribution of the data. The best predictor, i.e., the one that minimizes the mean-squared prediction error, is given in (5.1), the conditional expectation of $Z(\mathbf{s}_0)$ given the observed data. When the data follow a multivariate Gaussian distribution, this expectation is linear in the data and is equivalent to the simple kriging predictor (5.10). For other distributions, this conditional expectation may not be linear and so linear predictors may be poor approximations to this optimal conditional expectation. Statisticians often cope with such problems by transforming the data, so that the transformed data follow a Gaussian distribution and then performing analyses with the transformed data. In this section, we discuss several approaches to constructing nonlinear predictors based on transformations of the data.

### *5.6.1 Lognormal Kriging*

Suppose the logarithm of the random function $Z(\mathbf{s})$ is a Gaussian random field so that $Y(\mathbf{s}) = \log\{Z(\mathbf{s})\}$ follows a multivariate Gaussian distribution. For the development here, assume that $Y(\cdot)$ is intrinsically stationary with mean $\mu_Y$ and semivariogram $\gamma_Y(\mathbf{h})$. Simple kriging of $Y(\mathbf{s}_0)$ using data $Y(\mathbf{s}_1), \cdots, Y(\mathbf{s}_n)$ gives $p_{sk}(\mathbf{Y}; \mathbf{s}_0)$ from (5.10) and $\sigma^2_{sk}(\mathbf{Y}; \mathbf{s}_0)$ from (5.11). This suggests using $p(\mathbf{Z}; \mathbf{s}_0) = \exp\{p_{sk}(\mathbf{Y}; \mathbf{s}_0)\}$ as a predictor of $Z(\mathbf{s}_0)$. Unfortunately, this predictor is biased for $Z(\mathbf{s}_0)$. However, David (1988, pp. 117–118) and Cressie (1993, pp. 135–136) show how the properties of the lognormal distribution can be used to construct an unbiased predictor. First we draw on the results in Aitchison and Brown (1957). If

$$\mathbf{Y} = [Y_1, Y_2]' \sim G(\boldsymbol{\mu}, \Sigma)$$
$$\boldsymbol{\mu} = [\mu_1, \mu_2]'; \qquad \Sigma = \sigma_{ij},\ i, j = 1, 2$$

then $[\exp\{Y_1\}, \exp\{Y_2\}]'$ has mean $\nu$ and covariance matrix $T$, where

$$\nu = (\nu_1, \nu_2)' = [\exp\{\mu_1 + \sigma_{11}/2\}, \exp\{\mu_2 + \sigma_{22}/2\}]'$$

and

$$T = \begin{bmatrix} \nu_1^2(\exp\{\sigma_{11}\} - 1) & \nu_1\nu_2(\exp\{\sigma_{12}\} - 1) \\ \nu_1\nu_2(\exp\{\sigma_{21}\} - 1) & \nu_2^2(\exp\{\sigma_{22}\} - 1) \end{bmatrix}.$$

To appreciate this result and its implications for prediction/estimation consider the following, simple case.

**Example 5.10** Let $Y_1, \cdots, Y_n$ be independent and identically distributed Gaussian variables with unknown mean $\mu$ and known variance $\sigma^2$. We are interested in estimating

$$Q = \mathrm{E}[\exp\{Y\}] = \exp\{\mu + \sigma^2/2\}.$$

We first consider how to estimate $\mathrm{E}[Y_i]$, and then how to involve this estimator to derive an unbiased estimator of $Q$. The natural estimator for $\mu$ is the arithmetic average $\overline{Y} = \sum_{i=1}^n Y_i$. So, consider using $\widehat{Q}_1 = \exp\{\overline{Y}\}$ as your estimator of $Q$. Because the $Y_i$ are Gaussian, so is $\overline{Y}, \overline{Y} \sim G(\mu, \sigma^2/n)$. Applying the result of Aitchison and Brown yields

$$\mathrm{E}[\exp\{\overline{Y}\}] = \exp\{\mu + \mathrm{Var}[\overline{Y}]/2\} = \exp\{\mu + \sigma^2/(2n)\}.$$

$\widehat{Q}_1$ is biased for $Q$, but we can correct it. Some simple manipulations lead to

$$\begin{aligned} \mathrm{E}[\exp\{\overline{Y}\}] &= \exp\{\mu + \sigma^2/(2n)\} \\ &= \exp\{\mu + \sigma^2/2 - \sigma^2/2 + \sigma^2/(2n)\} \\ &= Q\exp\{-\sigma^2/2 + \sigma^2/(2n)\}. \end{aligned}$$

Consequently,

$$\exp\{\overline{Y}\}\exp\{\sigma^2/2 - \sigma^2/(2n)\} = \exp\{\overline{Y} + 0.5(\mathrm{Var}[Y] - \mathrm{Var}[\overline{Y}])\}$$

is an unbiased estimator of $Q$. ☐

Returning to the problem of finding a predictor $p(\mathbf{Z}; \mathbf{s}_0)$ based on data $\mathbf{Y}(\mathbf{s})$, we ask that the predictor be unbiased in the sense that previous predictors have been unbiased; its mean should equal $\mathrm{E}[Z(\mathbf{s})] = \mu_Z$. Applying the result of Aitchison and Brown (1957) twice, as in the example, first to $p(\mathbf{Z}; \mathbf{s}_0) = \exp\{p_{sk}(\mathbf{Y}; \mathbf{s}_0)\}$, and then inversely to $\mu_Y$ gives

$$\mathrm{E}[p(\mathbf{Z}; \mathbf{s}_0)] = \mathrm{E}[\exp\{p_{sk}(\mathbf{Y}; \mathbf{s}_0)\}] = \mu_Z \exp\left\{-\sigma_Y^2/2 + \mathrm{Var}[p_{sk}(\mathbf{Y}; \mathbf{s}_0)]/2\right\},$$

where $\sigma_Y^2 = \mathrm{Var}[Y(\mathbf{s}_i)]$. The bias-corrected predictor of $Z(\mathbf{s}_0)$ is then

$$\begin{aligned} p_{slk}(\mathbf{Z}; \mathbf{s}_0) &= \exp\left\{p_{sk}(\mathbf{Y}; \mathbf{s}_0) + \sigma_Y^2/2 - \mathrm{Var}[p_{sk}(\mathbf{Y}; \mathbf{s}_0)]/2\right\} \\ &= \exp\left\{p_{sk}(\mathbf{Y}; \mathbf{s}_0) + \sigma_{sk}^2(\mathbf{Y}; \mathbf{s}_0)/2\right\}. \end{aligned} \qquad (5.54)$$

Since $Y(\mathbf{s})$ is a Gaussian random field, $p_{sk}(\mathbf{Y};\mathbf{s}_0) = \mathrm{E}[Y(\mathbf{s}_0)|\mathbf{Y}]$ and so

$$p_{slk}(\mathbf{Z};\mathbf{s}_0) = \mathrm{E}[\exp\{Y(\mathbf{s}_0)\}|\mathbf{Y}] = \mathrm{E}[Z(\mathbf{s}_0)|\mathbf{Z}].$$

Thus, $p_{slk}(\mathbf{Z};\mathbf{s}_0)$ is the optimal predictor of $Z(\mathbf{s}_0)$. The corresponding conditional variance, which is also the minimized mean-squared prediction error (MSPE), is (Chilès and Delfiner, 1999, p. 191)

$$\mathrm{Var}[(p_{slk}(\mathbf{Z};\mathbf{s}_0) - Z(\mathbf{s}_0))|\mathbf{Z}] = (p_{slk}(\mathbf{Z};\mathbf{s}_0))^2[\exp\{\sigma^2_{sk}(\mathbf{Y};\mathbf{s}_0)\} - 1].$$

When ordinary kriging is used to predict $Y(\mathbf{s}_0)$, the properties of the lognormal distribution can be used as before (see Cressie, 1993, pp. 135–136) and in this case the bias-corrected predictor of $Z(\mathbf{s}_0)$ is

*Ordinary Lognormal Kriging Predictor*

$$\begin{aligned} p_{olk}(\mathbf{Z};\mathbf{s}_0) &= \exp\left\{p_{ok}(\mathbf{Y};\mathbf{s}_0) + \sigma^2_Y/2 - \mathrm{Var}[p_{ok}(\mathbf{Y};\mathbf{s}_0)]/2\right\} \\ &= \exp\left\{p_{ok}(\mathbf{Y};\mathbf{s}_0) + \sigma^2_{ok}(\mathbf{Y};\mathbf{s}_0)/2 - m_Y\right\}, \end{aligned} \tag{5.55}$$

where $m_Y$ is the Lagrange multiplier obtained with ordinary kriging of $Y(\mathbf{s}_0)$ based on data $\mathbf{Y}$. The bias-corrected MSPE (see e.g., Journel, 1980; David, 1988, p. 118) is

$$\begin{aligned} E\left[(p_{olk}(\mathbf{Z};\mathbf{s}_0) - Z(\mathbf{s}_0))^2\right] &= \exp\{2\mu_Y + \sigma^2_Y\}\exp\{\sigma^2_Y\} \\ &\quad \times \left\{1 + \left(\exp\{-\sigma^2_{ok}(\mathbf{Y};\mathbf{s}_0) + m_Y\}\right)\right. \\ &\quad \left. \times \ (\exp\{m_Y\} - 2)\right\}. \end{aligned}$$

Thus, unlike ordinary kriging, we need to estimate $\mu_Y$ and $\sigma^2_Y(\cdot)$ as well as $\gamma_Y(\cdot)$ in order to use lognormal kriging. Moreover, the optimality properties of $p_{olk}(\mathbf{Z};\mathbf{s}_0)$ are at best unclear. Finding a predictor that minimizes $\mathrm{Var}[p(\mathbf{Y};\mathbf{s}_0) - Y(\mathbf{s}_0)]$ within the class of linear, unbiased predictors of $Y(\mathbf{s}_0)$ (which in this case is $p_{ok}(\mathbf{Y};\mathbf{s}_0)$), does not imply that $p_{olk}(\mathbf{Z};\mathbf{s}_0)$ minimizes $\mathrm{Var}[(p(\mathbf{Z};\mathbf{s}_0) - Z(\mathbf{s}_0))|\mathbf{Z}]$ within the class of linear, unbiased predictors of $Z(\mathbf{s}_0)$.

The bias correction makes the ordinary lognormal kriging predictor sensitive to departures from the lognormality assumption and to fluctuations in the semivariogram (a criticism that applies to many nonlinear prediction methods and not just to lognormal kriging). Thus, some authors (e.g., Journel, 1980) have recommended calibration of $p_{olk}(\mathbf{Z};\mathbf{s}_0)$, forcing the mean of kriged predictions to equal the mean of the original $Z$ data. This may be a useful technique, but it is difficult to determine the properties of the resulting predictor. Others (e.g., Chilès and Delfiner, 1999, p. 191) seem to regard *mean* unbiasedness as unnecessary, noting that $\exp\{p_{ok}(\mathbf{Y};\mathbf{s}_0)\}$ is *median* unbiased (i.e., $\Pr(\exp\{p_{ok}(\mathbf{Y};\mathbf{s}_0)\} > Z(\mathbf{s}_0)) = \Pr(\exp\{p_{ok}(\mathbf{Y};\mathbf{s}_0)\} < Z(\mathbf{s}_0)) = 0.5$). Marcotte and Groleau (1997) propose an interesting approach that works around these problems. Instead of transforming the data, predicting $Y(\mathbf{s}_0)$, and then transforming back, they suggest predicting $Z(\mathbf{s}_0)$ using the original data $\mathbf{Z}(\mathbf{s})$ to obtain $p_{ok}(\mathbf{Z};\mathbf{s}_0)$, and then transforming via $p(\mathbf{Y};\mathbf{s}_0) = \log\{p_{ok}(\mathbf{Z};\mathbf{s}_0)\}$. Marcotte and Groleau (1997) suggest using $\mathrm{E}[Z(\mathbf{s}_0)|p(\mathbf{Y};\mathbf{s}_0)]$ as a predictor of $Z(\mathbf{s}_0)$. Using the properties of the lognormal distribution from Aitchison and Brown (1957) given above, Marcotte and Groleau (1997) derive a computational expression for this conditional expectation that depends on $\mu_Z$ and $\gamma_Z(\mathbf{h})$, and is relatively robust to departures from the lognormality assumption and to mis-specification of the semivariogram.

Although the theory of lognormal kriging has been developed and revisited by many authors including Rendu (1979), Journel (1980), Dowd (1982), and David (1988), problems with its practical implementation persist. David (1988) gives several examples that provide some advice on how to detect and correct problems with lognormal kriging and more modifications are provided in Chilès and Delfiner (1999). Nonlinear spatial prediction is an area of active research in geostatistics, and the last paragraph in Boufassa and Armstrong (1989) seems to summarize the problems and the frustration: "The user of geostatistics therefore is faced with the difficult task of choosing the most appropriate stationary model for their data. This choice is difficult to make given only information from a single realization. It would be helpful if statisticians could devise a way of testing this."

### *5.6.2 Trans-Gaussian Kriging*

The lognormal distribution is nice, mathematically speaking, since its moments can be written in terms of the moments of an underlying Gaussian distribution. In many other applications, the transformation required to achieve normality may not be the natural logarithm, and in such instances, it is difficult to obtain exact expressions relating the moments of the original data to those of the transformed variable. Trans-Gaussian kriging, suggested by Cressie (1993), is a more general approach to developing optimal predictors for non-Gaussian spatial data.

Assume $Z(\mathbf{s}) = \varphi(Y(\mathbf{s}))$, where $Y(\mathbf{s})$ follows a multivariate Gaussian distribution, and the function $\varphi$ is known. Again we assume that $Y(\cdot)$ is intrinsically stationary with mean $\mu_Y$ and semivariogram $\gamma_Y(\mathbf{h})$. We assume that $\mu_Y$ is unknown and use $p_{ok}(\mathbf{Y}; \mathbf{s}_0)$ in (5.21) as the predictor of $Y(\mathbf{s}_0)$, although analogous derivations can be done using simple kriging. In this context, a natural predictor of $Z(\mathbf{s}_0)$ is $p(\mathbf{Z}; \mathbf{s}_0) = \varphi(p_{ok}(\mathbf{Y}; \mathbf{s}_0))$, but we need to determine its expected value in order to correct for any bias, and then derive the variance of the resulting bias-corrected predictor. Cressie (1993, pp. 137–138) uses the delta method to derive these properties. In what follows, we provide the details underlying this development.

Expand $\varphi(p_{ok}(\mathbf{Y}; \mathbf{s}_0)) \equiv \varphi(\widehat{Y}_0)$ around $\mu_Y$ in a second-order Taylor series:

$$\varphi(p_{ok}(\mathbf{Y}; \mathbf{s}_0)) \approx \varphi(\mu_Y) + \varphi'(\mu_Y)(\widehat{Y}_0 - \mu_Y) + \frac{\varphi''(\mu_Y)}{2}(\widehat{Y}_0 - \mu_Y)^2.$$

Taking expectations we find

$$\mathrm{E}[p(\mathbf{Z}; \mathbf{s}_0)] = \mathrm{E}[\varphi(p_{ok}(\mathbf{Y}; \mathbf{s}_0))] \approx \varphi(\mu_Y) + \frac{\varphi''(\mu_Y)}{2}\mathrm{E}[(\widehat{Y}_0 - \mu_Y)^2]. \quad (5.56)$$

For unbiasedness, this should be equal to $\mathrm{E}[Z(\mathbf{s}_0)]$, which can be obtained by applying the same type of expansion to $\varphi(Y(\mathbf{s}_0))$ giving

$$\mathrm{E}[Z(\mathbf{s}_0)] = \mathrm{E}[\varphi(Y(\mathbf{s}_0))] \approx \varphi(\mu_Y) + \frac{\varphi''(\mu_Y)}{2}\mathrm{E}[(Y(\mathbf{s}_0) - \mu_Y)^2]. \qquad (5.57)$$

To make (5.56) equal to (5.57), we need to add

$$\frac{\varphi''(\mu_Y)}{2}\mathrm{E}[(Y(\mathbf{s}_0) - \mu_Y)^2] - \frac{\varphi''(\mu_Y)}{2}\mathrm{E}[(\widehat{Y}_0 - \mu_Y)^2]$$
$$= \frac{\varphi''(\mu_Y)}{2}(\sigma^2_{ok}(\mathbf{Y};\mathbf{s}_0) - 2m_Y)$$

to $p(\mathbf{Z};\mathbf{s}_0)$. Thus, the trans-Gaussian predictor of $Z(\mathbf{s}_0)$ is

*Trans-Gaussian Predictor*

$$p_{tg}(\mathbf{Z};\mathbf{s}_0) = \varphi(p_{ok}(\mathbf{Y};\mathbf{s}_0)) + \frac{\varphi''(\mu_Y)}{2}(\sigma^2_{ok}(\mathbf{Y};\mathbf{s}_0) - 2m_Y). \qquad (5.58)$$

The mean-squared prediction error of $p_{tg}(\mathbf{Z};\mathbf{s}_0)$, based on just a first-order Taylor series expansion, is

$$\mathrm{E}[(p_{tg}(\mathbf{Z};\mathbf{s}_0) - Z(\mathbf{s}_0))^2] \approx \left[\varphi'(\mu_Y)\right]^2 \sigma^2_{ok}(\mathbf{Y};\mathbf{s}_0). \qquad (5.59)$$

Statisticians are well aware of many different families of transformations that can be used to transform a given set of data to a Gaussian distribution. Common examples include variance stabilizing transformations such as the square-root for the Poisson distribution and the arcsine for the Binomial distribution (see e.g., Draper and Smith, 1981, pp. 237–238), the Box-Cox family of power transformations (Box and Cox, 1964), the Freeman-Tukey transformation (Freeman and Tukey, 1950), and monotone mappings such as the logit or probit for Binomial proportions and the log-transform for counts. There is also an empirical way to infer $\varphi$ that is commonly-used in geostatistics. This transformation, called the **anamorphosis** function (Rivoirard, 1994, pp. 46–48), is obtained by matching the percentiles of the data to those of a standard Gaussian distribution. If we consider obtaining $\varphi^{-1}$ instead of $\varphi$ (arguably, the more natural procedure), this transformation is called the normal scores transformation (Deutsch and Journel, 1992, p. 138).

Let $\Phi(y)$ be the standard Gaussian cumulative distribution function and suppose $F(z)$ is the cumulative distribution function of the data. Then, the $p$-quantile of $F(z)$ is that value $z_p$ such that $F(z_p) = p$. Given $p$, the corresponding p-quantile from the standard Gaussian distribution is $y_p$, defined as $\Phi(y_p) = p$. Thus, equating these for each $p$ gives

$$\begin{aligned} z_p &= F^{-1}(\Phi(y_p)) \equiv \varphi(y_p) \quad \text{anamorphosis} \\ y_p &= \Phi^{-1}(F(z_p)) \equiv \varphi^{-1}(z_p) \quad \text{normal scores.} \end{aligned} \qquad (5.60)$$

**Example 5.11 Trans-Gaussian kriging of rainfall amounts.** In this example, we investigate the effects of departures from a Gaussian assumption and the effects of plug-in estimators of the semivariogram model parameters on the quality of predictions obtained from kriging. We begin by following the example given in De Oliveira, Kedem, and Short (1997) concerning rainfall measurements near Darwin, Australia. The data are rainfall amounts measured in millimeters (mm) accumulated over a 7-day period from the $76^{th}$–$82^{nd}$ day of 1991 at 24 rainfall collection stations (Figure 5.12). A histogram

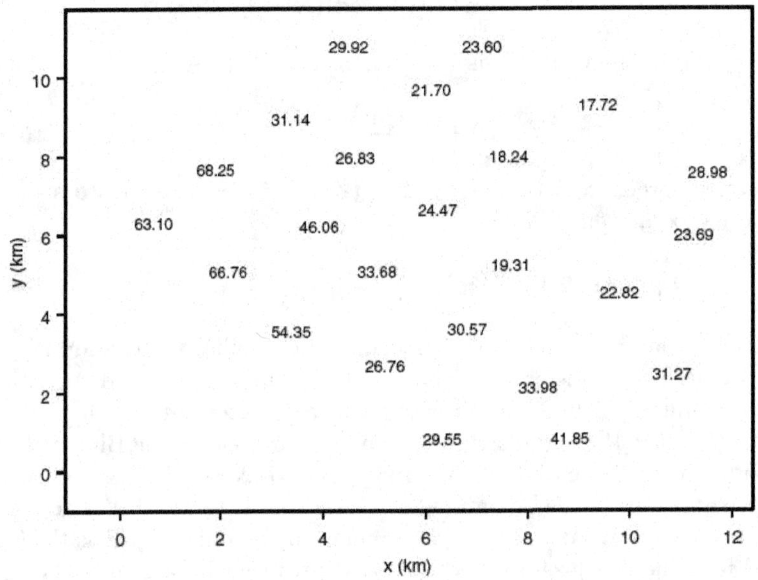


Figure 5.12 *Locations of the 24 rainfall monitoring stations and associated weekly rainfall amounts (mm). Data kindly provided by Dr. Victor De Oliveira, Department of Mathematical Sciences, University of Arkansas.*

of the weekly rainfall amounts (Figure 5.13) suggests that linear methods of spatial prediction may not be the best choice for interpolating the rainfall amounts. De Oliveira et al. (1997) consider a Box-Cox transformation

$$g_\lambda(z) = \begin{cases} \dfrac{z^\lambda - 1}{\lambda} & \text{if } \lambda \neq 0 \\ \log(z) & \text{if } \lambda = 0, \end{cases}$$

and estimate $\lambda$ using maximum likelihood. The distribution of the Box-Cox transformed data using $\hat{\lambda} = -0.486$ is shown in Figure 5.14. This transformation appears to have over-compensated for the skewness, and so we also consider the normal scores transformation given in (5.60). A histogram of the normal scores is shown in Figure 5.15. To use Trans-Gaussian kriging de-

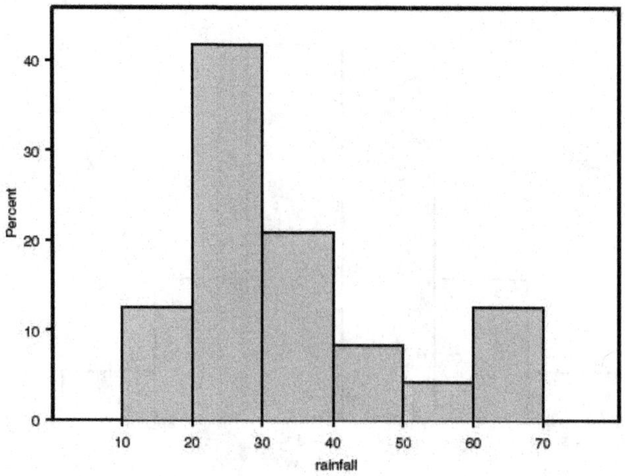


Figure 5.13 *Histogram of weekly rainfall amounts (mm).*

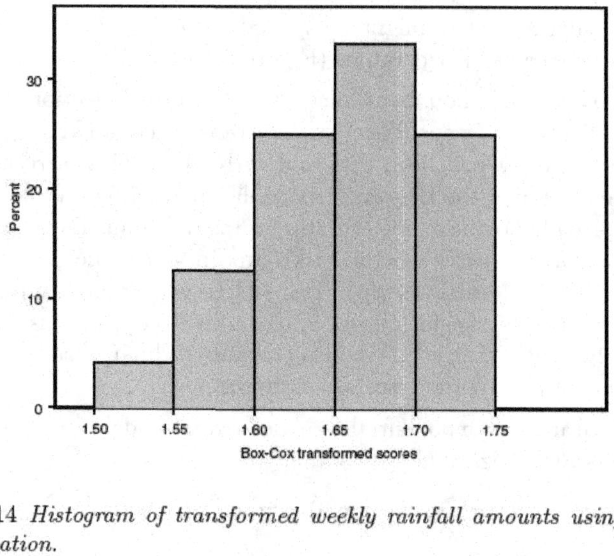


Figure 5.14 *Histogram of transformed weekly rainfall amounts using the Box-Cox transformation.*

scribed in §5.6.2, we need to compute the empirical semivariogram of the data on the transformed scale. The empirical semivariogram of the normal scores is shown in Figure 5.16. The empirical semivariograms of the Box-Cox transformed values and the original rainfall values (not shown here) have similar shapes, but different sills reflecting their differing scales. While it is difficult to

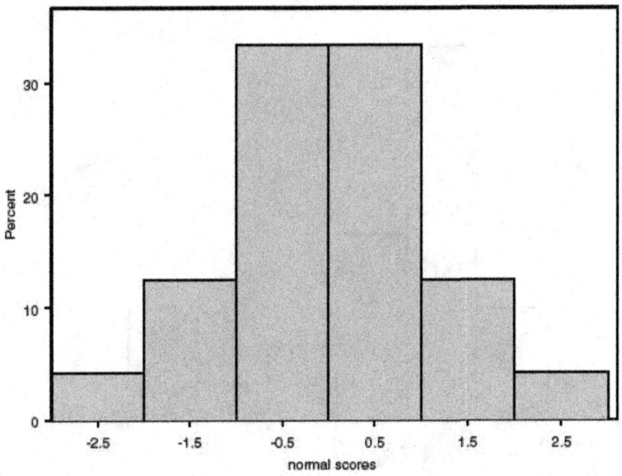


Figure 5.15 *Histogram of transformed weekly rainfall amounts using the normal scores transformation.*

adequately estimate a semivariogram with only 24 data values, the empirical semivariogram suggests there might be spatial autocorrelation in the rainfall amounts and their transformed values (Figure 5.16).

Predictions of rainfall amounts at locations without a collection station can be obtained in a variety of ways. We consider Trans-Gaussian kriging using the Box-Cox transformed values, both with and without the bias correction factor in (5.58), since obtaining the Lagrange multiplier needed for this correction is not routinely provided by most software programs that implement kriging. We also consider a normal scores version of kriging in which the data are transformed and back-transformed as with Trans-Gaussian kriging, but given the nature of $\varphi^{-1}(z)$, back-transformation is more involved and bias corrections cannot be easily made (if at all). We also consider ordinary kriging since this is the easiest spatial prediction method to implement.

We chose a parametric exponential semivariogram model (without a nugget effect) in the parameterization

$$C(\sigma^2, \rho, ||\mathbf{s}_i - \mathbf{s}_j||) = \sigma^2 \rho^{||\mathbf{s}_i - \mathbf{s}_j||}$$

to describe the shape of the spatial autocorrelation depicted in Figure 5.16. Since using plug-in estimates of the autocorrelation parameters may introduce bias in the kriging standard errors, we adjust the predictions using the Prasad-Rao MSE estimator in (5.53). To provide some insight into how the different methods perform, we considered a cross validation study as in De Oliveira et al. (1997). For each data value (either a transformed value or a given rainfall amount), we delete this value and use the remaining 23 values to predict it and

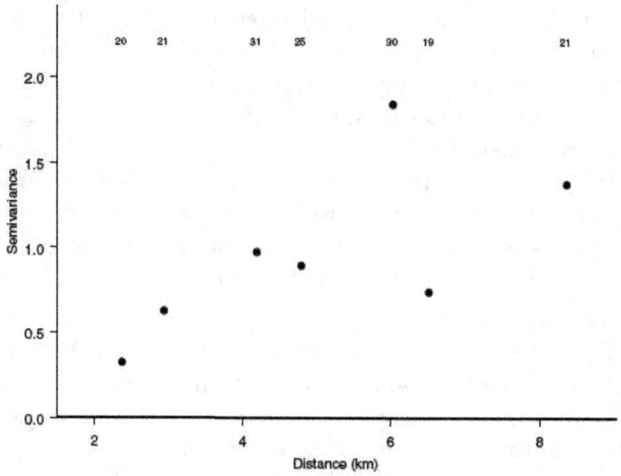


Figure 5.16 *Empirical semivariogram of normal score transformed values. Values across top represent number of pairs in lag class.*

construct a prediction interval (PI) for the omitted, true value. Thus, we can compare each predicted value to the true value omitted from the analysis and obtain measures of bias and the percentage of prediction intervals containing the true values. The covariance parameters and the overall mean were re-estimated for each of the 24 cross validation data sets.

With ordinary kriging, we describe the spatial autocorrelation in the original rainfall amounts using both an exponential semivariogram in the above parameterization and a spherical semivariogram model to examine the impact of the semivariogram model.

Predictions for Trans-Gaussian kriging based on the Box-Cox transformed scores were obtained using (5.58) and the standard errors were obtained using (5.59). We also obtained a biased Trans-Gaussian predictor obtained by ignoring the second term in (5.58) that depends on the Lagrange multiplier, $m_Y$. Note that $\varphi(\cdot)$ in §5.6.2 pertains to the original rainfall amounts, $Z(\mathbf{s}_i)$. The transformed amounts are $Y(\mathbf{s}) = \varphi^{-1}(Z(\mathbf{s}))$ so that $g_\lambda(z) = \varphi^{-1}(z)$.

The normal scores transformation matches the cumulative probabilities that define the distribution function of $Z(\mathbf{s})$ to those of a standard normal distribution. Thus, the transformed scores are just the corresponding percentiles of the associated standard normal distribution. However, transforming back after prediction *at a new location* is tricky, since the predicted values will not coincide with actual transformed scores for which there is a direct link back to the original rainfall amounts. Thus, for predictions that lie within two consecutively-ranked values, predictions on the original scale are obtained through interpolation. Rather than back-transforming the standard errors and creating prediction intervals on the original scale as with Trans-Gaussian kriging, we back-transformed the prediction intervals constructed from kriging with normal scores. When back-transforming prediction intervals based on normal scores to obtain prediction intervals on the original scale, the prediction interval bounds may be less than the minimum and exceed the maximum of the normal score transformed values. In such cases, we used linear extrapolation in the tails. Deutsch and Journel (1992, pp. 131–135) provide a detailed discussion of this issue and suggest more sophisticated approaches. We deliberately chose a conservative approach that does not assume long tails at either end of the distribution of the original rainfall amounts.

Restricted maximum likelihood was used to estimate the spatial autocorrelation parameters. The mean, $\mu_Y$, needed for Trans-Gaussian kriging was estimated using generalized least squares. Prediction intervals (PIs) were based on the $(1-\alpha/2)$ percentage point from the standard normal distribution (1.96 for a 95% PI). Since there are only 24 measurements available for estimation and prediction, we also used percentage points from a t-distribution on 22 degrees of freedom to construct the prediction intervals. In addition, since the use of plug-in estimates for the variance and the spatial autocorrelation parameters can affect these intervals, we also combined the Prasad-Rao MSE estimator with adjusted degrees of freedom according to Kenward and Roger (1997). The Kenward-Roger adjustment is described in §6.2.3.1.

To assess the performance of each prediction method, we computed the following:

- The average relative bias:

$$\text{avg}\left\{\frac{\widehat{Z}(\mathbf{s}_0) - Z(\mathbf{s}_0)}{Z(\mathbf{s}_0)}\right\} \times 100\%.$$

  This should be close to zero for methods that are unbiased.

- The average relative absolute error

$$\text{avg}\left\{\frac{|\widehat{Z}(\mathbf{s}_0) - Z(\mathbf{s}_0)|}{Z(\mathbf{s}_0)}\right\} \times 100\%.$$

  This should be relatively small for methods that are accurate (precise as well as unbiased).

- The average percentage increase in prediction standard errors due to the Prasad-Rao adjustment;

- The percentage of 95% prediction intervals containing the true value. This should be close to 95%.

The results of the cross validation study are shown in Table 5.4.

From Table 5.4 we see that all empirical PI coverages were surprisingly close to nominal. Also, surprisingly, the biased-corrected version of the Trans-

Table 5.4 *Results from cross-validation of predictions of rainfall amounts.*

| Method | Average Relative Bias | Average Absolute Error | % Increase in Std. Error from PR-adjustment | Gaussian 95% PI |
|---|---|---|---|---|
| OK (EXP) | 3.53 | 15.57 | 0.04 | 95.8 |
| OK (SPH) | 3.42 | 15.12 | 1.75 | 95.8 |
| TGK (no b.c.) | 0.57 | 15.27 | 0.55 | 95.8 |
| TGK (b.c.) | 5.13 | 16.73 | 0.55 | 95.8 |
| NS | -0.94 | 13.36 | N/A | 91.7 |

Gaussian kriging predictor has the largest bias and the largest absolute relative error. Predictions from the bias-corrected version of Trans-Gaussian kriging are all higher than those obtained without this bias correction. Cressie (1993, p. 137) notes that the approximations underlying the derivation of the biased-corrected Trans-Gaussian predictor rely on the kriging variance of the transformed variables being small. On the average, for the cross-validation study presented here, this variance was approximately 0.035, so perhaps this is not "small" enough. Also, the Box-Cox transformation tends to over-correct the skewness in the distribution of the rainfall amounts and this may be impacting the accuracy of the predictions from Trans-Gaussian kriging (Figure 5.14). The predictions from the normal scores transformation were relatively (and surprisingly) accurate, considering no effort was made to adjust for bias and the amount of information lost during back-transformation. The Prasad-Rao/Kenward-Roger adjustment did change the standard errors slightly, but not enough to result in different inferences. However, in this example, the spatial autocorrelation is very strong ($\widehat{\rho}$ is between 0.92 and 0.98). Following the recommendation of Zimmerman and Cressie (1992), on page 267, the adjustment may not be needed in this case.

The empirical probability of coverage from the normal scores was below nominal, but we used a fairly conservative method for back transformation in the tails of the distribution that may be affecting these results. Using a percentage point from a t-distribution to construct these intervals had little impact on their probability of coverage. Ordinary kriging seems to do as well as the other techniques and certainly requires far less computational (and cerebral) effort. However, the distribution of the rainfall amounts (Figure 5.13) is not tremendously skewed or long-tailed, so the relative accuracy of ordinary kriging in this example may be somewhat misleading. On the other hand, perhaps the effort spent in correcting departures from assumptions should be directly proportional to their relative magnitude and our greatest efforts should be spent on situations that show gross departures from assumptions. □

### 5.6.3 Indicator Kriging

The spatial prediction techniques described in previous sections are geared towards finding good approximations to $E[Z(s_0)|\mathbf{Z}(s)]$. Alternatively, if we can estimate $\Pr(Z(s_0) \leq z|Z(s_1), \ldots, Z(s_n)) \equiv G(s_0, z\ |\mathbf{Z}(s))$, the conditional probability distribution at each location, we can also obtain $E[Z(s_0)|\mathbf{Z}(s)]$ and $E[g(Z(s_0))|\mathbf{Z}(s)]$. Switzer (1977) proposed the idea of using indicator functions of the data to estimate a stationary, univariate distribution function, $F(z) = \Pr(Z(s) \leq z)$. This idea was then extended by Journel (1983) for nonparametric estimation and mapping of $\Pr(Z(s_0) \leq z|Z(s_1), \cdots, Z(s_n))$ and is now known as **indicator kriging**.

*Indicator Transform*

Assume that the process $\{Z(s) : s \in D \subset \mathbb{R}^d\}$ is strictly stationary and consider an indicator transform of $Z(s)$

$$I(s, z) = \begin{cases} 1 & \text{if } Z(s) \leq z \\ 0 & \text{otherwise.} \end{cases} \tag{5.61}$$

This function transforms $Z(s)$ into a binary process whose values are determined by the threshold $z$. Since $E[I(s_0, z)] = \Pr(Z(s_0) \leq z) = F(z)$ is unknown, we can use ordinary kriging to predict $I(s_0, z)$ from the indicator data $\mathbf{I}(s, z) = [I(s_1, z), \cdots, I(s_n, z)]'$. This gives

$$p_{ik}(\mathbf{I}(s, z); I(s_0, z)) = \boldsymbol{\lambda}(z)'\mathbf{I}(s, z), \tag{5.62}$$

where the weight vector $\boldsymbol{\lambda}(z)$ is obtained from (5.19) using the indicator semivariogram, $\gamma_I(\mathbf{h}) = \frac{1}{2}\mathrm{Var}[I(s+\mathbf{h}, z) - I(s, z)]$. Since ordinary kriging of $Z(s_0)$ provides an approximation to $E[Z(s_0)|Z(s_1), \cdots, Z(s_n)]$, indicator kriging provides an approximation to

$$E[I(s_0, z)|I(s_1, z) \cdots, I(s_n, z)] = \Pr(Z(s_0) \leq z|I(s_1, z) \cdots, I(s_n, z)).$$

Note, however, that $\Pr(Z(s_0) \leq z|I(s_1, z) \cdots, I(s_n, z))$ is not the same as $\Pr(Z(s_0) \leq z|Z(s_1), \cdots, Z(s_n))$; much information is lost by making the indicator transform. Thus, in theory, the indicator kriging predictor in (5.62) is a crude estimator of the conditional probability of interest. However, indicator semivariograms contain more information than may be apparent at first glance. First, they contain information about the bivariate distributions since $\mathrm{Var}[I(s + \mathbf{h}, z) - I(s, z)] = \Pr(Z(s + \mathbf{h}) \leq z, Z(s) \leq z)$. Second, the indicator functions do contain some information from one threshold to the next since they are defined cumulatively. Finally, Goovaerts (1994, 1997) has shown that, in practice, more complex indicator methods that more directly account for information available across all thresholds (e.g., indicator cokriging, see Goovaerts, 1997, pp. 297–299) offer little improvement over the more simple indicator kriging approach.

In many applications, such as environmental remediation and risk analysis, we are interested in **exceedance** probabilities, e.g., $\Pr(Z(s_0) > z|\mathbf{Z}(s))$. This probability can be estimated using indicator kriging based on the compliment of (5.61), $I^c(s, z) = 1 - I(s, z)$. In other applications, the data may already be binary (e.g., presence/absence records). In applications such as these, mapping the estimates of $\Pr(Z(\mathbf{s}_0) > z|\mathbf{Z}(\mathbf{s}))$ is often the ultimate inferential goal. However, indicator kriging is also used to provide nonparametric predictions of any functional $g(Z(\mathbf{s}_0))$ by using $K$ different indicator variables defined at thresholds $z_k$, $k = 1, \cdots, K$. This produces an estimate of the entire conditional distribution at each location $\mathbf{s}_0$. Thus, for each threshold $z_k$, indicator kriging based on the corresponding indicator data $I(\mathbf{s}_1, z_k), \cdots, I(\mathbf{s}_n, z_k)$ gives an approximation to $\Pr(Z(\mathbf{s}_0) \leq z_k|Z(\mathbf{s}_1), \cdots, Z(\mathbf{s}_n))$. Given this approximate conditional distribution, denoted here as $\hat{F}(\mathbf{s}_0, z\ |\mathbf{Z}(\mathbf{s}))$, a predictor of $g(Z(\mathbf{s}_0))$ is

$$p(\mathbf{Z}, g(Z(\mathbf{s}_0))) = \int g(z)\, d\widehat{F}(\mathbf{s}_0, z\ |\mathbf{Z}(\mathbf{s})).$$

When $g(Z(\mathbf{s}_0)) = Z(\mathbf{s}_0)$, this is called the "E-type estimate" of $Z(\mathbf{s}_0)$ (Deutsch and Journel, 1992, p. 76). A measure of uncertainty is given by

$$\sigma^2(\mathbf{s}_0) = \int [p(\mathbf{Z}, g(Z(\mathbf{s}_0))) - g(z)]^2\, d\widehat{F}(\mathbf{s}_0, z\ |\mathbf{Z}(\mathbf{s})).$$

Computation of this nonparametric predictor of $g(Z(\mathbf{s}_0))$ requires that $K$ semivariograms be estimated and modeled. *Median indicator kriging* (Journel, 1983) alleviates this tedious chore by using a common semivariogram model based on the median threshold value, $\Pr(Z(\mathbf{s}) < z_M) = 0.5$. However, this is only valid if all the indicator semivariograms are proportional to one another (Matheron, 1982; Goovaerts, 1997, p. 304). A more troublesome issue is the fact that $\widehat{F}(\mathbf{s}_0, z\ |\mathbf{Z}(\mathbf{s}))$ need not satisfy the theoretical properties of a cumulative distribution function: it may be negative, exceed one, and is not necessarily monotonic. These "order-relation problems" are circumvented in practice by using a variety of "fix-ups" ranging from clever modeling strategies to brute-force alteration of any offending estimates (Deutsch and Journel, 1992, p. 77–81). While the cause of these problems is often blamed on negative indicator kriging weights or the lack of data between two thresholds, the basic problem is two-fold. First, the estimator is not constrained to satisfy these properties, and second, there is no guarantee that any joint probability distribution exists with the specified marginal and bivariate distributions (see §5.8).

### *5.6.4 Disjunctive Kriging*

Disjunctive kriging is a method for nonlinear spatial prediction proposed by Matheron (1976) to make more use of the information contained in indicator variables. The method is based on a technique known as *disjunctive coding.* Let $\{R_k\}$ be a partition of $\Re$, i.e., $R_i \cap R_j = \emptyset$ for $i \neq j$ and $\cup_k R_k = \Re$. Now define indicator variables

$$I_k(\mathbf{s}) = \begin{cases} 1 & \text{if } Z(\mathbf{s}) \in R_k \\ 0 & \text{otherwise.} \end{cases}$$

If the partition is sufficiently fine, i.e., there are many subsets $R_k$, then any function $g(Z(\mathbf{s}_0))$ can be approximated by a linear combination of these indicator functions

$$g(Z(\mathbf{s}_0)) = g_1 I_1(\mathbf{s}_0) + g_2 I_2(\mathbf{s}_0) + g_3 I_3(\mathbf{s}_0) + \cdots.$$

In the situation we describe here, each $I_k(\mathbf{s}_0)$ is unknown, but we can obtain a predictor of any $I_k(\mathbf{s}_0)$ using indicator kriging of the data associated with the $k^{th}$ set $\{I_k(\mathbf{s}_i), i = 1, \cdots, n\}$. However, as discussed above, this does not make optimal use of all the indicator information. Another approach is to obtain a predictor of $I_k(\mathbf{s}_0)$ using not only the data associated with the $k^{th}$ set, but also the data associated with all of the other sets, i.e., use data $\{I_1(\mathbf{s}_i), i = 1, \cdots, n\}, \cdots, \{I_k(\mathbf{s}_i), i = 1, \cdots, n\}, \cdots$. Thus, if we use a linear combination of all available indicator data to predict each $I_k(\mathbf{s}_0)$, the predictor can be written as

$$\widehat{I}_k(\mathbf{s}_0) = \sum_i \sum_k \lambda_{ik} I_k(\mathbf{s}_i)$$

and a predictor of $g(Z(\mathbf{s}_0))$ is then

$$p(\mathbf{Z}, g(Z(\mathbf{s}_0)) = \sum_i \sum_k g_{ki} I_k(\mathbf{s}_i) \equiv \sum_i g_i(Z(\mathbf{s}_i)). \qquad (5.63)$$

This is the general form of the *disjunctive kriging* predictor.

The key to disjunctive kriging is the determination of the functions $g_i(Z(\mathbf{s}_i))$. These can be obtained using all the indicator information as described above, but in any practical application, this requires estimation and modeling of covariances and cross-covariances of all the indicator variables. Thus, in practice, disjunctive kriging relies on models that expand a given function in terms of other functions that are uncorrelated. In geostatistics, this is called **factorizing** (Rivoirard, 1994, p. 11). Vector space theory, and Hilbert space theory in particular, gives us a very elegant way to obtain such functions through the use of orthogonal polynomials.

### *5.6.4.1 Orthonormal Polynomials for the Gaussian Distribution*

Consider a system of functions $\{\chi_p(x), p = 0, 1, \cdots\}$ that forms an orthonormal basis in the space $L^2_w$ associated with the nonnegative weight function $w(x)$. Then, any measurable function $g(x)$ for which $\int g(x)^2 w(x) dx < \infty$ can be expressed as a linear combination of these basis functions, i.e.,

$$g(x) = \sum_{p=0}^{\infty} a_p \chi_p(x).$$

The basis functions have the property

$$\int \chi_p(x) \chi_m(x) w(x) dx = \delta_{pm},$$

where $\delta_{pm} = 1$ if $p = m$ and is equal to 0 otherwise.

The weight function determines the system $\{\chi_p(x)\}$. For example, taking $w(x) = f(x) = \frac{1}{\sqrt{2\pi}} e^{-x^2/2}$ over the interval $(-\infty, \infty)$, the standard Gaussian density function, gives the system of *Chebyshev-Hermite polynomials* (Stuart and Ord, 1994, p. 226-228)

*Chebyshev-Hermite Poly-nomials*

$$H_p(x) = (-1)^p e^{x^2/2} \frac{d^p}{dx^p} e^{-x^2/2}. \tag{5.64}$$

They are called polynomials since $H_p(x)$ is in fact a polynomial of degree $p$ :

$$\begin{aligned} H_0(x) &= 1 \\ H_1(x) &= x \\ H_2(x) &= x^2 - 1 \\ \vdots & \quad \vdots \quad \vdots \\ H_{p+1}(x) &= xH_p(x) - pH_{p-1}(x). \end{aligned}$$

Chebyshev-Hermite polynomials are orthogonal, but not orthonormal

$$\frac{1}{\sqrt{2\pi}} \int_{-\infty}^{\infty} H_p(x) H_m(x) e^{-x^2/2} dx = \begin{cases} 0 & m \neq p \\ p! & m = p. \end{cases}$$

In terms of statistical concepts such as expectation and variance, the standardized Chebyshev-Hermite polynomials, $\eta_p(x) = H_p(x)/\sqrt{(p!)}$ satisfy

$$\mathrm{E}[\eta_p(x)] = \int \eta_p(x) f(x) dx = 0$$

$$\mathrm{Var}[\eta_p(x)] = \int (\eta_p(x))^2 f(x) dx = 1$$

$$\mathrm{E}[\eta_p(x)\eta_m(x)] = \begin{cases} 0 & m \neq p \\ 1 & m = p. \end{cases}$$

Thus, the polynomials $\eta_p(x)$ form an orthonormal basis on $L^2$ with respect to the standard Gaussian density and we can expand any measurable function $g(x)$ as

$$g(x) = \sum_{p=0}^{\infty} b_p \eta_p(x).$$

The coefficients $\{b_p\}$ can be obtained from the orthogonal properties of the polynomials. Since $\mathrm{E}[g(x)\eta_m(x)] = \sum_{p=0}^{\infty} b_p \mathrm{E}[\eta_p(x)\eta_m(x)] \equiv b_p$,

$$b_p = \int g(x) \eta_p(x) f(x) dx. \tag{5.65}$$

*5.6.4.2 Disjunctive Kriging with Gaussian Data*

Suppose that we now want to predict $g(Z(\mathbf{s}_0))$ using the predictor given in (5.63). Then, from the above development,

$$g(Z(\mathbf{s}_0)) = \sum_{p=0}^{\infty} b_{0p}\eta_p(Z(\mathbf{s}_0)), \tag{5.66}$$

and

$$g_i(Z(\mathbf{s}_i)) = \sum_{p=0}^{\infty} b_{ip}\eta_p(Z(\mathbf{s}_i)).$$

If we could predict $\eta_p(Z(\mathbf{s}_0))$ from the available data, then we would have a predictor of $g(Z(\mathbf{s}_0))$. Predicting $\eta_p(Z(\mathbf{s}_0))$ from $\eta_p(Z(\mathbf{s}_i))$ is now an easier task since $\eta_p(Z(\mathbf{s}))$ and $\eta_p(Z(\mathbf{s}))$ are uncorrelated. Thus, we can use ordinary kriging based on the data $\eta_p(Z(\mathbf{s}_1)), \cdots, \eta_p(Z(\mathbf{s}_n))$ to predict $\eta_p(Z(\mathbf{s}_0))$. To obtain the kriging equations, we need the covariance between $\eta_p(Z(\mathbf{s}+\mathbf{h}))$ and $\eta_p(Z(\mathbf{s}))$. Matheron (1976) showed that if $Z(\mathbf{s}+\mathbf{h})$ and $Z(\mathbf{s})$ are bivariate Gaussian with correlation function $\rho(\mathbf{h})$, then for $p \geq 1$ this covariance is

$$\mathrm{Cov}[\eta_p(Z(\mathbf{s}+\mathbf{h})), \eta_p(Z(\mathbf{s}))] = [\rho(\mathbf{h})]^p.$$

Thus, the optimal predictor of $\eta_p(Z(\mathbf{s}_0))$ is (Chilès and Delfiner, 1999, p. 393)

$$\widehat{\eta}_p(Z(\mathbf{s}_0)) = p(\{\eta_p(Z(\mathbf{s}_i)\}; \eta_p(Z(\mathbf{s}_0)) = \sum_{i=1}^{p} \lambda_{pi}\eta_p(Z(\mathbf{s}_i)),\ p = 1, 2, \cdots, \tag{5.67}$$

where the $\lambda_{pi}$ satisfy

$$\boldsymbol{\lambda}\mathbf{R} = \boldsymbol{\rho}. \tag{5.68}$$

Here, $\mathbf{R}$ is an $n \times n$ matrix with elements $[\rho_{ij}(\mathbf{h})]^p$ and $\boldsymbol{\rho}$ is an $n \times 1$ vector with elements $[\rho_{0i}(\mathbf{h})]^p$. The kriging variance associated with $\widehat{\eta}_p(Z(\mathbf{s}_0))$ is (Chilès and Delfiner, 1999, p. 394).

$$\sigma_\eta^2(\mathbf{s}_0) = 1 - \boldsymbol{\lambda}'\boldsymbol{\rho}. \tag{5.69}$$

The disjunctive kriging predictor of $g(Z(\mathbf{s}_0))$ is obtained by substituting for $\eta_p(Z(\mathbf{s}_0))$ in (5.66) $\widehat{\eta}_p(Z(\mathbf{s}_0))$ from (5.67) using weights from (5.68):

$$p_{dk}(\mathbf{Z}; g(Z(\mathbf{s}_0))) = \sum_{p=0}^{\infty} b_{0p}\widehat{\eta}_p(Z(\mathbf{s}_0)). \tag{5.70}$$

The kriging variance associated with the disjunctive kriging predictor is

$$\sigma_{dk}^2(\mathbf{Z}; g(Z(\mathbf{s}_0))) = \sum_{p=1}^{\infty} b_{0p}^2\sigma_\eta^2(\mathbf{s}_0). \tag{5.71}$$

Often, the coefficients $b_p$ will be zero in the Hermetian expansion (5.66). Also, the correlation function $[\rho(\mathbf{h})]^p$ tends to that of an uncorrelated white noise process as $p$ becomes large. Thus, in practice only a few (usually less than a dozen, Rivoirard, 1994, p. 43) Hermite polynomials need to be predicted for disjunctive kriging.

**Example 5.12** The key to disjunctive kriging is obtaining the expansion in (5.66). This is easier than it first appears. For example, suppose we are interested in $g(Z(\mathbf{s}_0)) = Z(\mathbf{s}_0)$. The Hermetian expansion of $Z(\mathbf{s}_0)$ is

$$Z(\mathbf{s}_0) = \sum_{p=0}^{\infty} b_p \frac{H_p(Z(\mathbf{s}_0))}{\sqrt{p!}},$$

where the coefficients satisfy (5.65). Thus $b_0 = \int z H_0(z) f(z) dz = 0$, since $H_0(x) = 1$ and $\mathrm{E}[Z(\mathbf{s}_0)] = 0$. Similarly, $b_1 = \int z H_1(z) f(z) dz = \mathrm{Var}[Z(\mathbf{s})] = 1$. For $p \geq 2$,

$$b_p = \int z H_p(z) f(z) dz = \int H_1(z) H_p(z) f(z) dz = 0.$$

Thus, the Hermitian expansion of $Z(\mathbf{s}_0)$ is just $H_1(Z(\mathbf{s}_0))$.

Now consider $g(Z(\mathbf{s}_0)) = I(\mathbf{s}_0, z_k)$. It is easy to determine the first coefficient as

$$b_0 = \int_{-\infty}^{z_k} H_0(z) f(z) dz = F(z_k).$$

To determine the other coefficients, note that from (5.64), the Hermite polynomials also satisfy

$$H_p(x) f(x) = (-1)^p \frac{d}{dx}(H_{p-1}(x) f(x)).$$

Using this to determine $b_p$ gives (Chilès and Delfiner, 1999, p. 641)

$$\begin{aligned} b_p &= \int_{-\infty}^{z_k} \frac{H_p(z)}{\sqrt{p!}} f(z) dz \\ &= \int_{-\infty}^{z_k} \frac{(-1)^p}{\sqrt{p!}} \frac{d}{dx}(H_{p-1}(x) f(x)) dx \\ &= (-1)^p \frac{H_{p-1}(z_k) f(z_k)}{\sqrt{p!}}. \end{aligned}$$

Thus,

$$I(\mathbf{s}_0, z_k) = F(z_k) + f(z_k) \sum_{p=1}^{\infty} (-1)^p \frac{H_{p-1}(z_k) H_p(Z(\mathbf{s}_0))}{p!}. \tag{5.72}$$

Rivoirard (1994) gives some excellent elementary examples that can be performed with a calculator to show how these expansions and disjunctive kriging work in practice with actual data. □

*5.6.4.3 Disjunctive Kriging with Non-Gaussian Data*

If the data do not follow a Gaussian distribution, they can be transformed using the anamorphosis or normal scores transformations described in §5.6.2. Unfortunately, not all data can be transformed to have a Gaussian distribution. For example, if the data are discrete or their distribution is very skewed, a Gaussian assumption even on the transformed scale may not be realistic. Thus, we may want to consider other probability measures, not just the standard Gaussian, and thus, other orthonormal polynomials. For example, if our weight function is $w(x) \propto e^{-x}x^{\alpha}$ on $(0, \infty)$, we may use the generalized Laguerre polynomials that form an orthogonal basis in $L^2$ with respect to the Gamma distribution. More generally, Matheron (1984) developed isofactorial models that have the following general form

$$F_{i,j}(dx_i, dx_j) = \sum_{m=0}^{\infty} T_m(i,j)\chi_m(x_i)\chi_m(x_j)F(dx_i)F(dx_j),$$

where $F_{i,j}(dx_i, dx_j)$ is a bivariate distribution with marginals $F(dx_i)$ and $F(dx_j)$, and the $\chi_m(z)$ are orthonormal polynomials with respect to some probability measure $F(dx)$. In kriging the polynomials, the covariances needed for the kriging equations are given by the $T_m(i,j)$. These are inferred from assumptions pertaining to the bivariate distribution of the pairs $(Z(s_i), Z(s_j))$. For example, as we noted above, if $(Z(s_i), Z(s_j))$ is bivariate Gaussian with correlation function $\rho(\mathbf{h})$, then $T_m(i,j) = [\rho(||i-j||)]^m$. However, to actually predict the factors, we need to know (and parametrically model) $T_m(\mathbf{h})$. The general form of $T_m(\mathbf{h})$ has been worked out in special cases (see Chilès and Delfiner, 1999, pp. 398–413), but many of the models seem contrived, or there are undesirable constraints on the form of the $T_m(i,j) = [\rho(||i-j||)]^m$ needed to ensure a valid bivariate distribution. Thus, Gaussian disjunctive kriging remains the isofactorial model that is most commonly used in practice.

## 5.7 Change of Support

In the previous sections we have assumed that the data were located at "points" within a spatial domain $D$ and that the inferential goal was prediction at another "point" in $D$. However, spatial data come in many forms. Instead of measurements associated with point locations, we could have measurements associated with lines, areal regions, surfaces, or volumes. In geology and mining, observations often pertain to rocks, stratigraphic units, or blocks of ore that are three dimensional. The inferential goal may also not be limited to point predictions. We may want to predict the grade of a volume of ore or estimate the probability of contamination in a volume of soil. Data associated with areal regions are particularly common in geographical studies where counts or rates are obtained as aggregate measures over geopolitical regions such as counties, Census tracts, and voting districts. In many instances, spatial aggregation is necessary to create meaningful units for analysis. This latter aspect was perhaps best described by Yule and Kendall (1950, p. 312), when they stated "... geographical areas chosen for the calculation of crop yields are modifiable units and necessarily so. Since it is impossible (or at any rate agriculturally impractical) to grow wheat and potatoes on the same piece of ground simultaneously we must, to give our investigation any meaning, consider an area containing both wheat and potatoes and this area is modifiable at choice." This problem is now known as the **modifiable areal unit problem** (MAUP) (Openshaw and Taylor, 1979).

The MAUP is comprised of two interrelated problems. The first occurs when different inferences are obtained when the same set of data is grouped into increasingly larger areal units. This is often referred to as the **scale effect** or **aggregation effect**. The second, often termed the **grouping effect** or the **zoning effect**, arises from the variability in results due to alternative formations of the areal units that produce units of different shape or orientation at the same or similar scales (Openshaw and Taylor, 1979; Openshaw, 1984; Wong, 1996). We illustrated the implications of the zoning effect using the Woodpecker data in §3.3.4. These situations, and the MAUP in general, are special cases of what is known as the **change of support problem** (COSP) in geostatistics. The term **support** includes the geometrical size, shape, and spatial orientation of the units or regions associated with the measurements (see e.g., Olea, 1991). Changing the support of a variable (typically by averaging or aggregation) creates a new variable. This new variable is related to the original one, but has different statistical and spatial properties. For example, average values are not as variable as point measurements. When we have statistically independent data, deriving the variance of their average is relatively easy. When the data are spatially dependent, inferring this variance is more difficult. It depends on both the block itself and on the variability in the point measurements. The problem of how the spatial variation in one variable associated with a given support relates to that of the other variable with a different support is called the **change of support problem.** Many of the statistical solutions to the change of support problem can be traced back to Krige's "regression effect" and subsequent corrections used in mining blocks of ore in the 1950's (Krige, 1951). From the beginning, the field of geostatistics has incorporated solutions to change of support problems, beginning with the early work of Matheron (1963). In the following sections, we describe some common change of support problems and their solutions. A more detailed description of the change of support problem and recent statistical solutions can be found in Gotway and Young (2002).

### *5.7.1 Block Kriging*

Consider the process $\{Z(\mathbf{s}) : \mathbf{s} \in D \subset \mathbb{R}^d\}$, where $Z(\mathbf{s})$ has mean $\mu$ and covariance function $\mathrm{Cov}[Z(\mathbf{u}), Z(\mathbf{v})] = C(\mathbf{u}, \mathbf{v})$ for $\mathbf{u}, \mathbf{v}$ in $D$. Suppose that instead of predicting $Z(\mathbf{s}_0)$ from data $Z(\mathbf{s}_1), \cdots, Z(\mathbf{s}_n)$, we are interested in predicting the average value in a particular region B (block) of volume $|B|$,

$$g(Z(\mathbf{s}_0)) = Z(B) = \frac{1}{|B|}\int_B Z(\mathbf{s})\, d\mathbf{s}. \tag{5.73}$$

The spatial region or block associated with the data, $B$, is called the **support** of the variable $Z(B)$. To adapt the ideas of ordinary kriging to the prediction of $Z(B)$, we consider predictors of the form $p(\mathbf{Z}; Z(B)) = \sum_{i=1}^{n} \lambda_i Z(\mathbf{s}_i)$, where the weights are chosen to minimize the mean-squared prediction error $E[(p(\mathbf{Z}; Z(B)) - Z(B))^2]$. Since $E[Z(\mathbf{s})] = \mu$, $E[Z(B)] = \mu$, and the same ideas used in the development of the ordinary kriging predictor in §5.2.2 can be applied to the prediction of $Z(B)$. This leads to the **block kriging** predictor

*Block Kriging Predictor*

$$p(\mathbf{Z}; Z(B)) = \sum_{i=1}^{n} \lambda_i Z(\mathbf{s}_i),$$

where optimal weights $\{\lambda_i\}$ are obtained by solving (Journel and Huijbregts, 1978; Chilès and Delfiner, 1999)

$$\begin{aligned} \sum_{k=1}^{n} \lambda_k C(\mathbf{s}_i, \mathbf{s}_k) - m &= \text{Cov}[Z(B), Z(\mathbf{s}_i)] \quad i = 1, \cdots, n; \\ \sum_{i=1}^{n} \lambda_i &= 1. \end{aligned} \tag{5.74}$$

In matrix terms, the optimal weights satisfy

$$\begin{aligned} \boldsymbol{\lambda}' &= \left( \boldsymbol{\sigma}(B, \mathbf{s}) + \mathbf{1} \frac{1 - \mathbf{1}'\boldsymbol{\Sigma}^{-1}\boldsymbol{\sigma}(B, \mathbf{s})}{\mathbf{1}'\boldsymbol{\Sigma}^{-1}\mathbf{1}} \right)' \boldsymbol{\Sigma}^{-1} \\ m &= \frac{1 - \mathbf{1}'\boldsymbol{\Sigma}^{-1}\boldsymbol{\sigma}(B, \mathbf{s})}{\mathbf{1}'\boldsymbol{\Sigma}^{-1}\mathbf{1}}, \end{aligned}$$

where the elements of the vector $\boldsymbol{\sigma}(B, \mathbf{s})$ are $\text{Cov}[Z(B), Z(\mathbf{s}_i)]$ (see Cressie, 1993). These "point-to-block" covariances can be derived from the covariance function $C(\cdot)$ of the underlying $Z(\mathbf{s})$ process as

$$\text{Cov}[Z(B), Z(\mathbf{s})] = \frac{1}{|B|} \int_B C(\mathbf{u}, \mathbf{v}) d\mathbf{u} d\mathbf{v}. \tag{5.75}$$

(see Journel and Huijbregts, 1978; Chilès and Delfiner, 1999). Thus, the weights needed for block kriging are the same as those used in ordinary kriging with $\boldsymbol{\sigma}$ replaced by $\boldsymbol{\sigma}(B, \mathbf{s})$.

The minimized mean-squared prediction error, the block kriging variance, is

*Block Kriging Variance*

$$\begin{aligned} \sigma^2_{ok}(B) &= \sigma(B, B) - \boldsymbol{\lambda}'\boldsymbol{\sigma}(B, \mathbf{s}) + m \\ &= \sigma(B, B) - \boldsymbol{\sigma}(B, \mathbf{s})'\boldsymbol{\Sigma}^{-1}\boldsymbol{\sigma}(B, \mathbf{s}) + \frac{\left(1 - \mathbf{1}'\boldsymbol{\Sigma}^{-1}\boldsymbol{\sigma}(B, \mathbf{s})\right)^2}{\mathbf{1}'\boldsymbol{\Sigma}^{-1}\mathbf{1}}, \end{aligned}$$

where

$$\sigma(B, B) = \text{Cov}[Z(B), Z(B)] = \frac{1}{|B|^2} \int_B \int_B C(\mathbf{u}, \mathbf{v}) d\mathbf{u} d\mathbf{v}. \tag{5.76}$$

The covariance function, $C(\mathbf{u}, \mathbf{v})$ (here a point-to-point covariance), is assumed known for theoretical derivations, but is then estimated and modeled with a valid positive definite function based on the data as in ordinary point kriging. In practice, the integrals in (5.75) and (5.76) are computed by discretizing $B$ into points, $\{\mathbf{u}'_j\}$, so that (5.75) is approximated using

$$\text{Cov}[Z(B), Z(\mathbf{s})] \approx 1/N \sum_{j=1}^{N} C(\mathbf{u}'_j, \mathbf{s}),$$

and (5.76) is approximated using

$$\text{Cov}[Z(B), Z(B)] \approx 1/N^2 \sum_{i=1}^{N} \sum_{j=1}^{N} C(\mathbf{u}'_i, \mathbf{u}'_j).$$

Block kriging can also be carried out using the semivariogram. The relationship between the semivariogram associated with $Z(B)$ and that associated with the underlying process of point support $Z(\mathbf{s})$ is given by (Cressie, 1993, p. 16),

$$\begin{aligned} 2\gamma(B_i, B_j) = & - \frac{1}{|B_i||B_i|} \int_{B_i} \int_{B_i} \gamma(\mathbf{u} - \mathbf{v}) \, d\mathbf{u}d\mathbf{v} \\ & - \frac{1}{|B_j||B_j|} \int_{B_j} \int_{B_j} \gamma(\mathbf{u} - \mathbf{v}) \, d\mathbf{u}d\mathbf{v} \\ & + \frac{2}{|B_i||B_j|} \int_{B_i} \int_{B_j} \gamma(\mathbf{u} - \mathbf{v}) \, d\mathbf{u}d\mathbf{v} \, , \end{aligned} \qquad (5.77)$$

where $2\gamma(\mathbf{u} - \mathbf{v}) = \text{Var}[Z(\mathbf{u}) - Z(\mathbf{v})]$ is the variogram of the point-support process $\{Z(\mathbf{s})\}$.

**Example 5.13** It is illustrative to examine the effects of integrating a covariance function, although in practice approximations are invoked. Imagine a one-dimensional (transect) process with point-to-point semivariogram

$$\gamma(s_i, s_j) = 1 - \exp\left\{ -\frac{3|s_i - s_j|}{5} \right\}.$$

We wish to predict $Z(B)$, where $B$ is the line segment from $s = 2$ to $s = 4$. The point-to-block semivariogram is then

$$\gamma(s_i, B) = \frac{1}{2} \int_2^4 1 - \exp\left\{ -\frac{3|u - s_j|}{5} \right\} du.$$

Figure 5.17 shows the point-to-point and the point-to-block semivariances in the neighborhood of the prediction location $s = 3$. Whereas $\gamma(s_i, s_0 = 3)$ approaches 0 as $s_i$ nears the prediction location, the same is not true for the point-to-block semivariances. The semivariogram is one half of the variance between a point datum, $Z(\mathbf{s})$, and an aggregate $Z(B) = \int Z(\mathbf{u})\, du$. This difference is not zero at the prediction location.

□

The point-to-block change of support problem is one that routinely arises in geostatistics, but the same idea can be used to solve other change of support problems. Suppose the data are $Z(A_1), \cdots, Z(A_n)$, with $Z(A_i)$ defined

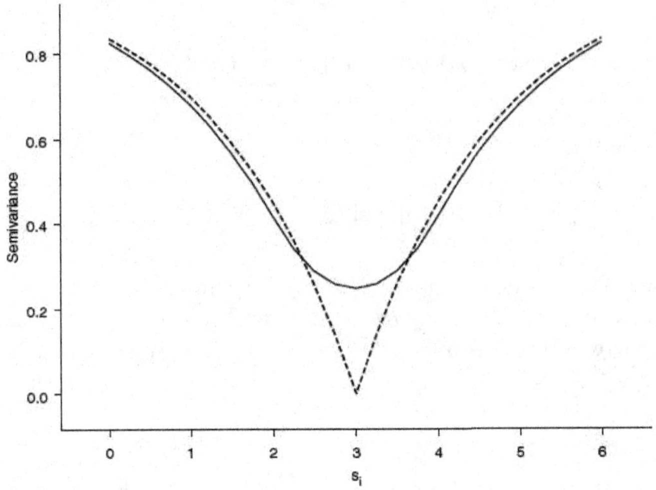


Figure 5.17 *Point-to-point (dashed line) and point-to-block semivariances (solid line) near the prediction location at $s = 3$.*

through (5.73) and prediction of $Z(B)$ is desired. The optimal linear predictor of $Z(B)$ based on data $\{Z(A_i)\}$ is $\widehat{Z}(B) = \sum_{i=1}^{n} \lambda_i Z(A_i)$, where the optimal weights $\{\lambda_i\}$ are solutions to the equations obtained by replacing the point-to-point covariances $C(\mathbf{s}_i, \mathbf{s}_k)$ and the point-to-block covariances $\text{Cov}[Z(B), Z(\mathbf{s}_i)]$ in (5.74) with

$$\text{Cov}[Z(A_i), Z(A_k)] = \int_{A_k}\int_{A_i} C(\mathbf{u}, \mathbf{v}) d\mathbf{u} d\mathbf{v} / (|A_i|\,|A_k|), \tag{5.78}$$

and

$$\text{Cov}[Z(B), Z(A_i)] = \int_{B}\int_{A_i} C(\mathbf{u}, \mathbf{v}) d\mathbf{u} d\mathbf{v} / (|B|\,|A_i|).$$

Because data on any support can be built from data with point-support, these relationships can be used for the case when $|A_i| < |B|$ (aggregation), the case when $|B| < |A_i|$ (disaggregation), and also the case of overlapping units on essentially the same scale. However, unlike the previous situation, where we observed point-support data and could easily estimate the point-support covariance function $C(\mathbf{u}, \mathbf{v})$, in practice this function is more difficult to infer from aggregate data. If we assume a parametric model, $\gamma(\mathbf{u} - \mathbf{v}; \boldsymbol{\theta})$, for $\gamma(\mathbf{u} - \mathbf{v})$, a generalized estimating equations (GEE) approach can be used to estimate $\boldsymbol{\theta}$ (see McShane et al., 1997). Consider the squared differences

$$Y_{ij}^{(1)} = (Z(B_i) - Z(B_j))^2. \tag{5.79}$$

Note that $\text{E}[Z(B_i) - Z(B_j)] = 0$ and $\text{E}[Y_{ij}] = 2\gamma(B_i, B_j; \boldsymbol{\theta})$. Taking an identity working variance-covariance matrix for the $Y_{ij}^{(1)}$, the generalized estimating equations (see §4.5.3 in Chapter 4) can be written as

$$U(\boldsymbol{\theta}; \{Y_{ij}^{(1)}\}) = 2\sum_{i=1}^{n-1}\sum_{j=i+1}^{n} \frac{\partial\gamma(B_i, B_j; \boldsymbol{\theta})}{\partial\boldsymbol{\theta}'} \left(Y_{ij}^{(1)} - 2\gamma(B_i, B_j; \boldsymbol{\theta})\right) \equiv \mathbf{0}. \quad (5.80)$$

Mockus (1998) considers a very similar approach based on least-squares fitting of a parametric covariance function.

### 5.7.2 *The Multi-Gaussian Approach*

In many cases, $\mathrm{E}[Z(B)|\mathbf{Z}(\mathbf{s})]$ is not linear in the data $\mathbf{Z}(\mathbf{s})$ and, in others, prediction of a nonlinear function of $Z(B)$ is of interest. These problems require more information about the conditional distribution of $Z(B)$ given the data, $F_B(z|\mathbf{Z}(\mathbf{s})) = \Pr(Z(B) \leq z|\mathbf{Z}(\mathbf{s}))$, than that used for linear prediction. Moreover, in many cases, such as mining and environmental remediation, the quantity $\Pr(Z(B) > z|\mathbf{Z}(\mathbf{s}))$ has meaning in its own right (e.g., proportion of high-grade blocks available in mining evaluation or the risk of contamination in a volume of soil). Nonlinear geostatistics offers solutions to COSPs that arise in this context. The multi-Gaussian approach (Verly, 1983) to nonlinear prediction in the point-to-block COSP assumes that available point data $Z(\mathbf{s}_1), \cdots, Z(\mathbf{s}_n)$ can be transformed to Gaussian variables, $\{Y(s)\}$, by $Z(s) = \varphi(Y(s))$. The block $B$ is discretized into points $\{\mathbf{u}_j', j = 1, \cdots, N\}$, and $Z(B)$ is approximated as

$$Z(B) \approx \frac{1}{N}\sum_{j=1}^{N} Z(\mathbf{u}_j'). \quad (5.81)$$

Then

$$\begin{aligned} F_B(z|\mathbf{Z}(\mathbf{s})) &\approx \Pr\left(\frac{1}{N}\sum_{j=1}^{N} Z(\mathbf{u}_j') < z|\mathbf{Z}(\mathbf{s})\right) \\ &= \Pr\left(\sum_{j=1}^{N} \phi(Y(\mathbf{u}_j')) < Nz|Y(\mathbf{s}_1), Y(\mathbf{s}_2), \cdots, Y(\mathbf{s}_n)\right). \end{aligned}$$

This probability is estimated through simulation (see Chapter 7). The vector $\mathbf{Y}(\mathbf{u}) = [Y(\mathbf{u}_1), \cdots, Y(\mathbf{u}_N)]'$ is simulated from the conditional distribution of $\mathbf{Y}(\mathbf{u})|\mathbf{Y}(\mathbf{s})$. Since $\mathbf{Y}$ is Gaussian, this conditional distribution can be obtained by kriging and simulation is straightforward. Then, $F_B(z|\mathbf{Z}(\mathbf{s}))$ is estimated as the proportion of vectors satisfying $\sum_{j=1}^{N} \varphi(Y(\mathbf{u}_j')) < Nz$.

If, instead of point support data, data $Z(A_1), \cdots, Z(A_n)$, $|A_i| < |B|$, are available, this approach can still be used provided an approximation similar to that of equation (5.81) remains valid. More general COSP models based on the multi-Gaussian approximation may be possible by building models from data based on point support as described in §5.7.1.

### 5.7.3 The Use of Indicator Data

Consider again indicator data $\mathbf{I}(\mathbf{s}, z) = [I(\mathbf{s}_1, z), \cdots, I(\mathbf{s}_n, z)]'$, derived from the indicator transform in (5.61). From §5.6.3, indicator kriging provides an estimate of $F_{s_0}(z\,|\mathbf{Z}(\mathbf{s})) = \Pr(Z(\mathbf{s}_0) < z|\mathbf{Z}(\mathbf{s}))$. For nonlinear prediction in the point-to-block COSP, it is tempting to use block kriging, described in §5.7.1, with the indicator data. However, this will yield a predictor of

$$I^*(B) = \frac{1}{|B|}\int_B I(Z(\mathbf{s}) \leq z)\, d\mathbf{s},$$

which is the proportion of $B$ consisting of points where $Z(s)$ is at or below $z$. This quantity is clearly not the same as

$$I(B) = \begin{cases} 1 & \text{if } Z(B) \leq z \\ 0 & \text{otherwise,} \end{cases} \tag{5.82}$$

which would provide an estimate of $\Pr(Z(B) \leq z|\mathbf{Z}(\mathbf{s}))$, the probability that the average value of $Z(\cdot)$ is at or below $z$. This latter quantity is the one of interest in COSPs. The problem arises with any nonlinear function of $Z(\mathbf{s})$, because the mean of block-support data will not be the same as the block average of the point support data. This is also true in the more general COSP based on data with supports $A_i$ that differ from support $B$.

Goovaerts (1997) suggests a solution to nonlinear block prediction based on simulation. The block is discretized and data $Z(\mathbf{u}'_j)$ are simulated at each discretized node. Simulated block values are then obtained via equation (5.81). Based on these simulated block values, block indicator values are constructed using equation (5.82), and $\Pr(Z(B) \leq z|\mathbf{Z}(\mathbf{s}))$ is then estimated as the average of these block indicator values. Goovaerts (1997) recommends LU decomposition for the simulation of the $Z$-values, but any conditional simulation technique (i.e., one that forces the realizations to honor the available data) could be used (see §7.2.2).

### 5.7.4 Disjunctive Kriging and Isofactorial Models

The ideas underlying the development of the disjunctive kriging predictor and the use of isofactorial models can also be used for COSPs. For example, suppose all pairs $(Z(\mathbf{s}), Z(\mathbf{u}))$ are bivariate Gaussian and we want to predict $I(B)$ in equation (5.82). This function can be expanded in terms of Hermite polynomials using (5.64):

$$I(B) = F(z) + f(z)\sum_{p=1}^{\infty}(-1)^p \frac{H_{p-1}(z)H_p(Z(B))}{p!}.$$

Then, the disjunctive kriging predictor of $I(B)$ is obtained by replacing each $H_p(Z(B))$ with its predictor obtained by kriging based on the equations

$$\sum_{i=1}^{n} \lambda_{pi} \, [\mathrm{Cov}[Z(\mathbf{s}_i), Z(\mathbf{s}_j)]]^p = [\mathrm{Cov}[Z(\mathbf{s}_j), Z(B)]]^p \,, \quad j = 1, \cdots n.$$

These are analogous to those §5.6.4, but adapted to the point-block COSP through the term $[\mathrm{Cov}[Z(\mathbf{s}_j), Z(B)]]^p$. They also have a more general form in the case of isofactorial models (§5.6.4.3):

$$\sum_{i=1}^{n} \lambda_{pi}[T_p(i,j)] = T_p(B,j), \; j = 1, \cdots n.$$

The covariances $T_p(i,j)$ (point-to-point), $T_p(B,j)$ (point-to-block), and $T_p(B, B)$ (block-to-block) needed for disjunctive kriging and for calculation of the prediction standard errors must be derived for the particular system of orthonormal polynomials being used. In practice, this has only been done in special cases, e.g., using what is called the **discrete Gaussian model** (Rivoirard, 1994; Chilès and Delfiner, 1999).

### *5.7.5 Constrained Kriging*

Dissatisfied by the solutions to the change of support problem described above, Cressie (1993b) proposes **constrained kriging**, which uses $g(\boldsymbol{\lambda}'\mathbf{Z}(\mathbf{s}))$ to predict $g(Z(B))$. If ordinary kriging is used to obtain the weights, $\boldsymbol{\lambda}$, the corresponding predictor, $g(\boldsymbol{\lambda}'\mathbf{Z}(\mathbf{s}))$, will be too smooth. In constrained kriging, a variance constraint is added to compensate for this undesirable smoothness. Thus, the weights are chosen to minimize the mean-squared prediction error of $\boldsymbol{\lambda}'\mathbf{Z}(\mathbf{s})$ as a predictor of $Z(B)$, subject to both an unbiasedness constraint as in ordinary kriging and a variance constraint $\mathrm{Var}[\boldsymbol{\lambda}'\mathbf{Z}(\mathbf{s})] = \mathrm{Var}[Z(B)]$. Thus, we choose $\boldsymbol{\lambda}$ by minimizing $\mathrm{E}[Z(B) - \boldsymbol{\lambda}'\mathbf{Z}(\mathbf{s})]$, subject to $\mathrm{E}[\boldsymbol{\lambda}'\mathbf{Z}(\mathbf{s})] = \mathrm{E}[Z(B)] = \mu$ and

$$\mathrm{Var}[\boldsymbol{\lambda}'\mathbf{Z}(\mathbf{s})] = \sigma(B,B) = \int_B \int_B C(\mathbf{u},\mathbf{v}) \, d\mathbf{u} d\mathbf{v}.$$

Cressie (1993b) shows that the optimal weights are given by

$$\boldsymbol{\lambda}' = (1/m_2)\boldsymbol{\sigma}(B,\mathbf{s})'\boldsymbol{\Sigma}^{-1} - (m_1/m_2)\mathbf{1}'\boldsymbol{\Sigma}^{-1}$$

$$m_1 = -\frac{m_2}{\mathbf{1}'\boldsymbol{\Sigma}^{-1}\mathbf{1}} + \frac{\boldsymbol{\sigma}(B,\mathbf{s})'\boldsymbol{\Sigma}^{-1}\mathbf{1}}{\mathbf{1}'\boldsymbol{\Sigma}^{-1}\mathbf{1}}$$

$$m_2 = \frac{[(\boldsymbol{\sigma}(B,\mathbf{s})'\boldsymbol{\Sigma}^{-1}\boldsymbol{\sigma}(B,\mathbf{s}))(\mathbf{1}'\boldsymbol{\Sigma}^{-1}\mathbf{1}) - (\boldsymbol{\sigma}(B,\mathbf{s})'\boldsymbol{\Sigma}^{-1}\mathbf{1})^2]^{1/2}}{[(\mathbf{1}'\boldsymbol{\Sigma}^{-1}\mathbf{1})\sigma(B,B) - 1]^{1/2}},$$

where $m_1$ and $m_2$ are Lagrange multipliers from the constrained minimization, the vector $\boldsymbol{\sigma}(B,\mathbf{s})$ has elements $\mathrm{Cov}[Z(B), Z(\mathbf{s}_i)]$ given in (5.75) and $\boldsymbol{\Sigma}$ has elements $C(\mathbf{u},\mathbf{v})$.

The weights obtained from this constrained minimization are optimal for

$\boldsymbol{\lambda}'$ as a predictor of $Z(B)$, i.e., for linear prediction. Thus, $g(\boldsymbol{\lambda}'\mathbf{Z}(\mathbf{s}))$ will not be optimal for $g(Z(B))$, but the advantage of constrained kriging is that the weights depend only on $C(\mathbf{u}, \mathbf{v})$, the point-point covariance and the range of $g(\boldsymbol{\lambda}'\mathbf{Z}(\mathbf{s}))$ exactly matches that of $g(Z(B))$. Simulations in Cressie (1993b) and Aldworth and Cressie (1999) indicate that accurate nonlinear predictions of aggregate data can be made using this approach. An extension of this, **covariance-matching constrained kriging**, has been shown to have even better mean-squared prediction error properties (Aldworth and Cressie, 2003).

## 5.8 On the Popularity of the Multivariate Gaussian Distribution

We showed in sections §5.1 and §5.2 that it was possible to determine the best predictor, $\mathrm{E}[Z(\mathbf{s}_0)|\,\mathbf{Z}(\mathbf{s})]$, when the data follow a multivariate Gaussian distribution. When the assumption of a Gaussian distribution is relaxed, we must either re-define the notion of an optimal predictor by imposing additional criteria such as linearity and unbiasedness, or transform the data to a Gaussian distribution in order to make use of its nice theoretical properties. This led us to ask the question: Why is there such a dependence on the multivariate Gaussian distribution in spatial statistics? This section explores several answers to this crucial question. Additional discussion is given in §6.3.3 and §7.4.

For simplicity, we begin by considering bivariate distributions. Let $Z_1$ and $Z_2$ be two random variables with bivariate distribution function $F_{12}(z_1, z_2) = \Pr(Z_1 \leq z_1, Z_2 \leq z_2)$. The marginal distributions $F(z_1)$ and $F(z_2)$ can be obtained from the bivariate distribution $F(z_1, z_2)$ as

$$F_1(z_1) = F_{12}(z_1, \infty); \quad F_2(z_2) = F_{12}(\infty, z_2).$$

A well-known example is that of the bivariate Gaussian distribution where

$$\begin{aligned} F_{12}(z_1, z_2) &= \frac{1}{2\pi\sigma_1\sigma_2\sqrt{1-\rho^2}} \int_{-\infty}^{z_1}\int_{-\infty}^{z_2} \exp\left\{\frac{-1}{2(1-\rho^2)}\left[\left(\frac{z_1-\mu_1}{\sigma_1}\right)^2 - \right.\right. \\ &\quad \left.\left. 2\rho\left(\frac{z_1-\mu_1}{\sigma_1}\right)\left(\frac{z_2-\mu_2}{\sigma_2}\right) + \left(\frac{z_2-\mu_2}{\sigma_2}\right)^2\right]\right\}, \end{aligned}$$

with $-1 < \rho < 1$, $F_1(z_1) \sim G(\mu_1, \sigma_1^2)$ and $F_2(z_2) \sim G(\mu_2, \sigma_2^2)$. The question of interest in this section is: Can we go the other way, i.e., given $F_1(z_1)$ and $F_2(z_2)$ can we construct $F_{12}(z_1, z_2)$ such that its marginals are $F_1(z_1)$ and $F_2(z_2)$ and $\mathrm{Corr}[Z_1, Z_2] = \rho$? The answer is "yes" but as we might expect, there may be some caveats depending on the particular case of interest.

There are several different ways to construct bivariate (and multivariate) distributions (Johnson and Kotz, 1972; Johnson, 1987):

1. through generalizations of Pearson's system of distributions derived from differential equations;

2. by series expansions like the isofactorial models described in §5.6.4.3 or the well-known Edgeworth expansions;
3. by transformation;
4. through the use of sequential conditional distributions.

However, it is important to consider the characteristics required of the resulting bivariate (and multivariate) distribution. Several important considerations include the nature of the marginals, the nature of the conditional distributions, the permissible correlation structure, whether the distribution factors appropriately under independence, and the limiting distribution. We consider several examples, focusing on bivariate distributions for simplicity.

Consider the bivariate Cauchy distribution (see, e.g., Mardia, 1970, p. 86)

$$F(z_1, z_2) = (2\pi)^{-1}c[c^2 + z_1^2 + z_2^2]^{-3/2}, \ -\infty < z_1, z_2, < \infty, \quad c > 0.$$

The marginal distributions of $Z_1$ and $Z_2$ are both Cauchy, but $\mathrm{E}[Z_2|Z_1]$ does not exist. When working with Poisson distributions, the result is more severe: no bivariate distribution exists having both marginal and conditional distributions of Poisson form (Mardia, 1970). Even if the conditional moments do exist, other constraints arise. For example, consider the Farlie-Gumbel-Morganstern system of bivariate distributions (Johnson and Kotz, 1975)

$$F_{12}(z_1, z_2) = F_1(z_1)F_2(z_2)\{1 + \alpha[1 - F_1(z_1)][1 - F_2(z_2)]\},$$

constructed from specified marginals $F_1(z_1)$ and $F_2(z_2)$. Here, $\alpha$ is a real number so that $F_{12}(z_1, z_2)$ satisfies the theoretical properties of a distribution function. The bivariate exponential distribution is a member of this class, obtained by taking both $F_1(z_1)$ and $F_2(z_2)$ to be exponential. With this system, the conditional moments exist, but $\mathrm{Corr}[Z_1, Z_2] \leq 1/3$ (Huang and Kotz, 1984). Many bivariate distributions that are not constructed from an underlying bivariate Gaussian distribution have a similar constraint on the correlation; some like the bivariate Binomial permit only negative correlations (Mardia, 1970). Plackett's family of bivariate distribution functions (Plackett, 1965) overcomes this difficulty. This family is given by the function $F_{12}(z_1, z_2)$ that satisfies

$$\psi = \frac{F_{12}(1 - F_1 - F_2 + F_{12})}{(F_1 - F_{12})(F_1 - F_{12})} \quad \psi \in (0, \infty).$$

If $\psi = 1$, then $Z_1$ and $Z_2$ are independent. By varying $\psi$, we can generate bivariate distributions with different strengths of dependence. Mardia (1967) shows that $\mathrm{Corr}[Z_1, Z_2] = [(\psi - 1)(1 + \psi) - 2\psi \log(\psi)]/(1 - \psi)^2$, and that $0 < \mathrm{Corr}[Z_1, Z_2] < 1$ if $\psi > 1$ and $\mathrm{Corr}[Z_1, Z_2] < 0$ if $0 < \psi < 1$. Unfortunately, Plackett's family does not seem to have a nice multivariate generalization. As a final example, consider the class of isofactorial models described in §5.6.4.3. This is a factorization of an *existing* bivariate distribution (Lancaster, 1958). While it can be used to build a bivariate distribution from specified marginals, we must verify that the resulting bivariate distribution is a valid distribution function and then we must derive the range of permissible correlations.

These difficulties are compounded when generalizing these results to multivariate distributions. Perhaps the most commonly-used non-Gaussian multivariate distributions are the elliptically contoured distributions (see, e.g., Johnson, 1987, pp. 106–124). These are defined through affine transformations of the class of spherically symmetric distributions with a density of the form

$$F(d\mathbf{z}) = k_n|\boldsymbol{\Sigma}|^{-1/2}h\left((\mathbf{z}-\boldsymbol{\mu})'\boldsymbol{\Sigma}^{-1}(\mathbf{z}-\boldsymbol{\mu})\right),$$

where $k_n$ is a proportionality constant and $h$ is a one-dimensional real-valued function (Johnson, 1987, p. 107). Clearly, the multivariate Gaussian distribution is in the class, obtained by taking $k_n = (2\pi)^{-n/2}$ and $h(z) = \exp\{-z/2\}$. These distributions share many of their properties with the multivariate Gaussian distribution, making them almost equally as nice from a theoretical viewpoint. However, this also means we are not completely free of some sort of multivariate Gaussian assumption, and many members of this family have limiting marginal distributions that are normal (e.g., Pearson's Type II, Johnson, 1987, p. 114). Generating multivariate distributions sequentially from specified conditionals overcomes some of these difficulties. In this approach we specify $Z_1 \sim F_1$, then $Z_2|Z_1 \sim F_{2|1}$, and $Z_3|Z_1, Z_2 \sim F_{3|21}$, and so on. This approach is routinely used in Bayesian hierarchical modeling and it can allow us to generate fairly complex multivariate distributions. However, we may not always be certain of some of the properties of the resulting distribution. Consider the following example. Suppose $Z_1(\mathbf{s})$ is a second-order stationary process with $\mathrm{E}[Z_1(\mathbf{s})] = 1$ and $\mathrm{Cov}[Z_1(\mathbf{u}), Z_1(\mathbf{u}+\mathbf{h})] = \sigma^2\rho_1(\mathbf{h})$. Conditional on $Z_1(\mathbf{s})$, suppose $Z_2(\mathbf{s})$ is a white noise process with mean and variance given by

$$\mathrm{E}[Z_2(\mathbf{s})|Z_1(\mathbf{s})] = \exp\{\mathbf{x}(\mathbf{s})'\boldsymbol{\beta}\}Z_1(\mathbf{s}) \equiv \mu(\mathbf{s}), \quad \mathrm{Var}[Z_2(\mathbf{s})|Z_1(\mathbf{s})] = \mu(\mathbf{s}).$$

This is a simplified version of a common model used for modeling and inference with count data (see., e.g., Zeger, 1988 and McShane et al., 1997). This model is attractive since the marginal mean $\mathrm{E}[Z_2(\mathbf{s})] = \exp\{\mathbf{x}(\mathbf{s})'\boldsymbol{\beta}\}$, depends only on the unknown parameter $\boldsymbol{\beta}$, and the marginal variance, $\mathrm{Var}[Z_2(\mathbf{s})] = \mu(\mathbf{s}) + \sigma^2\mu(\mathbf{s})^2$, allows overdispersion in the data $Z_2(\mathbf{s})$. Now consider the marginal correlation of $Z_2(\mathbf{s})$

$$\mathrm{Corr}[Z_2(\mathbf{s}), Z_2(\mathbf{s}+\mathbf{h})] = \frac{\rho_1(\mathbf{h})}{\left[\left(1+\dfrac{1}{\sigma^2\mu(\mathbf{s})}\right)\left(1+\dfrac{1}{\sigma^2\mu(\mathbf{s}+\mathbf{h})}\right)\right]^{1/2}}.$$

If $\sigma^2$, $\mu(\mathbf{s})$, and $\mu(\mathbf{s}+\mathbf{h})$ are small, $\mathrm{Corr}[Z_2(\mathbf{s}), Z_2(\mathbf{s}+\mathbf{h})] << \rho_1(\mathbf{h})$. For example, taking $\sigma^2 = 1$ and $\mu(\mathbf{s}) = \mu(\mathbf{s}+\mathbf{h}) = 1$, $\mathrm{Corr}[Z_2(\mathbf{s}), Z_2(\mathbf{s}+\mathbf{h})] = \rho_1(\mathbf{h})/2$. Thus, while the conditioning induces both overdispersion and autocorrelation in the $Z_2$ process, the marginal correlation has a definite upper bound and so may not be a good model for highly correlated data.

Given all of this discussion (and entire books reflecting almost 50 years of research in this area), it is now easy to see why the multivariate Gaussian distribution is so popular: it has a closed form expression, permits pairwise correlations in $(-1, 1)$, each $(Z_i, Z_j)$ has a bivariate Gaussian distribution whose moments can be easily derived, all marginal distributions are Gaussian, and all conditional distributions are Gaussian. Moreover, (almost) equally tractable multivariate distributions can be derived from the multivariate Gaussian (e.g., the multivariate lognormal and the multivariate $t$-distribution) and these play key roles in classical multivariate analysis. Thus, the multivariate Gaussian distribution has earned its truly unique place in statistical theory.

Note that in geostatistical modeling, we are working with multivariate data, i.e., rather than just considering $F_{ij}(z_i, z_j)$ we must be concerned with $F_{1,2,\cdots,n}(z_1, z_2, \cdots, z_n)$ and the relationships permitted under this multivariate distribution. Herein lies the problem with the nonparametric indicator approaches and non-Gaussian disjunctive kriging models: they attempt to build a multivariate distribution from bivariate distributions. With indicator kriging this is done through indicator semivariograms, and with disjunctive kriging it is done through isofactorial models. From the above discussion, we have to wonder if there is indeed a multivariate distribution that gives rise to these bivariate distributions. Sometimes, this consideration may seem like just a theoretical nuisance. However, in some practical applications it can cause difficulties, e.g., "covariance" matrices that are not positive definite, numerical instability, and order-relations problems. These ideas are important to keep in mind as we go on to consider more complex models for spatial data in subsequent chapters.

## 5.9 Chapter Problems

**Problem 5.1** The prediction theorem states that for any random vector $\mathbf{U}$ and any random variable $Y$ we have either of the following

- For every function $g$, $\mathrm{E}[(Y - g(\mathbf{U})^2] = \infty$;
- $\mathrm{E}[(Y - \mathrm{E}[Y \mid \mathbf{U}])^2] \leq \mathrm{E}[(Y - g(\mathbf{U})^2]$ for every $g$ with equality only if $g(\mathbf{U}) = \mathrm{E}[Y \mid \mathbf{U}]$.

Hence, the conditional expectation is the best predictor under squared error loss. Prove this theorem.

**Problem 5.2** Consider prediction under squared error loss. Let $p^0(\mathbf{Z}; \mathbf{s}_0) = \mathrm{E}[Z(\mathbf{s}_0)|\mathbf{Z}(\mathbf{s})]$. Establish that

$$\mathrm{E}[(Z(\mathbf{s}_0) - p^0(\mathbf{Z}; \mathbf{s}_0))^2] = \mathrm{Var}[Z(\mathbf{s}_0)] - \mathrm{Var}[p^0(\mathbf{Z}; \mathbf{s}_0)].$$

**Problem 5.3** Consider a standard linear model with uncorrelated errors, $\mathbf{Y} = \mathbf{X}\beta + \mathbf{e}$, $\mathbf{e} \sim (\mathbf{0}, \sigma^2\mathbf{I})$. It is easy to establish that the ordinary least squares estimator of $\beta$ requires only assumption about the first two moments of $\mathbf{Y}$ and is identical to the maximum likelihood estimator of $\beta$ when $\mathbf{e}$ is Gaussian. Use this correspondence to revisit the link between best prediction under squared-error loss for Gaussian data and best linear unbiased prediction under squared-error loss.

**Problem 5.4** Let the random variables $X$ and $Y$ have joint density $f(x, y)$. Show that the best linear unbiased predictor of Y based on $X$ is given by the linear regression function

$$BLUP(Y|X) = \alpha + \beta X,$$

where $\beta = \text{Cov}[X, Y]/\text{Var}[X]$ and $\alpha = \text{E}[Y] - \beta\text{E}[X]$.

**Problem 5.5** (adapted from Goldberger, 1991, p. 55) Consider the joint mass function $p(x, y)$ in the following table.

| | $X$ | | |
|---|---|---|---|
| $Y$ | $x = 1$ | $x = 2$ | $x = 3$ |
| $y = 0$ | 0.15 | 0.10 | 0.30 |
| $y = 1$ | 0.15 | 0.30 | 0.00 |

(i) Find the conditional expectation function $\text{E}[Y|X]$.
(ii) Find the $BLUP(Y|X)$.
(iii) Compare the conditional expectation and the BLUP for $x = 1, 2, 3$.
(iv) Calculate $\text{E}[(Y - \text{E}[Y|X])^2]$ and $\text{E}[(Y - BLUP(Y|X))^2]$. Which one is smaller?

**Problem 5.6** Let random variables $X$ and $Y$ have joint probability density function

$$f(x, y) = \frac{6}{7}(x + y)^2, \; 0 \leq x \leq 1; \, 0 \leq y \leq 1.$$

Find $\text{E}[Y|X]$, $BLUP(Y|X)$, and compare the mean-squared prediction errors for the two predictors.

**Problem 5.7** The simple kriging predictor (5.10) on page 223 is the solution to the problem of best linear prediction with known mean. Show that $p_{sk}(\mathbf{Z}; \mathbf{s}_0)$ not only is an extremum of the mean squared prediction error $\text{E}[(p(\mathbf{Z}; \mathbf{s}_0) - Z(\mathbf{s}_0))^2]$, but that it is a minimum. Assume that $\text{Var}[\mathbf{Z}(\mathbf{s})]$ is positive definite.

**Problem 5.8** Show that (5.14) and (5.15) are the solution to the ordinary kriging problem in §5.2 (page 227). Verify the formula (5.16) for $\sigma^2_{ok}(\mathbf{s}_0)$.

**Problem 5.9** Verify that (5.17) is the ordinary kriging predictor, provided $\widehat{\mu}$ is the generalized least squares estimate of $\mu$.

**Problem 5.10** Verify (5.19)–(5.22).

**Problem 5.11** Refer to the seven point ordinary kriging Example 5.5, p. 229. Repeat the example with a prediction point that falls outside the hull of the observed points, for example, $\mathbf{s}_0 = [50, 20]$ or $\mathbf{s}_0 = [60, 10]$.

- Do the same conclusions as in Example 5.5 regarding the effects of nugget, sill, and range still hold?
- Which points are screening?
- What happens to the kriging variance?
- In which models does the predicted value change, in which models does it remain the same?

**Problem 5.12** Assume that you observe in the domain $A$ a spatial random field $\{\mathbf{Z}(\mathbf{s}) : \mathbf{s} \in A \subset \mathbb{R}^2\}$ with covariance function $C(\mathbf{h})$. Further let $B_i$ and $B_j$ be some regions in $A$ with volumes $|B_i|$ and $|B_j|$, respectively. Show that

$$\mathrm{Cov}[Z(B_i), Z(B_j)] = \frac{1}{|B_i||B_j|} \int_{B_i} \int_{B_j} C(\mathbf{u}, \mathbf{v})\, d\mathbf{u} d\mathbf{v}.$$

**Problem 5.13** The table below shows 31 observations of two variables, a response variable $y$ and a regressor variable $x$.

- Prepare a scatter plot of these data and suggest a regression to model the relationship between $y$ and $x$.
- Obtain the ordinary least squares fit of your model of choice and of a simple linear regression.
- Compute predicted values for a localized simple linear regression model, where the weights are given by the Gaussian kernel function (page 240) with bandwidths $\lambda = 0.1, 0.25, 0.5, 1, 2$, and 5. How would you choose the value of $\lambda$?

| $x$ | $y$ | $x$ | $y$ | $x$ | $y$ | $x$ | $y$ | $x$ | $y$ |
|---|---|---|---|---|---|---|---|---|---|
| 3.03 | 0.03 | 3.10 | −0.20 | 3.23 | −0.44 | 3.34 | 0.48 | 3.42 | −0.79 |
| 3.51 | −0.35 | 3.62 | −0.21 | 3.72 | −1.02 | 3.80 | −1.49 | 3.91 | −1.78 |
| 4.04 | −0.43 | 4.15 | −0.91 | 4.23 | −1.33 | 4.32 | −1.41 | 4.44 | −1.14 |
| 4.53 | −1.67 | 4.65 | −1.36 | 4.73 | −0.96 | 4.85 | −1.15 | 4.92 | −1.27 |
| 5.02 | −0.37 | 5.12 | −1.25 | 5.24 | −0.96 | 5.31 | −0.44 | 5.44 | −2.08 |
| 5.51 | −0.68 | 5.63 | −1.23 | 5.72 | −0.82 | 5.81 | −0.51 | 5.92 | −0.45 |
| 6.01 | −0.40 | | | | | | | | |

CHAPTER 6

# Spatial Regression Models

In §2.4.1 we introduced the operational decomposition

$$\begin{aligned} Z(\mathbf{s}) &= \mu(\mathbf{s}) + e(\mathbf{s}) \\ e(\mathbf{s}) &= W(\mathbf{s}) + \eta(\mathbf{s}) + \epsilon(\mathbf{s}) \end{aligned}$$

of data from a random field process into large-scale trend $\mu(\mathbf{s})$, smooth, small-scale variation $W(\mathbf{s})$, micro-scale variation $\eta(\mathbf{s})$, and measurement error $\epsilon(\mathbf{s})$. This decomposition was also used to formulate statistical models for spatial prediction in the previous chapter. For example, the ordinary kriging predictor was obtained for $\mu(\mathbf{s}) = \mu$, the universal kriging predictor for $\mu(\mathbf{s}) = \mathbf{x}'(\mathbf{s})\boldsymbol{\beta}$. The focus in the previous chapter was on spatial prediction; predicting $Z(\mathbf{s})$ or the noiseless $S(\mathbf{s}) = \mu(\mathbf{s}) + W(\mathbf{s}) + \eta(\mathbf{s})$ at observed or unobserved locations. Developing best linear unbiased predictors ultimately required best linear unbiased estimators of $\mu$ and $\boldsymbol{\beta}$. The fixed effects $\boldsymbol{\beta}$ were important in that they need to be properly estimated to account for a spatially varying mean and to avoid bias. The fixed effects were not the primary focus of the analysis, however. They were essentially nuisance parameters. The covariance parameters $\boldsymbol{\theta}$ were arguably of greater importance than the parameters of the mean function, as $\boldsymbol{\theta}$ drives the various prediction equations and the precision of the predictors along with the model chosen for $\boldsymbol{\Sigma} = \mathrm{Var}[\mathbf{e}(\mathbf{s})]$.

Statistical practitioners are accustomed to the exploration of relationships among variables, modeling these relationships with regression and classification (ANOVA) models, testing hypotheses about regression and treatment effects, developing meaningful contrasts, and so forth. When first exposed to spatial statistics, the practitioner often appears to abandon these classical lines of data inquiry—that focus on aspects of the mean function—in favor of spatial prediction and the production of colorful maps. What happened? When you analyze a field experiment with spatially arranged experimental units, for example, you can rely on randomization theory or on a spatial model as the framework for statistical inference (more on the distinction below). In either case, the goal is to make decisions about the effects of the treatments applied in the experiment. And since the treatment structure is captured in the mean function—unless treatment levels are selected at random—we can not treat $\mu(\mathbf{s})$ as a nuisance. It is central to the inquiry.

In this chapter we discuss models for spatial data analysis where the focus is on modeling and understanding the mean function. In a reversal from Chapter 5, the covariance parameters may, at times, take on the role of the nuisance parameters. The models differ in the degree to which spatial dependence is incorporated and whether it is modeled directly or indirectly. For example, just as there are methods of spatial prediction that assume uncorrelated errors (see §5.3.1 and §5.3.2), some spatial regression models account for all spatial variation through the mean function. In practice you will find that your needs with respect to prediction and mean function inference are varied. Often we are interested in both estimation of fixed effects and covariance parameters. Neither $\boldsymbol{\beta}$ nor $\boldsymbol{\theta}$ are nuisance parameters; you need to pay attention to both. In §6.1–6.2 we assume that the spatial model has a linear mean structure,

$$\begin{aligned} \mathbf{Z}(\mathbf{s}) &= \mathbf{X}(\mathbf{s})\boldsymbol{\beta} + \mathbf{e}(\mathbf{s}), \\ \mathbf{e}(\mathbf{s}) &\sim (\mathbf{0}, \boldsymbol{\Sigma}(\boldsymbol{\theta})). \end{aligned} \tag{6.1}$$

In addition, we may assume $\mathbf{e}(\mathbf{s}) \sim G(\mathbf{0}, \boldsymbol{\Sigma}(\boldsymbol{\theta}))$, when necessary to obtain confidence intervals or to construct hypothesis tests. In §6.3 we consider a more general formulation based on generalized linear models (GLMs) and relax this distributional assumption and the linearity of the mean function.

All models in this chapter have in common that spatial structure is incorporated, either as part of the mean function, the error process, or both. None of the models *ignore* spatial structure, they accommodate it in different ways. Naturally, this begs the question about the respective merits and demerits of the various approaches. An interesting case in point is the comparison of models for analyzing data from designed (field) experiments. Recently, random-field methods based on model (6.1), with spatial autocorrelation modeled through $\boldsymbol{\Sigma}(\boldsymbol{\theta})$, have received considerable attention (see, for example, Zimmerman and Harville, 1991; Brownie, Bowman, and Burton, 1993; Stroup, Baenziger, and Mulitze, 1994; Brownie and Gumpertz, 1997; Gotway and Stroup, 1997). It has generally been reported that modeling such data by including spatial autocorrelation yields a more powerful analysis than the traditional design-based analysis of variance (based on randomization theory, e.g., the CRD or RCBD analysis) when the spatial structure is pronounced. In the cases examined, it is typically the case that the design structure involves some form of blocking scheme with a fairly large number of treatments, e.g., variety or plant breeding trials. A short-sighted comparison of the two approaches that considers only the precision of parameter estimates is dangerous since the methods can represent completely different philosophies. From a randomization perspective (Kempthorne, 1955), the error-control and treatment design lead to a linear model that usually has the form of an analysis of variance model with independent errors. The statistical model is not something hypothesized or assumed, it is the result of executing a particular design. Spatial autocorrelation is addressed under randomization at two stages. Large-scale trends are ostensibly *eliminated* through blocking, small-scale trends that operate at the scale of the experimental units are *neutralized* through randomization. Situations where spatial random field models with correlated errors have seemingly out-performed design-based analyses are basically dramatic examples of inefficient blocking that failed to eliminate spatial trends. Making poor design choices does not affect the validity of cause-and-effect inferences in design-based analyses under randomization. It only makes it difficult to detect treatment differences because of a large experimental error variance. When experimental data are subjected to modeling, it is possible to increase the statistical precision of treatment contrasts. The ability to draw cause-and-effect conclusions has been lost, however, unless it can be established that the model is correct.

Some statisticians take general exception with the modeling of experimental data, whether its focus is on the mean or the covariance structure of the data, because it is not consistent with randomization inference. Any deviation from the statistical model that reflects the execution of the particular design is detrimental in their view. We agree that you should "analyze 'em the way you randomize 'em," whenever possible; this is the beauty of designed-based inference. Nevertheless, we also know from experience that things can go wrong and that scientists want to make the most of the data they have worked so hard to collect. Thus, modeling of experimental data should also be a choice, provided we attach the important caveat that modeling experimental data does not lend itself to cause-and-effect inferences. If, for example, blocking has been carried out too coarsely to provide a reduction in experimental error variance substantial enough to yield smaller standard errors of treatment contrasts than an analysis that accounts for heterogeneity outside of the error-control design, why not proceed down that road?

## 6.1 Linear Models with Uncorrelated Errors

When model errors are uncorrelated, the standard linear model battery can be brought to bear and analyses are particularly simple. In designed experiments, design-based analysis based on randomization theory does not need to explicitly model spatial dependencies, since these are neutralized through randomization at the spatial scale of the experimental unit. If spatial structure is present at scales smaller or larger than the unit, or if data are observational, or if one does not want to adopt a randomization framework for inference, uncorrelated errors are justified if all of the spatial variation is captured with, accounted for, or explained by the mean function. This leads to the addition of terms in the mean function to account for spatial configuration and structure, or to the transformation of the regressor space.

The significance of regression or ANCOVA models with uncorrelated errors for spatial data is twofold. First, we want to discuss their place in spatial analysis. Many statisticians have been led to believe that these models are inadequate for use with spatial data and that ordinary least squares estimates of fixed effects are biased or inefficient. This is not always true. Second, these models provide an excellent introduction to more complex models with correlated errors that follow later. For example, models with nearest neighbor adjustments, such as the Papadakis analysis (Papadakis, 1937), are forerunners of spatial autoregressive models. The danger of assuming that the salient spatial structure can be captured through the mean function alone is that there is "little room for error." In the words of Zimmerman and Harville (1991), a correlated error structure can "soak up" spatial heterogeneity. In longitudinal data analyses it is not uncommon to model completely unstructured covariance matrices, in part to protect against omitted covariates. In the spatial case unstructured covariance matrices are impractical, but even a highly parameterized covariance structure can provide an important safety net, protecting you against the danger of a misspecified mean function.

**Example 6.1 Soil carbon regression.** This application is a continuation of Example 4.2, which considered the spatial variation of soil C/N ratios on an agricultural field. In contrast to the analyses in Chapter 4 we are now concerned with modeling the relationship between soil carbon percentage and the soil nitrogen percentage.

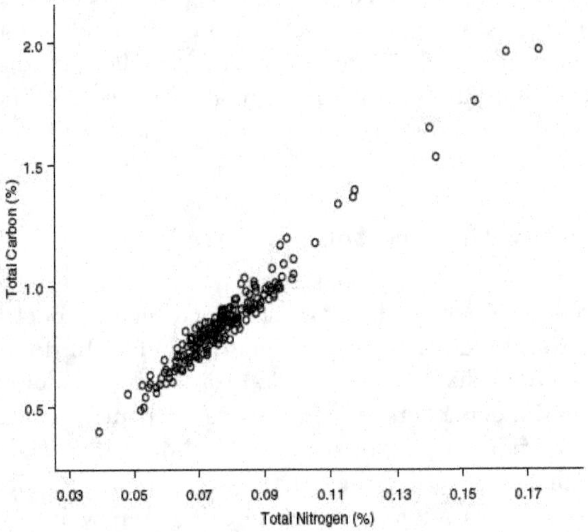


Figure 6.1 *Scatterplot of soil carbon (%) versus soil nitrogen (%). Spatial configuration of samples is shown in Figure 4.6 on page 156. Data kindly provided by Dr. Thomas G. Mueller, Department of Agronomy, University of Kentucky.*

Previously we established that the empirical semivariogram of the C/N ratios can be modeled with an exponential semivariogram; there appears to be spatial autocorrelation in the C/N measurements. This finding does not tell us anything about the relationship between C and N, however. For example, it is possible that the C/N ratios exhibit spatial dependency because C% depends on N% and it is the latter attribute that varies spatially. Could spatial autocorrelation in the N-process induce the spatial variation in C? Also, an exponential semivariogram may not be appropriate to model the spatial dependency in soil carbon adjusted for soil nitrogen. A scatterplot between the two variables certainly argues for the fact that a linear relationship exists between the two soil attributes (Figure 6.1). A regression analysis with mean function $\mathrm{E}[C(\mathbf{s})|N(\mathbf{s})] = \beta_0 + \beta_1 N(\mathbf{s})$ seems appropriate. An ordinary least squares analysis yields a highly significant relationship with $R^2 = 0.89$. The important question is, however, whether the simple linear regression structure is sufficient to explain the (co-)variation in soil C. In other words, how do we model the errors in the model

$$C(\mathbf{s}_i)|N(\mathbf{s}_i) = \beta_0 + \beta_1 N(\mathbf{s}_i) + e(\mathbf{s}_i)?$$

□

### 6.1.1 *Ordinary Least Squares—Inference and Diagnostics*

If we are satisfied with our choice of a particular model, parameter estimation is followed by confirmatory inference, that is, the testing of hypothesis, the computation of confidence intervals, and so forth. While we are in the process of model building, we need to raise questions about the model-data agreement, that is, diagnose the extent to which the data conform to model assumptions, the extent to which observations are influential in the analysis, and the extent to which model components inter-relate. In the case of linear models with uncorrelated errors, there is a large battery of tools to choose from to perform these diagnostic tasks as well as confirmatory inference. In this subsection we briefly re-iterate some of these well-known tools. We will see in the next section how considerably more complicated diagnostic and inferential tasks can become when correlated errors are introduced. It is particularly important to us to address the properties of OLS residuals and their use in diagnosing the fit of a spatial regression model.

#### 6.1.1.1 *Estimators and Their Properties*

Assume that you are fitting a linear model with uncorrelated, homoscedastic errors to spatial data,

$$\mathbf{Z}(\mathbf{s}) = \mathbf{X}(\mathbf{s})\boldsymbol{\beta} + \mathbf{e}(\mathbf{s}), \qquad \mathbf{e}(\mathbf{s}) \sim (\mathbf{0}, \sigma^2\mathbf{I}). \tag{6.2}$$

The ordinary least squares estimator of the fixed effects and the customary estimator of the residual variance are

$$\begin{aligned} \widehat{\boldsymbol{\beta}}_{ols} &= \left(\mathbf{X}(\mathbf{s})'\mathbf{X}(\mathbf{s})\right)^{-1}\mathbf{X}(\mathbf{s})'\mathbf{Z}(\mathbf{s}) \\ \widehat{\sigma}^2 &= \frac{1}{n - \mathrm{rank}\{\mathbf{X}(\mathbf{s})\}}\left(\mathbf{Z}(\mathbf{s}) - \mathbf{X}(\mathbf{s})\widehat{\boldsymbol{\beta}}_{ols}\right)'\left(\mathbf{Z}(\mathbf{s}) - \mathbf{X}(\mathbf{s})\widehat{\boldsymbol{\beta}}_{ols}\right). \end{aligned}$$

These estimators are based on the least squares criterion: find $\boldsymbol{\beta}$ that minimizes the residual sum of squares:

$$\left(\mathbf{Z}(\mathbf{s}) - \mathbf{X}(\mathbf{s})'\boldsymbol{\beta}\right)'\left(\mathbf{Z}(\mathbf{s}) - \mathbf{X}(\mathbf{s})'\boldsymbol{\beta}\right).$$

Using ideas from calculus theory of minimization (e.g., differentiate, equate to zero, solve) leads to *normal equations* and then to the OLS estimator of $\boldsymbol{\beta}$. The minimized error variance, corrected for bias, is then the OLS estimator of $\sigma^2$. If we are willing to make a distributional assumption about $\mathbf{Z}(\mathbf{s})$, then we can also use maximum likelihood estimation. Assume that $\mathbf{Z}(\mathbf{s}) \sim G(\mathbf{0}, \sigma^2\mathbf{I})$. The maximum likelihood (ML) estimators of $\boldsymbol{\beta}$ and $\sigma^2$ maximize the joint likelihood of the data, or, equivalently, minimize the negative of twice the Gaussian log likelihood given by

$$\varphi(\boldsymbol{\beta}, \sigma^2; \mathbf{Z}(\mathbf{s})) = n\ln\{2\pi\} + n\ln\{\sigma^2\} + \frac{(\mathbf{Z}(\mathbf{s}) - \mathbf{X}(\mathbf{s})\boldsymbol{\beta})'(\mathbf{Z}(\mathbf{s}) - \mathbf{X}(\mathbf{s})\boldsymbol{\beta})}{\sigma^2}.$$

Here, ideas from calculus lead to *score equations* which are then solved to obtained the ML estimators. For linear models, least squares and ML produce equivalent estimators. Thus, the ML estimator of $\boldsymbol{\beta}$ and the bias-corrected ML estimator of $\sigma^2$ (the REML estimator) are equivalent to the OLS estimators given above.

To establish a few important properties of $\widehat{\boldsymbol{\beta}}_{ols}$, let us define the "hat" matrix $\mathbf{H} = \mathbf{X}(\mathbf{s})(\mathbf{X}(\mathbf{s})'\mathbf{X}(\mathbf{s}))^{-1}\mathbf{X}(\mathbf{s})'$ and $\mathbf{M} = \mathbf{I} - \mathbf{H}$. Obviously, $\mathbf{H}$ and $\mathbf{M}$ are projection matrices (symmetric and idempotent), and estimates of the mean are obtained as $\widehat{\mathbf{Z}}(\mathbf{s}) = \mathbf{X}(\mathbf{s})\widehat{\boldsymbol{\beta}}_{ols} = \mathbf{HZ}(\mathbf{s})$. The OLS residuals are then simply $\widehat{\mathbf{e}}_{ols}(\mathbf{s}) = \mathbf{Z}(\mathbf{s}) - \widehat{\mathbf{Z}}(\mathbf{s}) = \mathbf{MZ}(\mathbf{s})$. Because $\mathbf{H}$ is a projector, $\mathbf{H}$ and $\mathbf{M}$ project onto orthogonal subspaces. The expression $\mathbf{Z}(\mathbf{s}) = \mathbf{HZ}(\mathbf{s}) + \mathbf{MZ}(\mathbf{s})$ states that the $(n \times 1)$ vector $\mathbf{Z}(\mathbf{s})$ is decomposed into a component projected onto that subspace of $\mathbb{R}^n$ which can be generated by the columns of $\mathbf{X}(\mathbf{s})$, and a component projected onto the orthogonal complement of this space.

*OLS Properties*

If model (6.2) is correct, then

$$\begin{aligned}
\mathrm{E}[\widehat{\boldsymbol{\beta}}_{ols}] &= \boldsymbol{\beta} \\
\mathrm{E}[\widehat{\mathbf{e}}_{ols}(\mathbf{s})] &= \mathbf{0} \\
\mathrm{Var}[\widehat{\boldsymbol{\beta}}_{ols}] &= \sigma^2\left(\mathbf{X}(\mathbf{s})'\mathbf{X}(\mathbf{s})\right)^{-1} \\
\mathrm{Var}[\widehat{\mathbf{e}}_{ols}(\mathbf{s})] &= \sigma^2\mathbf{M} \\
\mathbf{1}'\widehat{\mathbf{e}}_{ols}(\mathbf{s}) &= 0.
\end{aligned}$$

The first and second properties follow from the assumption that the model errors $\mathbf{e}(\mathbf{s})$ have zero mean. The third and fourth properties are based on the assumption that the variance of the model errors is $\sigma^2\mathbf{I}$. The last property of the OLS residuals (sum-to-zero) applies when the $\mathbf{X}(\mathbf{s})$ matrix contains a constant column (an intercept). (We typically assume in this text that an intercept is present.)

We can also derive the expected value and variance of $\widehat{\sigma}^2$. One of the advantages of maximum likelihood estimation is that the inverse of the information matrix provides the variance-covariance matrix of the maximum likelihood estimators. For the linear model considered here, this matrix can be obtained as a special case of that given in (5.47).

*6.1.1.2 Testing Linear Hypotheses*

A linear hypothesis is a hypothesis of the form $H_0\colon \mathbf{L}\boldsymbol{\beta} = \mathbf{l}_0$, where $\mathbf{L}$ is an $l \times p$ matrix of coefficients and $\mathbf{l}_0$ is a specified $l \times 1$ vector. For example, $H_0\colon \beta_j = 0$ can be obtained by choosing $\mathbf{L}$ to be a $p$ vector of zeros with a one in the $j$th position, and taking $\mathbf{l}_0 = 0$. If $\mathbf{X}(\mathbf{s})$ is deficient in rank, then we need to worry about whether the hypothesis $\mathbf{L}\boldsymbol{\beta} = \mathbf{l}_0$ is testable. This is the case if $\boldsymbol{\alpha}_i\left(\mathbf{X}(\mathbf{s})'\mathbf{X}(\mathbf{s})\right)^{-}\mathbf{X}(\mathbf{s})'\mathbf{X}(\mathbf{s}) = \boldsymbol{\alpha}_i$, where $\boldsymbol{\alpha}_i$ is the $i$th row of $\mathbf{L}$.

The standard test statistic for the testable hypothesis $H_0$ is

*F-Test for Linear Hypothesis*

$$F = \frac{\left(\mathbf{L}\widehat{\boldsymbol{\beta}}_{ols} - \mathbf{l}_0\right)' \left[\mathbf{L}\left(\mathbf{X}(\mathbf{s})'\mathbf{X}(\mathbf{s})\right)^{-1}\mathbf{L}'\right]^{-1} \left(\mathbf{L}\widehat{\boldsymbol{\beta}}_{ols} - \mathbf{l}_0\right)}{\widehat{\sigma}^2 \text{rank}\{\mathbf{L}\}}. \qquad (6.3)$$

If $\mathbf{e}(\mathbf{s})$ is Gaussian distributed, then $F$ follows an $F$-distribution with rank$\{\mathbf{L}\}$ numerator and $n - \text{rank}\{\mathbf{X}(\mathbf{s})\}$ denominator degrees of freedom. The customary $t$-test for $H_0\colon \beta_j = 0$ is a special case of (6.3). In this case, $F = \widehat{\beta}_j^2/\text{ese}(\widehat{\beta}_j)^2$, where ese denotes the estimated standard error. Since $F_{\alpha,1,\nu} = t_{\alpha/2,\nu}$ the statistic for the $t$-test is

$$t = \text{sign}\left(\widehat{\beta}_j\right) \times \sqrt{F},$$

and a $(1-\alpha) \times 100\%$ confidence interval for $\beta_j$ is given by

$$\widehat{\beta}_j \pm t_{\alpha/2, n-\text{rank}\{\mathbf{X}(\mathbf{s})\}} \times \text{ese}\left(\widehat{\beta}_j\right).$$

In a model with uncorrelated errors, the test statistic (6.3) has another, appealing interpretation, in terms of reductions of sums of squares. Let

$$SSR = \left(\mathbf{Z}(\mathbf{s}) - \mathbf{X}(\mathbf{s})'\widehat{\boldsymbol{\beta}}_{ols}\right)'\left(\mathbf{Z}(\mathbf{s}) - \mathbf{X}(\mathbf{s})'\widehat{\boldsymbol{\beta}}_{ols}\right)$$

denote the residual sum of squares of the model and let $SRR_r$ denote the same sum of squares subject to the linear constraints $H_0\colon \mathbf{L}\boldsymbol{\beta} = \mathbf{l}_0$. Then

*Sum of Squares Reduction Test*

$$F = \frac{SSR_r - SSR}{\text{rank}\{\mathbf{L}\}\widehat{\sigma}^2}.$$

*6.1.1.3 Residual and Influence Diagnostics*

The ordinary least squares residuals $\widehat{\mathbf{e}}_{ols}(\mathbf{s})$ have zero mean and variance $\sigma^2\mathbf{M} = \sigma^2(\mathbf{I} - \mathbf{H})$. If $h_{ii}$ denotes the $i$th diagonal element of $\mathbf{H}$, then the $i$th residual can be *standardized* as

*Standardization*

$$\frac{\widehat{e}_{ols}(\mathbf{s}_i)}{\sqrt{\text{Var}[\widehat{e}_{ols}(\mathbf{s}_i)]}} = \frac{\widehat{e}_{ols}(\mathbf{s}_i)}{\sigma\sqrt{1-h_{ii}}}.$$

Since $\sigma$ is unknown, one can compute instead the *studentized* residual

*Studentized Residual*

$$r_i = \frac{\widehat{e}_{ols}(\mathbf{s}_i)}{\sqrt{\widehat{\text{Var}}[\widehat{e}_{ols}(\mathbf{s}_i)]}} = \frac{\widehat{e}_{ols}(\mathbf{s}_i)}{\widehat{\sigma}\sqrt{1-h_{ii}}}.$$

The studentized residual is important in diagnostic work because (i) it avoids some of the pitfalls of raw residuals, and (ii) many other diagnostic measures can be expressed in terms of the $r_i$. Among the pitfalls of raw residuals are that they are correlated (M is not a diagonal matrix), they are not homoscedastic (the diagonals of M are not of the same value), and they are rank-deficient (M is a $(n \times n)$ matrix of rank $(n - \text{rank}\{\mathbf{X}(\mathbf{s})\})$. Studentization at least takes care of one of these issues, namely the unequal variance of the residuals (see the following sub-section).

The quantity $r_i$ is also referred to as an *internally* studentized residual, because the estimate $\widehat{\sigma}$ is based on all $n$ data points. If $\widehat{\sigma}^2_{-i}$ is the estimator of $\sigma^2$ obtained with the $i$th data point removed from the analysis, then

$$t_i = \frac{\widehat{e}_{ols}(\mathbf{s}_i)}{\widehat{\sigma}_{-i}\sqrt{1 - h_{ii}}}$$

is called the *externally* studentized residual.

The quantity $h_{ii}$ is also called the **leverage**; it expresses how unusual an observation is in the "X"-space. Data points with high leverage have the potential to be influential on the analysis, but are not necessarily so. In a linear model with uncorrelated errors and an intercept, the leverages are bounded, $1/n \leq h_{ii} \leq 1$, and the sum of the leverages equals the rank of $\mathbf{X}(\mathbf{s})$. Note that

*Leverage Matrix*

$$\mathbf{H} = \frac{\partial \widehat{\mathbf{Z}}(\mathbf{s})}{\partial \mathbf{Z}(\mathbf{s})}.$$

Diagnostic measures based on the sequential removal of data points are particularly simple to compute for linear models with uncorrelated errors. At the heart of the matter is the following powerful result. If $\mathbf{X}(\mathbf{s})_{-i}$ is the $((n-1) \times p)$ regressor matrix with the $i$th row removed, then the estimates of the fixed effects—if the $i$th data point is not part of the analysis—are

$$\widehat{\boldsymbol{\beta}}_{ols,-i} = \left(\mathbf{X}(\mathbf{s})'_{-i}\mathbf{X}(\mathbf{s})_{-i}\right)^{-1}\mathbf{X}(\mathbf{s})_{-i}\mathbf{Z}(\mathbf{s})_{-i}.$$

The difficult component in this calculation is the matrix inversion. Fortunately, we can compute this inverse easily because of the following result (known as the Sherman-Morrison-Woodbury theorem)

$$\begin{aligned}\left(\mathbf{X}(\mathbf{s})'_{-i}\mathbf{X}(\mathbf{s})_{-i}\right)^{-1} &= \left(\mathbf{X}(\mathbf{s})'\mathbf{X}(\mathbf{s})\right)^{-1} \\ &\quad + \frac{(\mathbf{X}(\mathbf{s})'\mathbf{X}(\mathbf{s}))^{-1}\mathbf{x}(\mathbf{s}_i)\mathbf{x}(\mathbf{s}_i)'(\mathbf{X}(\mathbf{s})'\mathbf{X}(\mathbf{s}))^{-1}}{1 - h_{ii}}.\end{aligned}$$

After some algebra, we are led to a simple expression for the "leave-one-out" estimate of the fixed effects,

*Leave-One-Out OLS Estimates*

$$\begin{aligned}\widehat{\boldsymbol{\beta}}_{ols,-i} &= \widehat{\boldsymbol{\beta}}_{ols} - \frac{(\mathbf{X}(\mathbf{s})'\mathbf{X}(\mathbf{s}))^{-1}\mathbf{x}(\mathbf{s}_i)\widehat{e}(\mathbf{s}_i)}{1 - h_{ii}} \\ &= \widehat{\boldsymbol{\beta}}_{ols} - (\mathbf{X}(\mathbf{s})'\mathbf{X}(\mathbf{s}))^{-1}\mathbf{x}(\mathbf{s}_i)\widehat{e}(\mathbf{s}_i)_{-i}.\end{aligned}$$

The quantity $\widehat{e}(\mathbf{s}_i)_{-i} = Z(\mathbf{s}_i) - \widehat{Z}(\mathbf{s}_i)_{-i}$ is known as the PRESS residual (Allen,

1974). The importance of these results for diagnosing the fit of linear models is that statistics can be computed efficiently based only on the fit of the model to the full data and that many statistics depend on only a fairly small number of elementary measures such as leverages and raw residuals. For example, a PRESS residual is simply

*PRESS Residual*

$$\widehat{e}(\mathbf{s}_i)_{-i} = Z(\mathbf{s}_i) - \widehat{Z}(\mathbf{s}_i)_{-i} = \frac{\widehat{e}(\mathbf{s}_i)}{1-h_{ii}},$$

and Cook's $D$ (Cook, 1977, 1979), a measure for the influence of an observation on the estimate $\widehat{\beta}$, can be written as

*Cook's D*

$$D = \frac{r_i^2 h_{ii}}{k(1-h_{ii})},$$

where $k = \text{rank}\{\mathbf{X}(\mathbf{s})\}$. The DFFITS statistic of Belsely, Kuh, and Welsch (1980) measures the change in fit in terms of standard error units, and can be written as

$$DFFITS_i = t_i\sqrt{\frac{h_{ii}}{1-h_{ii}}}.$$

These and many other influence statistics are discussed in the monographs by Belsely, Kuh, Welsch (1980) and Cook and Weisberg (1982).

### 6.1.2 *Working with OLS Residuals*

Fitted residuals in a statistical model are commonly used to examine the underlying assumptions about the model. For example, a QQ-plot or histogram of the $\widehat{\mathbf{e}}_{ols}(\mathbf{s})$ is used to check whether it is reasonable to assume a Gaussian distribution, scatter plots of the residuals are used to test a constant variance assumption or the appropriateness of the mean function. In spatial models, whether the errors are assumed to be correlated or uncorrelated, an important question is whether the covariance structure of the model has been chosen properly. It seems natural, then, to use the fitted residuals to judge whether the assumed model $\text{Var}[\mathbf{e}(\mathbf{s})] = \mathbf{\Sigma}(\boldsymbol{\theta})$ appears adequate. When, as in this section, it is assumed that $\mathbf{\Sigma}(\boldsymbol{\theta}) = \sigma^2\mathbf{I}$ and the model is fit by ordinary least squares, one would use the $\widehat{\mathbf{e}}_{ols}(\mathbf{s})$ to inquire whether there is any residual spatial autocorrelation. Common devices are a test for autocorrelation based on Moran's $I$ with regional data and estimation of semivariograms of the residuals with geostatistical data. To proceed with such analyses in a meaningful way, the properties of residuals need to be understood.

Recall from the previous section that the "raw" residuals from an OLS fit are

$$\widehat{\mathbf{e}}_{ols}(\mathbf{s}) = \mathbf{Z}(\mathbf{s}) - \widehat{\mathbf{Z}}(\mathbf{s}) = \mathbf{Z}(\mathbf{s}) - \mathbf{HZ}(\mathbf{s}) = \mathbf{MZ}(\mathbf{s}), \tag{6.4}$$

where $\mathbf{H} = \mathbf{X}(\mathbf{s})(\mathbf{X}(\mathbf{s})'\mathbf{X}(\mathbf{s}))^{-1}\mathbf{X}(\mathbf{s})'$ is the "hat" (leverage) matrix. Since we aim to use $\widehat{\mathbf{e}}_{ols}(\mathbf{s})$ to learn about the unobservable $\mathbf{e}(\mathbf{s})$, let us compare their features. First, the elements of $\mathbf{e}(\mathbf{s})$ have zero mean, are non-redundant, uncorrelated, and homoscedastic. By comparison, the elements of $\widehat{\mathbf{e}}_{ols}(\mathbf{s})$ are

- **rank deficient**: If $\mathbf{X}(\mathbf{s})$ is $(n \times p)$ of rank $k$, then
$$\mathbf{H} = \mathbf{X}(\mathbf{s})(\mathbf{X}(\mathbf{s})'\mathbf{X}(\mathbf{s}))^{-1}\mathbf{X}(\mathbf{s})'$$
is of rank $k$. The fitted values $\widehat{\mathbf{Z}}(\mathbf{s})$ are a projection of $\mathbf{Z}(\mathbf{s})$ onto a $k$-dimensional subspace of $\mathbb{R}^n$. As a consequence, only $n-k$ of the OLS residuals carry information about the model disturbances $\mathbf{e}(\mathbf{s})$. The remaining $k$ residuals are redundant.
- **correlated**: The variance of the OLS residual vector is
$$\text{Var}[\widehat{\mathbf{e}}_{ols}(\mathbf{s})] = \sigma^2\mathbf{M},$$
a non-diagonal matrix. The residuals are correlated because they result from fitting a model to data and thus obey certain constraints. For example, $\mathbf{X}(\mathbf{s})'\widehat{\mathbf{e}}_{ols}(\mathbf{s}) = \mathbf{0}$. Fitted residuals exhibit more negative correlations than the model errors.
- **heteroscedastic**: The leverage matrix $\mathbf{H}$ is the gradient of the fitted values with respect to the observed data. A diagonal element $h_{ii}$ reflects the weight of an observation in determining its predicted value. With the exception of some balanced classification models, the $h_{ii}$ are not of equal value and the residuals are thus not equi-dispersed. A large residual in a plot of the $\widehat{e}_{ols}(\mathbf{s}_i)$ against the fitted values may not convey a model breakdown or an outlying observation. A large residual is more likely if $\text{Var}[\widehat{e}_{ols}(\mathbf{s}_i)]$ is large ($h_{ii}$ small). Furthermore, $\sigma^2(1-h_{ii}) < \sigma^2$, since in an OLS model with intercept $1/n \le h_{ii} \le 1$. Thus, the sill of a semivariogram computed from the $\widehat{e}_{ols}(\mathbf{s}_i)$ does not reflect the variability of the process ($\sigma^2$).

The one (only) property the residuals have in common with the model errors is a zero mean. Note that these discrepancies between $\widehat{e}_{ols}(\mathbf{s})$ and $\mathbf{e}(\mathbf{s})$ have nothing to do with fitting a model to spatial data or with spatial autocorrelation. They are a mere consequence of fitting a model to data by ordinary least squares. It is now becoming clear that computing a semivariogram from residuals in order to understand the autocorrelation pattern of the data can be a problematic undertaking. This empirical semivariogram estimates the semivariogram of the residuals, not the semivariogram of $\mathbf{e}(\mathbf{s})$. Furthermore, the "data" for this analysis, $\widehat{e}_{ols}(\mathbf{s})$, fail to meet an important condition for a semivariogram analysis: the residual process is not second-order stationary. Moreover, negative correlations among the residuals may confound spatial autocorrelation. We should have similar reservations about a Moran's $I$ statistic based on the $\widehat{e}_{ols}(\mathbf{s}_i)$.

At this point, three courses of action emerge:

(i) proceed with the raw residuals but understand the implications and (try to) interpret the results accordingly;

(ii) manipulate (adjust, transform) the raw residuals in such a way that an autocorrelation analysis is more meaningful;

(iii) use spatial statistics that take into account the fact that you are working with quantities that result from a model fit, rather than actual, observed data.

Most practitioners adopt (i), usually without the caveat and understandably so: it is difficult to interpret the results once we realize all the problems that can arise when working with raw residuals. One approach to (ii) is to derive a set of $n-k$ "new" quantities that overcome the problems inherent in working with raw residuals. This approach is termed **error recovery**. As for (iii), spatial statistics typically used for spatial autocorrelation analysis can be modified for use with residuals. We describe error recovery and some modifications to spatial autocorrelation statistics for use in OLS residual analysis in subsequent paragraphs.

### *6.1.2.1 Error Recovery*

Error recovery is the process of generating "variance-covariance" standardized residuals that have zero mean, unit variance, and are uncorrelated. Two methods can be distinguished for recovering uncorrelated errors, depending on whether they draw on the projection properties in the fitted model or on the sequential forecasting of observations. The latter approach gives rise to *recursive* or *sequential* residuals (Brown et al., 1975; Kianifard and Swallow, 1996). The process of residual recursion starts with fitting the model to $k = \text{rank}(\mathbf{X})$ data points. The remaining $n-k$ observations are entered sequentially and the $j$th recursive residual is the scaled difference between $Z(\mathbf{s}_j)$ and $\widehat{Z}_{-j}(\mathbf{s}_j)$, the predicted value based on the previous observations. The $n-k$ scaled differences so obtained are the recursive residuals and are useful in detecting outliers, changes in regression coefficients, heteroscedasticity, and serial correlation (Galpin and Hawkins, 1984; Kianifard and Swallow, 1996).

Error recovery based on projections is also known as Linearly Unbiased Scaled (LUS) estimation and is due to Theil (1971). If $\widehat{\mathbf{e}}(\mathbf{s})$ is an $(n \times 1)$ residual vector with variance $\text{Var}[\widehat{\mathbf{e}}(\mathbf{s})] = \mathbf{A}$, then to transform the residuals so that they are uncorrelated with unit variance we need to find an $(n \times n)$ matrix $\mathbf{Q}$ such that

*Linearly Unbiased Scaled Estimates*

$$\text{Var}[\mathbf{Q}'\widehat{\mathbf{e}}(\mathbf{s})] = \mathbf{Q}'\mathbf{A}\mathbf{Q} = \begin{bmatrix} \mathbf{I}_{(n-k)} & \mathbf{0}_{((n-k)\times k)} \\ \mathbf{0}_{(k\times(n-k))} & \mathbf{0}_{(k\times k)} \end{bmatrix}.$$

The first $n-k$ elements of $\mathbf{Q}'\widehat{\mathbf{e}}(\mathbf{s})$ are the LUS estimates of $\mathbf{e}(\mathbf{s})$. Jensen and Ramirez (1999) call these elements the **linearly recovered errors** (which we refer to as LREs).

In the case of OLS residuals based on a linear model with uncorrelated, homoscedastic errors, $\mathbf{A} = \sigma^2\mathbf{M}$, with $\mathbf{M}$ defined on page 305. Since $\mathbf{M}$ is a projection matrix (symmetric and idempotent), it has a spectral decomposition $\mathbf{M} = \mathbf{P}\boldsymbol{\Delta}\mathbf{P}'$, where $\mathbf{P}$ is orthogonal and $\boldsymbol{\Delta}$ is a diagonal matrix containing the eigenvalues of $\mathbf{M}$. Because of these properties,

$$\mathbf{\Delta} = \begin{bmatrix} \mathbf{I}_{(n-k)} & \mathbf{0}_{((n-k)\times k)} \\ \mathbf{0}_{(k\times(n-k))} & \mathbf{0}_{(k\times k)} \end{bmatrix}.$$

Hence, the matrix $\mathbf{Q} = \mathbf{P}/\sigma$ will transform the OLS residuals into LREs that have variance-covariance matrix $\mathbf{\Delta}$.

The process of error recovery by the aforementioned techniques produces $n-k$ recovered residuals with identity variance-covariance matrix and $k$ "residuals" whose variance is zero. If $n$ is not much larger than $k$, only a few independent residuals can be recovered. In spatial regression models this is usually not of concern, since the number of observations usually far exceeds the rank of the $\mathbf{X}(\mathbf{s})$ matrix. The linearly recovered errors also do not correspond to any particular one of the OLS residuals, they are linear combinations of them. Linearly recovered errors are thus not suited for the identification of outliers (recursive residuals do not have this drawback). Furthermore, the recovered errors are not unique. Any orthogonal rotation of an LUS estimate is also an LUS estimate. And residuals recovered by means of a Cholesky decomposition depend on the ordering of the data. Residuals recovered by spectral or singular value decomposition depend on the ordering of the eigenvalues and corresponding eigenvectors. Note that in the OLS case, where $\mathbf{M}$ is a projection matrix, all non-zero eigenvalues of $\mathbf{M}$ are equal to one and the ordering of the columns of $\mathbf{Q}$ corresponding to the non-zero eigenvalues is arbitrary. In the correlated error case an ordering of the eigenvalues by decreasing size is common. But there is no reason why recovered errors computed with this convention in mind should be *best* in any sense. Jensen and Ramirez (1999) use linearly recovered errors to test for normality of the model disturbances. These authors settle the uniqueness issue of the LUS estimates by finding an orthogonal rotation $\mathbf{BQ}'$ that maximizes the kurtosis of the recovered errors, making them as "ugly as possible," to slant diagnostic tests away from normality. Very little is known about using an analogous approach for assessing whether there is spatial autocorrelation in the recovered errors.

#### *6.1.2.2 Semivariogram Estimation using OLS Residuals*

When working with spatial data, we need to assess whether there is any *spatial* variation that has not been accounted for by the model. Such variation can be due to an omitted spatially-varying covariate or to spatial autocorrelation in the data, or both. The only real information we have for this assessment comes from the OLS residuals. If the OLS residuals exhibit spatial patterns, then the OLS regression model is not adequate to describe the spatial variation in the data. We note in passing that a simple map of the residuals can be extremely informative as can other spatial visualization techniques.

One of the most common tools for assessing spatial autocorrelation is the empirical semivariogram. Unfortunately, the empirical semivariogram computed from the OLS residuals (which we refer to as the residual semivariogram) may not be a good estimate of the semivariogram of the error process, $\mathbf{e}(\mathbf{s})$. As noted previously, the statistical properties of the two processes are very different. It is difficult to determine whether any structure (or lack thereof) apparent in the residual semivariogram is due to spatial autocorrelation in the error process or to artifacts induced by the rank deficiency, correlation, and heteroscedasticity among the residuals.

**Example 6.1 (Soil carbon regression. Continued)** Figure 6.2 displays the empirical semivariograms of the soil carbon percentages and of the OLS residuals in a simple linear regression of C% on N%. Both panels display some structure. In the left panel we do not know whether the spatial structure could be due to the fact that the average C percentage changes systematically with soil N%. In other words, the degree to which the empirical semivariogram is determined by large-scale trend is not discernible. In the right-hand panel it is not clear to what extent the structure is induced by the properties of OLS residuals, and to what extent it reflects spatial autocorrelation in the (unobservable) model errors. Note the smaller "sill" of the residual semivariogram, the decrease in the sill reflects the systematic variation in C% removed by the regression on N%.

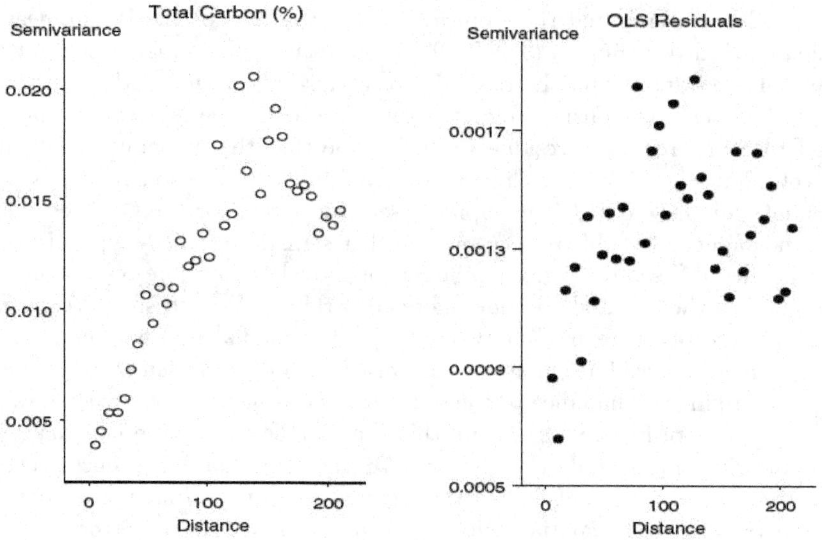


Figure 6.2 *Empirical semivariograms of soil carbon (%) (left panel) and OLS residuals of simple linear regression of C% on N%.*

□

There are several options for reducing the potential for such artifacts in the residual semivariogram. First, we could use the empirical semivariogram computed from the studentized residuals to assess residual spatial autocorrelation. Although studentized residuals remain correlated and rank-deficient, as far as computing a semivariogram is concerned, studentized residuals at least meet the needed stationary properties since they have mean zero and constant variance. This approach is fairly simple since most statistical software packages compute studentized residuals. Another, more comprehensive, solution is to compute the empirical semivariogram of the recursive residuals. Recursive residuals remove all the problems inherent in the OLS residuals. Moreover, Grondona (1989) and Grondona and Cressie (1995) show that for the linear model with uncorrelated errors, the covariance function computed from recursive residuals is an unbiased estimator of the white noise (or nugget) covariance function $C(h) = \sigma^2 I[h = 0]$. Thus, the empirical semivariogram computed from recursive residuals should be that of a white noise process unless there is residual spatial variation. If $k = \text{rank}(\mathbf{X})$ is small, it may be difficult to fit the model to the first $k$ data points needed to begin the recursive process. Thus, alternatively, we can compute the empirical semivariogram based on the LREs described above. As with the recursive residuals, the empirical semivariogram of the LREs should also resemble that of a white noise process.

**Example 6.1 (Soil carbon regression. Continued)** For the C/N regression model we computed the empirical semivariograms of the Best Linearly Unbiased Scaled estimates (Theil, 1971), the recursive residuals, the studentized OLS residuals, and the raw OLS residuals (Figure 6.3). All four types of residuals exhibit considerable structure. Based on the semivariograms of the BLUS and recursive residuals we conclude that the model errors should be considered spatially correlated. A simple linear regression model is not adequate for these data. The empirical semivariogram for the OLS residuals does not appear too different from the other semivariograms. Typically, the bias in the OLS semivariogram is larger for models with more regression coefficients, as the residuals become more constrained. With a single regressor and an intercept, there are only two redundant residuals and the leverages are fairly homogeneous. The empirical semivariogram of the studentized residuals bears a striking resemblance to the semivariograms of the recovered errors in the top panels of Figure 6.3. Again, this is partly helped by the fact that the $\mathbf{X}(\mathbf{s})$ matrix contains only two columns. On the other hand, it is encouraging that the simple process of scaling the residuals to equal variance provides as clear a picture of the spatial autocorrelation as the recovered errors, whose construction is more involved.

A possible disadvantage of recovered errors is their dependence on data order. What is important is that the residual semivariogram conveys the presence of spatial structure and the need for a spatial analysis. Figure 6.4 displays empirical semivariograms of the recursive residuals for 10 random permutations of the data set. To show the variation among the semivariograms, they are shown as series plots rather than as scatter plots. Any of the sets is equally well suited to address the question of residual spatial dependency. □

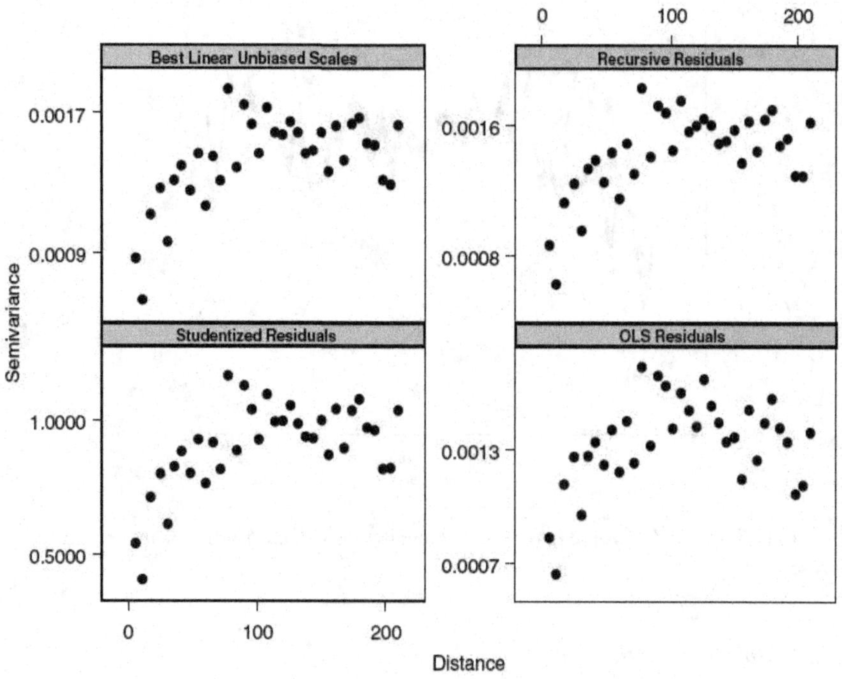


Figure 6.3 *Empirical semivariograms of various residual measures in C–N simple linear regression.*

Relatively little, practical work has been done on residual diagnostics for spatial models. In particular, the practical impacts of rank deficiency, correlation and heteroscedasticity among the OLS residuals on inferences drawn from residual variography are not clearly understood. Also, it is not clear to what extent the solutions described above are truly effective, and whether or not they may introduce other problems (e.g., lack of uniqueness in LREs). Some additional research has been done for more general linear models with correlated errors and this is discussed in §6.2.3.

Earlier we mentioned three courses of action when working with residuals: to proceed with an analysis of raw residuals, to compute transformed residuals, and to use adjusted spatial statistics. The careful interpretation of the semivariogram of OLS residuals is an example of the first action. The semivariograms for the recursive, studentized, and BLUS residuals in Figure 6.3 represent the second course. An example of taking into account the fact that statistics are computed form fitted, rather than observed, quantities is the test for autocorrelation in lattice data based on Moran's $I$ for OLS residuals.

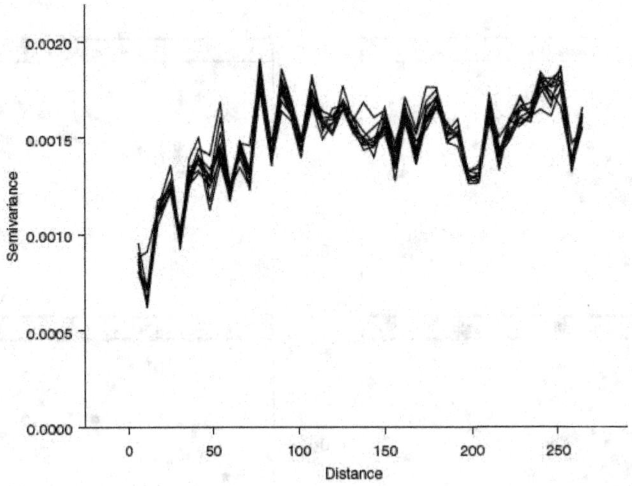


Figure 6.4 *Empirical semivariograms of recursive residuals for 10 random orderings of the data set.*

### 6.1.2.3 Moran's I with OLS Residuals

In the case of lattice data, it is tempting to compute measures of spatial autocorrelation—such as Moran's $I$ or Geary's $c$—from the OLS residuals. The Moran statistic based on an OLS fit is

*Moran's I for OLS Residuals*

$$I_{res} = \frac{n}{w_{..}} \frac{\widehat{\mathbf{e}}_{ols}(\mathbf{s})'\mathbf{W}\widehat{\mathbf{e}}_{ols}(\mathbf{s})}{\widehat{\mathbf{e}}_{ols}(\mathbf{s})'\widehat{\mathbf{e}}_{ols}(\mathbf{s})}, \tag{6.5}$$

where $w_{..} = \mathbf{1}'\mathbf{W}\mathbf{1} = \sum_{i=1}^{n}\sum_{j=1}^{n} w_{ij}$ and $\mathbf{W}$ is the spatial connectivity matrix (see §1.3.2). Note that $I_{res}$ can also be written as

$$I_{res} = \frac{n}{w_{..}} \frac{\mathbf{e}(\mathbf{s})'\mathbf{M}\mathbf{W}\mathbf{M}\mathbf{e}(\mathbf{s})}{\mathbf{e}(\mathbf{s})'\mathbf{M}\mathbf{e}(\mathbf{s})}.$$

So far, these expressions are a simple application of the Moran's $I$ statistic to $\widehat{\mathbf{e}}_{ols}(\mathbf{s})$, no adjustment has been made yet for the fact that the "data" are fitted quantities. We noted earlier that one path to OLS residual analysis is to apply methods that have been specifically devised to account for the properties of fitted residuals. This accounting takes place in the next step. Since the error terms, $e(\mathbf{s}_i)$, are assumed to be *iid* Gaussian—under the null hypothesis of no spatial autocorrelation—Cliff and Ord (1981, pp. 202–203) show that $I_{res}$ is asymptotically Gaussian (for $n \to \infty$) with mean

*Mean and Variance of $I_{res}$*

$$\mathrm{E}[I_{res}] = -\left[\frac{n}{(n-k)w_{..}}\right] \mathrm{tr}\left\{(\mathbf{X}(\mathbf{s})'\mathbf{X}(\mathbf{s}))^{-1}\mathbf{X}(\mathbf{s})'\mathbf{W}\mathbf{X}(\mathbf{s})\right\}$$

$$= \frac{n}{(n-k)w_{..}}\text{tr}\{\mathbf{MW}\} \tag{6.6}$$

and variance

$$\begin{aligned}\text{Var}[I_{res}] &= \left\{\frac{n^2}{w_{..}^2(n-k)(n-k+2)}\right\}\\ &\times\left\{S_1 + 2\text{tr}\left\{\mathbf{G}^2\right\} - \text{tr}\{\mathbf{F}\} - \frac{2[\text{tr}\{\mathbf{G}\}]^2}{(n-k)}\right\},\end{aligned} \tag{6.7}$$

where

$$\begin{aligned}S_1 &= \frac{1}{2}\sum_{i=1}^{n}\sum_{j=1}^{n}(w_{ij}+w_{ji})^2,\\ \mathbf{F} &= (\mathbf{X}(\text{s})'\mathbf{X}(\text{s}))^{-1}\mathbf{X}(\text{s})'(\mathbf{W}+\mathbf{W}')^2\mathbf{X}(\text{s}), \quad \text{and}\\ \mathbf{G} &= (\mathbf{X}(\text{s})'\mathbf{X}(\text{s}))^{-1}\mathbf{X}(\text{s})'\mathbf{W}\mathbf{X}(\text{s}).\end{aligned}$$

Cliff and Ord (1981, p. 200) note that randomizing the residuals does not provide the appropriate reference set for a permutation test of autocorrelation based on OLS residuals. They consider only the asymptotic test based on the asymptotic results in equations (6.6) and (6.7), and an assumption that the data are Gaussian. Thus, an approximate test of the null hypothesis of no spatial autocorrelation can be made by comparing the observed value of

$$z = \frac{I_{res} - \text{E}[I_{res}]}{\sqrt{\text{Var}[I_{res}]}} \tag{6.8}$$

to the appropriate percentage point of the standard Gaussian distribution. Note that this test is different than the one described in §1.3.2. For example, the comparable test for Moran's $I$ described in (§1.3.2) uses a mean of $-(n-1)^{-1}$ to construct the test statistic. If this test statistic is used with OLS residuals, which can be easily accomplished by importing the residuals into a software program that computes Moran's $I$, the resulting test will be incorrect; the correct mean is given in (6.6). The same argument applies to the variance. The computer will not know that you are working with residuals and so cannot adjust the mean and variance to give the proper test statistic.

As with the semivariogram, we can also compute Moran's $I$ using the recursive residuals or the LREs from LUS estimation described above. Since the recursive residuals and the LREs are free of the problems inherent in OLS residuals, a permutation test can be used. In addition, Cliff and Ord (1981, p. 204) give the moments of $I$ computed from the LREs so that an approximate $z$-test can be constructed. However, Cliff and Ord (1981, p. 204) note the problems with determining which $n-k$ recovered errors to use and the same concern applies to the use of recursive residuals in this context. They tentatively suggest using the most well-connected observations, those for which $\sum_j w_{ij}$ is the largest. Their discussion and recommendation clearly implies the potential inferential perils that can occur when using recovered errors to assess spatial autocorrelation.

### 6.1.3 Spatially Explicit Models

Consider again the linear regression model with uncorrelated errors given in (6.2). It is a spatial regression model since the dependent variable $\mathbf{Z}(\mathbf{s})$, and the independent variables comprising $\mathbf{X}(\mathbf{s})$, are recorded at spatial locations $\mathbf{s}_1, \ldots, \mathbf{s}_n$. However, for most independent variables, the spatial aspect of the problem serves only to link $\mathbf{Z}(\mathbf{s})$ and $\mathbf{X}(\mathbf{s})$. Once the dependent and independent variables are linked through location, there is nothing in the analysis that explicitly considers spatial pattern or spatial relationships. In fact, if we give you $\mathbf{Z}(\mathbf{s})$ and $\mathbf{X}(\mathbf{s})$, but simply refer to them as $\mathbf{Z}$ and $\mathbf{X}$, you could apply any and all tools from regression analysis to understand the effect of $\mathbf{X}$ on $\mathbf{Z}$. Moreover, you could move them around in space and still get the same results (provided you move $Z(\mathbf{s})$ and its corresponding covariates together). Fotheringham, Brunsdon, and Charlton (2002) refer to such analyses as **aspatial**, a term we find informative. The field of spatial statistics is far from aspatial, and even in the simple linear model case, there is more that can be done to use spatial information more explicitly.

#### 6.1.3.1 Local Polynomial and Geographically Weighted Regression

One of the easiest ways to make more use of spatial information and relationships is to use covariates that are polynomial functions of the spatial coordinates $\mathbf{s}_i = [x_i, y_i]'$. The trend surface models described in §5.3.1 are an example of this approach. For example, a linear trend surface uses a first degree polynomial in [x,y] to describe the spatial variation in the response, e.g.,

$$Z(\mathbf{s}_i) = \beta_0 + \beta_1 x_i + \beta_2 y_i + \epsilon_i, \quad \epsilon_i \sim iid\,(0, \sigma^2).$$

Ordinary least squares estimation inference can be used to estimate the $\beta$ parameters. However, such an analysis is not aspatial; $\mathbf{X}$ is clearly completely tied to the spatial locations. Although in §5.3.1, the parameter estimates were simply a means of obtaining a response surface, the $\beta$ coefficients themselves have a spatial interpretation, measuring the strength of large-scale spatial trends in $Z(\mathbf{s}_i)$.

The parameter estimates from a trend surface analysis provide a fairly broad, large-scale interpretation of the spatial variation in $Z(\mathbf{s}_i)$. However, they are essentially aspatial, since the model has one set of parameters that apply everywhere, regardless of spatial location. As discussed in §5.3.2, we can adapt traditional local polynomial regression to the spatial case by fitting a polynomial model at any specified spatial location $\mathbf{s}_0$. This model is essentially a spatial version of the local estimation procedures commonly referred to as LOESS or nonparametric regression, where the covariates are polynomial functions of the spatial coordinates. In traditional applications of LOESS and nonparametric regression methods where general covariates form $\mathbf{X}$, the term *local* refers to the attribute or $X$-space and not to spatial location. The weights are functions of $x_i - x_0$, differences in covariate values, and the analysis is aspatial (Fotheringham et al., 2002, pp. 3–4). What is really needed is a model that is fit locally in the spatial sense, but allows general covariates that are not necessarily polynomial functions of the spatial coordinates. The same general model as that described in §5.3.2 can be used, but with general covariates

$$\mathbf{Z}(\mathbf{s}) = \mathbf{X}(\mathbf{s})\boldsymbol{\beta}(\mathbf{s}_0) + \mathbf{e}_0, \quad \mathbf{e}_0 \sim (\mathbf{0}, \sigma^2\mathbf{W}(\mathbf{s}_0)^{-1}), \tag{6.9}$$

where the weights forming the elements of $\mathbf{W}(\mathbf{s}_0)$, $W(\mathbf{s}_i, \mathbf{s}_0)$, determine the degree to which $Z(\mathbf{s}_i)$ is allowed to affect the estimate of the relationship between $\mathbf{X}(\mathbf{s})$ and $\mathbf{Z}(\mathbf{s})$ at the point of interest $\mathbf{s}_0$. The covariates in $\mathbf{X}(\mathbf{s})$ do not have to be polynomial functions of spatial coordinates. They can be, e.g., price, income, elevation, race, etc. However, in contrast to traditional kernel regression approaches, the local aspect of the model is based on the spatial locations, not on the covariate values. In contrast to the models for local estimation described in §5.3.2, the interest lies not in a smoothed response surface, but in smoothed maps of the $\beta_j$ that describe the spatial variation in the relationship between $x_j(\mathbf{s})$ and $Z(\mathbf{s})$. In geography, this model is known as **geographically weighted regression** (Fotheringham et al., 2002).

To avoid unnatural discontinuities in the parameter estimates, the spatial weight function has to be chosen in such a way that the influence of an observation decreases with increasing distance from $\mathbf{s}_0$. The kernel weight functions given in §5.3.2 are some examples of weights that satisfy this condition. Cressie (1998) and Fotheringham et al. (2002) provide more general discussions on the choice of spatial function and additional alternatives.

Once the spatial weight function is selected, $\boldsymbol{\beta}(\mathbf{s}_0)$ can be estimated by weighted least squares

$$\widehat{\boldsymbol{\beta}}(\mathbf{s}_0) = \left(\mathbf{X}(\mathbf{s})'\mathbf{W}(\mathbf{s}_0)\mathbf{X}(\mathbf{s})\right)^{-1}\mathbf{X}(\mathbf{s})'\mathbf{W}(\mathbf{s}_0)\mathbf{Z}(\mathbf{s}),$$

and the variance of the parameter estimates is

$$\mathrm{Var}[\widehat{\boldsymbol{\beta}}(\mathbf{s}_0)] = \sigma^2\mathbf{C}\mathbf{C}', \quad \mathbf{C} = \left(\mathbf{X}(\mathbf{s})'\mathbf{W}(\mathbf{s}_0)\mathbf{X}(\mathbf{s})\right)^{-1}\mathbf{X}(\mathbf{s})'\mathbf{W}(\mathbf{s}_0).$$

If $\boldsymbol{\beta}$ is estimated at the sample locations, $\mathbf{s}_i$, then the fitted values are given by $\widehat{\mathbf{Z}}(\mathbf{s}) = \mathbf{L}\mathbf{Z}(\mathbf{s})$, where the $i^{th}$ row of $\mathbf{L}$ is $\mathbf{x}(\mathbf{s}_i)\left(\mathbf{X}(\mathbf{s})'\mathbf{W}(\mathbf{s}_i)\mathbf{X}(\mathbf{s})\right)^{-1}\mathbf{X}(\mathbf{s})'\mathbf{W}(\mathbf{s}_i)$. The variance component, $\sigma^2$, can be estimated from the residuals of this fit using (Cressie, 1998)

$$\widehat{\sigma}^2 = \frac{(\mathbf{Z}(\mathbf{s}) - \widehat{\mathbf{Z}}(\mathbf{s}))'(\mathbf{Z}(\mathbf{s}) - \widehat{\mathbf{Z}}(\mathbf{s}))}{\mathrm{tr}\{(\mathbf{I} - \mathbf{L})(\mathbf{I} - \mathbf{L})'\}}.$$

Cressie (1998) gives more details on how this local model can be used in geostatistical analysis and Fotheringham et al. (2002) provide many practical examples of how this model, and various extensions of it, can be used in geographical analysis.

#### 6.1.3.2 Models with Neighbor Adjustments—Papadakis Analysis

Models with neighbor adjustments originated in agricultural experimentation to cope with the problem of spatial trends at scales different from the size of blocks. In particular, to obtain valid inferences in the presence of soil fertility trends in large variety trials. Starting from a classification model that captures the treatment and design structure—often the design-based model—the model is changed to an analysis of covariance. The new covariates are functions of residuals between neighboring sites, assumed to capture the salient spatial and treatment effects, so that unbiased (and precise) comparisons of treatments can be made. The process can be illustrated nicely with an important representative in this class of models, the Papadakis model (Papadakis, 1937).

Suppose an experimental area is a two-dimensional layout of $r$ rows and $c$ columns. The $rc$ experimental units are grouped into $b$ blocks. The treatments $T_1, \cdots, T_t$ are assigned at random to the units in each block so that treatment $T_i$ is replicated $n_i$ times within a block. Most common is the randomized complete block design for which $n_i = 1, \forall i$. The design-based linear model for the analysis of this experiment can be written as

$$
\begin{aligned}
Z(i,j)_{kl} &= \mu + \rho_k + \tau_l + \epsilon_{kl} \\
i &= 1,\cdots,r \quad j = 1,\cdots,c \\
k &= 1,\cdots,b \quad l = 1,\cdots,t,
\end{aligned}
\tag{6.10}
$$

where $\rho_k$ is the effect of the $k$th block and $\tau_l$ is the effect of the $l$th treatment (Hinkelmann and Kempthorne, 1994). The somewhat unusual notation is used to identify blocks and treatments (indices $k$ and $l$), as well as lattice positions.

The Papadakis analysis is essentially an analysis of covariance where the block effects in (6.10) are replaced by functions of OLS residuals in the model $Z(i,j)_{kl} = \mu + \tau_l + \epsilon_{kl}$. Because the residuals are based on a different model, the analysis involves several steps. In the first step the model with only treatment effects is fit and the residuals

$$
\widehat{\epsilon}(i,j) = Z(i,j)_{kl} - \widehat{\mathrm{E}}[Z(i,j)_{kl}] = Z(i,j)_{kl} - \overline{Z}_k
$$

are computed. Here, $\overline{Z}_k$ denotes the arithmetic average of the treatment $k$, the treatment that was assigned to the plot in row $i$, column $j$. Based on these residuals, and a definition of which plots constitute spatial neighbors, covariates are computed. For example, Stroup, Baenziger, and Mulitze (1994) define the East-West and North-South differences

$$
\begin{aligned}
x_1(i,j) &= \frac{1}{2}(\widehat{\epsilon}(i,j-1) + \widehat{\epsilon}(i,j+1)) \\
x_2(i,j) &= \frac{1}{2}(\widehat{\epsilon}(i-1,j) + \widehat{\epsilon}(i+1,j))
\end{aligned}
$$

These adjustments then replace the block effects in (6.10). A Papadakis model for these data could be

$$Z(i,j)_{kl} = \beta_0 + \tau_l + \beta_1 x_1(i,j) + \beta_2 x_2(i,j) + \epsilon^*_{kl}.$$

There are many ambiguities in this analysis for which the user needs to make a determination. In a two-dimensional layout, you can define a single covariate adjusting in both directions, or separate covariates in each direction. You need to decide on how to handle edge-effects, EUs near the boundary of the experimental area. You need to decide whether to include immediate neighbors into the adjustments or extend to second-or higher-order differences. The analysis can be non-iterative (as described above) or iterative, the residuals from the analysis of covariance model are then used to recompute the covariates and the process continues until changes are sufficiently small.

Since the covariates in the Papadakis analysis are linear functions of the responses $\mathbf{Z}(\mathbf{s})$, the analysis essentially accounts for local linear trends. This is one of the reasons why mean adjustments with this analysis tend to be more moderate than in models that add trend surface components, which tend to have a higher degree of the polynomial. Overfit trend surface models can produce large local adjustments to the treatment means (Brownie, Bowman, and Burton, 1993).

It is instructive to view the Papadakis analysis in a slightly different light. The OLS residuals from which the covariates are constructed, are linear functions of the data in the model $\mathbf{Z}(\mathbf{s}) = \mathbf{X}\boldsymbol{\tau} + \boldsymbol{\epsilon}$, namely $\widehat{\boldsymbol{\epsilon}} = \mathbf{MZ}(\mathbf{s})$, $\mathbf{M} = \mathbf{I} - \mathbf{X}(\mathbf{X}'\mathbf{X})^{-1}\mathbf{X}'$. Consider the case of a single Papadakis covariate. Because $\widehat{\boldsymbol{\tau}}_{ols} = (\mathbf{X}'\mathbf{X})^{-1}\mathbf{X}'\mathbf{Z}(\mathbf{s})$, the resulting model can be written as

$$\begin{aligned}\mathbf{Z}(\mathbf{s}) &= \mathbf{X}\boldsymbol{\tau} + \beta\mathbf{A}\widehat{\boldsymbol{\epsilon}} + \boldsymbol{\epsilon}^* \\ &= \mathbf{X}\boldsymbol{\tau} + \beta\mathbf{AMZ}(\mathbf{s}) + \boldsymbol{\epsilon}^* \\ &= \mathbf{X}\boldsymbol{\tau} + \beta\mathbf{A}(\mathbf{Z}(\mathbf{s}) - \mathbf{X}\widehat{\boldsymbol{\tau}}_{ols}) + \boldsymbol{\epsilon}^*.\end{aligned}$$

The matrix $\mathbf{A}$ determines how the averages of the residuals are determined. One of the shortcomings of the Papadakis analysis is that the covariates $\mathbf{MZ}(\mathbf{s})$ are assumed fixed, only $\boldsymbol{\epsilon}^*$ is treated as a random component on the right hand side. Furthermore, the elements of $\boldsymbol{\epsilon}^*$ are considered uncorrelated and homoscedastic. If the variation in $\mathbf{Z}(\mathbf{s})$ were taken into account on the right hand side, a very different correlation model would result. But if we can accommodate $\mathbf{Z}(\mathbf{s})$ on the right hand side as a random variable, then there is no need to rely on the fitted OLS residuals $\widehat{\boldsymbol{\tau}}_{ols}$ in the first place. We could then fit a model of the form

$$\mathbf{Z}(\mathbf{s}) = \mathbf{X}\boldsymbol{\tau} + \beta\mathbf{W}(\mathbf{Z}(\mathbf{s}) - \mathbf{X}\boldsymbol{\tau}) + \boldsymbol{\epsilon}^*. \tag{6.11}$$

The matrix $\mathbf{A}$ has been replaced by the matrix $\mathbf{W}$, because how you involve the model errors may be different from how you use fitted residuals for adjustments. The important point is that in equation (6.11) the response is regressed on its own residual. This model has autoregressive form, it is a simultaneous autoregressive model (§6.2.2.1), a special case of a correlated error model. Studies as those by Zimmerman and Harville (1991), Brownie,

Bowman, and Burton (1993), Stroup, Baenziger, and Mulitze (1994), and Brownie and Gumpertz (1997) have shown that correlated error models typically outperform trend surface models and neighbor-adjusted models with uncorrelated errors. The importance of models with neighbor adjustments lies in their connection to autoregressive models, and thus as stepping stones to other correlated error models.

#### *6.1.3.3 First Difference Model*

The idea of adding polynomial trends in spatial coordinates or nearest neighbor residuals in the Papadakis analyses is that suitable adjustments to the mean function justify a model with uncorrelated errors. A different route, also leading to a model with uncorrelated errors, was taken by Besag and Kempton (1986), who described a model for spatially arranged experimental designs. They consider field experiments arranged in adjacent columns, where the number of rows is much larger than the number of columns (which may be 1). In other words, these layouts are rectangular and elongated. In this case, adjustments for neighboring plots are likely to be made within the row only, partly because plants within a row are often planted more closely than plants across rows. We focus on a single column for the time being, the extension to multiple rows and columns is straightforward. The first difference model is formed from

$$Z(s) = \mathbf{X}\boldsymbol{\tau} + \mathbf{e}(\mathbf{s}), \qquad \mathbf{e}(\mathbf{s}) \sim (\mathbf{0}, \sigma^2\boldsymbol{\Sigma}). \tag{6.12}$$

If $Z(i)$ is the response value in row $i$, then it is assumed that $\text{Var}[Z(i) - Z(i+1)] = \sigma^2$ and $\text{Cov}[Z(i) - Z(i+1), Z(k) - Z(k+1)] = 0, i \neq k$. In other words, the first differences between observations in the same column are homoscedastic and uncorrelated. The model (6.12) can then be transformed to a first-difference model by multiplying both sides of the model (from the left) with the matrix

$$\boldsymbol{\Delta} = \begin{bmatrix} 1 & -1 & 0 & \cdots & 0 \\ 0 & 1 & -1 & \cdots & 0 \\ \vdots & \vdots & & & \vdots \\ 0 & 0 & \cdots & 1 & -1 \end{bmatrix}, \tag{6.13}$$

The resulting model is an OLS model:

$$\begin{aligned} \boldsymbol{\Delta}\mathbf{Z}(\mathbf{s}) &= \boldsymbol{\Delta}\mathbf{X}\boldsymbol{\tau} + \boldsymbol{\Delta}\mathbf{e}(\mathbf{s}) \\ \mathbf{Z}(\mathbf{s})^* &= \mathbf{X}^*\boldsymbol{\tau} + \mathbf{e}(\mathbf{s})^* \\ \mathbf{e}(\mathbf{s})^* &\sim (\mathbf{0}, \sigma^2\mathbf{I}), \end{aligned}$$

from which the treatment effects and their precision can be estimated as

$$\begin{aligned} \widehat{\boldsymbol{\tau}} &= (\mathbf{X}^{*\prime}\mathbf{X}^*)^{-1}\mathbf{X}^{*\prime}\mathbf{Z}(\mathbf{s})^* \\ \widehat{\text{Var}}[\widehat{\boldsymbol{\tau}}] &= \widehat{\sigma}^2(\mathbf{X}^{*\prime}\mathbf{X}^*)^{-1} \\ \widehat{\sigma}^2 &= \frac{1}{n - \text{rank}\{\mathbf{X}\}}(\mathbf{Z}(\mathbf{s})^* - \mathbf{X}^*\widehat{\boldsymbol{\tau}})'(\mathbf{Z}(\mathbf{s})^* - \mathbf{X}^*\widehat{\boldsymbol{\tau}}). \end{aligned}$$

The importance of the Besag-Kempton model lies in its connection to generalized least squares estimation in a correlated error model. In the initial model (6.12) the errors are correlated, but a matrix $\mathbf{\Delta}$ is known a priori, such that $\mathbf{\Delta\Sigma\Delta}' = \mathbf{I}$. To bring out the connection more clearly, take a correlated error model with general linear mean function,

$$\mathbf{Z}(\mathbf{s}) = \mathbf{X}(\mathbf{s})\boldsymbol{\beta} + \mathbf{e}(\mathbf{s}), \qquad \mathbf{e}(\mathbf{s}) \sim (\mathbf{0}, \sigma^2\mathbf{\Sigma}).$$

If the (correlation) matrix $\mathbf{\Sigma}$ is known, the generalized least squares estimates of $\boldsymbol{\beta}$ can be obtained by applying ordinary least squares to a suitably transformed model. Let $\mathbf{L}$ be a matrix such that $\mathbf{LL}' = \mathbf{\Sigma}$. If $\mathbf{\Sigma}$ is positive definite, for example, $\mathbf{L}$ can be chosen as its lower triangular Cholesky root. Or, $\mathbf{L}$ can be constructed based on a singular value decomposition. Then, $\mathbf{L}^{-1}\mathbf{\Sigma}\mathbf{L}'^{-1} = \mathbf{I}$, and the model can be transformed to

$$\begin{aligned}
\mathbf{L}^{-1}\mathbf{Z}(\mathbf{s}) &= \mathbf{L}^{-1}\mathbf{X}(\mathbf{s})\boldsymbol{\beta} + \mathbf{L}^{-1}\mathbf{e}(\mathbf{s}) \\
\mathbf{Z}(\mathbf{s})^* &= \mathbf{X}(\mathbf{s})^*\boldsymbol{\beta} + \mathbf{e}(\mathbf{s})^* \\
\mathbf{e}(\mathbf{s})^* &\sim (\mathbf{0}, \sigma^2\mathbf{I}).
\end{aligned}$$

The ordinary least squares estimates in the transformed model are the generalized least squares estimates in the correlated error model:

$$\begin{aligned}
\widehat{\boldsymbol{\beta}}_{gls} &= (\mathbf{X}(\mathbf{s})^{*\prime}\mathbf{X}(\mathbf{s})^*)^{-1}\mathbf{X}(\mathbf{s})^{*\prime}\mathbf{Z}(\mathbf{s})^* \\
&= (\mathbf{X}(\mathbf{s})'\mathbf{L}^{-1\prime}\mathbf{L}^{-1}\mathbf{X}(\mathbf{s}))^{-1}\mathbf{X}(\mathbf{s})'\mathbf{L}^{-1\prime}\mathbf{L}^{-1}\mathbf{Z}(\mathbf{s}) \\
&= (\mathbf{X}(\mathbf{s})'\mathbf{\Sigma}^{-1}\mathbf{X}(\mathbf{s}))^{-1}\mathbf{X}(\mathbf{s})'\mathbf{\Sigma}^{-1}\mathbf{Z}(\mathbf{s}).
\end{aligned}$$

The first difference matrix $\mathbf{\Delta}$, when applied to model (6.12), plays the role of the inverse "square root" matrix $\mathbf{L}^{-1}$. It transforms the model into one with uncorrelated errors. In the correlated error model, the transformation matrix $\mathbf{L}^{-1}$ is known because $\mathbf{\Sigma}$ is known. In the first-difference approach we presume knowledge about the transformation directly, at least up to a multiplicative constant. Note also, that the differencing process produces a model for $n-1$, rather than $n$, observations. The reality of fitting models with correlated errors is that $\mathbf{\Sigma}$ is unknown, at least it is unknown up to some parameter vector $\boldsymbol{\theta}$. Thus, the square root matrix is also unknown.

## 6.2 Linear Models with Correlated Errors

In this section, we continue to assume a linear relationship between the outcome variable and the covariates, but in addition to the spatially-varying covariates, the components of the error term may now be spatially-correlated. Thus, we assume the more general model of equation (6.1).

In addition to there being numerous estimation techniques for the covariance parameters $\boldsymbol{\theta}$, there are different modeling approaches that can be used to describe the dependence of $\mathbf{\Sigma}(\boldsymbol{\theta})$ on $\boldsymbol{\theta}$. We discuss two approaches in this chapter. The first is based on the models for the covariance function described in §4.3 and applies to spatially-continuous (geostatistical) data. The second approach is based on the spatial proximity measures described in §1.3 and applies to regional data. In what follows, we first focus on geostatistical models for $\mathbf{\Sigma}(\boldsymbol{\theta})$ and their use in linear regression models with spatially correlated errors. The approach for regional data will be described in §6.2.2. Our concern now is primarily in estimates of $\boldsymbol{\theta}$ for inference about $\boldsymbol{\beta}$, rather than in prediction of the $\mathbf{Z}(\mathbf{s})$ process.

If $\boldsymbol{\theta}$ is known, the generalized least estimator

$$\widehat{\boldsymbol{\beta}}_{gls} = \left(\mathbf{X}(\mathbf{s})'\mathbf{\Sigma}(\boldsymbol{\theta})^{-1}\mathbf{X}(\mathbf{s})\right)^{-1}\mathbf{X}(\mathbf{s})'\mathbf{\Sigma}(\boldsymbol{\theta})^{-1}\mathbf{Z}(\mathbf{s}) \tag{6.14}$$

can be used to estimate $\boldsymbol{\beta}$. In order to use this estimator for statistical inference, we need to be sure it is a consistent estimator of $\boldsymbol{\beta}$ and then determine its distributional properties. If we assume the errors, $\mathbf{e}(\mathbf{s})$, follow a Gaussian distribution, then the maximum likelihood estimator of $\boldsymbol{\beta}$ is equivalent to the generalized least squares estimator. Thus, the generalized least squares estimator is consistent for $\boldsymbol{\beta}$ and $\widehat{\boldsymbol{\beta}}_{gls} \sim G(\boldsymbol{\beta}, (\mathbf{X}(\mathbf{s})\mathbf{\Sigma}(\boldsymbol{\theta})^{-1}\mathbf{X}(\mathbf{s}))^{-1})$. However, if the errors are not Gaussian, these properties are not guaranteed. Consistency of $\widehat{\boldsymbol{\beta}}_{gls}$ depends on the $\mathbf{X}(\mathbf{s})$ matrix and the covariance matrix $\mathbf{\Sigma}(\boldsymbol{\theta})$. One condition that will ensure the consistency of $\widehat{\boldsymbol{\beta}}_{gls}$ for $\boldsymbol{\beta}$ is

$$\lim_{n\to\infty} \frac{(\mathbf{X}(\mathbf{s})'\mathbf{\Sigma}(\boldsymbol{\theta})^{-1}\mathbf{X}(\mathbf{s}))}{n} = \mathbf{Q}, \tag{6.15}$$

where $\mathbf{Q}$ is a finite, nonsingular matrix (see, e.g., Judge et al., 1985, p. 175). The asymptotic properties of $\widehat{\boldsymbol{\beta}}_{gls}$ are derived from the asymptotic properties of

$$\sqrt{n}(\widehat{\boldsymbol{\beta}}_{gls} - \boldsymbol{\beta}) = \left(\frac{\mathbf{X}(\mathbf{s})'\mathbf{\Sigma}(\boldsymbol{\theta})^{-1}\mathbf{X}(\mathbf{s})}{n}\right)^{-1} \frac{\mathbf{X}(\mathbf{s})'\mathbf{\Sigma}(\boldsymbol{\theta})^{-1}\mathbf{e}(\mathbf{s})}{\sqrt{n}}.$$

Most central limit theorems given in basic statistics books are not directly relevant to this problem since $\mathbf{X}(\mathbf{s})'\mathbf{\Sigma}(\boldsymbol{\theta})^{-1}\mathbf{e}(\mathbf{s})$ is not a sum of independent and identically distributed random variables. One condition that can be applied here is called the Lindberg-Feller central limit theorem and this condition can be used to show that (Schmidt, 1976)

$$\sqrt{n}(\widehat{\boldsymbol{\beta}}_{gls} - \boldsymbol{\beta}) \xrightarrow{d} G(\mathbf{0}, \mathbf{Q}^{-1}).$$

Thus, if the condition in (6.15) and some added regularity conditions for central limit theorems are satisfied, then

$$\widehat{\boldsymbol{\beta}}_{gls} \stackrel{\cdot}{\sim} G(\boldsymbol{\beta}, (\mathbf{X}(\mathbf{s})'\mathbf{\Sigma}(\boldsymbol{\theta})^{-1}\mathbf{X}(\mathbf{s}))^{-1}).$$

If $\boldsymbol{\theta}$ is unknown, iteratively reweighted generalized least squares can be used to estimate $\boldsymbol{\beta}$ and $\boldsymbol{\theta}$ in the general model of equation (6.1) (see §5.5.1). This gives an estimated generalized least squares estimator of the fixed effects $\boldsymbol{\beta}$,

*Estimated GLS Estimator*

$$\widetilde{\boldsymbol{\beta}}_{egls} = \left(\mathbf{X}(\mathbf{s})'\mathbf{\Sigma}(\widetilde{\boldsymbol{\theta}})^{-1}\mathbf{X}(\mathbf{s})\right)^{-1}\mathbf{X}(\mathbf{s})'\mathbf{\Sigma}(\widetilde{\boldsymbol{\theta}})^{-1}\mathbf{Z}(\mathbf{s}), \tag{6.16}$$

where $\widetilde{\boldsymbol{\theta}}$ is an estimator of the covariance parameters, obtained by the methods in §5.5 in the case of a spatially varying mean (and by the methods in §4.5–§4.6 in the case of a constant mean). As in Chapter 5, we need to be concerned with the effects of using estimated covariance parameters in plug-in expressions such as (6.16).

Derivation of the distributional properties of $\widetilde{\boldsymbol{\beta}}_{egls}$ is difficult because $\boldsymbol{\Sigma}(\widetilde{\boldsymbol{\theta}})$ and $\mathbf{e}(\mathbf{s})$ will be correlated. Thus, we again turn to asymptotic results. If, in addition to the condition in (6.15), the conditions

$$\lim_{n\to\infty} n^{-1}\mathbf{X}(\mathbf{s})'[\boldsymbol{\Sigma}(\widetilde{\boldsymbol{\theta}}) - \boldsymbol{\Sigma}(\boldsymbol{\theta})]\mathbf{X}(\mathbf{s}) \xrightarrow{p} 0$$

and

$$\lim_{n\to\infty} n^{-1/2}\mathbf{X}(\mathbf{s})'[\boldsymbol{\Sigma}(\widetilde{\boldsymbol{\theta}}) - \boldsymbol{\Sigma}(\boldsymbol{\theta})]\mathbf{e}(\mathbf{s}) \xrightarrow{p} \mathbf{0}$$

hold, then $\widetilde{\boldsymbol{\beta}}_{egls}$ has the same limiting distribution as $\widehat{\boldsymbol{\beta}}_{gls}$ and so

$$\sqrt{n}(\widetilde{\boldsymbol{\beta}}_{egls} - \boldsymbol{\beta}) \xrightarrow{d} G(\mathbf{0}, \mathbf{Q}^{-1})$$

(Theil, 1971; Schmidt, 1976; Judge et al., 1985, p. 176). Thus, if these conditions hold,

$$\widetilde{\boldsymbol{\beta}}_{egls} \stackrel{\cdot}{\sim} G(\boldsymbol{\beta}, (\mathbf{X}(\mathbf{s})'\boldsymbol{\Sigma}(\boldsymbol{\theta})^{-1}\mathbf{X}(\mathbf{s}))^{-1}).$$

In many cases, these conditions are met if $\widetilde{\boldsymbol{\theta}}$ is a consistent estimator of $\boldsymbol{\theta}$. However, the consistency of $\widetilde{\boldsymbol{\theta}}$ is not sufficient; sufficient conditions are given in Fuller and Battese (1973). Fuller and Battese show that these conditions are met for some commonly nested error structure models and Theil (1971) gives conditions for the AR(1) time series process.

We can bypass the checking of these conditions by using likelihood methods to estimate $\boldsymbol{\beta}$ and $\boldsymbol{\theta}$ simultaneously. We discussed ML and REML estimation in §5.5.2, and the variance-covariance matrix of these estimators can be obtained from the information matrix given in (5.47). However, even if we use ML or REML, or we can be satisfied that the asymptotic properties of the estimated generalized least squares estimator are valid, in practice inference will require estimation of

$$\text{Var}(\widetilde{\boldsymbol{\beta}}) = (\mathbf{X}(\mathbf{s})'\boldsymbol{\Sigma}(\boldsymbol{\theta})^{-1}\mathbf{X}(\mathbf{s}))^{-1}.$$

The consequences of using the plug-in expression $(\mathbf{X}(\mathbf{s})'\boldsymbol{\Sigma}(\widetilde{\boldsymbol{\theta}})^{-1}\mathbf{X}(\mathbf{s}))^{-1})$ as an estimator of $\text{Var}(\widetilde{\boldsymbol{\beta}})$ are discussed in §6.2.3.1.

**Example 6.1 (Soil carbon regression. Continued)** By studying empirical semivariograms of various types of residuals, we convinced ourselves that the model errors in the regression of soil C% on soil N% are correlated (see Figure 6.3 on page 313). Based on these semivariograms we now propose the following model:

$$\begin{aligned} C(\mathbf{s}_i)|N(\mathbf{s}_i) &= \beta_0 + \beta_1 N(\mathbf{s}_i) + e(\mathbf{s}_i) \\ e(\mathbf{s}_i) &\sim G(0, \sigma^2) \\ \text{Cov}[e(\mathbf{s}_i), e(\mathbf{s}_j)] &= \sigma^2 \exp\{-||\mathbf{s}_i - \mathbf{s}_j||/\theta\}, \end{aligned} \quad (6.17)$$

a spatial regression model with soil N% as the regressor and an exponential covariance structure. Later in this chapter we will diagnose whether this is the appropriate spatial covariance structure for these data. Table 6.1 displays the estimates for the fixed effects and covariance parameters under the assumption that the errors are uncorrelated and in the spatial models. At first glance the estimates of the fixed effects and their standard errors do not appear to differ much between a model with uncorrelated errors (OLS in Table 6.1) and the REML/ML fits of the correlated error model. The estimates of the mean soil carbon percentage will be similar. This should not be too surprising, since the OLS estimates are unbiased as long as the model errors have zero mean. You do not incur bias in OLS estimation because of correlations of the model errors. However, the OLS estimates can be highly inefficient if data are correlated.

Table 6.1 *Estimated parameters for C–N regression. Covariance structure for REML and ML estimation is exponential. Independent error assumption for OLS estimation.*

| | OLS | | REML | | ML | |
|---|---|---|---|---|---|---|
| Parameter | Est. | Std. Err. | Est. | Std. Err. | Est. | Std. Err. |
| $\beta_0$ | −0.0078 | 0.0202 | 0.0064 | 0.0228 | 0.0058 | 0.0226 |
| $\beta_1$ | 10.900 | 0.2652 | 10.733 | 0.2960 | 10.738 | 0.2936 |
| $\theta$ | — | — | 14.128 | 3.0459 | 13.661 | 2.9172 |
| $\sigma^2$ | 0.0016 | 0.0001 | 0.0017 | 0.0002 | 0.0016 | 0.0002 |

Furthermore, a cursory glance at parameter estimates for the fixed effects and their standard errors does not tell us whether the inclusion of correlated errors improved the model. We need a formal method to compare the models. We are revisiting the data in Table 6.1 in §6.2.3.1 where these formal methods are discussed.

Note that the standard errors of the OLS estimates in Table 6.1 are smaller than those for the correlated error model. This is one indication of the bias in estimating the precision of the estimates. OLS tends to overstate the precision of the estimates.

An important difference between the model with uncorrelated and with uncorrelated data is in prediction of soil carbon values. In the spatial model a prediction of soil carbon at one of the sampled locations reproduces the observed value. In the OLS model the prediction equals the estimate of the mean. Predictions of soil C% are spatially sensitive in the correlated error model. Predictions in the OLS model are spatially sensitive only to the extent that the regressor, soil N%, varies spatially. □

### 6.2.1 Mixed Models

In the previous section, any variation not explained by the parametric mean function, $\mathbf{X}(\mathbf{s})\boldsymbol{\beta}$, was assumed to be unstructured, random, spatial variation. However, in many applications, the variation reflected in $\mathbf{e}(\mathbf{s})$ may have a systematic component. For example, in a randomized complete block design, it can be advantageous to separate out the variation explained by the blocking and not simply lump this variation into a general error term. Thus, we can consider **mixed models** that contain both fixed and random effects.

*General Linear Mixed Model*

The general form of a linear mixed model (LMM) is

$$\mathbf{Z}(\mathbf{s}) = \mathbf{X}(\mathbf{s})\boldsymbol{\beta} + \mathbf{U}(\mathbf{s})\boldsymbol{\alpha} + \boldsymbol{\epsilon}(\mathbf{s}), \qquad (6.18)$$

where $\boldsymbol{\alpha}$ is a $(K \times 1)$ vector of random effects with mean $\mathbf{0}$ and variance $\mathbf{G}$. The vector of model errors $\boldsymbol{\epsilon}(\mathbf{s})$ is independent of $\boldsymbol{\alpha}$ and has mean $\mathbf{0}$ and variance $\mathbf{R}$. Our inferential goal is now more complicated. In addition to estimators of the fixed effects, $\boldsymbol{\beta}$, and any parameters characterizing $\mathbf{R}$, we will also need a predictor of the random effects, $\boldsymbol{\alpha}$, as well as estimators of any parameters characterizing $\mathbf{G}$.

The mixed model (6.18) can be related to a signal model (see §2.4.1),

$$\begin{aligned} \mathbf{Z}(\mathbf{s}) &= \mathbf{S}(\mathbf{s}) + \boldsymbol{\epsilon}(\mathbf{s}) \\ \mathbf{S}(\mathbf{s}) &= \mathbf{X}(\mathbf{s})\boldsymbol{\beta} + \mathbf{W}(\mathbf{s}) + \boldsymbol{\eta}(\mathbf{s}), \end{aligned}$$

so that $\mathbf{U}(\mathbf{s})\boldsymbol{\alpha}$ corresponds to $\mathbf{W}(\mathbf{s}) + \boldsymbol{\eta}(\mathbf{s})$, the smooth-scale and micro-scale components. For the subsequent discussion, we combine these two components into $\boldsymbol{\upsilon}(\mathbf{s})$, so that (6.18) is a special case of $\mathbf{Z}(\mathbf{s}) = \mathbf{X}(\mathbf{s})\boldsymbol{\beta} + \boldsymbol{\upsilon}(\mathbf{s}) + \boldsymbol{\epsilon}(\mathbf{s})$. The various approaches to spatial modeling that draw on linear mixed model technology, differ in how $\mathbf{U}(\mathbf{s})\boldsymbol{\alpha}$ is constructed, and in their assumptions regarding $\mathbf{G}$ and $\mathbf{R}$.

But first, let us return to the general case and assume that $\mathbf{G}$ and $\mathbf{R}$ are known. The mixed model equations of Henderson (1950) are a system of equations whose solution yield $\widehat{\boldsymbol{\beta}}$ and $\widehat{\boldsymbol{\alpha}}$, the estimates of the fixed effects and predictors of the random effects. The mixed model equations can be derived using a least squares criterion and augmenting the traditional $\boldsymbol{\beta}$ vector with the random effects vector $\boldsymbol{\alpha}$. Another derivation of the mixed model equations which we present here commences by specifying the joint likelihood of $[\boldsymbol{\alpha}, \boldsymbol{\epsilon}(\mathbf{s})]$ and maximizing it with respect to $\boldsymbol{\beta}$ and $\boldsymbol{\alpha}$. Under a Gaussian assumption for both random components, this joint density is

$$\begin{aligned} f(\boldsymbol{\alpha}, \boldsymbol{\epsilon}(\mathbf{s})) &= \frac{1}{(2\pi)^{(n+K)/2}} \left| \begin{array}{cc} \mathbf{G} & \mathbf{0} \\ \mathbf{0} & \mathbf{R} \end{array} \right|^{-1/2} \\ &\quad \times \exp\left\{ -\frac{1}{2} \left[ \begin{array}{c} \boldsymbol{\alpha} \\ \mathbf{Z}(\mathbf{s}) - \mathbf{X}(\mathbf{s})\boldsymbol{\beta} - \mathbf{U}(\mathbf{s})\boldsymbol{\alpha} \end{array} \right] \left[ \begin{array}{cc} \mathbf{G} & \mathbf{0} \\ \mathbf{0} & \mathbf{R} \end{array} \right]^{-1} \right. \\ &\quad \left. \times \left[ \begin{array}{c} \boldsymbol{\alpha} \\ \mathbf{Z}(\mathbf{s}) - \mathbf{X}(\mathbf{s})\boldsymbol{\beta} - \mathbf{U}(\mathbf{s})\boldsymbol{\alpha} \end{array} \right] \right\}, \end{aligned}$$

and it is maximized when

$$Q(\beta,\alpha) = (\mathbf{Z}(\mathbf{s}) - \mathbf{X}(\mathbf{s})\beta - \mathbf{U}(\mathbf{s})\alpha)'\mathbf{R}^{-1}(\mathbf{Z}(\mathbf{s}) - \mathbf{X}(\mathbf{s})\beta - \mathbf{U}(\mathbf{s})\alpha) + \alpha'\mathbf{G}^{-1}\alpha \quad (6.19)$$

is minimized. The resulting system of equations is known as Henderson's mixed model equations

*Mixed Model Equations*

$$\begin{bmatrix} \mathbf{X}(\mathbf{s})'\mathbf{R}^{-1}\mathbf{X}(\mathbf{s}) & \mathbf{X}(\mathbf{s})'\mathbf{R}^{-1}\mathbf{U}(\mathbf{s}) \\ \mathbf{U}(\mathbf{s})'\mathbf{R}^{-1}\mathbf{X}(\mathbf{s}) & \mathbf{U}(\mathbf{s})'\mathbf{R}^{-1}\mathbf{U}(\mathbf{s}) + \mathbf{G}^{-1} \end{bmatrix} \begin{bmatrix} \widehat{\beta} \\ \widehat{\alpha} \end{bmatrix} = \begin{bmatrix} \mathbf{X}(\mathbf{s})'\mathbf{R}^{-1}\mathbf{Z}(\mathbf{s}) \\ \mathbf{U}(\mathbf{s})'\mathbf{R}^{-1}\mathbf{Z}(\mathbf{s}) \end{bmatrix}.$$

The solutions are

$$\widehat{\beta} = (\mathbf{X}(\mathbf{s})'\mathbf{V}^{-1}\mathbf{X}(\mathbf{s}))^{-1}\mathbf{X}(\mathbf{s})'\mathbf{V}^{-1}\mathbf{Z}(\mathbf{s}) \quad (6.20)$$

$$\widehat{\alpha} = \mathbf{G}\mathbf{U}(\mathbf{s})'\mathbf{V}^{-1}(\mathbf{Z}(\mathbf{s}) - \mathbf{X}(\mathbf{s})\widehat{\beta}) \quad (6.21)$$

$$\mathbf{V} = \mathbf{U}(\mathbf{s})\mathbf{G}\mathbf{U}(\mathbf{s})' + \mathbf{R}.$$

The estimator for the fixed effects given in equation (6.20) is just the generalized least squares estimator of $\beta$ obtained using $\text{Var}[\mathbf{Z}(\mathbf{s})] = \mathbf{V}$. Also, it can be shown that $\widehat{\alpha}$ is the best linear unbiased predictor (BLUP) of the random effects, $\alpha$, under squared-error loss. Note that when there are no random effects, the linear mixed model reduces to that given in equation (6.1) and discussed earlier.

In practical applications, $\mathbf{V}$ is parameterized as $\mathbf{V}(\theta)$ and the elements of $\theta$ are estimated by one of the methods discussed previously. It is common in mixed model analyses to follow the assumption of Gaussian distributed errors and a Gaussian distribution for $\alpha$ and to estimate $\theta$ by maximum or restricted maximum likelihood. The estimators for the fixed effects then take on the EGLS form

$$\widehat{\beta}_{ml} = \Omega(\widehat{\theta}_{ml})\mathbf{X}(\mathbf{s})\mathbf{V}(\widehat{\theta}_{ml})^{-1}\mathbf{Z}(\mathbf{s})$$

$$\widehat{\beta}_{reml} = \Omega(\widehat{\theta}_{reml})\mathbf{X}(\mathbf{s})\mathbf{V}(\widehat{\theta}_{reml})^{-1}\mathbf{Z}(\mathbf{s})$$

and the predictors become

$$\widehat{\alpha}_{ml} = \Sigma_S(\widehat{\theta}_{ml})\mathbf{V}(\widehat{\theta}_{ml})^{-1}(\mathbf{Z}(\mathbf{s}) - \mathbf{X}(\mathbf{s})\widehat{\beta}_{ml})$$

$$\widehat{\alpha}_{reml} = \Sigma_S(\widehat{\theta}_{reml})\mathbf{V}(\widehat{\theta}_{reml})^{-1}(\mathbf{Z}(\mathbf{s}) - \mathbf{X}(\mathbf{s})\widehat{\beta}_{reml}).$$

However, it is much easier to write this than it is to accomplish this modeling in practice. Typically, statisticians think of modeling $\mathbf{G}$ and $\mathbf{R}$, rather than modeling $\mathbf{V}$. In the spatial case, it will generally be difficult to model both of these as parametric functions of a smaller number of parameters; it is difficult for any statistical model fitting algorithm to distinguish between spatially structured random effects ($U(\mathbf{s})\alpha$, with $\text{Var}[\alpha] \equiv \mathbf{G}(\theta_\alpha)$) and spatially autocorrelated errors ($\text{Var}[\epsilon(\mathbf{s})] = \mathbf{R}(\theta_\epsilon)$). In subsequent sections, we describe several special cases of the general linear mixed model that can be used for modeling spatial data.

*6.2.1.1 Penalized Splines and Low-rank Smoothers*

An important special case of the linear mixed model arises when $\mathbf{R} = \sigma_\epsilon^2\mathbf{I}$ and $\mathbf{G} = \sigma^2\mathbf{I}$, a **variance component model**. This is a particularly simple mixed model, and fast algorithms are available to estimate the parameters $\sigma^2$ and $\sigma_\epsilon^2$; for example, the modified W-transformation of Goodnight and Hemmerle (1979). The linear mixed model criterion (6.19) now becomes

*Mixed Model Criterion*

$$\begin{aligned} Q(\boldsymbol{\beta}, \boldsymbol{\alpha}) &= \sigma_\epsilon^{-2}(\mathbf{Z}(\mathbf{s}) - \mathbf{X}(\mathbf{s})\boldsymbol{\beta} - \mathbf{U}(\mathbf{s})\boldsymbol{\alpha})'(\mathbf{Z}(\mathbf{s}) - \mathbf{X}(\mathbf{s})\boldsymbol{\beta} - \mathbf{U}(\mathbf{s})\boldsymbol{\alpha}) \\ &\quad + \sigma^{-2}\boldsymbol{\alpha}'\boldsymbol{\alpha} \\ &= \sigma_\epsilon^{-2}||\mathbf{Z}(\mathbf{s}) - \mathbf{X}(\mathbf{s})\boldsymbol{\beta} - \mathbf{U}(\mathbf{s})\boldsymbol{\alpha}||^2 + \sigma^{-2}||\boldsymbol{\alpha}||^2. \end{aligned} \tag{6.22}$$

This expression is very closely related to the objective function minimized in spline smoothing. In this subsection we elicit this connection and provide details on the construction of the spatial design or regressor matrix $\mathbf{U}(\mathbf{s})$. The interested reader is referred to the text by Ruppert, Wand, and Carroll (2003), on which the exposition regarding splines and radial smoothers is based.

The problem of modeling the mean function in $Y_i = f(x_i) + \epsilon_i$ has many statistical solutions. Among the nonparametric ones are scatterplot smoothers based on splines. A spline model essentially decomposes $f(x)$ into an "overall" mean component and a linear combination of piecewise functions. For example, define the truncated line function

$$(x-t)_+ = \begin{cases} x-t & x > t \\ 0 & \text{otherwise.} \end{cases}$$

The functions $1, x, (x-t_1), \cdots, (x-t_K)$ are linear spline basis functions, and their linear combinations are called *splines* (Ruppert, Wand, and Carroll, 2003, p. 62). The points $t_1, \cdots, t_K$ are the **knots** of the spline. Ruppert et al. (2003) term a smoother as *low-rank*, if the number of knots is considerably less than the number of data points, if $K \approx n$, then the smoother is termed *full-rank*.

A linear spline model, for example, uses the linear spline basis functions

$$f(x) = \beta_0 + \beta_1 x + \sum_{j=1}^{K} \alpha_j (x - t_j)_+. \tag{6.23}$$

This model and its basis functions are a special case of the more general power spline of degree $p$

$$f(x) = \beta_0 + \beta_1 x + \cdots \beta_p x^p + \sum_{j=1}^{K} \alpha_j ((x-t_j)_+)^p.$$

We have used the symbol $\alpha$ for the spline coefficients to initiate the connection between spline smoothers and the mixed model in equation (6.18). We have not decided to treat the coefficients as random quantities, however. Whether $\alpha$ is fixed or random, we can write the observational model for (6.23) as

$$\mathbf{y} = \mathbf{X}\boldsymbol{\beta} + \mathbf{U}\boldsymbol{\alpha} + \boldsymbol{\epsilon},$$

where

$$\mathbf{X} = \begin{bmatrix} 1 & x_1 \\ \vdots & \vdots \\ 1 & x_n \end{bmatrix} \qquad \mathbf{U} = \begin{bmatrix} (x_1 - t_1) & \cdots & (x_1 - t_K) \\ \vdots & \ddots & \vdots \\ (x_n - t_1) & \cdots & (x_n - t_K) \end{bmatrix}.$$

In order to fit a spline model to data, penalties are imposed that prevent the fit from being too variable. A penalty criterion that restricts the variation of the spline coefficients leads to the minimization of

*Spline Criterion*

$$Q^*(\boldsymbol{\beta}, \boldsymbol{\alpha}) = ||\mathbf{y} - \mathbf{X}\boldsymbol{\beta} - \mathbf{U}\boldsymbol{\alpha}||^2 + \lambda^2||\boldsymbol{\alpha}||^2, \tag{6.24}$$

where $\lambda > 0$ is the smoothing parameter.

The connection between spline smoothing and linear mixed models is becoming clear. If $\widehat{\boldsymbol{\beta}}$, $\widehat{\boldsymbol{\alpha}}$ minimize (6.24), then they also minimize

$$Q^*(\boldsymbol{\beta}, \boldsymbol{\alpha})/\sigma_\epsilon^2 = Q(\boldsymbol{\beta}, \boldsymbol{\alpha}),$$

the mixed model criterion (6.19), that led to Henderson's mixed model equation, with $\lambda^2 = \sigma_\epsilon^2/\sigma^2$.

You can use mixed model software to perform spline smoothing: construct the random effects regressor matrix $\mathbf{U}$ from the spline basis functions, and assume a variance component model for their dispersion. This correspondence has another, far reaching, consequence. The smoothing parameter does not have to be determined by cross-validation or a comparison of information criteria. If $\widehat{\sigma}_\epsilon^2$ and $\widehat{\sigma}^2$ are the (restricted) maximum likelihood estimates of the variance components, then the smoothing parameter is determined as

$$\lambda = \left(\frac{\widehat{\sigma}_\epsilon^2}{\widehat{\sigma}^2}\right)^{1/2}$$

in a linear spline model. In a power spline of degree $p$, this expression is taken to the $p$th power (Ruppert, Wand, and Carroll, 2003, p. 113). This automated, model-driven estimation of the smoothing parameter, replaces the smoothing parameter selection that is a matter of much controversy in nonparametric modeling. In fact, the actual value of the smoothing parameter in these models may not even be of interest. It is implied by the estimates of the variance components, it has been "de-mystified."

There is another important consequence of adopting the mixed model framework and ML/REML estimation of the smoothing parameter. In order for there to be a variance component $\sigma^2$, the spline coefficients $\boldsymbol{\alpha}$ have to be random variables. If you use the "mixed model crank" to obtain the solutions $\widehat{\boldsymbol{\beta}}$ and $\widehat{\boldsymbol{\alpha}}$, this is one thing. If you are interested in the precision of predicted values such as $\mathbf{x}'\widehat{\boldsymbol{\beta}} + \mathbf{u}'\widehat{\boldsymbol{\alpha}}$, then the appropriate estimate of precision depends on whether $\boldsymbol{\alpha}$ is fixed or random. We argue that $\boldsymbol{\alpha}$ is a random component and should be treated accordingly. Recall the correspondence $\mathbf{U}(\mathbf{s})\boldsymbol{\alpha} = \upsilon(\mathbf{s})$

in the initial formulation of the linear mixed model. The spline coefficients are part of the spatial process $\upsilon(\mathbf{s})$, a random process. The randomness of $\upsilon(\mathbf{s})$ cannot be induced by $\mathbf{U}(\mathbf{s})$, it is a matrix of constants. The randomness must be induced by $\boldsymbol{\alpha}$, hence the coefficients are random variables.

So far, the spline model has the form of a scatterplot smoother in a single dimension. In order to cope with multiple dimensions, e.g., two spatial coordinates, we need to make some modifications. Again, we follow Ruppert, Wand, and Carroll (2003, Ch. 13.4–13.5). For a process in $\mathbb{R}^d$, the first modification is to use radial basis functions as the spline basis. Define $\mathbf{h}_{ik}$ as the Euclidean distance between the location $\mathbf{s}_i$ and the spline knot $\mathbf{t}_k$, $\mathbf{h}_{ik} = ||\mathbf{s}_i - \mathbf{t}_k||$. Similarly, define $\mathbf{c}_{kl} = ||\mathbf{t}_k - \mathbf{t}_l||$, the distance between the knots $\mathbf{t}_k$ and $\mathbf{t}_l$, and $p = 2m - d$, $p > 0$. Then, the $(n \times K)$ matrix $\mathbf{U}^*(\mathbf{s})$ and $(K \times K)$ matrix $\boldsymbol{\Omega}(\mathbf{t})$ are defined to have typical elements

$$
\begin{aligned}
\mathbf{U}^*(\mathbf{s}) &= \begin{cases} [\mathbf{h}_{ik}^p] & d \text{ odd} \\ [\mathbf{h}_{ik}^p] \log\{\mathbf{h}_{ik}\} & d \text{ even} \end{cases} \\
\boldsymbol{\Omega}(\mathbf{t}) &= \begin{cases} [\mathbf{c}_{kl}^p] & d \text{ odd} \\ [\mathbf{c}_{kl}^p] \log\{\mathbf{c}_{kl}\} & d \text{ even.} \end{cases}
\end{aligned}
$$

From a singular value decomposition of $\boldsymbol{\Omega}(\mathbf{t})$ you can obtain the square root matrix $\boldsymbol{\Omega}(\mathbf{t})^{1/2}$. Define the mixed linear model

*Multivariate Radial Smoother*

$$
\begin{aligned}
\mathbf{Z}(\mathbf{s}) &= \mathbf{X}(\mathbf{s})\boldsymbol{\beta} + \mathbf{U}(\mathbf{s})\boldsymbol{\alpha} + \boldsymbol{\epsilon} \qquad (6.25) \\
\mathbf{U}(\mathbf{s}) &= \mathbf{U}^*(\mathbf{s})\boldsymbol{\Omega}(\mathbf{t})^{-1/2} \\
\boldsymbol{\alpha} &\sim (\mathbf{0}, \sigma^2\mathbf{I}) \\
\boldsymbol{\epsilon} &\sim (\mathbf{0}, \sigma_\epsilon^2\mathbf{I}) \\
\mathrm{Cov}[\boldsymbol{\alpha}, \boldsymbol{\epsilon}] &= \mathbf{0}.
\end{aligned}
$$

The transformation of $\mathbf{U}^*(\mathbf{s})$ to $\mathbf{U}(\mathbf{s})$ is made so that this low-rank radial smoother approximates a (full-rank) thin-plate spline, see Ruppert et al. (2003, p. 253). This is also the reason behind the logarithmic factor in the case of $d$ even. The number $m$ in the exponent relates to the order of the derivative penalized in the spline. For one- and two-dimensional processes, $m = 2$ is an appropriate choice.

Model (6.25), the low-rank radial smother for spatial data, has several remarkable aspects. For example, you can fit a spatial model by numerical optimization of a single parameter, $\sigma^2$. As discussed in §5.5.2 and §5.5.3, in (restricted) maximum likelihood estimation the fixed effects as well as the residual variance component can be profiled from the objective function. The profiled (restricted) log-likelihood is a function of the remaining covariance parameters. In model (6.25), only $\sigma^2$, the variance of the spline coefficients, remains. Estimates of the fixed effects and predictions of the spline coefficients are computed after $\widehat{\sigma}^2$ has been determined; either directly from (6.20), (6.21), or from sweeping the augmented mixed model equations (Goodnight, 1979). Also, the mixed model has an "independent error" structure, $\text{Var}[\mathbf{Z}(\mathbf{s})|\gamma] = \sigma_\epsilon^2\mathbf{I}$, permitting the computationally fast modified W-transformation of Goodnight and Hemmerle (1979).

The degree of smoothing is determined again from the relative size of the two variance components. Note that $\sigma_\epsilon^2$ is the variance of observations about the noise-free random function. Its magnitude scales with the response variable $Z(\mathbf{s})$. On the other hand, the magnitude of $\sigma^2$ depends on the smoothness of the process, as well as on the coordinate metric. Rescaling the spatial coordinates will affect $\sigma^2$. When this variance component is near zero, convergence difficulties may arise. Dividing the coordinates by a constant factor will move the solution away from the boundary of the parameter space.

The computational simplicity of model (6.25) has been achieved at some cost. Recall the more general model $\mathbf{Z}(\mathbf{s}) = \mathbf{X}(\mathbf{s})\boldsymbol{\beta} + \boldsymbol{\upsilon}(\mathbf{s}) + \boldsymbol{\epsilon}(\mathbf{s})$. Assuming $\text{Var}[\boldsymbol{\epsilon}(\mathbf{s})] = \sigma_\epsilon^2\mathbf{I}$, the spatial dependency in the error-free signal is expressed through $\text{Var}[\boldsymbol{\upsilon}(\mathbf{s})] = \boldsymbol{\Sigma}_\upsilon$. On the contrary, the low-rank radial smoother replaces $\boldsymbol{\upsilon}(\mathbf{s})$ with $\mathbf{U}(\mathbf{s})\boldsymbol{\alpha}$ and assumes a very simple covariance structure for $\boldsymbol{\alpha}$, $\text{Var}[\boldsymbol{\alpha}] = \sigma^2\mathbf{I}$. Where did the spatial information go? It is contained, of course, in the "spline basis" $\mathbf{U}(\mathbf{s}) = \mathbf{U}^*(\mathbf{s})\boldsymbol{\Omega}(\mathbf{t})^{-1/2}$, which depends on the spatial configuration of data *and* knots. Because both $\mathbf{U}^*(\mathbf{s})$ and $\boldsymbol{\Omega}(\mathbf{t})$ are known, given the data and knot locations, the matrix $\mathbf{U}(\mathbf{s})$ does not depend on any unknowns. If, however, $\mathbf{U}(\mathbf{s})$ contains additional spatial dependence parameters, for example, because $\boldsymbol{\Omega}(\mathbf{t})$ is chosen as a spatial covariance model, then computational demand increases quickly. In the general model, the spatial information is contained in the random process itself, that is, $\mathbf{U}(\mathbf{s}) \equiv \mathbf{I}$, $\boldsymbol{\alpha} \equiv \boldsymbol{\upsilon}(\mathbf{s})$.

Radial smoothing based on mixed models takes an interesting intermediate position between correlated error models for geostatistical data in which the spatial correlation structure is specified directly (§6.2.1.3) and correlated error models for lattice (regional) data in which the error structure is determined indirectly (§6.2.2). Conditional on the random spline knots $\boldsymbol{\alpha}$, the radial smoothing model is a model with uncorrelated errors. Likelihood-based estimation proceeds, however, from a marginal distribution. The marginal variance is that of a correlated error model, $\text{Var}[\mathbf{Z}(\mathbf{s})] = \sigma^2\mathbf{U}(\mathbf{s})\mathbf{U}(\mathbf{s})' + \sigma_\epsilon^2\mathbf{I}$. The marginal correlations are a consequence of $\mathbf{U}(\mathbf{s})$ not being a diagonal matrix and are determined by the structure of $\mathbf{U}(\mathbf{s})$. The number of spline knots and their location in the domain indirectly determine the correlation "structure" of the resulting model. This makes the selection of the knots an important aspect of radial smoothing, also, because the computational burden (the size of the mixed model equations) increases with the number of knots.

Limiting the number of knots to a fraction of the total sample size—a feature of low-rank smoothers—brings with it tremendous computational savings. If the number of knots and their placement is selected properly, these savings do not come at a loss in quality of fit. It does, however, raise the important question as to how knots should be selected, and how many one should choose ($K$). In the one-dimensional case ($d = 1$), Ruppert et al. (2003, p. 126) recommend for automatic smoothers to choose $K$ as one fourth of the (unique) coordinate locations, but not more than 35. The issue of uniqueness rarely comes into play with spatial data, but for other applications of scatterplot smoothing, replications may be common. The $t$th knot is then placed at the $(t+1)/(K+2)$th sample quantile of the (unique) coordinates. In the two-dimensional, spatial case, selecting the number of knots and their placement is a much more complicated problem. A regular grid of values is not necessarily a good choice. The spatial domain may be irregularly shaped, and there may be pockets with higher and lower data frequency. Ruppert et al. (2003, p. 257) select no less than 20 knots, but no more than $n/4$, with an upper limit of $K = 150$ and recommend placement of the knots according to a space-filling design. These designs are obtained by optimizing a stress criterion, selecting either $K$ of the observed data points (discrete spatial design) or $K$ arbitrary locations in the domain (continuous spatial design). For large data sets, the optimization involved is numerically challenging and may turn out to be the computationally costliest aspect of the entire analysis. This, to some extent, defeats the formulation of the low-rank smoother as a variance component model for which fast estimation algorithms are available.

#### *6.2.1.2 Filtering Measurement Error*

As another special case, consider again the general linear model (see also §5.4.3)

$$Z(\mathbf{s}) = S(\mathbf{s}) + \epsilon(\mathbf{s}), \quad \mathbf{s} \in D,$$

where $\mathrm{E}[\epsilon(\mathbf{s})] = 0$, $\mathrm{Var}[\epsilon(\mathbf{s})] = \sigma^2_\epsilon$, $\mathrm{Cov}[\epsilon(\mathbf{s}_i), \epsilon(\mathbf{s}_j)] = 0$, for all $i \neq j$ and $S(\mathbf{s})$ and $\epsilon(\mathbf{s})$ are independent. We again assume that the unobserved signal $S(\mathbf{s})$ can be described by a general linear model with autocorrelated errors

$$S(\mathbf{s}) = \mathbf{x}(\mathbf{s})'\boldsymbol{\beta} + \upsilon(\mathbf{s}),$$

where $\mathrm{E}[\upsilon(\mathbf{s})] = 0$, and $\mathrm{Cov}[\upsilon(\mathbf{s}_i), \upsilon(\mathbf{s}_j)] = \mathrm{Cov}[S(\mathbf{s}_i), S(\mathbf{s}_j)] = C_S(\mathbf{s}_i, \mathbf{s}_j)$. Note that this model has a linear mixed model formulation,

*Mixed Model Formulation*

$$\begin{aligned}
\mathbf{Z}(\mathbf{s}) &= \mathbf{X}(\mathbf{s})\boldsymbol{\beta} + \boldsymbol{\upsilon}(\mathbf{s}) + \boldsymbol{\epsilon}(\mathbf{s}) \\
\boldsymbol{\upsilon} &\sim (\mathbf{0}, \boldsymbol{\Sigma}_S) \\
\boldsymbol{\epsilon}(\mathbf{s}) &\sim (\mathbf{0}, \sigma^2_\epsilon \mathbf{I}) \\
\mathrm{Cov}[\boldsymbol{\upsilon}, \boldsymbol{\epsilon}(\mathbf{s})] &= \mathbf{0},
\end{aligned}$$

and (6.18) is a special case, obtained by taking $\mathbf{U}(\mathbf{s})\boldsymbol{\alpha} \equiv \boldsymbol{\upsilon}(\mathbf{s})$, $R \equiv \sigma^2_\epsilon \mathbf{I}$, and $\mathbf{G} \equiv \boldsymbol{\Sigma}_S$. The matrix $\boldsymbol{\Sigma}_S$ contains the covariance terms $C_S(\mathbf{s}_i, \mathbf{s}_j)$. In practice, $\boldsymbol{\Sigma}_S$ is modeled parametrically as $\boldsymbol{\Sigma}(\boldsymbol{\theta}_S)$. Note that the moments are

$$\begin{aligned}
\mathrm{E}[\mathbf{Z}(\mathbf{s})] &= \mathbf{X}(\mathbf{s})\boldsymbol{\beta} && (6.26) \\
\mathrm{Var}[\mathbf{Z}(\mathbf{s})] &= \boldsymbol{\Sigma}_S + \sigma^2_\epsilon \mathbf{I} = \mathbf{V}, && (6.27)
\end{aligned}$$

which are those of a linear model with correlated errors of the form of (6.1). Thus, as far as estimation of the fixed effects is concerned, the two models are equivalent. Because of this equivalency, we can also obtain solutions based on Henderson's mixed model equations:

$$\begin{aligned}\begin{bmatrix}\widehat{\boldsymbol{\beta}}\\ \widehat{\boldsymbol{\upsilon}}(\mathbf{s})\end{bmatrix} &= \begin{bmatrix}\sigma_\epsilon^{-2}\mathbf{X}(\mathbf{s})'\mathbf{X}(\mathbf{s}) & \sigma_\epsilon^{-2}\mathbf{X}(\mathbf{s})' \\ \sigma_\epsilon^{-2}\mathbf{X}(\mathbf{s}) & \sigma_\epsilon^{-2}\mathbf{I}+\boldsymbol{\Sigma}_S^{-1}\end{bmatrix}^{-}\begin{bmatrix}\sigma_\epsilon^{-2}\mathbf{X}(\mathbf{s})'\mathbf{Z}(\mathbf{s})\\ \sigma_\epsilon^{-2}\mathbf{Z}(\mathbf{s})\end{bmatrix}\\ &= \mathbf{C}\begin{bmatrix}\sigma_\epsilon^{-2}\mathbf{X}(\mathbf{s})'\mathbf{Z}(\mathbf{s})\\ \sigma_\epsilon^{-2}\mathbf{Z}(\mathbf{s})\end{bmatrix}\end{aligned}$$

It is left as an exercise to verify that

$$\mathbf{C} = \begin{bmatrix}\boldsymbol{\Omega} & -\boldsymbol{\Omega}\mathbf{X}(\mathbf{s})\mathbf{V}^{-1}\boldsymbol{\Sigma}_S \\ -\boldsymbol{\Sigma}_S\mathbf{V}^{-1}\mathbf{X}(\mathbf{s})'\boldsymbol{\Omega} & (\sigma_\epsilon^2\mathbf{I}+\boldsymbol{\Sigma}_S^{-1})^{-1}+\boldsymbol{\Sigma}_S\mathbf{V}^{-1}\mathbf{X}(\mathbf{s})\boldsymbol{\Omega}\mathbf{X}(\mathbf{s})\mathbf{V}^{-1}\boldsymbol{\Sigma}_S\end{bmatrix}, \tag{6.28}$$

where $\boldsymbol{\Omega} = (\mathbf{X}(\mathbf{s})'\mathbf{V}^{-1}\mathbf{X}(\mathbf{s}))^{-}$. The solutions to the mixed model equations are then

$$\widehat{\boldsymbol{\beta}} = \boldsymbol{\Omega}\mathbf{X}(\mathbf{s})'\mathbf{V}^{-1}\mathbf{Z}(\mathbf{s}) \tag{6.29}$$

$$\widehat{\boldsymbol{\upsilon}}(\mathbf{s}) = \boldsymbol{\Sigma}_S\mathbf{V}^{-1}(\mathbf{Z}(\mathbf{s})-\mathbf{X}(\mathbf{s})\widehat{\boldsymbol{\beta}}). \tag{6.30}$$

Note that (6.29) is the GLS estimator of $\boldsymbol{\beta}$ and that (6.30) is the best linear unbiased predictor of $\boldsymbol{\upsilon}(\mathbf{s})$ under squared-error loss.

Since $\mathbf{Z}(\mathbf{s}) = \mathbf{S}(\mathbf{s})+\boldsymbol{\epsilon}(\mathbf{s}) = \mathbf{X}(\mathbf{s})\boldsymbol{\beta}+\boldsymbol{\upsilon}(\mathbf{s})+\boldsymbol{\epsilon}(\mathbf{s})$, it is natural to consider as a predictor of $Z(\mathbf{s}_i)$ the quantity

$$\begin{aligned}Z(\mathbf{s}_i) &= \mathbf{x}(\mathbf{s}_i)'\widehat{\boldsymbol{\beta}}+\widehat{\upsilon}(\mathbf{s}_i)\\ &= \mathbf{x}(\mathbf{s}_i)'\widehat{\boldsymbol{\beta}}+\boldsymbol{\sigma}(\mathbf{s}_i)'\mathbf{V}^{-1}(\mathbf{Z}(\mathbf{s})-\mathbf{X}(\mathbf{s})\widehat{\boldsymbol{\beta}}),\end{aligned} \tag{6.31}$$

where $\boldsymbol{\sigma}(\mathbf{s}_i) = \mathrm{Cov}[\upsilon(\mathbf{s}_i), \boldsymbol{\upsilon}(\mathbf{s})]$. This is equivalent to the universal kriging predictor with filtered measurement error (see Chapter problems and §5.3.3).

**Example 6.1 (Soil carbon regression. Continued)** For the C/N data we fitted two conditional models. A mixed model

$$\mathbf{Z}(\mathbf{s}) = \mathbf{X}(\mathbf{s})\boldsymbol{\beta}+\boldsymbol{\upsilon}(\mathbf{s})+\boldsymbol{\epsilon}(\mathbf{s}),$$

where $\boldsymbol{\upsilon}(\mathbf{s})$ is a zero mean random field with exponential covariance structure, and a low-rank radial smoother with 49 knots, placed on an evenly spaced grid throughout the domain. The large-scale trend structure is the same in the two models, as before, C% is modeled as a linear function of total soil nitrogen (N%). The low-rank smoother has two covariance parameters, the variance of the random spline coefficients, and the residual variance, $\sigma_\epsilon^2 = \mathrm{Var}[\epsilon(\mathbf{s})]$. The spatial mixed model has three covariance parameters, the variance and range of the exponential covariance structure $\boldsymbol{\Sigma}(\boldsymbol{\theta}) = \mathrm{Var}[\boldsymbol{\upsilon}(\mathbf{s})]$ and the nugget variance $\sigma_\epsilon^2 = \mathrm{Var}[\epsilon(\mathbf{s})]$.

Because of the presence of the variance component $\sigma^2$, predictions at the observed locations are filtered and do not honor the data. The left-hand panel of Figure 6.5 shows the adjustments that are made to the estimate of the mean to predict the C% at the observed locations. The right-hand panel of the figure compares the C% predictions in the two conditional models. They are generally very close. The computational effort to fit the radial smoother model is considerably smaller, however. The low-rank smoother has a simpler variance structure, $\sigma^2\mathbf{U}(\mathbf{s})'\mathbf{U}(\mathbf{s})+\sigma_\epsilon^2\mathbf{I}$ compared to $\mathbf{\Sigma}(\boldsymbol{\theta})+\sigma_\epsilon^2\mathbf{I}$, that does not have second derivatives. Representing the spatial random structure requires $K = 49$ random components in the smoothing model and $n = 195$ components in the spatial mixed model.

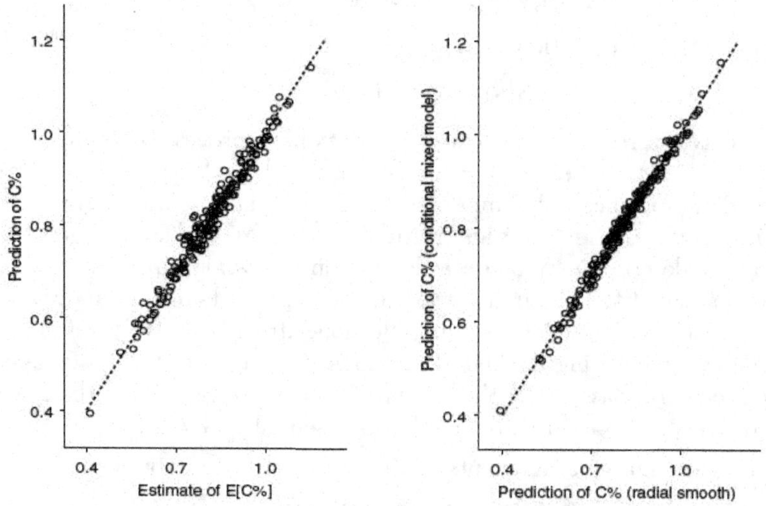


Figure 6.5 *Comparison of fitted and predicted values in two conditional models for the C–N spatial regression. Left-hand panel compares predictions of the C-process and estimates of the mean carbon percentage in the radial smoother model. The right-hand panel compares the C-predictions in the two conditional formulations.*

□

#### 6.2.1.3 *The Marginal and the Conditional Specifications*

It is often convenient to specify parametric models by just considering their first two moments. For example, we might assume that the mean and variance under consideration are

*Marginal Formulation*

$$\begin{aligned} \mathrm{E}[\mathbf{Z}(\mathbf{s})] &= \mathbf{X}(\mathbf{s})\boldsymbol{\beta} \\ \mathrm{Var}[\mathbf{Z}(\mathbf{s})] &= \sigma_\epsilon^2\mathbf{I} + \mathbf{\Sigma}(\boldsymbol{\theta}), \end{aligned} \tag{6.32}$$

where $\sigma_\epsilon^2$ represents the measurement error variance and $\boldsymbol{\theta}$ includes other variance components as well as spatial autocorrelation parameters. Estimation of $\boldsymbol{\beta}$ and $\boldsymbol{\theta}$ can be done using iteratively reweighted generalized least squares (§5.5.1) and universal kriging (§5.3.3) can be used to make predictions of $Z(\mathbf{s}_0)$. There are no random effects and all the spatial autocorrelation is modeled through what is essentially $\mathbf{R} = \sigma_\epsilon^2 \mathbf{I} + \mathbf{\Sigma}(\boldsymbol{\theta})$.

*Conditional Formulation*

Modeling these moments directly can be a difficult task and it is often easier to hypothesize a spatially-varying latent process, $S(\mathbf{s})$, and to construct models around the moments of this process. For example, we could assume that, given an unobservable spatial process $S(\mathbf{s})$,

$$\begin{aligned} \mathrm{E}[Z(\mathbf{s})|S] &\equiv S(\mathbf{s}), \\ \mathrm{Var}[Z(\mathbf{s})|S] &= \sigma_\epsilon^2, \\ \mathrm{Cov}[Z(\mathbf{s}_i), Z(\mathbf{s}_j)|S) &= 0 \qquad \text{for } \mathbf{s}_i \neq \mathbf{s}_j. \end{aligned}$$

To complete the specification we assume

$$S(\mathbf{s}) = \mathbf{x}(\mathbf{s})'\boldsymbol{\beta} + \upsilon(\mathbf{s}),$$

where $\upsilon(\mathbf{s})$ is a zero mean, second-order stationary process with covariance function $C_\upsilon(\mathbf{s}_i, \mathbf{s}_j) = C_S(\mathbf{s}_i, \mathbf{s}_j) = \mathrm{Cov}[\upsilon(\mathbf{s}_i), \upsilon(\mathbf{s}_j)] = \mathrm{Cov}[S(\mathbf{s}_i), S(\mathbf{s}_j)]$, and corresponding variance-covariance matrix $\mathbf{\Sigma}(\boldsymbol{\theta}_S)$. This is an example of a **conditionally-specified** or **hierarchical** model. At the first stage of the hierarchy, we describe how the data depend on the random process $S(\mathbf{s})$. At the second stage of the hierarchy, we model the moments of the random process. Estimation of $\boldsymbol{\beta}$ and $\boldsymbol{\theta}$ can also be done using iteratively reweighted least squares, or by using likelihood methods if we assume $S(\mathbf{s})$ is Gaussian (as described earlier as part of the theory of mixed models). All of the spatial autocorrelation is modeled through what is essentially $\mathbf{G} = \mathbf{\Sigma}(\boldsymbol{\theta}_S)$.

Note that the marginal moments of $\mathbf{Z}(\mathbf{s})$ are then easily obtained as

$$\begin{aligned} \mathrm{E}[\mathbf{Z}(\mathbf{s})] &= \mathrm{E}_S[\mathrm{E}[\mathbf{Z}(\mathbf{s})|\mathbf{S}(\mathbf{s})]] = \mathrm{E}_S[\mathrm{E}[\mathbf{Z}(\mathbf{s})|\boldsymbol{\upsilon}(\mathbf{s})]] = \mathbf{X}(\mathbf{s})\boldsymbol{\beta} \\ \mathrm{Var}[\mathbf{Z}(\mathbf{s})] &= \mathrm{Var}_S[\mathrm{E}[\mathbf{Z}(\mathbf{s})|\mathbf{S}(\mathbf{s})]] + \mathrm{E}_S[\mathrm{Var}[\mathbf{Z}(\mathbf{s})|\mathbf{S}(\mathbf{s})] \\ &= \mathrm{Var}_S[\mathbf{S}(\mathbf{s})] + \mathrm{E}[\sigma_\epsilon^2 \mathbf{I}] \\ &= \mathbf{\Sigma}(\boldsymbol{\theta}_S) + \sigma_\epsilon^2 \mathbf{I} = \mathbf{V}. \end{aligned}$$

Thus, our conditionally-specified, or hierarchical, model is a linear mixed model and is marginally equivalent to a general linear model with autocorrelated errors where the moments are given in (6.32). Because $\mathrm{Var}[Z(\mathbf{s})|S(\mathbf{s})] = \sigma_\epsilon^2$, this variance component represents the nugget effect of the marginal spatial covariance structure. For identifiability of the variance components, the covariance structure of $\upsilon(\mathbf{s})$ is free of a nugget effect, or it is assumed that the overall nugget effect can be decomposed in micro-scale and measurement error variation in known proportions.

With a linear model and only 2 levels in the hierarchy, the use of the conditional specification over the marginal specification is basically one of personal preference done for ease of understanding. The two approaches give equivalent inference if the marginal moments coincide. However, if we want to consider additional levels in the hierarchy or nonlinear models, the conditional specification can offer some advantages and we revisit this formulation again later in this chapter.

### 6.2.2 *Spatial Autoregressive Models*

With regional data, a direct specification of the variance-covariance matrix $\mathbf{\Sigma}$ limits our measure of spatial proximity to the distances among point locations assumed to represent each region (e.g., intercentroid distances). In this subsection, we describe other approaches to modeling autocorrelation in spatial regression models that can incorporate the neighborhood structures often used when modeling regional data.

In time series, autoregressive models represent the data at time $t$ as a linear combination of *past* values. The spatial analog represents the data at location $\mathbf{s}$ as a linear combination of *neighboring* values. This autoregression induces spatial dependence in the data. Thus, instead of specifying the spatial autocorrelation structure directly, spatial autoregressive models induce spatial autocorrelation in the data through the autoregression and the spatial proximity measure used to define the neighborhood structure among the data.

#### 6.2.2.1 *Simultaneous Autoregressive (SAR) Models*

We begin by applying the idea of spatial autoregression to the vector of residual errors, $\mathbf{e}(\mathbf{s})$, in the spatial linear regression model with Gaussian data. That is, we regress $\epsilon(\mathbf{s}_i)$ on all the other error terms giving

*Spatial Autoregression*

$$\begin{aligned}\mathbf{Z}(\mathbf{s}) &= \mathbf{X}(\mathbf{s})\boldsymbol{\beta} + \mathbf{e}(\mathbf{s}) \\ \mathbf{e}(\mathbf{s}) &= \mathbf{B}\mathbf{e}(\mathbf{s}) + \boldsymbol{\upsilon},\end{aligned} \tag{6.33}$$

where $\mathbf{B}$ is a matrix of spatial dependence parameters with $b_{ii} = 0$ (so we do not regress $\epsilon(\mathbf{s}_i)$ on itself). We assume the residual errors from the autoregression, $\upsilon_i, i = 1, \cdots, n$, have mean zero and a diagonal variance-covariance matrix $\mathbf{\Sigma}_\upsilon = \text{diag}[\sigma_1^2, \ldots \sigma_n^2]$. If all the $b_{ij}$ are zero, there is no autoregression and the model reduces to the traditional linear regression model with uncorrelated errors.

We can also express this autoregressive model as

$$(\mathbf{I} - \mathbf{B})(\mathbf{Z}(\mathbf{s}) - \mathbf{X}(\mathbf{s})\boldsymbol{\beta}) = \boldsymbol{\upsilon}, \tag{6.34}$$

and from this expression we can derive the variance-covariance matrix of $\mathbf{Z}(\mathbf{s})$ as

$$\mathbf{\Sigma}_{SAR} = \text{Var}[\mathbf{Z}(\mathbf{s})] = (\mathbf{I} - \mathbf{B})^{-1}\mathbf{\Sigma}_\upsilon(\mathbf{I} - \mathbf{B}')^{-1}, \tag{6.35}$$

assuming $(\mathbf{I} - \mathbf{B})^{-1}$ exists. Thus, the autoregression induces a particular model for the general covariance structure of the data $\mathbf{Z}(\mathbf{s})$ that is defined by the parameters $b_{ij}$; the covariance structure is determined indirectly through $\mathbf{B}$ and our choice of $\mathbf{\Sigma}_\upsilon$.

The model in (6.34) was introduced by Whittle (1954), and often appears in the literature as the *simultaneous autoregressive (SAR)* model, where the adjective "simultaneous" describes the $n$ autoregressions that occur simultaneously at each data location in this formulation. It further serves to distinguish this type of spatial model from the class of *conditional* autoregressive models defined in §6.2.2.2.

Clearly, the matrix of spatial dependence parameters, $\mathbf{B}$, plays an important role in SAR models. To make progress with estimation and inference, we will need to reduce the number of spatial dependence parameters through the use of a parametric model for the $\{b_{ij}\}$ and, for interpretation, we would like to relate them to the ideas of proximity and autocorrelation we have described previously. One way to do this is to take $\mathbf{B} = \rho\mathbf{W}$, where $\mathbf{W}$ is one of the spatial proximity matrices discussed in §1.3.2. For example, if the spatial locations form a regular lattice, $\mathbf{W}$ may be a matrix of 0's and 1's based on a rook, bishop, or queen move. In geographical analysis, $\mathbf{W}$ may be based on the length of shared borders, centroid distances, or other measures of regional proximity. With this parameterization of $\mathbf{B}$, the SAR model can be written as

*Simultaneous Auto-regressive Model (SAR)*

$$\begin{aligned} \mathbf{Z}(\mathbf{s}) &= \mathbf{X}(\mathbf{s})\boldsymbol{\beta} + \mathbf{e}(\mathbf{s}) \\ \mathbf{e}(\mathbf{s}) &= \rho\mathbf{W}\mathbf{e}(\mathbf{s}) + \boldsymbol{\upsilon}, \end{aligned} \tag{6.36}$$

where $\mathbf{B} = \rho\mathbf{W}$. We can manipulate this model in a variety of ways, and it is often intuitive to write it as

$$\mathbf{Z}(\mathbf{s}) = \mathbf{X}(\mathbf{s})\boldsymbol{\beta} + (\mathbf{I} - \rho\mathbf{W})^{-1}\boldsymbol{\upsilon} \tag{6.37}$$

$$= \mathbf{X}(\mathbf{s})\boldsymbol{\beta} - \rho\mathbf{W}\mathbf{X}(\mathbf{s})\boldsymbol{\beta} + \rho\mathbf{W}\mathbf{Z}(\mathbf{s}) + \boldsymbol{\upsilon}. \tag{6.38}$$

From equation (6.37) we can see how the autoregression induces spatial autocorrelation in the linear regression model through the term $(\mathbf{I} - \rho\mathbf{W})^{-1}\boldsymbol{\upsilon}$. From equation (6.38) we obtain a better appreciation for what this means in terms of a linear regression model with uncorrelated errors: we now have two additional terms in the regression model: $\rho\mathbf{W}\mathbf{X}\boldsymbol{\beta}$ and $\rho\mathbf{W}\mathbf{Z}(\mathbf{s})$. These terms are called **spatially lagged** variables.

For a well-defined model, we require $(\mathbf{I}-\rho\mathbf{W})$ to be non-singular (invertible). This restriction imposes conditions on $\mathbf{W}$ and also on $\rho$, best summarized through the eigenvalues of the matrix $\mathbf{W}$. If $\vartheta_{max}$ and $\vartheta_{min}$ are the largest and smallest eigenvalues of $\mathbf{W}$, and if $\vartheta_{min} < 0$ and $\vartheta_{max} > 0$, then $1/\vartheta_{min} < \rho < 1/\vartheta_{max}$ (Haining, 1990, p. 82). For a large set of identical square regions, these extreme eigenvalues approach $-4$ and 4, respectively, as the number of regions increases, implying $|\rho| < 0.25$, but actual constraints on $\rho$ may be more severe, especially when the sites are irregularly spaced. Often, the row sums of $\mathbf{W}$ are standardized to 1 by dividing each entry in $\mathbf{W}$ by its row sum, $\sum_j w_{ij}$. Then, $\vartheta_{max} = 1$ and $\vartheta_{min} \leq -1$, so $\rho < 1$ but may be less than $-1$ (see Haining, 1990, §3.2.2).

*Estimation and Inference in SAR Models.* The one-parameter SAR model described by equations (6.36), (6.37) and (6.38), with $\boldsymbol{\Sigma}_\upsilon = \sigma^2\mathbf{I}$, is by far the most commonly-used SAR model in practical applications. Probably the main reason for the popularity of this model is the difficulty of actually fitting models with more parameters. In what follows we describe both least squares and maximum likelihood methods for estimation and inference with this one-parameter SAR model.

If $\rho$ is known, then generalized least squares can be used to estimate $\boldsymbol{\beta}$ and to obtain an estimate of $\sigma^2$. Thus,

$$\widehat{\boldsymbol{\beta}}_{gls} = \left(\mathbf{X}(\mathrm{s})'\boldsymbol{\Sigma}_{SAR}^{-1}\mathbf{X}(\mathrm{s})\right)^{-1}\mathbf{X}(\mathrm{s})'\boldsymbol{\Sigma}_{SAR}^{-1}\mathbf{Z}(\mathrm{s}),$$

and

$$\widehat{\sigma}^2 = \frac{(\mathbf{Z}(\mathrm{s}) - \mathbf{X}(\mathrm{s})\widehat{\boldsymbol{\beta}}_{gls})'\boldsymbol{\Sigma}_{SAR}^{-1}(\mathbf{Z}(\mathrm{s}) - \mathbf{X}(\mathrm{s})\widehat{\boldsymbol{\beta}}_{gls})}{n-k},$$

with $\boldsymbol{\Sigma}_{SAR}$ given in equation (6.35) using $\mathbf{B} = \rho\mathbf{W}$ and $\boldsymbol{\Sigma}_{\upsilon} = \sigma^2\mathbf{I}$. Of course $\rho$ is not known, and we might be tempted to use iteratively re-weighted least squares to estimate both $\rho$ and $\boldsymbol{\beta}$ simultaneously through iteration. However, $\mathbf{Z}(\mathrm{s})$ and $\boldsymbol{\upsilon}$ are, in general, not independent, making the least squares estimator of $\rho$ inconsistent (Whittle, 1954; Ord, 1975). Based on the work of Ord (1975) and Cliff and Ord (1981, p. 160), Haining (1990, p. 130) suggested the use of a modified least squares estimator of $\rho$ that is consistent, although inefficient

$$\widehat{\rho} = \frac{\mathbf{Z}(\mathrm{s})'\mathbf{W}'\mathbf{W}\mathbf{Z}(\mathrm{s})}{\mathbf{Z}(\mathrm{s})'\mathbf{W}'\mathbf{W}^2\mathbf{Z}(\mathrm{s})}.$$

However, given this problem with least squares estimation and the unifying theory underlying maximum likelihood, parameters of SAR models are usually estimated by maximum likelihood.

Suppose the data are multivariate Gaussian with the general SAR model defined in (6.36) and (6.35), i.e., the data are multivariate Gaussian with mean $\mathbf{X}(\mathrm{s})\boldsymbol{\beta}$ and variance-covariance matrix given in equation (6.35),

$$\mathbf{Z}(\mathrm{s}) \sim G\left(\mathbf{X}(\mathrm{s})\boldsymbol{\beta}, (\mathbf{I}-\mathbf{B})^{-1}\boldsymbol{\Sigma}_{\upsilon}(\mathbf{I}-\mathbf{B}')^{-1}\right).$$

We consider a more general variance-covariance structure for $\boldsymbol{\upsilon}$, by reparameterizing the diagonal matrix $\boldsymbol{\Sigma}_{\upsilon}$ as $\boldsymbol{\Sigma}_{\upsilon} = \sigma^2\mathbf{V}_{\upsilon}$, so the variance covariance matrix of a SAR can be written as

$$\boldsymbol{\Sigma}_{SAR}(\boldsymbol{\theta}) = \sigma^2(\mathbf{I}-\mathbf{B})^{-1}\mathbf{V}_{\upsilon}(\mathbf{I}-\mathbf{B}')^{-1} = \sigma^2\mathbf{V}_{SAR}(\boldsymbol{\theta}), \qquad (6.39)$$

where the vector $\boldsymbol{\theta}$ contains the spatial dependence parameters $b_{ij}$, and the parameters of $V_{\upsilon}$. Under this model, the maximum likelihood procedure described in §5.5.2 can be used to estimate all unknown parameters. Specifically, estimates of $\boldsymbol{\beta}$ and $\boldsymbol{\theta}$ are obtained by minimizing the negative of twice the Gaussian likelihood, $\varphi(\boldsymbol{\beta};\boldsymbol{\theta};\mathbf{Z}(\mathrm{s}))$, given in equation (5.41) with $\boldsymbol{\Sigma}(\boldsymbol{\theta})$ replaced with $\boldsymbol{\Sigma}_{SAR}(\boldsymbol{\theta})$. The parameters $\sigma^2$ and $\boldsymbol{\beta}$ can be profiled from the log likelihood as in §5.5.2. If $\widehat{\boldsymbol{\theta}}_{ml}$ is the maximum likelihood estimator of $\boldsymbol{\theta}$, then

$$\widehat{\boldsymbol{\beta}}_{ml} = \left(\mathbf{X}(\mathrm{s})'\boldsymbol{\Sigma}_{SAR}(\widehat{\boldsymbol{\theta}}_{ml})^{-1}\mathbf{X}(\mathrm{s})\right)^{-1}\mathbf{X}(\mathrm{s})'\boldsymbol{\Sigma}_{SAR}(\widehat{\boldsymbol{\theta}}_{ml})\mathbf{Z}(\mathrm{s}), \qquad (6.40)$$

and

$$\widehat{\sigma}^2_{ml} = \frac{(\mathbf{Z}(\mathrm{s}) - \mathbf{X}(\mathrm{s})\widehat{\boldsymbol{\beta}}_{ml})'\boldsymbol{\Sigma}_{SAR}(\widehat{\boldsymbol{\theta}}_{ml})^{-1}(\mathbf{Z}(\mathrm{s}) - \mathbf{X}(\mathrm{s})\widehat{\boldsymbol{\beta}}_{ml})}{n}. \qquad (6.41)$$

In most cases, there is no closed form expression for $\widehat{\boldsymbol{\theta}}_{ml}$ and it is obtained using numerical methods.

In this general development, there are many model parameters contained in the vector $\boldsymbol{\theta}$, and minimizing $\varphi(\boldsymbol{\beta};\boldsymbol{\theta};\mathbf{Z}(\mathbf{s}))$ may be difficult or impossible. Thus, in practice, we usually assume the parameters in $\mathbf{V}_\upsilon$ are known (e.g., the $(i,i)^{th}$ element is equal to 1 or equal to $1/n_i$ to account for differing population sizes in a geographical analysis) and parameterize $\mathbf{B}$ as a parametric function of a spatial proximity matrix $\mathbf{W}$. For example, if we take $\mathbf{B} = \rho\mathbf{W}$ and $\mathbf{V}_\upsilon = \sigma^2\mathbf{I}$, the only unknown parameter is $\rho$, and in this case, minimizing $\varphi(\boldsymbol{\beta};\boldsymbol{\theta};\mathbf{Z}(\mathbf{s}))$ is typically straightforward. The information matrix—assuming $\boldsymbol{\beta}$ is not profiled—then has a much simpler form (Ord, 1975; Cliff and Ord, 1981, p. 242):

*SAR Information Matrix*

$$\begin{aligned} I(\boldsymbol{\beta},\sigma^2,\rho) &= \begin{bmatrix} \sigma^{-2}(\mathbf{X}(\mathbf{s})'\mathbf{A}(\rho)^{-1}\mathbf{X}(\mathbf{s})) & \mathbf{0} \\ \mathbf{0}' & I(\sigma^2,\rho) \end{bmatrix} \qquad (6.42)\\ I(\sigma^2,\rho) &= \begin{bmatrix} n\sigma^{-4}/2 & \sigma^{-2}\mathrm{tr}\{\mathbf{G}\} \\ \sigma^{-2}\mathrm{tr}\{\mathbf{G}\} & \sigma^{-2}(\alpha + \mathrm{tr}\{\mathbf{G}'\mathbf{G}\}) \end{bmatrix} \\ \mathbf{A}(\rho) &= (\mathbf{I} - \rho\mathbf{W})^{-1}(\mathbf{I} - \rho\mathbf{W}')^{-1} \\ \mathbf{G} &= \mathbf{W}(\mathbf{I} - \rho\mathbf{W})^{-1} \\ \alpha &= \sum(\vartheta_i^2/(1-\rho\vartheta_i)^2), \end{aligned}$$

where $\vartheta_i$ are the eigenvalues of $\mathbf{W}$. The inverse of the information matrix provides the asymptotic standard errors of the estimators.

We note, without providing further details, that restricted maximum likelihood estimation is of course a possible alternative in this setup. The needed adjustments to the foregoing formulas are straightforward, see §5.5.3.

### 6.2.2.2 *Conditional Autoregressive (CAR) Models*

The simultaneous approach to spatial autoregression provides one multivariate model that describes the spatial interactions among the data. We may find it more intuitive to follow the approach adopted in the analysis of time series and specify models for the set of *conditional* probability distributions of each observation, $Z(\mathbf{s}_i)$, given the observed values of all of the other observations. That is, we model $f(Z(\mathbf{s}_i)|\mathbf{Z}(\mathbf{s})_{-i})$, where $\mathbf{Z}(\mathbf{s})_{-i}$ denotes the vector of all observations except $Z(\mathbf{s}_i)$, and we do this for each observation in turn.

*Markov Property*

In time series, a sequence of random variables $Y_1, Y_2, \cdots, Y_T$ is said to have the *Markov property* if the conditional distribution of $Y_{t+1}$ given $Y_1, Y_2, \cdots, Y_t$ is the same as the conditional distribution of $Y_{t+1}$ given $Y_t$. That is, the value at time $t+1$ depends only on the previous value. A sequence of random variables having the Markov property is also called a *Markov process.* We can extend these ideas to the spatial domain by assuming $Z(\mathbf{s}_i)$ depends only on a set of neighbors, i.e., $Z(\mathbf{s}_i)$ depends on $Z(\mathbf{s}_j)$ only if location $\mathbf{s}_j$ is in the neighborhood set, $\mathcal{N}_i$, of $\mathbf{s}_i$. In this special case, the process $Z(\mathbf{s})$ is called a *Markov random field.* Thus, with the conditional autoregressive approach, we construct models for $f(Z(\mathbf{s}_i)|Z(\mathbf{s}_j), \mathbf{s}_j \in \mathcal{N}_i)$. For example, if we assume each of these conditional distributions is Gaussian, then we might model them using:

*Markov Random Field*

$$\mathrm{E}[Z(\mathbf{s}_i)|\mathbf{Z}(\mathbf{s})_{-i}] = \mathbf{x}(\mathbf{s}_i)'\boldsymbol{\beta} + \sum_{j=1}^{n} c_{ij}(Z(\mathbf{s}_j) - \mathbf{x}(\mathbf{s}_i)'\boldsymbol{\beta}), \quad (6.43)$$

$$\mathrm{Var}[Z(\mathbf{s}_i)|\mathbf{Z}(\mathbf{s})_{-i}] = \sigma_i^2, \; i = 1, \cdots, n, \quad (6.44)$$

where the $c_{ij}$ denote spatial dependence parameters that are nonzero only if $\mathbf{s}_j \in \mathcal{N}_i$, and $c_{ii} = 0$.

For estimation and inference, we need to ensure the existence of the joint distribution. Building a valid joint distribution from a collection of marginal or conditional distributions is difficult and this type of "upwards" construction almost always results in constraints on the interactions among the data (see also §5.8 and §6.3.3). In the case we consider here, the Hammersley-Clifford theorem (first proved in Besag (1974)), describes the conditions necessary for a set of conditional distributions, $\{f(Z(\mathbf{s}_i)|Z(\mathbf{s}_j), \mathbf{s}_j \in \mathcal{N}_i)\}$, to define a valid joint distribution, $f(Z(\mathbf{s}_1), Z(\mathbf{s}_2), Z(\mathbf{s}_n))$. In order to contrast CAR models with SAR models (the latter of which are really only defined for a multivariate Gaussian distribution), we begin with Gaussian CAR models and address the implications resulting from the Hammersley-Clifford theorem for non-Gaussian models in the continuation of Example 1.4 on page 377.

If we assume that the conditional distributions are Gaussian, with conditional means and variances given by equations (6.43) and (6.44), respectively, the conditions required by the Hammersley-Clifford theorem are not too restrictive and Besag (1974) shows (see also Cliff and Ord, 1981, p. 180) that these conditional distributions generate a valid joint multivariate Gaussian distribution with mean $\mathbf{X}(\mathbf{s})\boldsymbol{\beta}$ and variance

$$\boldsymbol{\Sigma}_{CAR} = (\mathbf{I} - \mathbf{C})^{-1}\boldsymbol{\Sigma}_c, \quad (6.45)$$

where $\boldsymbol{\Sigma}_c = \mathrm{diag}[\sigma_1^2, \cdots \sigma_n^2]$. To ensure that this variance-covariance matrix is symmetric, the constraints

$$\sigma_j^2 c_{ij} = \sigma_i^2 c_{ji} \quad (6.46)$$

must be imposed.

Clearly, this model is similar to a SAR model, but with a different variance-covariance matrix. In fact, if we set $\boldsymbol{\Sigma}_c = \sigma^2\mathbf{I}$ in the development of the CAR models and $\boldsymbol{\Sigma}_v = \sigma^2\mathbf{I}$ in the development of the SAR models, then a comparison of the variances in equations (6.35) and (6.45) shows that any SAR model with spatial dependence matrix $\mathbf{B}$ can be expressed as a CAR model with spatial dependence matrix $\mathbf{C} = \mathbf{B} + \mathbf{B}' - \mathbf{B}\mathbf{B}'$. Any CAR model can also be expressed as a SAR model, but the relationships between $\mathbf{B}$ and $\mathbf{C}$ are somewhat contrived (see Haining, 1990, p. 81), and the neighborhood structure of the two models may not be the same (Cressie, 1993, p. 409).

*Estimation and Inference with Gaussian CAR models.* Following the ideas regarding SAR models, we consider the case where $\mathbf{\Sigma}_c = \sigma^2\mathbf{I}$ and the spatial dependence parameters can be written as functions of a single spatial autocorrelation parameter, e.g., $\mathbf{C} = \rho\mathbf{W}$.

Unlike the SAR model, the least squares estimator of the autocorrelation parameter $\rho$ in a one-parameter CAR model is consistent. Thus, iteratively re-weighted generalized least squares (described in §5.5.1) can be used to estimate all of the parameters of this CAR model. As before, $\boldsymbol{\beta}$ is estimated using generalized least squares with $\mathbf{\Sigma}(\boldsymbol{\theta}) = \mathbf{\Sigma}_{CAR} = \mathbf{\Sigma}_{CAR}(\rho)$, and $\rho$ is estimated using (Haining, 1990, p. 130)

$$\widehat{\rho}_{OLS} = \frac{\widehat{\boldsymbol{\epsilon}}'\mathbf{W}\widehat{\boldsymbol{\epsilon}}}{\widehat{\boldsymbol{\epsilon}}'\mathbf{W}^2\widehat{\boldsymbol{\epsilon}}},$$

where $\widehat{\boldsymbol{\epsilon}}$ is the residual vector from OLS regression.

To perform maximum likelihood estimation, we consider a slightly more general formulation for $\mathbf{\Sigma}_c$, as in the case of the SAR models. Specifically, we reparameterize it as $\mathbf{\Sigma}_c = \sigma^2\mathbf{V}_c$, with $\mathbf{V}_c$ known. Thus, with $\mathbf{C} = \rho\mathbf{W}$, the variance-covariance matrix of this CAR model can be written as

$$\mathbf{\Sigma}_{CAR} = \sigma^2(\mathbf{I} - \mathbf{C})^{-1}\mathbf{V}_c = \sigma^2\mathbf{V}_{CAR}(\rho). \tag{6.47}$$

Maximizing the Gaussian likelihood based on this variance-covariance structure is usually straightforward, and the information matrix has a form very similar to that associated with the one-parameter SAR model (see (6.42)): (Cliff and Ord, 1981, p. 242)

$$I(\boldsymbol{\beta}, \sigma^2, \rho) = \begin{bmatrix} \sigma^{-2}(\mathbf{X}(\mathbf{s})'\mathbf{A}(\rho)^{-1}\mathbf{X}(\mathbf{s})) & \mathbf{0} & \mathbf{0} \\ \mathbf{0}' & \frac{n}{2}\sigma^{-4} & \frac{1}{2}\sigma^{-2}tr(\mathbf{G}) \\ \mathbf{0}' & \frac{1}{2}\sigma^{-2}tr(\mathbf{G}) & \frac{1}{2}\alpha \end{bmatrix}, \tag{6.48}$$

where $\mathbf{A}(\rho) = (I - \rho\mathbf{W})^{-1}$, $\mathbf{G} = \mathbf{W}(I - \rho\mathbf{W})^{-1}$, $\alpha = \sum(\vartheta_i^2/(1 - \rho\vartheta_i)^2)$, and $\vartheta_i$ are the eigenvalues of $\mathbf{W}$.

#### *6.2.2.3 Spatial Prediction with Autoregressive Models*

Spatial autoregressive models were developed for use with regional data and in most applications the data are aggregated over a set of finite areal regions. Thus, prediction at a new location (which is actually a new region) is usually not of interest unless there is missing data. Assume that in the Gaussian case the SAR model $\mathbf{Z}(\mathbf{s}) \sim G(\mathbf{X}(\mathbf{s})\boldsymbol{\beta}, \mathbf{\Sigma}_{SAR})$, with $\mathbf{\Sigma}_{SAR}$ given in (6.35), or the CAR model $\mathbf{Z}(\mathbf{s}) \sim G(\mathbf{X}(\mathbf{s})\boldsymbol{\beta}, \mathbf{\Sigma}_{CAR})$ with $\mathbf{\Sigma}_{CAR}$ given in (6.45), hold for the observed data as well at the new location. Then universal kriging can be used to predict at the new location by jointly modeling the data and the unobservables as in equations (5.31) and (5.32) and using the predictor in equation (5.33).

While prediction at a new location is usually not of interest in spatial autoregressive models, predictions from the models for the given set of regions can be important. If predictions are made using the same covariate values, these predictions represent smoothed values of the original data that are adjusted for the covariate effects. The adjustment also accounts for spatial autocorrelation as measured by the spatial dependence parameters in the model and these are usually assumed to be determined by the neighborhood structure imposed on the lattice system. Recall that universal kriging honors the data, and so one way to obtain these smoothed values is to use the filtered version described in §5.3.3.

### *6.2.3 Generalized Least Squares—Inference and Diagnostics*

#### *6.2.3.1 Hypothesis Testing*

*Linear Hypotheses About Fixed Effects.* The correlated error models of this section, whether marginal models, mixed models, or autoregressive models, have an estimated generalized least squares solution for the fixed effects. As for models with uncorrelated errors, we consider linear hypotheses involving $\boldsymbol{\beta}$ of the form

$$\begin{aligned} H_0&: \quad \mathbf{L}\boldsymbol{\beta} = \mathbf{l}_0 \\ H_1&: \quad \mathbf{L}\boldsymbol{\beta} \neq \mathbf{l}_0, \end{aligned} \tag{6.49}$$

where $\mathbf{L}$ is a $l \times p$ matrix of contrast coefficients and $\mathbf{l}_0$ is a specified $l \times 1$ vector. The equivalent statistic to (6.3) is the Wald $F$ statistic

*Wald F Statistic*

$$\widetilde{F} = \frac{(\mathbf{L}\widetilde{\boldsymbol{\beta}} - \mathbf{l}_0)'[\mathbf{L}(\mathbf{X}(\mathbf{s})'\boldsymbol{\Sigma}(\widetilde{\boldsymbol{\theta}})^{-1}\mathbf{X}(\mathbf{s}))^{-1}\mathbf{L}']^{-1}(\mathbf{L}\widetilde{\boldsymbol{\beta}} - \mathbf{l}_0)}{\text{rank}(\mathbf{L})}, \tag{6.50}$$

taking $\widetilde{\boldsymbol{\beta}}$ as the EGLS estimator based on $\widetilde{\boldsymbol{\theta}}$, the latter being either the IRWGLS, ML, or REML estimators. For uncorrelated, Gaussian distributed errors, the regular $F$ statistic ((6.3), page 305) followed an $F$ distribution.

In the correlated error case, the distributional properties of $\widetilde{F}$ are less clear-cut. If $\widetilde{\boldsymbol{\theta}}$ is a consistent estimator, then $\widetilde{F}$ has an approximate Chi-square distribution with rank$\{\mathbf{L}\}$ degrees of freedom. This is also the distribution of $\widetilde{F}$ if $\boldsymbol{\theta}$ is known and $\mathbf{Z}(\mathbf{s})$ is Gaussian. Consequently, $p$-values computed from the Chi-square approximation tend to be too small, the test tends to be liberal, Type-I error rates tend to exceed the nominal level. A better approximation to the nominal Type-I error level is achieved when $p$-values for $\widetilde{F}$ are computed from an $F$ distribution with rank$\{\mathbf{L}\}$ numerator and $(n - \text{rank}\{\mathbf{X}(\mathbf{s})\})$ denominator degrees of freedom.

**Example 6.2** A similar phenomenon can be observed in *iid* random samples.

If $Y_1, \cdots, Y_n$ is a random sample of size $n$ from a distribution with mean $\mu$ and finite variance $\sigma^2$, and $\overline{Y}$, $s^2$, denote the sample mean and sample variance, respectively, then the distribution of

$$T = \frac{\overline{Y} - \mu}{s/\sqrt{n}} \tag{6.51}$$

approaches a $G(0,1)$ distribution as $n \to \infty$ (by the Central Limit Theorem). Because $s^2$ is a consistent estimator of $\sigma^2$, the asymptotic distribution is the same as that of

$$T^* = \frac{\overline{Y} - \mu}{\sigma/\sqrt{n}}.$$

In order to test the hypothesis $H_0{:}\, \mu = \mu_0$ with test statistic $T$, one can draw on the standard Gaussian distribution to compute $p$-values, critical values, etc. Since $n$ is finite, one may obtain better adherence to nominal Type-I error levels, if the distribution applied to $T$ is one that takes "into account" the fact that $s$ is a random variable for any finite $n$. This is the rationale for using a $t$-distribution with $n-1$ degrees of freedom in tests based on (6.51), although $T \sim t_{n-1}$ only if the $Y_i$ are Gaussian distributed. □

When $\mathbf{L}$ is a $1 \times p$ vector whose only nonzero entry is a 1 for the $i^{th}$ element and $\mathbf{l}_0 = 0$, the test based on $\widetilde{F}$ reduces to the familiar $t$-test of $\beta_i = 0$. Denoting the diagonal elements of $(\mathbf{X}(\mathbf{s})'\mathbf{\Sigma}(\widetilde{\boldsymbol{\theta}})^{-1}\mathbf{X}(\mathbf{s}))^{-1}$ by $s^{ii}$, then the estimated standard error for a single $\widehat{\beta}_i$ is $\sqrt{s^{ii}}$, and a (1-$\alpha$)% confidence interval for $\beta_i$ is

$$\widehat{\beta}_i \pm (t_{\alpha/2,n-k})\sqrt{s^{ii}}, \tag{6.52}$$

where $t_{\alpha/2,n-k}$ is the $\alpha/2$ percentage point from a $t$-distribution with $n-k$ degrees of freedom.

Using an $F$-distribution ($t$-distribution) instead of the asymptotic Chi-square (Gaussian) distribution improves the properties of the test of $H_0{:}\, \mathbf{L}\boldsymbol{\beta} = \mathbf{l}_0$, but it does not address yet the essential problem that

$$\mathbf{\Omega}(\widetilde{\boldsymbol{\theta}}) = \left(\mathbf{X}(\mathbf{s})'\mathbf{\Sigma}(\widetilde{\boldsymbol{\theta}})^{-1}\mathbf{X}(\mathbf{s})\right)^{-1}$$

is not the variance-covariance matrix of $\widetilde{\boldsymbol{\beta}}$. It is an estimate of $\mathrm{Var}[\widetilde{\boldsymbol{\beta}}]$ but not necessarily a good one at that. Indeed, it is an estimate of the variance of the GLS estimator. We saw in §5.5.4 that plugging-in covariance parameter estimates can affect the prediction standard errors. By the same token, plugging-in affects the variance of the EGLS estimator relative to the GLS estimator. The test statistic $\widetilde{F}$ should be adjusted in possibly two ways. First,

$$\mathbf{L}\left(\mathbf{X}(\mathbf{s})'\mathbf{\Sigma}(\widetilde{\boldsymbol{\theta}})^{-1}\mathbf{X}(\mathbf{s})\right)^{-1}\mathbf{L}'$$

is supposed to be an estimate of $\mathrm{Var}[\mathbf{L}\widetilde{\boldsymbol{\beta}}]$. This requires a different estimate of $\mathrm{Var}[\widetilde{\boldsymbol{\beta}}]$, one that accounts for the fact that uncertainty is associated with

$\widetilde{\boldsymbol{\theta}}$ (regardless of how $\boldsymbol{\theta}$ was estimated). If this adjusted test statistic is denoted $\widetilde{F}^*$, then the appropriate null distribution may not be the same as that used in a test based on $\widetilde{F}$. The adjustment for the variance estimator of the EGLS estimator flows directly from the work of Kackar and Harville (1984), Prasad and Rao (1990), and Harville and Jeske (1992); summarized in §5.5.4. Kenward and Roger (1997) gave computationally simple expressions for a bias-corrected estimator of the variance of $\widehat{\boldsymbol{\beta}}$ based on (5.53) and REML estimation. They furthermore gave expressions for adjusting the degrees of freedom of the reference $F$-distribution.

Let $\widehat{\boldsymbol{\theta}}$ denote the REML estimator of $\boldsymbol{\theta}$. The adjusted Kenward-Roger variance estimator then takes the form

*Bias-adjusted Variance Estimator*

$$\boldsymbol{\Upsilon}(\widehat{\boldsymbol{\theta}}) = \boldsymbol{\Omega}(\widehat{\boldsymbol{\theta}}) + 2\boldsymbol{\Omega}(\widehat{\boldsymbol{\theta}}) \left\{ \sum_{i=1}^{q} \sum_{j=1}^{q} a_{ij} \mathbf{D}_{ij} \right\} \boldsymbol{\Omega}(\widehat{\boldsymbol{\theta}}), \tag{6.53}$$

where $a_{ij}$ is the $(i,j)$th element of $\text{Var}[\widehat{\boldsymbol{\theta}}]$, $q$ is the dimension of $\boldsymbol{\theta}$ and

$$\begin{aligned}
\mathbf{D}_{ij} &= \mathbf{X}'\frac{\partial \boldsymbol{\Sigma}(\boldsymbol{\theta})^{-1}}{\partial \theta_i}\boldsymbol{\Sigma}(\widehat{\boldsymbol{\theta}})\frac{\partial \boldsymbol{\Sigma}(\boldsymbol{\theta})^{-1}}{\partial \theta_i}\mathbf{X} - \\
&\quad \mathbf{X}'\frac{\partial \boldsymbol{\Sigma}(\boldsymbol{\theta})^{-1}}{\partial \theta_i}\mathbf{X}\boldsymbol{\Omega}(\widehat{\boldsymbol{\theta}})\mathbf{X}'\frac{\partial \boldsymbol{\Sigma}(\boldsymbol{\theta})^{-1}}{\partial \theta_j}\mathbf{X} - \\
&\quad \frac{1}{4}\mathbf{X}'\boldsymbol{\Sigma}(\widehat{\boldsymbol{\theta}})^{-1}\frac{\partial^2 \boldsymbol{\Sigma}(\boldsymbol{\theta})}{\partial \theta_i \partial \theta_j}\boldsymbol{\Sigma}(\widehat{\boldsymbol{\theta}})^{-1}\mathbf{X}.
\end{aligned}$$

The derivatives in this expression are evaluated at $\widehat{\boldsymbol{\theta}}$. When $\boldsymbol{\Sigma}$ depends linearly on $\boldsymbol{\theta}$, the last term vanishes. This is the case in certain mixed models, but usually not when the covariance structure has spatial components.

In the next step, Kenward and Roger (1997) replace $\boldsymbol{\Omega}(\widehat{\boldsymbol{\theta}})$ in (6.50) with (6.53) and adjust the test statistic and its degrees of freedom. The Kenward-Roger $F$-test has test statistic

*Kenward-Roger Wald Test*

$$\widetilde{F}^* = \lambda \frac{(\mathbf{L}\widehat{\boldsymbol{\beta}} - \mathbf{l}_0)'[\mathbf{L}\boldsymbol{\Upsilon}(\widehat{\boldsymbol{\theta}})\mathbf{L}']^{-1}(\mathbf{L}\widehat{\boldsymbol{\beta}} - \mathbf{l}_0)}{\text{rank}(\mathbf{L})}, \tag{6.54}$$

its distribution under $H_0\colon \mathbf{L}\boldsymbol{\beta} = \mathbf{l}_0$ is taken to be $F$ with rank$\{\mathbf{L}\}$ numerator and $m$ denominator degrees of freedom. The interested reader is referred to Kenward and Roger (1997) for computational details regarding $\lambda$ and $m$.

*Testing Hypotheses About Covariance Parameters.* Since maximum likelihood estimators are asymptotically Gaussian, we can obtain approximate confidence intervals or associated hypothesis tests for the elements of $\boldsymbol{\theta}$ using

$$\widehat{\theta}_{i(ml)} \quad \pm \quad (z_{\alpha/2})\sqrt{\text{Var}[\widehat{\theta}_{i(ml)}]} \tag{6.55}$$

$$\widehat{\theta}_{i(reml)} \quad \pm \quad (z_{\alpha/2})\sqrt{\text{Var}[\widehat{\theta}_{i(reml)}]} \tag{6.56}$$

where the variance of the estimators derives from the inverse of the observed or expected information matrix, substituting (plugging-in) ML (REML) estimates for any unknown parameters. Alternatively, we can use likelihood ratio tests of hypotheses about $\boldsymbol{\theta}$ and compare two nested parametric models to see whether a subset of the covariance parameters fits the data as well as the full set. Consider comparing two models of the same form, one based on parameters $\boldsymbol{\theta}_1$ and a larger model based on $\boldsymbol{\theta}_2$, with $\dim(\boldsymbol{\theta}_2) > \dim(\boldsymbol{\theta}_1)$. That is, $\boldsymbol{\theta}_1$ is obtained by constraining some parameters in $\boldsymbol{\theta}_2$, usually setting them to zero, and $\dim(\boldsymbol{\theta})$ denotes the number of free parameters. Then a test of $H_0: \boldsymbol{\theta} = \boldsymbol{\theta}_1$ against the alternative $H_1: \boldsymbol{\theta} = \boldsymbol{\theta}_2$ can be carried out by comparing

*Likelihood Ratio Test Statistic*

$$\varphi(\boldsymbol{\beta}; \boldsymbol{\theta}_1; \mathbf{Z}(\mathbf{s})) - \varphi(\boldsymbol{\beta}; \boldsymbol{\theta}_2; \mathbf{Z}(\mathbf{s})) \tag{6.57}$$

to a $\chi^2$ distribution with $\dim(\boldsymbol{\theta}_2) - \dim(\boldsymbol{\theta}_1)$ degrees of freedom. (Recall that $\varphi(\boldsymbol{\beta}; \boldsymbol{\theta}; \mathbf{Z}(\mathbf{s}))$ denotes twice the negative of the log-likelihood, so $\varphi(\boldsymbol{\beta}; \boldsymbol{\theta}_1; \mathbf{Z}(\mathbf{s})) > \varphi(\boldsymbol{\beta}; \boldsymbol{\theta}_2; \mathbf{Z}(\mathbf{s}))$.)

**Example 6.3** Consider a linear spatial regression model $\mathbf{Z}(\mathbf{s}) = \mathbf{X}(\mathbf{s})\boldsymbol{\beta} + \mathbf{e}(\mathbf{s})$ with

$$\text{Cov}[e(\mathbf{s}_i), e(\mathbf{s}_j)] = c_0 + \sigma^2 \exp\{-||\mathbf{s}_i - \mathbf{s}_j||/\alpha\},$$

an exponential model with practical range $3\alpha$ and nugget $c_0$. Let $\boldsymbol{\theta}_2 = [c_0, \sigma^2, \alpha]'$. The alternative covariance functions are

1. $\text{Cov}[e(\mathbf{s}_i), e(\mathbf{s}_j)] = \sigma^2 \exp\{-||\mathbf{s}_i - \mathbf{s}_j||/\alpha\}$, $\boldsymbol{\theta}_1 = [\sigma^2, \alpha]'$.
2. $\text{Cov}[e(\mathbf{s}_i), e(\mathbf{s}_j)] = \sigma^2 \exp\{-||\mathbf{s}_i - \mathbf{s}_j||^2/\alpha^2\}$

Model 1 is an exponential model without nugget effect. Since it can be obtained from the original model by setting $c_0 = 0$, the two models are nested. A likelihood ratio test of $H_0: c_0 = 0$ is possible. The test statistic is $\varphi(\boldsymbol{\beta}; \boldsymbol{\theta}_1; \mathbf{Z}(\mathbf{s})) - \varphi(\boldsymbol{\beta}; \boldsymbol{\theta}_2; \mathbf{Z}(\mathbf{s}))$. Model 2 is a gaussian covariance structure and is not of the same form as the original model or model 1. Constraining parameters of the other models does not yield model 2. A likelihood ratio test comparing the two models is not possible. □

An important exception to the distributional properties of likelihood ratio statistics occurs when parameters lie on the boundary of the parameter space. In that case, the test statistic in equation (6.57) is a mixture of $\chi^2$ distributions. Such boundary exceptions arise in practice when variance components (e.g., the nugget or the variance of the low-rank smoother's spline coefficients) are tested against zero. If we are testing only one of these variance components against 0, the test statistic has a distribution that is a mixture of a degenerate distribution giving probability 1 to the value zero, and a $\chi^2$ distribution with $\dim(\boldsymbol{\theta}_2) - \dim(\boldsymbol{\theta}_1)$ degrees of freedom (Self and Liang, 1987; Littell, Milliken, Stroup and Wolfinger, 1996). Thus, to make the test, simply divide the $p$-value obtained from a $\chi^2$ with $\dim(\boldsymbol{\theta}_2) - \dim(\boldsymbol{\theta}_1)$ degrees of freedom by 2.

The likelihood ratio test can also be carried out for REML estimation, provided the hypothesis is about the covariance parameters. In that case we are comparing

$$\varphi_R(\boldsymbol{\theta}_1; \mathbf{KZ}(\mathbf{s})) - \varphi_R(\boldsymbol{\theta}_2; \mathbf{KZ}(\mathbf{s})) \qquad (6.58)$$

to a $\chi^2$ distribution with $\dim(\boldsymbol{\theta}_2) - \dim(\boldsymbol{\theta}_1)$ degrees of freedom. While likelihood ratio tests can be formulated for models that are nested with respect to the mean structure and/or the covariance structure, tests based on $\varphi_R(\boldsymbol{\theta}; \mathbf{KZ}(\mathbf{s}))$ can only be carried out for models that are nested with respect to the covariance parameters and have the same fixed effects structure (same $\mathbf{X}(\mathbf{s})$ matrix).

**Example 6.4** Consider a partitioned matrix $\mathbf{X}(\mathbf{s}) = [\mathbf{X}(\mathbf{s})_1 \; \mathbf{X}(\mathbf{s})_2]$ and the following models:

1. $\mathbf{Z}(\mathbf{s}) = \mathbf{X}(\mathbf{s})[\boldsymbol{\beta}_1', \boldsymbol{\beta}_2']' + \mathbf{e}(\mathbf{s})$ with $\text{Cov}[\mathbf{e}(\mathbf{s}_i), \mathbf{e}(\mathbf{s}_j)] = c_0 + \sigma^2 \exp\{|\mathbf{s}_i - \mathbf{s}_j||/\alpha\}$
2. $\mathbf{Z}(\mathbf{s}) = \mathbf{X}(\mathbf{s})_2\boldsymbol{\beta}_2 + \mathbf{e}(\mathbf{s})$ with $\text{Cov}[\mathbf{e}(\mathbf{s}_i), \mathbf{e}(\mathbf{s}_j)] = c_0 + \sigma^2 \exp\{|\mathbf{s}_i - \mathbf{s}_j||/\alpha\}$
3. $\mathbf{Z}(\mathbf{s}) = \mathbf{X}(\mathbf{s})[\boldsymbol{\beta}_1', \boldsymbol{\beta}_2']' + \mathbf{e}(\mathbf{s})$ with $\text{Cov}[\mathbf{e}(\mathbf{s}_i), \mathbf{e}(\mathbf{s}_j)] = \sigma^2 \exp\{|\mathbf{s}_i - \mathbf{s}_j||/\alpha\}$
4. $\mathbf{Z}(\mathbf{s}) = \mathbf{X}(\mathbf{s})_2\boldsymbol{\beta}_2 + \mathbf{e}(\mathbf{s}))$ with $\text{Cov}[\mathbf{e}(\mathbf{s}_i), \mathbf{e}(\mathbf{s}_j)] = \sigma^2 \exp\{|\mathbf{s}_i - \mathbf{s}_j||/\alpha\}$

To test the hypothesis $H_0\colon \boldsymbol{\beta}_2 = \mathbf{0}$, a likelihood ratio test is possible, based on the log likelihoods from models 1 and 2 (or models 3 and 4). The REML log likelihoods from these two models can not be compared because $\mathbf{K}_1\mathbf{Z}(\mathbf{s})$ and $\mathbf{K}_2\mathbf{Z}(\mathbf{s})$ are two different sets of "data."

To test the hypothesis $H_0\colon c_0 = 0$, models 1 and 3 (or models 2 and 4) can be compared, since they are nested. The test can be based on either the $-2$ log likelihood $(\varphi(\boldsymbol{\beta}; \boldsymbol{\theta}; \mathbf{Z}(\mathbf{s})))$ or on the $-2$ residual log likelihood $(\varphi_R(\boldsymbol{\theta}; \mathbf{KZ}(\mathbf{s})))$. The test based on $\varphi_R$ is possible because both models use the same matrix of error contrasts.

Finally, a test of $H_0\colon c_0 = 0, \boldsymbol{\beta}_2 = \mathbf{0}$ is possible by comparing models 1 and 4. Since the models differ in their regressor matrices (mean model), only a test based on the log likelihood is possible. □

**Example 6.1 (Soil carbon regression. Continued)** Early on in §6.2 we compared OLS estimates in the soil carbon regression model to maximum likelihood and restricted maximum likelihood estimates in Table 6.1 on page 324. At that point we deferred a formal test of significance of the spatial autocorrelation.

We can compare the models

$$\begin{aligned} C(\mathbf{s}_i)|N(\mathbf{s}_i) &= \beta_0 + \beta_1 N(\mathbf{s}_i) + e(\mathbf{s}_i) \\ e(\mathbf{s}_i) &\sim G(0, \sigma^2) \\ \text{Cov}[e(\mathbf{s}_i), e(\mathbf{s}_j)] &= \sigma^2 \exp\{-||\mathbf{s}_i - \mathbf{s}_j||/\theta\} \end{aligned}$$

and

$$\begin{aligned} C(\mathbf{s}_i)|N(\mathbf{s}_i) &= \beta_0 + \beta_1 N(\mathbf{s}_i) + e(\mathbf{s}_i) \\ e(\mathbf{s}_i) &\sim G(0, \sigma^2) \\ \text{Cov}[e(\mathbf{s}_i), e(\mathbf{s}_j)] &= 0 \end{aligned}$$

Table 6.2 *Minus two times (restricted) log likelihood values for C–N regression analysis. See Table 6.1 on page 324 for parameter estimates*

| Fit Statistic | Covariance Structure<br>Independent | Exponential |
|---|---|---|
| $\varphi(\boldsymbol{\beta}; \boldsymbol{\theta}_1; \mathbf{C}(\mathbf{s}))$ | −706 | −747 |
| $\varphi_R(\boldsymbol{\theta}; \mathbf{KC}(\mathbf{s}))$ | −695 | −738 |

with a likelihood ratio test, provided that the models are nested; that is, the independence model can be derived from the correlated error model by a restriction of the parameter space. The exponential covariance structure $\exp\{-||\mathbf{s}_i - \mathbf{s}_j||/\theta\}$ approaches the independence model only asymptotically, as $\theta \to 0$. For $\theta = 0$ the likelihood cannot be computed, so it appears that the models cannot be compared. There are two simple ways out of this "dilemma." We can reparameterize the covariance model so that the OLS model is nested. For example,

$$\exp\left\{-\frac{||\mathbf{s}_i - \mathbf{s}_j||}{\theta}\right\} = \rho^{||\mathbf{s}_i - \mathbf{s}_j||}$$

for $\rho = \exp\{-1/\theta\}$. The likelihood ratios statistic for $H_0\colon \rho = 0$ is $-706 - (-747) = 41$ with $p$-value $\Pr(\chi^2_1 > 41) < 0.0001$. The addition of the exponential covariance structure significantly improves the model. The test based on the restricted likelihood ratio test yields the same conclusion ($-695-(-738) = 43$). It should be obvious that $-747$ is also the $-2$ log likelihood for the model in the $\exp\{-||\mathbf{s}_i - \mathbf{s}_j||/\theta\}$ parameterization, so we could have used the log likelihood from the exponential model directly. The second approach is to evaluate the log likelihood at a value for the range parameter far enough away from zero to prevent floating point exceptions and inaccuracies, but small enough so that the correlation matrix is essentially the identity.

For a stationary process, $\rho = 0$ falls on the boundary of the parameter space. The $p$-value of the test could be adjusted, but it would not alter our conclusion (the adjustment shrinks the $p$-value). The correlated error model fits significantly better than a model assuming independence. □

To compare models that are not nested, we can draw on likelihood-based information criteria. For example, Akaike's Information Criterion (AIC, Akaike, 1974), a penalized-log-likelihood ratio, is defined by

$$\mathrm{AIC}(\boldsymbol{\beta}, \boldsymbol{\theta}) = \varphi(\boldsymbol{\beta}; \boldsymbol{\theta}; \mathbf{Z}(\mathbf{s})) + 2(k + q) \quad (6.59)$$

$$\mathrm{AIC}_R(\boldsymbol{\theta}) = \varphi_R(\boldsymbol{\theta}; \mathbf{Z}(\mathbf{s})) + 2q. \quad (6.60)$$

A finite-sample corrected version of this information criterion is AICC (Hurvich and Tsai, 1989; Burnham and Anderson, 1998)

$$\begin{aligned} \mathrm{AICC}(\boldsymbol{\beta}, \boldsymbol{\theta}) &= \varphi(\boldsymbol{\beta}; \boldsymbol{\theta}; \mathbf{Z}(\mathbf{s})) + 2\frac{n(k+q)}{n-k-q-1} && (6.61)\\ \mathrm{AICC}_R(\boldsymbol{\theta}) &= \varphi_R(\boldsymbol{\theta}; \mathbf{Z}(\mathbf{s})) + 2\frac{(n-k)q}{n-k-q-1}. && (6.62) \end{aligned}$$

In theory, we select the model with the smallest AIC or AICC value. In practice, one should observe the following guidelines, however.

- When fitting several competing models, for example, models that differ in their covariance structure, one often finds that the information criteria group models into sets of models that are clearly inferior and a set of model with fairly similar values. One should then not necessarily adopt the model with the smallest information criterion, but a model in the group of plausible model that is parsimonious and interpretable.
- The presence of unusual observations (outliers, influential data points), can substantively affect the covariance model. For example, in a simple one-way classification model, a single extreme value in one of the groups may suggest that a heterogeneous variance model may be appropriate. Strict reliance on information criteria alone may then lead to a model that is more complex than necessary. The complexity of the model tries to adjust to a few unusual observations.
- In the case of residual maximum likelihood estimation, these information criteria should not be used for model comparisons, if the models have different mean structures (whether the mean models are nested or not). The criteria are based on residual log likelihoods that apply to different sets of data when the mean model changes. Comparisons of non-nested models based on information criteria that are derived from $\varphi_R(\boldsymbol{\theta}; \mathbf{Z}(\mathbf{s}))$ should be restricted to models that are non-nested only with respect to the covariance structure.

*Hypothesis Testing in Autoregressive Models.* Since the SAR model can be viewed as a linear model with spatially autocorrelated errors, all of the methods of hypothesis testing involving $\boldsymbol{\beta}$ or $\boldsymbol{\theta}$ described previously can be used with SAR models. Specifically, hypotheses involving $\boldsymbol{\beta}$ can be tested using the Wald test in (6.50) or the Kenward-Roger test in (6.54). However, one of the most common uses of the one-parameter SAR model is to provide an alternative model for a test of residual spatial autocorrelation in OLS residuals. That is, we consider the following two models, a traditional linear regression model with independent errors,

$$\mathbf{Z}(\mathbf{s}) = \mathbf{X}(\mathbf{s})\boldsymbol{\beta} + \boldsymbol{\epsilon}, \qquad \boldsymbol{\Sigma} = \sigma^2\mathbf{I} \tag{6.63}$$

and the one parameter SAR model:

$$\mathbf{Z}(\mathbf{s}) = \mathbf{X}(\mathbf{s})\boldsymbol{\beta} + (I - \rho\mathbf{W})^{-1}\boldsymbol{\upsilon}, \qquad \boldsymbol{\Sigma} = \sigma^2(\mathbf{I} - \rho\mathbf{W})^{-1}(\mathbf{I} - \rho\mathbf{W}')^{-1}. \tag{6.64}$$

Setting $\rho = 0$ in the SAR model (6.64), we obtain the traditional linear regression model with independent errors (6.63). This nesting of the two models, together with a traditional null model for which estimation is rather straightforward (i.e., OLS), provides several approaches for constructing tests of $H_0: \rho = 0$ vs. $H_1: \rho \neq 0$ (or a corresponding one-sided alternative). Two of the most common are the likelihood ratio test described earlier in §6.2.3.1 and the use of Moran's $I$ with OLS residuals (see §6.1.2). In order to carry out the likelihood ratio test for $H_0: \rho = 0$ in a SAR model, we are comparing the alternative model in equation (6.64) with parameters $\boldsymbol{\theta}_2 = [\boldsymbol{\beta}', \sigma^2, \rho]'$ to the null model defined by equation (6.63) with parameters $\boldsymbol{\theta}_1 = [\boldsymbol{\beta}', \sigma^2]'$, with $\dim(\boldsymbol{\theta}_2) - \dim(\boldsymbol{\theta}_1) = 1$. A test of $H_0: \boldsymbol{\theta} = \boldsymbol{\theta}_1$ against the alternative $H_1: \boldsymbol{\theta} = \boldsymbol{\theta}_2$ is also a test of $H_0: \rho = 0$ vs. $H_1: \rho \neq 0$ and it can be done by comparing

$$\varphi(\boldsymbol{\beta}; \boldsymbol{\theta}_1; \mathbf{Z}(\mathbf{s})) - \varphi(\boldsymbol{\beta}; \boldsymbol{\theta}_2; \mathbf{Z}(\mathbf{s})) \tag{6.65}$$

to a $\chi^2$ distribution with a single degree of freedom. The $\chi^2$ distribution is a large sample approximation to the likelihood ratio statistic, and thus will give approximately valid inference only for large $n$ (the number of points on the lattice). The size of the lattice necessary for the approximation to work well depends on several factors including the particular structure of the spatial proximity matrix $\mathbf{W}$. In general, for small lattices, the $\chi^2_1$ distribution can provide a very poor approximation to the distribution of the likelihood ratio statistic.

The primary difference between hypothesis testing for CAR and SAR models based on Gaussian data is the definition of $\boldsymbol{\Sigma}(\boldsymbol{\theta})$. In the simplest one-parameter cases, $\boldsymbol{\Sigma}(\boldsymbol{\theta}) = \sigma^2(I - \rho\mathbf{W})^{-1}$ for a CAR model and $\boldsymbol{\Sigma}(\boldsymbol{\theta}) = \sigma^2(I - \rho\mathbf{W})^{-1}(I - \rho\mathbf{W}')^{-1}$ for a SAR model. Thus, the methods for testing $H_0: \rho = 0$ vs. $H_1: \rho \neq 0$ apply to SAR and CAR models alike.

#### *6.2.3.2 GLS Residual Analysis*

The goals of residual analysis in the correlated error model

$$\mathbf{Z}(\mathbf{s}) = \mathbf{X}(\mathbf{s})\boldsymbol{\beta} + \mathbf{e}(\mathbf{s}), \qquad \mathbf{e}(\mathbf{s}) \sim (\mathbf{0}, \boldsymbol{\Sigma}(\boldsymbol{\theta}))$$

are the same as in standard linear models. We would like to derive quantities whose properties instruct us about the behavior of $\mathbf{e}(\mathbf{s})$. We want to use such residual-type quantities to detect, for example, whether important terms have been omitted from $\mathbf{X}(\mathbf{s})$ and whether the assumed model for $\boldsymbol{\Sigma}(\boldsymbol{\theta})$ is correct. Except, we are no longer working with ordinary least squares residuals, but with (E)GLS residuals,

$$\widehat{\mathbf{e}}_{egls}(\mathbf{s}) = \mathbf{Z}(\mathbf{s}) - \mathbf{X}(\mathbf{s})\widehat{\boldsymbol{\beta}}_{egls}.$$

To contrast residual analysis in the correlated error model with the analysis of OLS residuals (§6.1.2), it is important to understand the consequences of engaging a non-diagonal variance-covariance matrix in estimation. We will focus on this aspect here, and therefore concentrate on GLS estimation. A second layer of complications can be added on top of these differences because $\boldsymbol{\theta}$ is being estimated.

The first important result is that raw GLS residuals are about as useful as raw OLS residuals in diagnosing properties of the model errors $\mathbf{e}(\mathbf{s})$. We can write

$$\begin{aligned}
\widehat{\mathbf{e}}_{gls}(\mathbf{s}) &= \mathbf{Z}(\mathbf{s}) - \widehat{\mathbf{Z}}(\mathbf{s}) \\
&= \mathbf{Z}(\mathbf{s}) - \mathbf{X}(\mathbf{s})\left(\mathbf{X}(\mathbf{s})'\boldsymbol{\Sigma}(\boldsymbol{\theta})^{-1}\mathbf{X}(\mathbf{s})\right)^{-1}\mathbf{X}(\mathbf{s})'\boldsymbol{\Sigma}(\boldsymbol{\theta})^{-1}\mathbf{Z}(\mathbf{s}) \\
&= \left(\mathbf{I} - \mathbf{X}(\mathbf{s})\left(\mathbf{X}(\mathbf{s})'\boldsymbol{\Sigma}(\boldsymbol{\theta})^{-1}\mathbf{X}(\mathbf{s})\right)^{-1}\mathbf{X}(\mathbf{s})'\boldsymbol{\Sigma}(\boldsymbol{\theta})^{-1}\right)\mathbf{Z}(\mathbf{s}) \\
&= (\mathbf{I} - \mathbf{H}(\boldsymbol{\theta}))\,\mathbf{Z}(\mathbf{s}) = \mathbf{M}(\boldsymbol{\theta})\mathbf{Z}(\mathbf{s}). \qquad (6.66)
\end{aligned}$$

The matrix $\mathbf{H}(\boldsymbol{\theta})$ is the gradient of the fitted values with respect to the observed data,

*Leverage Matrix*

$$\mathbf{H}(\boldsymbol{\theta}) = \frac{\partial \widehat{\mathbf{Z}}(\mathbf{s})}{\partial \mathbf{Z}(\mathbf{s})}$$

and it is thus reasonable to consider it the "leverage" matrix of the model. As with the model with uncorrelated errors, the leverage matrix is not diagonal, and its diagonal entries are not necessarily the same. So, as with OLS residuals, GLS residuals are correlated, and heteroscedastic, and since $\text{rank}\{\mathbf{H}(\boldsymbol{\theta})\} = k$, the GLS residuals are also rank-deficient. These properties stem from the fitting the model to data, and have nothing to do with whether the application is spatial or not. Since $\widehat{\mathbf{e}}_{gls}(\mathbf{s})$ is afflicted with the same problems as $\widehat{\mathbf{e}}_{ols}(\mathbf{s})$, the same caveats apply. For example, a semivariogram based on the GLS residuals is a biased estimator of the semivariogram of $\mathbf{e}(\mathbf{s})$. In fact, the semivariogram of the GLS residuals may not be any more reliable or useful in diagnosing the correctness of the covariance structure as the semivariogram of the OLS residuals. From (6.66) it follows immediately that

$$\text{Var}[\widehat{\mathbf{e}}_{gls}(\mathbf{s})] = \mathbf{M}(\boldsymbol{\theta})\boldsymbol{\Sigma}(\boldsymbol{\theta})\mathbf{M}(\boldsymbol{\theta})', \qquad (6.67)$$

which can be quite different from $\boldsymbol{\Sigma}(\boldsymbol{\theta})$. Consequently, the residuals will not be uncorrelated, even if the model fits well. If you fit a model by OLS but the true variance-covariance matrix is $\text{Var}[\mathbf{e}(\mathbf{s})] = \boldsymbol{\Sigma}(\boldsymbol{\theta})$, then the variance of the OLS residuals is

$$\text{Var}[\widehat{\mathbf{e}}_{ols}(\mathbf{s})] = \mathbf{M}_{ols}\boldsymbol{\Sigma}(\boldsymbol{\theta})\mathbf{M}_{ols},$$

where $\mathbf{M}_{ols} = \mathbf{I} - \mathbf{X}(\mathbf{s})(\mathbf{X}(\mathbf{s})'\mathbf{X}(\mathbf{s}))^{-1}\mathbf{X}(\mathbf{s})'$.

There is more. Like the OLS residuals, the GLS residuals have zero mean, provided that $\text{E}[\mathbf{e}(\mathbf{s})] = \mathbf{0}$. The OLS residuals also satisfy very simple constraints, for example $\mathbf{X}(\mathbf{s})'\widehat{\mathbf{e}}_{ols}(\mathbf{s}) = \mathbf{0}$. Thus, if $\mathbf{X}(\mathbf{s})$ contains an intercept column, the OLS residuals not only will have zero mean in expectation. They will sum to zero in every sample. The GLS residuals do not satisfy $\mathbf{X}(\mathbf{s})'\widehat{\mathbf{e}}_{gls}(\mathbf{s}) = \mathbf{0}$.

The question, then, is how to diagnose whether the covariance model has been chosen properly? Brownie and Gumpertz (1997) recommend a procedure that uses OLS residuals and the fit of a correlated error model in tandem. Their procedure is to

1. Fit the models $\mathbf{Z}(\mathbf{s}) = \mathbf{X}(\mathbf{s})\boldsymbol{\beta} + \mathbf{e}(\mathbf{s})$, $\mathbf{e}(\mathbf{s}) \sim (\mathbf{0}, \boldsymbol{\Sigma}(\boldsymbol{\theta}))$ and $\mathbf{Z}(\mathbf{s}) = \mathbf{X}(\mathbf{s})\boldsymbol{\beta} + \boldsymbol{\epsilon}$, $\boldsymbol{\epsilon} \sim (\mathbf{0}, \sigma^2\mathbf{I})$. The first model can be fit by REML, for example, yielding the estimate $\widehat{\boldsymbol{\theta}}_{reml}$. The second model is fit by ordinary least squares.
2. Compute $\widehat{\gamma}_e(h)$, the empirical semivariogram of the OLS residuals $\widehat{\mathbf{e}}_{ols}(\mathbf{s})$.
3. Compute an estimate of $\text{Var}[\widehat{\mathbf{e}}_{ols}(\mathbf{s})]$ under the assumption that $\boldsymbol{\Sigma}(\boldsymbol{\theta})$ is the correct model for the covariance structure. The "natural" estimate is the plug-in estimate
$$\widehat{\text{Var}}[\widehat{\mathbf{e}}_{ols}(\mathbf{s})] = \mathbf{M}_{ols}\boldsymbol{\Sigma}(\widehat{\boldsymbol{\theta}}_{reml})\mathbf{M}_{ols} = \mathbf{A}.$$
Construct the "semivariogram" of the OLS residuals under the assumed model for $\boldsymbol{\Sigma}$ from the matrix $\mathbf{A}$: for all pairs of observations in a particular lag class $h$, let $g_e(h)$ be the average value of $\frac{1}{2}([\mathbf{A}]_{ii} + [\mathbf{A}]_{jj}) - [\mathbf{A}]_{ij}$.
4. Graphically compare the empirical semivariogram $\widehat{\gamma}_e(h)$ and the function $g_e(h)$. If $\boldsymbol{\Sigma}(\boldsymbol{\theta})$ is the correct covariance function, the two should agree quite well. $\widehat{\gamma}_e(h)$ should scatter about $g_e(h)$ in a non-systematic fashion.

**Example 6.1 (Soil carbon regression. Continued)** Based on Figure 6.3 on page 313 we assumed that the model errors in the spatial linear regression of C% on N% are spatially correlated. REML and ML parameter estimates for a model with exponential covariance structure were then presented in Table 6.1 on page 324. Was the exponential model a reasonable choice? Following the steps of the Brownie and Gumpertz (1997) method, Figure 6.6 displays the comparisons of the empirical semivariograms of OLS residuals and $g_e(h)$. The exponential model (left-hand panel) is not appropriate. The corresponding plot is shown in the right-hand panel for an anisotropic covariance structure,
$$\text{Cov}[Z(\mathbf{s}_i), Z(\mathbf{s}_j)] = \sigma^2 \exp\left\{-\theta_x|x_i - x_j|^{\phi_x} - \theta_y|y_i - y_j|^{\phi_y}\right\},$$
that provides a much better fit. The estimates of the covariance parameters for this model are $\widehat{\theta}_x = 0.73$, $\widehat{\theta}_y = 0.40$, $\widehat{\phi}_x = 0.2$, $\widehat{\phi}_y = 0.2$, $\widehat{\sigma}^2 = 0.0016$. Because this model is not nested within the isotropic exponential model, a quantitative comparison can be made based on information criteria. The bias-corrected AIC criteria for the models are $-733.9$ for the isotropic model and $-755.3$ for the anisotropic model. The latter is clearly preferable. □

Notice that the matrix $\mathbf{M}(\boldsymbol{\theta})$ is transposed on the right hand side of equation (6.67). The leverage matrix $\mathbf{H}(\boldsymbol{\theta})$ is an oblique projector onto the column space of $\mathbf{X}(\mathbf{s})$ (Christensen, 1991). It is not a projection matrix in the true sense, since it is not symmetric. It is idempotent, however:

$$\begin{aligned}\mathbf{H}(\boldsymbol{\theta})\mathbf{H}(\boldsymbol{\theta}) &= \mathbf{X}(\mathbf{s})\left(\mathbf{X}(\mathbf{s})'\boldsymbol{\Sigma}(\boldsymbol{\theta})^{-1}\mathbf{X}(\mathbf{s})\right)^{-1}\mathbf{X}(\mathbf{s})'\boldsymbol{\Sigma}(\boldsymbol{\theta})^{-1} \times \\ &\quad \mathbf{X}(\mathbf{s})\left(\mathbf{X}(\mathbf{s})'\boldsymbol{\Sigma}(\boldsymbol{\theta})^{-1}\mathbf{X}(\mathbf{s})\right)^{-1}\mathbf{X}(\mathbf{s})'\boldsymbol{\Sigma}(\boldsymbol{\theta})^{-1}\end{aligned}$$

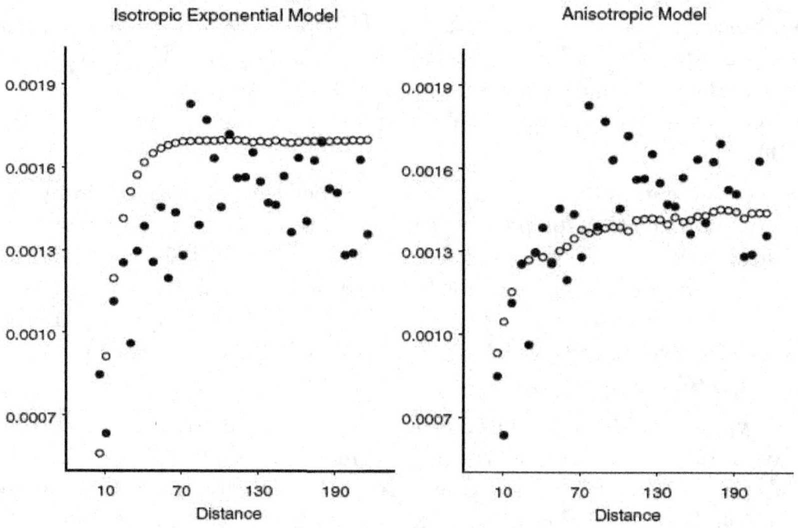


Figure 6.6 *Comparison of $g_e(h)$ (open circles) and empirical semivariogram of OLS residual for exponential models with and without anisotropy.*

$$= \mathbf{X}(\mathbf{s})\left(\mathbf{X}(\mathbf{s})'\boldsymbol{\Sigma}(\boldsymbol{\theta})^{-1}\mathbf{X}(\mathbf{s})\right)^{-1}\mathbf{X}(\mathbf{s})'\boldsymbol{\Sigma}(\boldsymbol{\theta})^{-1}.$$

It also shares other properties of the leverage matrix in an OLS model (see Chapter problems), and has the familiar interpretation as a gradient. Because diagonal elements of $\mathbf{H}(\boldsymbol{\theta})$ can be negative, other forms of leverage matrices in correlated error models have been proposed. For example, $\boldsymbol{\Sigma}(\widehat{\boldsymbol{\theta}})^{-1}\mathbf{H}(\boldsymbol{\theta})$ and $\mathbf{L}^{-1}\mathbf{H}(\boldsymbol{\theta})\mathbf{L}$, where $\mathbf{L}$ is the lower-triangular Cholesky root of $\boldsymbol{\Sigma}(\boldsymbol{\theta})$ (Martin, 1992; Haining, 1994).

In order to remedy shortcomings of the GLS residuals as diagnostic tools, we can apply similar techniques as in §6.1.2. For example, the plug-in estimate of the estimated variance of the EGLS residuals is

$$\begin{aligned} \widehat{\mathrm{Var}}[\widehat{\mathbf{e}}_{egls}(\mathbf{s})] &= (\mathbf{I} - \mathbf{H}(\widehat{\boldsymbol{\theta}}))\boldsymbol{\Sigma}(\widehat{\boldsymbol{\theta}})(\mathbf{I} - \mathbf{H}(\widehat{\boldsymbol{\theta}})') \\ &= \boldsymbol{\Sigma}(\widehat{\boldsymbol{\theta}}) - \mathbf{H}(\widehat{\boldsymbol{\theta}})\boldsymbol{\Sigma}(\widehat{\boldsymbol{\theta}}), \end{aligned} \tag{6.68}$$

and the $i$th studentized EGLS residual is computed by dividing $\widehat{e}_{egls}(\mathbf{s}_i)$ by the square root of the $i$th diagonal element of (6.68). The Kenward-Roger variance estimator (6.53) can be used to improve this studentization.

We can also apply ideas of error recovery to derive a set of $n-k$ variance-covariance standardized, uncorrelated, and homoscedastic residuals. Recall from §6.1.2 that if the variance of the residuals is $\mathbf{A}$, then we wish to find an $(n \times n)$ matrix $\mathbf{Q}$ such that

$$\mathrm{Var}[\mathbf{Q}'\widehat{\mathbf{e}}_{egls}(\mathbf{s})] = \mathbf{Q}'\mathbf{A}\mathbf{Q} = \begin{bmatrix} \mathbf{I}_{(n-k)} & \mathbf{0}_{((n-k)\times k)} \\ \mathbf{0}_{(k\times(n-k))} & \mathbf{0}_{(k\times k)} \end{bmatrix}.$$

The first $n-k$ elements of $\mathbf{Q}'\widehat{\mathbf{e}}_{egls}(\mathbf{s})$ are the linearly recovered errors (LREs) of $\mathbf{e}(\mathbf{s})$. Since $\mathbf{A} = \text{Var}[\widehat{\mathbf{e}}_{egls}(\mathbf{s})] = \mathbf{\Sigma}(\widehat{\boldsymbol{\theta}}) - \mathbf{H}(\widehat{\boldsymbol{\theta}})\mathbf{\Sigma}(\widehat{\boldsymbol{\theta}})$ is real, symmetric, and positive semi-definite, it has a spectral decomposition $\mathbf{P}\mathbf{\Delta}\mathbf{P}' = \mathbf{A}$. Let $\mathbf{D}$ denote the diagonal matrix whose $i$th diagonal element is $1/\sqrt{\delta_i}$ if $\delta_i > 0$ and zero otherwise. The needed matrix to transform the residuals $\widehat{\mathbf{e}}_{egls}(\mathbf{s})$ is $\mathbf{Q} = \mathbf{PD}$.

You can also compute a matrix $\mathbf{Q}$ with the needed properties by way of a Cholesky decomposition for positive semi-definite matrices. This decomposition yields a lower triangular matrix $\mathbf{L}$ such that $\mathbf{LL}' = \mathbf{A}$. This decomposition is obtained row-wise and elements of $\mathbf{L}$ corresponding to singular rows are replaced with zeros. Then choose $\mathbf{Q} = \mathbf{L}^-$, where the superscript $-$ denotes a generalized inverse; obtained, for example, by applying the sweep operator to all rows of $\mathbf{L}$ (Goodnight, 1979).

Other types of residuals can be considered in correlated error models. Haslett and Hayes (1998), for example, define marginal and conditional (prediction) residuals. Houseman, Ryan, and Coull (2004) define rotated residuals based on the Cholesky root of the inverse variance matrix of the data, rather than a root constructed from the variance of the residuals.

## 6.3 Generalized Linear Models

### *6.3.1 Background*

A linear model may not always be appropriate, particularly for discrete data that might be assumed to follow a Poisson or Binomial distribution. Generalized linear models (comprehensively described and illustrated in the treatise by McCullagh and Nelder, 1989) are one class of statistical models developed specifically for such situations. These models are now routinely used for modeling non-Gaussian longitudinal data, usually using a "GEE" approach for inference. The GEE approach was adapted for time series count data by Zeger (1988), and in the following sections we show how these ideas can be applied to non-Gaussian spatial data.

In all of the previous sections we have assumed that the mean response is a linear function of the explanatory covariates, i.e., $\boldsymbol{\mu} = \text{E}[\mathbf{Z}(\mathbf{s})] = \mathbf{X}(\mathbf{s})\boldsymbol{\beta}$. We also implicitly assumed that the variance and covariance of observations does not depend on the mean. Note that this is a separate assumption from mean stationarity. The implicit assumption was that the mean $\mu$ does not convey information about the variation of the data. For non-Gaussian data, these assumptions are usually no longer tenable. Suppose that $Y_1, \cdots, Y_n$ denote uncorrelated binary observations whose mean depends on some covariate $x$. If $\text{E}[Y_i] = \mu(x_i)$, then

- $\text{Var}[Y_i] = \mu(x_i)(1-\mu(x_i))$. Knowing the mean of the data provides complete knowledge about the variation of the data.

- Relating $\mu$ and $x$ linearly is not a reasonable proposition on (at least) two grounds. The mean $\mu(x_i)$ must be bounded between 0 and 1 and this constraint is difficult to impose on a linear function such as $\beta_0 + \beta_1 x$. The effect of the regressor $x$ may not be linear on the scale of the mean, which is a probability. Linearity is more likely on a transformed scale, for example, the logit scale $\log\{\mu(x_i)/(1-\mu(x_i))\}$.

The specification of a generalized linear model (GLM) entails three components: the linear predictor, the link function, and the distributional specification. The linear predictor in a GLM expresses the model parameters as a linear function of covariates, $\mathbf{x}(\mathbf{s})'\boldsymbol{\beta}$. We now assume that some monotonic function of the mean, called the *link function*, is linearly related to the covariates, i.e.,

$$g(\boldsymbol{\mu}) = g(\mathrm{E}[\mathbf{Z}(\mathbf{s})]) = \mathbf{X}(\mathbf{s})\boldsymbol{\beta}, \tag{6.69}$$

where $\boldsymbol{\mu}$ is the vector of mean values $[\mu(\mathbf{s}_1), \mu(\mathbf{s}_2), \cdots \mu(\mathbf{s}_n)]'$ of the data vector $\mathbf{Z}(\mathbf{s})$. This enables us to model nonlinear relationships between the data and the explanatory covariates that often arise with binary data, count data, or skewed continuous data. In a GLM, the distribution of the data is a member of the exponential family. This family is very rich and includes continuous distributions such as the Gaussian, Inverse Gaussian, Gamma, and Beta distributions. Among the discrete members of the exponential family are the binary (Bernoulli), geometric, Binomial, negative Binomial, and Poisson distributions. The name *exponential* family stems from the fact that the densities or mass functions in this family can be written in a particular exponential form. The identification of a link function and the mean-variance relationship is straightforward in this form. A density or mass function in the exponential family can be written as

*Exponential Family Density*

$$f(y) = \exp\left\{\frac{y\phi - b(\phi)}{\psi} + c(y, a(\psi))\right\} \tag{6.70}$$

for some functions $b()$ and $c()$. The parameter $\psi$ is a scale parameter and not all exponential family distributions have this parameter (for those that do not, we may simply assume $\psi \equiv 1$).

**Example 6.5** Let $Y$ be a binary random variable so that

$$\Pr(Y = y) = \begin{cases} \mu & y = 1 \\ 1-\mu & y = 0, \end{cases}$$

or more concisely, $p(y) = \mu^y(1-\mu)^{1-y}, y \in \{0,1\}$. Exponentiating the mass function you obtain

$$p(y) = \exp\left\{y \log\left\{\frac{\mu}{1-\mu}\right\} + \log\{1-\mu\}\right\}.$$

The natural parameter is the logit of the mean, $\phi = \log\{\mu/(1-\mu)\}$. The other elements in the exponential form (6.70) are $b(\phi) = \log\{1+e^{\phi}\} = -\log\{1-\mu\}$, $\psi = 1$, $c(y, \psi) = 1$. □

To specify a statistical model in the exponential family of distributions you can draw on some important relationships. For example, the mean and variance of the response relate to the first and second derivative of $b(\phi)$ as follows: $\mathrm{E}[Y] = \mu = b'(\phi)$, $\mathrm{Var}[Y] = \psi b''(\phi)$. If we express the natural (=canonical) parameter $\phi$ as a function of the mean $\mu$, the second derivative $v(\mu) = b''(\phi(\mu))$ is called the **variance function** of the GLM. In the binary example above you can easily verify that $v(\mu) = \mu(1-\mu)$.

In (6.69), the linear predictor $\mathbf{x}(\mathbf{s})'\boldsymbol{\beta}$ was related to the *linked mean* $g(\mu)$. Since the link function is monotonic, we can also express the mean as a function of the inverse linked linear predictor, $\mu = g^{-1}(\mathbf{x}(\mathbf{s})'\boldsymbol{\beta})$. Compare this to the relationship between $b(\phi)$ and the mean, $\mu = b'(\phi)$. If we substitute $\phi = \mathbf{x}(\mathbf{s})'\boldsymbol{\beta}$, then the first derivative of $b()$ could be the inverse link function. In other words, every exponential family distribution implies a link function that arises naturally from the relationship between the natural parameter $\phi$ and the mean of the data $\mu$. Because $\phi$ is also called the canonical parameter, this link is often referred to as the canonical link. The function $b'(\phi)$ is then the *inverse canonical link* function. For Poisson data, for example, the canonical link is the log link, for binary and binomial data it is the logit link, for Gaussian data it is the identity link (no transformation). You should feel free to explore other link functions than the canonical ones. Although they are good starting points in most cases, non-canonical links are preferable for some distributions. For example, data following a Gamma distribution are non-negative, continuous, and right-skewed. The canonical link for this distribution is the reciprocal link, $1/\mu = \mathbf{x}(\mathbf{s})'\boldsymbol{\beta}$. This link does not guarantee non-negative predicted values. Instead, the log link ensures positivity, $\mu = \exp\{\mathbf{x}(\mathbf{s})'\boldsymbol{\beta}\}$.

### 6.3.2 *Fixed Effects and the Marginal Specification*

In traditional applications of GLMs (e.g., in the development of Dobson, 1990), the data are assumed to be independent, but with heterogeneous variances given by the variance function. Thus, the variance-covariance matrix of the data $\mathbf{Z}(\mathbf{s})$ is

$$\boldsymbol{\Sigma} = \mathrm{Var}[\mathbf{Z}(\mathbf{s})] = \psi \mathbf{V}_{\boldsymbol{\mu}}, \tag{6.71}$$

where $\mathbf{V}_{\boldsymbol{\mu}}$ is an $(n \times n)$ diagonal matrix with the variance function terms $v(\mu(\mathbf{s}_i))$ on the diagonal.

To adapt GLMs for use with spatial data, we need to modify the traditional GLM variance-covariance matrix in equation (6.71) to reflect small-scale spatial autocorrelation. In §6.2 we incorporated spatial autocorrelation by allowing a more general variance-covariance matrix $\boldsymbol{\Sigma}(\boldsymbol{\theta})$, where $\boldsymbol{\theta}$ is a $q \times 1$ vector of unknown parameters with $q << n$. We extend this idea here by modifying $\boldsymbol{\Sigma}(\boldsymbol{\theta})$ to include the variance-to-mean relationships inherent in a GLM specification. Based on the ideas in Wolfinger and O'Connell (1993) and Gotway and Stroup (1997), one such approach is to model the variance-covariance matrix of the data as

$$\mathrm{Var}[\mathbf{Z}(\mathbf{s})] = \boldsymbol{\Sigma}(\boldsymbol{\mu}, \boldsymbol{\theta}) = \sigma^2 \mathbf{V}_{\boldsymbol{\mu}}^{1/2} \mathbf{R}(\boldsymbol{\theta}) \mathbf{V}_{\boldsymbol{\mu}}^{1/2}, \tag{6.72}$$

where $\mathbf{R}(\boldsymbol{\theta})$ is a correlation matrix with elements $\rho(\mathbf{s}_i - \mathbf{s}_j; \boldsymbol{\theta})$, the spatial correlogram defined in §1.4.2. The diagonal matrix $\mathbf{V}_{\boldsymbol{\mu}}^{1/2}$ has elements equal to the square root of the variance functions, $\sqrt{v(\mu)}$.

If $\mathbf{R}(\boldsymbol{\theta}) = \mathbf{I}$, then (6.72) reduces to $\sigma^2 \mathbf{V}_{\boldsymbol{\mu}}$, which is not quite the same as (6.71). The parameter $\sigma^2$ obviously equals the scale parameter $\psi$ for those exponential family distributions that possess a scale parameter. In cases where $\psi \equiv 1$, for example for binary, Binomial, or Poisson data, the parameter $\sigma^2$ measures *overdispersion* of the data. Overdispersion is the phenomenon by which the data are more dispersed than is consistent with a particular distributional assumption. Adding the multiplicative scale factor $\sigma^2$ in (6.72) is a basic method to account for the "inexactness" in the variance-to-mean relationship.

Recall that the correlogram or autocorrelation function is directly related to the covariance function and the semivariogram, and so many of the ideas concerning parametric modeling of covariance functions and semivariograms described in §4.3 apply here as well. Often we take $q = 1$ and $\theta$ to equal the range of spatial autocorrelation, since the sill of a correlogram is 1. A nugget effect can be included by using

$$\mathrm{Var}[\mathbf{Z}(\mathbf{s})] = c_0 \mathbf{V}_{\boldsymbol{\mu}} + \sigma_1^2 \mathbf{V}_{\boldsymbol{\mu}}^{1/2} \mathbf{R}(\boldsymbol{\theta}) \mathbf{V}_{\boldsymbol{\mu}}^{1/2}.$$

In this case, $\mathrm{Var}[Z(\mathbf{s}_i)] = (c_0 + \sigma^2) v(\mu(\mathbf{s}_i))$, and the covariance between any two variables is $\mathrm{Cov}[Z(\mathbf{s}_i), Z(\mathbf{s}_j)] = \sigma^2 \sqrt{v(\mu(\mathbf{s}_i)) v(\mu(\mathbf{s}_j))} \rho(\mathbf{s}_i - \mathbf{s}_j)$.

### 6.3.3 A Caveat

Before we proceed further with spatial models for non-Gaussian data, an important caveat of the presented model needs to be addressed. In the Gaussian case you can always construct a multivariate distribution with mean $\boldsymbol{\mu}$ and variance $\boldsymbol{\Sigma}$. This also leads to valid models for the marginal distributions, these are of course Gaussian with respective means and variances $\mu_i$ and $[\boldsymbol{\Sigma}]_{ii}$. The model

$$\mathbf{Y} = \boldsymbol{\mu} + \boldsymbol{\epsilon}, \qquad \boldsymbol{\epsilon} \sim G(\mathbf{0}, \boldsymbol{\Sigma})$$

is a generalization of

$$\mathbf{Y} = \boldsymbol{\mu} + \boldsymbol{\epsilon}, \qquad \boldsymbol{\epsilon} \sim G(\mathbf{0}, \mathbf{I}).$$

For non-Gaussian data, such generalizations are not possible. In the independence case, there is a known marginal distribution for each observation based on the exponential family. There is also a valid joint likelihood, the product of the individual likelihoods. Estimation in generalized linear models can largely be handled based on a specification of the first two moments of the responses alone. Expressions (6.69) and (6.72) extend these moment specifications to the non-Gaussian, spatially correlated case. There is, however, no claim made at this point that the underlying joint distribution may be a "multivariate Binomial" distribution, or some such thing. We may even have to step back from the assumption that a joint distribution in which

- the mean is defined by (6.69),
- the variance is given by (6.72) with $\mathbf{R} \neq \mathbf{I}$,
- the marginal distributions are Binomial, Poisson, etc.,

exists at all.

Moreover, in addition to the difficulties inherent in building non-Gaussian multivariate distributions described in §5.8, the mean-variance relationship also imposes constraints on the possible covariation between the responses. In the simple case of $n$ binary outcomes with common mean (success probability) $\mu$, Gilliland and Schabenberger (2001) examine the constraints placed on the association parameter $\rho$ in a model of equi-correlation (compound symmetry), when the joint distribution on $\{0,1\}^n$ is symmetric with respect to permutation of coordinates. While the association parameter is not restricted from above, it is severely restricted from below (Figure 6.7). The absolute minimum in each case is of course $\rho_{\min} = -1/(n-1)$. This lower bound applies to any $n$ equi-correlated random variables, regardless of their distribution. But in the binary case—especially for small $n$—the lower bound can be substantially larger than $\rho_{\min}$. The bound is achieved for $\mu = 0.5$ if $n$ is even and for $\mu = \pm 1/n$. The valid parameter space depends on the sample size and on the unknown success probability.

If one places further restrictions on the joint distributions, the valid combinations of $(\mu, \rho)$ are further constrained. If one sets third- and higher-order correlations to zero as in Bahadur (1961), then the lower bounds are even more restrictive than shown in Figure (6.7) and the range of $\rho$ is now also bounded from above (see also Kupper and Hasemann, 1978). Such higher-order restrictions may be necessary to facilitate a particular parameterization or estimation technique, e.g., second-order generalized estimating equations. Prentice (1988) notes that without higher-order effects, models for correlated binomial data sacrifice a desired marginalization and flexibility.

### *6.3.4 Mixed Models and the Conditional Specification*

In the development in §6.3.2, we assumed that $\boldsymbol{\beta}$ is a vector of fixed, unknown parameters. The literature refers to this approach as the *marginal* specification since the marginal mean, $\mathrm{E}[\mathbf{Z}(\mathbf{s})]$, is modeled as a function of fixed, non-random (but unknown) parameters. An alternative specification defines the distribution of each $Z(\mathbf{s})$ conditional on an unobserved (latent) spatial process. We considered this type of specification with linear models in §6.2.1.3. With linear models, the marginal and conditional specifications give the same inference, but in a GLM, the two approaches generally lead to models with

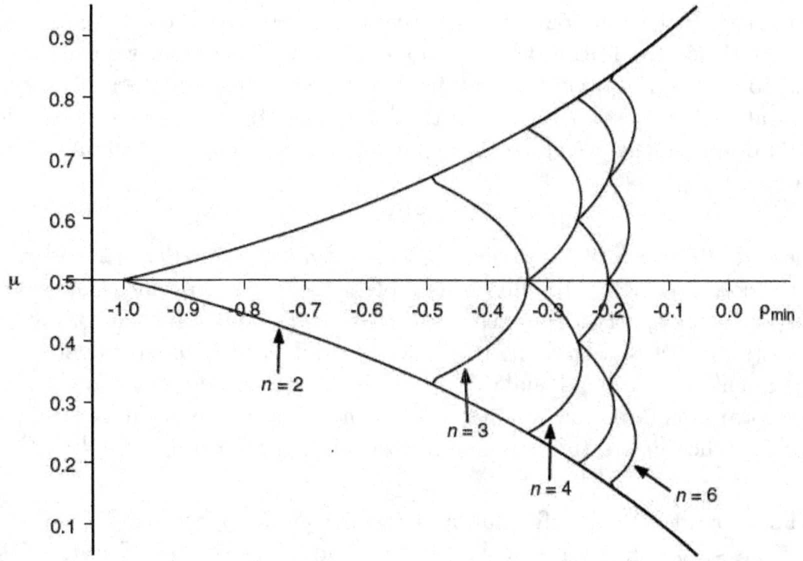


Figure 6.7 *Lower bounds of the correlation parameter of equi-correlated binary observation with mean* $\mu$ *so that the n-variate joint distribution is permutation symmetric* $(n = 2, 3, 4, 5, 6)$. *For a given n, the region to the right of the respective curve is that of possible combinations of* $(\mu, \rho)$ *under this model. After Gilliland and Schabenberger (2001).*

very different interpretations. In a GLM, the conditional approach incorporates the unobserved spatial process through the use of random effects within the mean function and models the conditional mean and variance of $Z(\mathbf{s})$ as a function of both fixed covariate effects and these random effects deriving from the unobserved spatial process. The conditional formulation leads to a generalized linear mixed model (GLMM).

We assume the data are conditionally dependent on an underlying, smooth, spatial process $\{S(\mathbf{s}) : \mathbf{s} \in D\}$. Given $\mathbf{S}(\mathbf{s})$, $Z(\mathbf{s})$ has distribution in the exponential family. Instead of relating the marginal mean $\mathrm{E}[Z(\mathbf{s})]$ to the covariates, we now consider the conditional mean

$$\mathrm{E}[Z(\mathbf{s})|\mathbf{S}] \equiv \mu(\mathbf{s}).$$

The link function relates this conditional mean to the explanatory covariates and also to the underlying spatial random field so that

$$g[\mu(\mathbf{s})] = \mathbf{x}(\mathbf{s})'\boldsymbol{\beta} + S(\mathbf{s}). \quad (6.73)$$

Note that $S(\mathbf{s})$, a random effect at location $\mathbf{s}$, enters the linear component of the GLM as an addition to the intercept. At any location, we can consider $S(\mathbf{s})$ to represent a *random intercept* that varies with spatial location (or more accurately, a random addition to the intercept). In tandem with traditional GLM development, we allow the conditional variance to depend on the mean via

$$\mathrm{Var}[Z(\mathbf{s})|\mathbf{S}] = \sigma^2 v(\mu(\mathbf{s})), \quad (6.74)$$

where the function $v(\cdot)$ is the variance function described earlier and $\sigma^2$ is a dispersion parameter. Finally, to complete the model specification, we need to specify the spatial dependence structure in the data. Instead of modeling $\mathrm{Var}[\mathbf{Z}(\mathbf{s})]$ directly, we assume the data are conditionally independent (i.e., independent given $\mathbf{S}$) and that $\{S(\mathbf{s})\}$ is a Gaussian random field with mean 0 and covariance function $\sigma_S^2\rho_S(\mathbf{s}_i - \mathbf{s}_j)$. Thus, the assumption of conditional independence defers treatment of spatial autocorrelation to the $\{S(\mathbf{s})\}$ process.

This conditional specification and the marginal specification given in the previous section are different in structure and interpretation. Because $S(\mathbf{s})$ is a component of the conditional mean and because the link function $g(\cdot)$ is a nonlinear function, we have,

$$\begin{aligned} \mathrm{E}[Z(\mathbf{s})] &= \mathrm{E}_S[\mathrm{E}[Z(\mathbf{s})|S]] \\ &= \mathrm{E}_S[g^{-1}(\mathbf{x}(\mathbf{s})'\boldsymbol{\beta} + S(\mathbf{s}))] \\ &\neq g^{-1}(\mathbf{x}(\mathbf{s})'\boldsymbol{\beta}). \end{aligned}$$

Taking expectations is a linear operation and does not carry through in case of a nonlinear link function. Evaluating the inverse link function at $\mathbf{x}(\mathbf{s})'\boldsymbol{\beta}$ does not yield the marginal mean. In some cases, such as the log link, the marginal mean can be derived, or equivalently, corrections can be applied to $g^{-1}(\mathbf{x}(\mathbf{s})'\boldsymbol{\beta})$; see, for example, Zeger, Liang, and Albert (1988).

**Example 6.6** Consider a model with canonical link, $g(\mu) = \log\{\mu\}$, identity variance function $v(\mu) = \mu$, and let $m(\mathbf{s}) = \exp\{\mathbf{x}(\mathbf{s})'\boldsymbol{\beta}\}$. Such a construction arises, for example, in Poisson regression. Using results from statistical theory relating marginal and conditional means and variances, together with the mean and variance relationships of the lognormal distribution (see §5.6.1), the marginal moments of the data can be derived as

$$\begin{aligned} \mathrm{E}[Z(\mathbf{s})] &= \mathrm{E}_S[\mathrm{E}[Z(\mathbf{s})|S]] = \mathrm{E}_S[m(\mathbf{s})\exp\{S\}] \\ &= m(\mathbf{s})\exp\{\sigma_S^2/2\}, \\ \mathrm{Var}[Z(\mathbf{s})] &= m(\mathbf{s})\Big[\sigma^2\exp\{\sigma_S^2/2\} \\ &\quad + m(\mathbf{s})\exp\{\sigma_S^2\}(\exp\{\sigma_S^2\} - 1)\Big], \end{aligned}$$

$$\text{Cov}[Z(\mathbf{s}_i), Z(\mathbf{s}_j)] = m(\mathbf{s}_i)m(\mathbf{s}_j)\exp\{\sigma_S^2\} \times (\exp\{\sigma_S^2\rho(\mathbf{s}_i - \mathbf{s}_j)\} - 1).$$

Note $\text{Var}[Z(\mathbf{s})] > \text{E}[Z(\mathbf{s})]$, even if $\sigma^2 = 1$. Also, both the overdispersion and the autocorrelation induced by the latent process $\{S(\mathbf{s})\}$ depend on the mean, so the conditional model can be used with non-stationary spatial processes. □

As with the marginal spatial GLMs described in §6.3.2, the correlation function $\rho_S(\mathbf{s}_i - \mathbf{s}_j)$, can be modeled as a function of a $q \times 1$ vector of unknown parameters $\boldsymbol{\theta}_S$ that completely characterizes the spatial dependence in the underlying process, i.e., $\text{Corr}[S(\mathbf{s}_i), S(\mathbf{s}_j)] = \rho_S(\mathbf{s}_i - \mathbf{s}_j; \boldsymbol{\theta}_S)$.

### *6.3.5 Estimation in Spatial GLMs and GLMMs*

Traditional GLMs allow us to move away from the Gaussian distribution and utilize other distributions that allow mean-variance relationships. However, the likelihood-based inference typically used with these models requires a *multivariate* distribution, and when the data are spatially-autocorrelated we cannot easily build this distribution as a product of marginal likelihoods as we do when the data are independent.

We can bypass this problem by using conditionally-specified generalized linear mixed models (GLMMs), since these models assume that, conditional on random effects, the data are independent. The conditional independence and hierarchical structure allow us to build a multivariate distribution, although we cannot always be sure of the properties of this distribution. However, the hierarchical structure of GLMMs poses problems of its own. As noted in Breslow and Clayton (1993), exact inference for first-stage model parameters (e.g., the fixed covariate effects) typically requires integration over the distribution of the random effects. This necessary multi-dimensional integration is difficult and can often result in numerical instabilities, requiring approximations or more computer-intensive estimation procedures.

There are several ways to avoid all of these problems (although arguably they all introduce other problems). To address the situation where we can define means and variances, but not necessarily the entire likelihood, Wedderburn (1974) introduced the notion of *quasi-likelihood* based on the first two moments of a distribution, and the approach sees wide application for GLMs based on independent data (McCullagh and Nelder, 1989). This leads to an iterative *estimating equation* based on only the first two moments of a distribution that can be used with spatial data. Another solution is based on an initial Taylor series expansion that then allows *pseudo-likelihood* methods for spatial inference similar to those described in §5.5.2. A similar approach, called *penalized quasi-likelihood* by Breslow and Clayton (1993), uses a Laplace approximation to the log-likelihood. If combined with a Fisher-scoring algorithm, the estimating equations for the fixed effects parameters and predictors of random effects in the model are of the same form as the mixed model equations in the pseudo-likelihood approach of Wolfinger and O'Connell (1993). Finally, Bayesian hierarchical models and the simulation methods used for inference with them offer another popular alternative.

#### 6.3.5.1 Generalized Estimating Equations

As noted briefly above, quasi-likelihood builds inference in terms of the first two moments only, rather than the entire joint distribution, but it has many of the familiar properties of a joint likelihood. For example, estimates derive from maximizing the quasi-likelihood function using score equations (partial derivatives set equal to zero) and have nice asymptotic properties such as limiting Gaussian distributions. While most fully developed for independent data, McCullagh and Nelder (1989) extend quasi-likelihood estimation to dependent data, and Gotway and Stroup (1997) cast the approach in a spatial GLM context, which we summarize here.

*Quasi-Log Likelihood*

The quasi-likelihood (or quasi-log-likelihood, to be more exact) function $Q(\boldsymbol{\mu}; \mathbf{z})$ is defined by the relationship

$$\frac{\partial Q(\boldsymbol{\mu}; \mathbf{z})}{\partial \boldsymbol{\mu}} = \mathbf{V}^{-1}(\mathbf{Z}(\mathbf{s}) - \boldsymbol{\mu}(\boldsymbol{\beta})),$$

where $\mathbf{V}$ represents the variance-covariance matrix capturing the spatial correlation, and $\boldsymbol{\mu}(\boldsymbol{\beta}) \equiv \boldsymbol{\mu} = (\mu_1, \cdots, \mu_n)'$ is the mean vector. We use the notation $\boldsymbol{\mu}(\boldsymbol{\beta})$ to emphasize the dependence of the mean on $\boldsymbol{\beta}$ that arises through the link function. Note that in most GLMs, the elements of $\mathbf{V}$ are also functions of $\boldsymbol{\mu}(\boldsymbol{\beta})$. Differentiating $Q(\boldsymbol{\mu}; \mathbf{z})$ with respect to each element of $\boldsymbol{\beta}$ yields the set of quasi-likelihood score equations

*Quasi-Likelihood Score*

$$\boldsymbol{\Delta}'\mathbf{V}^{-1}(\mathbf{Z}(\mathbf{s}) - \boldsymbol{\mu}(\boldsymbol{\beta})) = \mathbf{0}$$

where $\boldsymbol{\Delta}$ denotes the matrix with elements $[\boldsymbol{\Delta}]_{ij} = \partial \mu_i / \partial \beta_j$ and $j = 1, \cdots, p$ indexes the parameters in the linear portion of the GLM. Solving the score equations for $\boldsymbol{\beta}$ yields the quasi-likelihood estimates of the model parameters.

McCullagh and Nelder (1989, pp. 333–335) note that the inverse of the variance-covariance matrix $\mathbf{V}^{-1}$ must satisfy several conditions to guarantee a solution to the score equations, some not easily verified in practice. One problem in practical applications is that $\mathbf{V}^{-1}$ is assumed to depend only on $\boldsymbol{\beta}$ and not on any unknown spatial autocorrelation parameters $\boldsymbol{\theta}$. As a result, Wolfinger and O'Connell (1993) and Gotway and Stroup (1997) follow Liang and Zeger (1986) and Zeger and Liang (1986) and limit attention to variance-covariance matrices of the form introduced in equation (6.72),

$$\mathbf{V} \equiv \boldsymbol{\Sigma}(\boldsymbol{\mu}, \boldsymbol{\theta}) = \sigma^2 \mathbf{V}_{\boldsymbol{\mu}}^{1/2} \mathbf{R}(\boldsymbol{\theta}) \mathbf{V}_{\boldsymbol{\mu}}^{1/2},$$

where $\mathbf{R}(\boldsymbol{\theta})$ denotes a matrix of correlations among the observations parameterized by the vector $\boldsymbol{\theta}$, and $\mathbf{V}_{\boldsymbol{\mu}}^{1/2}$ a diagonal matrix of variance functions. With $\mathbf{V}$ now also depending on unknown autocorrelation parameters, the quasi-likelihood score equations are known as *generalized estimating equations.* Liang and Zeger (1986) and Zeger and Liang (1986) show in the context of longitudinal data, that, under mild regularity conditions, these equations generate consistent estimators of $\boldsymbol{\beta}$, even with mis-specified correlation matrices. Zeger (1988) shows that the same conditions are met for a single time series replicate, provided the covariance function is "well behaved" in the sense that $\Sigma(\boldsymbol{\mu}, \boldsymbol{\theta})$ breaks into independent blocks for large $n$. This same result holds for spatial data as well (McShane et al., 1997). In the spatial setting, to avoid any distributional assumptions in a completely marginal analysis, Gotway and Stroup (1997) suggest using an iteratively re-weighted generalized least squares approach to solving the generalized estimating equations for $\boldsymbol{\beta}$ and $\boldsymbol{\theta}$. With $\boldsymbol{\eta} = \mathbf{X}(\mathrm{s})\boldsymbol{\beta}$, the matrix $\boldsymbol{\Delta}$ can be written as $\boldsymbol{\Delta} = \boldsymbol{\Psi}\mathbf{X}(\mathrm{s})$, where $\boldsymbol{\Psi} = \mathrm{diag}[\partial\mu_i/\partial\eta_i]$, and the generalized estimating equations can be written as

$$(\mathbf{X}(\mathrm{s})'\mathbf{A}(\boldsymbol{\theta})\mathbf{X}(\mathrm{s}))^{-1}\boldsymbol{\beta} = \mathbf{X}(\mathrm{s})'\mathbf{A}(\boldsymbol{\theta})\mathbf{Z}(\mathrm{s})^*, \tag{6.75}$$

where $\mathbf{A}(\boldsymbol{\theta}) = \boldsymbol{\Psi}\Sigma(\boldsymbol{\mu}, \boldsymbol{\theta})^{-1}\boldsymbol{\Psi}$ and $\mathbf{Z}(\mathrm{s})^* = \mathbf{X}\boldsymbol{\beta} + \boldsymbol{\Psi}^{-1}(\mathbf{Z}(\mathrm{s}) - \boldsymbol{\mu})$.

Assume for the moment that $\boldsymbol{\theta}$ is known. The solution to the estimating equation (6.75) is

$$\widehat{\boldsymbol{\beta}} = (\mathbf{X}(\mathrm{s})'\mathbf{A}(\boldsymbol{\theta})\mathbf{X}(\mathrm{s}))^{-1}\mathbf{X}(\mathrm{s})'\mathbf{A}(\boldsymbol{\theta})\mathbf{Z}(\mathrm{s})^*,$$

a generalized least squares estimator in the model

$$\mathbf{Z}(\mathrm{s})^* = \mathbf{X}(\mathrm{s})\boldsymbol{\beta} + \mathbf{e}(\mathrm{s}), \quad \mathbf{e}(\mathrm{s}) \sim (\mathbf{0}, \mathbf{A}(\boldsymbol{\theta})^{-1}).$$

In fact, this model can be derived from a first-order Taylor series of the mean $\boldsymbol{\mu} = g^{-1}(\mathbf{X}(\mathrm{s})\boldsymbol{\beta})$ in the model

$$\mathbf{Z}(\mathrm{s}) = g^{-1}(\mathbf{X}(\mathrm{s})\boldsymbol{\beta}) + \mathbf{e}^*(\mathrm{s}), \quad \mathbf{e}^*(\mathrm{s}) \sim (\mathbf{0}, \mathbf{V}). \tag{6.76}$$

This marginal formulation is somewhat "unnatural" in generalized linear models, since it assumes a zero mean error vector, but it suffices for estimating techniques that require only the first two moments. Now, expanding about some value $\widetilde{\boldsymbol{\beta}}$, we obtain

$$\begin{aligned}\boldsymbol{\mu} &\doteq g^{-1}(\mathbf{X}(\mathrm{s})\widetilde{\boldsymbol{\beta}}) + \frac{\partial g^{-1}(\mathbf{X}(\mathrm{s})\boldsymbol{\beta})}{\partial \boldsymbol{\beta}}\bigg|_{\widetilde{\boldsymbol{\beta}}} \mathbf{X}(\mathrm{s})(\boldsymbol{\beta} - \widetilde{\boldsymbol{\beta}}) \\ &= \widetilde{\boldsymbol{\mu}} + \widetilde{\boldsymbol{\Psi}}\mathbf{X}(\mathrm{s})(\boldsymbol{\beta} - \widetilde{\boldsymbol{\beta}}).\end{aligned}$$

Substituting the approximated mean into (6.76) and re-arranging yields the pseudo-model

*Linearized Pseudo-Model*

$$\begin{aligned}\mathbf{Z}(\mathrm{s})^* &= \widetilde{\boldsymbol{\Psi}}^{-1}(\mathbf{Z}(\mathrm{s}) - \widetilde{\boldsymbol{\mu}}) + \mathbf{X}(\mathrm{s})\widetilde{\boldsymbol{\beta}} \\ &= \mathbf{X}(\mathrm{s})\boldsymbol{\beta} + \widetilde{\boldsymbol{\Psi}}^{-1}\mathbf{e}^*(\mathrm{s}). \end{aligned} \tag{6.77}$$

This model is the linearized form of a nonlinear model with spatial correlated error structure. The pseudo-response $\mathbf{Z}(\mathrm{s})^*$ is also called the "working" outcome variable, but the attribute should not be confused with the concept of a "working correlation structure" in GEE estimation. The working outcome variable earns the name because its values change every time we "work through" the generalized least squares problem.

Solving (6.75) is an iterative process (even if $\boldsymbol{\theta}$ is known), because in order to evaluate the pseudo-response $\mathbf{Z}(\mathbf{s})^*$, we need to know the mean, hence, $\boldsymbol{\beta}$ must be available. In order to start the process, we assume some starting value for $\boldsymbol{\beta}$. Another way of looking at the iterative nature is through the Taylor series expansion (6.77). The initial expansion locus (the starting value $\widetilde{\boldsymbol{\beta}}$) is arbitrary. Hence, the GLS solution to (6.75) depends on that choice. Thus, after a solution has been obtained, we update the pseudo-response $\mathbf{Z}(\mathbf{s})^*$ and repeat the fitting process. The procedure stops if the estimates of $\boldsymbol{\beta}$ do not change between two model fits.

The formulation in terms of the linearization of a nonlinear model with correlated error is also helpful because it points us to methods for estimating the covariance parameters. If $\boldsymbol{\theta}$ is unknown—as is typically the case—we form residuals after each fit of model (6.77) and apply the geostatistical techniques in Chapter 4. It is recommended to perform a studentization first, to remedy the heteroscedasticity problem. See (6.68) on page 351 on the studentization of GLS residuals. Probably the two most frequent current practices are to

(i) estimate $\boldsymbol{\theta}$ by fitting a semivariogram model to the raw GLS residuals, and to repeat this process after each GLS fit;

(ii) estimate $\boldsymbol{\theta}$ by initially fitting a semivariogram model to the raw residuals from a generalized linear model without any spatial correlation structure ($\mathbf{R}(\boldsymbol{\theta}) = \mathbf{I}$). Then, the GLS fit is performed once, assuming that $\widehat{\boldsymbol{\theta}}$ so obtained equals the true values. In other words, no further updates of the covariance parameters are obtained.

You could also formulate a second set of pseudo-responses based on the GLS residuals and apply the generalized estimating equation or composite likelihood techniques of §4.5.3.

#### *6.3.5.2 Pseudo-likelihood Estimation*

Instead of quasi-likelihood, Wolfinger and O'Connell (1993) suggest an approach termed *pseudo-likelihood* (PL) as a flexible and efficient way of estimating the unknown parameters in a generalized linear mixed model (GLMM). (We note that this "pseudo-likelihood" is different from the one developed by Besag (1975) for inference with non-Gaussian auto models.

The pseudo-likelihood approach of Wolfinger and O'Connell (1993) differs from the quasi-likelihood approach above in that at any step of the iterative process, a function that is a true joint likelihood is used to estimate unknown parameters. In the case we consider here, a pseudo-likelihood approach assumes $\boldsymbol{\beta}$ is known and estimates $\sigma^2$ and $\boldsymbol{\theta}$ using ML (or REML, as described in §5.5.2–5.5.3), and then assumes $\boldsymbol{\theta}$ is known and estimates $\boldsymbol{\beta}$ using ML (or EGLS), and iterates until convergence. We can apply this approach to marginal GLMs as special cases of a GLMM.

The idea behind the pseudo-likelihood approach is to linearize the problem so we can use the approach in §5.5.2–5.5.3 for estimation and inference. This is done using a first-order Taylor series expansion of the link function to give what Wolfinger and O'Connell (1993) call "pseudo data" (similar to the "working" outcome variable used in the generalized estimating equations) $\mathbf{Z}(\mathbf{s})^*$). As before, the pseudo-data is constructed as

$$\nu_i = g(\widehat{\mu}_i) + g'(\widehat{\mu}_i)(Z(\mathbf{s}_i) - \widehat{\mu}_i), \tag{6.78}$$

where $g'(\widehat{\mu}_i)$, is the first derivative of the link function with respect to $\mu$, evaluated at the current estimate $\widehat{\mu}$. To apply the methods in §5.5.2–5.5.3, we need the mean and variance-covariance matrix of the pseudo data, $\boldsymbol{\nu}$. Conditioning on $\boldsymbol{\beta}$ and $\mathbf{S}$, assuming $\mathrm{Var}[\mathbf{Z}(\mathbf{s})|\mathbf{S}]$ has the form of equation (6.72), and using some approximations described in Wolfinger and O'Connell (1993), these can be derived in almost the traditional fashion as

$$\begin{aligned} \mathrm{E}[\boldsymbol{\nu}|\boldsymbol{\beta}, \mathbf{S}] &= \mathbf{X}\boldsymbol{\beta} + \mathbf{S} \\ \mathrm{Var}[\boldsymbol{\nu}|\boldsymbol{\beta}, \mathbf{S}] &= \boldsymbol{\Sigma}_{\boldsymbol{\mu}}, \end{aligned}$$

with

$$\boldsymbol{\Sigma}_{\widehat{\boldsymbol{\mu}}} = \sigma^2 \widehat{\boldsymbol{\Psi}}^{-1} \mathbf{V}^{1/2}_{\widehat{\boldsymbol{\mu}}}\, \mathrm{R}(\theta)\, \mathbf{V}^{1/2}_{\widehat{\boldsymbol{\mu}}} \widehat{\boldsymbol{\Psi}}^{-1} = \widehat{\boldsymbol{\Psi}}^{-1} \boldsymbol{\Sigma}(\widehat{\mu}, \theta) \widehat{\boldsymbol{\Psi}}^{-1}. \tag{6.79}$$

Recall that the matrix $\widehat{\boldsymbol{\Psi}}$ is an $(n \times n)$ diagonal matrix with typical element $[\partial\mu(\mathbf{s}_i))/\partial\eta(\mathbf{s}_i)]$ and is evaluated at $\widehat{\mu}$. The marginal moments of the pseudo-data are:

$$\begin{aligned} \mathrm{E}[\boldsymbol{\nu}] &= \mathbf{X}\boldsymbol{\beta} \\ \mathrm{Var}[\boldsymbol{\nu}] &= \boldsymbol{\Sigma}_S + \boldsymbol{\Sigma}_{\widehat{\boldsymbol{\mu}}} \equiv \boldsymbol{\Sigma}_{\boldsymbol{\nu}}, \end{aligned}$$

and $\boldsymbol{\Sigma}_S$ has $(i,j)$th element $\sigma^2_S \rho_S(\mathbf{s}_i - \mathbf{s}_j)$. This can be considered as a general linear regression model with spatially autocorrelated errors as described in §6.2, since the mean (of the pseudo-data $\boldsymbol{\nu}$) is linear in $\boldsymbol{\beta}$. Thus, if we are willing to assume $\boldsymbol{\Sigma}_{\widehat{\boldsymbol{\mu}}}$ is known (or at least does not depend on $\boldsymbol{\beta}$) when we want to estimate $\boldsymbol{\beta}$, and that $\boldsymbol{\beta}$ is known when we want to estimate $\boldsymbol{\theta}$, we can maximize the log-likelihood analytically yielding the least squares equations

$$\widehat{\boldsymbol{\beta}} = (\mathbf{X}'\boldsymbol{\Sigma}_{\boldsymbol{\nu}}^{-1}\mathbf{X})^{-1}\mathbf{X}'\boldsymbol{\Sigma}_{\boldsymbol{\nu}}^{-1}\boldsymbol{\nu}, \tag{6.80}$$

$$\widehat{\mathbf{S}} = \boldsymbol{\Sigma}_S\boldsymbol{\Sigma}_{\boldsymbol{\nu}}^{-1}(\boldsymbol{\nu} - \mathbf{X}\widehat{\boldsymbol{\beta}}), \tag{6.81}$$

$$\widehat{\sigma}^2 = \frac{(\boldsymbol{\nu} - \mathbf{X}\widehat{\boldsymbol{\beta}})'(\boldsymbol{\Sigma}_{\boldsymbol{\nu}}^*)^{-1}(\boldsymbol{\nu} - \mathbf{X}\widehat{\boldsymbol{\beta}})}{n}. \tag{6.82}$$

(The matrix $\boldsymbol{\Sigma}_{\boldsymbol{\nu}}^*$ in (6.82) is obtained by factoring the residual variance from $\boldsymbol{\Sigma}_{\boldsymbol{\mu}}$ and $\boldsymbol{\Sigma}_S$.) However, because $\boldsymbol{\Sigma}_{\widehat{\boldsymbol{\mu}}}$ does depend on $\boldsymbol{\beta}$ we iterate as follows:

1. Obtain an initial estimate of $\widehat{\mu}$ from the original data. An estimate from the non-spatial generalized linear model often works well;

2. Compute the pseudo-data from equation (6.78);
3. Using ML (or REML) with the pseudo-data, obtain estimates of the spatial autocorrelation parameters, $\boldsymbol{\theta}$, and $\sigma_S^2$ in $\boldsymbol{\Sigma}_{\nu}$;
4. Use these estimates to compute generalized least squares estimates (which are also the maximum pseudo-likelihood estimators) of $\boldsymbol{\beta}$ and $\sigma^2$ from equations (6.80) and (6.82) and to predict $\mathbf{S}$ from equation (6.81);
5. Update the estimate of $\boldsymbol{\mu}$, using $\widehat{\boldsymbol{\mu}} = g^{-1}(\mathbf{X}\widehat{\boldsymbol{\beta}} + \widehat{\mathbf{S}})$;
6. Repeat these steps until convergence.

Approximate standard errors for the fixed effects derive from

$$\widehat{\text{Var}}(\widehat{\boldsymbol{\beta}}) = (\mathbf{X}'\widehat{\boldsymbol{\Sigma}}_{\nu}^{-1}\mathbf{X})^{-1}$$

using the converged parameter estimates of $\boldsymbol{\theta}$ in $\boldsymbol{\Sigma}_{\nu}$ to define the estimator $\widehat{\boldsymbol{\Sigma}}_{\nu}$. We can construct approximate $p$-values using $t$-tests analogous to those described in §6.2.3.1.

As for the marginal model and the generalized estimating equation approach, we can cast the pseudo-likelihood approach in terms of a linearization and a pseudo-model. In the marginal case, the pseudo-model had the form of a general linear model with correlated errors. In the conditional case, we arrive at a linear mixed model. First, take a first-order Taylor series of $\boldsymbol{\mu}$ about some values $\widehat{\boldsymbol{\beta}}$ and $\widehat{\mathbf{S}}(\mathbf{s})$,

$$\boldsymbol{\mu} \doteq \widehat{\boldsymbol{\mu}} + \widehat{\boldsymbol{\Psi}}\mathbf{X}(\mathbf{s})(\boldsymbol{\beta} - \widehat{\boldsymbol{\beta}}) + \widehat{\boldsymbol{\Psi}}(\mathbf{S}(\mathbf{s}) - \widehat{\mathbf{S}}(\mathbf{s})).$$

Rearranging terms we can write

$$\widehat{\boldsymbol{\Psi}}^{-1}(\boldsymbol{\mu} - \widehat{\boldsymbol{\mu}}) + \mathbf{X}(\mathbf{s})\widehat{\boldsymbol{\beta}} + \widehat{\mathbf{S}}(\mathbf{s}) \doteq \mathbf{X}(\mathbf{s})\boldsymbol{\beta} + \mathbf{S}(\mathbf{s}).$$

Notice that the left-hand side of this expression is the conditional expectation, given $\mathbf{S}(\mathbf{s})$), of

$$\boldsymbol{\nu} = \widehat{\boldsymbol{\Psi}}^{-1}(\mathbf{Z}(\mathbf{s}) - \widehat{\boldsymbol{\mu}}) + \mathbf{X}(\mathbf{s})\widehat{\boldsymbol{\beta}} + \widehat{\mathbf{S}}(\mathbf{s}), \tag{6.83}$$

since we assume that the expansion locus in the Taylor series is fixed. The conditional variance is given by (6.79). We can now consider a linear mixed model $\boldsymbol{\nu} = \mathbf{X}(\mathbf{s})\boldsymbol{\beta} + \mathbf{S}(\mathbf{s}) + \mathbf{e}(\mathbf{s})$, where $\text{Var}[\mathbf{e}(\mathbf{s})] = \boldsymbol{\Sigma}_{\widehat{\boldsymbol{\mu}}}$, and $\text{Var}[\mathbf{S}(\mathbf{s})] = \boldsymbol{\Sigma}_S$. In order to estimate $\boldsymbol{\theta}$, we further assume that the pseudo-data $\boldsymbol{\nu}$ follows a Gaussian distribution and apply ML or REML estimation. For example, minus twice the negative restricted log-likelihood for the linearized pseudo-model is

$$\begin{aligned} \varphi_R(\boldsymbol{\theta}; \mathbf{K}\boldsymbol{\nu}) &= \ln\{|\boldsymbol{\Sigma}_{\nu}|\} + \ln\{|\mathbf{X}(\mathbf{s})'\boldsymbol{\Sigma}_{\nu}^{-1}\mathbf{X}(\mathbf{s})|\} \\ &\quad + \mathbf{r}'\boldsymbol{\Sigma}_{\nu}^{-1}\mathbf{r} + (n-k)\ln\{2\pi\}, \end{aligned} \tag{6.84}$$

where $\mathbf{r} = \boldsymbol{\nu} - \mathbf{X}(\mathbf{s})(\mathbf{X}(\mathbf{s})\boldsymbol{\Sigma}_{\nu}^{-1}\mathbf{X}(\mathbf{s}))^{-1}\mathbf{X}(\mathbf{s})\boldsymbol{\Sigma}_{\nu}^{-1}\boldsymbol{\nu}$.

While in the marginal model (with generalized estimating equations) approach we are repeatedly fitting a general linear model, in the pseudo-likelihood approach we are repeatedly fitting a linear mixed model. One of the advantages of this approach is that steps 3 and 4 in the algorithm above can be accommodated within the mixed model framework. Estimates of $\boldsymbol{\theta}$ are obtained by minimizing (6.84), and updated estimates of $\boldsymbol{\theta}$ and predictions of $\mathbf{S}(\mathbf{s})$ are obtained by solving the mixed model equations,

$$\left[\begin{array}{cc} \mathbf{X}(\mathbf{s})'\boldsymbol{\Sigma}_{\widehat{\mu}}\mathbf{X}(\mathbf{s}) & \mathbf{X}(\mathbf{s})'\boldsymbol{\Sigma}_{\widehat{\mu}} \\ \boldsymbol{\Sigma}_{\widehat{\mu}}\mathbf{X}(\mathbf{s}) & \boldsymbol{\Sigma}_{\widehat{\mu}} + \boldsymbol{\Sigma}_S^{-1} \end{array}\right] \left[\begin{array}{c} \widehat{\boldsymbol{\beta}} \\ \widehat{\mathbf{S}}(\mathbf{s}) \end{array}\right] = \left[\begin{array}{c} \mathbf{X}(\mathbf{s})'\boldsymbol{\Sigma}_{\widehat{\mu}}\boldsymbol{\nu} \\ \mathbf{S}(\mathbf{s})\boldsymbol{\Sigma}_{\widehat{\mu}}\boldsymbol{\nu} \end{array}\right]. \tag{6.85}$$

The solutions to the mixed model equations are (6.80) and (6.81).

We have a choice as to how to model any spatial autocorrelation: through $\mathbf{R}$ or through $\boldsymbol{\Sigma}_S$. Marginal models let $\mathbf{R}$ be a spatial correlation matrix, $\mathbf{R}(\boldsymbol{\theta})$, and set $\mathbf{S} = \mathbf{0}$. Conditional models are specified through $\mathbf{S}$ and $\boldsymbol{\Sigma}_S$, with spatial dependence incorporated in $\boldsymbol{\Sigma}_S = \sigma_s^2\mathbf{V}(\boldsymbol{\theta}_S)$, and $\mathbf{R}$ equal to an identity matrix. Once we have determined which type of model we want to use, we use the iterative approach just described to estimate any unknown parameters ($\sigma^2, \boldsymbol{\beta}, \boldsymbol{\theta}$ for a marginal model and $\sigma^2, \sigma_s^2, \boldsymbol{\beta}, \boldsymbol{\theta}_S$ for a conditional model).

This distinction between marginal and conditional approaches should not be confused with another issue of "marginality" that comes about in the pseudo-likelihood approach. Because we have started with a Taylor series in a conditional model about $\widehat{\boldsymbol{\beta}}$ and $\widehat{\mathbf{S}}(\mathbf{s})$, the question is how to choose these expansion loci. In case of the random field $\mathbf{S}(\mathbf{s})$, there are two obvious choices. We can expand about some current predictor of $\mathbf{S}(\mathbf{s})$, for example, the solution to the mixed model equations, or about the mean $\mathrm{E}[\mathbf{S}(\mathbf{s})] = \mathbf{0}$. The latter case is termed a marginal or population-averaged expansion, a term borrowed from longitudinal data analysis. The former expansion is also referred to as a conditional or subject-specific expansion. A marginal expansion also yields a pseudo-model of the linear mixed model variety, however, the pseudo-data in this case is

$$\boldsymbol{\nu}_m = \widehat{\boldsymbol{\Psi}}^{-1}(\mathbf{Z}(\mathbf{s}) - \widehat{\boldsymbol{\mu}}) + \mathbf{X}(\mathbf{s})\widehat{\boldsymbol{\beta}}.$$

Notice that this is the same pseudo-data as in (6.77), but the right-hand side of the pseudo-model remains a mixed model. Changing the expansion locus from the conditional to the marginal form has several important implications:

- predictions of the random field $\mathbf{S}(\mathbf{s})$ are not required in order to compute the pseudo-data. You can hold the computation of $\widehat{\mathbf{S}}(\mathbf{s})$ until the overall, doubly iterative algorithm has converged. This can save substantial computing time.
- the expansion about $\widehat{\mathbf{S}}(\mathbf{s})$ typically produces a better approximation, but it is also somewhat more sensitive to model mis-specification.
- the interpretation of $\widehat{\mathbf{S}}(\mathbf{s})$—computed according to (6.81) after the algorithm converged—is not clear.

The conditional, mixed model approach also provides another important advantage. Our formulation of the conditioning random field $\mathbf{S}(\mathbf{s})$ was entirely general. As in §6.2.1 we can provide additional structure. For example, let $\mathbf{S}(\mathbf{s}) = \mathbf{U}(\mathbf{s})\boldsymbol{\alpha}$, and determine $\mathbf{U}(\mathbf{s})$ as a low-rank radial smoother. Then,

$\text{Var}[\alpha] = \sigma^2\mathbf{I}$, and the mixed model equations are particularly easy to solve. This provides a technique, to spatially smooth, for example, counts and rates.

#### *6.3.5.3 Penalized Quasi-likelihood*

The pseudo-likelihood technique is by no means the only approach to fit generalized linear mixed models. It is in fact only one representative of a particular class of methods, the linearization methods. For the case of clustered (longitudinal and repeated measures) data, Schabenberger and Gregoire (1996) described more than ten algorithms representing different assumptions and approaches to the problem. Among the strengths of linearization methods are their flexibility, the possibility to accommodate complex random structures and high-dimensional random effects, and to cope with correlated errors. The derivation of the pseudo-model rests solely on the first two moments of the data. In the marginal GEE approach, no further distributional assumptions were made. In the pseudo-likelihood approach we also assumed that the pseudo-response follows a Gaussian distribution, which led to the objective function (6.84) (in the case of REML estimation). To be perfectly truthful, it is not necessary to make a Gaussian assumption for the pseudo-response (6.83). The REML objective function is a reasonable objective function for the estimation of $\boldsymbol{\theta}$, regardless of the distribution of $\boldsymbol{\nu}$. From this vantage point, the solutions (6.80) and (6.81) are simply a GLS estimator and a best linear unbiased predictor. We prefer, however, to think of $\boldsymbol{\nu}$ as a Gaussian random variable, and of (6.84) as minus twice its restricted log likelihood, even if it is only a vehicle to arrive at estimates and predictions. The analysis to some extent depends on our choice of vehicle.

The pseudo-likelihood approach is so eminently feasible, because it yields a Gaussian linear mixed model in which the marginal distribution can be easily obtained. It is this marginal distribution on which the optimization is based, but because it is dependent on the expansion locus, the optimization must be repeated; the process is doubly iterative. In other words, the linearization provides a solution to the problem of obtaining the marginal distribution

$$f(\mathbf{Z}(\mathbf{s})) = \int f(\mathbf{Z}(\mathbf{s})|\mathbf{S}(\mathbf{s}))f_s(\mathbf{S}(\mathbf{s}))\,d\mathbf{S}(\mathbf{s}), \tag{6.86}$$

by approximating the model. The disadvantage of the approach is that you are not working with the likelihood of $\mathbf{Z}(\mathbf{s})$, but with the likelihood of $\boldsymbol{\nu}$, the pseudo-response.

A different class of methods approaches the problem of obtaining the marginal distribution of the random process by numerically solving (6.86). The methods abound to accomplish the integration numerically, ranging from quadrature methods and Laplace approximations, to Monte Carlo integration, importance sampling, and Markov chain Monte Carlo (MCMC) methods. We mention this class of methods here only in passing. Their applicability to spatial data analysis is limited by the dimensionality of the integral in (6.86). Note that this is an integral over the random variables in $\mathbf{S}(\mathbf{s})$, an $n$-dimensional problem. A numerical integration via quadrature or other multi-dimensional technique is computationally not feasible, unless the random effects structure is much simplified (which is the case for the radial smoothing models). However, an integral approximation that does not require high-dimensional integration and assumes conditional independence ($\mathbf{R} = \sigma_\epsilon^2 \mathbf{I}$) is an option. In the case of longitudinal data, where the marginal covariance matrix is block-diagonal, Breslow and Clayton (1993) applied a Laplace approximation in an approach that they termed penalized quasi-likelihood.

Assume that you wish to compute $\int p(\boldsymbol{\tau})d\boldsymbol{\tau}$, that $\boldsymbol{\tau}$ is $(m \times 1)$ and that the function $p(\boldsymbol{\tau})$ can be written as $\exp\{nf(\boldsymbol{\tau})\}$. Then the Laplace approximation of the target integral (Wolfinger, 1993) is

$$\int p(\boldsymbol{\tau})\, d\boldsymbol{\tau} = \int \exp\{nh(\boldsymbol{\tau})\} \doteq \left(\frac{2\pi}{n}\right)^{m/2} \exp\{nh(\widehat{\boldsymbol{\tau}})\} \left|-h''(\widehat{\boldsymbol{\tau}})\right|^{-1/2}.$$

The quantity $\widehat{\boldsymbol{\tau}}$ in this approximation is not just any value, it is the value that maximizes $\exp\{nh(\boldsymbol{\tau})\}$, or equivalently, maximizes $\log\{h(\boldsymbol{\tau})\}$. The term $|-h''(\widehat{\boldsymbol{\tau}})|$ is the determinant of the second derivative matrix, evaluated at that maximizing value.

In the case of a GLMM, we have $p(\boldsymbol{\tau}) = f(\mathbf{Z}(\mathbf{s})|\mathbf{S}(\mathbf{s}))\, f_s(\mathbf{S}(\mathbf{s}))$, $\boldsymbol{\tau} = [\boldsymbol{\beta}', \mathbf{S}(\mathbf{s})']'$, and $h(\boldsymbol{\beta}, \boldsymbol{\theta}) = h(\boldsymbol{\beta}, \mathbf{S}(\mathbf{s})) = n^{-1}\log\{f(\mathbf{Z}(\mathbf{s})|\mathbf{S}(\mathbf{s}))\, f_s(\mathbf{S}(\mathbf{s}))\}$. Note that this approximation involves the random field $\mathbf{S}(\mathbf{s})$, rather than the covariance parameters. Breslow and Clayton (1993) use a Fisher-scoring algorithm, which amounts to replacing $-h''(\widehat{\boldsymbol{\tau}})$ with the expected value of

$$-\frac{\partial^2}{\partial\boldsymbol{\tau}\partial\boldsymbol{\tau}'} \log\left\{f(\mathbf{Z}(\mathbf{s})|\mathbf{S}(\mathbf{s}))\, f_s(\mathbf{S}(\mathbf{s}))\right\},$$

which turns out to be

$$\begin{bmatrix} \mathbf{X}(\mathbf{s})'\boldsymbol{\Sigma}_{\widehat{\mu}}^{-1}\mathbf{X}(\mathbf{s}) & \mathbf{X}(\mathbf{s})'\boldsymbol{\Sigma}_{\widehat{\mu}}^{-1} \\ \boldsymbol{\Sigma}_{\widehat{\mu}}^{-1}\mathbf{X}(\mathbf{s}) & \boldsymbol{\Sigma}_{\widehat{\mu}}^{-1} + \boldsymbol{\Sigma}_S^{-1} \end{bmatrix}.$$

You will recognize this matrix as one component of the mixed model equations (6.85). From the first order conditions of the problem we find the values that maximize $h(\boldsymbol{\beta}, \mathbf{S}(\mathbf{s}))$ as

$$\begin{aligned} \mathbf{X}(\mathbf{s})\boldsymbol{\Psi}\boldsymbol{\Sigma}_{\widehat{\mu}}^{-1}(\mathbf{Z}(\mathbf{s}) - g^{-1}(\mathbf{X}(\mathbf{s})\boldsymbol{\beta} + \mathbf{S}(\mathbf{s}))) &= \mathbf{0} \\ \boldsymbol{\Psi}\boldsymbol{\Sigma}_{\widehat{\mu}}^{-1}(\mathbf{Z}(\mathbf{s}) - g^{-1}(\mathbf{X}(\mathbf{s}) + \mathbf{S}(\mathbf{s}))) - \boldsymbol{\Sigma}_S^{-1}\mathbf{S}(\mathbf{s}) &= \mathbf{0}. \end{aligned}$$

The solutions for $\boldsymbol{\beta}$ and $\mathbf{S}(\mathbf{s})$ are (6.80) and (6.81). Substituting into the Laplace approximation yields the objective function that is maximized to obtain the estimates of the covariance parameters. One can show that this objective function differs from the REML log likelihood in the pseudo-likelihood method only by a constant amount. The two approaches will thus yield the same estimates.

*6.3.5.4 Inference and Diagnostics*

The marginal as well as the conditional formulation of the generalized linear model for spatial data arrive at a doubly iterative fitting algorithm. After producing the initial pseudo-data, the fit alternates between estimation of the covariance parameters and the estimation of the fixed effects. The final estimate of $\boldsymbol{\theta}$ takes on the form of an estimated generalized least squares estimate in either case. Thus, hypothesis tests, confidence intervals, and other inferential procedures for $\boldsymbol{\beta}$ can be based on the methods in §6.2.3. Essentially, this applies the formulas for a general linear model with correlated errors to the last linearized model that was fit. In other words, this produces inferences on the linear scale. In a model for Binomial rates with a logit link, the inferences apply to the logit scale. This is generally meaningful, since the model states that covariate effects are additive (linear) on the linked scale. In order to produce estimates of the mean, associated standard errors, and confidence intervals, however, the predictions on the linearized scale can be transformed. Standard errors are derived by the delta method.

**Example 6.7** Assume that a model has been fit with a log link, $\mathrm{E}[Y] = \mu = \exp\{\eta\} = \exp\{\mathbf{x}'\boldsymbol{\beta}\}$ and we have at our disposal the estimates $\widehat{\boldsymbol{\beta}}$ as well as an estimate of their variability, $\mathrm{Var}[\widehat{\boldsymbol{\beta}}]$. Then the plug-in estimate of $\mu$ is $\widehat{\mu} = \exp\{\widehat{\eta}\}$ and the variance of the linear predictor is $\mathrm{Var}[\widehat{\eta}] = \mathbf{x}'\mathrm{Var}[\widehat{\boldsymbol{\beta}}]\mathbf{x}$. To obtain an approximate variance of $\widehat{\mu}$, expand it in a first-order Taylor series about $\eta$,

$$\widehat{\mu} \doteq \mu + \frac{\partial \widehat{\mu}}{\partial \widehat{\eta}}_{|\eta} (\widehat{\eta} - \eta).$$

In our special case,

$$\widehat{\mu} = \exp\{\eta\} + \exp\{\eta\}(\widehat{\eta} - \eta),$$

so that $\mathrm{Var}[\widehat{\mu}] \doteq \exp\{\eta\}\mathrm{Var}[\widehat{\eta}]$. The estimate of this approximate variance is then

$$\widehat{\mathrm{Var}}[\widehat{\mu}] = \exp\{\widehat{\eta}\}\widehat{\mathrm{Var}}[\widehat{\eta}].$$

□

Since link functions are monotonic, confidence limits for $\widehat{\eta} = \mathbf{x}(\mathbf{s})\widehat{\boldsymbol{\beta}}$ can be transformed to confidence limits for $\mu(\mathbf{s})$, by using the upper and lower limits as arguments of the inverse link function $g^{-1}(\cdot)$.

Confirmatory statistical inference about the covariance parameters $\boldsymbol{\theta}$ in generalized linear mixed models is more difficult compared to their linear mixed model counterparts. Likelihood-ratio or restricted likelihood-ratio tests are not immediately available in methods based on linearization. The obvious reason is that adding or removing columns of the $\mathbf{X}(\mathbf{s})$ matrix or changing the structure of $\mathbf{S}(\mathbf{s})$ changes the linearization. When the linearization changes, so does the pseudo-data, and the pseudo objective functions such as (6.84) are not comparable. Information based criteria such as AIC or AICC also should not be used for model comparisons, unless the linearizations are the same; this holds for ML as well as REML estimation. The fact that two models are nested with respect to their large-scale trend structure ($\mathbf{X}(\mathbf{s})_2$ is a subset of $\mathbf{X}(\mathbf{s})_1$), and we perform ML pseudo-likelihood estimation does not change this fact. An exception to this rule, where models can be compared based on their pseudo objective functions, occurs when their large-scale trend structures are the same, their formulation of $\mathbf{S}(\mathbf{s})$ is the same, and they are nested with respect to the $\mathbf{R}(\boldsymbol{\theta})$ structure.

### 6.3.6 *Spatial Prediction in GLMs*

The marginal as well as the conditional generalized spatial models have a pseudo-data formulation that corresponds to either a linear model with correlated errors or a linear mixed model. The goal of prediction at unobserved locations is the process $Z(\mathbf{s})$ in the marginal formulation or $g^{-1}(\mathbf{x}'(\mathbf{s}_0)\boldsymbol{\beta} + S(\mathbf{s}_0))$ in the conditional formulation (filtering). In either case, these are predictions on the scale of the data, not the scale of the pseudo-data. It is illustrative, however, to approach the prediction problem in spatial generalized linear models from the pseudo-data. Consider a marginal formulation. Since the model for the pseudo-data (6.77) is linear with mean $\mathbf{X}(\mathbf{s})\boldsymbol{\beta}$ and variance $\boldsymbol{\Sigma}_{\mu} = \boldsymbol{\Psi}^{-1}\boldsymbol{\Sigma}(\boldsymbol{\mu},\boldsymbol{\theta})\boldsymbol{\Psi}^{-1}$, we can apply universal kriging and obtain the UK predictor for the pseudo-data as

$$p(\boldsymbol{\nu};\nu(\mathbf{s}_0)) = \mathbf{x}'(\mathbf{s}_0)\widehat{\boldsymbol{\beta}}_{gls} + \boldsymbol{\sigma}'\boldsymbol{\Sigma}_{\mu}^{-1}\left(\boldsymbol{\nu} - g^{-1}\left(\mathbf{X}(\mathbf{s})\widehat{\boldsymbol{\beta}}_{gls}\right)\right),$$

where $\boldsymbol{\sigma}$ is the covariance vector between the pseudo-data for a new observation and the "observed" vector $\boldsymbol{\nu}$. Plug-in estimation replaces the GLS estimates with EGLS estimates and evaluates $\boldsymbol{\sigma}$ and $\boldsymbol{\Sigma}_{\mu}$ at the estimated covariance parameters. The mean-squared prediction error $\mathrm{E}[(p(\boldsymbol{\nu};\nu(\mathbf{s}_0)) - \nu(\mathbf{s}_0))^2] = \sigma^2_{\nu}(\mathbf{s}_0)$ is computed as in §5.3.3, equation (5.30). To convert this prediction into one for the original data, you can apply the inverse link function,

$$\widehat{Z}(\mathbf{s}_0) = g^{-1}\left(p(\boldsymbol{\nu};\nu(\mathbf{s}_0))\right) \tag{6.87}$$

and apply the Delta method to obtain a measure of prediction error,

$$\left(\frac{\partial g^{-1}\left(p(\boldsymbol{\nu};\nu(\mathbf{s}_0))\right)}{\partial p(\boldsymbol{\nu};\nu(\mathbf{s}_0))}\right)^2 \sigma^2_{\nu}(\mathbf{s}_0) = \left(\frac{\partial \mu}{\partial \eta}_{|p}\right)^2 \sigma^2_{\nu}(\mathbf{s}_0). \tag{6.88}$$

However, expression (6.88) is not the mean-squared prediction error of the inverse linked predictor (6.87). It is the prediction error of a different predictor of the original data, which we derive as follows, see Gotway and Wolfinger (2003).

To predict the original data (and not the pseudo data), assume $\nu(\mathbf{s}_0)$ and the new observation to be predicted, $Z(\mathbf{s}_0)$, and their associated predictors,

$\widehat{\nu}(\mathbf{s}_0)$ and $\widehat{Z}(\mathbf{s}_0)$ also satisfy equation (6.78), so that

$$\widehat{\nu}(\mathbf{s}_0) = g(\widehat{\mu}(\mathbf{s}_0)) + g'(\widehat{\mu}(\mathbf{s}_0))(\widehat{Z}(\mathbf{s}_0) - \widehat{\mu}(\mathbf{s}_0)), \tag{6.89}$$

where $g'(\widehat{\mu}(\mathbf{s}_0))$ denotes the derivative of $g(\mu) = \eta$ with respect to $\mu$, evaluated at $\widehat{\mu}(\mathbf{s}_0)$. Note that this derivative corresponds to the reciprocal diagonal elements of the matrix $\boldsymbol{\Psi}$ defined earlier. Solving for $\widehat{\nu}_0(\mathbf{s}_0)$ in equation (6.89) yields the predictor

$$\widehat{Z}(\mathbf{s}_0) = \widehat{\mu}(\mathbf{s}_0) + \left(g'(\widehat{\mu}(\mathbf{s}_0))\right)^{-1}(\widehat{\nu}(\mathbf{s}_0) - g(\widehat{\mu}(\mathbf{s}_0))). \tag{6.90}$$

The mean-squared prediction error associated with this predictor is $\mathrm{E}[(\widehat{Z}(\mathbf{s}_0) - Z(\mathbf{s}_0))^2]$, which can be obtained from the mean-prediction squared error of $\widehat{\nu}_0$ obtained from universal kriging by noting that

$$\widehat{Z}(\mathbf{s}_0) - Z(\mathbf{s}_0) = \left(g'(\widehat{\mu}(\mathbf{s}_0))\right)^{-1}(\widehat{\nu}_0 - \nu_0),$$

and then

$$\begin{aligned}
\sigma^2_Z(\mathbf{s}_0) &= \mathrm{E}[(\widehat{Z}(\mathbf{s}_0) - Z(\mathbf{s}_0))^2] \\
&= [(g'(\widehat{\mu}(\mathbf{s}_0)))^{-1}]^2 \mathrm{E}[(\widehat{\nu}_0 - \nu_0)^2] \\
&= [(g'(\widehat{\mu}(\mathbf{s}_0)))^{-1}]^2 \sigma^2_\nu(\mathbf{s}_0) \\
&= \left[\frac{\partial \mu}{\partial \eta}_{|\widehat{\mu}(\mathbf{s}_0)}\right]^2 \sigma^2_\nu(\mathbf{s}_0).
\end{aligned} \tag{6.91}$$

The root mean-squared prediction error, $\sqrt{\sigma^2_Z(\mathbf{s}_0)}$, is usually reported as a prediction standard error.

In the case of a marginal GLMM where the mean function is defined as in equation (6.69), the predictor based on pseudo-likelihood estimation (equation 6.90) is very similar in nature to that proposed by Gotway and Stroup (1997) without the explicit linearization of pseudo data. Similarly, the mean-squared prediction errors are comparable to those derived by Vijapurkar and Gotway (2001) by expanding the nonlinear mean function instead of the link function; see also Vallant (1985).

**Example 1.4 (Blood lead levels in children, Virginia 2000. Continued)** To illustrate the use of GLMs and GLMMs in the analysis of spatial data, we again consider Example 1.4, first introduced on page 10. The outcome data of interest are the percent of children under 6 years of age with elevated blood lead levels for each Virginia county in 2000 (shown in Figure 1.5, page 10). As discussed in Example 1.4, the primary source of elevated blood lead levels in children is dust from lead-based paint in older homes in impoverished areas. Thus, if data on home age and household income (and perhaps other variables as well) were available, we could use these variables to explain (at least some of) the variation in the percent of children with elevated blood lead levels. However, the only data available to us are aggregated to the county level. Since we do not have data on individual homes, we will use instead the median housing value per county as a surrogate for housing age and maintenance quality (Figure 1.6). We also do not have data on whether or not an individual child in the study was living in poverty in 2000. Thus, we obtained from the U.S. Census Bureau the number of children in each county under 17 years of age living in poverty in 2000, and will use this variable as a surrogate measure of impoverishment (Figure 6.8).

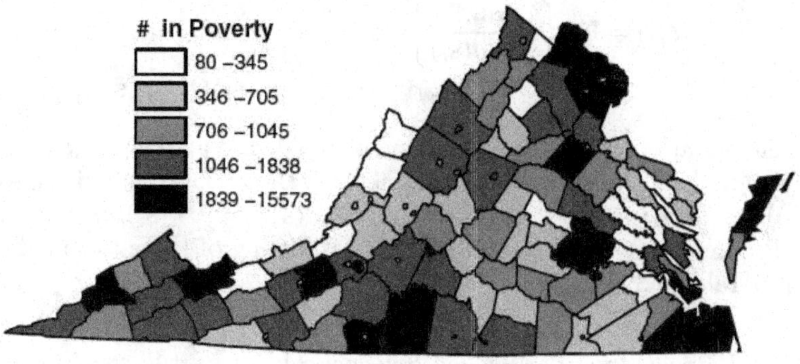


Figure 6.8 *Number of children under 17 years of age living in poverty in Virginia in 2000. Source: U.S. Census Bureau.*

Thus, in terms of the building blocks of a statistical model we have the following:

$Z(\mathbf{s}_i) \equiv Z_i$ = the number of children under 6 years of age with elevated blood lead levels in county $i$, $i = 1, \cdots, N = 133$;

$n(\mathbf{s}_i) \equiv n_i$ = the number of children under 6 years of age tested for elevated blood lead levels in county $i$;

$p(\mathbf{s}_i) \equiv p_i = Z_i/n_i \times 100$ the percent of children under 6 years of age with elevated blood lead levels in county $i$;

$x_1(\mathbf{s}_i) \equiv x_{1i}$ = the median housing value in \$100,000 (i.e., the actual value from the Census, divided by 100,000);

$x_2(\mathbf{s}_i) \equiv x_{2i}$ = the number of children under 17 years of age living in poverty in 2000, per 100,000 children at risk (i.e., the actual value from the Census, divided by 100,000).

Since the outcome data are percentages, it seems reasonable to fit a generalized linear regression model to these data and we could use either of the following models:

- A Poisson distribution with $Z(\mathbf{s}_i) \sim \text{Poisson}(\mu(\mathbf{s}_i) \equiv n(\mathbf{s}_i)\lambda(\mathbf{s}_i))$,

$$\begin{aligned} \log\{\lambda(\mathbf{s}_i)\} &= \beta_0 + \beta_1 x_1(\mathbf{s}_i) + \beta_2 x_2(\mathbf{s}_i) \\ g(\mu(\mathbf{s}_i)) = \log\{\mu(\mathbf{s}_i)\} &= \log\{n_i\} + \beta_0 + \beta_1 x_1(\mathbf{s}_i) + \beta_2 x_2(\mathbf{s}_i) \\ v(\mu(\mathbf{s}_i)) &= \mu(\mathbf{s}_i), \end{aligned}$$

so that

$$\begin{aligned} \mathrm{E}[p(\mathbf{s}_i)] &= \lambda(\mathbf{s}_i) \\ \mathrm{Var}[p(\mathbf{s}_i)] &= \lambda(\mathbf{s}_i)/n(\mathbf{s}_i); \end{aligned}$$

- A Binomial distribution with $Z(\mathbf{s}_i) \sim \mathrm{Binomial}(n(\mathbf{s}_i), \mu(\mathbf{s}_i))$, so that

$$\begin{aligned} \mathrm{E}[p(\mathbf{s}_i)] &= \mu(\mathbf{s}_i) \\ g(\mu(\mathbf{s}_i)) = \log\left\{\frac{\mu(\mathbf{s}_i)}{1-\mu(\mathbf{s}_i)}\right\} &= \beta_0 + \beta_1 x_1(\mathbf{s}_i) + \beta_2 x_2(\mathbf{s}_i) \\ v(\mu(\mathbf{s}_i)) &= n(\mathbf{s}_i)\mu(\mathbf{s}_i)(1-\mu(\mathbf{s}_i)). \end{aligned}$$

To accommodate possible overdispersion of the data—at this stage of modeling—we add a multiplicative scale parameter to the variance functions, so that $v(\mu(\mathbf{s}_i)) = \sigma^2\mu(\mathbf{s}_i)$ in the Poisson case and $v(\mu(\mathbf{s}_i)) = \sigma^2 n(\mathbf{s}_i)\mu(\mathbf{s}_i)(1-\mu(\mathbf{s}_i))$ in the Binomial case. The results from REML fitting of these two models are shown in Table 6.3.

Table 6.3 *Results from Poisson and Binomial regressions with overdispersion.*

Poisson-Based Regression

| Effect | Estimate | Std. Error. | t-value | p-value |
|---|---|---|---|---|
| Intercept ($\beta_0$) | −2.2781 | 0.1606 | −14.18 | $< 0.0001$ |
| Median Value ($\beta_1$) | −0.6228 | 0.2120 | −2.94 | 0.0039 |
| Poverty ($\beta_2$) | 0.6293 | 1.4068 | 0.45 | 0.6554 |
| $\hat{\sigma}^2$ | 6.7825 | 0.8413 | | |

Binomial-Based Regression

| Effect | Estimate | Std. Error. | t-value | p-value |
|---|---|---|---|---|
| Intercept ($\beta_0$) | −2.1850 | 0.1712 | −12.76 | $< 0.0001$ |
| Median Value ($\beta_1$) | −0.6552 | 0.2241 | −2.92 | 0.0041 |
| Poverty ($\beta_2$) | 0.6566 | 1.5086 | 0.44 | 0.6641 |
| $\hat{\sigma}^2$ | 7.2858 | 0.9037 | | |

There is very little difference in the results obtained from the two models. There is also very little to guide us in choosing between them. Comparing the value of the AIC criterion from both models may be misleading since the models are based on two different distributions whose likelihoods differ. Moreover, GLMs are essentially weighted regressions, and when different distributions and different variance functions are used, the weights are different as well. Thus, comparing traditional goodness of fit measures for these two models (and for the other GLMs and GLMMs considered below) may not be valid.

Instead, we rely on some general observations to help us choose between these two models.

Typically, when there is overdispersion in the data, the Poisson-based model often appears to fit better than one based on the Binomial distribution, since the Poisson distribution has more variability relative to the Binomial. As a consequence, it appears to have less overdispersion. To see this, suppose $Z|n \sim \text{Binomial}(n, \pi)$, and $n \sim \text{Poisson}(\lambda)$. Then, $Z$ is a Poisson random variable with mean $\lambda\pi$ and $\text{Var}[Z] > \text{E}[n]\pi(1-\pi)$. Also, when $\pi$ is small, and $n\pi \to \lambda$, the Binomial distribution converges to a Poisson distribution. For the Virginia blood lead level data, $\overline{Z} = 12.63$ and $s_Z^2 = 1378.64$, $\overline{p} = 7.83$ and $s_p^2 = 110.25$, reflecting a great deal of overdispersion in the outcome data and also indicating that perhaps elevated blood lead levels in children are unusual (and hopefully so). These observations might lead us toward choosing the Poisson-based model over the Binomial-based model and, for the rest of this example, we will adopt the Poisson-based model. Repeating the analysis using the Binomial-based model is left as Exercise 6.17.

One key feature of the results in Table 6.3 is the large amount of overdispersion in the data (measured by $\widehat{\sigma}^2 = 6.7825$). This may just be indicating that the data do not follow a Poisson (or a Binomial) distribution, but this variation could also be spatial. Thus, if we make the model more spatially explicit, e.g., by using spatial GLMs and GLMMs, we might be able to account for some of the overdispersion and obtain a better fit. To determine whether the data are spatially autocorrelated and an adjustment of the covariance structure of the is necessary, we follow an analogous procedure to that used for the C/N ratio data (continuation of Example 6.1 on page 311). Since the current situation is more complicated—as we are using generalized linear, and not linear models—we use traditional residuals from the Poisson-based GLM, rather than recovered errors. Thus, we compute the empirical semivariogram of the standardized Pearson residuals based on Euclidean distances between county centroids, measured in kilometers after converting the centroids to Virginia state plane coordinates. The standardized Pearson residuals are

*Standardized Pearson Residuals*

$$r_i = \frac{Z_i - \widehat{\mu}_i}{\widehat{\sigma}^2 v(\widehat{\mu}_i)(1 - l_{ii})},$$

where $l_{ii}$ is the *ith* diagonal element of

$$\widehat{\Delta}^{1/2}\mathbf{X}(\mathbf{s})(\mathbf{X}(\mathbf{s})'\widehat{\Delta}\mathbf{X}(\mathbf{s}))^{-1}\mathbf{X}(\mathbf{s})'\widehat{\Delta}^{1/2},$$

and $\widehat{\Delta}$ is a diagonal matrix with *ith* element $(\widehat{\sigma}^2 v(\widehat{\mu}_i))^{-1}(g'(\widehat{\mu}_i))^{-2}$.

The empirical semivariogram based on these residuals is shown in Figure 6.9. We use it only as a rough guide to help indicate whether or not we need a model with a more complicated variance structure and, if so, to suggest a possible parametric form. This semivariogram is probably not definitive of the spatial variability in our data. For example, it does not lend itself to an interpretation of the spatial dependency in a conditional model, where the covariance structure of a latent spatial random field is modeled on the logarithmic (linked) scale. The empirical semivariogram in Figure 6.9 suggests that there might be some spatial autocorrelation in the standardized Pearson residuals and that perhaps a spherical model might be an adequate parametric representation.

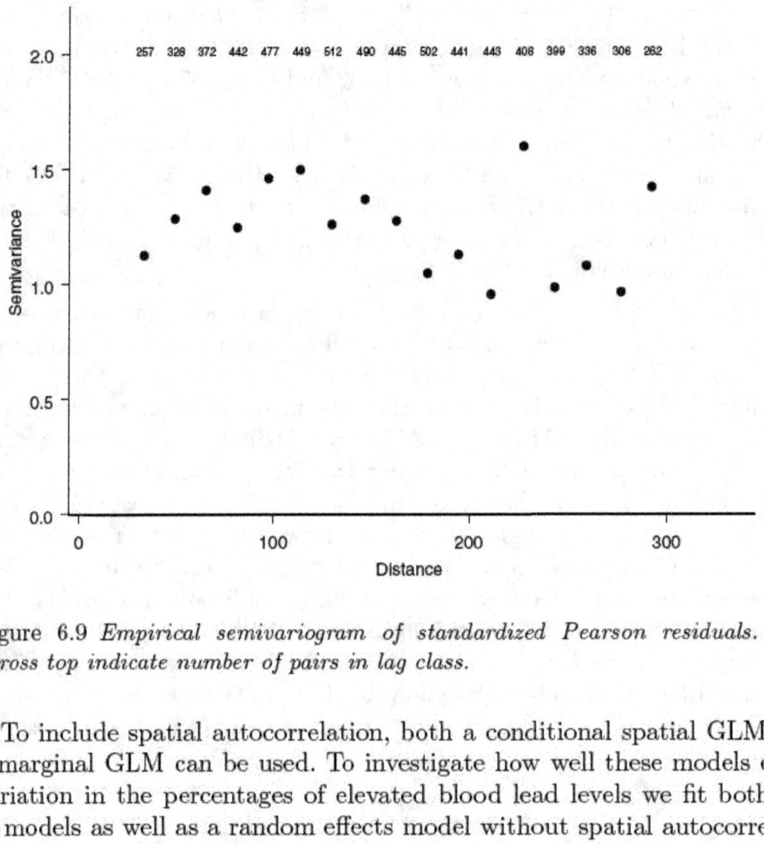


Figure 6.9 *Empirical semivariogram of standardized Pearson residuals. Values across top indicate number of pairs in lag class.*

To include spatial autocorrelation, both a conditional spatial GLMM and a marginal GLM can be used. To investigate how well these models explain variation in the percentages of elevated blood lead levels we fit both types of models as well as a random effects model without spatial autocorrelation. Specifically, we considered the following models, with $g(\mu(\mathbf{s})) = \log\{\mu(\mathbf{s})\}$ in all cases:

- Random Effects Model:

$$\begin{aligned} Z(\mathbf{s}_i)|S(\mathbf{s}_i) &\overset{ind}{\sim} \text{Poisson}(\mu(\mathbf{s}_i) \equiv n(\mathbf{s}_i)\lambda(\mathbf{s}_i)), \\ \log\{\lambda(\mathbf{s}_i)\} &= \beta_0 + \beta_1 x_1(\mathbf{s}_i) + \beta_2 x_2(\mathbf{s}_i) + S(\mathbf{s}_i), \\ \text{Var}[\mathbf{Z}(\mathbf{s})|\mathbf{S}(\mathbf{s})] &= \sigma^2 \mathbf{V}_{\boldsymbol{\mu}}, \\ \mathbf{S}(\mathbf{s}) &\sim G(\mathbf{0}, \ \sigma_S^2 \mathbf{I}); \end{aligned}$$

- Conditional Spatial GLMM:

$$Z(\mathbf{s}_i)|S(\mathbf{s}_i) \overset{ind}{\sim} \text{Poisson}(\mu(\mathbf{s}_i) \equiv n(\mathbf{s}_i)\lambda(\mathbf{s}_i)),$$

$$\log\{\lambda(\mathbf{s}_i)\} = \beta_0 + \beta_1 x_1(\mathbf{s}_i) + \beta_2 x_2(\mathbf{s}_i) + S(\mathbf{s}_i),$$
$$\text{Var}[\mathbf{Z}(\mathbf{s})|\mathbf{S}(\mathbf{s})] = \sigma^2 \mathbf{V}_{\boldsymbol{\mu}},$$
$$\mathbf{S}(\mathbf{s}) \sim G(\mathbf{0},\ \sigma_S^2 \mathbf{R}_S(\alpha_s));$$

- Marginal Spatial GLM

$$\text{E}[Z(\mathbf{s})] = \mu(\mathbf{s}) \equiv n(\mathbf{s})\lambda(\mathbf{s})),$$
$$\log\{\lambda(\mathbf{s})\} = \beta_0 + \beta_1 x_1(\mathbf{s}) + \beta_2 x_2(\mathbf{s}),$$
$$\text{Var}[\mathbf{Z}(\mathbf{s})] = \sigma_0^2 \mathbf{V}_{\boldsymbol{\mu}} + \sigma_1^2 \mathbf{V}_{\boldsymbol{\mu}}^{1/2} \mathbf{R}_S(\alpha_m) \mathbf{V}_{\boldsymbol{\mu}}^{1/2}.$$

For the spatial models, spatial autocorrelation was modeled using the spherical model defined in (4.13). The parameters of the models were estimated using the pseudo-likelihood estimation described in §6.3.5.2. The results are shown in Table 6.4.

From Table 6.4, some tentative conclusions emerge:

- The results from the random effects model indicate that the relationship between median housing value and the percentage of children with elevated blood lead levels is not significant. They also show an inverse relationship between the percentage of children with elevated blood lead levels and poverty, although the relationship is not significant. The results from all other models indicate just the opposite: a significant association with median housing value and a nonsignificant poverty coefficient that has a positive sign. Thus, some of spatial variation in the data arising from spatial autocorrelation (incorporated through $\mathbf{R}(\alpha)$) is perhaps being attributed to median housing value and poverty in the random effects model. In the conditionally-specified model choices about $\mathbf{R}(\alpha)$ affect $S(\mathbf{s})$ which in turn can affect the estimates of the fixed effects; This just reflects the fact that the general decomposition of "data=f(fixed effects, random effects, error)" is not unique. The random effects model does not allow any spatial structure in the random effects or the errors. The conditional and marginal spatial models do allow spatial structure in these components, with the conditional model assigning spatial structure to the random effects, and the marginal model assigning the same sort of structure to "error." Thus, while the conditional and marginal spatial GLMs accommodate variation in different ways, incorporating spatially-structured variation in spatial regression models can be important.
- The marginal spatial model and the traditional Poisson-based model (Table 6.3) give similar results. However, the standard errors from the marginal spatial GLM are noticeably higher than for the traditional Poisson-based model. Thus, with the "data=f(fixed effects, error)" decomposition in marginal models, the marginal spatial GLMs incorporate spatial structure through the "error" component.
- The majority of the models suggests that median housing value is significantly associated with the percentage of children with elevated blood lead

Table 6.4 *Results from spatial GLMs and GLMMs.*

Random Effects Model

| Effect | Estimate | Std. Error | t-value | p-value |
|---|---|---|---|---|
| Intercept ($\beta_0$) | −2.6112 | 0.1790 | −14.59 | $< 0.0001$ |
| Median Value ($\beta_1$) | −0.2672 | 0.2239 | −1.19 | 0.2349 |
| Poverty ($\beta_2$) | −0.9688 | 2.900 | −0.33 | 0.7389 |
| $\hat{\sigma}_S^2$ | 0.5047 | 0.1478 | | |
| $\hat{\sigma}^2$ | 1.1052 | 0.3858 | | |

Conditional Spatial GLMM

| Effect | Estimate | Std. Error | t-value | p-value |
|---|---|---|---|---|
| Intercept ($\beta_0$) | −2.4246 | 0.3057 | −7.93 | $< 0.0001$ |
| Median Value ($\beta_1$) | −0.7835 | 0.3610 | −2.17 | 0.0318 |
| Poverty ($\beta_2$) | 3.5216 | 2.5933 | 1.36 | 0.1768 |
| $\hat{\sigma}_S^2$ | 0.8806 | 0.1949 | | |
| $\hat{\sigma}^2$ | 0.7010 | 0.1938 | | |
| $\hat{\alpha}_s$ (km) | 78.81 | 6.51 | | |

Marginal Spatial GLM

| Effect | Estimate | Std. Error | t-value | p-value |
|---|---|---|---|---|
| Intercept ($\beta_0$) | −2.2135 | 0.2010 | −11.01 | $< 0.0001$ |
| Median Value ($\beta_1$) | −0.8096 | 0.2713 | −2.98 | 0.0034 |
| Poverty ($\beta_2$) | 1.3172 | 1.5736 | 0.84 | 0.4041 |
| $\hat{\sigma}_0^2$ | 6.1374 | 0.8846 | | |
| $\hat{\sigma}_1^2$ | 0.9633 | 0.7815 | | |
| $\hat{\alpha_m}$ (km) | 186.65 | 70.04 | | |

levels, and that counties with higher median housing values tend to have a lower percentage of children with elevated blood lead levels. Also, there appears to be a positive relationship between the percentage of children with elevated blood lead levels and poverty, although the relationship is not significant (and may not be well estimated).

The spatial GLM and GLMM considered thus far in this example use a geostatistical approach to model autocorrelation in the data. However, the regional nature of the data suggests that using a different measure of spatial proximity might be warranted. For linear regression models, the spatial autoregressive models described in §6.2.2 can incorporate such measures of spatial proximity. However, the SAR models described in §6.2.2.1 are only defined for multivariate Gaussian data and extending the CAR models described in §6.2.2.2 is fraught with problems. Besag (1974) and Cressie (1993, Section 6.4) discuss the conditions necessary to construct a joint likelihood from a conditional specification. Central to these conditions is the Hammersley-Clifford theorem (Besag, 1974) that derives the general form of the joint likelihood. This theorem shows that the joint likelihood allows interaction among the data only for data observed at locations that form a **clique**, a set of locations that are all neighbors of each other. The practical impact of this occurs when we want to go the other way and specify a neighborhood structure together with a set of conditional distributions. For many of the distributions in the exponential family guaranteeing a valid joint likelihood results in excessive restrictions on the neighborhood structure (cf., §5.8). For example, even assuming just pairwise dependence among locations, so that no clique contains more than two locations, a conditional Poisson models permits only negative spatial dependence.

However, adopting a modeling approach allows us to circumvent some of these difficulties. For example, in a conditionally-specified spatial model, we could describe the Gaussian process $S(\mathbf{s})$ using either a SAR or a CAR model. This basically amounts to different choices for $\mathbf{\Sigma}_S$. For example, in the simplest one-parameter case, we could choose $\mathbf{\Sigma}_S = \sigma^2(I - \rho\mathbf{W})^{-1}$ for a CAR model and $\mathbf{\Sigma}_S = \sigma^2(I - \rho\mathbf{W})^{-1}(I - \rho\mathbf{W}')^{-1}$ for a SAR model, where $\mathbf{W}$ is a specified spatial proximity matrix and $\rho$ is the spatial dependence parameter. With a marginal GLM—since it makes only assumptions about the first two moments—we could model $\sigma^2\mathbf{R}$ as either $\sigma^2(I - \rho\mathbf{W})^{-1}$ corresponding to the variance-covariance matrix from a CAR model, or as $\sigma^2(I - \rho\mathbf{W})^{-1}(I - \rho\mathbf{W}')^{-1}$ corresponding to the variance-covariance matrix from a SAR model. While this strategy does not model the *data* using a CAR or SAR model, it does allow us to consider more general parametric models for Var$[\mathbf{Z}(\mathbf{s})]$ that may be more desirable when working with regional data. Moreover, we are not limited to the forms arising from CAR and SAR models; we can create any parametric representation as long as the resulting matrices are positive definite and the parameters are estimable.

To illustrate this idea, we used the binary connectivity weights defined in equation (1.6) of §1.3.2 to measure spatial similarity in the percent of children under 6 years of age with elevated blood lead levels for each Virginia county. Using this measure of spatial proximity, the value of Moran's $I$ computed from the standardized Pearson residuals obtained from the traditional Poisson-based model is $I = 0.0884$, (the eigenvalues of $\mathbf{W}$ indicate this value can range from $-0.3275$ to $0.1661$) with a z-score of $z = 1.65$. Thus, there may be some evidence of spatial autocorrelation in the residuals, although the assumptions underlying any test, and even the definition and interpretation of Moran's $I$ computed from standardized Pearson residuals, are extremely questionable. Thus, we considered instead the following marginal spatial GLM using $g(\mu(\mathbf{s})) = \log\{\mu(\mathbf{s})\}$:

$$\mathrm{E}[Z(\mathbf{s})] = \mu(\mathbf{s}) \equiv n(\mathbf{s})\lambda(\mathbf{s}),$$
$$\log\{\lambda(\mathbf{s})\} = \beta_0 + \beta_1 x_1(\mathbf{s}) + \beta_1 x_2(\mathbf{s}),$$
$$\mathrm{Var}[\mathbf{Z}(\mathbf{s})] = \sigma_0^2 \mathbf{V}_{\boldsymbol{\mu}} + \sigma_1^2 \mathbf{V}_{\boldsymbol{\mu}}^{1/2}(I - \rho\mathbf{W})^{-1}\mathbf{V}_{\boldsymbol{\mu}}^{1/2},$$

where $\mathbf{W}$ is a spatial proximity matrix constructed from the binary connectivity weights. The results are shown in Table 6.5.

Table 6.5 *Results from a marginal spatial GLM with a CAR autocorrelation structure.*

| Effect | Estimate | Std. Error | t-value | p-value |
|---|---|---|---|---|
| Intercept ($\beta_0$) | −2.1472 | 0.1796 | −11.95 | < 0.00001 |
| Median Value ($\beta_1$) | −0.7532 | 0.2471 | −3.05 | 0.0014 |
| Poverty ($\beta_2$) | 0.5856 | 1.4417 | 0.40 | 0.6574 |
| $\hat{\sigma}_0^2$ | 0.0000 | | | |
| $\hat{\sigma}_1^2$ | 6.2497 | | | |
| $\hat{\rho}$ | 0.0992 | | | |

These results are quite similar to those from the marginal spatial GLM given in Table 6.4, although it is difficult to generalize this conclusion. Often, adjusting for spatial autocorrelation is more important than the parametric models used to do the adjustment, although in some applications, different models for the spatial autocorrelation can lead to different results and conclusions.

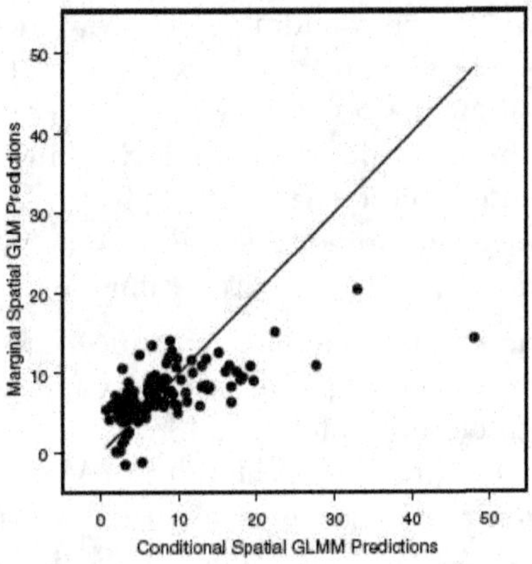


Figure 6.10 *Predictions from marginal spatial GLM vs. predictions from conditional spatial GLMM.*

The fitted mean from the regression, $\widehat{\mu}$, can be interpreted as the predicted percentage of children with elevated blood lead levels, adjusted for median income and poverty. For the models with random effects, this mean is $g^{-1}(\mathbf{X}(\mathbf{s})\widehat{\beta} + \widehat{\mathbf{S}}(\mathbf{s}))$, where $\widehat{\mathbf{S}}(\mathbf{s})$ is obtained from (6.81). For the marginal spatial models, this mean should be based on prediction of $Z(\mathbf{s}_i)$ at the data locations, using the methods described in §6.3.6 (cf., §5.4.3).

Figure 6.10 compares the adjusted percentages for the conditional spatial GLMM and the marginal spatial GLM with a geostatistical variance structure. There is a strong relationship between them, although there is more smoothing in the marginal model. This is dramatically evident from the adjusted maps shown in Figure 6.12.

The maps of the adjusted percentages obtained from the conditional spatial GLMM are more similar to the raw rates shown in Figure 1.5, since this model tends to shrink the percentages to local means (through $S(\mathbf{s})$) rather than to a global mean as in the marginal models. Figure 6.11 shows the adjusted percentages from the two marginal spatial GLMs, one with a geostatistical variance, and one with the CAR variance. The adjusted percentages are more different than the maps in Figure 6.12 reflect. The model with the CAR variance smooths the data much more than the model with the geostatistical variance. This may result from our choice of adjacency weights since with these weights adjacent counties are assumed to be spatially similar and so the contribution from $\sigma_0^2$ which mitigates local spatial similarity is reduced.

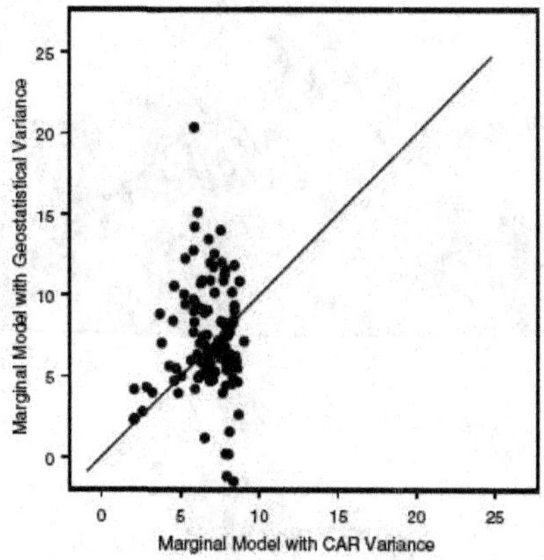


Figure 6.11 *Predictions from marginal GLM using geostatistical variance structure vs. predictions from marginal GLM using CAR variance structure.*

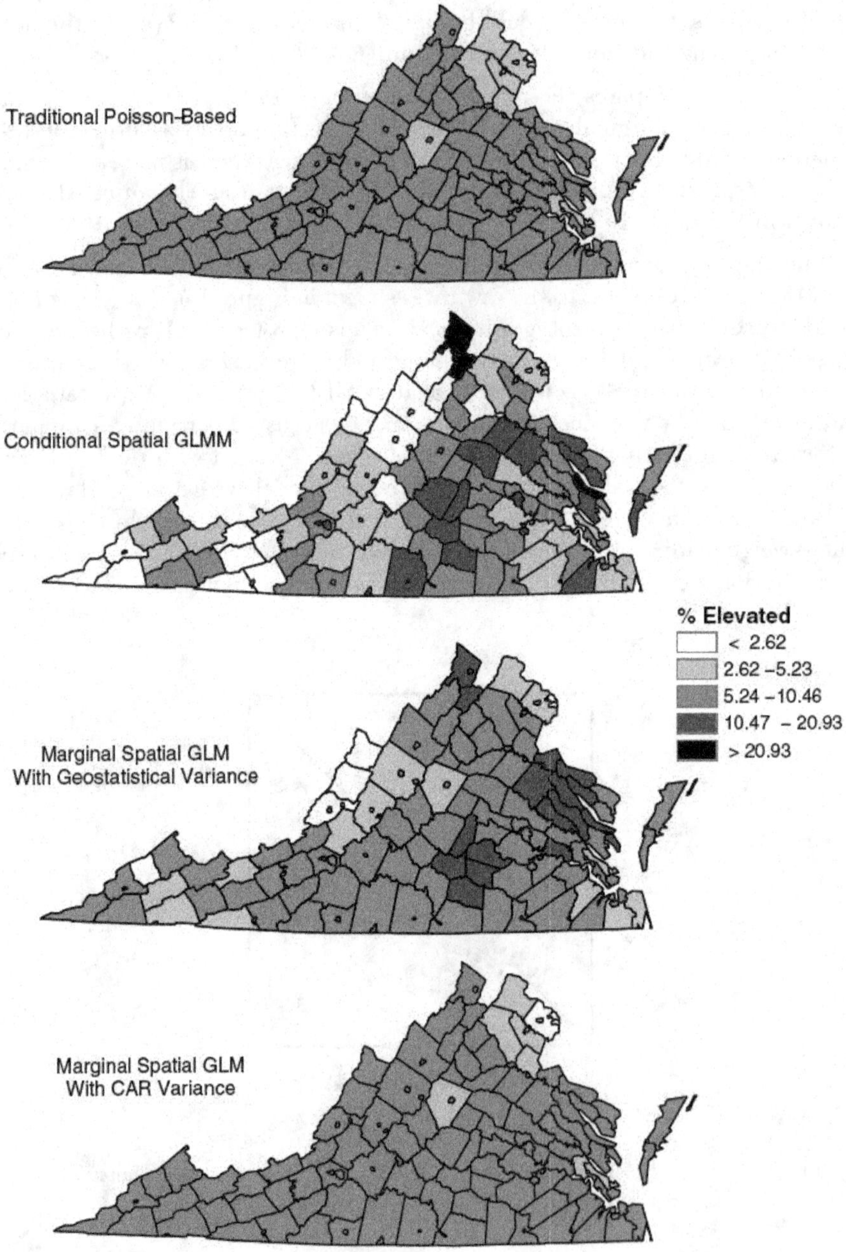


Figure 6.12 *Predictions from regression fits.*

Both the conditional spatial GLMM and the marginal spatial GLM indicate three areas where blood lead levels in children remain high, after adjusting for median housing value and poverty level: one, comprised of Frederick and Warren counties, in the north; another in south-central Virginia near Farmville; and another in the northeast in the Rappahannock river basin. To investigate these further, a more thorough analysis that includes other potentially important covariates (e.g., other demographic variables, more refined measures of known lead exposures), more local epidemiological and environmental information, and subject-matter experts, is needed.

Moreover, these analyses all assume that the relationship between the percentage of children with elevated blood lead level and median housing value and poverty level is the same across all of the Virginia. We can relax this assumption by considering a Poisson version of geographically weighted regression (§6.1.3.1, Fotheringham et al., 2002). Instead of assuming a linear relationship between the data and the covariates as in (6.9), we assume

$$\mathrm{E}[Z(\mathbf{s})] = \mu(\mathbf{s}) \equiv n(\mathbf{s})\lambda(\mathbf{s}),$$

$$\log\{\lambda(\mathbf{s})\} = \beta_0(\mathbf{s}) + \beta_1(\mathbf{s})x_1(\mathbf{s}) + \beta_2(\mathbf{s})x_2(\mathbf{s}),$$

$$\mathrm{Var}[\mathbf{Z}(\mathbf{s})] = \sigma_1^2 \mathbf{V}_{\boldsymbol{\mu}}^{1/2}\mathbf{W}(\mathbf{s}_0)^{-1}\mathbf{V}_{\boldsymbol{\mu}}^{1/2}.$$

We chose weights defined by the Gaussian kernel, $w_{ij} = \exp\{-\frac{1}{2}d_{ij}/b\}$ where $d_{ij}$ is the Euclidean distance between regional centroids and $b$ is a bandwidth parameter which controls the degree of local smoothing. Following Fotheringham et al. (2002), we estimated $b$ by cross validation and used local versions of the generalized estimating equations derived in §6.3.5.1 to estimate $\boldsymbol{\beta}_0$. Since the data are aggregated, we obtained estimates of $\boldsymbol{\beta}_0$ and standard errors at each county centroid. A map of the effect of median housing value on the percentage of children with elevated blood lead levels is shown in Figure 6.13 and the corresponding standard error map is shown in Figure 6.14.

Figure 6.13 clearly shows that the effect of median housing value on the percentage of children with elevated blood lead levels is not constant across Virginia. There is a rather large area in south central Virginia where this effect is as we would expect; as median housing value increases, blood lead levels increases. Referring back to Figure 1.5, this is also an area with some of the largest percentages of elevated blood lead levels. It is also a fairly rural region with relatively low median housing values (cf., Figure 1.6). There is another area of interest near West Virginia, where the relationship between median housing value and the percentage of elevated blood lead levels is not what we would expect; as median housing value increases, blood lead levels also tend to increase. This area also has relatively low median housing values, but is also one that has some of the lowest percentages of elevated blood lead levels in Virginia. This suggests that the homes in this area might be different from those in south central Virginia; perhaps many of the older homes are not painted at all or only those with relatively high median values for this region are painted with lead-based paint. Conjectures like these are one of the

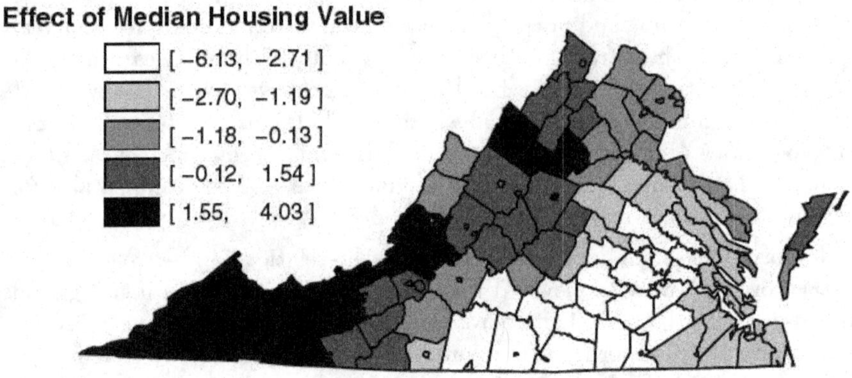


Figure 6.13 *Estimates of the effect of median housing value from GWR.*

major advantages of this type of spatial analysis: it allows us to fine tune our hypotheses and isolate more specific exposure sources that may affect blood lead levels in children.

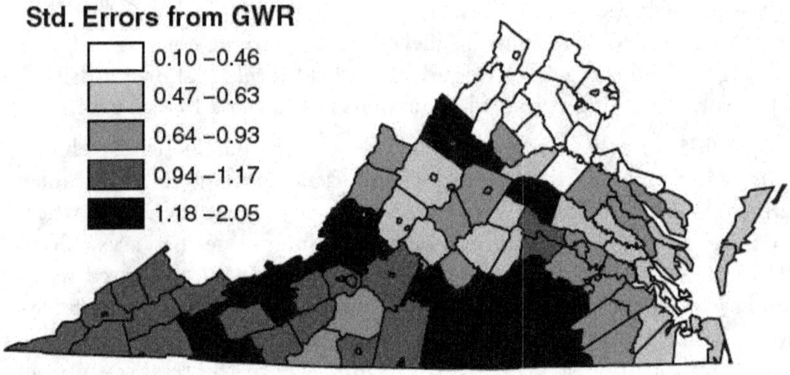


Figure 6.14 *Standard errors of the effect of median housing value estimated from GWR.*

□

## 6.4 Bayesian Hierarchical Models

The conditionally specified generalized linear mixed models discussed in the previous section are examples of hierarchical models. At the first stage of the hierarchy, we describe how the data depend on the random effects, $S(\mathbf{s})$, and on other covariate values deemed fixed and not random. At the second stage of the hierarchy, we model the distribution of the random effects. Since the data are assumed to be conditionally independent, constructing the likelihood function (the likelihood of the data given unknown parameters) is relatively easy and forms the basis for inference about the unknown parameters. Thus, in the abstract, we have data $\mathbf{Z} = (Z_1, \cdots, Z_n)'$ whose distribution depends on unknown quantities $\phi$ (these may be parameters, missing data, or even other variables). We assume that $\mathbf{Z}|\phi \sim f(\mathbf{z}|\phi)$, where, borrowing from traditional statistical terminology, $f(\mathbf{z}|\phi)$ is called the likelihood function. The distribution of $\phi$ is specified using $\pi(\phi)$ and is called the **prior** distribution. The key to Bayesian inference is the **posterior distribution** of the unknown quantities given the data that reflects the uncertainty about $\phi$ after observing the data. From Bayes' theorem, the posterior distribution is proportional to the product of the likelihood function and the prior distribution

*Posterior Distribution*

$$h(\phi|\mathbf{Z}) \propto f(\mathbf{z}|\phi)\pi(\phi), \tag{6.92}$$

where the constant of proportionality is $\int f(\mathbf{z}|\phi)\pi(\phi)d\phi$, so that ensures posterior distribution integrates to one.

**Example 6.8** Suppose that given $\phi$, the data $Z_i$ are independent Poisson variables, $Z_i|\phi \overset{ind}{\sim} \text{Poisson}(\phi)$, and suppose $\phi \sim \Gamma(\alpha, \beta)$. Then, the likelihood function is

$$f(\mathbf{z}|\phi) = \prod_{i=1}^{n} e^{-\phi}\frac{\phi^{z_i}}{z_i!},$$

and the posterior is

$$h(\phi|\mathbf{z}) = \prod_{i=1}^{n} e^{-\phi}\frac{\phi^{z_i}}{z_i!} \times \frac{\beta^{\alpha}}{\Gamma(\alpha)}\phi^{\alpha-1}e^{-\beta\phi}.$$

Combining terms, the posterior can be written as

$$h(\phi|\mathbf{z}) \propto \phi^{\alpha+\sum z_i - 1}e^{-(\beta+n)\phi},$$

which is $\Gamma(\alpha + \sum z_i, \beta + n)$.

Given the posterior distribution, we can use different summary statistics from this distribution to provide inferences about $\phi$. One that is commonly used is the posterior mean, which in this case is

$$\mathrm{E}[\phi|\mathbf{z}] = \frac{\alpha + \sum z_i}{\beta + n} = \bar{z}\left(\frac{n}{n+\beta}\right) + \frac{\alpha}{\beta}\left(1 - \frac{n}{n+\beta}\right).$$

Thus, the Bayes estimate is a linear combination of the maximum likelihood estimate $\bar{z}$, and the prior mean, $\alpha/\beta$. □

**Example 6.9** Consider the linear mixed model

$$Z_i = \mu_i + e_i,$$

where the $\mu_i$ are random effects and the errors are *iid* with a common known variance $\sigma^2$. One Bayesian hierarchical formulation might be

$$\begin{aligned}[\mathbf{Z}|\boldsymbol{\mu},\sigma^2] &\sim G(\boldsymbol{\mu},\sigma^2\mathbf{I})\\ [\boldsymbol{\mu}|\alpha,\tau^2] &\sim G(\alpha\mathbf{1},\tau^2\mathbf{I})\\ [\alpha] &\sim G(a,b^2).\end{aligned}$$

At the final stage of the hierarchy, numerical values must be specified for all unknowns, in this case $\tau^2$, $a$, and $b^2$. The lack of information on these parameters has caused much controversy in statistics leading many to adopt an **empirical Bayes** approach. An empirical Bayes version of this model is

$$\begin{aligned}[\mathbf{Z}|\boldsymbol{\mu},\sigma^2] &\sim G(\boldsymbol{\mu},\sigma^2\mathbf{I})\\ [\boldsymbol{\mu}|\alpha,\tau^2] &\sim G(\alpha\mathbf{1},\tau^2\mathbf{I})\end{aligned}$$

where $\alpha$ and $\tau^2$ are unknown and then estimated from the data using the likelihood

$$f(\mathbf{z}|\sigma^2,\alpha,\tau^2) = \int f(\mathbf{z}|\boldsymbol{\mu},\sigma^2)\pi(\boldsymbol{\mu}|\alpha,\tau^2)d\boldsymbol{\mu}.$$

In this case, because of the conditional independence,

$$f(\mathbf{z}|\sigma^2,\alpha,\tau^2) = \prod_{i=1}^{n}\int f(z_i|\mu_i,\sigma^2)\pi(\mu_i|\alpha,\tau^2)d\mu_i,$$

which is $G(\alpha,\sigma^2+\tau^2)$. Thus, $\alpha$ and $\tau^2$ can be estimated by maximum likelihood giving $\widehat{\alpha} = \bar{Z}$, and $\widehat{\tau}^2 = \max(0, s^2 - \sigma^2)$, where $s^2 = \sum(Z_i - \bar{Z})^2/n$. □

A more general Bayesian hierarchical model—corresponding to the GLMMs discussed in the previous section—is slightly more complicated. Suppose we model spatial autocorrelation through the random effects, so that $\mathbf{R} = \sigma^2\mathbf{I}$, with $\sigma^2$ known, and $\text{Var}[\mathbf{S}(\mathbf{s})] = \boldsymbol{\Sigma}(\boldsymbol{\theta})$. For a hierarchical specification, we must specify how the data depend on both the fixed and the random effects, and then how the distribution of the random effects depends on $\boldsymbol{\theta}$. Thus, the joint posterior distribution may look something like

*Joint Posterior Distribution*

$$h(\boldsymbol{\theta},\mathbf{S}(\mathbf{s}),\boldsymbol{\beta}|\mathbf{Z}(\mathbf{s})) \propto f(\mathbf{Z}(\mathbf{s})|\boldsymbol{\beta},\mathbf{S}(\mathbf{s}))\,\pi(\mathbf{S}(\mathbf{s})|\boldsymbol{\theta})\,\pi(\boldsymbol{\beta})\,\pi(\boldsymbol{\theta}). \tag{6.93}$$

Note that a fully Bayesian analysis treats all model parameters as random variables rather than fixed but unknown values. However, we still refer to "fixed effects" as parameters pertaining to all experimental units and "random effects" as parameters that vary among experimental units. This distinction remains a key component of mixed models regardless of the inferential approach. There is another important distinction. The covariance parameters $\boldsymbol{\theta}$ enter the model at a second stage of the hierarchy and are also assigned a prior distribution. They are often referred to as **hyperparameters**. Finally, to construct (6.93), we assume statistical independence between the fixed effects parameters $\boldsymbol{\beta}$ and the hyperparameters $\boldsymbol{\theta}$, yielding a product of prior distributions for the different model parameters.

### *6.4.1 Prior Distributions*

In principle, the prior distribution is a statistical rendition of the information known about a parameter of interest. Most parameters are not completely unknown; we usually have some information about them. For example, we know that variance components are positive, and from previous experience we might expect their distribution to be skewed. We also know that probabilities lie between 0 and 1. There might be historical data from other studies that provides information that can be used to construct a prior distribution. Theoretically, the prior distribution may be elicited from subject-matter experts, but in practice this is often done only in complex risk analysis studies. A more intriguing philosophical approach to prior solicitation advocates eliciting, not a prior distribution, but plausible data (Lele, 2004).

To make progress analytically, a **conjugate prior**, one that leads to a posterior distribution belonging to the same family as the prior, is often used. With a conjugate prior, the prior parameter can often be interpreted as a prior sample, with the posterior distribution being just an updated version based on new data. In the first example given above, the $\Gamma(\alpha, \beta)$ prior for $\phi$ is a conjugate prior.

From the beginning, Bayesian analysis was meant to be subjective. As the question "Where do you get the prior?" became more persistent, attempts were made to make Bayesian analysis more objective by defining **noninformative** priors. For example, suppose the parameter space is discrete, taking on $m$ values. Then the uniform prior $\pi(\phi) = 1/m$ is a noninformative prior in that every value of $\phi$ is equally likely. In this case, the term "noninformative" is somewhat misleading; this prior does indeed provide information about $\phi$, namely the fact that every value of $\phi$ is equally likely. Other noninformative priors include flat priors (e.g., the prior distribution is uniform on $(-\infty, \infty)$), vague priors (e.g., $\phi \sim G(0, 10,000)$ or $1/\sigma^2 \sim \Gamma(0.001, 0.001)$), and the Jeffrey's prior ($\pi(\phi) \propto I(\phi)^{1/2}$, where $I(\phi)$ is the Fisher information in the model). In the multi-parameter case, the situation is more complex since we need to specify prior distributions that are joint distributions, e.g., with $\phi = (\phi_1, \phi_2)$ the prior distribution of $\phi$ must define the joint distribution of $\phi_1$ and $\phi_2$.

Unfortunately, many practical applications of Bayesian analysis seem to specify noninformative priors when very little information is known about the prior distribution or the impact of the prior distribution is to be minimized. We have to wonder why one would choose a Bayesian analysis in such cases. and caution against the use of the automatic inference for complex models such choices permit. Moreover, many prior distributions, particularly those used in complex hierarchical modeling situations, are chosen for convenience (e.g., multivariate Gaussian, the inverse Gamma) and often, in addition to conditional independence, the hyperparameters are also assumed to be independent. Thus, while there are now many more answers to the question "Where do you get the prior?" the question itself still remains.

### *6.4.2 Fitting Bayesian Models*

Suppose $\phi = (\theta, \tau)'$ with joint posterior distribution $h(\theta, \tau|\mathbf{z})$. Inference about $\theta$ is made from the marginal posterior distribution of $\theta$ obtained by integrating $\tau$ out of the joint posterior distribution. For most realistic models, the posterior distribution is complex and high-dimensional and such integrals are difficult to evaluate. However, in some cases, it is often possible to simulate realizations from the desired distribution. For example, suppose $h(\theta, \tau|\mathbf{z}) = f(\theta|\tau, \mathbf{z})g(\tau|\mathbf{z})$, and suppose it is easy to simulate realizations from $f(\theta|\tau, \mathbf{z})$ and $g(\tau|\mathbf{z})$. Then

$$h(\theta|\mathbf{z}) = \int f(\theta|\tau, \mathbf{z})g(\tau|\mathbf{z})d\tau,$$

and we can obtain a sample $\theta_1, \cdots, \theta_m$ from $h(\theta|\mathbf{z})$ as follows:

1. Given observed data $\mathbf{z}$, generate $\tau^*$ from $g(\tau|\mathbf{z})$;
2. Generate $\theta^*$ from $f(\theta|\tau, \mathbf{z})$;
3. repeat $m$ times.

The pairs $(\theta_1, \tau_1), \cdots, (\theta_m, \tau_m)$ are then a random sample from joint posterior distribution $h(\theta, \tau|\mathbf{z}) = f(\theta|\tau, \mathbf{z})g(\tau|\mathbf{z})$ and $\theta_1, \cdots, \theta_m$ are a random sample from the marginal posterior distribution $h(\theta|\mathbf{z}) = \int f(\theta|\tau, \mathbf{z})g(\tau|\mathbf{z})d\tau$ (see, e.g., Tanner and Wong, 1987; Tanner, 1993). When this is possible, it makes Bayesian inference fairly easy. However, in many applications, the distributions involved are non-standard and sampling from them is difficult at best. Sometimes, it is possible to approximate the desired integrals without having to directly simulate from a specified distribution, e.g., importance sampling (Ripley, 1987; Robert and Casella, 1999) and rejection sampling (Ripley, 1987; Smith and Gelfand, 1992; Tanner, 1993). Taking this one step further, it is also possible to generate a sample, say $\theta_1, \cdots, \theta_m \sim g$, without directly simulating from $g$. However, generating $\theta_1, \cdots, \theta_m$ *independently* is difficult, so attention has focused on generating a *dependent* sample from a specified distribution indirectly, without having to calculate the density or determine an adequate approximation. This is the essence of **Markov chain Monte Carlo (MCMC)**.

A **Markov chain** is a sequence of random variables $\{X_m; m \geq 0\}$ governed by the transition probabilities $\Pr(X_{m+1} \in A|X_1 \ldots X_m) = \Pr(X_{m+1} \in A|X_m) \equiv P(x_m, A)$, so that the distribution of the next value depends only on the present "state" or value. This is called the **Markov property**. An invariant distribution $\pi(x)$ for the Markov chain is a density satisfying $\pi(A) =$

$\int P(x, A)\pi(x)dx$. Markov chain Monte Carlo (MCMC) is a sampling based simulation technique for generating a dependent sample from a specified distribution of interest, $\pi(x)$. Under certain conditions, if we "run" a Markov chain, i.e., we generate $X^{(1)}, X^{(2)}, \cdots$, then

$$X^m \xrightarrow{d} X \sim \pi(x) \text{ as } m \to \infty.$$

Thus, if a long sequence of values is generated, the chain will converge to a stationary distribution, i.e., after convergence, the probability of the chain being in any particular "state" (or taking any particular value) at any particular time remains the same. In other words, after convergence, any sequence of observations from the Markov chain represents a sample from the stationary distribution. To start the chain, a starting value, $X_0$ must be specified. Although the chain will eventually converge to a stationary distribution that does not depend on $X_0$, usually the first $m^*$ iterations are discarded (called the "burn-in") to minimize any potential impact of the starting value on the remaining values in the sequence. Note that the values in the remaining, post-convergence sequence are dependent since each new value still depends on the previous value. Thus, in practice, one selects an independent random sample by using only every $k^{th}$ value appearing in the sequence, where $k$ is large enough to ensure the resulting sample mimics that of a purely random process. Tierney (1994) and Robert and Casella (1999) provide much more rigorous statistical treatments of MCMC sampling and convergence, and Gilks et al. (1996) provide a practical discussion.

### *6.4.2.1 The Gibbs Sampler*

One of the most popular approaches to construct a Markov chain with a specified target density $\pi(x)$ is the **Gibbs sampler**, introduced by Geman and Geman (1984) in image analysis and popularized in statistics by Gelfand and Smith (1990). The Gibbs sampler is perhaps best explained by an example. Suppose we have a model for data vector $\mathbf{Z}$ that has two parameters $\phi_1$ and $\phi_2$. Suppose that the desired posterior distribution, $f(\phi_1, \phi_2|\mathbf{Z})$, is intractable, but we can derive and sample from the full conditional distributions

*Full Conditional Distributions*

$$f(\phi_1|\phi_2, \mathbf{Z})$$
$$f(\phi_2|\phi_1, \mathbf{Z}).$$

Armed with a starting values $\phi_2^{(0)}$ we iterate through the full conditional as follows:

$$\begin{aligned}
&\text{generate } \phi_1^{(1)} \text{ from } f(\phi_1|\phi_2^{(0)}, \mathbf{Z}),\\
&\text{generate } \phi_2^{(1)} \text{ from } f(\phi_2|\phi_1^{(1)}, \mathbf{Z}),\\
&\text{generate } \phi_1^{(2)} \text{ from } f(\phi_1|\phi_2^{(1)}, \mathbf{Z}),\\
&\text{generate } \phi_2^{(2)} \text{ from } f(\phi_2|\phi_1^{(2)}, \mathbf{Z}),\\
&\qquad\vdots
\end{aligned}$$

As we continue to sequentially update the values of $\phi_1$ and $\phi_2$, they will eventually become indistinguishable from samples from the joint posterior distribution $f(\phi_1, \phi_2|\mathbf{Z})$, *provided* such a stationary distribution (Gelfand and Smith, 1990). Casella and George (1992) provide an excellent, elementary treatment of the Gibbs sampler. More theoretical treatments are given in Gelfand and Smith (1990), Smith and Roberts (1993), and Robert and Casella (1999) and practical guidelines for implementation can be found in Gilks (1996).

Besag (1974) showed that knowledge of all full conditional distributions uniquely determines the joint distribution, provided that the joint distribution exists and is proper. Note, however, a collection of proper full conditional distributions does not guarantee the existence of a proper joint distribution (see, e.g., Casella and George, 1992, for a simple, analytical example). Schervish and Carlin (1992) provide general convergence conditions needed for the Gibbs sampler. If Gibbs sampling is used with a set of full conditionals that do not determine a proper joint distribution, the sampler may fail to converge or give nonsensical results. Unfortunately, such situations may not always manifest themselves in such obvious ways, and determining convergence or lack thereof can be tricky; in particularly with complex spatial models.

**Example 6.10** Consider a hierarchical model that builds on the model given in Example 6.8 above. Suppose that

$$\begin{aligned} Z_i|\lambda_i &\overset{ind}{\sim} \text{Poisson}(\lambda_i) \\ \lambda_i &\overset{ind}{\sim} \Gamma(\alpha, \beta_i) \\ \beta_i &\overset{ind}{\sim} \Gamma(a, b) \end{aligned}$$

with $\alpha, a$, and $b$ known. Then, the posterior distribution is

$$h(\boldsymbol{\lambda}, \boldsymbol{\beta}|\mathbf{z}) = \prod_{i=1}^{n} \frac{e^{-\lambda_i}\lambda_i^{z_i}}{z_i!} \times \frac{\beta_i^{\alpha}}{\Gamma(\alpha)}\lambda_i^{\alpha-1}e^{-\beta_i\lambda_i} \times \frac{a^b}{\Gamma(a)}\beta_i^{a-1}e^{-b\beta_i}.$$

Gilks (1996) demonstrates that full conditionals can be determined by simply collecting the terms of the joint posterior that correspond to the parameter of interest. So, to obtain the full conditional of $\lambda_i$, collect the terms involving $\lambda_i$ and this gives

$$h(\lambda_i|\beta_i, \mathbf{z}, \alpha, a, b) \sim \Gamma(z_i + \alpha, 1 + \beta_i),$$

and similarly the full conditional for $\beta_i$ is

$$h(\beta_i|\lambda_i, \mathbf{z}, \alpha, a, b) \sim \Gamma(\alpha + a, \lambda_i + b).$$

Thus, given a starting value for $\lambda_i$ and values for $\alpha, a$, and $b$, $\beta_i$ can be generated from $\Gamma(\alpha + a, \lambda_i + b)$ and then $\lambda_i$ can be generated from $\Gamma(z_i + \alpha, 1 + \beta_i)$ to obtain a sample $(\lambda_i, \beta_i)_1, \cdots, (\lambda_i, \beta_i)_m$. □

### *6.4.2.2 The Metropolis-Hastings Algorithm*

Another popular approach to construct a Markov chain with a specified target density $\pi(x)$ is the **Metropolis-Hastings algorithm.** It was originally developed by Metropolis, Rosenbluth, Teller, and Teller (1953) and generalized by Hastings (1970), whence the name was coined. It is frequently used in geostatistical simulation where it forms the basis of simulated annealing (§7.3, Deutsch and Journel, 1992). Tierney (1994) was among the first to demonstrate its utility in Bayesian computing. The Gibbs sampling algorithm discussed in §6.4.2.1 is a special case of the Metropolis-Hastings algorithm.

The algorithm depends on the creation of what Chib and Greenberg (1995) call a **candidate generating density** $q(x^{(t)}, y)$ that depends on the current state, $x^{(t)}$, and can generate a new sample $x$, e.g., $q(x^{(t)}, y) = G(y - x^{(t)}, \sigma^2\mathbf{I})$. Then, a Markov chain is generated as follows:

1. Repeat for $j = 1, \ldots m$;
2. Draw $y \sim q(\cdot, x^{(j)})$ and $u \sim U(0, 1)$;
3. If $u \leq \frac{\pi(y)}{\pi(x^{(j)})}$ then set $x^{(j+1)} = y$;
4. Otherwise set $x^{(t+1)} = x^{(t)}$.
5. This gives $x^{(1)}, x^{(2)}, \cdots, x^{(m)}$.

Eventually—after the burn-in and under certain mild regularity conditions—the distribution of samples generated by this algorithm will converge to the target distribution, $\pi(x)$.

The key to the algorithm is selection of the candidate generating density, $q(x, y)$. Often, this density will depend on unknown location or scale parameters that must be tuned during the burn-in period. The choice of scale parameter is important since it affects the rate of convergence and the region of the sample space covered by the chain. Typically, this choice is made by specifying an acceptance rate, the percentage of times a move to a new state is made. This rate is often set to 25–50%, but depends on the dimensionality of the problem and the candidate generating density. More implementational details are given in Chib and Greenberg (1995) and Robert and Casella (1999).

### *6.4.2.3 Diagnosing Convergence*

At convergence, the MCMC values should represent a sample from the target distribution of interest. Thus, for large $m$ they should randomly fluctuate around a stable mean value. So, a first diagnostic is to plot the sequence $\phi_t$ against $t$. Another approach is to use multiple chains with different starting values. This allows us to compare results from different replications; for example, Gelman and Rubin (1992) suggest convergence statistics based on the ratio of between-chain variances to within-chain variances. However, convergence to a stationary distribution may not be the only concern. If we are interested in functions of the parameters, we might be more interested in convergence of averages of the form $1/m \sum g(\phi_t)$. Another consideration is the dependence in the samples. If independent samples are required for inference, we can obtain these by subsampling the chain, but then we need to assess how close the resulting sample is to being independent. There have been many different convergence diagnostics proposed in the literature and Cowles and Carlin (1996) provide a comprehensive, comparative review. Many of these, and others proposed more recently, are described and illustrated in Robert and Casella (1999).

#### *6.4.2.4 Summarizing Posterior Distributions*

Summaries of the post-convergence MCMC samples provide posterior inference for model parameters. For instance, the sample mean of the (post-convergence) sampled values for a particular model parameter provides an estimate of the marginal posterior mean and a point estimate of the parameter itself. Other measures of central tendency such as the median might also be used as point estimates. The utility of the simulation approach is that it provides an estimate of the entire posterior distribution which allows us to obtain interval estimates as well. For example, the 2.5th and 97.5th quantiles of the (post-convergence) sampled values for a model parameter provides a 95% interval estimate of the parameter. In Bayesian inference, such an interval is termed a **credible set** to distinguish it from the confidence interval in frequentist statistics. Unlike confidence intervals, credibility intervals have a direct probabilistic interpretation: A 95% credible set defines an interval having a 0.95 posterior probability of containing the parameter of interest.

A comprehensive discussion of the theory or the practice of Bayesian hierarchical modeling and MCMC is beyond the scope of this text. Instead, we have provided a summary that will allow us to understand how such models might be used in spatial data analysis. A discussion of the details, nuances, considerations in implementation, and illustrative examples can be found in Gilks et al. (1996), Robert and Casella (1999), Carlin and Louis (2000), Congdon (2001, 2003), Banerjee, Carlin and Gelfand (2003), and Gelman et al. (2004).

### *6.4.3 Selected Spatial Models*

At this point, there has been nothing explicitly spatial about our discussion and the examples provided were basically aspatial. It is difficult to give a general treatment of Bayesian hierarchical models, in general, and their use in spatial data analysis, in particular, since each application can lead to a unique model. Instead, we give an overview of several general types of models that have been useful in spatial data analysis. More concrete examples can be found in e.g., Carlin and Louis (2000) and Banerjee, Carlin, and Gelfand (2003).

*6.4.3.1 Linear Regression and Bayesian Kriging*

Consider the simple linear model

$$\mathbf{Z}(\mathbf{s}) = \mathbf{X}(\mathbf{s})\boldsymbol{\beta} + \boldsymbol{\epsilon}(\mathbf{s}),$$

with $\mathrm{E}[\boldsymbol{\epsilon}(\mathbf{s})] = \mathbf{0}$ and $\mathrm{Var}[\boldsymbol{\epsilon}(\mathbf{s})] = \mathbf{R}$, where $\mathbf{R}$ is assumed known. A simple hierarchical rendition is

$$\begin{aligned} [\mathbf{Z}(\mathbf{s})|\boldsymbol{\beta}] &\sim G(\mathbf{X}(\mathbf{s})\boldsymbol{\beta}, \mathbf{R}) \\ [\boldsymbol{\beta}] &\sim G(\mathbf{m}, \mathbf{Q}), \end{aligned} \tag{6.94}$$

with $\mathbf{m}$ and $\mathbf{Q}$ known. We assume that $\mathbf{Q}$ and $(\mathbf{X}(\mathbf{s})'\mathbf{R}^{-1}\mathbf{X}(\mathbf{s}))$ are both of full rank. The posterior distribution, $h(\boldsymbol{\beta}|\mathbf{Z}(\mathbf{s}))$, is

$$\begin{aligned} h(\boldsymbol{\beta}|\mathbf{Z}(\mathbf{s})) &\propto f(\mathbf{Z}(\mathbf{s})|\boldsymbol{\beta})\pi(\boldsymbol{\beta}) \\ &\propto \exp\left\{-\frac{1}{2}\left[(\mathbf{Z}(\mathbf{s}) - \mathbf{X}(\mathbf{s})\boldsymbol{\beta})'\mathbf{R}^{-1}(\mathbf{Z}(\mathbf{s}) - \mathbf{X}(\mathbf{s})\boldsymbol{\beta})\right.\right. \\ &\quad \left.\left. + (\boldsymbol{\beta} - \mathbf{m})'\mathbf{Q}^{-1}(\boldsymbol{\beta} - \mathbf{m})\right]\right\} \\ &= \exp\left\{-\frac{1}{2}\left[\boldsymbol{\beta}'(\mathbf{Q}^{-1} + \mathbf{X}(\mathbf{s})'\mathbf{R}^{-1}\mathbf{X}(\mathbf{s}))\boldsymbol{\beta}\right.\right. \\ &\quad - \boldsymbol{\beta}'(\mathbf{Q}^{-1}\mathbf{m} + \mathbf{X}(\mathbf{s})'\mathbf{R}^{-1}\mathbf{Z}(\mathbf{s})) \\ &\quad - (\mathbf{Q}^{-1}\mathbf{m} + \mathbf{X}(\mathbf{s})'\mathbf{R}^{-1}\mathbf{Z}(\mathbf{s}))'\boldsymbol{\beta} \\ &\quad \left.\left. + \mathbf{m}'\mathbf{Q}^{-1}\mathbf{m} + \mathbf{Z}(\mathbf{s})\mathbf{R}^{-1}\mathbf{Z}(\mathbf{s})\right]\right\} \\ &\propto \exp\left\{-\frac{1}{2}(\boldsymbol{\beta} - \mathbf{m}^*)'(\mathbf{Q}^*)^{-1}(\boldsymbol{\beta} - \mathbf{m}^*)\right\} \end{aligned}$$

with

$$\begin{aligned} \mathbf{m}^* &= (\mathbf{Q}^{-1} + \mathbf{X}(\mathbf{s})'\mathbf{R}^{-1}\mathbf{X}(\mathbf{s}))^{-1}(\mathbf{X}(\mathbf{s})'\mathbf{R}^{-1}\mathbf{Z}(\mathbf{s}) + \mathbf{Q}^{-1}\mathbf{m}) \\ \mathbf{Q}^* &= (\mathbf{Q}^{-1} + \mathbf{X}(\mathbf{s})'\mathbf{R}^{-1}\mathbf{X}(\mathbf{s}))^{-1}. \end{aligned}$$

Thus, $h(\boldsymbol{\beta}|\mathbf{Z}(\mathbf{s})) \propto G(\mathbf{m}^*, \mathbf{Q}^*)$. Note that if $\mathbf{Q}^{-1} = \mathbf{0}$, $\mathbf{m}^*$ reduces to the generalized least squares estimator of (the now fixed effects) $\boldsymbol{\beta}$ with variance-covariance matrix $\mathbf{Q}^*$. This is also true if we assume independent vague or uniform priors for the elements of $\boldsymbol{\beta}$. If, on the other hand, $\mathbf{Q}^{-1}$ is much larger than $(\mathbf{X}(\mathbf{s})'\mathbf{R}^{-1}\mathbf{X}(\mathbf{s}))$, the prior information dominates the data and $\mathbf{m}^* \doteq \mathbf{m}$, and $\mathbf{Q}^* \doteq \mathbf{Q}$.

Now consider the prediction of a $k \times 1$ vector of new observations, $\mathbf{Z}(\mathbf{s}_0)$. For this we assume $\mathbf{Z}(\mathbf{s}_0) = \mathbf{X}(\mathbf{s}_0)\boldsymbol{\beta} + \boldsymbol{\epsilon}(\mathbf{s}_0)$ and that

$$\begin{bmatrix} \mathbf{Z}(\mathbf{s}) \\ \mathbf{Z}(\mathbf{s}_0) \end{bmatrix} \sim G\left(\begin{bmatrix} \mathbf{X}(\mathbf{s}) \\ \mathbf{X}(\mathbf{s}_0) \end{bmatrix}\boldsymbol{\beta}, \begin{bmatrix} \mathbf{R}_{zz} & \mathbf{R}_{z0} \\ \mathbf{R}'_{0z} & \mathbf{R}_{00} \end{bmatrix}\right). \tag{6.95}$$

For spatial prediction with the hierarchical model in (6.94), we require the **predictive distribution** of $\mathbf{Z}(\mathbf{s}_0)$, the marginal distribution of $\mathbf{Z}(\mathbf{s}_0)$ given both the data and the prior information,

$$p(\mathbf{Z}(\mathbf{s}_0)|\mathbf{Z}(\mathbf{s})) = \int_{\beta} f(\mathbf{Z}(\mathbf{s}_0)|\mathbf{Z}(\mathbf{s}), \beta) f(\beta)\, d\beta.$$

*Predictive Distribution*

The mean of this distribution can be obtained as

$$\begin{aligned} \mathrm{E}[\mathbf{Z}(\mathbf{s}_0)|\mathbf{Z}(\mathbf{s})] &= \mathrm{E}_{\beta}[\mathrm{E}(\mathbf{Z}(\mathbf{s}_0)|\mathbf{Z}(\mathbf{s}); \beta] \\ &= \int_{\beta} [\mathbf{X}(\mathbf{s}_0)\beta + \mathbf{R}_{0z}\mathbf{R}_{zz}^{-1}(\mathbf{Z}(\mathbf{s}) - \mathbf{X}(\mathbf{s})\beta)] f(\beta)\, d\beta, \end{aligned}$$

and the variance can be obtained using

$$\mathrm{Var}[\mathbf{Z}(\mathbf{s}_0)|\mathbf{Z}(\mathbf{s})] = \mathrm{E}_{\beta}[\mathrm{Var}(\mathbf{Z}(\mathbf{s}_0)|\mathbf{Z}(\mathbf{s}); \beta)] + \mathrm{Var}_{\beta}[\mathrm{E}(\mathbf{Z}(\mathbf{s}_0)|\mathbf{Z}(\mathbf{s}); \beta)].$$

For the model in (6.94) these results give (Kitanidis, 1986)

$$\begin{aligned} \mathrm{E}[\mathbf{Z}(\mathbf{s}_0)|\mathbf{Z}(\mathbf{s})] &= (\mathbf{X}(\mathbf{s}_0) - \mathbf{R}_{0z}\mathbf{R}_{zz}^{-1}\mathbf{X}(\mathbf{s}))(\mathbf{Q}^{-1} + \mathbf{X}(\mathbf{s})'\mathbf{R}_{zz}^{-1}\mathbf{X}(\mathbf{s}))^{-1}\mathbf{Q}^{-1}\mathbf{m}' \\ &\quad + [\mathbf{R}_{0z}\mathbf{R}_{zz}^{-1} + (\mathbf{X}(\mathbf{s}_0) - \mathbf{R}_{0z}\mathbf{R}_{zz}^{-1}\mathbf{X}(\mathbf{s})) \\ &\quad \times (\mathbf{Q}^{-1} + \mathbf{X}(\mathbf{s})'\mathbf{R}_{zz}^{-1}\mathbf{X}(\mathbf{s}))^{-1}\mathbf{X}(\mathbf{s})'\mathbf{R}_{zz}^{-1}]\mathbf{Z}(\mathbf{s}), \end{aligned}$$

and

$$\begin{aligned} \mathrm{Var}[\mathbf{Z}(\mathbf{s}_0)|\mathbf{Z}(\mathbf{s})] &= \mathbf{R}_{00} - \mathbf{R}_{0z}\mathbf{R}_{zz}^{-1}\mathbf{R}_{z0} + (\mathbf{X}(\mathbf{s}_0) - \mathbf{R}_{0z}\mathbf{R}_{zz}^{-1}\mathbf{X}(\mathbf{s})) \\ &\quad \times (\mathbf{Q}^{-1} + \mathbf{X}(\mathbf{s})'\mathbf{R}_{zz}^{-1}\mathbf{X}(\mathbf{s}))^{-1}(\mathbf{X}(\mathbf{s}_0) - \mathbf{R}_{0z}\mathbf{R}_{zz}^{-1}\mathbf{X}(\mathbf{s}))'. \end{aligned}$$

If $\mathbf{Q}^{-1} = \mathbf{0}$, (or we assume independent vague or uniform priors for the elements of $\beta$) these reduce to the universal kriging predictor and its variance, given in (5.29) ad (5.30), respectively.

Consider slightly relaxing the assumption that $\mathbf{R}$ is completely known and assume $\mathbf{R} = \sigma^2\mathbf{V}$, where $\mathbf{V}$ is known, but $\sigma^2$ is unknown. Now there are two types of unknown parameters to consider, $\beta$ and $\sigma^2$. The natural conjugate family of prior distributions is the **normal-inverse-Gamma** distribution (O'Hagan, 1994, p. 246), defined as

$$\begin{aligned} \pi(\beta, \sigma^2) &= \frac{(a/2)^{d/2}}{(2\pi)^{p/2}|\mathbf{Q}|^{1/2}\Gamma(d/2)}(\sigma^2)^{-(d+p+2)/2} \\ &\quad \times \exp\left\{-\frac{1}{2\sigma^2}\left[(\beta - \mathbf{m})'\mathbf{Q}^{-1}(\beta - \mathbf{m}) + a\right]\right\}, \end{aligned}$$

with hyperparameters $a, d, \mathbf{m}$, and $\mathbf{Q}$. This distribution is denoted $NIG(a, d, \mathbf{m}, \mathbf{Q})$. The hierarchical model is now

$$\begin{aligned} [\mathbf{Z}(\mathbf{s})|\beta, \sigma^2] &\sim G(\mathbf{X}(\mathbf{s})\beta, \sigma^2\mathbf{V}) \\ [\beta, \sigma^2] &\sim NIG(a, d, \mathbf{m}, \mathbf{Q}), \end{aligned} \tag{6.96}$$

with $\mathbf{V}, a, d, \mathbf{m}$, and $\mathbf{Q}$ known. Another way of writing this model is

$$\begin{aligned} [\mathbf{Z}(\mathbf{s})|\beta, \sigma^2] &\sim G(\mathbf{X}(\mathbf{s})\beta, \sigma^2\mathbf{V}) \\ [\beta|] &\sim G(\mathbf{m}, \sigma^2\mathbf{Q}) \\ [\sigma^2] &\sim IG(a, d), \end{aligned}$$

where $IG(a, d)$ denotes the inverse Gamma distribution with density

$$\pi(\sigma^2) = \frac{(a/2)^{d/2}}{\Gamma(d/2)}(\sigma^2)^{-(d+2)/2} \exp\left\{-a/(2\sigma^2)\right\},$$

and mean $a/(d-2)$. Then $\sigma^{-2}$ has a Gamma distribution with parameters $(d/2, a/2)$ and mean $d/a$. The joint posterior distribution is $h(\boldsymbol{\beta}, \sigma^2|\mathbf{Z}(\mathbf{s})) = f(\mathbf{Z}(\mathbf{s})|\boldsymbol{\beta}, \sigma^2)\pi(\boldsymbol{\beta}, \sigma^2)$, but inferences about $\boldsymbol{\beta}$ must be made from $h(\boldsymbol{\beta}|\mathbf{Z}(\mathbf{s}))$, obtained by integrating out $\sigma^2$ from the joint posterior. We can obtain the moments of this distribution using results on conditional expectations. Conditioning on $\sigma^2$ is the same as assuming $\sigma^2$ is known, so this mean is equivalent to that derived above. Thus, the posterior mean is $\mathrm{E}[\boldsymbol{\beta}|\mathbf{Z}(\mathbf{s})] = \mathrm{E}[\mathrm{E}(\boldsymbol{\beta}|\sigma^2, \mathbf{Z}(\mathbf{s}))] = \mathbf{m}^*$. However, the variance is $\mathrm{Var}[\boldsymbol{\beta}|\mathbf{Z}(\mathbf{s})] = \mathrm{E}[\sigma^2|\mathbf{Z}(\mathbf{s})]\mathbf{Q}^*$. The posterior mean of $\sigma^2$, $\mathrm{E}[\sigma^2|\mathbf{Z}(\mathbf{s})]$, can be derived using similar arguments (and much more algebra). O'Hagan (1994, p. 249) shows that it is a linear combination of three terms: the prior mean, $\mathrm{E}(\sigma^2)$, the usual residual sum of squares, and a term that compares $\widehat{\boldsymbol{\beta}}_{gls}$ to $\mathbf{m}$. The real impact of the prior distribution, $\pi(\boldsymbol{\beta}, \sigma^2)$, is that the posterior density function of $\boldsymbol{\beta}$ follows a multivariate $t$-distribution with mean $\mathbf{m}^*$, scale $a^*\mathbf{Q}^*$ and $d+n$ degrees of freedom, where $a^* = a + \mathbf{m}'\mathbf{Q}^{-1}\mathbf{m} + \mathbf{Z}(\mathbf{s})'\mathbf{V}^{-1}\mathbf{Z}(\mathbf{s}) - (\mathbf{m}^*)'\mathbf{Q}^{*-1}\mathbf{m}^*$ (Judge et al., 1985; O'Hagan, 1994).

We chose the $NIG(a, d, \mathbf{m}, \mathbf{Q})$ as a prior distribution for $(\boldsymbol{\beta}, \sigma^2)$ since the conjugacy leads to tractable (although messy) analytical results. However, this distribution specifies a particular relationship between $\boldsymbol{\beta}$ and $\sigma^2$. Often, we will have no reason to assume these two parameters are related. Thus, instead we could assume $f(\boldsymbol{\beta}, \sigma^2) = f(\boldsymbol{\beta})f(\sigma^2)$, (independence), and then we are free to independently choose priors for $\boldsymbol{\beta}$ and $\sigma^2$. Common choices include an improper uniform prior for $\boldsymbol{\beta}$ and an inverse Gamma distribution for $\sigma^2$. If $\mathbf{Q}^{-1} = \mathbf{0}$ (corresponding to an improper uniform prior for $\boldsymbol{\beta}$), and we choose $a = 0$, then $\mathbf{m}^* = \widehat{\boldsymbol{\beta}}_{gls}$, and $\mathrm{E}[\sigma^2|\mathbf{Z}(\mathbf{s})] = \{(n-p)/(d+n-2)\widehat{\sigma}^2\}$, with $\widehat{\sigma}^2 = (\mathbf{Z}(\mathbf{s}) - \mathbf{X}(\mathbf{s})\widehat{\boldsymbol{\beta}}_{gls})'\mathbf{V}^{-1}(\mathbf{Z}(\mathbf{s}) - \mathbf{X}(\mathbf{s})\widehat{\boldsymbol{\beta}}_{gls})/(n-p)$. Thus, the posterior distribution of $\boldsymbol{\beta}$ is $\widehat{\boldsymbol{\beta}}_{gls}$ gives the classical inference for $\boldsymbol{\beta}$.

The predictive distribution of $\mathbf{Z}(\mathbf{s}_0)$ given $\mathbf{Z}(\mathbf{s})$ is derived analogously to that given above, and Kitanidis (1986) and Le and Zidek (1992) show that it has a multivariate $t$-distribution with mean

$$\begin{aligned}
\mathrm{E}[\mathbf{Z}(\mathbf{s}_0)|\mathbf{Z}(\mathbf{s})] &= (\mathbf{X}(\mathbf{s}_0) - \mathbf{V}_{0z}\mathbf{V}_{zz}^{-1}\mathbf{X}(\mathbf{s}))(\mathbf{Q}^{-1} + \mathbf{X}(\mathbf{s})'\mathbf{V}_{zz}^{-1}\mathbf{X}(\mathbf{s}))^{-1}\mathbf{Q}^{-1}\mathbf{m}' \\
&\quad + [\mathbf{V}_{0z}\mathbf{V}_{zz}^{-1} + (\mathbf{X}(\mathbf{s}_0) - \mathbf{V}_{0z}\mathbf{V}_{zz}^{-1}\mathbf{X}(\mathbf{s})) \\
&\quad \times (\mathbf{Q}^{-1} + \mathbf{X}(\mathbf{s})'\mathbf{V}_{zz}^{-1}\mathbf{X}(\mathbf{s}))^{-1}\mathbf{X}(\mathbf{s})'\mathbf{V}_{zz}^{-1}]\mathbf{Z}(\mathbf{s}), \qquad (6.97)
\end{aligned}$$

and variance

$$\begin{aligned}
\mathrm{Var}[\mathbf{Z}(\mathbf{s}_0)|\mathbf{Z}(\mathbf{s})] &= \left[\mathbf{V}_{00} - \mathbf{V}_{0z}\mathbf{V}_{zz}^{-1}\mathbf{V}_{z0} + (\mathbf{X}(\mathbf{s}_0) - \mathbf{V}_{0z}\mathbf{V}_{zz}^{-1}\mathbf{X}(\mathbf{s}))\right. \\
&\quad \times (\mathbf{Q}^{-1} + \mathbf{X}(\mathbf{s})'\mathbf{V}_{zz}^{-1}\mathbf{X}(\mathbf{s}))^{-1} \\
&\quad \left.\times (\mathbf{X}(\mathbf{s}_0) - \mathbf{V}_{0z}\mathbf{V}_{zz}^{-1}\mathbf{X}(\mathbf{s}))'\right]\nu^*, \qquad (6.98)
\end{aligned}$$

where

$$\nu^* = \frac{a + (n-p)\hat{\sigma}^2 + (\hat{\boldsymbol{\beta}}_{gls} - \mathbf{m})'(\mathbf{X}(\mathbf{s})'\mathbf{V}_{zz}^{-1}\mathbf{X}(\mathbf{s}))(\hat{\boldsymbol{\beta}}_{gls} - \mathbf{m})}{n + d - 2}.$$

Finally, consider the more general case with $\mathbf{R} = \sigma^2 \mathbf{V}(\boldsymbol{\theta})$, where $\boldsymbol{\theta}$ characterizes the spatial autocorrelation in the data. Now we now need to assume something for the prior distribution $\pi(\boldsymbol{\beta}, \sigma^2, \boldsymbol{\theta})$ For convenience, Handcock and Stein (1993) choose $\pi(\boldsymbol{\beta}, \sigma^2, \boldsymbol{\theta}) \propto \pi(\boldsymbol{\theta})/\sigma^2$ and use the Matérn class of covariance functions to model the dependence of $\mathbf{V}$ on $\boldsymbol{\theta}$, choosing uniform priors for both the smoothness and the range parameters in this class. In this case, and for other choices of $\pi(\boldsymbol{\beta}, \sigma^2, \boldsymbol{\theta})$, analytical derivation of the predictive distribution becomes impossible. Computational details for the use of importance sampling or MCMC methods for a general $\pi(\boldsymbol{\theta})$ and any parametric model for $\mathbf{V}(\boldsymbol{\theta})$ are given in Gaudard et al. (1999).

Unfortunately, and surprisingly, many (arguably most) of the noninformative priors commonly used for spatial autocorrelation (including the one used by Handcock and Stein described above) can lead to improper posterior distributions (Berger, De Oliveira, and Sansó, 2001). Berger et al. (2001) provide tremendous insight into how such choices should be made and provide a flexible alternative (based on a reference prior approach) that always produces proper prior distributions. However, this prior has not been widely used in spatial applications, so its true impacts on Bayesian modeling of spatial data are unknown.

#### *6.4.3.2 Smoothing Disease Rates*

One of the most popular uses of Bayesian hierarchical modeling of spatial data is in disease mapping. In this context, for each of $i = 1, \cdots, N$ geographic regions, let $Z(\mathbf{s}_i)$ be the number of incident cases of disease, where $\mathbf{s}_i$ is a generic spatial index for the $i^{th}$ region. Suppose that we also have available the number of incident cases expected in each region based on the population size and demographic structure within region $i$ $(E(\mathbf{s}_i), i = 1, \ldots, N)$; often the $E(\mathbf{s}_i)$ reflect age-standardized values. The $Z(\mathbf{s}_i)$ are random variables and we assume the $E(\mathbf{s}_i)$ are fixed values. In many applications where information on $E(\mathbf{s}_i)$ is unknown, $n(\mathbf{s}_i)$, the population at risk in region $i$ is used instead, as in Example 1.4.

To allow for the possibility of region-specific risk factors in addition to those defining each $E(\mathbf{s}_i)$, Clayton and Kaldor (1987) propose a set of region-specific relative risks $\zeta(\mathbf{s}_i), i = 1, \cdots, n$, and define the first stage of a hierarchical model through

$$[Z(\mathbf{s}_i)|\zeta(\mathbf{s}_i)] \overset{ind}{\sim} \text{Poisson}(E(\mathbf{s}_i)\zeta(\mathbf{s}_i)). \tag{6.99}$$

Thus $\mathrm{E}[Z(\mathbf{s}_i)|\zeta(\mathbf{s}_i)] = E(\mathbf{s}_i)\zeta(\mathbf{s}_i)$, so that the $\zeta(\mathbf{s}_i)$ represent an additional multiplicative risk associated with region $i$, not already accounted for in the calculation of $E(\mathbf{s}_i)$.

The ratio of observed to expected counts, $Z(\mathbf{s}_i)/E(\mathbf{s}_i)$ corresponds to the local standardized mortality ratio (SMR) and is the maximum likelihood estimate of the relative risk, $\zeta(\mathbf{s}_i)$, experienced by individuals residing in region $i$. Instead of SMR, the same type of model can be used with rates, $Z(\mathbf{s}_i)/n_i$, where $n_i$ is the population at risk in region $i$.

We can incorporate fixed and random effects within the relative risk parameter if we include covariates and let

$$\log\{\zeta(\mathbf{s}_i)\} = \mathbf{x}(\mathbf{s}_i)'\boldsymbol{\beta} + \psi(\mathbf{s}_i), \tag{6.100}$$

so that the model is

$$[Z(\mathbf{s}_i)|\boldsymbol{\beta}, \psi(\mathbf{s}_i)] \overset{ind}{\sim} \text{Poisson}(E(\mathbf{s}_i)\exp\{\mathbf{x}(\mathbf{s}_i)'\boldsymbol{\beta} + \psi(\mathbf{s}_i)\}). \tag{6.101}$$

Note that this is the same model used in Example 1.4, with $\psi(\mathbf{s}_i)$ representing $S(\mathbf{s}_i)$.

The next step is to specify distributions for the parameters $\boldsymbol{\beta}$ and $\psi$. For the fixed effects $\boldsymbol{\beta}$, noninformative priors are often used (e.g., uniform, or Gaussian with very wide prior variance). Since the elements comprising $\boldsymbol{\beta}$ are continuous parameters potentially taking values anywhere in $[-\infty, \infty]$ the uniform prior distribution is an improper distribution, i.e., its probability density function does not integrate to one. In many cases when the joint likelihood is sufficiently well-defined, the posterior distribution arising from an improper prior distribution will be a proper distribution, but care must be taken when using improper priors, in general, and in spatial modeling in particular.

We also have several choices for the prior distribution of the random effects $\psi(\mathbf{s}_i)$. A common assumption is

$$\psi(\mathbf{s}_i) \overset{ind}{\sim} G(0, \sigma^2_\psi), \quad i = 1, \cdots, N. \tag{6.102}$$

With this choice of prior, the $\psi(\mathbf{s}_i)$ do not depend on location $i$ and are said to be *exchangeable*. The effect of this choice is to add excess (but not spatially structured) variation to the model in equation (6.100), and hence to the model in equation (6.101). As we saw in §6.3.4, even though the prior distributions are essentially aspatial, they do induce some similarity among the observations. Exploiting this similarity to obtain better predictions of the relative risks is the concept of "borrowing strength" across observations commonly referred to in discussions of smoothing disease rates.

Estimating the remaining model parameter $\sigma^2_\psi$ (or the parameters from other choices of the prior for $\psi$) from the data, leads to empirical Bayes inference often used in smoothing disease rates prior to mapping (e.g., Clayton and Kaldor, 1987; Marshall, 1991; Cressie, 1992; Devine et al., 1994). For fully Bayes inference, we complete the hierarchy by defining a hyperprior distribution for $\sigma^2_\psi$. This is often taken to be a conjugate prior; for the variance parameter of a Gaussian distribution, the inverse Gamma distribution provides the conjugate family.

Most applications complete the model at this point, assigning fixed values to the two parameters of the inverse Gamma hyperprior. There has been much recent discussion on valid noninformative choices for these parameters since the inverse Gamma distribution is only defined for positive values and zero is a degenerate value (see e.g., Kelsall and Wakefield, 1999 and Gelman et al., 2004, Appendix C). Ghosh et al. (1999) and Sun et al. (1999) define conditions on the inverse Gamma parameters to ensure a proper posterior. In our experience, convergence of MCMC algorithms can be very sensitive to choices for these parameters, so they must be selected carefully.

To more explicitly model spatial autocorrelation, we can use the prior distribution for $\psi$ to allow a spatial pattern among the $\psi(\mathbf{s}_i)$, e.g., through a parametric covariance function linking pairs $\psi(\mathbf{s}_i)$ and $\psi(\mathbf{s}_j)$ for $j \neq i$. Thus, we could consider a joint multivariate Gaussian prior distribution for $\psi$ with spatial covariance matrix $\mathbf{\Sigma}_\psi$, i.e.,

$$\psi \sim G(\mathbf{0}, \mathbf{\Sigma}_\psi), \tag{6.103}$$

and, following the geostatistical paradigm, define the elements of $\mathbf{\Sigma}_\psi$ through a parametric covariance function, e.g., such as in Handcock and Stein (1993) discussed above. A more popular approach is to use a conditionally specified prior spatial structure for $\psi$ similar to the conditional autoregressive models introduced in §6.2.2.2. With a CAR prior, we define the distribution of $\psi$ conditionally, i.e.,

$$\psi(\mathbf{s}_i)|\psi(\mathbf{s}_{j\neq i}) \sim G\left(\sum_{i=1}^{N} c_{ij}\psi(\mathbf{s}_j), \tau_i^2\right), i = 1, \cdots, N,$$

which is equivalent to

$$\psi \sim G(\mathbf{0}, (\mathbf{I}-\mathbf{C})\tau^2\mathbf{M}), \tag{6.104}$$

where $\mathbf{M} = \text{diag}(\tau_1^2, \cdots, \tau_N^2)$. As in §6.2.2.2, the $c_{ij}$s denote spatial dependence parameters and we set $c_{ii} = 0$ for all $i$. Typical applications take $\mathbf{M} = \tau^2\mathbf{I}$ and consider adjacency-based weights where $c_{ij} = \phi$ if region $j$ is adjacent to region $i$, and $c_{ij} = 0$, otherwise. Clayton and Kaldor (1987) propose priors in an empirical Bayes setting, and Breslow and Clayton (1993) apply CAR priors as random effects distributions within likelihood approximations for generalized linear mixed models.

For a fully Bayesian approach, we need to specify a prior for $[\phi, \tau^2]$, e.g.,

$$\begin{aligned}
\pi(\phi, \tau^2) &= \pi(\phi)\pi(\tau^2) \\
\pi(\tau^2) &\sim IG(a, b) \\
\pi(\phi) &\sim U(\gamma_{min}^{-1}, \gamma_{max}^{-1}),
\end{aligned}$$

where $\gamma_{min}^{-1}$ and $\gamma_{max}^{-1}$ are the smallest and largest eigenvalues of $\mathbf{W}$, where $\mathbf{C} = \phi\mathbf{W}$ and $\mathbf{W}$ is a spatial proximity matrix. (e.g., Besag et al., 1991; Stern and Cressie, 1999).

There are many issues surrounding the use of CAR priors, e.g., choices for $c_{ij}$ that lead to desired interpretations (cf., intrinsic autoregression of Besag and Kooperberg, 1995), the "improperness" that results from the pairwise construction of CAR models and the use of constraints on $\psi$ to assure proper posterior distributions, etc. (see, e.g., Besag et al., 1991; Besag et al., 1995).

#### *6.4.3.3 Mapping Non-Gaussian Data*

When the data are not Gaussian, the optimal predictor of the data at new locations, $E[\mathbf{Z}(s_0)|\mathbf{Z}(s)]$, may not be linear in the data and so linear predictors (like the universal kriging predictor described in §5.3.3 and the Bayesian kriging predictors described above in §6.4.3.1) may be poor approximations to this optimal conditional expectation. Statisticians typically address this problem in one of two ways, either through transformation as described in §5.6 or through the use of generalized linear models described in §6.3. Just as there is a Bayesian hierarchical model formulation for universal kriging, there are also Bayesian hierarchical model formulations for trans-Gaussian (§5.6.2) and indicator kriging (§5.6.3) and for spatial GLMMs (§6.3.4).

De Oliveira et al. (1997) extend the ideas given in §6.4.3.1 to the case of transformed Gaussian fields. Consider data $\mathbf{Z}(s)$, and a parametric family of monotone transformations, $\{g_\lambda(\cdot)\}$, for which $g_\lambda(\mathbf{Z}(s))$ is Gaussian. In contrast to the methods described in §5.6.2, De Oliveira et al. (1997) assume $\lambda$ is unknown. On the transformed scale, $(g_\lambda(\mathbf{Z}(s)), g_\lambda(\mathbf{Z}(s_0))'$ is assumed to follow the same multivariate Gaussian distribution as in (6.95), with $\mathbf{R} = \sigma^2\mathbf{V}(\boldsymbol{\theta})$. The idea is to build a hierarchical model that will incorporate uncertainty in the sampling distribution through a prior distribution on $\lambda$. Following ideas in Box and Cox (1964), De Oliveira et al. (1997) choose the prior

$$\pi(\boldsymbol{\beta}, \sigma^2, \boldsymbol{\theta}, \lambda) \propto \sigma^2 \pi(\boldsymbol{\theta}) / J_\lambda^{p/n},$$

where $J_\lambda = \prod |g'_\lambda(z(s_i))|$, and then assume (bounded) uniform priors for $\lambda$ and each component of $\boldsymbol{\theta}$.

The predictive distribution is obtained from

$$p(\mathbf{Z}(s_0)|\mathbf{Z}(s)) = \int f(\mathbf{Z}(s_0)|\mathbf{Z}(s), \boldsymbol{\phi}) f(\boldsymbol{\phi}|\mathbf{Z}(s)) d\boldsymbol{\phi}, \qquad (6.105)$$

where $\boldsymbol{\phi} = (\boldsymbol{\beta}, \sigma^2, \boldsymbol{\theta}, \lambda)'$. By first fixing $\boldsymbol{\theta}$ and $\lambda$, De Oliveira et al. (1997) make repeated use of the results for linear models given in §6.4.3.1 to construct $f(\boldsymbol{\phi}|\mathbf{Z}(s))$, and to integrate out $\boldsymbol{\beta}$ and $\sigma^2$ from (6.105), using the fact that $p(g_\lambda(\mathbf{Z}(s_0))|\mathbf{Z}(s), \boldsymbol{\theta}, \lambda)$ has a multivariate $t$-distribution with mean and variance similar to those given in equations (6.97) and (6.98). This leaves (6.105) as a function of $\boldsymbol{\theta}$ and $\lambda$ and De Oliveira et al. (1997) construct a Monte Carlo algorithm to integrate out these parameters.

De Oliveira et al. (1997) compare their approach, which they call the Bayesian transformed Gaussian model (BTG), to the more traditional trans-Gaussian kriging approach (TGK). They use a cross-validation study based on rainfall amounts measured at 24 stations near Darwin, Australia. Their results indicate that both approaches were about as accurate, but the empirical probability coverages based on 95% prediction intervals from the BTG model were much closer to nominal than those from trans-Gaussian kriging. BTG intervals covered 91.6% of the 24 true rainfall values, while TGK covered 75%, although the kriging prediction intervals were based on $z$-scores rather than $t$-scores. Thus, the BTG model is another approach to adjust for use of the plug-in variance in kriging in small samples (in addition to that of Kackar and Harville and Kenward and Roger discussed in §5.5.4).

Kim and Mallick (2002) adopt a slightly different approach to the problem of spatial prediction for non-Gaussian data. They utilize the multivariate skew-Gaussian (SG) distribution developed by Azzalini and Capitanio (1999). The SG distribution is based on the multivariate Gaussian distribution, but includes an extra parameter for shape. Kim and Mallick (2002) construct a hierarchical model very similar to the most general model described in §6.4.3.1, but with the likelihood based on a skew-Gaussian distribution rather than a Gaussian distribution. Because the multivariate Gaussian distribution forms the basis for the SG distribution, the SG distribution has many similar properties and the ideas and results in §6.4.3.1 can be used to obtain all full conditionals needed for Gibbs sampling.

These approaches assume the data are continuous and that there is no relationship between the mean and the variance. In a generalized linear modeling context, Diggle et al. (1998) consider the model in §6.3.4 where given an underlying, smooth, spatial process $\{S(\mathbf{s}) : \mathbf{s} \in D\}$, $Z(\mathbf{s})$ has distribution in the exponential family. Thus,

$$\begin{aligned} \mathrm{E}[Z(\mathbf{s})|\mathbf{S}] &\equiv \mu(\mathbf{s}), \\ g[\mu(\mathbf{s})] &= \mathbf{x}(\mathbf{s})'\boldsymbol{\beta} + S(\mathbf{s}), \end{aligned} \tag{6.106}$$

and we assume $S(\mathbf{s})$ is a Gaussian random field with mean 0 and covariance function $\sigma_S^2 \rho_S(\mathbf{s}_i - \mathbf{s}_j; \boldsymbol{\theta})$.

Thus, in a hierarchical specification

$$\begin{aligned} [\mathbf{Z}(\mathbf{s})|\boldsymbol{\beta}, \mathbf{S}] &\overset{ind}{\sim} f(g^{-1}(\mathbf{X}(\mathbf{s})\boldsymbol{\beta} + \mathbf{S})) \\ [\mathbf{S}|\boldsymbol{\theta}] &\sim G(\mathbf{0}, \boldsymbol{\Sigma}^S(\boldsymbol{\theta})) \\ [\boldsymbol{\beta}, \boldsymbol{\theta}] &\sim \pi(\boldsymbol{\beta}, \boldsymbol{\theta}), \end{aligned}$$

where $f(\cdot)$ belongs to the exponential family, and has mean $g^{-1}(\mathbf{X}(\mathbf{s})\boldsymbol{\beta} + \mathbf{S})$.

Diggle et al. (1998) take a two step approach, first constructing an algorithm to generate sample from the full conditionals of $\boldsymbol{\theta}$, $\boldsymbol{\beta}$ and $\mathbf{S}(\mathbf{s})$, and then considering prediction of new values $\mathbf{S}(\mathbf{s}_0)$. For prediction of $\mathbf{S}(\mathbf{s}_0)$, again assume that, conditional on $\boldsymbol{\beta}$ and $\boldsymbol{\theta}$, $\mathbf{S}(\mathbf{s})$ and $\mathbf{S}(\mathbf{s}_0)$ follow a multivariate normal distribution of the form of equation (6.95), where

$$\begin{bmatrix} \mathbf{S}(\mathbf{s}) \\ \mathbf{S}(\mathbf{s}_0) \end{bmatrix} \sim G\left(\mathbf{0}, \begin{bmatrix} \boldsymbol{\Sigma}_{ss}^S(\boldsymbol{\theta}) & \boldsymbol{\Sigma}_{s0}^S(\boldsymbol{\theta}) \\ \boldsymbol{\Sigma}_{0s}^S(\boldsymbol{\theta})' & \boldsymbol{\Sigma}_{00}^S(\boldsymbol{\theta}) \end{bmatrix}\right).$$

Diggle et al. (1998) assume that the data $\mathbf{Z}(\mathbf{s})$ are independent of $\mathbf{S}(\mathbf{s}_0)$ (which we note in passing seems counterintuitive to the geostatistical paradigm) and so predicting $\mathbf{S}(\mathbf{s}_0)$ given values of $\mathbf{Z}(\mathbf{s})$, $\boldsymbol{\beta}$, and $\boldsymbol{\theta}$ can be done by using ordinary kriging of $\mathbf{S}(\mathbf{s}_0)$ assuming data $\mathbf{S}(\mathbf{s})$.

Once the entire two step algorithm is complete, prediction of data $\mathbf{Z}(\mathbf{s}_0)$ is done using $g^{-1}(\mathbf{X}(\mathbf{s})\boldsymbol{\beta} + \mathbf{S}(\mathbf{s}_0))$. Diggle et al. (1998) called their approach "model based geostatistics," a name we disagree with for reasons that should be obvious from Chapter 5.

As with other Bayesian hierarchical formulations for modeling spatial data, care must be taken when choosing prior distributions, particulary those for spatial autocorrelation parameters. The cautions in Berger et al. (2001) apply here as well, and recent work by Zhang (2004) indicates that parameters of the Matérn class are not estimable in model-based geostatistics.

Moyeed and Papritz (2002) compare this approach (MK) to ordinary kriging (OK), lognormal kriging (LK), and disjunctive kriging (DK) using a validation study based on 2649 measurements on copper (Cu) and cobalt (Co) concentrations. They predicted values at each of 2149 locations, using predictions from each model constructed from 500 observations. Their results are similar to those of De Oliveira et al. (1997):

1. The accuracy of the predictions was essentially the same for all methods.
2. The empirical probability coverages for OK were not nominal; OK gave smaller coverage at the tails and higher coverage for central values than nominal.
3. The degree to which the empirical probability coverages departed from nominal depends on the strength of autocorrelation, supporting a similar finding by Zimmerman and Cressie (1992) (§5.5.4). For Co, with relatively weak spatial dependence, the empirical probability coverages for OK were about the same as those for MK with a spherical covariance structure. For Cu, with stronger spatial dependence, the empirical probability coverages for OK were much worse, while those for MK were far closer to nominal.
4. The MK method is sensitive to the choice of the parametric covariance function used to construct $\boldsymbol{\Sigma}^S(\boldsymbol{\theta})$. For some choices, the empirical probability coverages were close to nominal; for other choices the results were similar to OK or LK.
5. The empirical probability coverages for LK and DK were as good or better as those from MK, even though LK and DK do not account for the uncertainty in estimated autocorrelation parameters.

While most Bayesian methods for spatial prediction are compared to traditional kriging approaches, this is not the most enlightening comparison. Better comparisons would use as a comparison OK using a t-statistic to account for the small sample size, OK or universal kriging with the Kackar and Harville-type adjustments, or results from kriging-based geostatistical simulation methods (Chapter 7), invented primarily for the purpose of adjusting for the smoothness in predictions from kriging.

## 6.5 Chapter Problems

**Problem 6.1** Consider a Gaussian linear mixed model for longitudinal data. Let $\mathbf{Z}_i$ denote the $(n_i \times 1)$ response vector for subject $i = 1, \cdots, s$. If $\mathbf{X}_i$ and $\mathbf{Z}_i$ are $(n_i \times p)$ and $(n_i \times q)$ matrices of known constants and $\mathbf{b}_i \sim G(\mathbf{0}, \mathbf{G})$, then the conditional and marginal distributions in the linear mixed model are given by

$$\begin{aligned} \mathbf{Y}_i|\mathbf{b}_i &\sim G(\mathbf{X}_i\boldsymbol{\beta} + \mathbf{Z}_i\mathbf{b}_i, \mathbf{R}_i) \\ \mathbf{Y}_i &\sim G(\mathbf{X}_i\boldsymbol{\beta}, \mathbf{V}_i) \\ \mathbf{V}_i &= \mathbf{Z}_i\mathbf{G}\mathbf{Z}_i' + \mathbf{R}_i \end{aligned}$$

(i) Derive the maximum likelihood estimator for $\boldsymbol{\beta}$ based on the marginal distribution of $\mathbf{Y} = [\mathbf{Y}_1', \cdots, \mathbf{Y}_s']'$.

(ii) Formulate a model in which to perform local estimation. Assume that you want to localize the conditional mean function $\mathrm{E}[\mathbf{Y}_i|\mathbf{b}_i]$.

**Problem 6.2** In the first-difference approach model (6.12), describe the correlation structure among the observations in a column that would lead to uncorrelated observations when the differencing matrix (6.13) is applied.

**Problem 6.3** Show that the OLS residuals (6.4) on page 307 have mean zero, even if the assumption that $\mathrm{Var}[\mathbf{e}(\mathbf{s})] = \sigma_\epsilon^2\mathbf{I}$ is not correct.

**Problem 6.4** In model (5.24) show that

$$\mathrm{Var}[Z(\mathbf{s}_i)] - \mathbf{h}(\boldsymbol{\theta})_i'\mathrm{Cov}[Z(\mathbf{s}_i), \mathbf{Z}(\mathbf{s})] \geq 0,$$

where $\mathbf{h}(\boldsymbol{\theta})_i'$ is the $i$th row of

$$\mathbf{H}(\boldsymbol{\theta}) = \mathbf{X}(\mathbf{s})(\mathbf{X}(\mathbf{s})'\boldsymbol{\Sigma}(\boldsymbol{\theta})^{-1}\mathbf{X}(\mathbf{s}))^{-1}\mathbf{X}(\mathbf{s})'\boldsymbol{\Sigma}(\boldsymbol{\theta})^{-1}.$$

**Problem 6.5** Consider a statistical model of the following form

$$\mathbf{Y} = \mathbf{X}\boldsymbol{\beta} + \mathbf{e} \qquad \mathbf{e} \sim (\mathbf{0}, \mathbf{V}).$$

Assume that $\mathbf{V}$ is a known, positive definite matrix and that the model is fit by (a) ordinary least squares and (b) generalized least squares. The leverage matrices of the two fits are

(a) $\mathbf{H}_{ols} = \mathbf{X}(\mathbf{X}'\mathbf{X})^{-}\mathbf{X}'$

(b) $\mathbf{H}_{gls} = \mathbf{X}(\mathbf{X}'\mathbf{V}^{-1}\mathbf{X})^{-1}\mathbf{X}'\mathbf{V}^{-1}$

(i) If $\mathbf{X}$ contains an intercept, find lower and upper bounds for the diagonal elements of $\mathbf{H}_{ols}$. Can you find similar bounds for $\mathbf{H}_{gls}$?

(ii) Find the lower bound for the diagonal elements of $\mathbf{H}_{gls}$ if $\mathbf{X} = \mathbf{1}$.

(iii) Compare $\mathrm{tr}\{\mathbf{H}_{ols}\}$ and $\mathrm{tr}\{\mathbf{H}_{gls}\}$.

**Problem 6.6** Consider a statistical model of the following form

$$\mathbf{Y} = \mathbf{X}\boldsymbol{\beta} + \mathbf{e} \qquad \mathbf{e} \sim (\mathbf{0}, \mathbf{V}).$$

Assume that $\mathbf{V}$ is positive definite so there exists a lower triangular matrix $\mathbf{L}$ such that $\mathbf{V} = \mathbf{L}\mathbf{L}'$. An alternative model to the one given is

$$\begin{aligned} \mathbf{L}^{-1}\mathbf{Y} &= \mathbf{L}^{-1}\mathbf{X}\boldsymbol{\beta} + \mathbf{L}^{-1}\mathbf{e} \\ &= \mathbf{X}^*\boldsymbol{\beta} + \mathbf{e}^* \\ \mathbf{e}^* &\sim (\mathbf{0}, \mathbf{I}). \end{aligned}$$

(i) Derive the estimators for $\boldsymbol{\beta}$ in both models and compare.

(ii) Find the projector in the second model and compare its properties to the oblique projector $\mathbf{H}_{gls}$ from the previous problem.

**Problem 6.7** In a standard regression model $\mathbf{Y} = \mathbf{X}\boldsymbol{\beta} + \mathbf{e}$, $\mathbf{e} \sim (\mathbf{0}, \sigma^2\mathbf{I})$, where $\mathbf{X}$ contains an intercept, the OLS residuals

$$\widehat{e}_i = y_i - \widehat{y}_i = y_i - \mathbf{x}_i'(\mathbf{X}'\mathbf{X})^{-1}\mathbf{X}'\mathbf{Y}$$

have expectation zero. Furthermore, $\sum_{i=1}^{n} \widehat{e}_i = 0$. Now assume that $\mathrm{Var}[\mathbf{e}] = \mathbf{V}$ and that $\boldsymbol{\beta}$ is estimated by GLS (EGLS). Which of the mentioned properties of the OLS residuals are shared by the GLS (EGLS) residuals?

**Problem 6.8** Suppose that $t$ treatments are assigned to field plots arranged in a single long row so that each treatment appears in exactly $r$ plots. To account for soil fertility gradients as well as measurement error, the basic model is

$$\mathbf{Z}(\mathbf{s}) = \mathbf{X}\boldsymbol{\tau} + \mathbf{S}(\mathbf{s}) + \boldsymbol{\epsilon},$$

where $\mathbf{S}(\mathbf{s}) \sim (\mathbf{0}, \sigma^2\boldsymbol{\Sigma})$ and $\boldsymbol{\epsilon} \sim (\mathbf{0}, \sigma_\epsilon^2\mathbf{I})$. Assume it is known that the random field $\{\mathbf{S}(\mathbf{s})\}$ is such that the $(rt-1) \times rt$ matrix of first differences ($\boldsymbol{\Delta}$, see (6.13) on page 320) transforms $\boldsymbol{\Sigma}$ as follows: $\boldsymbol{\Delta}\boldsymbol{\Sigma}\boldsymbol{\Delta}' = \mathbf{I}$.

(i) Describe the estimation of treatment effects $\boldsymbol{\tau}$ and the variance components $\sigma^2$ and $\sigma_\epsilon^2$ in the context of a linear mixed model, §6.2.1.3.

(ii) Is this a linear mixed model with or without correlations in the conditional distribution? Hint: After the transformation matrix $\boldsymbol{\Delta}$ is applied, what do you consider to be the random effects in the linear mixed model?

**Problem 6.9** In the mixed model formulation of the linear model with autocorrelated errors (§6.2.1.3), verify the expression for $\mathbf{C}$ in (6.28) and the expressions for the mixed model equation solutions, (6.29)–(6.30).

**Problem 6.10** In the mixed model formulation of §6.2.1.3, show that $\mathbf{C}$ in (6.28) is the variance matrix of $[\widehat{\boldsymbol{\beta}}, (\widehat{\boldsymbol{\upsilon}}(\mathbf{s}) - \boldsymbol{\upsilon}(\mathbf{s})']$. Verify that the element in the lower right corner of $\mathbf{C}$ is the prediction mean squared error of $\widehat{\boldsymbol{\upsilon}}(\mathbf{s})$.

**Problem 6.11** Show that the mixed model predictor (6.31) has the form of a filtered, universal kriging predictor.

**Problem 6.12** Imagine a lattice process on a $2 \times 3$ rectangle. The sites $s_1, s_2, s_3$ make up the first row, the remaining sites make up the second row. Assume that a spatial connectivity matrix is given by

$$\mathbf{W} = \begin{bmatrix} 0 & 1 & 0 & 1 & 0 & 0 \\ 1 & 0 & 1 & 0 & 1 & 0 \\ 0 & 1 & 0 & 0 & 0 & 1 \\ 1 & 0 & 0 & 0 & 1 & 0 \\ 0 & 1 & 0 & 1 & 0 & 1 \\ 0 & 0 & 1 & 0 & 1 & 0 \end{bmatrix}.$$

For a simultaneous and a conditional autoregressive scheme with $\text{Var}[\mathbf{Z}(\mathbf{s})] = \sigma^2(\mathbf{I} - \rho\mathbf{W})^{-1}(\mathbf{I} - \rho\mathbf{W}')^{-1}$ and $\text{Var}[\mathbf{Z}(\mathbf{s})] = \sigma^2(\mathbf{I} - \rho\mathbf{W})^{-1}$, respectively, perform the following tasks for $\sigma^2 = 1, \rho = 0.25$.

(i) Derive the variance-covariance matrix for the SAR and CAR scheme.

(ii) Determine which of the processes is second-order stationary.

(iii) Describe the correlation pattern that results. Are observations equicorrelated that are the same distance apart? Do correlations decrease with increasing lag distance?

**Problem 6.13** Consider the following exponential family distributions:

(a) Binomial$(n, \pi)$:

$$\Pr(Y = y) = \binom{n}{y} \pi^y (1-\pi)^{n-y}, \quad y = 0, 1, \cdots, n$$

(b) Gaussian, $G(\mu, \sigma^2)$:

$$f(y) = \frac{1}{\sigma\sqrt{2\pi}} \exp\{-0.5(y-\mu)^2/\sigma^2\}, \quad -\infty < y < \infty$$

(c) Negative Binomial, NegBin$(k, \mu)$:

$$\Pr(Y = y) = \frac{\Gamma(y + 1/k)}{\Gamma(y+1)\Gamma(1/k)} \frac{(k\mu)^y}{(1+k\mu)^{y+1/k}}, \quad y = 0, 1, 2, \cdots$$

(d) Inverse Gaussian, IG$(\mu, \sigma)$:

$$f(y) = \frac{1}{\sigma\sqrt{2\pi y^3}} \exp\left\{-\frac{1}{2y}\left(\frac{y-\mu}{\mu\sigma}\right)^2\right\}, \quad y > 0$$

(i) Write the densities or mass functions in exponential form

(ii) Identify the canonical link, variance function, and scale parameter.

(iii) Is the canonical link a reasonable link for modeling? If not, suggest an appropriate link.

(iv) Assume a random sample of size $m$ from each of these distributions. For the distributions with a scale parameter, can you profile this parameter from the likelihood?

**Problem 6.14** If $Y = \log\{X\}$ has a Gaussian distribution, then $X = \exp\{Y\}$ as a lognormal distribution.

(i) Derive the density of $X$, $f(x)$.

(ii) Prove or disprove, whether the lognormal distribution is a member of the exponential family.

(iii) Name a distribution that is a member of the exponential family and has a similar shape and the same support as the lognormal distribution.

**Problem 6.15** If $Z|n$ Binomial$(n, \pi)$, and $n \sim$ Poisson$(\lambda)$, find the distribution of $Z$. Is it more or less dispersed than the distribution of $Z|n$?

**Problem 6.16** Repeat the previous exercise where $n$ is a truncated Poisson variate, that is,

$$\Pr(n = m) = \frac{\lambda^m}{m!}\frac{\exp\{-\lambda\}}{1 - \exp\{-\lambda\}}, \qquad m = 1, 2, \cdots.$$

**Problem 6.17** For the data on blood lead levels in children in Virginia, 2000, repeat the analysis in Example 1.4 using models based on the Binomial distribution.

CHAPTER 7

# Simulation of Random Fields

Simulating spatial data is important on several grounds.

- The worth of a statistical method for georeferenced data can often be established convincingly only if the method has exhibited satisfactory long-run behavior.
- Statistical inference for spatial data often relies on randomization tests, e.g., tests for spatial point patterns. The ability to simulate realizations of a hypothesized process quickly and efficiently is important to allow a sufficient number of realizations to be produced.
- The absence of replication in most spatial data sets requires repeated *observation* of a phenomenon to obtain empirical estimates of mean, variation, and covariation.

Constructing a realization of a random field $\{Z(\mathbf{s}) : \mathbf{s} \in D \subset \mathbf{R}^d\}$ is not a trivial task, since it requires knowledge of the spatial distribution

$$\Pr(Z(\mathbf{s}_1) < z_1, Z(\mathbf{s}_2) < z_2, \cdots, Z(\mathbf{s}_n) < z_n).$$

From a particular set of data we can only infer, under stationarity assumptions, the first and second moment structure of the field. Inferring the joint distribution of the $Z(\mathbf{s}_i)$ from the mean and covariance function is not possible unless the random field is a Gaussian random field (GRF). Even if the spatial distribution were known, it is usually not possible to construct a realization via simulation from it. The best we may be able to do—unless the random field is Gaussian—is to produce realizations whose marginal or maybe bivariate distributions agree with the respective distributions of the target field. In other instances, for example, the simulation of correlated counts, even this condition may be difficult to attain. Generating a random field in which the $Z(\mathbf{s})$ are marginally Bernoulli($\pi$) random variables with a particular covariance function may be impossible, because the model itself is vacuous. What we may be able to accomplish is to generate random deviates with known autocovariance function whose marginal moments (mean and variance) "behave like" those of Bernoulli($\pi$) variables.

In this Chapter we review several methods for generating spatial data with the help of computers and random number generators. Some methods will generate spatial data with known spatial distribution, for example, the methods to generate Gaussian random fields (§7.1). Other methods generate data that comply only to first and second moment assumptions, that is, data whose mean, variance, and covariance function are known. Particularly important among the latter methods are those that generate data behaving like counts or proportions. Convolution theory (§7.4) can be used to simulate such data as well as simulated annealing methods (§7.3).

## 7.1 Unconditional Simulation of Gaussian Random Fields

The Gaussian random field holds a core position in the theory of spatial data analysis much in the same vein as the Gaussian distribution is key to many classical approaches of statistical inferences. Best linear unbiased kriging predictors are identical to conditional means in Gaussian random fields, establishing their optimality beyond the class of linear predictors. Second-order stationarity in a Gaussian random field implies strict stationarity. The statistical properties of estimators derived from Gaussian data are easy to examine and test statistics usually have a known and simple distribution. Hence, creating realizations from GRFs is an important task. Fortunately, it is comparatively simple.

Chilès and Delfiner (1999, p. 451) discuss an instructive example that highlights the importance of simulation. Imagine that observations are collected along a transect at 100-meter intervals measuring the depth of the ocean floor. The goal is to measure the length of the profile. One could create a continuous profile by kriging and then obtain the length as the sum of the segments between the observed transect locations. Since kriging is a smoothing of the data in-between the observed locations, this length would be an underestimate of the profile length. In order to get a realistic estimate, we need to generate values of the ocean depth in-between the 100-meter sampling locations that are consistent with the stochastic variation we would have seen, had the sampling interval been shorter.

In this example it is reasonable that the simulated profile passes through the observed data points. After all, these were the values which were observed, and barring measurement error, reflect the actual depth of the ocean. A simulation method that *honors the data* in the sense that the simulated value at an observed location agrees with the observed value is termed a **conditional** simulation. Simulation methods that do not honor the data, for example, because no data has yet been collected, are called **unconditional** simulations.

Several methods are available to simulate GRFs unconditionally, some are more brute-force than others. The simplest—and probably crudest— method relies on the reproductive property of the (multivariate) Gaussian distribution and the fact that a positive-definite matrix $\boldsymbol{\Sigma}$ can be represented as

$$\boldsymbol{\Sigma} = \boldsymbol{\Sigma}^{1/2}\boldsymbol{\Sigma}'^{1/2}.$$

If $\mathbf{Y} \sim G(\boldsymbol{\mu}, \boldsymbol{\Sigma})$, and $\mathbf{X} \sim G(\mathbf{0}, \mathbf{I})$, then

$$\boldsymbol{\mu} + \boldsymbol{\Sigma}^{1/2}\mathbf{X}$$

has a $G(\boldsymbol{\mu}, \boldsymbol{\Sigma})$ distribution. Two of the elementary ways of obtaining a *square root* matrix of the variance-covariance matrix $\boldsymbol{\Sigma}$ are the Cholesky decomposition and the spectral decomposition.

### *7.1.1 Cholesky (LU) Decomposition*

If $\boldsymbol{\Sigma}_{n\times n}$ is a positive definite matrix, then there exists an upper triangular matrix $\mathbf{U}_{n\times n}$ such that $\boldsymbol{\Sigma} = \mathbf{U}'\mathbf{U}$. The matrix $\mathbf{U}$ is called the Cholesky root of $\boldsymbol{\Sigma}$ and is unique up to sign (Graybill, 1983). Since $\mathbf{U}'$ is *lower*-triangular and $\mathbf{U}$ is *upper*-triangular, the decomposition is often referred to as the *lower-upper* or LU decomposition. Many statistical packages can calculate a Cholesky root, for example, the `root()` function of the SAS\IML® module. Since Gaussian random number generators are also widely available, this suggests a simple method of generating data from a $G_n(\boldsymbol{\mu}, \boldsymbol{\Sigma})$ distribution. Generate $n$ independent standard Gaussian random deviates and store them in vector $\mathbf{x}$. Calculate the Cholesky root $\mathbf{U}'$ of the variance-covariance matrix $\boldsymbol{\Sigma}$ and a $(n \times 1)$ vector of means $\boldsymbol{\mu}$. Return $\mathbf{y} = \boldsymbol{\mu} + \mathbf{U}'\mathbf{x}$ as a realization from a $G(\boldsymbol{\mu}, \boldsymbol{\Sigma})$. It works well for small to moderate sized problems. As $n$ grows large, however, calculating the Cholesky decomposition is numerically expensive.

### *7.1.2 Spectral Decomposition*

A second method of generating a *square root* matrix of $\boldsymbol{\Sigma}$ relies on the spectral decomposition of a real symmetric matrix. If $\mathbf{A}_{p\times p}$ is a real symmetric matrix, then there exists a $(p \times p)$ orthogonal matrix $\mathbf{P}$ such that

$$\mathbf{A} = \mathbf{P}\boldsymbol{\Delta}\mathbf{P}',$$

where $\boldsymbol{\Delta}$ is a diagonal matrix containing the eigenvalues of $\mathbf{A}$. Since $\mathbf{P}'\mathbf{P} = \mathbf{I}$, the matrix

$$\boldsymbol{\Sigma}^{1/2} = \mathbf{P}\boldsymbol{\Delta}^{1/2}\mathbf{P}'$$

has the needed properties to function as the *square root* matrix to generate $G(\boldsymbol{\mu}, \boldsymbol{\Sigma})$ deviates by

$$\mathbf{y} = \boldsymbol{\mu} + \boldsymbol{\Sigma}^{1/2}\mathbf{x}.$$

The spectral decomposition can be calculated in The SAS® System with the `eigen` function of the SAS\IML® module.

## 7.2 Conditional Simulation of Gaussian Random Fields

A conditional simulation is a realization $S(\mathbf{s})$ of a random field $Z(\mathbf{s})$ that honors the observed values of $Z(\mathbf{s})$. Let $Z(\mathbf{s}_1), Z(\mathbf{s}_2), \cdots, Z(\mathbf{s}_m)$ denote the observed values of a random field. A conditional simulation produces $n = m + k$ values

$$\mathbf{S}(\mathbf{s}) = [Z(\mathbf{s}_1), Z(\mathbf{s}_2), \cdots, Z(\mathbf{s}_m), S(\mathbf{s}_{m+1}), \cdots, S(\mathbf{s}_{m+k})]'.$$

If $m = 0$, the simulation is unconditional. Some methods for conditional simulation condition on the data directly, while others start with an unconditional simulation which is then conditioned.

### 7.2.1 Sequential Simulation

The idea of sequential simulation is simple. For the general case consider simulating a $(n \times 1)$ random vector $\mathbf{Y}$ with known distribution $F(y_1, \cdots, y_n) = \Pr(Y_1 \leq y_1, \cdots, Y_n \leq y_n)$. The joint cdf can be decomposed into conditional distributions

$$F(y_1, \cdots, y_n) = F(y_1)F(y_2|y_1) \times \cdots \times F(y_n|y_1, \cdots, y_{n-1}).$$

If the conditional distributions are accessible, the joint distribution can be generated from the conditionals. The name of the method stems from the sequential nature of the decomposition. In the spatial case, the conditioning sequence can be represented as

$$\begin{aligned}
\Pr\left(S(\mathbf{s}_{m+1}) \leq s_{m+1}, \cdots, S(\mathbf{s}_{m+k}) \leq s_{m+k} | z(\mathbf{s}_1), \cdots, z(\mathbf{s}_m)\right) \\
= \Pr\left(S(\mathbf{s}_{m+1}) \leq s_{m+1} | z(\mathbf{s}_1), \cdots, z(\mathbf{s}_m)\right) \\
\times \Pr\left(S(\mathbf{s}_{m+2}) \leq s_{m+2} | z(\mathbf{s}_1), \cdots, z(\mathbf{s}_m), s(\mathbf{s}_{m+1})\right) \\
\vdots \\
\times \Pr\left(S(\mathbf{s}_{m+k}) \leq s_{m+k} | z(\mathbf{s}_1), \cdots, z(\mathbf{s}_m), s(\mathbf{s}_{m+1}), \cdots, s(\mathbf{s}_{m+k-1})\right)
\end{aligned}$$

The advantage of sequential simulations is that it produces a random field not only with the correct covariance structure, but the correct spatial distribution. The disadvantage is having to work out the conditional distributions. In one particular case, when $Z(\mathbf{s})$ is a GRF, the conditional distributions are simple. If

$$\begin{bmatrix} Z(\mathbf{s}_0) \\ \mathbf{Z}(\mathbf{s}) \end{bmatrix} = G\left( \begin{bmatrix} \mu_0 \\ \boldsymbol{\mu} \end{bmatrix}, \begin{bmatrix} \sigma^2 & \mathbf{c}_0' \\ \mathbf{c}_0 & \boldsymbol{\Sigma} \end{bmatrix} \right),$$

then $Z(\mathbf{s}_0)|\mathbf{Z}(\mathbf{s})$ is Gaussian distributed with mean $\mathrm{E}[Z(\mathbf{s}_0)|\mathbf{Z}(\mathbf{s})] = \mu_0 + \mathbf{c}'\boldsymbol{\Sigma}^{-1}(\mathbf{Z}(\mathbf{s}) - \boldsymbol{\mu})$ and variance $\mathrm{Var}[Z(\mathbf{s}_0)|\mathbf{Z}(\mathbf{s})] = \sigma^2 - \mathbf{c}'\boldsymbol{\Sigma}^{-1}\mathbf{c}$. If the mean of the random field is known, this is the simple kriging predictor and the corresponding kriging variance. Thus we can calculate the conditional distributions of $S(\mathbf{s}_{m+i})$ given $z(\mathbf{s}_1), \cdots, z(\mathbf{s})_m, s(\mathbf{s}_{m+1}), \cdots, s(\mathbf{s}_{m+i-1})$ as Gaussian. The mean of the distribution is the simple kriging predictor of $S(\mathbf{s}_{m+i})$ based on the data $Z(\mathbf{s}_1), \cdots, Z(\mathbf{s}_m)$ and $S(\mathbf{s}_m), \cdots, S(\mathbf{s}_{m+i-1})$. Notice that if $m = 0$, this leads to an unconditional simulation of a GRF where successive values are random draws from Gaussian distributions. The means and variances of these distributions are updated sequentially. Fortunately, the stochastic properties are independent of the order in which the $S(\mathbf{s})$ values are being generated.

### *7.2.2 Conditioning a Simulation by Kriging*

Consider a random field with covariance function $C(\mathbf{h})$, sampled at locations $\mathbf{s}_1, \cdots, \mathbf{s}_m$ and vector of realizations $\mathbf{Z}(\mathbf{s}) = [Z(\mathbf{s}_1), \cdots, Z(\mathbf{s}_m)]'$. We want to simulate a random field with the same mean and covariance structure as $Z(\mathbf{s})$, but ensure that the realization passes through the observed values $Z(\mathbf{s}_1), \cdots, Z(\mathbf{s}_m)$. This can be accomplished based on an unconditional simulation $S(\mathbf{s})$ of the random field with the same covariance function,

$$\mathrm{Cov}[S(\mathbf{h}), S(\mathbf{s}+\mathbf{h})] = \mathrm{Cov}[Z(\mathbf{s}), Z(\mathbf{s}+\mathbf{h})]$$

as follows. The decomposition

$$Z(\mathbf{s}) = p_{sk}(\mathbf{s}; \mathbf{Z}) + Z(\mathbf{s}) - p_{sk}(\mathbf{s}; \mathbf{Z})$$

holds always, but the simple kriging residual $Z(\mathbf{s}) - p_{sk}(\mathbf{s}; \mathbf{Z})$ is not observable. Instead we can substitute for it $S(\mathbf{s}) - p_{sk}(\mathbf{s}; \mathbf{S}_m)$, where $p_{sk}(\mathbf{s}; \mathbf{S}_m)$ denotes the simple kriging predictor at location $\mathbf{s}$ based on the values of the unconditional simulation at the locations $\mathbf{s}_1, \cdots, \mathbf{s}_m$ where $Z$ was observed. That is, $\mathbf{S}_m(\mathbf{s}) = [S(\mathbf{s}_1), \cdots, S(\mathbf{s}_m)]'$. We define the conditional realization as

$$\begin{aligned} Z_c(\mathbf{s}) &= p_{sk}(\mathbf{s}; \mathbf{Z}) + S(\mathbf{s}) - p_{sk}(\mathbf{s}; \mathbf{S}_m) \\ &= S(\mathbf{s}) + \mathbf{c}'\mathbf{\Sigma}^{-1}(\mathbf{Z} - \mathbf{S}_m). \end{aligned} \tag{7.1}$$

Note that (7.1) corrects the unconditional simulation at $S(\mathbf{s})$ by the residual between the observed data and the values from the unconditional simulation. Also, it is not necessary that $S(\mathbf{s})$ is simulated with the same mean as $Z(\mathbf{s})$. Any mean will do, for example, $\mathrm{E}[S(\mathbf{s})] = 0$. It is left as an exercise to show that (7.1) has the needed properties. In particular,

(i) For $\mathbf{s}_0 \in \{\mathbf{s}_1, \cdots, \mathbf{s}_m\}$, $Z_c(\mathbf{s}_0) = Z(\mathbf{s}_0)$; the realization honors the data;

(ii) $\mathrm{E}[Z_c(\mathbf{s})] = \mathrm{E}[Z(\mathbf{s})]$, i.e., the conditional simulation is (unconditionally) unbiased;

(iii) $\mathrm{Cov}[Z_c(\mathbf{s}), Z_c(\mathbf{s}+\mathbf{h})] = \mathrm{Cov}[Z(\mathbf{s}), Z(\mathbf{s}+\mathbf{h})] \quad \forall \mathbf{h}$.

The idea of a conditional simulation is to reproduce data where it is known but not to smooth the data in-between. The kriging predictor is a best linear unbiased predictor of the random variables in a spatial process that smoothes in-between the observed data. A conditional simulation of a random field will exhibit more variability between the observed points than the kriging predictor. In fact, it is easy to show that

$$\mathrm{E}\left[(Z_c(\mathbf{s}) - Z(\mathbf{s}))^2\right] = 2\sigma^2_{sk},$$

where $\sigma^2_{sk}$ is the simple kriging variance.

## 7.3 Simulated Annealing

On occasion we may want to constrain the simulations more than having the proper covariance on average or honoring the data points. For example, we may want *all* realizations have the same empirical covariance function than an observed process. Imagine that $m$ observations have been collected and the empirical semivariogram $\widehat{\gamma}(\mathbf{h}_1), \cdots, \widehat{\gamma}(\mathbf{h}_k)$ has been calculated for a set of $k$ lag classes. A conditional realization of the random field is to be simulated that honors the data $Z(\mathbf{s}_1), \cdots, Z(\mathbf{s}_m)$ and whose semivariogram agrees completely with the empirical semivariogram $\widehat{\gamma}(\mathbf{h})$.

To place additional constraints on the realizations, the optimization method of simulated annealing (SA) can be used. It is a heuristic method that is used in operations research, for example, to find solutions in combinatorial optimization when standard mathematical programming tools fail. The monograph by Laarhoven and Aarts (1987) gives a wonderful account of SA, its history, theory, and applications.

The name of the method reveals a metallurgic connection. If a molten metal is cooled slowly, the molecules can move freely, attaining eventually a state of the solid with low energy and little stress. If cooling occurs too quickly, however, the system will not be able to find the low-energy state because the movement of the molecules is obstructed. The solid will not reach thermal equilibrium at a given temperature, defects are frozen into the solid.

Despite the name, simulated annealing originated in statistical physics. The probability density function $f(a)$ of state $a$ in a system with energy $U(a)$ and absolute temperature $T > 0$, when the system reaches thermal equilibrium, is given by the Boltzmann distribution,

$$f(a) = g(T)^{-1} \exp\{-U(a)/(k_B T)\}, \tag{7.2}$$

where $k_B$ is the Boltzmann constant, $g(T)^{-1}$ is a normalizing constant, and $\exp\{-U(a)/(k_B T)\}$ is known as the Boltzmann factor.

The probability that the system is in a high energy state at temperature $T$ tends to zero because the density function $f(a)$ concentrates on the set of states with minimum energy. The idea of simulated annealing is to continue to generate realizations from the state space as the temperature $T$ is reduced. If the process converges, uniform sampling from the states that have minimum energy is achieved. In statistical applications, the energy function $U(a)$ represents a measure of the discrepancy between features of the current state (= current realization) and the intended features of the simulated process.

Simulated annealing is related to iterative improvement (II) algorithms (Laarhoven and Aarts, 1987, p. 4). An II algorithm presupposes a state space of configurations, a cost function, and a mechanism to generate a transition from one state to another by means of small perturbations. Starting from an initial configuration (state) of the system, transitions from the current to a new state are selected if a new configuration lowers the cost function. If a configuration is found such that no neighboring state has lower energy (cost), the configuration is adopted as the solution. Unfortunately, II algorithms will find a local minimum that depends on the initial configuration. Simulated annealing avoids this pitfall by allowing the algorithm to accept a transition if it increases the energy in the system and not only configurations that lower the system's energy. The key is to allow these transitions to higher energy states with the appropriate frequency. In addition, the probability to accept higher-energy states will gradually decrease in the process.

The connection of simulated annealing to optimization and an illustration on why this procedure works, is given in the following example.

**Example 7.1** Consider that we want to find the global maximum of the function

$$f(x) = \exp\left\{-\frac{1}{2}U(x)\right\} = \exp\left\{-\frac{1}{2}(x-\mu)^2\right\}.$$

Obviously, the function is maximized for $x = \mu$ when the energy function $U(x)$ takes on its minimum. In order to find the maximum of $f(x)$ by simulated annealing we sample from a system with probability distribution $g_T^*(x) \propto f(x)^{1/T}$, where $T$ denotes the absolute *temperature* of the system. For example, we can sample from

$$g_T(x) = \frac{1}{\sqrt{2\pi T}} f(x)^{1/T} = \frac{1}{\sqrt{2\pi T}} \exp\left\{-\frac{1}{2}\frac{(x-\mu)^2}{T}\right\}.$$

Note that $g(x)$ is the Gaussian density function with mean $\mu$ and variance $T$. The process commences by drawing a single sample from $g_T(x)$ at high temperatures and samples are drawn continuously while the temperature is dropped. As $T \to 0$, $x \to \mu$ with probability one. From this simple example we can see that the final realization will not be identical to the abscissa where $f(x)$ has its maximum, but should be close to it. □

The connection of annealing a metal to the simulation of thermal equilibrium was drawn by the Monte Carlo approach of Metropolis et al. (1953). First, we wish to generate a sequence of configurations of the solid in thermal equilibrium at temperature $T$. Assume that the solid has energy $U_u$ at the $u$th step of the procedure. A randomly chosen particle is slightly displaced. This perturbation will lead to a new energy state of the solid. If $U_{u+1} < U_u$, the new configuration is accepted. Otherwise, the new configuration is accepted with probability $\exp\{-(U_{u+1} - U_u)/(k_B T)\}$. After a large number of random perturbations, the system should achieve thermal equilibrium at temperature $T$. The application in optimization problems involves—in addition to the described process—a controlled reduction of the temperature $T$. After the system has reached thermal equilibrium, the temperature is reduced, and a new equilibrium is sought. The entire process continues until, for some small temperature, no perturbations lead to configurations with lower energy. The acceptable random perturbations are then a random sampling from the uniform distribution of the set of configurations with minimal energy.

The application of this iterative simulated annealing process to the simulation of spatial random fields requires the following:

(i) An initial realization of the random field. This can be a conditional or unconditional simulation. We call the realization at the $u$th iteration of the process the $u$th image.

(ii) A rule by which to perturb the image to reallocate its *molecules*.

(iii) A rule by which to determine whether the perturbed image at iteration $u+1$ should be accepted or rejected compared to the image at iteration $u$.

(iv) A rule by which to change the *temperature* of the system, the cooling schedule.

Denote by $\mathbf{S}_u(\mathbf{s})$ the realization of the random field at stage $u$ of the iterative process; $\mathbf{S}_0(\mathbf{s})$ is the initial image. The most common rule of perturbing the image is to randomly select two locations $\mathbf{s}_i$ and $\mathbf{s}_j$ and to swap (exchange) their values. If the simulation is conditional such that the eventual image honors an observed set of data $Z(\mathbf{s}_1), \cdots, Z(\mathbf{s}_m)$, we choose

$$\mathbf{S}_0(\mathbf{s}) = [Z(\mathbf{s}_1), Z(\mathbf{s}_2), \cdots, Z(\mathbf{s}_m), S_0(\mathbf{s}_{m+1}), \cdots, S_0(\mathbf{s}_{m+k})]',$$

and perturb only the values at the unobserved locations.

After swapping values we decide whether to accept the new image. $\mathbf{S}_{u+1}(\mathbf{s})$ is accepted if $U_{u+1} < U_u$. Otherwise the new image is accepted with probability $f_{u+1}/f_u$, where

$$f_u \propto \exp\{-U/T_u\}.$$

Chilès and Delfiner (1999, p. 566) point out that when the temperature $T_u$ is reduced according to the cooling schedule $T_u = -c\log\{u+1\}$, the process converges to a global minimum energy state. Here, $c$ is chosen as the minimum increase in energy that moves the process out of a local energy minimum—that is not the global minimum—into a state of lower energy. In practice, the cooling schedule is usually governed by simpler functions such as $T_u = T_0\lambda$, $\lambda < 1$. The important issues here are the magnitude of the temperature reduction ($\lambda$) and the frequency with which the temperature is reduced. It is by no means necessary or prudent to reduce the temperature every time a new image is accepted. At each temperature, the system should be allowed to achieve thermal equilibrium. Goovaerts (1997, p. 420) discusses adjusting the temperature after a sufficient number of perturbations have been accepted ($2n$—$10n$) or too many have been tried unsuccessfully ($10n$—$100n$); $n$ denotes the number of sites on the image.

Simulated annealing is a heuristic, elegant, intuitive, brute-force method. It starts with an initial image of $n$ atoms and tries to improve the image by rearranging the atoms in pairs. Improvement is measured by a user-defined discrepancy function, the energy function $U$. Some care should be exercised when simulating spatial random fields by this method.

- If several simulations are desired one should not start them from the same initial image $\mathbf{S}_0(\mathbf{s})$, but from different initial images. Otherwise the realizations are too similar to each other.

- Care must be exercised in choosing and monitoring the objective function. If $U$ measures the discrepancy between a theoretical and the empirical semivariogram for the current image and a set of $k$ lags,

$$U_u(\mathbf{h}) = \sum_{s=1}^{k} (\widehat{\gamma}(\mathbf{h}_s) - \gamma(\mathbf{h}_s))^2$$

  it is not necessarily desirable to achieve $U = 0$. This would imply that all realizations in the set of realizations with minimum energy have the same empirical semivariogram. If we sample a surface of a random field with semivariogram $\gamma(\mathbf{h})$, even a sampling conditional on $m$ known values of the surface, we expect the empirical semivariogram to agree with the theoretical semivariogram within the limits of sampling variation. We do not expect perfect agreement. If states with zero energy exist, it may be advisable to stop the annealing algorithm before such states are reached to allow the simulated system to represent uncertainty realistically.

- Since successive steps of the algorithm involve swapping values of the current image, the quality of the initial image is important. An initial image with high energy will require many iterations to achieve a low energy configuration. The minimum energy configuration that is achievable must be viewed in light of the initial image that is being used. Simulated annealing—if the process converges—finds one of the configurations of the sites that has lowest energy among the states that can be achieved starting from the initial image. This lowest energy configuration is not necessarily a good representation of the target random field if the initial image was chosen poorly.

- Simulated annealing is a computer intensive method. Hundreds of thousands of iterations are often necessary to find a low energy configuration. It is thus important to be able to update the objective function between perturbations quickly. For example, if the objective function monitors the empirical semivariogram of the realization, at stage $u$ we can calculate the new semivariogram at lag $\mathbf{h}$ by subtracting the contribution of the swapped values from the previous semivariogram and adding their contributions at that lag to the new semivariogram. A complete re-calculation of the empirical semivariogram is not needed (see Chapter problems).

## 7.4 Simulating from Convolutions

Simulating Gaussian random fields via the LU decomposition (§7.1) exploits two important facts: linear combinations of Gaussian random variables are Gaussian, and the mean and variance of a Gaussian random variable can be determined independently. The only obstacle is the size of the spatial data set being generated, calculating the Cholesky decomposition of the variance-covariance matrix of a $n$-dimensional Gaussian distribution is computer-intensive for large $n$.

The difficulties encountered in generating spatial data with non-Gaussian distributions are formidable in comparison, however. In recent years, statistical methods for analyzing correlated data have flourished. The generalized estimating equation (GEE) approach of Liang and Zeger (1986) for longitudinal data and Zeger and Liang (1986), for example, incorporates the correlation with clustered data by choosing a working correlation matrix and estimating the parameters of this matrix by the method of moments based on model residuals. In essence, correlation structures common for Gaussian processes are paired with models for the mean that draw on generalized linear models. One appealing feature of the GEE method for longitudinal data is that the estimates of the parameters in the mean model are consistent, provided the estimators of the working correlation matrix are consistent and the variance matrix satisfies certain limiting conditions. This does not require that the working variance-covariance matrix converges to the true variance-covariance matrix of the response; it requires that the working matrix converges to *something*. In addition, a sandwich estimator allows "robust" estimation of the variance of the estimated coefficients, down-weighing the impact of having employed a working correlation model, rather than the true one.

As a result of the popularity of the GEE approach, which requires only assumptions about the first two moments, it may be tempting to combine models for mean and covariance structure that maybe should not be considered in the same breath. The resulting model may be vacuous. For example, when modeling counts correlated in time, you may aim for Poisson random variables with an AR(1) correlation structure. A valid joint distribution with that correlation structure in which all marginal distributions are Poisson may not exist. The essential problem is the linkage of the mean and variance for many distributions, which imposes constraints on the joint distributions, see §5.8 and §6.3.3. The acid test to determine whether pairing a certain marginal behavior with a particular correlation model is appropriate is of course whether a valid joint probability distribution exists. The *sloppy* test is to ask: "is there a way to simulate such behavior?" It should give us pause if we intend to model data as marginally Poisson with an exponential covariance function and no mechanism comes to mind that could generate this behavior.

In many instances, statistical models for correlated data are based on first and second moments. When modeling counts, for example, it is thus natural to draw on distributions such as the Binomial, Poisson, or Negative Binomial, to suggest a mean-variance relationship for the model. What is usually required is not that the data have marginally a known distribution, but that they exhibit certain mean-variance behavior that concurs with models used for uncorrelated data. Such data can be generated in many instances. For example, is a model for $Z(\mathbf{s})$ with $\mathrm{E}[Z(\mathbf{s})] = \pi$, $\mathrm{Var}[Z(\mathbf{s})] = \pi(1-\pi)$, $\mathrm{Cov}[Z(\mathbf{s}), Z(\mathbf{s}+\mathbf{h})] = \sigma^2 \exp\{-||\mathbf{h}||^2/\alpha\}$ vacuous? Most likely not, if a data-generating mechanism can be devised. The convolution representation is helpful in the construction (and simulation) of this mechanism.

Recall that a spatial random field $\{Z(\mathbf{s}) : \mathbf{s} \in D \subset \mathbf{R}^d\}$ can be represented as the convolution of a kernel function $K(\mathbf{u})$ and an excitation field $X(\mathbf{s})$,

$$Z(\mathbf{s}) = \int_{\mathbf{u}} K(\mathbf{s}-\mathbf{u})X(\mathbf{u})\,d\mathbf{u}.$$

$Z(\mathbf{s})$ will be second-order stationary, if $X(\mathbf{s})$ is second-order stationary. In addition, if the excitation field has $\mathrm{E}[X(\mathbf{s})] = \mu_x$, $\mathrm{Var}[X(\mathbf{s})] = \sigma_x^2$, and $\mathrm{Cov}[X(\mathbf{s}), X(\mathbf{s}+\mathbf{h})] = 0, \mathbf{h} \neq \mathbf{0}$, then

(i) $\mathrm{E}[Z(\mathbf{s})] = \mu_x \int_{\mathbf{u}} K(\mathbf{u})\,d\mathbf{u}$;

(ii) $\mathrm{Var}[Z(\mathbf{s})] = \sigma_x^2 \int_{\mathbf{u}} K^2(\mathbf{u})\,d\mathbf{u}$;

(iii) $\mathrm{Cov}[Z(\mathbf{s}), Z(\mathbf{s}+\mathbf{h})] = \sigma_x^2 \int_{\mathbf{u}} K(\mathbf{u})K(\mathbf{u}+\mathbf{h})\,d\mathbf{u}$.

Since the kernel function governs the covariance function of the $Z(\mathbf{s})$ process, an intuitive approach to generating random fields is as follows. Generate a dense field of the $X(\mathbf{s})$, choose a kernel function $K(\mathbf{u})$, and convolve the two. This ensures that we understand (at least) the first and second moment structure of the generated process. Unless the marginal distribution of $X(\mathbf{s})$ has some reproductive property, however, the distribution of $Z(\mathbf{s})$ is difficult to determine by this device. A notable exception is the case where $X(\mathbf{s})$ is Gaussian; $Z(\mathbf{s})$ then will also be Gaussian.

If it is required that the marginal distribution of $Z(\mathbf{s})$ is characterized by a certain mean-variance relationship or has other moment properties, then these can be constructed by matching moments of the desired random field with the excitation process. We illustrate with an example.

**Example 7.2** Imagine it is desired to generate a random field $Z(\mathbf{s})$ that exhibits the following behavior:

$$\mathrm{E}[Z(\mathbf{s})] = \lambda, \quad \mathrm{Var}[Z(\mathbf{s})] = \lambda, \quad \mathrm{Cov}[Z(\mathbf{s}), Z(\mathbf{s}+\mathbf{h})] = C(\mathbf{h}).$$

In other words, the $Z(\mathbf{s})$ are marginally supposed to behave somewhat like Poisson($\lambda$) variables. From $\lambda = \mu_x \int K(\mathbf{u})\,d\mathbf{u}$ and $\lambda = \sigma_x^2 \int K^2(\mathbf{u})\,d\mathbf{u}$ we obtain the condition

$$\sigma_x^2 = \mu_x \frac{\int K(\mathbf{u})d\mathbf{u}}{\int K^2(\mathbf{u})d\mathbf{u}}. \tag{7.3}$$

For a given kernel function $K(\mathbf{u})$ we could choose a Gaussian white noise field with mean $\mu_x = \lambda / \int K(\mathbf{u})\,d\mathbf{u}$ and variance $\sigma_x^2 = \lambda / \int K^2(\mathbf{u})\,d\mathbf{u}$. But that does not guarantee that the $Z(\mathbf{s})$ are non-negative. The near-Poisson($\lambda$) behavior should be improved if the stochastic properties of the excitation field are somewhat close to the target. Relationship (7.3) suggests a count variable whose variance is proportional to the mean. The Negative Binomial random variable behaves this way. If $X \sim \mathrm{NegBinomial}(a, b)$, then $\mathrm{E}[X] = ab$, $\mathrm{Var}[X] = ab(b+1) = \mathrm{E}[X](b+1)$. By choosing the parameters $a$ and $b$ of the Negative Binomial excitation field carefully, the convolved process will have mean $\lambda$, variance $\lambda$, and covariance function $C(\mathbf{h}) = \int K(\mathbf{u})K(\mathbf{u}+\mathbf{h})\,d\mathbf{u}$.

Matching moments we obtain

$$\lambda = ab \int_{\mathbf{u}} K(\mathbf{u})\, d\mathbf{u} \Leftrightarrow ab = \frac{\lambda}{\int_{\mathbf{u}} K(\mathbf{u})\, d\mathbf{u}} \tag{7.4}$$

$$\lambda = ab(b+1) \int_{\mathbf{u}} K^2(\mathbf{u})\, d\mathbf{u}. \tag{7.5}$$

Substituting (7.4) into (7.5) leads to

$$b = \frac{\int K(\mathbf{u})\, d\mathbf{u}}{\int K^2(\mathbf{u})\, d\mathbf{u}} - 1, \tag{7.6}$$

and back-substitution yields

$$a = \frac{\lambda}{\int K(\mathbf{u})d\mathbf{u}} \left[ \frac{\int K^2(\mathbf{u})d\mathbf{u}}{\int K(\mathbf{u})d\mathbf{u} - \int K^2(\mathbf{u})d\mathbf{u}} \right]. \tag{7.7}$$

Although these expressions are not pretty, they are straightforward to calculate, since $K(\mathbf{u})$ is a known function and hence the integrals are constants. Additional simplifications arise when $\int K(\mathbf{u})d\mathbf{u} = 1$, since then the range of values in the excitation field equals the range of values in the convolved process. It is also obvious that the kernel function must be chosen so that $\int K(\mathbf{u})d\mathbf{u} > \int K^2(\mathbf{u})d\mathbf{u}$ in this application for $a$ and $b$ to be non-negative.

Figure 7.1 shows (average) semivariograms of random fields generated by this device. The domain of the process was chosen as a transect of length 500 with observations collected at unit intervals. A Gaussian kernel function

$$K(s-u,h) = \frac{1}{h\sqrt{(2\pi)}} \exp\left\{ -\frac{1}{2}\frac{(s-u)^2}{h^2} \right\}$$

was used and the range of integration was extended outside of the domain to avoid edge effects. The goal was to generate a random field in which the realizations have mean $\lambda = 5$, variance $\lambda = 5$ and are autocorrelated. The parameters $a$ and $b$ of the Negative Binomial excitation field were determined by (7.6) and (7.7). The Gaussian kernel implies a gaussian covariance function, other covariance functions can be generated by choosing the kernel accordingly (§2.4.2). For each of $h = 5, 10, 15, 20$, ten random convolutions were generated and their empirical semivariograms calculated. Figure 7.1 shows the average semivariogram over the ten repetitions for the different kernel bandwidths. With increasing $h$, the smoothness of the semivariogram as well as its range increases. The sample means and sample variances averaged over the ten realizations are shown in Table 7.1. Even with only 10 simulations, the first two moments of $Z(\mathbf{s})$ match the target closely.

□

Another important example is the generation of a random field in which the $Z(\mathbf{s})$ behave like Bernoulli($\pi$) random variables, i.e., $\mathrm{E}[Z(\mathbf{s})] = \pi$, $\mathrm{Var}[Z(\mathbf{s})] = \pi(1-\pi)$. Since $\pi$ is the success probability in a Bernoulli experiment, it is convenient to consider excitation fields with support $(0,1)$. For example, let

Table 7.1 *Average sample mean and sample variance of simulated convolutions whose semivariograms are shown in Figure 7.1*

| Bandwidth $h$ | $\overline{\overline{Z}}$ | $\overline{s^2}$ |
|---|---|---|
| 5 | 4.98 | 4.29 |
| 10 | 5.30 | 4.67 |
| 15 | 5.23 | 4.02 |
| 20 | 5.07 | 4.72 |

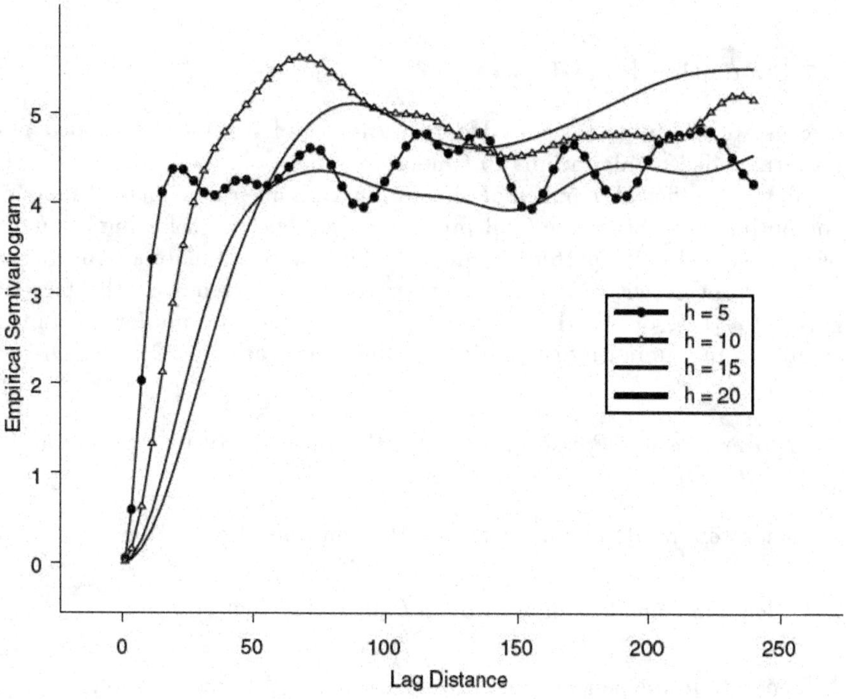


Figure 7.1 *Average semivariograms for Negative Binomial excitation fields convolved with Gaussian kernel functions of different bandwidth h (=standard deviation).*

$X(\mathbf{s})$ be a Beta$(\alpha, \beta)$ random variable with $\mathrm{E}[X(\mathbf{s})] = \alpha/(\alpha+\beta)$, $\mathrm{Var}[X(\mathbf{s})] = \alpha\beta(\alpha+\beta)^{-2}/(\alpha+\beta+1)$. Matching the moments of the convolution with the target yields the equations

$$\pi = \frac{\alpha}{\alpha+\beta}\int_{\mathbf{u}} K(\mathbf{u})\,d\mathbf{u}$$

$$\pi(1-\pi) \quad = \quad \frac{\alpha\beta}{(\alpha+\beta)^2(\alpha+\beta+1)} \int_{\mathbf{u}} K^2(\mathbf{u})\,d\mathbf{u}.$$

It is left as an exercise to establish that the parameters $\alpha$ and $\beta$ of the Beta excitation field must be chosen to satisfy

$$\alpha \quad = \quad \frac{\pi}{c}\left\{\frac{d(c-\pi)}{(1-\pi)c^2} - 1\right\} \tag{7.8}$$

$$\beta \quad = \quad \frac{c-\pi}{c}\left\{\frac{d(c-\pi)}{(1-\pi)c^2} - 1\right\}, \tag{7.9}$$

where $c = \int K(\mathbf{u})d\mathbf{u}$ and $d = \int K^2(\mathbf{u})d\mathbf{u}$. Notice that if the kernel is chosen such that $\int K(\mathbf{u})d\mathbf{u} = 1$, then the expressions simplify considerably; $\alpha = \pi(d-1)$, $\beta = (1-\pi)(d-1)$.

## 7.5 Simulating Point Processes

Because of the importance of Monte Carlo based testing for mapped point patterns, efficient algorithms to simulate spatial point processes are critical to allow a sufficient number of simulation runs in a reasonable amount of computing time. Since the CSR process is the benchmark for initial analysis of an observed point pattern, we need to be able to simulate a homogeneous Poisson process with intensity $\lambda$. A simple method relies on the fact that if $N(A) \sim \text{Poisson}(\lambda)$, then, given $N(A) = n$, the $n$ events form a random sample from a uniform distribution (a Binomial process, §3.2).

### *7.5.1 Homogeneous Poisson Process on the Rectangle $(0,0) \times (a,b)$ with Intensity $\lambda$*

1. Generate a random number from a Poisson($\lambda ab$) distribution $\rightarrow n$.
2. Order $n$ independent draws from a $U(0,a)$ distribution $\rightarrow x_1 < x_2 < \cdots < x_n$.
3. Generate $n$ independent realizations from a $U(0,b)$ distribution $\rightarrow y_1, \cdots, y_n$.
4. Return $(x_1, y_1), \cdots, (x_n, y_n)$ as the coordinates of the homogeneous Poisson process.

Extensions of this algorithm to processes in $\mathbf{R}^d$ are immediate. Comparing an observed pattern versus simulated ones, the number of events in the simulated patterns typically equal those in the observed pattern. In that case, step 1 of the algorithm is omitted and $n$ for step 2 is set equal to the number of observed points. If the study region is of irregular shape, a Poisson process can be generated on a bounding rectangle that encloses the study region. Points that fall outside the area of interest are removed.

### *7.5.2 Inhomogeneous Poisson Process with Intensity $\lambda(s)$*

Lewis and Shedler (1979) suggested the following acceptance/rejection algorithm to simulate a Poisson process on $A$ with spatially varying intensity.

1. Simulate with intensity $\lambda_0 \geq \max\{\lambda(\mathbf{s})\}$ a homogeneous Poisson process on $A$
   $\rightarrow (x_1, y_1), \cdots, (x_m, y_m)$.
2. Generate a $U(0, 1)$ realization (independently of the others) for each event in $A$ after step 1
   $\rightarrow u_1, \cdots, u_m$.
3. Retain event $i$ if $u_i \leq \lambda(\mathbf{s})/\lambda_0$, otherwise remove the event.

The initial step of the Lewis-Shedler algorithm generates a homogeneous Poisson process whose intensity dominates that of the inhomogeneous process everywhere. Steps 2 and 3 are thinning steps that remove excessive events with the appropriate frequency. The algorithm works for any $\lambda_0 \geq \max\{\lambda(\mathbf{s})\}$, but if $\lambda_0$ is chosen too large many events will need to be thinned in steps 2 and 3. It is sufficient to choose $\lambda_0 = \max\{\lambda(\mathbf{s})\}$.

## 7.6 Chapter Problems

**Problem 7.1** Generate data on a $10 \times 10$ regular lattice by drawing *iid* observations from a $G(\mu, \sigma^2)$ distribution and assign the realizations at random to the lattice positions. Given a particular nearest-neighbor definition (rook, bishop, queen move), produce a rearrangement of the observations by simulated annealing such that the sample correlation between $Z(\mathbf{s}_i)$ and the average of its neighbors takes on some value. Compare the empirical semivariogram between the lattice arrangements so achieved and the original arrangement for various magnitudes of the sample correlations.

*Things to consider: you will have to decide on a cooling schedule, the number of successful perturbation attempts before a new temperature is selected, and the number of unsuccessful attempts. How do you select the convergence criterion? How close is the actual sample correlation to the desired one upon convergence? How do your choices affect the end result?*

**Problem 7.2** Consider the Matheron estimator of the empirical semivariogram (see §4.4.1). Swap two values, $Z(\mathbf{s}_i)$ and $Z(\mathbf{s}_j)$. Give an update formula for the empirical semivariogram that avoids re-calculation of the entire semivariogram.

**Problem 7.3** Consider a bivariate and a trivariate Gaussian distribution. Develop explicit formulas to generate a realization by sequential simulation. Given an input vector for the mean $\boldsymbol{\mu}$ and the variance matrix $\boldsymbol{\Sigma}$ of the Gaussian distribution, write the software to generate realizations based on a standard Gaussian random number generator.

**Problem 7.4** In a conditional simulation by kriging, verify that the conditional realization (7.1) satisfies the following requirements: (i) $Z_c(\mathrm{s})$ honors the data, (ii) $Z_c(\mathrm{s})$ is unbiased in the sense that its mean equals $\mathrm{E}[Z(\mathrm{s})]$, (iii) its variance and covariance function are identical to that of $Z(\mathrm{s})$. Furthermore, establish that $\mathrm{E}[(Z_c(\mathrm{s}) - Z(\mathrm{s}))^2] = 2\sigma^2_{sk}$.

**Problem 7.5** In a conditional simulation by kriging, establish whether the conditional distribution of $Z_c(\mathrm{s})$ given the observed data $Z(\mathrm{s}_1), \cdots, Z(\mathrm{s}_m)$ is stationary. *Hint: consider the conditional mean and variance.*

**Problem 7.6** Verify formulas (7.8) and (7.9), that is, show that a convolution where $X(\mathrm{s}) \sim \mathrm{Beta}(\alpha, \beta)$ satisfies $\mathrm{E}[Z(\mathrm{s})] = \pi$, $\mathrm{Var}[Z(\mathrm{s})] = \pi(1-\pi)$.

**Problem 7.7** Use the convolution method to devise a mechanism to generate $Z(\mathrm{s})$ such that $\mathrm{E}[Z(\mathrm{s})] = k\pi$ and $\mathrm{Var}[Z(\mathrm{s})] = k\pi(1-\pi)$.

CHAPTER 8

# Non-Stationary Covariance

## 8.1 Types of Non-Stationarity

Throughout this text (second-order) stationarity of the stochastic process was an important assumption, without which there was little hope to make progress with statistical inference based on a sample of size one (§2.1). Recall that a process $\{Z(\mathbf{s}) : \mathbf{s} \in D \subset \mathbf{R}^d\}$ is second-order (weakly) stationary, if $\mathrm{E}[Z(\mathbf{s})] = \mu$ and $\mathrm{Cov}[Z(\mathbf{s}), Z(\mathbf{s}+\mathbf{h})] = C(\mathbf{h})$. A non-stationary process is any random field for which these conditions do not hold; some aspect of the spatial distribution is not translation invariant, it depends on the spatial location.

A non-constant mean function and variance heterogeneity are two frequent sources of non-stationarity. Mean and variance non-stationarity are not the focus of this chapter. Changes in the mean value can be accommodated in spatial models by parameterizing the mean function in terms of spatial coordinates and other regressor variables. Handling non-stationarity through fixed-effects structure of the model was covered in Chapter 6. Variance heterogeneity can sometimes be allayed by transformations of the response variables.

Non-stationarity is a common feature of many spatial processes, in particular those observed in the earth sciences. It can also be the result of operations on stationary random fields. For example, let $X(\mathbf{s})$ be a white noise random field in $\mathbf{R}^2$ with mean $\mu$ and variance $\sigma^2$. The domain consists of subregions $S_1, \cdots, S_k$ and we consider modeling the block aggregates $Z(S_i) = \int_{S_i} X(\mathbf{u})\, d\mathbf{u}$. Obviously, the $Z(S_i)$ are neither mean constant nor homoscedastic, unless $\nu(S_i) = \nu(S_j), \forall i \neq j$. Lattice processes with unequal areal units are typically not stationary and the formulation of lattice models takes the variation and covariation structure into account. The type of non-stationarity that is of concern in this section is covariance non-stationarity, the absence of translation invariance of the covariance function in geostatistical applications.

When the covariance function varies spatially, $\mathrm{Cov}[Z(\mathbf{s}), Z(\mathbf{s}+\mathbf{h})] = C(\mathbf{s}, \mathbf{h})$, two important consequences arise. One, the random field no longer "replicates itself" in different parts of the domain. This implication of stationarity enabled us to estimate the spatial dependency from pairs of points that shared the same distance but without regard to their absolute coordinates. Two, the covariogram or semivariogram models considered so far no longer apply.

The approaches to model covariance non-stationarity can be classified coarsely into global and local methods. A global method considers the entire domain, a local method assumes that a globally non-stationarity process can be represented as a combination of locally stationary processes. Parametric non-stationary covariance models and space deformation methods are examples of global modeling. The employ of convolutions, weighted averages, and moving windows is typical of local methods. The "replication" mechanisms needed to estimate dispersion and covariation from the data are different in the two approaches. Global methods, such as space deformation, require multiple observations for at least a subset of the spatial locations. Spatio-temporal data, where temporal stationarity can be assumed, can work well in this case. Local methods do not require actual replicate observations at a site, but local stationarity in a neighborhood of a given site. Estimates of variation and spatial covariation then can be based on observations within the neighborhood.

It is sometimes stated that spatial prediction is not possible for non-stationary processes. That assertion is not correct. Consider the ordinary kriging predictor for predicting at a new location $\mathbf{s}_0$,

$$p_{ok}(\mathbf{Z}; \mathbf{s}_0) = \widehat{\mu} + \mathbf{c}'\mathbf{\Sigma}^{-1}(\mathbf{Z}(\mathbf{s}) - \mathbf{1}\mu),$$

where $\mathbf{\Sigma}$ is the variance matrix of the random field and $\mathbf{c} = \text{Cov}[Z(\mathbf{s}_0), Z(\mathbf{s})]$. It is perfectly acceptable for the covariance matrix $\mathbf{\Sigma}$ to be that of a non-stationary process. We require $\mathbf{\Sigma}$ to be positive definite, a condition entirely separate from stationarity. There *is*, however, a problem with spatial prediction for non-stationary processes. The elements of $\mathbf{\Sigma}$ are unknown and must be estimated from the data. If $\mathbf{\Sigma}$ is parameterized, then the parameters of $\mathbf{\Sigma}(\boldsymbol{\theta})$ must be estimated. In either case, we require valid semivariogram or covariance function models for non-stationary data *and* a mechanism for estimation.

## 8.2 Global Modeling Approaches

### *8.2.1 Parametric Models*

If one understands the mechanisms that contribute to covariance non-stationarity, these can be incorporated in a model for the covariance structure. The parameters of the non-stationary covariance model can subsequently be estimated based on (restricted) maximum likelihood. Hughes-Oliver et al. (1998a) present a correlation model for stochastic processes driven by one or few point sources. Such a point source can be an industrial plant emitting air pollution or waste water, or the center of a wafer in semiconductor processing (Hughes-Oliver et al., 1998b). Their model for a single point source at location $\mathbf{c}$ is

*Point Source Correlation Model, Hughes-Oliver et al.*

$$\text{Corr}[Z(\mathbf{s}_i), Z(\mathbf{s}_j)] = \exp\left\{-\theta_1||\mathbf{s}_i - \mathbf{s}_j|| \exp\left\{\theta_2|c_i - c_j| + \theta_3 \min[c_i, c_j]\right\}\right\}, \tag{8.1}$$

where $c_i = ||\mathbf{s}_i - \mathbf{c}||$ and $c_j = ||\mathbf{s}_j - \mathbf{c}||$ are the distances between sites and the point source. This model is non-stationary because the correlation between sites $\mathbf{s}_i$ and $\mathbf{s}_j$ depends on the site distances from the point source through $c_i$ and $c_j$.

The Hughes-Oliver model is a clever generalization of the exponential correlation model. First note that when $\theta_2 = \theta_3 = 0$, (8.1) reduces to the exponential correlation model with practical range $\alpha = 3/\theta_1$. In general, the correlation between two sites $\mathbf{s}_i$ and $\mathbf{s}_j$ is that of a process with exponential correlation model and practical range

$$\alpha(\mathbf{s}_i, \mathbf{s}_j) = 3\frac{\exp\{-\theta_2|c_i - c_j| - \theta_3 \min[c_i, c_j]\}}{\theta_1}.$$

Consider two sites equally far from the point source, so that $c_i = c_j$. Then,

$$\alpha(\mathbf{s}_i, \mathbf{s}_j) = 3\frac{\exp\{-\theta_3||\mathbf{s}_i - \mathbf{c}||\}}{\theta_1}.$$

The spatial correlation will be small if $\theta_3$ large and/or the site is far removed from the point source.

The correlation model (8.1) assumes that the effects of the point source are circular. Airborne pollution, for example, does not evolve in circular pattern. Point source anisotropy can be incorporated by correcting distances for geometric anisotropy,

$$\begin{aligned}\mathrm{Corr}[Z(\mathbf{s}_i), Z(\mathbf{s}_j)] &= \exp\{-\theta_1 h^*_{ij} \\ &\quad \times \exp\{\theta_2|c^*_i - c^*_j| + \theta_3 \min[c^*_i, c^*_j]\}\},\end{aligned}$$

where $h^*_{ij} = ||\mathbf{A}(\mathbf{s}_i - \mathbf{s}_j)||$, $c^*_i = ||\mathbf{A}_c(\mathbf{s}_i - \mathbf{c})||$ and $\mathbf{A}$, $\mathbf{A}_c$ are anisotropy matrices.

To establish whether a non-stationary correlation model such as (8.1) is valid can be a non-trivial task. Hughes-Oliver et al. (1998a) discuss that the obvious requirements $\theta_1 > 0$, $\theta_2, \theta_3 \geq 0$ are necessary but not sufficient for a covariance matrix defined by (8.1) to be positive semi-definite. If it is not possible to derive conditions on the model parameters that imply positive semi-definiteness, then one must examine the eigenvalues of the estimated covariance or correlation matrix to ensure that at least the estimated model is valid.

### *8.2.2 Space Deformation*

If a random process does not have the needed attributes for statistical inference, it is common to employ a transformation of the process that leads to the desired properties. Lognormal kriging (§5.6.1) or modeling a geometrically anisotropic covariance structure (§4.3.7) are two instances where transformations in spatial statistics are routinely employed. An important difference between the two types of transformations is whether they transform the response variable (lognormal kriging) or the coordinate system (anisotropic modeling). Recall from §4.3.7 that if iso-correlation contours are elliptical, a linear transformation $\mathbf{s}^* = f(\mathbf{s}) = \mathbf{A}\mathbf{s}$ achieves the rotation and scaling of the coordinate system so that covariances based on the $\mathbf{s}^*$ coordinates are isotropic. If $g()$ is a variogram, then

$$\text{Var}[Z(\mathbf{s}_i) - Z(\mathbf{s}_j)] = g(||f(\mathbf{s}_i) - f(\mathbf{s}_j)||), \tag{8.2}$$

and $\widehat{g}(||f(\mathbf{s}_i) - f(\mathbf{s}_j)||)$ is its estimate, assuming $f$ is known.

This general idea can be applied to spatial processes with non-stationary covariance function. The result is a method known as "space deformation:"

- find a function $f$ that transforms the space in such a way that the covariance structure in the transformed space is stationary and isotropic;
- find a function $g$ such that $\text{Var}[Z(\mathbf{s}_i) - Z(\mathbf{s}_j)] = g(||f(\mathbf{s}_i) - f(\mathbf{s}_j)||)$;
- if $f$ is unknown, then a natural estimator is

$$\widehat{\text{Var}}[Z(\mathbf{s}_i) - Z(\mathbf{s}_j)] = \widehat{g}(||\widehat{f}(\mathbf{s}_i) - \widehat{f}(\mathbf{s}_j)||).$$

The work by Sampson and Guttorp (1992) marks an important milestone for this class of non-stationary models. We only give a rough outline of the approach, the interested reader is referred to the paper by Sampson and Guttorp (1992) for details and to Mardia, Kent, and Bibby (1979, Ch. 14). The technique of multidimensional scaling, key in determining the space deformation, was discussed earlier (see §4.8.2).

In the non-stationary case, neither $g$ nor $f$ are known and the Sampson-Guttorp approach determines both in a two-step process. First, the "spatial dispersions"

$$d_{ij}^2 = \widehat{\text{Var}}[Z(\mathbf{s}_i) - Z(\mathbf{s}_j)]$$

are computed and multidimensional scaling is applied to the matrix $\mathbf{D} = [d_{ij}]$. Starting from an initial configuration of sites, the result of this operation is a set of points $\{\mathbf{s}_1^*, \cdots, \mathbf{s}_n^*\}$ and a monotone function $m$, so that

$$m(d_{ij}) \approx ||\mathbf{s}_i^* - \mathbf{s}_j^*||.$$

Since $m$ is monotone we can solve $d_{ij} = m^{-1}(||\mathbf{s}_i^* - \mathbf{s}_j^*||)$ and obtain $d_{ij}^2 = g(||\mathbf{s}_i^* - \mathbf{s}_j^*||)$. The approach of Sampson and Guttorp (1992) in analyzing spatio-temporal measurements of solar radiation is to assume temporal stationarity and to estimate the $d_{ij}^2$ from sample moments computed at each spatial location across the time points. Once multidimensional scaling has been applied to the $d_{ij}$, a model $g()$ is fit to the scatterplot of the $d_{ij}^2$ versus distances $h_{ij}^* = ||\mathbf{s}_i^* - \mathbf{s}_j^*||$ in the deformed space. This fit can be a parametric model or a nonparametric alternative as discussed in §4.6. Sampson and Guttorp chose a nonparametric approach.

The space deformation method presents several challenges to the analyst.

- An empirical semivariogram is needed as input for the multidimensional scaling algorithm. Sampson and Guttorp (1992) call the $d_{ij}^2$ the spatial dispersions instead of the sample variogram to disassociate the quantities from notions of stationarity. How to obtain estimates of the $d_{ij}^2$ without stationarity assumptions or replication is not clear. Repeated measurements over time of each spatial location and an assumption of temporal stationarity enable the estimation of the spatial dispersions in Sampson and Guttorp's case.

- In order to compute the covariance between arbitrary points $\mathbf{s}_i$ and $\mathbf{s}_j$ in the domain of measurement, a smooth and injective mapping is required that yields the corresponding coordinates of the points in the deformed space in which the Euclidean distances $h_{ij}$ are computed. In other words, one needs a function $f(\mathbf{s}) = \mathbf{s}^*$. Sampson and Guttorp (1992) determine $f$ as a smoothing spline.
- The result of the multidimensional scaling algorithm depends on the initial configuration of sites. The observation sites are a logical choice.

## 8.3 Local Stationarity

A function $f(x)$ that changes in $x$ can be viewed as constant or approximately constant in an interval $[a, a+\epsilon]$, provided $|\epsilon| > 0$ is sufficiently small. A similar idea can be applied to non-stationary random fields; even if the mean and/or covariance function are non-stationary throughout the domain, it may be reasonable to assume that

(i) the process is stationary in smaller subregions;

(ii) the process can be decomposed into simpler, stationary processes.

We label methods for non-stationary data making assumptions (i) or (ii) as "local" methods. Three important representatives are moving window techniques, convolutions, and weighted stationary processes.

### *8.3.1 Moving Windows*

This technique was developed by Haas (1990) to perform spatial prediction in non-stationary data and extended to the spatio-temporal case in Haas (1995). To compute the ordinary kriging predictor

$$p_{ok}(\mathbf{Z}; \mathbf{s}_0) = \widehat{\mu} + \mathbf{c}'\mathbf{\Sigma}^{-1}(Z(\mathbf{s}) - \mathbf{1}\mu)$$

for a set of prediction locations $\mathbf{s}_0{}^{(1)}, \cdots, \mathbf{s}_0{}^{(m)}$, one only needs to recompute the vector of covariances between the observed and prediction location, $\mathbf{c}$. For large data sets, however, the inversion of the variance-covariance matrix $\mathbf{\Sigma}$ is a formidable computational problem, even if performed only once. In addition, observations far removed from the prediction location may contribute only little to the predicted value at $\mathbf{s}_0$, their kriging weights are close to zero. It is thus a commonplace device to consider for prediction at $\mathbf{s}_0{}^{(i)}$ only those observed sites within a pre-defined neighborhood of $\mathbf{s}_0{}^{(i)}$. This kriging window changes with prediction location and points outside of the window have kriging weight 0 (see §5.4.2). The local kriging approach has advantages and disadvantages. The predictor that excludes observed sites is no longer best and the analyst must decide on the size and shape of the kriging neighborhood. As points are included and excluded in the neighborhoods with changing prediction location, spurious discontinuities can be introduced. On the upside, local kriging is computationally less involved than solving the kriging equations based on all $n$ data points for every prediction location. Also, if the mean is non-stationary, it may be reasonable to assume that the mean is constant within the kriging window and to re-estimate $\mu$ based on the observations in the neighborhood.

Whether the mean is estimated globally or locally, the spatial covariation in local kriging is based on the same global model. Assume that the covariances are determined based on some covariance or semivariogram model with parameter vector $\boldsymbol{\theta}$. Local kriging can then be expressed as

$$p_{ok}(\mathbf{Z}; \mathbf{s}_0^{(i)}) = \widehat{\mu} + \mathbf{c}^{(i)}(\widehat{\boldsymbol{\theta}})'\boldsymbol{\Sigma}^{(i)}(\widehat{\boldsymbol{\theta}})^{-1}(\mathbf{Z}(\mathbf{s})^{(i)} - \mathbf{1}\mu),$$

where $\mathbf{Z}(\mathbf{s})^{(i)}$ denotes the subset of points in the kriging neighborhood, $\boldsymbol{\Sigma}^{(i)} = \text{Var}[\mathbf{Z}(\mathbf{s})^{(i)}]$, and $\mathbf{c}^{(i)} = \text{Cov}[Z(\mathbf{s}_0{}^{(i)}), \mathbf{Z}(\mathbf{s})^{(i)}]$. All $n$ data points contribute to the estimation of $\boldsymbol{\theta}$ in local kriging.

The moving window approach of Haas (1990, 1995) generalizes this idea by re-estimating the semivariogram or covariance function locally within a circular neighborhood (window). A prediction is made at the center of the window using the local estimate of the semivariogram parameters,

$$p_{ok}(\mathbf{Z}; \mathbf{s}_0^{(i)}) = \widehat{\mu} + \mathbf{c}^{(i)}(\widehat{\boldsymbol{\theta}}^{(i)})'\boldsymbol{\Sigma}^{(i)}(\widehat{\boldsymbol{\theta}}^{(i)})^{-1}(\mathbf{Z}(\mathbf{s})^{(i)} - \mathbf{1}\mu).$$

The neighborhood for local kriging could conceivably be different from the neighborhood used to derive the semivariogram parameters $\boldsymbol{\theta}^{(i)}$, but the neighborhoods are usually the same. Choosing the window size must balance the need for a sufficient number of pairs to estimate the semivariogram parameters reliably (large window size), and the desire to make the window small so that a stationarity assumption within the window is tenable. Haas (1990) describes a heuristic approach to determine the size of the local neighborhood: enlarge a circle around the prediction site until at least 35 sites are included, then include five sites at a time until there is at least one pair of sites at each lag class and the nonlinear least squares fit of the local semivariogram converges.

### *8.3.2 Convolution Methods*

Constructing non-stationary processes from convolutions is an elegant and powerful approach with great promise. We include convolution methods in the class of local modeling approaches because of the presence of a kernel function that can be viewed as a local weighing function, decreasing with the distance from the target point $\mathbf{s}$, and because of local window techniques used at the estimation stage. Two illustrative referenced for this approach are Higdon (1998) and Higdon, Swall, and Kern (1999).

Consider a zero-mean white noise process $X(\mathbf{s})$ such that $\text{E}[X(\mathbf{s})] = \mu_x = 0$, $\text{Var}[X(\mathbf{s})] = \phi_x$, and $\text{E}[X(\mathbf{s})X(\mathbf{s}+\mathbf{h})] = 0, \mathbf{h} \neq \mathbf{0}$. Then a weakly stationary random field $Z(\mathbf{s})$ can be constructed by convolving $X(\mathbf{s})$ with a kernel function $K_{\mathbf{s}}(\mathbf{u})$, centered at s. The random field

$$Z(\mathbf{s}) = \int_{\text{all } \mathbf{u}} K_{\mathbf{s}}(\mathbf{u})X(\mathbf{u})\,d\mathbf{u}$$

has mean $\mathrm{E}[Z(\mathbf{s})] = \mu_x \int_{\mathbf{u}} K_{\mathbf{s}}(\mathbf{u})\,d\mathbf{u} = 0$ and covariance function

$$\begin{aligned}\mathrm{Cov}[Z(\mathbf{s}), Z(\mathbf{s}+\mathbf{h})] &= \mathrm{E}[Z(\mathbf{s})Z(\mathbf{s}+\mathbf{h})] \\ &= \int_{\mathbf{u}}\int_{\mathbf{v}} K_{\mathbf{s}}(\mathbf{u})K_{\mathbf{s}+\mathbf{h}}(\mathbf{v})\mathrm{E}[X(\mathbf{u})X(\mathbf{v})]\,d\mathbf{u}\,d\mathbf{v} \\ &= \phi_x \int_{\mathbf{v}} K_{\mathbf{s}}(\mathbf{v})K_{\mathbf{s}+\mathbf{h}}(\mathbf{v})\,d\mathbf{v}.\end{aligned}$$

Since the covariance function depends on the choice of the kernel function, a non-stationary covariance function can be constructed by varying the kernel spatially. Following Higdon, Swall, and Kern (1999), consider the following progression for a process in $\mathbf{R}^2$ and points $\mathbf{u} = [u_x, u_y], \mathbf{s} = [s_x, s_y]$.

1. $K_{\mathbf{s}}(\mathbf{u}, \sigma^2) = (2\pi\sigma^2)^{-1}\exp\{-0.5[(u_x - s_x)^2 + (u_y - s_y)^2]\}$.

   This bivariate normal kernel is the product kernel of two univariate Gaussian kernels with equal variances. Note that $\sigma^2$ is the dispersion of the kernel functions, not the variance of the excitation process $X(\mathbf{s})$. The result of the convolution with $X(\mathbf{s})$ is a stationary, isotropic random field with gaussian covariance structure. The kernel function has a single parameter, $\sigma^2$.

   *Isotropic Convolution*

2. $K_{\mathbf{s}}(\mathbf{u}, \boldsymbol{\theta}) = |\boldsymbol{\Sigma}|^{-1/2}(2\pi)^{-1}\exp\{-0.5(\mathbf{u}-\mathbf{s})'\boldsymbol{\Sigma}^{-1}(\mathbf{u}-\mathbf{s})\}$,

   $$\boldsymbol{\Sigma} = \begin{bmatrix} \sigma_x^2 & \rho\sigma_x\sigma_y \\ \rho\sigma_x\sigma_y & \sigma_y^2 \end{bmatrix}.$$

   *Anisotropic Convolution*

   The bivariate Gaussian kernel with correlation and unequal variances yields a random field with geometric anisotropy and gaussian covariance structure. The parameters of this kernel function are $\boldsymbol{\theta} = [\sigma_x, \sigma_y, \rho]$. The isotropic situation in 1. is a special case, $\rho = 0, \sigma_x = \sigma_y$.

3. $K_{\mathbf{s}}(\mathbf{u}, \boldsymbol{\theta}(\mathbf{s})) = |\boldsymbol{\Sigma}(\mathbf{s})|^{-1/2}(2\pi)^{-1}\exp\{-0.5(\mathbf{u}-\mathbf{s})'\boldsymbol{\Sigma}(\mathbf{s})^{-1}(\mathbf{u}-\mathbf{s})\}$.

   The parameters of the kernel function depend on the spatial location, $\boldsymbol{\theta}(\mathbf{s}) = [\sigma_x(\mathbf{s}), \sigma_y(\mathbf{s}), \rho(\mathbf{s})]$. The resulting process has non-stationary covariance function

   *Non-stationary Convolution*

   $$\mathrm{Cov}[Z(\mathbf{s}_i), Z(\mathbf{s}_j)] = \phi_x \int_{\mathbf{v}} K_{\mathbf{s}_i}(\mathbf{v}, \boldsymbol{\theta}(\mathbf{s}_i))K_{\mathbf{s}_j}(\mathbf{v}, \boldsymbol{\theta}(\mathbf{s}_j))\,d\mathbf{v}.$$

   Higdon et al. (1999) give a closed-form expression for this integral.

For a process in $\mathbf{R}^d$ and $n$ sites, the non-stationary convolution model has $nd(d+1)/2$ kernel parameters, a large number. To make estimation feasible, Higdon (1998) draws on the local window idea and the connection between the gaussian semivariogram and the gaussian convolution kernel. First, estimate the kernel parameters only for a subset of sites. Second, at each site in the subset estimate the semivariogram parameters in a local neighborhood and convert the parameter estimates to parameters of the kernel function. Third, interpolate the kernel function parameters between the estimation sites so that the kernel functions vary smoothly throughout the domain.

### *8.3.3 Weighted Stationary Processes*

The method of weighted stationary processes is closely related to convolution methods and many models in this class have a convolution representation. The important difference between this and the previously discussed approach is the assumption about which model component is spatially evolving. The convolution method of Higdon (1998) and Higdon et al. (1999) varies parameters of the kernel function spatially. The weighted stationary process approach of Fuentes (2001) varies the stationary processes but not the kernel.

Fuentes (2001) assumes that the non-stationary process $Z(\mathbf{s})$ can be written as the weighted sum of stationary processes $Z_1(\mathbf{s}), \cdots, Z_k(\mathbf{s})$,

$$Z(\mathbf{s}) = \sum_{i=1}^{k} w_i(\mathbf{s}) Z_i(\mathbf{s}). \tag{8.3}$$

The local processes are uncorrelated, $\mathrm{Cov}[Z_i(\mathbf{s}), Z_j(\mathbf{s})] = 0, i \neq j$ and have covariance functions $\mathrm{Cov}[Z_i(\mathbf{s}), Z_i(\mathbf{s}+\mathbf{h})] = C(\mathbf{h}, \boldsymbol{\theta}_i)$. The resulting, non-stationary covariance function of the observed process is

$$\begin{aligned}
\mathrm{Cov}[Z(\mathbf{s}), Z(\mathbf{s}+\mathbf{h})] &= \mathrm{Cov}\left[\sum_{i=1}^{k} Z_i(\mathbf{s}) w_i(\mathbf{s}), \sum_{i=1}^{k} Z_i(\mathbf{s}+\mathbf{h}) w_i(\mathbf{s}+\mathbf{h})\right] \\
&= \sum_{i=1}^{k}\sum_{j=1}^{k} \mathrm{Cov}[Z_i(\mathbf{s}), Z_j(\mathbf{s}+\mathbf{h})] w_i(\mathbf{s}) w_j(\mathbf{s}+\mathbf{h}) \\
&= \sum_{i=1}^{k} C(\mathbf{h}, \boldsymbol{\theta}_i) w_i(\mathbf{s}) w_i(\mathbf{s}+\mathbf{h}).
\end{aligned}$$

To see how mixing locally stationary processes leads to a model with non-stationary covariance, we demonstrate the Fuentes model with the following, simplified example.

**Example 8.1** Consider a one-dimensional stochastic process on the interval $(0, 10)$. The segment is divided into four intervals of equal widths, $S_1, \cdots, S_4$. The covariance function in segment $i = 1, \cdots, 4$ is

$$\mathrm{Cov}[Z_i(t_1), Z_j(t_2)] = \exp\{-|t_2 - t_1|/\theta_i\},$$

with $\theta_1 = 10, \theta_2 = 2, \theta_3 = 2, \theta_4 = 10$. The weight functions $w_i(t)$ are Gaussian kernels with standard deviation $\sigma = 1.5$, centered in the respective segment, so that

$$w_i(t) = \frac{1}{\sqrt{2\pi}\sigma} \exp\left\{-0.5(t - c_i)^2/\sigma^2\right\},$$

$c_1 = 1.25, c_2 = 3.75, c_3 = 6.25, c_4 = 8.75$. Figure 8.1 shows a single realization of this process along with the kernel functions. Although the covariance function changes discontinuously between segments, the smooth, overlapping kernels create a smooth process.

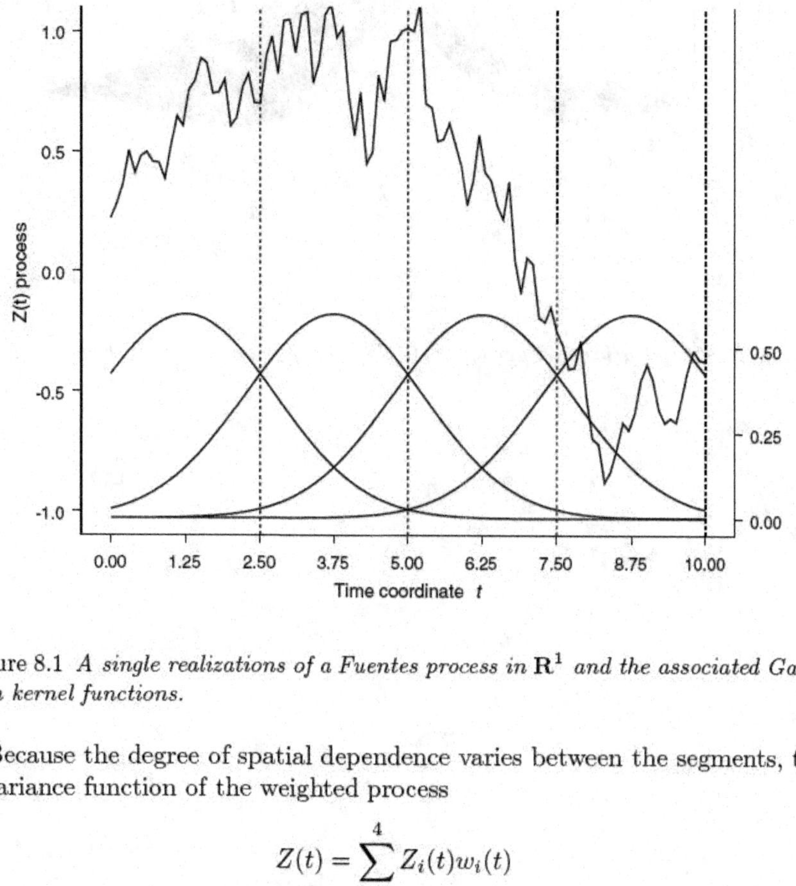


Figure 8.1 *A single realizations of a Fuentes process in* $\mathbf{R}^1$ *and the associated Gaussian kernel functions.*

Because the degree of spatial dependence varies between the segments, the covariance function of the weighted process

$$Z(t) = \sum_{i=1}^{4} Z_i(t) w_i(t)$$

is non-stationary (Figure 8.2). The variances are not constant, and the correlations between locations decrease slower for data points in the first and last segment and faster in the intermediate segments. □

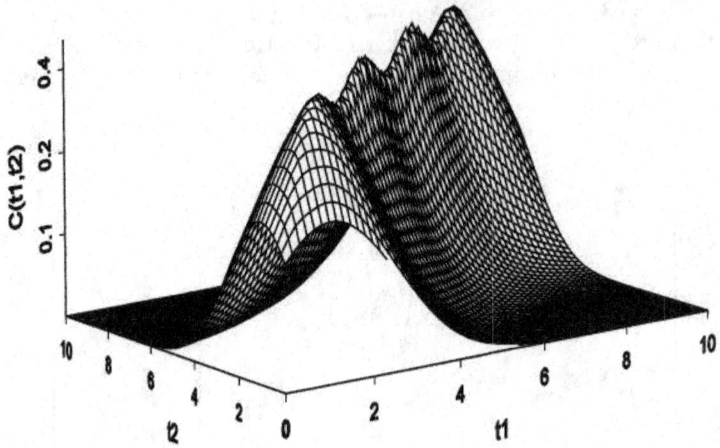


Figure 8.2 *Covariance function of the Fuentes process shown in Figure 8.1.*

CHAPTER 9

# Spatio-Temporal Processes

## 9.1 A New Dimension

The significant advance in the statistical analysis of spatial data is to acknowledge the fact that the configuration of observations carries important information about the relationship of data points.

We made the argument early on that incorporating the spatial context into the statistical analysis is a need and a benefit when "space matters." By the same token, we must argue now that addressing the time component in space-time processes must not be overlooked and that many—if not most—spatial processes change over time. Unfortunately, statistical tools for the analysis of spatio-temporal processes are not (yet) as fully developed as methods for time series or spatial data alone. Also, there is paucity of commercial software solutions for such data. The temptation arises thus naturally to proceed along one of the following lines:

1. separate spatial analyses for each time point;

2. separate temporal analyses for each location;

3. spatio-temporal data analysis with methods for random fields in "$\mathbf{R}^{d+1}$".

The first two approaches can be considered *conditional* methods because they isolate a particular time point or location and apply standard techniques for the type of data that results. A two-stage variation on the theme is to combine the results from the conditional analyses in a second stage. Two-stage approaches are common in statistical application where multiple sources of variation are at work but methodology and/or software (and computing power) are unavailable for joint modeling. A case in point are nonlinear mixed model applications for clustered data. Two important sources of variation there are the changes in response as a function of covariates for each cluster and cluster-to-cluster heterogeneity. The first source of variation is captured by a nonlinear regression model. The second source is expressed by varying at random model coefficients among clusters. The obvious two-stage approach is to fit the nonlinear model separately to each cluster and to combine the regression coefficients into a set of overall (population-averaged) coefficients in a second stage. The reader interested in multi-stage and modern, mixed model based approaches to nonlinear mixed models is referred to Davidian and Giltinan (1995).

Two-stage approaches are appealing because of simplicity but have also serious shortcomings.

- If it is not possible to analyze the data spatially for a time point $t_j$, then data collected at that time will not contribute in the second stage when the spatial analyses are combined. Data that are sparse in time or space can present difficulties in this regard. Integrating observations over time if data are sparse temporally, may enable a spatial analysis but can confound temporal effects. When data "locations" are unique in time and space, a two-stage analysis fails.
- There are many ways to summarize the results from the first-stage analysis. One can examine the temporal distribution of semivariogram ranges and sills, for example. To reasonably combine those statistics into some overall measure—if that is desired—requires information about the temporal correlation between the statistics.
- Interpolation of observations in a continuous space-time process should take into account the interactions between the spatial and temporal components and allow for predictions in time and space. Separate analyses in time (space) allow predictions in space (time) only.

Joint analyses of spatio-temporal data are preferable to separate analyses. In the process of building a joint model, separate analyses are valuable tools, however. Again, we can draw a parallel to nonlinear mixed modeling. Fitting the nonlinear response model separately to all those clusters that provide a sufficient number of observations and examining the variation of the regression coefficients enables one to make better choices about which coefficients to vary at random. Examining semivariogram parameter estimates over time and auto-regressive estimates for each location leads to more informed model selection for a joint analysis.

To express the space-time nature of the process under study, we expand our previous formulation of a stochastic process in the following way:

$$\{Z(\mathbf{s},t) : \mathbf{s} \in D(t) \subset \mathbf{R}^2, t \in T\}. \tag{9.1}$$

The observation of attribute $Z$ at time $t$ and location $\mathbf{s}$ is denoted as $Z(\mathbf{s},t)$ to emphasize that the temporal "location" $t$ is not just an added spatial coordinate. The dependence of the domain $D$ on time symbolizes the condition where the spatial domain changes over time. For simplicity, we may assume that $D(t) \equiv D$. The spatial component in (9.1) can be discrete or continuous, fixed or random. The same applies to the temporal component of the process. The distribution of particles in the atmosphere, for example, can be seen as a spatio-temporal process with continuous spatial domain and continuous time. Data on the disposable household incomes by counties on April 15 of every year, comprises a spatial lattice process and a discrete time process. This creates a dizzying array of spatio-temporal data structures.

The third approach mentioned above, to treat spatio-temporal data as spatial data with an extra dimension is not encouraged. Time and space are not directly comparable. Space has no past, present, and future and the spatial coordinate units are not comparable to temporal units. If certain assumptions are met, it is acceptable to model spatio-temporal data with separable covariance structures. This is not the same as modeling spatio-temporal data as "3-D" data. A spatio-temporal process with a spatial component in $\mathbf{R}^2$ may be separable, but it is not a process in $\mathbf{R}^3$. Assume we would treat the spatio-temporal data as the realization of a process in $\mathbf{R}^3$ and denote the coordinates as $\mathbf{s}_i = [x_i, y_i, t_i]'$. Let $\mathbf{s}_i - \mathbf{s}_j = [\mathbf{h}_{ij}, t_i - t_j]$. An exponential correlation model results in

$$\begin{aligned} \text{Corr}[Z(\mathbf{s}_i), Z(\mathbf{s}_j)] &= \exp\{-\theta((x_i - x_j)^2 + (y_i - y_j)^2 + (t_i - t_j)^2)^{1/2}\} \\ &= \exp\{-\theta(||\mathbf{h}_{ij}||^2 + (t_i - t_j)^2)^{1/2}\}. \end{aligned} \qquad (9.2)$$

This is a valid correlation function in $\mathbf{R}^3$ and two observations at the same point in time have correlation $\exp\{-\theta||\mathbf{h}_{ij}||\}$. Similarly, the serial correlation of two observations at the same spatial location is $\exp\{-\theta|t_i - t_j|\}$. Both covariance functions are valid in $\mathbf{R}^2$ and $\mathbf{R}^1$ but the "3-D" covariance does not make much practical sense. The range of the spatial process is different from the temporal process, the units are not comparable. The anisotropy of spatio-temporal models must be reflected in the covariance functions. Gneiting (2002) points out that from a mathematical point of view the space-time domain $\mathbf{R}^d \times \mathbf{R}$ is no different than the spatial domain $\mathbf{R}^{d+1}$. A valid spatio-temporal covariance function is also a valid spatial covariance function. He argues that through notation and construction, the difference between space and time must be brought to the fore.

There are two distances between points in a space-time process: the Euclidean distance between points in space and the Euclidean distance between points in time. The spatio-temporal lag between $Z(\mathbf{s}, t)$ and $Z(\mathbf{s} + \mathbf{h}, t + k)$ is $[\mathbf{h}; k]$. If time is considered an "added" dimension, then the anisotropy must be taken into account. A class of spatio-temporal covariance functions that uses this approach constructs models as follows. If $R$ is a valid correlation function in $\mathbf{R}^{d+1}$, then

$$\text{Corr}[Z(\mathbf{s}_i, t_i), Z(\mathbf{s}_j, t_j)] = R(\theta_s||\mathbf{h}_{ij}||^2 + \theta_t k^2). \qquad (9.3)$$

The parameters $\theta_s$ and $\theta_t$ are the spatial and temporal anisotropy parameters, respectively.

A more reasonable correlation structure than (9.2), drawing again on the exponential model, is

$$\text{Corr}[Z(\mathbf{s}_i, t_i), Z(\mathbf{s}_j, t_j)] = \exp\{-\theta_s||\mathbf{h}_{ij}||\} \times \exp\{-\theta_t|t_i - t_j|\}. \qquad (9.4)$$

This is an example of a separable covariance function. If the time points are evenly spaced, the temporal process has an AR(1) structure and the spatial component has an exponential structure. A model of this kind is considered in Mitchell and Gumpertz (2003) in modeling $CO_2$ released over time on rice plots. Also, it is easy to see that this type of model is a member of the anisotropic class (9.3), if one chooses the gaussian correlation model in (9.4).

## 9.2 Separable Covariance Functions

Whether separable or not, a spatio-temporal covariance function must be a valid covariance function in $\mathbf{R}^d \times \mathbf{R}$ (we restrict discussion to $d = 2$ in this section). Because mathematically there is no difference between that domain and $\mathbf{R}^{d+1}$, but our notation accommodates the special role of space and time, we re-state the conditions for stationarity and validity of a covariance function. A spatio-temporal covariance function is (second-order) stationary in space and time if

$$\mathrm{Cov}[Z(\mathbf{s}_i, t_i), Z(\mathbf{s}_j, t_j)] = C(\mathbf{s}_i - \mathbf{s}_j, t_i - t_j).$$

The covariance function is furthermore isotropic in space if

$$\mathrm{Cov}[Z(\mathbf{s}_i, t_i), Z(\mathbf{s}_j, t_j)] = C(||\mathbf{s}_i - \mathbf{s}_j||, |t_i - t_j|).$$

The functions $C(\mathbf{h}, 0)$ and $C(\mathbf{0}, k)$ are spatial and temporal covariance functions, respectively. The condition of positive definiteness must be met for $C$ to be a valid covariance function. This implies that for any set of spatio-temporal locations $(\mathbf{s}_i, t_i), \cdots, (\mathbf{s}_k, t_k)$ and real numbers $a_1, \cdots, a_k$

$$\sum_{i=1}^{k}\sum_{j=1}^{k} a_i a_j C(\mathbf{s}_i - \mathbf{s}_j, t_i - t_j) \geq 0. \tag{9.5}$$

These features are obvious "extensions" of the requirements for covariance functions in section §2.2. Two of the elementary properties of covariance functions reviewed there are especially important for the construction of spatio-temporal covariance functions:

If $C_j(\mathbf{h})$ are valid covariance functions, $j = 1, \cdots, k$, then $\sum_{j=1}^k b_j\, C_j(\mathbf{h})$ is a valid covariance function, if $b_j \geq 0 \forall j$.

If $C_j(\mathbf{h})$ are valid covariance functions, $j = 1, \cdots, k$, then $\prod_{j=1}^k C_j(\mathbf{h})$ is a valid covariance function.

*Separable Covariance Function*

Separability of spatio-temporal covariance functions can be defined based on the additive and multiplicative properties. A separable spatio-temporal covariance function decomposes $\mathrm{Cov}[Z(\mathbf{s}_i, t_i), Z(\mathbf{s}_j, t_j)]$ into a purely spatial and a purely temporal component. The components have usually different parameters to allow for space-time anisotropy. The final spatio-temporal covariance function is then the result of addition and or multiplication operations. For example, let $C_s(\mathbf{h}; \boldsymbol{\theta}_s)$ and $C_t(k; \boldsymbol{\theta}_t)$ be a spatial and temporal covariance function. Valid separable spatio-temporal covariance functions are

$$\mathrm{Cov}[Z(\mathbf{s}, t), Z(\mathbf{s} + \mathbf{h}, t + k)] = C_s(\mathbf{h}; \boldsymbol{\theta}_s)\, C_t(k; \boldsymbol{\theta}_t)$$

and

$$\mathrm{Cov}[Z(\mathbf{s}, t), Z(\mathbf{s} + \mathbf{h}, t + k)] = C_s(\mathbf{h}; \boldsymbol{\theta}_s) + C_t(k; \boldsymbol{\theta}_t)$$

These covariance models are sometimes referred to as the product and the sum covariance structures. The product-sum covariance structure

$$\mathrm{Cov}[Z(\mathbf{s}, t), Z(\mathbf{s} + \mathbf{h}, t + k)] = C_s(\mathbf{h}; \boldsymbol{\theta}_s)\, C_t(k; \boldsymbol{\theta}_t) + C_s(\mathbf{h}; \boldsymbol{\theta}_s) + C_t(k; \boldsymbol{\theta}_t)$$

of De Cesare, Myers, and Posa (2001) is generally nonseparable.

Separable covariance functions are easy to work with and valid, provided the components are valid covariance functions. Furthermore, existing commercial software for spatial data analysis can sometimes be coaxed into fitting separable spatio-temporal covariance models. Mitchell and Gumpertz (2003), for example, use a spatio-temporal covariance function with product separability in which the temporal process has a first-order autoregressive correlation structure (see equation (9.4)). This, in turn, enables the authors to rewrite the observational model in autoregressive form, which makes parameter estimation possible based on nonlinear mixed model tools in SAS.

Cressie and Huang (1999) note that separable models are often chosen for their convenience. We add that this is true for related models. Posa (1993) notes the invariance of kriging predictions to scaling of the covariance functions. That is, the kriging solutions in no-nugget models remain the same, regardless of changes in scale of the observations (the kriging variance, mind you, is not invariant). The spatio-temporal covariance models considered by Posa (1993) are of the form

$$\mathrm{Cov}[Z(\mathbf{s},t), Z(\mathbf{s}+\mathbf{h},t+k) = \sigma^2(t)\, C_s(\mathbf{h}). \tag{9.6}$$

Only the sill of the covariance function is time dependent to account for non-stationarity in time. If $\sigma^2(t) \equiv \sigma^2$, model (9.6) is a product separable model with a nugget-only model in time. Although the model is non-stationary, the kriging predictions at a given time point are not affected by the non-stationary temporal variance.

The primary drawback of separable models is not to incorporate space-time interactions. Consider a product separable structure

$$\mathrm{Cov}[Z(\mathbf{s},t), Z(\mathbf{s}+\mathbf{h},t+k)] = C_s(\mathbf{h};\boldsymbol{\theta}_s)\, C_t(k;\boldsymbol{\theta}_t)$$

The spatial covariances at time lag $k$ and $u \neq k$ have the same shape, they are proportional to each other. The temporal and spatial components represent dependencies in time (given spatial location) and space (given time). The processes do not act upon each other. To incorporate space-time interactions requires non-separability of the covariance function.

## 9.3 Non-Separable Covariance Functions

Non-separable, valid covariance functions for spatio-temporal data are typically more complicated than separable models but incorporate space-time interactions. Some non-separable models reduce to separable ones for particular values of model parameters and allow a test for separability. Methods for constructing valid covariance functions with spatio-temporal interactions include the monotone function approach of Gneiting (2002), the spectral method of Cressie and Huang (1999), the mixture approach, and the partial differential equation approach of Jones and Zhang (1997).

### 9.3.1 Monotone Function Approach

Gneiting (2002) presented a flexible and elegant approach to construct spatio-temporal covariance functions. The method is powerful because it does not require operations in the spectral domain and builds valid covariance functions from elementary components whose validity is easily checked. To fix ideas let $[\mathbf{h}; k]$ denote a lag vector in $\mathbf{R}^d \times \mathbf{R}^1$ and choose two functions $\phi(t), t \geq 0$ and $\psi(t), t \geq 0$ such that $\phi(t)$ is completely monotone and $\psi(t)$ is positive with a completely monotone derivative. The functions $\phi(t) = \exp\{-ct\}$, $c > 0$ and $\psi(t) = (at+1)$, $a > 0$, for example, satisfy the requirements. Tables 1 and 2 in Gneiting (2002) list a variety of functions and valid ranges of their parameters.

With these ingredients in place a valid covariance function in $\mathbf{R}^d \times \mathbf{R}^1$ is

$$\text{Cov}[Z(\mathbf{s},t), Z(\mathbf{s}+\mathbf{h}, t+k)] = C(\mathbf{h}, k) = \frac{\sigma^2}{\psi(|k|^2)^{d/2}} \phi\left(||\mathbf{h}||^2/\psi(|k|^2)\right). \quad (9.7)$$

Gneiting (2002) illustrates the construction with functions $\phi(t) = \exp\{-ct^\gamma\}$ and $\psi(t) = (at^\alpha + 1)^\beta$. They satisfy the requirements for

$$c, a > 0; \quad 0 < \gamma, \alpha \leq 1; \quad 0 \leq \beta \leq 1.$$

Substituting into (9.7) yields (for $d = 2$)

$$C(\mathbf{h}, k) = \frac{\sigma^2}{(a|k|^{2\alpha}+1)^\beta} \exp\left\{-\frac{c||\mathbf{h}||^{2\gamma}}{(a|k|^{2\alpha}+1)^{\beta\gamma}}\right\}. \quad (9.8)$$

For $\beta = 0$ the covariance function does not depend on the time lag. Multiplying (9.8) with a purely temporal covariance function leads to a separable model for $\beta = 0$. For $||\mathbf{h}|| = 0$, (9.8) reduces to the temporal covariance function $C_t(k) = (a|k|^{2\alpha} + 1)^{-\beta_t}$. The function

$$C_t(k) \times C(\mathbf{h}, k) = \frac{\sigma^2}{(a|k|^{2\alpha}+1)^{\beta_t+\beta}} \exp\left\{-\frac{c||\mathbf{h}||^{2\gamma}}{(a|k|^{2\alpha}+1)^{\beta\gamma}}\right\} \quad (9.9)$$

is a valid spatio-temporal covariance function. It is separable for $\beta = 0$ and non-separable otherwise. Since the separable and non-separable models are nested, a statistical test for $H_0$: $\beta = 0$ can be carried out. One technique would be to estimate the parameters of (9.9) by (restricted) maximum likelihood with and without the constraint $\beta = 0$ and to compare twice the negative (restricted) log likelihood. A correction for the fact that the null value of the test falls on the boundary of the parameter space can be applied (Self and Liang, 1987).

### 9.3.2 Spectral Approach

We noted in §4.3 (page 141) that by Bochner's theorem valid covariance functions have a spectral representation. For a process in $\mathbf{R}^d$ we can write

$$C(\mathbf{h}) = \int_{-\infty}^{\infty} \cdots \int_{-\infty}^{\infty} \exp\{i\boldsymbol{\omega}'\mathbf{h}\} s(\boldsymbol{\omega})\, d\boldsymbol{\omega},$$

which suggests the following method for constructing a valid covariance function: determine a valid spectral density and take its inverse Fourier transform. To acknowledge the physical difference between time and space, the spectral representation of a spatio-temporal covariance that satisfies (9.5) is written as

$$C(\mathbf{h}, k) = \int_{-\infty}^{\infty} \cdots \int_{-\infty}^{\infty} \exp\{i\boldsymbol{\omega}'\mathbf{h} + i\tau k\} s(\boldsymbol{\omega}, \tau)\, d\boldsymbol{\omega} d\tau, \qquad (9.10)$$

where $s(\boldsymbol{\omega}, \tau)$ is the spectral density. One could proceed with the construction of covariance functions as in the purely spatial case: determine a valid spatio-temporal spectral density and integrate. This is essentially the method applied by Cressie and Huang (1999), but these authors use two clever devices to avoid the selection of a joint spatio-temporal spectral density.

First, because of (9.10), the covariance function and the spectral density are a Fourier transform pair. Integration of the spatial and temporal components can be separated in the frequency domains:

$$\begin{aligned} s(\boldsymbol{\omega}, \tau) &= \frac{1}{(2\pi)^{d+1}} \int_{-\infty}^{\infty} \cdots \int_{-\infty}^{\infty} \exp\{-i\boldsymbol{\omega}'\mathbf{h} - i\tau k\} C(\mathbf{h}, k)\, d\mathbf{h} dk \\ &= \frac{1}{2\pi} \int_{-\infty}^{\infty} \exp\{-i\tau k\} \Bigg[ \frac{1}{(2\pi)^d} \int_{-\infty}^{\infty} \cdots \int_{-\infty}^{\infty} \\ &\qquad \exp\{-i\boldsymbol{\omega}'\mathbf{h}\} C(\mathbf{h}, k)\, d\mathbf{h} \Bigg] dk \\ &= \frac{1}{2\pi} \int_{-\infty}^{\infty} \exp\{-i\tau k\} h(\boldsymbol{\omega}, k)\, dk. \end{aligned} \qquad (9.11)$$

The function $h(\boldsymbol{\omega}, k)$ is the (spatial) spectral density for temporal lag $k$. What has been gained so far is that if we know the spectral density for a given lag $k$, the spatio-temporal spectral density is obtained with a one-dimensional Fourier transform. And it is presumably simpler to develop a model for $h(\boldsymbol{\omega}, k)$ than it is for $s(\boldsymbol{\omega}, \tau)$. To get the covariance function from there still requires complicated integration in (9.10).

The second device used by Cressie and Huang (1999) is to express $h(\boldsymbol{\omega}, k)$ as the product of two simpler functions. They put $h(\boldsymbol{\omega}, k) \equiv R(\boldsymbol{\omega}, k) r(\boldsymbol{\omega})$, where $R(\boldsymbol{\omega}, k)$ is a continuous correlation function, $r(\boldsymbol{\omega}) > 0$. If $\int R(\boldsymbol{\omega}, k) dk < \infty$ and $\int r(\boldsymbol{\omega}) d\boldsymbol{\omega} < \infty$, then substitution into (9.11) gives

$$s(\boldsymbol{\omega}, \tau) = \frac{1}{2\pi} r(\boldsymbol{\omega}) \int_{-\infty}^{\infty} \exp\{-i\tau k\} R(\boldsymbol{\omega}, k)\, dk$$

and

$$C(\mathbf{h}, k) = \int_{-\infty}^{\infty} \cdots \int_{-\infty}^{\infty} \exp\{-i\boldsymbol{\omega}'\mathbf{h}\} R(\boldsymbol{\omega}, k) k(\boldsymbol{\omega})\, d\boldsymbol{\omega}. \qquad (9.12)$$

Cressie and Huang (1999) present numerous examples for functions $R(\boldsymbol{\omega}, k)$ and $k(\boldsymbol{\omega})$ and the resulting spatio-temporal covariance functions. Gneiting (2002) establishes that some of the covariance functions in the paper by Cressie and Huang are not valid, because one of the correlation functions $R(\omega, k)$ used in their examples does not satisfy the needed conditions.

### 9.3.3 *Mixture Approach*

Instead of integration in the frequency domain, nonseparable covariance functions can also be constructed by summation or integration in the spatio-temporal domain. Notice that if $Z_s(\mathbf{s})$ and $Z_t(t)$ are purely spatial and temporal processes with covariance functions $C_s(\mathbf{h};\boldsymbol{\theta}_s)$ and $C_t(k;\boldsymbol{\theta}_t)$, respectively, then $Z(\mathbf{s},t) = Z_s(\mathbf{s})Z_t(t)$ has the separable product covariance function

$$C(\mathbf{h},k) = C_s(\mathbf{h};\boldsymbol{\theta}_s) \times C_t(k;\boldsymbol{\theta}_t),$$

if $Z_s(\mathbf{s})$ and $Z_t(t)$ are uncorrelated. Space-time interactions can be incorporated by mixing product covariance or correlation functions. Ma (2002) considers the probability mass functions $\pi_{ij}$, where $(i,j) \in \mathbf{Z}_+$. In other words, $\pi_{ij}$ is the mass function of $[U,V]$, a bivariate discrete random vector with support on the non-negative integers. If, conditional on $[U,V] = [i,j]$, the spatio-temporal process has correlation function

$$R(\mathbf{h},k\,|\,U=i, V=j) = R_s^i(\mathbf{h}) \times R_t^j(k),$$

then the correlation function of the unconditional process is the non-separable model

*Positive (bivariate) Power Mixture*

$$R(\mathbf{h},k) = \sum_{i=0}^{\infty}\sum_{j=0}^{\infty} R_s^i(\mathbf{h})R_t^j(k)\pi_{ij}. \tag{9.13}$$

Ma (2002) terms this model a *positive power mixture*, it makes use of the fact that if $R(\mathbf{u})$ is a correlation model in $\mathbf{R}^d$, then $R(\mathbf{u})^i$ is also a valid correlation model in $\mathbf{R}^d$ for any positive integer $i$. The method of power mixtures does not require a bivariate, discrete mass function. This is important, because such distributions are quite rare. A non-separable model can be constructed in the univariate case, too,

*Positive (univariate) Power Mixture*

$$R(\mathbf{h},k) = \sum_{i=0}^{\infty} \left(R_s(\mathbf{h})R_t(k)\right)^i \pi_i. \tag{9.14}$$

The right hand side of equation (9.14) bears a striking resemblance to the probability generating function (pgf) of a discrete random variable with support on the non-negative integers. If $U$ takes realizations in $\{0,1,\cdots\}$ with probability $\Pr(U=i) = \pi_i$, then its probability generating function is $G(w) = \sum_{i=0}^{\infty} w^i\pi_i$ for $0 \le w \le 1$. The correlation product in (9.14) takes the role of $w^i$ in the generating function. This provides a convenient method to construct spatio-temporal correlation models. Obtain the probability generating function and replace $w$ with $R_s(\mathbf{h})R_t(k)$. In the bivariate case, replace $w_1$ in the pgf with $R_s(\mathbf{h})$ and $w_2$ with $R_t(k)$. See Ma (2002) for further examples of the univariate and bivariate case.

**Example 9.1** Let $U \sim \text{Binomial}(n, \pi)$, so that its pgf is

$$G(w) = (\pi(w-1)+1)^n .$$

A valid spatio-temporal correlation model based on the power mixture of Binomial probabilities is

$$C(\mathbf{h}, k) = (\pi(R_s(\mathbf{h})R_t(k)-1)+1)^n .$$

If $U$ follows a Poisson distribution with mean $\lambda$, then its pgf is $G(w) = \exp\{\lambda(w-1)\}$ and $\exp\{\lambda(R_s(\mathbf{h})R_t(k)-1)\}$ is a spatio-temporal correlation function. □

A different approach to construct a non-separable covariance function from a product covariance function, is to make the spatial and temporal coordinates depend on one another. Ma (2002) terms this the **scale mixture approach.** Let $[U, V]$ be a bivariate random vector with distribution function $F(u, v)$, not necessarily discrete. If $[U, V]$ is independent of the purely spatial and temporal processes $Z_s(\mathbf{s})$ and $Z_t(t)$, which are independent of each other, then the scale mixture process

$$Z(\mathbf{s}, t) = Z_s(\mathbf{s}U)Z_t(tV)$$

has covariance function

$$C(\mathbf{h}, k) = \int C_s(\mathbf{h}u)C_t(kv)\, dF(u, v). \tag{9.15}$$

As with the power mixture approach, a univariate version follows easily:

$$C(\mathbf{h}, k) = \int C_s(\mathbf{h}u)C_t(ku)\, dF(u). \tag{9.16}$$

Covariance function (9.16) is a special case of the family of covariance functions of De Iaco, Myers, and Posa (2002). In their work, the distribution function $F(u)$ is replaced by a positive measure. They furthermore apply the mixing idea not only to separable product covariance functions, but also to product-sum functions and show the connection to the Cressie-Huang representation in the frequency domain.

Our notation in the preceding paragraphs may have suggested that the spatial and temporal covariance functions $C_s(\mathbf{h})$ and $C_t(k)$ are stationary. This is not necessarily the case. The development of the mixture models applies in the non-stationary situation as well. Most examples and applications start from stationary (and isotropic) covariance functions for the two components, however.

### *9.3.4 Differential Equation Approach*

This approach of constructing spatio-temporal covariance functions relies on the representation of the process as a stochastic differential equation. For example, a temporal process $Z(t)$ with exponential covariance function $C_t(k) =$

$\sigma^2 \exp\{-\theta k\}$ has representation

$$\left(\frac{d}{dt} + \theta\right) Z(t) = \epsilon(t),$$

where $\epsilon(t)$ is a white noise process with variance $\sigma_t^2$. In $\mathbf{R}^2$, Whittle (1954) considered the stochastic Laplace equation

$$\left(\frac{\partial^2}{\partial x^2} + \frac{\partial^2}{\partial y^2} - \theta^2\right) Z(\mathbf{s}) = \epsilon(\mathbf{s})$$

as describing the elementary process $Z(\mathbf{s}) = Z([x, y])$. This process has covariance function $C_s(||\mathbf{h}||) = \sigma_s^2\theta||\mathbf{h}||K_1(\theta||\mathbf{h}||)$.

A stochastic equation that combines the spatial and temporal components (Jones and Zhang, 1997),

$$\left(\frac{\partial^2}{\partial x^2} + \frac{\partial^2}{\partial y^2} - \theta_s^2\right)\left(\frac{\partial}{\partial t} + \theta_t\right) Z(\mathbf{s}, t) = \epsilon(\mathbf{s}, t),$$

leads to a product separable model with

$$C(\mathbf{h}, k) = C_t(k)C_s(\mathbf{h}) = \sigma_{st}^2 \exp\{-\theta_t k\} \times \theta_s||\mathbf{h}||K_1(\theta||\mathbf{h}||).$$

To construct non-separable models in $\mathbf{R}^d \times \mathbf{R}^1$, Jones and Zhang (1997) consider stochastic differential equations of the form

$$\left[\left(\sum_{i=1}^{d} \frac{\partial^2}{\partial s_i^2} - \theta^2\right)^p - c\frac{\partial}{\partial t}\right] Z(\mathbf{s}, t) = \epsilon(\mathbf{s}, t).$$

In this equation $s_1, \cdots, s_d$ denote the coordinates of a point $\mathbf{s}$. The parameter $p$ governs the smoothness of the process and must be greater than $\max\{1, d/2\}$. For $d = 2$ the spatio-temporal covariance function in the isotropic case is

$$C(h, k) = \frac{\sigma^2}{4c\pi} \int_0^\infty \frac{\tau \exp\{-(k/c)(\tau^2 + \theta^2)^p\}}{(\tau^2 + \theta^2)^p} J_0(\tau h)\, d\tau. \qquad (9.17)$$

The connection between (9.17) and purely spatial (isotropic) covariance models is interesting. Expressions (4.7)–(4.8) on page 141 expressed the covariance function in the isotropic case as a Hankel transformation. For $d = 2$ this is a Hankel transformation of zero order,

$$C_s(h) = \int_0^\infty J_0(h\omega)dH(\omega),$$

just as (9.17).

## 9.4 The Spatio-Temporal Semivariogram

To estimate the parameters of a spatio-temporal covariance function, the same basic methods can generally be applied as in Chapter 4. The model can be fit to the original data by maximum or restricted maximum likelihood, or using least squares or CL/GEE methods based on pseudo-data. The semivariogram for a stationary spatio-temporal process relates to the covariance function as before:

*Spatio-temporal Semivariogram*

$$\begin{aligned}\gamma(\mathbf{h}, k) &= \frac{1}{2}\mathrm{Var}[Z(\mathbf{s}, t) - Z(\mathbf{s}+\mathbf{h}, t+k)] \\ &= \mathrm{Var}[Z(\mathbf{s}, t)] - \mathrm{Cov}[Z(\mathbf{s}, t), Z(\mathbf{s}+\mathbf{h}, t+k)] \\ &= C(\mathbf{0}, 0) - C(\mathbf{h}, k).\end{aligned}$$

When the empirical spatio-temporal semivariogram is estimated from data, the implicit anisotropy of spatial and temporal dimensions must be accounted for. A graph of the spatio-temporal semivariogram (or covariogram) with an isotropic spatial component is a three-dimensional plot. The axes consist of temporal and spatial lags, the semivariogram values are the ordinates.

The empirical spatio-temporal semivariogram estimator that derives from the Matheron estimator (4.24), page 153, is

$$\widehat{\gamma}(\mathbf{h}, k) = \frac{1}{2|N(\mathbf{h}, k)|} \sum_{N(\mathbf{h},\mathbf{k})} \{Z(\mathbf{s}_i, t_i) - Z(\mathbf{s}_j, t_j)\}^2. \qquad (9.18)$$

The set $N(\mathbf{h}, k)$ consists of the points that are within spatial distance $\mathbf{h}$ and time lag $k$ of each other; $|N(\mathbf{h}, k)|$ denotes the number of distinct pairs in that set. When data are irregularly spaced in time and/or space, the empirical semivariogram may need to be computed for lag classes. The lag tolerances in the spatial and temporal dimensions need to be chosen differently, of course, to accommodate a sufficient number of point pairs at each spatio-temporal lag.

Note that (9.18) is an estimator of the joint space-time dependency. It is different from a conditional estimator of the spatial semivariogram at time $t$ which would be used in a two-stage method:

$$\widehat{\gamma}_t(\mathbf{h}) = \frac{1}{2|N_t(\mathbf{h})|} \sum_{N_t(\mathbf{h})} \{Z(\mathbf{s}_i, t) - Z(\mathbf{s}_j, t)\}^2. \qquad (9.19)$$

A weighted least squares fit of the joint spatio-temporal empirical semivariogram to a model $\gamma(\mathbf{h}, k; \boldsymbol{\theta})$ estimates $\boldsymbol{\theta}$ by minimizing

$$\sum_{j=1}^{m_s}\sum_{l=1}^{m_t} \frac{|N(\mathbf{h}_j, k_l)|}{2\gamma(\mathbf{h}_j, k_l; \boldsymbol{\theta})} \{\widehat{\gamma}(\mathbf{h}_j, k_l) - \gamma(\mathbf{h}_j, k_l; \boldsymbol{\theta})\}^2,$$

where $m_s$ and $m_t$ are the number of spatial and temporal lag classes, respectively. A fit of the conditional spatial semivariogram at time $t$ minimizes

$$\sum_{j=1}^{m_s} \frac{|N_t(\mathbf{h}_j)|}{2\gamma(\mathbf{h}_j, t; \boldsymbol{\theta})} \{\widehat{\gamma}(\mathbf{h}_j, t) - \gamma(\mathbf{h}_j, t; \boldsymbol{\theta})\}^2.$$

## 9.5 Spatio-Temporal Point Processes

### *9.5.1 Types of Processes*

A spatio-temporal point process is a spatio-temporal random field with a random spatial index $D$ and a temporal index $T$. As before, the temporal index can be either fixed or random, discrete or continuous. According to the nature of the temporal component we distinguish the following types of spatio-temporal point processes (Dorai-Raj, 2001).

- **Earthquake Process**
  Events are unique to spatial locations and time points, only one event can occur at a particular location and time. If a record indicates—in addition to time and location of the earthquake—the magnitude of the quake, the process is marked. The connection of this type of process to earthquakes is intuitive and it has been used in the study of seismic activity (Choi and Hall, 1999; Ogata, 1999). It plays an important role in many other applications. For example, the study of burglary patterns in a suburban area will invariably involve spatio-temporal point processes of this type, unless the data are temporally aggregated.
- **Explosion Process**
  The idea of an explosion process is the generation of a spatial point process at a time $t$ which itself is a realization in a stochastic process. The realization of an explosion process consists of locations $\mathbf{s}_{i1}, \cdots, \mathbf{s}_{in_i} \in D^*$ at times $t_i \in T^*$. Temporal events occur with intensity $\gamma(t)$ and produce point patterns with intensity $\lambda_t(\mathbf{s})$. An example of such a spatio-temporal process is the distribution of acorns around an oak tree. The time at which the (majority of the) acorns fall each year can be considered a temporal random process. The distribution of the acorns is a point process with some intensity, possibly spatially varying.
- **Birth-Death Process**
  This process is useful to model objects that are placed at a random location by *birth* at time $t_b$ and exist at that location for a random time $t_l$. Cressie (1993, p. 720) refers to such a process as a space-time survival point process. At time $t$, an event is recorded at location $\mathbf{s}$ if a *birth* occurred at $\mathbf{s}$ at time $t_b < t$ and the object has a lifetime of $t_b + t_l > t$. Rathbun and Cressie (1994) formulate the spatio-temporal distribution of longleaf pines in Southern Georgia through a birth-death process.

In the explosion process the time points at which the point patterns are observed are the realization of a stochastic process; they are a complete mapping of temporal events. If the observation times are not the result of a stochastic process, but chosen by the experimenter, the spatio-temporal pattern is referred to as a *point pattern sampled in time*. Even if sampling times are selected at random, they do not represent a mapping of temporal events, nor are they treated as a stochastic process. The realization of a birth-death process observed at fixed time points can be indistinguishable from a temporally sampled point pattern. Events observed at location s at time $t_i$ but not at time $t_{i+1}$ could be due to the death of a spatially stationary object or due to the displacement of a non-stationary object between the two time points.

### *9.5.2 Intensity Measures*

Recall from §3.4 that the first- and second-order intensities of a spatial point process are defined as the limits

$$\lambda(\mathbf{s}) = \lim_{|d\mathbf{s}|\to 0} \frac{\mathrm{E}[N(d\mathbf{s})]}{|d\mathbf{s}|}$$

$$\lambda_2(\mathbf{s}_i, \mathbf{s}_j) = \lim_{|d\mathbf{s}_i|,|d\mathbf{s}_j|\to 0} \frac{\mathrm{E}[N(d\mathbf{s}_i)N(d\mathbf{s}_j)]}{|d\mathbf{s}_i||d\mathbf{s}_j|},$$

where $d\mathbf{s}$ is an infinitesimal disk (ball) of area (volume) $|d\mathbf{s}|$. To extend the intensity measures to the spatio-temporal scenario, we define $N(d\mathbf{s}, dt)$ to denote the number of events in an infinitesimal cylinder with base $d\mathbf{s}$ and height $dt$ (Dorai-Raj, 2001). (Note that Haas (1995) considered cylinders in local prediction of spatio-temporal data.) The spatio-temporal intensity of the process $\{Z(\mathbf{s}, t) : \mathbf{s} \in D(t), t \in T\}$ is then defined as the average number of events per unit volume as the cylinder is shrunk around the point $(\mathbf{s}, t)$:

$$\lambda(\mathbf{s}, t) = \lim_{|d\mathbf{s}|,|dt|\to 0} \frac{\mathrm{E}[N(d\mathbf{s}, dt)]}{|d\mathbf{s}||dt|}. \tag{9.20}$$

To consider only a single component of the spatio-temporal process, the intensity (9.20) can be marginalized to obtain the marginal spatial intensity

$$\lambda(\mathbf{s}, \cdot) = \int_T \lambda(\mathbf{s}, v)\, dv, \tag{9.21}$$

or the marginal temporal intensity

$$\lambda(\cdot, t) = \int_D \lambda(\mathbf{u}, t)\, d\mathbf{u}. \tag{9.22}$$

If the spatio-temporal intensity can be marginalized, it can also be conditioned. The conditional spatial intensity at time $t$ is defined as

$$\lambda(\mathbf{s}|t) = \lim_{|d\mathbf{s}|\to 0} \frac{\mathrm{E}[N(d\mathbf{s}, t)]}{|d\mathbf{s}|},$$

and the conditional temporal intensity at location s as

$$\lambda(t|\mathbf{s}) = \lim_{|dt|\to 0} \frac{\mathrm{E}[N(\mathbf{s}, dt)]}{|dt|}.$$

In the case of an earthquake process, these conditional intensities are not meaningful and should be replaced by intensities constructed on intervals in time or areas in space (Rathbun, 1996).

Second-order intensities can also be extended to the spatio-temporal case by a similar device. Let $A_i = d\mathbf{s}_i \times dt_i$ be an infinitesimal cylinder containing point $(\mathbf{s}_i, t_i)$. The second-order spatio-temporal intensity is then defined as

$$\lambda_2(\mathbf{s}_i, \mathbf{s}_j, t_i, t_j) = \lim_{|A_i|,|A_j|\to 0} \frac{\mathrm{E}[N(A_i)N(A_j)]}{|A_i||A_j|}.$$

A large array of different marginal, conditional, and average conditional second-order intensities can be derived by similar arguments as for the first-order intensities.

### *9.5.3 Stationarity and Complete Randomness*

First- and second-order stationarity of a spatio-temporal point process can refer to stationarity in space, in time, or both. We thus consider an array of conditions.

(i) $Z(\mathbf{s}, t)$ is first-order stationary in space (FOS) if $\lambda(t|\mathbf{s}) = \lambda^*(t)$.

(ii) $Z(\mathbf{s}, t)$ is first-order stationary in time (FOT) if $\lambda(\mathbf{s}|t) = \lambda^{**}(\mathbf{s})$.

(iii) $Z(\mathbf{s}, t)$ is first-order stationary in space and time (FOST) if $\lambda(\mathbf{s}, t)$ does not depend on $\mathbf{s}$ or $t$.

Dorai-Raj (2001) shows that the spatio-temporal intensity (9.20) is related to the conditional intensities by

$$\lim_{|dt|\to 0} \int_{dt} \frac{\lambda(\mathbf{s}, v)}{|dt|}\, dv = \lim_{|d\mathbf{s}|\to 0} \int_{|d\mathbf{s}|} \frac{\lambda(t, \mathbf{u})}{|d\mathbf{s}|}\, d\mathbf{u}.$$

The following results are corollaries of this relationship.

(i) If a process is FOT, then $\lambda(\mathbf{s}, t) = \lambda^{**}(\mathbf{s})$ and $\lambda(\mathbf{s}, \cdot) = |T|\lambda^{**}(\mathbf{s})$.

(ii) If a process is FOS, then $\lambda(\mathbf{s}, t) = \lambda^*(t)$ and $\lambda(\cdot, t) = |A|\lambda^*(t)$.

Second-order stationarity in space and time requires that $\lambda(\mathbf{s}, t)$ does not depend on $\mathbf{s}$ or $t$, and that

$$\lambda_2(\mathbf{s}, \mathbf{s}+\mathbf{h}, t, t+k) = \lambda_2^*(\mathbf{h}, k).$$

Bartlett's complete covariance density function (§4.7.3.1) can be extended for FOST processes as

$$\lim_{|A_i|,|A_j|\to 0} \frac{\mathrm{Cov}[N(A_i), N(A_j)]}{|A_i||A_j|} = v(\mathbf{h}, k) + \lambda\delta(\mathbf{h}, k), \qquad (9.23)$$

Equipped with these tools, a spatio-temporal process can be defined as a completely spatio-temporally random (CSTR) process if it is a Poisson process in both space and time, that is, a process void of any temporal or spatial structure, so that $N(A, T) \sim \mathrm{Poisson}(\lambda|A \times T|)$. For this process, $\lambda(\mathbf{s}, t) = \lambda$ and $\lambda_2(\mathbf{s}, \mathbf{s}+\mathbf{h}, t, t+k) = \lambda^2$. If the CSR process is an unattainable standard for spatial point processes, then the CSTR process is even more so for spatio-temporal processes. Its purpose is to serve as the initial benchmark against which observed spatio-temporal patterns are tested, in much the same way as observed spatial point patterns are tested against CSR.

# References


Abramowitz, M. and Stegun, I.A. (1964) *Handbook of Mathematical Functions*, Applied Mathematics Series, Vol. 55. National Bureau of Standards, Washington, D.C (reprinted 1972 by Dover Publications, New York).

Aitchison, J. and Brown, J. (1957) *The Lognormal Distribution.* Cambridge University Press, London.

Akaike, H. (1974) A new look at the statistical model identification. *IEEE Transaction on Automatic Control*, AC-19, 716–723.

Aldworth, J. and Cressie, N. (1999) Sampling designs and prediction methods for Gaussian spatial processes. In: S. Ghosh (ed.), *Multivariate Analyses, Design of Experiments, and Survey Sampling.* Marcel Dekker, New York, 1–54.

Aldworth, J. and Cressie, N. (2003) Prediction of nonlinear spatial functionals. *Journal of Statistical Planning and Inference*, 112:3–41.

Allen, D.M. (1974) The relationship between variable selection and data augmentation and a method of prediction. *Technometrics*, 16:125–127.

Anselin, L. (1995) Local indicators of spatial association—LISA. *Geographic Analysis*, 27(2):93–115.

Armstrong, M. (1999) *Basic Linear Geostatistics.* Springer-Verlag, New York.

Azzalini, A. and Capitanio, A. (1999) Statistical applications of the multivariate skew normal distribution. *Journal of the Royal Statistical Society, Series B*, 61:579–602.

Baddeley, A. and Silverman, B.W. (1984) A cautionary example for the use of second order methods for analyzing point patterns. *Biometrics*, 40:1089–1094.

Bahadur, R.R. (1961) A representation of the joint distribution of responses to $n$ dichotomous items, In *Studies in Item Analysis and Prediction*, ed. H. Solomon, Stanford University Press, Stanford, CA, 158–165.

Banerjee, S., Carlin, B.P., and Gelfand, A.E. (2003) *Hierarchical Modeling and Analysis for Spatial Data.* Chapman and Hall/CRC, Boca Raton, FL.

Barry, R.P. and Ver Hoef, J.M. (1996) Blackbox kriging: spatial prediction without specifying variogram models. *Journal of Agricultural, Biological, and Environmental Statistics*, 1:297–322.

Bartlett, M.S. (1964) The spectral analysis of two-dimensional point processes. *Biometrika*, 51:299–311.

Bartlett, M.S. (1978) *Stochastic Processes. Methods and Applications.* Cambridge University Press, London.

Belsley, D.A., Kuh, E., and Welsch, R.E. (1980), *Regression Diagnostics; Identifying Influential Data and Sources of Collinearity.* John Wiley & Sons, New York.

Berger, J.O., De Oliveria, V., and Sansó, B. (2001) Objective Bayesian analysis of spatially correlated data. *Journal of the American Statistical Association*, 96: 1361–1374.

Besag, J. (1974) Spatial interaction and the statistical analysis of lattice systems. *Journal of the Royal Statistical Society, Series B*, 36:192–225.

Besag, J. (1975) Statistical analysis of non-lattice data. *The Statistician*, 24:179–195.

Besag, J., Green, P., Higdon, D., and Mengersen, K. (1995) Bayesian computation and stochastic systems (with discussion), *Statistical Science*, 10:3–66.

Besag, J. and Kempton, R. (1986) Statistical analysis of field experiments using neighbourng plots. *Biometrics*, 42:231–251.

Besag, J. and Kooperberg, C. (1995) On conditional and intrinsic autoregressions. *Biometrika*, 82:733–746.

Besag, J. and Newell, J. (1991) The detection of clusters in rare diseases. *Journal of the Royal Statistical Society, Series A*, 154:327–333.

Besag, J., York, J., and Mollié, A. (1991) Bayesian image restoration, with two applications in spatial statistics (with discussion). *Annals of the Institute of Statistical Mathematics*, 43:1–59.

Bloomfield, P. (1976) *Fourier Analysis of Time Series: An Introduction*. John Wiley & Sons, New York.

Boufassa, A. and Armstrong, M. (1989) Comparison between different kriging estimators. *Mathematical Geology*, 21:331–345.

Box, G.E.P. and Cox, D.R. (1964) An analysis of transformations. *Journal of the Royal Statistical Society, Series B*, 26:211–243.

Breslow, N.E. and Clayton, D.G. (1993) Approximate inference in generalized linear mixed models. *Journal of the American Statistical Association*, 88:9–25.

Breusch, T.S. (1980) Useful invariance results for generalized regression models. *Journal of Econometrics*, 13:327–340.

Brillinger, D.R. (1972) The spectral analysis of stationary interval functions. In *Proceedings of the 6th Berkeley Symposium on Mathematical Statistics and Probability*, 1:483–513.

Brown, R.L., Durbin, J., and Evans, J.M. (1975) Techniques for testing the constancy of regression relationships over time. *Journal of the Royal Statistical Society (B)*, 37:149–192.

Brownie, C., Bowman, D.T., and Burton, J.W. (1993) Estimating spatial variation in analysis of data from yield trials: a comparison of methods. *Agronomy Journal*, 85:1244–1253.

Brownie, C. and Gumpertz, M.L. (1997) Validity of spatial analyses of large field trials. *Journal of Agricultural, Biological, and Environmental Statistics*, 2(1):1–23.

Burnham, K.P. and Anderson, D.R. (1998) *Model Selection and Inference: A Practical Information-Theoretic Approach*. Springer-Verlag, New York.

Burt, W.H. (1943) Territoriality and homerange concepts as applied to mammals. *Journal of Wildlife Management*, 54:310–315.

Carlin, B.P. and Louis, T.A. (2000) *Bayes and Empirical Bayes Methods for Data Analysis*. 2nd Edition, Chapman and Hall/CRC, Boca Raton, FL.

Casella, G. and George, E.I. (1992) Explaining the Gibbs sampler. *The American Statistician*, 46:167–174.

Chatfield, C. (1996) *The Analysis of Time Series; An Introduction, 5th ed.* Chapman & Hall, London.

Cherry, S., Banfield, J., and Quimby, W. (1996) An evaluation of a nonparametric method of estimating semi-variograms of isotropic spatial processes. *Journal of Applied Statistics*, 23:435–449.

Chib, S. and Greenberg, E. (1995) Understanding the Metropolis-Hastings algorithm. *The American Statistician*, 49:327–335.

Chilès, J.P. and Delfiner, P. (1999) *Geostatistics. Modeling Spatial Uncertainty.* John Wiley & Sons, New York.

Choi, E. and Hall, P. (1999) Nonparametric approach to analysis of space-time data on earthquake occurrences. *Journal of Computational and Graphical Statistics*, 8:733–748.

Christensen, R. (1991) *Linear Models for Multivariate, Time Series, and Spatial Data.* Springer-Verlag, New York.

Clark, I. (1979) *Practical Geostatistics.* Applied Science Publishers, Essex, England.

Clayton, D.G. and Kaldor, J. (1987) Empirical Bayes estimates of age-standardized relative risks for use in disease mapping, *Biometrics*, 43:671–682.

Cleveland, W.S. (1979) Robust locally weighted regression and smoothing scatterplots. *Journal of the American Statistical Association*, 74:829–836.

Cliff, A.D. and Ord, J.K. (1981) *Spatial Processes; Models and Applications.* Pion Limited, London.

Cody, W.J. (1987) SPECFUN—A portable special function package. In *New Computing Environments: Microcomputers in Large-Scale Scientific Computing*, ed. A. Wouk, SIAM, Philadelphia, 1–12.

Congdon, P. (2001) *Bayesian Statistical Modelling.* John Wiley & Sons, Chichester.

Congdon, P. (2003) *Applied Bayesian Modelling.* John Wiley & Sons, Chichester.

Cook, R.D. (1977) Detection of influential observations in linear regression. *Technometrics*, 19:15–18.

Cook, R.D. (1979) Influential observations in linear regression. *Journal of the American Statistical Association*, 74:169–174.

Cook, R.D. and Weisberg, S. (1982) *Residuals and Influence in Regression.* Chapman and Hall, New York.

Cowles, M.K. and Carlin, B.P. (1996) Markov chain Monte Carlo convergence diagnostics: a comparative review. *Journal of the American Statistical Association*, 91:883–904.

Cressie, N. (1985) Fitting variogram models by weighted least squares. *Journal of the International Association for Mathematical Geology*, 17:563–586.

Cressie, N. (1990) The origins of kriging. *Mathematical Geology*, 22:239–252.

Cressie, N. (1992) Smoothing regional maps using empirical Bayes predictors. *Geographical Analysis*, 24:75–95.

Cressie, N.A.C. (1993) *Statistics for Spatial Data. Revised ed.* John Wiley & Sons, New York.

Cressie, N. (1993b) Aggregation in geostatistical problems. In: A. Soares (ed.), *Geostatistics Troia '92.* Kluwer Academic Publishers, Dordrecht, 25–35.

Cressie, N. (1998) Fundamentals of spatial statistics. pp. 9–33 in *Collecting Spatial Data: Optimum Design of Experiments for Random Fields.* W.G. Müller, (ed.) Physica-Verlag.

Cressie, N. and Grondona, M.O. (1992) A comparison of variogram estimation with covariogram estimation. In *The Art of Statistical Science*, ed. K.V. Mardia. John Wiley & Sons, New York, 191–208.

Cressie, N. and Hawkins, D.M. (1980) Robust estimation of the variogram, I. *Journal of the International Association for Mathematical Geology*, 12:115–125.

Cressie, N. and Huang, H.-C. (1999) Classes of nonseparable, spatio-temporal stationary covariance functions. *Journal of the American Statistical Association*, 94:1330–1340.

Cressie, N. and Lahiri, S.N. (1996) Asymptotics for REML estimation of spatial covariance parameters. *Journal of Statistical Planning and Inference*, 50:327–341.

Cressie, N. and Majure, J.J. (1997) Spatio-temporal statistical modeling of livestock waste in streams. *Journal of Agricultural, Biological, and Environmental Statistics*, 2:24–47.

Croux, C. and Rousseeuw, P.J. (1992) Time-efficient algorithms for two highly robust estimators of scale. In *Computational Statistics, Vol. 1*, eds. Y. Dodge, and J. Whittaker, Physika-Verlag, Heidelberg, 411–428.

Curriero, F.C. (1996) The use of non-Euclidean distances in geostatistics. Ph.D. thesis, Department of Statistics, Kansas State University, Manhattan, KS.

Curriero, F.C. (2004) Norm dependent isotropic covariogram and variogram models. *Journal of Statistical Planning and Inference*, In review.

Curriero, F.C. and Lele, S. (1999) A composite likelihood approach to semivariogram estimation. *Journal of Agricultural, Biological, and Environmental Statistics*, 4(1):9–28.

Cuzick, J. and Edwards, R. (1990) Spatial clustering for inhomogeneous populations (with discussion). *Journal of the Royal Statistical Society, Series B*, 52:73–104.

David, M. (1988) *Handbook of Applied Advanced Geostatistical Ore Reserve Estimation*. Elsevier, Amsterdam.

Davidian, M. and Giltinan, D.M. (1995) *Nonlinear Models for Repeated Measurement Data*. Chapman and Hall, New York.

De Cesare, L., Myers, D., and Posa, D. (2001) Estimating and modeling space-time correlation structures. *Statistics and Probability Letters*, 51(1):9–14.

De Iaco, S., Myers, D.E., and Posa, D. (2002) Nonseparable space-time covariance models: some parametric families. *Mathematical Geology*, 34(1):23–42.

De Oliveira, V., Kedem, B. and Short, D.A. (1997) Bayesian prediction of transformed Gaussian random fields. *Journal of the American Statistical Association*, 92:1422–1433.

Deutsch, C.V. and Journel, A.G. (1992) *GSLIB: Geostatistical Software Library and User's Guide*. Oxford University Press, New York.

Devine, O.J., Louis, T.A. and Halloran, M.E. (1994) Empirical Bayes methods for stabilizing incidence rates before mapping, *Epidemiology*, 5(6):622–630.

Diggle, P.J. (1983) *Statistical Analysis of Point Processes*. Chapman and Hall, New York.

Diggle, P.J. (1985) A kernel method for smoothing point process data. *Applied Statistics*, 34:138–147.

Diggle, P., Besag, J.E., and Gleaves, J.T. (1976) Statistical analysis of spatial patterns by means of distance methods. *Biometrics*, 32:659–667.

Diggle, P.J. and Chetwynd, A.G. (1991) Second-order analysis of spatial clustering for inhomogeneous populations. *Biometrics*, 47:1155–1163.

Diggle, P.J., Tawn, J.A., and Moyeed, R.A. (1998) Model-based geostatistics. *Applied Statistics*, 47:229–350.

Dobson, A.J. (1990) *An Introduction to Generalized Linear Models*. Chapman and Hall, London.

Dorai-Raj, S.S. (2001) *First- and Second-Order Properties of Spatiotemporal Point Processes in the Space-Time and Frequency Domains*. Ph.D. Dissertation, Dept. of Statistics, Virginia Polytechnic Institute and State University.

Dowd, P.A. (1982) Lognormal kriging—the general case. *Journal of the International Association for Mathematical Geology*, 14:475–499.

Draper, N. and Smith, H. (1981) *Applied Regression Analysis*, 2nd edition. John Wiley & Sons, New York.

Eaton, M.L. (1985) The Gauss-Markov theorem in multivariate analysis. In *Multivariate Analysis—VI*, ed. P.R. Krishnaiah, Amsterdam: Elsevier, 177–201.

Ecker, M.D. and Gelfand, A.E. (1997) Bayesian variogram modeling for an isotropic spatial process. *Journal of Agricultural, Biological, and Environmental Statistics*, 2:347–369.

Fedorov, V.V. (1974) Regression problems with controllable variables subject to error. *Biometrika*, 61:49–65.

Fisher, R.A., Thornton, H.G., and MacKenzie, W.A. (1922) The accuracy of the plating method of estimating the density of bacterial populations. *Annals of Applied Biology*, 9:325–359.

Fotheringham, A.S., Brunsdon, C. and Charlton, M. (2002) *Geographically Weighted Regression*. John Wiley & Sons, New York.

Freeman, M.F. and Tukey, J.W. (1950) Transformations related to the angular and the square root. *Annals of Mathematical Statistics*, 21:607–611.

Fuentes, M. (2001) A high frequency kriging approach for non-stationary environmental processes. *Environmetrics*, 12:469–483.

Fuller, W.A. and Battese, G.E. (1973) Transformations for estimation of linear models with nested error structure. *Journal of the American Statistical Association*, 68:626–632.

Galpin, J.S.and Hawkins, D.M. (1984) The use of recursive residuals in checking model fit in linear regression. *The American Statistician*, 38(2):94–105.

Gandin, L.S. (1963) *Objective Analysis of Meteorological Fields*. Gidrometeorologicheskoe Izdatel'stvo (GIMIZ), Leningrad (translated by Israel Program for Scientific Translations, Jerusalem, 1965).

Gaudard, M., Karson, M., Linder, E., and Sinha, D. (1999) Bayesian spatial prediction. *Environmental and Ecological Statistics*, 6:147–171.

Geary, R.C. (1954) The contiguity ratio and statistical mapping, *The Incorporated Statistician*, 5:115–145.

Gelfand, A.E. and Smith, A.F.M. (1990) Sampling-based approaches to calculating marginal densities. *Journal of the American Statistical Association*, 85:398–409.

Gelman, A., Carlin, J.B., Stern, H.S., and Rubin, D.B. (2004) *Bayesian Data Analysis*. Chapman and Hall/CRC, Boca Raton, FL.

Gelman, A. and Rubin, D.B. (1992) Inference from iterative simulation using multiple sequences (with discussion). *Statistical Science*, 7:457–511.

Geman, S. and Geman, D. (1984) Stochastic relaxation, Gibbs distributions and the Bayesian restoration of images. *IEEE Transactions on Pattern Analysis and Machine Intelligence*, 6:721–741.

Genton, M.G. (1998a) Highly robust variogram estimation. *Mathematical Geology*, 30:213–221.

Genton, M.G. (1998b) Variogram fitting by generalized least squares using an explicit formula for the covariance structure. *Mathematical Geology*, 30:323–345.

Genton, M.G. (2000) The correlation structure of Matheron's classical variogram estimator under elliptically contoured distributions. *Mathematical Geology*, 32(1):127–137.

Genton, M.G. (2001) Robustness problems in the analysis of spatial data. In *Spatial Statistics: Methodological Aspects and Applications*, ed. M. Moore, Springer-Verlag, New York, 21–38.

Genton, M.G. and Gorsich, D.J. (2002) Nonparametric variogram and covariogram estimation with Fourier-Bessel matrices. *Computational Statistics and Data Analysis*, 41:47–57.

Genton, M.G., He, L., and Liu, X. (2001) Moments of skew-normal random vectors and their quadratic forms. *Statistics & Probability Letters*, 51:319–325.

Gerrard, D.J. (1969) Competition quotient: a new measure of the competition affecting individual forest trees. *Research Bulletin No. 20*, Michigan Agricultural Experiment Station, Michigan State University

Ghosh, M., Natarajan, K., Waller, L.A., and Kim, D. (1999) Hierarchical GLMs for the analysis of spatial data: an application to disease mapping. *Journal of Statistical Planning and Inference*, 75:305–318.

Gilks, W.R. (1996) Full conditional distributions. In Gilks, W.R. Richardson, S. and Spiegelhalter, D.J. (eds.) *Markov Chain Monte Carlo in Practice* Chapman & Hall/CRC, Boca Raton, FL, 75–88.

Gilks, W.R., Richardson, S., and Spiegelhalter, D.J. (1996) Introducing Markov chain Monte Carlo. In Gilks, W.R. Richardson, S. and Spiegelhalter, D.J. (eds.) *Markov Chain Monte Carlo in Practice* Chapman & Hall/CRC, Boca Raton, FL, 1–19.

Gilliland, D. and Schabenberger, O. (2001) Limits on pairwise association for equicorrelated binary variables. *Journal of Applied Statistical Science*, 10:279–285.

Gneiting, T. (2002) Nonseparable, stationary covariance functions for space-time data. *Journal of the American Statistical Association*, 97:590–600.

Godambe, V.P. (1960) An optimum property of regular maximum likelihood estimation. *Annals of Mathematical Statistics*, 31:1208–1211.

Goldberger, A.S. (1962) Best linear unbiased prediction in the generalized linear regression model. *Journal of the American Statistical Association*, 57:369–375.

Goldberger, A.S. (1991) *A Course in Econometrics*. Harvard University Press, Cambridge, MA.

Goodnight, J.H. (1979) A tutorial on the sweep operator. *The American Statistician*, 33:149–158. (Also available as *The Sweep Operator: Its Importance in Statistical Computing*, SAS Technical Report R-106, SAS Institute, Inc. Cary, NC).

Goodnight, J.H. and Hemmerle, W.J. (1979) A simplified algorithm for the W-transformation in variance component estimation. *Technometrics*, 21:265–268.

Goovaerts, P. (1994) Comparative performance of indicator algorithms for modeling conditional probability distribution functions. *Mathematical Geology*, 26:389–411.

Goovaerts, P. (1997) *Geostatistics for Natural Resource Evaluation*. Oxford University Press, Oxford, UK.

Gorsich, D.J. and Genton, M.G. (2000) Variogram model selection via nonparametric derivative estimation. *Mathematical Geology*, 32(3):249–270.

Gotway, C.A. and Cressie, N. (1993) Improved multivariate prediction under a general linear model. *Journal of Multivariate Analysis*, 45:56–72.

Gotway, C.A. and Hegert, G.W. (1997) Incorporating spatial trends and anisotropy in geostatistical mapping of soil properties. *Soil Science Society of America Journal*, 61:298–309.

Gotway, C.A. and Stroup, W.W. (1997) A generalized linear model approach to spatial data analysis and prediction. *Journal of Agricultural, Biological and Environmental Statistics*, 2:157–178.

Gotway, C.A. and Wolfinger, R.D. (2003) Spatial prediction of counts and rates. *Statistics in Medicine*, 22:1415–1432.

Gotway, C.A. and Young, L.J. (2002) Combining incompatible spatial data. *Journal of the American Statistical Association*, 97:632–648.

Graybill, F.A. (1983) *Matrices With Applications in Statistics. 2nd ed.* Wadsworth International, Belmont, CA.

Greig-Smith, P. (1952) The use of random and contiguous quadrats in the study of the structure of plant communities. *Annals of Botany*, 16:293–316.

Grondona, M.O. (1989) Estimation and design with correlated observations. Ph.D. Dissertation, Iowa State University.

Grondona, M.O. and Cressie, N. (1995). Residuals based estimators of the covariogram. *Statistics*, 26:209–218.

Haas, T.C. (1990) Lognormal and moving window methods of estimating acid deposition. *Journal of the American Statistical Association*, 85:950–963.

Haas, T.C. (1995) Local prediction of a spatio-temporal process with an application to wet sulfate deposition. *Journal of the American Statistical Association*, 90:1189–1199.

Haining, R. (1990) *Spatial Data Analysis in the Social and Environmental Sciences.* Cambridge University Press, Cambridge.

Haining, R. (1994) Diagnostics for regression modeling in spatial econometrics. *Journal of Regional Science*, 34:325–341.

Hampel, F.R., Rochetti, E.M., Rousseeuw, P.J., and Stahel, W.A. (1986) *Robust Statistics, the Approach Based on Influence Functions.* John Wiley & Sons, New York.

Handcock, M.S. and Stein, M.L. (1993) A bayesian analysis of kriging. *Technometrics*, 35:403–410.

Handcock, M.S. and Wallis, J.R. (1994) An appproach to statistical spatial-temporal modeling of meteorological fields (with discussion). *Journal of the American Statistical Association*, 89:368–390.

Hanisch, K.-H. and Stoyan, D. (1979) Formulas for the second-order analysis of marked point processes. *Mathematische Operationsforschung und Statistik. Series Statistics*, 10:555–560.

Harville, D.A. (1974) Bayesian inference for variance components using only error contrasts. *Biometrika*, 61:383–385.

Harville, D.A. (1977) Maximum-likelihood approaches to variance component estimation and to related problems. *Journal of the American Statistical Association*, 72:320–340.

Harville, D.A. and Jeske, D.R. (1992) Mean squared error of estimation or Prediction under a general linear model. *Journal of the American Statistical Association*, 87:724–731.

Haslett, J. and Hayes, K. (1998) Residuals for the linear model with general covariance structure. *Journal of the Royal Statistical Society, Series B*, 60:201–215.

Hastings, W.K. (1970) Monte Carlo sampling methods using Markov chains and their applications. *Biometrika*, 57:97–109.

Hawkins, D.M. (1981) A cusum for a scale parameter. *Journal of Quality Technology*, 13:228–231.

Hawkins, D.M. and Cressie, N.A.C. (1984) Robust kriging—a proposal. *Journal of the International Association of Mathematical Geology*, 16:3–18.

Heagerty, P.J. and Lele, S.R. (1998) A composite likelihood approach to binary spatial data. *Journal of the American Statistical Association*, 93:1099–1111.

Henderson, C.R. (1950) The estimation of genetic parameters. *The Annals of Mathematical Statistics*, 21:309–310.

Heyde, C.C. (1997) *Quasi-Likelihood and Its Applications. A General Approach to Optimal Parameter Estimation.* Springer-Verlag, New York.

Higdon, D. (1998) A process-convolution approach to modeling temperatures in the North Atlantic Ocean. *Environmental and Ecological Statistics*, 5(2):173–190.

Higdon, D., Swall, J., and Kern, J. (1999) Non-stationary spatial modeling. *Bayesian Statistics*, 6:761–768.

Hinkelmann, K. and Kempthorne, O. (1994) *Design and Analysis of Experiments. Volume I. Introduction to Experimental Design.* John Wiley & Sons, New York.

Houseman, E.A., Ryan, L.M., and Coull, B.A. (2004) Cholesky residuals for assessing normal errors in a linear model with correlated outcomes. *Journal of the American Statistical Association*, 99:383–394.

Huang, J.S. and Kotz, S. (1984) Correlation structure in iterated Farlie-Gumble-Morgenstern distributions. *Biometrika*, 71:633–636.

Hughes-Oliver, J.M., Gonzalez-Farias, G., Lu, J.-C., and Chen, D. (1998) Parametric nonstationary correlation models. *Statistics & Probability Letters*, 40:267–278.

Hughes-Oliver, J.M., Lu, J.-C., Davis, J.C., and Gyurcsik, R.S. (1998) Achieving uniformity in a semiconductor fabrication process using spatial modeling. *Journal of the American Statistical Association*, 93:36–45.

Hurvich, C.M. and Tsai, C.-L. (1989) Regression and time series model selection in small samples. *Biometrika*, 76:297-307.

Isaaks, E.H. and Srivastava, R.M. (1989) *An Introduction to Applied Geostatistics.* Oxford University Press, New York.

Jensen, D.R. and Ramirez, D.E. (1999) Recovered errors and normal diagnostics in regression. *Metrica*, 49:107–119.

Johnson, M.E. (1987) *Multivariate Statistical Simulation.* John Wiley & Sons, New York.

Johnson, N.L. and Kotz, S. (1972) *Distributions in Statistics: Continuous Multivariate Distributions.* John Wiley & Sons, New York.

Johnson, N.L. and Kotz, S. (1975) On some generalized Farlie-Gumbel-Morganstern distributions. *Communications in Statistics*, 4:415–427.

Jones, R.H. (1993) *Longitudinal Data With Serial Correlation: A State-space Approach.* Chapman and Hall, New York.

Jones, R.H. and Zhang, Y. (1997) Models for continuous stationary space-time processes. In Gregoire, T.G. Brillinger, D.R. Diggle, P.J., Russek-Cohen, E., Warren, W.G., and Wolfinger, R.D. (eds.) *Modeling Longitudinal and Spatially Correlated Data*, Springer Verlag, New York, 289–298.

Journel, A.G. (1980) The lognormal approach to predicting local distributions of selective mining unit grades. *Journal of the International Association for Mathematical Geology*, 12:285–303.

Journel, A.G. (1983) Nonparametric estimation of spatial distributions. *Journal of the International Association for Mathematical Geology*, 15:445–468.

Journel, A.G. and Huijbregts, C.J. (1978) *Mining Geostatistics.* Academic Press, London.

Jowett, G.H. (1955a) The comparison of means of industrial time series. *Applied Statistics*, 4:32–46.

Jowett, G.H. (1955b) The comparison of means of sets of observations from sections of independent stochastic series. *Journal of the Royal Statistical Society, (B)*, 17:208–227.

Jowett, G.H. (1955c) Sampling properties of local statistics in stationary stochastic series. *Biometrika*, 42:160–169.

Judge, G.G., Griffiths, W.E., Hill, R.C., Lütkepohl, H., and Lee, T.-C. (1985) *The Theory and Practice of Econometrics*, John Wiley & Sons, New York.

Kackar, R.N. and Harville, D.A. (1984) Approximation for Standard errors of estimators of fixed and random effects in mixed linear models. *Journal of the American Statistical Association*, 79:853–862.

Kaluzny, S.P., Vega, S.C., Cardoso, T.P. and Shelly, A.A. (1998) *S+SpatialStats. User's Manual for Windows® and Unix®*. Springer Verlag, New York.

Kelsall, J.E. and Diggle, P.J. (1995) Non-parametric estimation of spatial variation in relative risk. *Statistics in Medicine*, 14: 2335–2342.

Kelsall, J.E. and Wakefield, J.C. (1999) Discussion of Best et al. 1999. In Bernardo, J.M. Berger, J.O. Dawid, A.P. and Smith, A.F.M. (eds.) *Bayesian Statistics 6*, Oxford University Press, Oxford, p. 151.

Kempthorne, O. (1955) The randomization theory of experimental inference. *Journal of the American Statistical Association*, 50:946–967.

Kenward, M.G. and Roger, J.H. (1997) Small sample inference for fixed effects from restricted maximum likelihood. *Biometrics*, 53:983–997.

Kern, J.C. and Higdon, D.M. (2000) A distance metric to account for edge effects in spatial analysis. In *Proceedings of the American Statistical Association, Biometrics Section*, Alexandria, VA, 49–52.

Kianifard, F. and Swallow, W.H. (1996) A review of the development and application of recursive residuals in linear models. *Journal of the American Statistical Association*, 91:391–400.

Kim, H. and Mallick, B.K. (2002) Analyzing spatial data using skew-Gaussian processes. In Lawson, A. B. and Denison, D. GT(eds.) *Spatial Cluster Modeling*, Chapman & Hall/CRC, Boca Raton, FL. pp. 163–173.

Kitanidis, P.K. (1983) Statistical estimation of polynomial generalized covariance functions and hydrological applications. *Water Resources Research*, 19:909–921.

Kitanidis, P.K. (1986) Parameter uncertainty in estimation of spatial functions: Bayesian analysis. *Water Resources Research*, 22:499–507.

Kitanidis, P.K. and Lane, R.W. (1985) Maximum likelihood parameter estimation of hydrological spatial processes by the Gauss-Newton method. *Journal of Hydrology*, 79:53–71.

Kitanidis, P.K. and Vomvoris, E.G. (1983) A geostatistical approach to the inverse problem in groundwater modeling (steady state) and one-dimensional simulations. *Water Resources Research*, 19:677–690.

Knox, G. (1964) Epidemiology of childhood leukemia in Northumberland and Durham. *British Journal of Preventative and Social Medicine*, 18:17–24.

Krige, D.G. (1951) A statistical approach to some basic mine valuation problems on the Witwatersrand. *Journal of Chemical, Metallurgical, and Mining Society of South Africa*, 52:119–139.

Krivoruchko, K. and Gribov, A. (2004) Geostatistical interpolation in the presence of barriers. In: *geoENV IV - Geostatistics for Environmental Applications*: Proceedings of the Fourth European Conference on Geostatistics for Environmental Applications 2002 (Quantitative Geology and Geostatistics), 331–342.

Kulldorff, M. (1997) A spatial scan statistic. *Communications in Statistics-Theory and Methods*, 26:1487–1496.

Kulldorff, M. and International Management Services, Inc. (2003) *SaTScan v. 4.0: Software for the spatial and space-time scan statistics*. National Cancer Institute, Bethesda, MD.

Kulldorff, M. and Nagarwalla, N. (1995) Spatial disease clusters: detection and inference. *Statistics in Medicine*, 14:799–810.

Kupper, L.L. and Haseman, J.K. (1978) The use of a correlated binomial model for the analysis of certain toxicological experiments, *Biometrics*, 34:69-76.

Laarhoven, P.J.M. van and Aarts, E.H.L. (1987) *Simulated Annealing: Theory and Applications.* Reidel Publishing, Dordrecht, Holland.

Lahiri, S.N., Lee, Y., Cressie, N. (2002) On asymptotic distribution and asymptotic efficiency of least squares estimators of spatial variogram parameters. *Journal of Statistical Planning and Inference*, 103:65–85.

Lancaster, H. O. (1958) The structure of bivariate distributions. *Annals of Mathematical Statistics*, 29:719–736.

Lawson, A.B. and Denison, D.G.T. (2002) *Spatial Cluster Modelling*, Chapman and Hall/CRC, Boca Raton, FL.

Le, N.D. and Zidek, J.V. (1992) Interpolation with uncertain spatial covariances: a Bayesian alternative to kriging. *Journal of Multivariate Analysis*, 43:351–374.

Lele, S. (1997) Estimating functions for semivariogram estimation, In: *Selected Proceedings of the Symposium on Estimating Functions*, eds. I.V. Basawa, V.P. Godambe, and R.L. Taylor, Hayward, CA: Institute of Mathematical Statistics, 381–396.

Lele, S. (2004) On using expert opinion in ecological analyses: A frequentist approach. *Environmental and Ecological Statistics*, to appear.

Lewis, P.A.W. and Shedler, G.S. (1979) Simulation of non-homogeneous Poisson processes by thinning. *Naval Research Logistics Quarterly*, 26:403–413.

Liang, K.-Y. and Zeger, S.L. (1986) Longitudinal data analysis using generalized linear models. *Biometrika*, 73:13–22.

Lindsay, B.G. (1988) Composite likelihood methods. *Contemporary Mathematics*, 80:221–239.

Littell, R. C., Milliken, G. A., Stroup, W. W., and Wolfinger, R. D. (1996) *The SAS System for Mixed Models.* SAS Institute, Inc., Cary, NC.

Little, L.S., Edwards, D. and Porter, D.E. (1997) Kriging in estuaries: as the crow flies or as the fish swims? *Journal of Experimental Marine Biology and Ecology*, 213:1–11.

Lotwick, H.W. and Silverman, B.W. (1982) Methods for analysing spatial processes of several types of points. *Journal of the Royal Statistical Society, Series B*, 44: 406–413.

Ma, C. (2002) Spatio-temporal covariance functions generated by mixtures. *Mathematical Geology*, 34(8):965–975.

Mantel, N. (1967) The detection of disease clustering and a generalized regression approach. *Cancer Research*, 27(2):209–220.

Marcotte, D. and Groleau, P. (1997) A simple and robust lognormal estimator. *Mathematical Geology*, 29:993–1009.

Mardia, K.V. (1967) Some contributions to contingency-type bivariate distributions. *Biometrika*, 54:235–249.

Mardia, K.V. (1970) *Families of Bivariate Distributions.* Hafner Publishing Company, Darien, CT.

Mardia, K.V., Kent, J.T., and Bibby, J.M. (1979) *Mutivariate Analysis.* Academic Press, London.

Mardia, K.V. and Marshall, R.J. (1984) Maximum likelihood estimation of models for residual covariance in spatial regression. *Biometrika*, 71:135–46.

Marshall, R.J. (1991) Mapping disease and mortality rates using empirical Bayes estimators. *Applied Statistics*, 40:283–294.

Martin, R.J. (1992) Leverage, influence and residuals in regression models when observations are correlated. *Communications in Statistics–Theory and Methods*, 21:1183–1212.

Matérn, B. (1960) Spatial variation. *Meddelanden fran Skogsforskningsinstitut*, 49(5).

Matérn, B. (1986) *Spatial Variation, 2nd ed.* Lecture Notes in Statistics, Springer-Verlag, New York.

Matheron, G. (1962) Traite de Geostatistique Appliquee, Tome I. *Memoires du Bureau de Recherches Geologiques et Minieres*, No. 14. Editions Technip, Paris.

Matheron, G. (1963) Principles of geostatistics. *Economic Geology*, 58:1246–1266.

Matheron, G. (1976) A simple substitute for conditional expectation: The disjunctive kriging. In: M. Guarascio, M. David, and C. Huijbregts (eds.), *Advanced Geostatistics in the Mining Industry*. Reidel, Dordrecht, 221–236.

Matheron, G. (1982) La destructuration des hautes teneures et le krigeage des indicatrices. Technical Report N-761, Centre de Géostatistique, Fontainebleau, France.

Matheron, G. (1984) Isofactorial models and change of support. In: G. Verly, M. David, A. Journel, A. Marechal (eds.), *Geostatistics for Natural Resources Characterization*. Reidel, Dordrecht, 449–467.

McCullagh, P. and Nelder, J.A. (1989) *Generalized Linear Models, Second Edition*. Chapman and Hall, New York.

McMillen D.P. (2003) Spatial autocorrelation or model misspecification? *International Regional Science Review*, 26:208–217.

McShane, L.M., Albert, P.S., and Palmatier, M.A. (1997) A latent process regression model for spatially correlated count data. *Biometrics*, 53:698–706.

Mercer, W.B. and Hall, A.D. (1911) The experimental error of field trials. *Journal of Agricultural Science*, 4:107–132.

Metropolis, N., Rosenbluth, M. N., Teller, A. H., and Teller, E. (1953) Equations of state calculations by fast computing machines. *Journal of Chemical Physics*, 21: 1087–1092.

Mitchell, M.W. and Gumpertz, M.L. (2003) Spatio-temporal prediction inside a free-air $CO_2$ enrichment system, *Journal of Agricultural, Biological, and Environmental Statistics*, 8(3):310–327.

Mockus, A. (1998) Estimating dependencies from spatial averages. *Journal of Computational and Graphical Statistics*, 7:501–513.

Møller, J. and Waagepetersen, R.P. (2003) *Statistical Inference and Simulation for Spatial Point Processes*, Chapman & Hall/CRC, Boca Raton, FL.

Moran, P.A.P. (1950) Notes on continuous stochastic phenomena, *Biometrika*, 37:17–23.

Moyeed, R.A. and Papritz, A. (2002) An emprical comparision of kriging methods for nonlinear spatial prediction. *Mathematical Geology*, 34:365–386.

Mugglestone, M.A. and Renshaw, E. (1996a) A practical guide to the spectral analysis of spatial point processes. *Journal of Computational Statistics & Data Analysis*, 21:43–65

Mugglestone, M.A. and Renshaw, E. (1996b) The exploratory analysis of bivariate spatial point patterns using cross-spectra. *Environmetrics*, 7:361–377.

Müller, W.G. (1999) Least-squares fitting form the variogram cloud. *Statistics & Probability Letters*, 43:93–98.

Nadaraya, E.A. (1964) On estimating regression. *Theory of Probability and its Applications*, 10:186–190.

Neyman, J. and Scott, E.L. (1972) Processes of clustering and applications. In:

P.A.W. Lewis, (ed.) *Stochastic Point Processes.* John Wiley & Sons, New York, 646–681.

Ogata, Y. (1999) Seismicity analysis through point-process modeling: a review. *Pure and Applied Geophysics*, 155:471–507.

O'Hagan, A. (1994) *Bayesian Inference. Kendall's Advanced Theory of Statistics, 2B*, Edward Arnold Publishers, London.

Olea, R. A. (ed.) (1991) *Geostatistical Glossary and Multilingual Dictionary.* Oxford University Press, New York.

Olea, R. A. (1999) *Geostatistics for Engineers and Earth Scientists.* Kluwer Academic Publishers, Norwell, Massachusetts.

Openshaw, S. (1984) *The Modifiable Areal Unit Problem.* Geobooks, Norwich, England.

Openshaw, S. and Taylor, P. (1979) A million or so correlation coefficients. In N. Wrigley (ed.), *Statistical Methods in the Spatial Sciences.* Pion, London, 127–144.

Ord, K. (1975) Estimation methods for models of spatial interaction. *Journal of the American Statistical Association*, 70:120–126.

Ord, K. (1990) Discussion of "Spatial Clustering for Inhomogeneous Populations" by J. Cuzick and R. Edwards. *Journal of the Royal Statistical Society, Series B*, 52:97.

Pagano, M. (1971) Some asymptotic properties of a two-dimensional periodogram. *Journal of Applied Probability*, 8:841–847.

Papadakis, J.S. (1937) Méthode statistique pour des expériences sur champ. *Bull. Inst. Amelior. Plant. Thessalonique*, 23.

Patterson, H.D. and Thompson, R. (1971) Recovery of inter-block information when block sizes are unequal. *Biometrika*, 58:545–554.

Percival, D.B. and Walden, A.T. (1993) *Spectral Analysis for Physical Applications. Multitaper and Conventional Univariate techniques.* Cambridge University Press, Cambridge, UK.

Plackett, R.L. (1965) A class of bivariate distributions. *Journal of the American Statistical Association*, 60:516–522.

Posa, D. (1993) A simple description of spatial-temporal processes. *Computational Statistics & Data Analysis*, 15:425–437.

Prasad, N.G.N. and Rao, J.N.K. (1990) The estimation of the mean squared error of small-area estimators. *Journal of the American Statistical Association*, 85:161–171.

Prentice, R.L. (1988) Correlated binary regression with covariates specific to each binary observation. *Biometrics*, 44:1033–1048.

Priestley, M.B. (1981) *Spectral analysis of time series. Volume 1: Univariate series.* Academic Press, New York.

Rathbun, S.L. (1996) Asymptotic properties of the maximum likelihood estimator for spatio-temporal point processes. *Journal of Statistical Planning and Inference*, 51:55–74.

Rathbun, S.L. (1998) Kriging estuaries. *Environmetrics*, 9:109–129.

Rathbun, S.L. and Cressie, N.A.C. (1994) A space-time survival point process for a longleaf pine forest in Southern Georgia. *Journal of the American Statistical Association*, 89:1164–1174.

Rendu, J.M. (1979) Normal and lognormal estimation. *Journal of the International Association for Mathematical Geology*, 11:407–422.

Renshaw, E. and Ford, E.D. (1983) The interpretation of process from pattern using two-dimensional spectral analysis: methods and problems of interpretation. *Applied Statistics*, 32:51–63.

Ripley, B.D. (1976) The second-order analysis of stationary point processes. *Journal of Applied Probability*, 13:255–266.

Ripley, B.D. (1977) Modeling spatial patterns. *Journal of the Royal Statistical Society (B)*, 39:172–192 (with discussion, 192–212).

Ripley, B.D. (1981) *Spatial Statistics*. John Wiley & Sons, New York.

Ripley, B.D. (1987) *Stochastic Simulation*. John Wiley & Sons, Chichester.

Ripley, B.D. and Silverman, B.W. (1978) Quick tests for spatial interaction. *Biometrika*, 65:641–642.

Rivoirard, J. (1994) *Introduction to Disjunctive Kriging and Nonlinear Geostatistics*. Clarendon Press, Oxford.

Robert, C.P. and Casella, G. (1999) *Monte Carlo Statistical Methods*. Springer-Verlag, New York.

Rogers, J.F., Thompson, S.J., Addy, C.L., McKeown, R.E., Cowen, D.J., and De-Coulfé, P. (2000) The association of very low birthweight with exposures to environmental sulfur dioxide and total suspended particulates. *American Journal of Epidemiology*, 151:602–613.

Rousseeuw, P.J. and Croux, C. (1993) Alternatives to the median absolute deviation. *Journal of the American Statistical Association*, 88:1273–1283.

Royaltey, H., Astrachan, E. and Sokal, R. (1975) Tests for patterns in geographic variation. *Geographic Analysis*, 7:369–396.

Ruppert, D., Wand, M.P., and Carroll, R.J. (2003) *Semiparametric Regression*, Cambridge University Press, Cambridge, UK.

Russo, D. and Bresler, E. (1981) Soil hydraulic properties as stochastic processes, 1. An analysis of field spatial variability. *Journal of the Soil Science Society of America*, 45:682–687.

Russo, D. and Jury, W.A. (1987) A theoretical study of the estimation of the correlation scale in spatially variable fields. 1. Stationary Fields. *Water Resources Research*, 7:1257–1268.

Sampson, P.D. and Guttorp, P. (1992) Nonparametric estimation of nonstationary spatial covariance structure. *Journal of the American Statistical Association*, 87:108–119.

Schabenberger, O. and Gregoire, T.G. (1996) Population-averaged and subject-specific approaches for clustered categorical data. *Journal of Statistical Computation and Simulation*, 54:231–253.

Schabenberger, O. and Pierce, F.J. (2002) *Contemporary Statistical Models for the Plant and Soil Sciences*. CRC Press, Boca Raton, FL.

Schervish, M.J. and Carlin, B.P. (1992) On the convergence of successive substitution sampling. *Journal of Computational and Graphical Statistics*, 1:111–127.

Schmidt, P. (1976) *Econometrics*, Marcel Dekker, New York.

Searle, S.R., Casella, G., and McCulloch, C.E. (1992) *Variance Components*. John Wiley & Sons, New York.

Self, S.G. and Liang, K.Y. (1987) Asymptotic properties of maximum likelihood estimators and likelihood ratio tests under nonstandard conditions. *Journal of the American Statistical Association*, 82:605–610.

Shapiro, A. and Botha, J.D. (1991) Variogram fitting with a general class of conditionally nonnegative definite functions. *Computational Statistics and Data Analysis*, 11:87–96.

Smith, A.F.M. and Gelfand, A.E. (1992) Bayesian statistics without tears: A sampling-resampling perspective. *The American Statistician*, 46:84–88.

Smith, A.F.M. and Roberts, G.O. (1993) Bayesian computation via the Gibbs sampler and related Markov chain Monte Carlo methods. *Journal of the Royal Statistical Society, Series B*, 55:3–24.

Solie, J.B., Raun, W.R., and Stone, M.L. (1999) Submeter spatial variability of selected soil and bermudagrass production variables. *Journal of the Soil Science Society of America*, 63:1724–1733.

Stein, M.L. (1999) *Interpolation of Spatial Data. Some Theory of Kriging.* Springer-Verlag, New York.

Stern, H. and Cressie, N. (1999) Inference for extremes in disease mapping. In A. Lawson et al. (eds.) *Disease Mapping and Risk Assessment for Public Health*, John Wiley & Sons, Chichester, 63–84.

Stoyan, D., Kendall, W.S. and Mecke, J. (1995) *Stochastic Geometry and its Applications. 2nd ed.* John Wiley & Sons, New York.

Stroup, D.F. (1990) Discussion of "Spatial Clustering for Inhomogeneous Populations" by J. Cuzick and R. Edwards. *Journal of the Royal Statistical Society, Series B*, 52:99.

Stroup, W.W., Baenziger, P.S., and Mulitze, D.K. (1994) Removing spatial variation from wheat yield trials: a comparison of methods. *Crop Science*, 86:62–66.

Stuart, A. and Ord, J.K. (1994) *Kendall's Advanced Theory of Statistics, Volume I: Distribution Theory*. Edward Arnold, London.

Sun, D., Tsutakawa, R.K., and Speckman, P. L. (1999) Posterior distribution of hierarchical models using CAR(1) distributions, *Biometrika*, 86:341–350.

Switzer, P. (1977) Estimation of spatial distributions from point sources with application to air pollution measurement. *Bulletin of the International Statistical Institute*, 47:123–137.

Szidarovsky, F., Baafi, E.Y., and Kim, Y.C. (1987). Kriging without negative weights, *Mathematical Geology*, 19:549–559.

Tanner, M.A. (1993) *Tools for Statistical Inference.* 2nd Edition, Springer-Verlag, New York.

Tanner, M.A. and Wong, W.H. (1987) The calculation of posterior distributions by data augmentation. *Journal of the American Statistical Association*, 82:528–540.

Theil, H. (1971) *Principles of Econometrics.* John Wiley & Sons, New York.

Thiébaux, H.J. and Pedder, M.A. (1987) *Spatial objective analysis with applications in atmospheric science.* Academic Press, London.

Thompson, H.R. (1955) Spatial point processes with applications to ecology. *Biometrika*, 42:102–115.

Thompson, H.R. (1958) The statistical study of plant distribution patterns using a grid of quadrats. *Australian Journal of Botany*, 6:322–342.

Tierney, L. (1994) Markov chains for exploring posterior distributions (with discussion). *Annals of Statistics*, 22:1701–1786.

Tobler, W. (1970) A computer movie simulating urban growth in the Detroit region. *Economic Geography*, 46:234–240.

Toutenburg, H. (1982) *Prior Information in Linear Models.* John Wiley & Sons, New York.

Turnbull, B.W., Iwano, E.J., Burnett, W.S., Howe, H.L., and Clark, L.C. (1990) Monitoring for clusters of disease: Application to leukemia incidence in upstate New York. *American Journal of Epidemiology*, 132:S136–S143.

Upton, G.J.G. and Fingleton, B. (1985) *Spatial Data Analysis by Example, Vol. 1: Point Pattern and Quantitative Data.* John Wiley & Sons, New York.

Vallant, R. (1985) Nonlinear prediction theory and the estimation of proportions in a finite population. *Journal of the American Statistical Association*, 80:631–641.

Vanmarcke, E. (1983) *Random Fields: Analysis and Synthesis.* MIT Press, Cambridge, MA.

Verly, G. (1983) The Multi-Gaussian approach and its applications to the estimation of local reserves. *Journal of the International Association for Mathematical Geology*, 15:259–286.

Vijapurkar, U.P. and Gotway, C.A. (2001) Assessment of forecasts and forecast uncertainty using generalized linear models for time series count data. *Journal of Statistical Computation and Simulation*, 68:321–349.

Waller, L.A. and Gotway, C.A. (2004) *Applied Spatial Statistics for Public Health Data.* John Wiley & Sons, New York.

Walters, J.R. (1990) Red-cockaded woodpeckers: a 'primitive' cooperative breeder. In: *Cooperative Breeding in Birds: Long-term Studies of Ecology and Behaviour.* Cambridge University Press, Cambridge, 67–101.

Wand, M.P. and Jones, M.C. (1995) *Kernel Smoothing.* Chapman and Hall/CRC Press, Boca Raton, FL.

Watson, G.S. (1964) Smooth regression analysis. *Sankhya (A)*, 26:359–372.

Webster, R. and Oliver, M.A. (2001) *Geostatistics for Environmental Scientists.* John Wiley & Sons, Chichester.

Wedderburn, R.W.M. (1974) Quasi-likelihood functions, generalized linear models and the Gauss-Newton method. *Biometrika*, 61:439–447.

Whittaker, E.T. and Watson, G.N. (1927) *A couse of modern analysis, 4th ed.*, Cambridge University Press, Cambridge, UK.

Whittle, P. (1954) On stationary processes in the plane. *Biometrika*, 41:434–449.

Wolfinger, R.D. (1993) Laplace's approximation for nonlinear mixed models. *Biometrika*, 80:791–795.

Wolfinger, R.D. and O'Connell, M. (1993) Generalized linear mixed models: a pseudo-likelihood approach. *Journal of Statistical Computing and Simulation*, 48:233–243.

Wolfinger, R., Tobias, R. and Sall, J. (1994) Computing gaussian likelihoods and their derivatives for general linear mixed models. *SIAM Journal on Scientific and Statistical Computing*, 15:1294–1310.

Wong, D.W.S. (1996) Aggregation effects in geo-referenced data. In D. Griffiths (ed.), *Advanced Spatial Statistics.* CRC Press, Boca Raton, Florida, 83–106.

Yaglom, A. (1987) *Correlation Theory of Stationary and Related Random Functions I.* Springer-Verlag, New York.

Yule, G. U. and Kendall, M. G. (1950) *An Introduction to the Theory of Statistics.* 14th Edition, Griffin, London.

Zahl, S. (1977) A comparison of three methods for the analysis of spatial pattern. *Biometrics*, 33:681–692.

Zeger, S.L. (1988) A regression model for time series of counts. *Biometrika*, 75:621–629.

Zeger, S.L. and Liang, K.-Y. (1986) Longitudinal data analysis for discrete and continuous outcomes. *Biometrics*, 42:121–130.

Zeger, S.K., Liang, K.-Y., and Albert, P.S. (1988) Models for longitudinal data: a generalized estimating equation approach. *Biometrics*, 44:1049–1006.

Zellner, A. (1986) Bayesian estimation and prediction using asymmetric loss functions. *Journal of the American Statistical Association*, 81:446–451.

Zhang, H. (2004) Inconsistent estimation and asymptotically equal interpolators in model-based geostatistics. *Journal of the American Statistical Association* 99: 250–261.

Zhao, L.P. and Prentice, R.L. (1990) Correlated binary regression using a quadratic exponential model. *Biometrika*, 77:642–648.

Zimmerman, D.L. (1989). Computationally efficient restricted maximum likelihood estimation of generalized covariance functions. *Mathematical Geology*, 21:655–672.

Zimmerman, D.L. and Cressie, N.A. (1992) Mean squared prediction error in the spatial linear model with estimated covariance parameters. *Annals of the Institute of Statistical Mathematics*, 32:1–15.

Zimmerman, D.L. and Harville, D.A. (1991) A random field approach to the analysis of field-plot experiments and other spatial experiments. *Biometrics*, 47:223–239.

Zimmerman, D.L. and Zimmerman, M.B. (1991) A comparison of spatial semivariogram estimators and corresponding kriging predictors. *Technometrics*, 33:77–91.

# Author Index

# Subject Index

Printed in the United States
by Baker & Taylor Publisher Services